VI European Conference on Neutron Scattering (ECNS2015)

Journal of Physics: Conference Series Volume 746

Zaragoza, Spain
30 August - 4 September 2015

ISBN: 978-1-5108-3105-6
ISSN: 1742-6588

Printed from e-media with permission by:

Curran Associates, Inc.
57 Morehouse Lane
Red Hook, NY 12571

Printed by Curran Associates, Inc. (2016)

For permission requests, please contact the Institute of Physics
at the address below.

Institute of Physics
Dirac House, Temple Back
Bristol BS1 6BE UK

Phone: 44 1 17 929 7481
Fax: 44 1 17 920 0979

techtracking@iop.org

Additional copies of this publication are available from:

Curran Associates, Inc.
57 Morehouse Lane
Red Hook, NY 12571 USA
Phone: 845-758-0400
Fax: 845-758-2633
Email: curran@proceedings.com
Web: www.proceedings.com

Table of contents

Volume 746

VI European Conference on Neutron Scattering (ECNS2015)
30 August to 4 September 2015, Zaragoza, Spain

Accepted papers received: 1 September 2016
Published online: 10 October 2016

Preface

Papers

Neutron Instrumentation, Optics, Sample Environment, Detectors and Software

Neutron Sources and Facilities

Engineering applications

Fundamental Science

Functional Materials

Magnetism, Superconductivity and other Electronic Phenomena

Zaragoza (Spain) hosted the VI European Conference on Neutron Scattering in 2015.

It was a great pleasure for the Materials Science Institute of Aragón (CSIC- University of Zaragoza) to host the VI European Conference on Neutron Scattering (ECNS) from the 30th of August until the 4th of September 2015. The meeting was held in Zaragoza, Spain, a prosperous and well communicated city founded by the Emperor Octavio Augustus over 2000 years ago. Zaragoza, a city where different cultures, Muslims, Jewish, and Christians have left their mark is famous for its landmarks such as the Basílica del Pilar, La Seo Cathedral and the Aljafería Palace as well as the local cuisine.

The conference is organized every four years as a forum for the European neutron scattering community to discuss recent developments and advances in all branches of science in which neutron scattering is, or eventually could be used. In 2015 the conference gathered more than 650 participants from 31 different countries from around the globe including Japan, United States of America, Taiwan, Republic of Korea, India, Argentina, China and Australia. This volume assembles the proceedings of the ECNS 2015.

Fig 1. The Spanish Secretary of State for Research, Development and Innovation, Carmen Vela; the Rector of the University of Zaragoza, Manuel José López Pérez; the Rector of the University of Cantabria, José Carlos Gómez Sal, and the President of the European Neutron Scattering Association, Christiane Alba-Simionesco; participated in the conference opening ceremony.

Journal of Physics: Conference Series **746** (2016) 011001 doi:10.1088/1742-6596/746/1/011001

ECNS2015
VI EUROPEAN CONFERENCE ON NEUTRON SCATTERING
August 30th, September 4th | ZARAGOZA (SPAIN)

Fig 2. Group photo in front of the ECNS2015 venue

The main topics in the conference were; Neutron Sources and Facilities, Neutron Instrumentation (Optics, Sample Environment, Detectors and Software), Fundamental Science, Chemistry of Materials, Magnetism, Superconductivity, Functional Materials, Glasses and Liquids, Thin Films and Interfaces, Soft Condensed Matter, Health and Life Sciences, Engineering applications, Cultural Heritage and Archaeometry.

Jean-Marie Tarascon, J. Manuel Perez-Mato, Roberto Caciuffo, Paul Schofield, Peter Fierlinger, Helmut Schober and Frank Gabel presented plenary talks. In addition, 33 keynote talks and 222 oral presentations were given during the four parallel sessions as well as 325 poster presentations. The poster sessions, which were held during the lunch and coffee breaks, were well attended and participants had time to visit not only the posters but also the exhibition.

The VI European Conference on Neutron Scattering (ECNS) was supported by our institutional sponsors (Government of Aragón, CSIC and University of Zaragoza) as well as one platinum sponsor (Consortium of German Neutron Sources), three gold sponsors (ILL, PSI, ESS), four silver sponsors (Swiss Neutronics, ISIS, ESS Bilbao, JINR-FLNP) and 12 bronze sponsors (NMi3, Airbus Defence & Space, Tessella, Oxford Instruments, General Electric, Laboratoire Léon Brillouin, HNF Technologies, AVS, ANTE, Mirrotron Ltd, Toshiba, and the ENATE winery). The Sponsors ranged from neutron sources, consortia and laboratories to industrial organisations.

Journal of Physics: Conference Series **746** (2016) 011001 doi:10.1088/1742-6596/746/1/011001

ECNS2015
VI EUROPEAN CONFERENCE
ON NEUTRON SCATTERING
August 30th, September 4th | ZARAGOZA (SPAIN)

Fig 3. Plenary talk on *Materials for Energy* given by Professor Jean-Marie Tarascon from *Collège de France*

A total of 19 graduate students and post-graduate students received the International Union of Crystallography fellowships (IUCR). Fellows came from Denmark, France, Hungary, Russia, UK, Australia, Japan and Spain.

ENSA has a long-standing culture of rewarding scientists for their work and at the ENSA session the 2015 Walter Hälg Prize was awarded to Prof Dr Helmut Rauch for his outstanding and seminal contributions to the fundamental aspects of neutron physics and optics and many related aspects of quantum physics.

The Seventh Erwin Felix Lewy Bertaut Prize of the European Crystallographic Association (ECA) and European Neutron Scattering Association (ENSA) was awarded to Dr. Giorgio Schirò for his work on movement of biological matter.

A number of satellite meetings were also held during the conference such as the Swiss Neutron Scattering Society General Assembly, the Neutron sources directors meeting as well as the European Neutron Scattering Association (ENSA) meeting. In particular, the Neutron sources directors meeting gathered directors from 14 Neutron sources from countries around the world. The ENSA meeting, which was held on the second day of the conference, hosted delegates from Holland, Norway, Denmark, Germany, Norway, Spain and France.

ECNS2015
VI EUROPEAN CONFERENCE ON NEUTRON SCATTERING
August 30th, September 4th | ZARAGOZA (SPAIN)

Fig 4. Prof Dr Helmut Rauch received the 2015 Walter Hälg Prize

Fig 5. Meeting of the ENSA Committee at the ECNS2015

 doi:10.1088/1742-6596/746/1/011001

ECNS2015
VI EUROPEAN CONFERENCE
ON NEUTRON SCATTERING
August 30th. September 4th | ZARAGOZA (SPAIN)

Several national facilities will close by 2020, which will cause the overall neutron capacity in Europe to be greatly reduced. Design studies for new, cost effective and energy efficient, accelerator-based sources must be initiated now in order to rejuvenate the ecosystem of European facilities by 2025. We hope that the high level of participation and the high quality of the science presented at ECNS in Zaragoza will transmit a clear sign to the European Authorities about the necessity to support our diverse community by maintaining the European capability and capacity, which is indispensable in both the short and long term.

Fig 6. Plenary talk on the *European Neutron Landscape* given by Professor Helmut Schober, Head of the Science Division at ILL.

The organizers would like to thank all the attendees for their active participation and enthusiasm, which ensured that the conference was a great success.

The different committees (Local Organising Committee, National Advisory Committee, International Advisory Board, and International Program Committee) appointed to advise us during the organization process and to develop the scientific program (which was not an easy task...) are also deeply recognised.

Javier Campo
Chair of ECNS 2015
Materials Science Institute of Aragón
(CSIC- University of Zaragoza)

ECNS2015
VI EUROPEAN CONFERENCE ON NEUTRON SCATTERING
August 30th, September 4th | ZARAGOZA (SPAIN)

Fig 7. Scientists enjoying the banquet.

Fig 8. "La Jota" The typical Aragonese folklore during the ECNS2015 welcome party.

Peer review statement

All papers published in this volume of *Journal of Physics: Conference Series* have been peer reviewed through processes administered by the proceedings Editors. Reviews were conducted by expert referees to the professional and scientific standards expected of a proceedings journal published by IOP Publishing.

Journal of Physics: Conference Series **746** (2016) 012001 doi:10.1088/1742-6596/746/1/012001

Neutron scattering at the OPAL research reactor

Garry J. McIntyre and Peter J. Holden

Bragg Institute, Australian Nuclear Science and Technology Organisation, New Illawarra Road, Lucas Heights NSW 2234, Australia

E-mail: garry.mcintyre@ansto.gov.au

Abstract. The current suite of 14 neutron scattering instruments at the multipurpose OPAL research reactor is described. All instruments have been constructed following best practice, using state-of-the-art components and in close consultation with the regional user base. First results from the most recently commissioned instruments match their design performance parameters. Selected recent scientific highlights illustrate some unique combinations of instrumentation and the regional flavour of topical applications.

1. Introduction

Australian neutron scattering leapt into the 21st century with the start up of the 20 MW OPAL nuclear research reactor at the Australian Nuclear Science and Technology Organisation (ANSTO) in 2006. OPAL is unique amongst modern high-power research reactors in that it is multipurpose, being used to produce medical isotopes and exploited for neutron transmutation doping and neutron activation analysis, as well as being used to provide beams for neutron scattering. While the multipurpose nature led to some compromises in the core design for the neutron-beam instruments, it brought with it valuable political and economic security. Both OPAL and its cold-neutron source, the running of which is not linked intimately to the operation of the reactor unlike at most research reactors, now operate consistently for around 300 days per year - an impressive statistic amongst modern research reactors.

The major part of the initial success at OPAL has been in crystallography, carrying on the excellent tradition established since the late 1950s at the HIFAR reactor at Lucas Heights. HIFAR had an illustrious career in itself. As well as providing a materials research capability for the (then) projected nuclear power programme in Australia, it was also the training ground for many young scientists that went on to distinguished careers in Australia and abroad. Hugo Rietveld and Alan Hewat of neutron powder diffraction (NPD) fame were just two of these, although it is noteworthy that their first neutron experiments were in single-crystal diffraction [1] and triple-axis spectrometry [2] respectively, since the power of NPD in materials discovery was yet to be realised, much less appreciated.

As befits a national facility, the neutron-scattering science at OPAL is largely driven by the external user programme, through a peer-review proposal system for access to the neutron-beam instruments. Fee-for-service commercial use is growing with the present target being 5% averaged over the full suite. Certain scientific themes, where neutrons offer distinct advantages as a research tool, are also promoted within the Bragg Institute, the entity that runs the neutron-scattering programme. These themes currently include energy materials, thermo-mechanical processes, magnetism, food science, biomolecules, and cultural heritage.

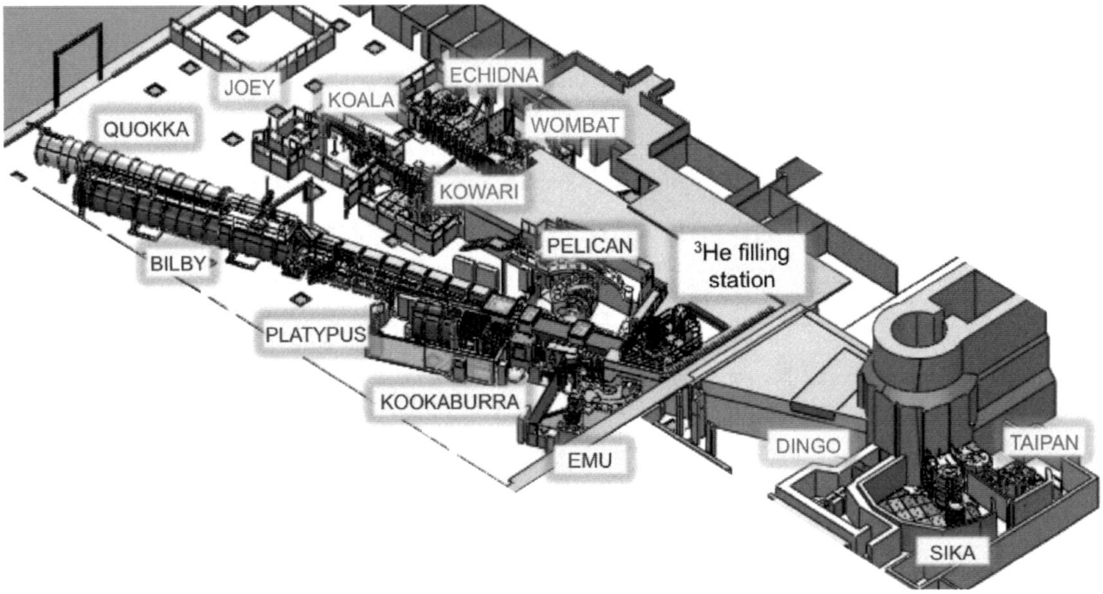

Figure 1: Schematic of the layout of the neutron-beam instruments at OPAL. DINGO, SIKA, and TAIPAN are located close to the reactor face, the other instruments are on neutron guides. Green text denotes the instruments on thermal neutron beams, blue text those on cold neutron beams. Also indicated is the convenient location of the ^{3}He spin-filter filling station atop the neutron guide bunker.

Here we introduce the OPAL neutron-beam facility, show the first results from the instruments most recently brought on-line, and describe recent scientific highlights allied closely to the new or unique instrumentation.

2. Neutron-beam instruments at OPAL

The construction of the neutron-beam instruments at OPAL has proceeded in three main tranches. The first, called *Neutron Beam Instrument Project - 1* (NBI-1), produced the first seven instruments [3] as part of the full funding for the reactor construction. The second, denoted *Major Capital Works* with funding from various internal and external sources, yielded four further instruments. Most recently, the *Neutron Beam Instrument Project - 2* (NBI-2) funded as part of the Super Science Initiative of 2009, has added three more [4], to bring the total number of instruments to 14. Figure 1 shows the layout of the current suite of neutron-beam instruments at OPAL, and table 1 describes each instrument with its key features and the year that it entered user operation.

Based largely on user wishes, the suite of instruments is fairly conventional and constructed using proven components following best practice elsewhere, to produce state-of-the-art instruments, but occasionally in unique combinations of components and with unique applications. One example is the time-of-flight spectrometer PELICAN, which has been designed from the outset to offer a full polarisation analysis capability using a supermirror polariser before the sample, a PASTIS insert [5] around the sample, and a wide-angle ^{3}He spin-filter cell after the sample, a worthy successor to the long-wavelength polarised-neutron spectrometer LONGPOL [6], which gave HIFAR a unique capability for several decades.

The cold-neutron triple-axis spectrometer, SIKA, constructed by National Central University, Taiwan, and just recently awarded its operating licence, has been presented elsewhere at

Journal of Physics: Conference Series **746** (2016) 012001 doi:10.1088/1742-6596/746/1/012001

Table 1: Neutron-scattering instruments at OPAL in 2015.

Instrument	Description	Features	In user program	Reference[a]
Diffraction				
ECHIDNA	High-resolution powder diffractometer	128-element multi-tube detector, robot sample changer	2008	[7]
JOEY[b]	Neutron Laue camera for single-crystal alignment	Fast-read-out CCD detector; precision sample table that supports 200 kg	-	[8]
KOALA	Thermal neutron Laue diffractometer	Sample volumes of typically 0.1 mm^3	2008	[9]
WOMBAT	High-intensity (powder) diffractometer	Large area detector, optional radial oscillating collimator	2008	[10]
Inelastic scattering				
EMU	High-resolution backscattering spectrometer	Nearly exact backscattering at low Q	2016[c]	
PELICAN	Time-of-flight spectrometer	Designed at the outset for optional polarisation analysis	2014	[11]
SIKA	Cold-neutron three-axis spectrometer	Choice of single or multi-analyser	2015	
TAIPAN	Thermal-neutron three-axis spectrometer	Optional Be-filter mode	2010	[12]
Small-angle scattering				
BILBY	Time-of-flight small-angle neutron-scattering (SANS) instrument	Large Q range for kinetic studies	2016[c]	
KOOKABURRA	Ultra-small-angle neutron-scattering instrument	Bonse-Hart geometry with choice of two wavelengths	2014	[13]
QUOKKA	Pinhole SANS instrument	Large area detector; refractive lens; polarised-neutron options	2009	[14]
Nano- to macro-scale imaging				
DINGO	Neutron imaging instrument	High-flux and high-resolution modes	2014	[15]
KOWARI	Strain scanner	Large-volume samples up to 1 t; texture studies	2008	[16]
PLATYPUS	Neutron reflectometer	Polarised-neutron options	2009	[17] [18]

[a]Additional details of each instrument can be found at [19]; [b]Accessible on an *ad hoc* basis; [c]Anticipated

ECNS 2015 by Dr Jason Gardner. SIKA is now operated by the National Synchrotron Radiation Research Center (NSRRC), Taiwan. The NSRRC team [20] at ANSTO is not just an outstation but is in many respects an integral part of the neutron-scattering program at ANSTO. A polarised-neutron capability using ^{3}He spin filters is currently being introduced on five instruments - PELICAN, QUOKKA, SIKA, TAIPAN, and WOMBAT - and added as an option to the capability already possible using polarising super-mirrors on PLATYPUS. Observation of the first scattered neutrons is an exciting time for the instrument construction teams, and figure 2 shows the first results from the most recently commissioned instruments or capabilities.

The high-resolution powder diffractometer, ECHIDNA, with 51 papers in 2014, has now achieved a publication rate amongst the world's best comparable instruments, and several other OPAL instruments are not far behind in such comparisons.

There is space for several additional instruments in the present guide hall, mostly by installation or extension of one each of additional thermal and cold neutron guides, the exact number of instruments depending on the types finally installed. There is also space directly opposite the present guide hall for a second guide hall, and for installation of a hot source if user demand warrants it. Just as for the first suite of instruments, the choice of additional instruments will be made in close consultation with the potential user base [21]. Such an approach has been fully vindicated by the consistent over-demand for beam time on the existing instruments.

The Bragg Institute currently includes the National Deuteration Facility (NDF), which produces a range of deuterated molecules, including lipids, proteins, biopolymer, aromatics, detergents, and sugars, principally for diffraction, SANS, and reflectometry experiments. The NDF also produces samples for other research techniques, such as NMR, and to study the effect of isotopic substitution itself on properties [24]. There is also a small computing cluster and limited atomistic-modelling support, as well as X-ray small-angle scattering and reflectometry instruments to complement the neutron user program, and most recently a Physical Properties Measurement System (Quantum Design Inc.).

The main commercial use of neutron beam time is for strain scanning of large pipes, rail components, and welds in general, with a growing interest in strain analysis of additive manufacturing, and radiography of cement samples. An interesting observation from our industrial program is that smaller companies are keen to advertise their use of the neutron-beam instruments to validate their work or process. The Bragg Institute Industrial Liaison Office [23] coordinates industrial activities, including products like deuteration at the NDF and software, in addition to neutron beam time.

3. Recent scientific highlights

Each set of authors of overviews of the OPAL neutron-beam instruments brings their own slant. Being the current research leaders in the Bragg Institute we focus on a small selection of the science that the new capabilities explore with emphasis on the instrumentation where OPAL is at the forefront and on topical regional applications.

3.1. 'Dynamic' sample environment

WOMBAT was the scene for the world's first *in situ operando* experiments on a functioning battery just five years ago [25], and ANSTO continues to play a leading role in *in situ* neutron diffraction on Li-ion batteries, producing 46% of the publications worldwide in this area since then.

'Dynamic' *in situ* experiments of a softer kind have also been pioneered by adapting a Rapid Visco^{TM} analyser to QUOKKA to study simultaneously structural and viscosity changes during cooking of starch[26], diffraction studies of crystallisation of the triglyceride, tripalmitin, under shear force using a rheometer on WOMBAT, and simultaneous differential scanning calorimetry

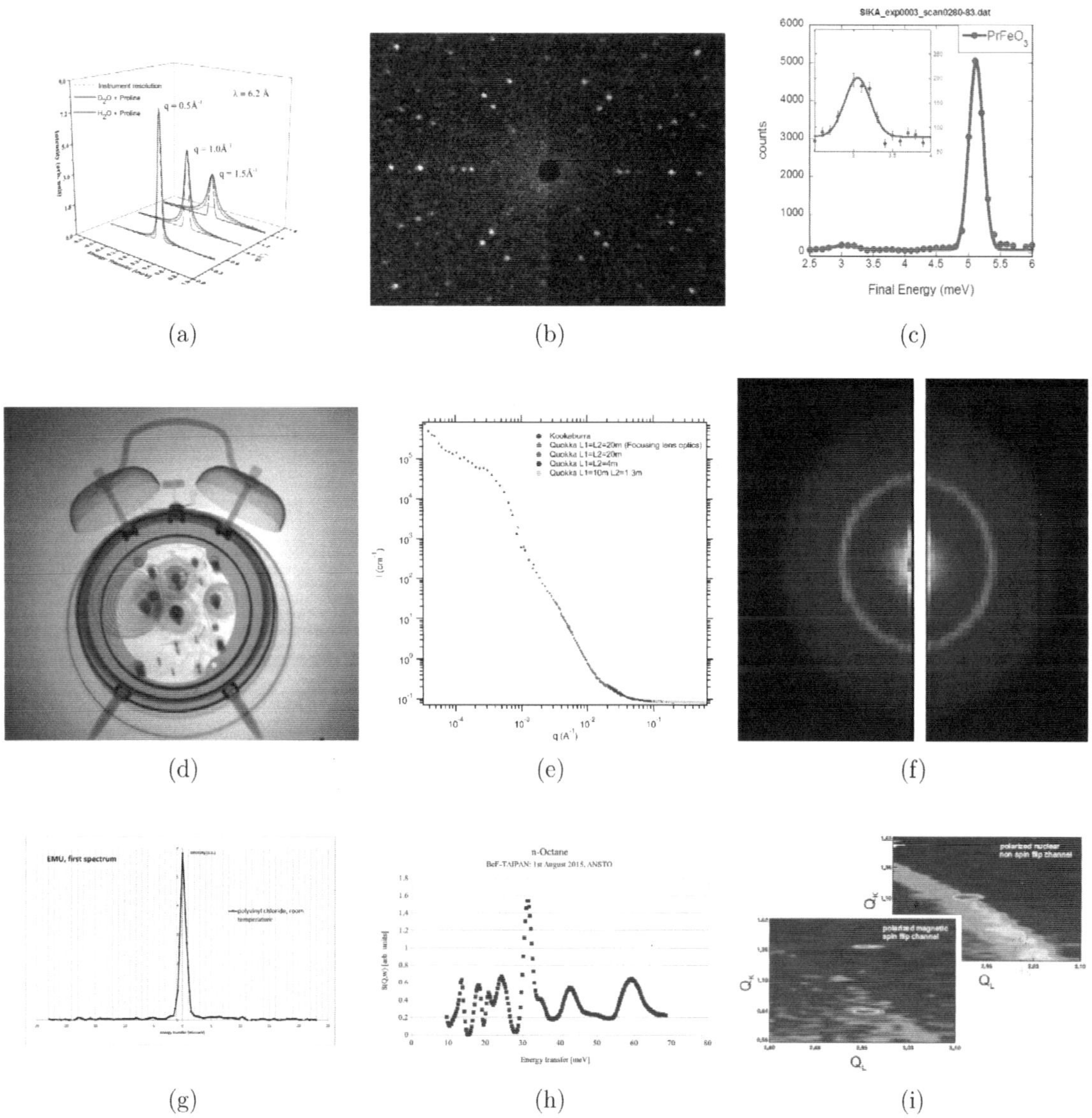

Figure 2: First results from the latest tranche of neutron-beam instruments or capabilities at OPAL: a) Quasi-elastic neutron scattering from 9.1% proline in H_2O/D_2O on PELICAN [22]; b) Laue pattern from $TbMn_2O_5$ on JOEY; c) Crystal-field excitation from $PrFeO_3$ recorded on SIKA; d) Radiograph of an alarm clock on DINGO; e) Extension of the Q range in small-angle scattering to 4 x 10^{-5} Å^{-1} enabled by KOOKABURRA; f) Small-angle scattering from silver behenate on BILBY; g) Energy-transfer resolution on EMU determined by an energy scan from polyvinyl chloride; h) Vibrational density-of states spectrum for octane from the Be-filter option on TAIPAN; i) Polarisation analysis in the magnetic phase of a multiferroic single-crystal sample using a ^{3}He spin filter on WOMBAT.

(DSC) and SANS on QUOKKA [27]. The signal in the last experiments, which monitored the periodicity of generally incommensurate lamellar structures of a binary blend of alkane molecules of different length, was enhanced by deuteration of one of the alkanes to give a considerable contrast advantage over similar studies with X-rays.

3.2. Chemical crystallography

With the regrettable withdrawal of the pioneering thermal-neutron Laue diffractometer, VIVALDI [28], from the user program at the Institut Laue-Langevin, KOALA, essentially a clone of VIVALDI, is now the sole reactor-based thermal-neutron Laue diffractometer accessible to external users, yet is leading the world in application of neutron diffraction to the study of challenging hydrides [9]. One surprising recent example is the 28 copper 15 hydride *Chinese puzzle molecule* found to form when certain copper compounds are mixed with a hydride of boron [29]. This new structure has alternating layers of hydride and copper wrapped in an outer shell of protecting dithiocarbamate molecules, and is much more stable than many other hydride compounds, which has promising implications for nanoparticle formation.

3.3. Spintronics

Not surprisingly, magnetism is a common theme on the majority of instruments, with transition-metal and rare-earth magnetism in bulk samples being strong favourites. There is also a growing effort in studies of thin-film magnetism, exploiting the tunability allowed by doping, strain, and proximity effects, of properties of importance for real applications. Reflectometry on PLATYPUS and diffraction on TAIPAN are the favoured techniques, often aided by structural characterisation using the in-house X-ray reflectometer. Recently, diffraction experiments on TAIPAN revealed that thin films of $SrCoO_{3-\delta}$ on a $DySrO_3$ substrate, which provides large epitaxial linear strain, undergo a strain-induced ferromagnetic-to-antiferromagnetic phase transition unlike similar films on the better-lattice-matched $SrTiO_3$ substrates [30]. Films as thin as 20 nm, a new record for successful neutron diffraction experiments, were necessary to maintain the strain through the film.

3.4. Granular materials

Granular materials are a unique form of matter that can act in a similar manner to a solid, liquid, or gas, depending on the circumstances. In quasi-static systems their ability to flow and support static shear stress provides for interesting physical behaviour. The constitutive response of granular materials is highly non-linear and incremental (historical) and has provided an active area of study for physicists and engineers for hundreds of years. Far from being of purely academic interest, the ability to model the response of granular materials is of high relevance to the handling of bulk commodities such as coal (4.72 tonnes 'flow' through Newcastle in New South Wales, Australia, each second), cereal grains, and mineral sands.

The full triaxial stress state within individual particles in a monodisperse spherical granular assembly has recently been measured for the first time [31]. This was made possible by neutron imaging and computed tomography on DINGO combined with neutron diffraction strain measurement techniques and associated stress reconstruction using KOWARI. The assembly in question consisted of 549 precision steel ball bearings under an applied load of 85 MPa in a cylindrical die. Clear evidence of force chains was observed in terms of both the shape of the probability distribution function for normal stresses and the network formed by highly loaded particles.

3.5. Imaging ancient Australian fauna

One could be forgiven for thinking that Australian fauna still hark back to the days of dinosaurs, but it is little appreciated that real dinosaurs did roam the southern continent and that there

Journal of Physics: Conference Series **746** (2016) 012001 doi:10.1088/1742-6596/746/1/012001

exist several quite important dig sites. The most significant of these are around Winton in outback Queensland, where the first bones were unearthed just 12 years ago. Ada Klinkhamer and David Elliot of the *Australian Age of Dinosaurs* Museum in Winton have been working with the DINGO scientist, Joseph Bevitt, to explore the possibilities of neutron radiography for non-invasive examination of rocks from the Winton area to detect ancient fossils. It can be hit or miss, but occasionally beautiful images like that in figure 3 can be teased out.

4. Instruments of communication: Public outreach

ANSTO is very conscious of the need to communicate our scientific and commercial activities to the general public. The external-communications group and our Discovery Centre are very proactive in outreach to the general public, having stalls at local fairs and celebration days, sponsoring groups from ANSTO in local charity sporting events, and conducting tours of the site including visits to the viewing platform overlooking the reactor guide hall. Figure 4 shows a recent billboard advertisement, and similar advertisements were posted on local bus shelters and on buses themselves.

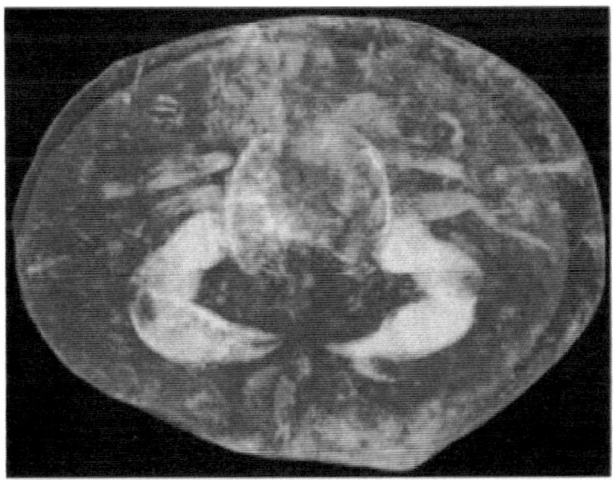

Figure 3: Non-invasive neutron radiography on DINGO reveals the outline of *Torynomma quadrata*, a cretaceous crab.

Figure 4: A Sutherland Shire billboard inviting the general public to visit ANSTO and the neutron-beam facilities.

Several groups of school classes, or community groups such as senior citizens, visit the site each week day, and the general public can register for guided tours on Saturdays. Annually, there are over 15,000 such visitors, with 11,000 in school excursions. The net result is that the nuclear-based activities at ANSTO are predominantly seen in a favourable light by the general public, and with considerable pride by the citizens of Sutherland Shire, where ANSTO is located.

The communications group also aids ANSTO scientists to create media and demonstration tools, such as the recent video that shows how EMU works [32] and a LEGO robotics model of TAIPAN [33], the latter in collaboration with Macquarie University.

5. Conclusions

The combination of state-of-the-art instrumentation, support facilities, expert scientific staff, and enthusiastic users of OPAL has yielded an impressive series of scientific results, as well as a fledgling industrial programme. The Bragg Institute's scientific projects are effective interfaces with the user community, have international impact, and often local public appeal. With currently 13 neutron-scattering instruments in the user programme, including a radiography/tomography instrument, all run by very expert, talented scientists and engineers, ANSTO is well on the way to achieving its commitment to make OPAL one of the top three research reactors for neutron scattering in the world.

Acknowledgments
We gratefully acknowledge the skill and commitment of all ANSTO staff, especially those in the Bragg Institute, that have led to the success of the neutron-scattering program at OPAL. We especially thank the researchers, whose work we have highlighted here, as well as the staff of the ANSTO external communications group, in particular Chris Wensrich (University of Newcastle, NSW), Wei Kong Pang (University of Wollongong), Rod Dowler (ANSTO communications), and Anna Paradowska and Joseph Bevitt (both ANSTO, Bragg Institute) who generously provided material for presentation at ECNS 2015. We also acknowledge partial funding of the NDF and part of the neutron instrument suite by the National Collaborative Research Infrastructure Strategy of the Australian Government.

References

[1] Clews C B J, Maslen E N, Rietveld H M and Sabine T M 1961 *Nature* **192** 154-5.
[2] Hewat A W 1969 *Solid State Comm.* **8** 187-9.
[3] Kennedy S J 2006 *Physica B* **385-6** 949-54.
[4] Klose F, Constantine P, Kennedy S J and Robinson R A 2014 *J. Phys: Conf. Ser.* **528** 012026.
[5] Stewart J R, Andersen K H, Babcock E and Schober H 2006 *Physica B* **385-6** 1142-5.
[6] Campbell S J, Ahmed N, Hicks T J, Ebdon F R and Wheeler D A 1974 *J. Phys. E* **7** 195-8.
[7] Liss K-D, Hunter B A, Hagen M E, Noakes T J and Kennedy S J 2006 *Physica B* **385-6** 1010-2.
[8] Ulrich C, Yethiraj M, Deng G, Callori S J, Brück S, Brule A, McIntyre G J and Klose F 2015 *Personal communication*
[9] Edwards A J 2011 *Aust. J. Chem.* **64** 869-72.
[10] Studer A J, Hagen M E and Noakes T J 2006 *Physica B* **385-6** 1013-5.
[11] Yu D-H, Mole R A, Noakes T, Kennedy S J and Robinson R A 2013 *J. Phys. Soc. Japan* **82** SA027.
[12] Danilkin S A, Horton G, Moore R, Braoudakis G and Hagen M E 2007 *J. Neutron Res.* **15** 55-60.
[13] Rehm C, Brule A, Freund A K and Kennedy S J 2013 *J. Appl. Cryst.* **46** 1699-704.
[14] Gilbert E P, Schulz J C and Noakes T J 2006 *Physica B* **385-6** 1180-2.
[15] Garbe U, Randall T and Hughes C 2011 *Nucl. Instrum. Methods Phys. Res. Sect. A* **651** 42-6.
[16] Brule A and Kirstein O 2006 *Physica B* **385-6** 1040-2.
[17] James M, Nelson A, Holt S A, Saerbeck T, Hamilton W A and Klose F 2011 *Nucl. Instr. Meth. Phys. Res. A* **632** 112-23.
[18] Saerbeck T, Klose F, Le Brun A P, Füzi J, Brule A, Nelson A, Holt S A and James M 2012 *Rev. Sci. Instrum.* **83** 081301.
[19] http://www.ansto.gov.au/ResearchHub/Bragg/Facilities/Instruments/index.htm
[20] http://www.nsrrc.org.tw/english/organizationDetail.aspx?DEPT_UID=52
[21] http://www.ansto.gov.au/cs/groups/corporate/documents/document/mdaw/mda2/~edisp/acs012397.pdf
[22] Yu D, Hennig M, Mole R A, Li J C, Wheeler C, Strässle T and Kearley G J 2013 *Phys. Chem. Chem. Phys.* **15** 20555-64.
[23] http://www.ansto.gov.au/ResearchHub/Bragg/Industry/index.htm
[24] Danos A, MacQueen R W, Cheng Y Y, Dvorák M, Darwish T A, McCamey D R and Schmidt T W 2015 *J. Phys. Chem. Lett.* **6** 3061-6.
[25] Sharma N, Peterson V K, Elcombe M M, Avdeev M, Studer A J, Blagojevic N, Yusoff R and Kamarulzaman N 2010 *Journal of Power Sources* **195** 8258-66.
[26] Doutch J, Bason M, Franceschini F, James K, Clowes D and Gilbert E P 2012 *Carbohydrate Polymers* **88** 1061-71.
[27] Pullen S A, Booth N, Olsen S R, Day B, Franceshini F, Mannicke D and Gilbert E P 2014 *Meas. Sci. Tech.* 25 055606.
[28] McIntyre G J, Lemée-Cailleau M-H and Wilkinson C 2006 *Physica B* **385-6** 1055-8.
[29] Edwards A J, Dhayal R S, Liao P-K, Liao J-H, Chiang M-H, Piltz R O, Kahlal S, Saillard J-Y and Liu C-W 2014 *Angew. Chem. Int. Ed. Engl.* **126** 7342-6.
[30] Callori S J, Hu S, Bertinshaw J, Yue Z J, Danilkin S, Wang X L, Nagarajan V, Klose F, Seidel J and Ulrich C 2015 *Phys. Rev. B* **91** 140405.
[31] Wensrich C M, Kisi E H, Luzin V, Garbe U, Kirstein O, Smith A L and Zhang J F 2014 *Phys. Rev. E* **90** 042203.
[32] http://www.ansto.gov.au/AboutANSTO/MediaCentre/News/ACS076204
[33] http://www.ansto.gov.au/AboutANSTO/MediaCentre/News/ACS076417

Journal of Physics: Conference Series **746** (2016) 012002 doi:10.1088/1742-6596/746/1/012002

A large-area double rotating-crystal monochromator for time-focusing neutron instruments

M Zanatta[1], A Orecchini[1], S Aisa[2], F Casinini[2], L Farnesini[3], P P Deen[4], A Paciaroni[1], C Petrillo[1], and F Sacchetti[1,2]

[1] Dipartimento di Fisica e Geologia, Università di Perugia, Perugia, Italy
[2] IOM-CNR c/o Dipartimento di Fisica e Geologia, Università di Perugia, Perugia, Italy
[3] INFN, Sezione di Perugia, Perugia, Italy
[4] European Spallation Source, Lund, Sweden

E-mail: andrea.orecchini@unipg.it

Abstract. We present the principle and the first prototypes of a double rotating-crystal monochromator, based on an assembly of smaller rotating elements. Such a device was developed as the key element to implement a parallel-beam modification of the time-focusing technique for neutron spectrometers. This concept is particularly promising for long-pulse sources and can bring specific advantages on continuous sources as well. Neutron tests performed on the first prototypes validate the mechanical reliability of the proposed design and the feasibility of a large-area double rotating-crystal monochromator based on this technology.

1. Introduction

The idea of a neutron monochromator based on a rotating crystal was first proposed by Brockhouse in the early age of neutron research [1]. An instrument based on this principle was the Rotating Crystal Spectrometer RXS, installed at Hanford (USA) [2, 3]. The use of rotating crystals was also considered for pulsed sources, an example is the rotating crystal analyzer of ROTAX at ISIS (UK) [4]. A major limitation to this design originates from the request of an accurate rotational stability, which prevents the use of large-area crystals. On the other hand, double crystal configurations have also been largely exploited in several instruments. An historical example is the hybrid spectrometer at the CP-5 reactor of the Argonne National Laboratory [5], which was employed by Copley and Rowe for their seminal researches on liquid metals [6]. Double crystal instruments are currently operational at several facilities, e.g. three axis spectrometers at LLB (Saclay, France) and BARC (Mumbay, India), and the imaging station ANTARES at FRMII (Garching, Germany). A combination of these two configurations, namely a double rotating-crystal monochromator (DRCM), was used at the IN4 beamline of the Institut Laue Langevin (Grenoble, France) in order to get rid of the unwanted Bragg diffraction orders [7].

In this paper, we discuss the construction and tests of a large-area rotating-crystal monochromator, based on an assembly of small rotating elements. This device was developed to build a DRCM with the aim to implement a modification of the well-known time-focusing technique for neutron spectrometers. In fact, a couple of properly phased rotating-crystal monochromators can be used to extract a wider wavelength band from a white beam of neutrons, in such a way that faster and slower neutrons reach the detector at the same time, i.e. are time-focused at the detector position. As compared to standard chopper instruments, this configuration exploits a longer extraction time, so that it is particularly effective for continuous or long-pulse sources. Moreover, the intensity gain due to time-focusing can be further boosted by using large-area beams. Finally, a DRCM has the advantage of shifting the monochromating beam well away from the primary white beam, thus avoiding a direct view of the neutron source from the sample position, with consequent background reduction. In addition, when installed in non-dispersive configuration, the DRCM fully preserves the incoming beam divergence. A more detailed conceptual description of a full time-focusing instrument based on this principle will be reported elsewhere [8].

2. Monochromator development

Our target is a rotating monochromator with ideally a large surface, of the order of 20×20 cm^2, that can rotate at frequencies up to about 100 Hz (6000 rpm). It is evident that the rotation of such a large-surface device implies facing considerable mechanical difficulties. In addition, it can introduce Doppler shift effects on neutrons diffracted by the peripheral regions of the monochromator. Although this can bring some advantages in specific conditions, for the time being we prefer to consider only cases where Doppler effects can be neglected. To this end, the large-area crystal can be partitioned into a suitable number of smaller rotating elements, thus reducing the peripheral speed of the single element. Each element would contain a subset of small crystals conveniently held and aligned by a proper mechanical support.

Having these considerations in mind, we designed a 2 cm-diameter cylindrical crystal holder where slab-shaped crystals lay on the cylinder axis and cut the cylinder along its diameter into two symmetric halves. A suitable number of such cylindrical elements can be easily disposed with parallel axes one next to each other, to cover the desired total monochromator area. An external supporting frame to hold the cylinder ends can be machined with high enough precision to obtain both thorough axial alignement and very small mechanical separations between adjacent cylinders. Spacings of few tenths of mm can be easily realized, thus covering most of the incoming beam area and minimising reflected beam losses. Finally, the cylindrical shape allows a great mechanical stability during rotation, which can thus be operated even in air, given the low frequencies into play. Proper phasing of all the rotating cylinders will result in simultaneous Bragg reflection from all the crystals of the array, thus reproducing the effect of a full large-area monochromator.

More specifically, each cylindrical holder is made of an aluminum alloy of the 6000 series. After cutting the cylinder along its diameter into two symmetric halves, most of the inner material is also removed until the cylinder external wall is reduced to less than 1 mm. In this way, neutron attenuation and background due to scattering by the supports are strongly reduced, while the necessary mechanical stability is preserved. The two halves are then assembled with the crystals in a sandwich structure and tightened together by M2 screws inserted along the cylinder diameter. Although the screws pierce the crystals perpendicularly to their surface, they do not constitute a real problem because the mounting system guarantees a negligible crystal damage during screws insertion.

As a first possible choice for the crystals, we considered Highly Oriented Pyrolitic Graphite (HOPG) that provides a high peak reflectivity even at relatively short wavelengths ($R_0 \simeq 0.7$ at about 1.3 Å [9]), with a clean reflected beam and no spurious components due to multiple reflections. In particular, a high reflectivity is crucial to minimize intensity losses due to the double monochromator configuration.

Fig. 1(a) shows a prototype assembly of three rotating elements. Each cylinder is 240 mm high with a diameter of 20 mm and hosts five $40 \times 20 \times 2$ mm^3 HOPG crystals. The overall weight of each rotating element is about 100 g, so that it can be easily rotated using a direct-drive small

DC motor or a stepper motor. The latter is preferable since it guarantees a digitally controlled velocity up to at least 112 Hz. It is worth noting that the present design exploits both faces of the rotating crystal for neutron reflection, hence the frequency f_M of the monochromatic neutron pulses is twice the *real* rotation frequency f_R. For special applications, it can be envisaged to realize a vertically focusing monochromator, by installing properly tilted crystals inside each cylinder. Of course, this implies the use of only one of the crystal faces, which yields a neutron pulse frequency $f_M = f_R$.

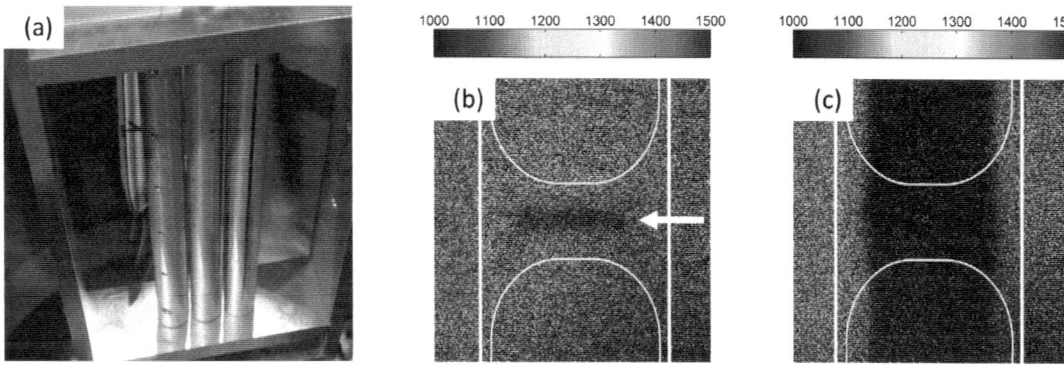

Figure 1. (a) Picture of one of the monochromator prototypes employed in the neutron tests, composed of three rotating elements. The thin 2 mm edge of the HOPG crystals appears as a vertical black line along the rightmost cylinder, whereas the heads of the horizontal screws holding the crystals are visible on the leftmost one. A neutron radiography of a 20×30 mm^2 portion of one cylinder shows the neutrons transmitted by the crystal when the latter is set off (b) and in Bragg reflection (c). White lines highlight the contours of the thickest parts of the cylinder aluminum structure. The white arrow in (b) marks the position of one of the brass screws.

3. Tests and results

In order to define the real performance of a DRCM, we produced and tested different prototypes using HOPG crystals with a typical mosaic spread of 0.5°. Neutron tests were performed at the Institut Laue Langevin (Grenoble, France) using the beamlines IN3 and T13A, and at the Laboratoire Léon Brillouin (Saclay, France) with the cold beam of the G4.3 three-axis spectrometer.

3.1. Transmission tests

Figures 1(b) and 1(c) show neutron radiographies obtained on IN3 using a CCD camera coupled to a thin (0.2 mm) lithium glass scintillator. Images were acquired in transmission geometry, using a monochromatic incident beam. The contours of the thickest parts of the aluminum cylindrical structure are highlighted by white lines: the vertical lines depict the external walls of the cylinder, the U-shaped lines delimit one of the inner parts hosting a screw. As expected, when the crystal is off Bragg reflection like in Fig. 1(b), the cylinder is almost completely transparent and the only absorbing detectable part is the horizontal brass screw marked by a white arrow. Conversely, when the graphite crystal is set into Bragg reflection, like in Fig. 1(c), the majority of the incident neutrons are scattered away and the transmitted intensity drops over all the crystal surface. Brass screws were used in this prototyping phase to allow frequent assembling and disassembling of the system. However, in a final design, the same aluminum alloy employed for the cylindrical holders would be used everywhere in the beam area, including screws. This is also useful to reduce the radiological activation of the device.

3.2. Crystal alignment

The alignment of the different crystals was checked using the test diffractometer T13A with an incident wavelength $\lambda_i = 2.27$ Å. Rocking curves were acquired along the cylinder vertical axis with a vertical step $\Delta h = 10$ mm. Results are reported in the intensity map of Fig. 2(a). The crystals appear well-aligned and the peak intensity shows only minor fluctuations that are compatible with the crystal size and with the presence of the brass screws. This qualitative impression is confirmed by the results of a Gaussian fit of the rocking curves. Fig. 2(b) and (c) show the peak position and intensity along the length of the holding cylinder. The average position is $108.63° \pm 0.02°$, that is a very good alignment compared to the typical rocking curve width of about $0.5°$. The same results were obtained for the other prototype devices, thus confirming that the holding system is valid and perfectly suitable to build a large-area device.

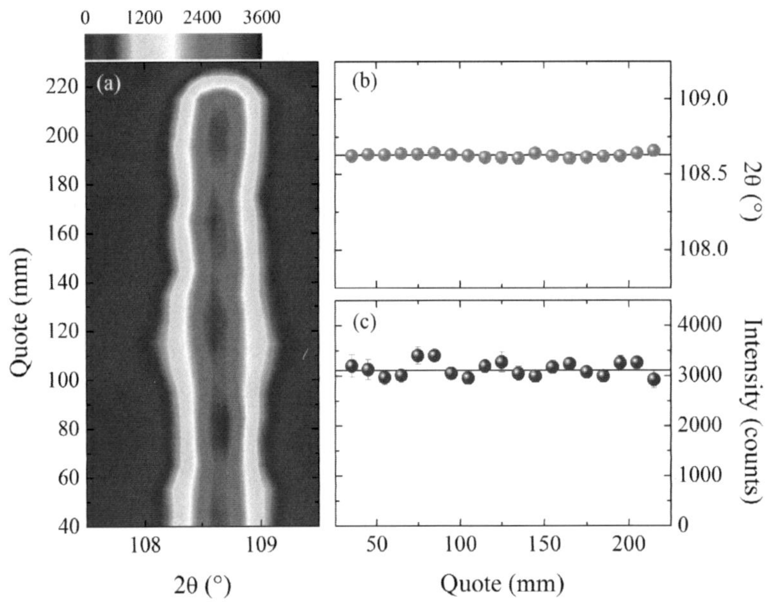

Figure 2. Intensity map produced from the rocking curves measured as a function of the vertical position along one of the cylinders (a). Measurements are performed with 10 mm steps. Peak angular position (b) and intensity (c) as a function of the height along the cylinder axis, as obtained by a Gaussian fit of the rocking curves.

3.3. Phasing and ToF measurements

An *in-operando* test of a complete DRCM was performed on the IN3 spectrometer, by acquiring time-of-flight (ToF) spectra. The test setup is shown in Fig. 3(a). It includes a single rotating monochromating element (M1), followed by an assembly of three rotating monochromating elements, like those depicted in Fig. 1(a), that constitutes the second monochromator (M2).

Each cylinder was rotated using a stepper motor having a step of 1.8° with a stability better than 0.09°. The rotation velocity was driven by a set of computer-controlled signals generated by means of a FPGA card (National Instrument single board RIO-9606) via the Labview environment. The same card was employed for the neutron ToF data acquisition. The use of the stepper motors coupled to the FPGA card has the advantage of a very precise control of the motor rotation, provided that the motors are correctly powered. The control clock of the system was operated at 40 MHz, thus allowing to drive the motor phase with time steps of 25 ns. Such short time steps ensure a more accurate phasing of all the rotating cylinders in the DRCM. Since the motors were controlled by a single clock, there was no relative difference in the velocity of the various motors and the overall phasing was well undemanding. This is a very important issue, as it allows a straightforward implementation of a device made up of several rotating elements.

The first monochromator was installed at the IN3 sample position, downstream a 0.67° collimator. Conversely, the second element of the DRCM replaced the spectrometer analyzer. In this configuration the distance between the two monochromators was about 1.12 m. Neutrons were acquired using the IN3 detector – a vertically-mounted 3 cm-diameter ^{3}He tube located at 2.02 m from the sample position – using a single-event ToF acquisition system [10]. As rotation control and acquisition exploited the same clock signal, there was no need for a ToF trigger from the motors to the acquisition system.

The fundamental wavelength selected by the IN3 graphite monochromator was $\lambda_1 = 4.26$ Å, that corresponds to a Bragg angle of 39.4°. Fig. 3(b) shows the ToF spectrum after the first rotating monochromator with $f_M = 28$ Hz. In addition to λ_1, three further harmonics are visible: $\lambda_2 = 2.13$ Å, $\lambda_3 = 1.42$ Å, and $\lambda_4 = 1.065$ Å.

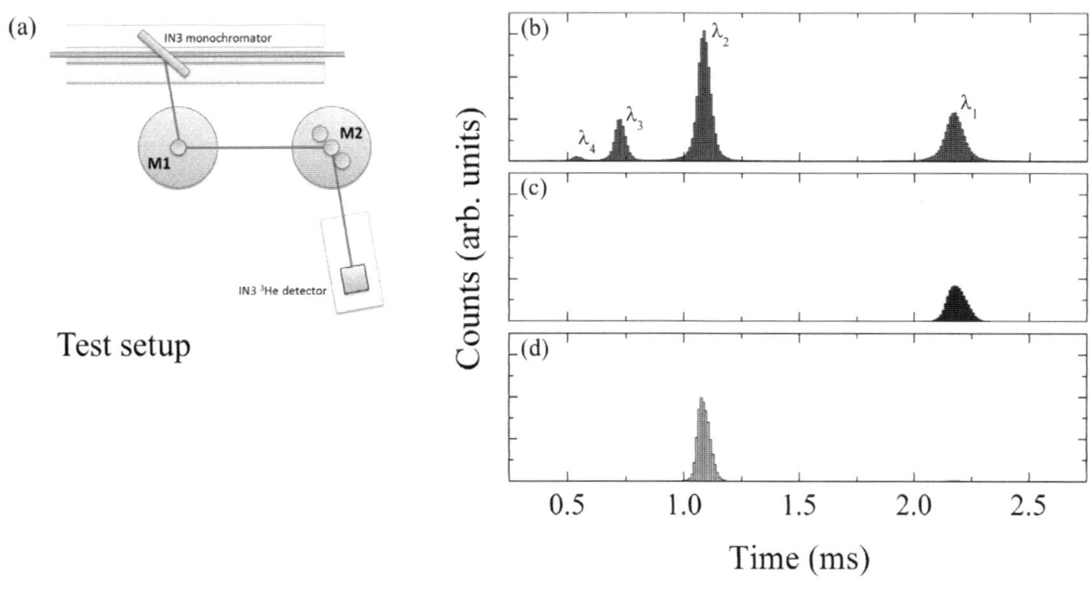

Figure 3. (a) Setup employed to test the DRCM. M1 and M2 indicate respectively the first and the second rotating-crystal monochromators, as explained in the text. ToF histograms measured at $f_M = 28$ Hz downstream M1 (b) and M2 ((c) and (d)). In (b) and (c), the second rotating-crystal monochromator was phased on λ_1 and λ_2 respectively.

Figs. 3(c) and (d) show two typical ToF spectra measured downstream the second rotating monochromator, that is using the whole DRCM. By phasing the first and the second monochromators an efficient selection of the different wavelengths can be achieved, even if the distance between the two monochromators is relatively short. All the data shown in Fig. 3 were acquired in fixed-event-number configuration, with the number of counts fixed to (2.5×10^5). The present acquisition system also allows operation in fixed-time mode, or in fixed-event-number mode controlled by a source other the acquisition detector itself, e.g. a beam monitor. ToF runs lasted several hours and showed that the phase stability is perfect. In fact, the relative phases of the various cylinders remained constant without any need of adjustment during each run. Similar tests, performed at the G4.3 beamline of the Laboratoire Léon Brillouin with wavelengths of 4.65 and 2.33 Å, fully confirmed the previous results of IN3.

4. Conclusions

The presented tests show the feasibility of a large-area rotating crystal monochromator. The flexibility of the proposed design allows to easily modify the size of the monochromator by adding or removing a rotating element. Similarly, the use of stepper motors coupled with a FPGA card allows to easily control and phase a large number of rotating devices, keeping their relative velocity constant up to frequencies of about 100 Hz (6000 rpm).

It is worth noting that HOPG is only one of the possible candidate materials for the crystals. Indeed, the HOPG does not allow for mosaic spreads narrower than 0.5°. Other possible candidates with this respect are Si, Cu, Ge and Pb. However, HOPG is known to provide the best performance in terms of reflectivity and quality of the beam.

Finally, the described tests also show that there are no fundamental mechanical problems in attempting to implement a time-focusing method, based on a large-area DRCM. On a long-pulse source, such as the forthcoming European Spallation Source, a spectrometer exploiting this concept could collect a large time slice of the source neutron pulse and thus allow significant intensity gains at the sample.

Acknowledgments

The presented results were part of the *Tempus Fugit* project, which has been supported by Elettra-Sincrotrone Trieste S.C.p.A. through a special grant of the Italian Ministry of Instruction, University and Research (MIUR). The Institut Laue Langevin and the Laboratoire Léon Brillouin are gratefully acknowledged for the time granted at their beamlines.

References

[1] Brockhouse B N, *Bull. Am. Phys. Soc.* 3,233 (1958).
[2] Harling O K, *Rev. Sci. Instrum.* **37**, 697 (1966).
[3] Harling O K, *J. Chem. Phys.* **50**, 5279 (1969).
[4] H. Tietze, W. Schmidt, R. Geick, *Physica B* **156**, 550 (1989).
[5] Kleb R, Ostrowski G E, Price D L, Rowe J M, *Nucl. Instr. and Meth.* **106**, 221-230 (1973).
[6] Copley J R D, Rowe J M, *Phys. Rev. Lett.* **32**, 49 (1974).
[7] Byrne J 2011 *Neutrons, Nuclei and Matter, an Exploration of the Physics of Slow Neutrons* (Mineola, New York: Dover publications Inc.), p 564.
[8] In preparation.
[9] Riste T, Otnes K, *Nucl. Instr. and Meth. A* **75**, 197 (1969).
[10] Casinini F, Petrillo C, Sacchetti F, *Nucl. Instr. and Meth. A* **675**, 1 (2012).

CASCADE - a multi-layer Boron-10 neutron detection system

M Köhli[1], M Klein[1,5], F Allmendinger[1], A-K Perrevoort[1], T Schröder[2], N Martin[2,3], C J Schmidt[4,5], and U Schmidt[1]

E-mail: koehli@physi.uni-heidelberg.de

[1] Physikalisches Institut, Heidelberg University, Im Neuenheimer Feld 226, D-69120 Heidelberg, Germany

[2] Heinz Maier-Leibnitz Zentrum (MLZ) and Physik Department, Technische Universität München, Lichtenbergstr. 1, 85748 Garching, Germany

[3] CEA, Centre de Saclay, DSM/IRAMIS/Laboratoire Léon Brillouin, 91191 Gif-sur-Yvette, France

[4] GSI Detector Laboratory, Planckstr. 1, 64291 Darmstadt, Germany

[5] CDT CASCADE Detector Technologies GmbH, Hans-Bunte-Str. 8–10, 69123 Heidelberg, Germany

Abstract. The globally increased demand for helium-3 along with the limited availability of this gas calls for the development of alternative technologies for the large ESS instrumentation pool. We report on the CASCADE Project - a novel detection system, which has been developed for the purposes of neutron spin echo spectroscopy. It features 2D spatially resolved detection of thermal neutrons at high rates. The CASCADE detector is composed of a stack of solid ^{10}B coated Gas Electron Multiplier foils, which serve both as a neutron converter and as an amplifier for the primary ionization deposited in the standard Argon-CO_2 counting gas environment. This multi-layer setup efficiently increases the detection efficiency and serves as a helium-3 alternative. It has furthermore been possible to extract the signal of the charge traversing the stack to identify the very thin conversion layer of about 1 micrometer. This allows the precise determination of the time-of-flight, necessary for the application in MIEZE spin echo techniques.

1. Introduction

The world of neutron detection has changed. Much of what once was established technology has been discarded. For them now substitutional ones have been presented. It began with production of tritium and peaked at the crisis of helium-3. Part of it was given to sciences for basic or applied research. Part for the industry, explorating oil deep in the rocks. And the largest part was given to homeland security, which above all else demanded for it for the protection against hazards. After the stockpile was nearly exhausted in 2008[1], concerns over the future supply[2] critical to perspectives of the European Spallation Source[3] led to developments of replacement technologies, most of them adapted from particle physics. The CASCADE neutron detection unit[4] is such a new generation system, which was designed specifically for the purposes of Neutron Spin Echo (NSE) spectroscopy[5]. This method and its successors, Neutron Resonance Spin Echo (NRSE)[6] and MIEZE (Modulation of IntEnsity by Zero Effort)[7], demanded a highly granular and time resolved detector to be operated efficiently at high rates. The CASCADE design is based on a combination of solid boron-10 coatings in

Journal of Physics: Conference Series **746** (2016) 012003 doi:10.1088/1742-6596/746/1/012003

several layers, modern gas amplification stages, a microstructured readout, multichannel ASICs (Application Specific Integrated Circuit) and FPGA (Field Programmable Gate Array) hardware triggered data acquisition. It is successfully in operation at the Forschungs-Neutronenquelle Heinz Maier-Leibnitz at the instruments RESEDA[8][9] and MIRA[10].

Figure 1. The CASCADE detector (without housing and shielding): The 20x20 cm^2 active detection volume behind the aluminum window is composed of 6 layers of boron-10 coatings and the central double sided readout. It is flushed by a gas mixture of Argon:CO_2. The 5 CIPix charge sensitive preamplifiers with 64 channels each are mounted on a Virtex-2 data acquisition board. Additionally pulse height spectra can be acquired from each channel. The measurements can be transmitted optically as compressed raw data or stored onboard as histograms by the event analysis firmware.

2. The CASCADE Detector

The CASCADE Detector consists of several layers of boron-10 coated Gas Electron Multipliers (GEM)[11]. Thermal neutrons convert to charged ions via the reaction $^{10}B + n \rightarrow ^7 Li + ^4 He$, which are emitted back to back at a total kinetic energy of up to 2.8 MeV. These particles ionize the counting gas Argon:CO_2 (70:30-90:10). Subsequently the electrical charge is then projected towards a common readout.

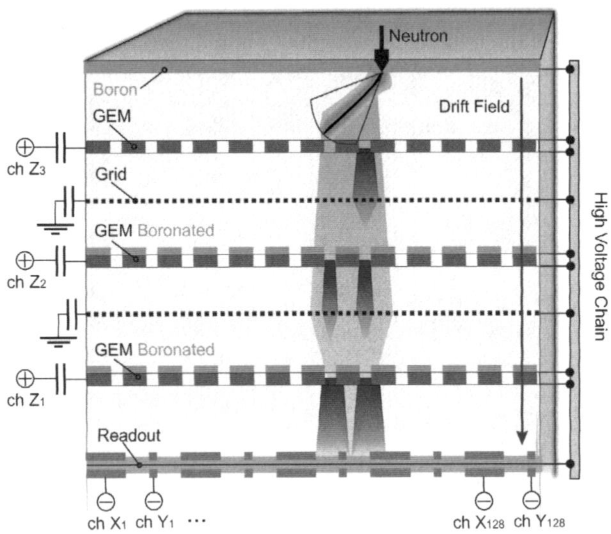

Figure 2. Schematic of the CASCADE detector of one half space: Thermal neutrons are converted in boron layers (orange). The conversion products ionize the counting gas and the electrons are projected towards a double sided readout of 128 channels in x and y each in crossed stripes of 1.57 mm width. Additionally the signal induced by the charge traversing the stack allows the identification of the conversion layer. In order to decouple the GEMs electrically and to limit the generated charge, grids are inserted in-between each stage.

The GEMs act as well as a substrate for the boron coating as they are 'charge transparent', which means that electron losses on the surface are compensated by gas amplification. GEMs are made of an insulator covered on both sides by copper. In etched holes of 0.05 mm diameter a high field gradient between the conducting surfaces causes Townsend avalanches leading to a high gas gain. The opacity of 88 % therefore requires at least a real gain of approximately 10. Further features of GEMs, as compared to wire based systems, the good ion backflow suppression, the high rate capability and the decoupling of drift field and gas gain. The standard spacing of the holes of 0.18 mm can be varied in the production process allowing for very high spatial resolution. In the CASCADE Detector presented here 6 layers divided into two half spaces are installed, see Fig 2, whereas in the top layer the drifts instead of the GEMs are coated. The coating thicknesses of 99 % enriched ^{10}B for the RESEDA configuration according to the measurements of CDT CASCADE Detector Technologies GmbH are from top to bottom in micrometers: 1.5, 0.8, 1.0, 0.95, 0.8, 1.0 (originally 2.0 before 07/2013). The mean remaining energy of the conversion products to ionize the gas depends on the layer thickness. The charge is projected towards a common double sided readout, which is a printed circuit board of meander shaped crossed stripes. Each side of 20 cm is divided into 128 segments of 1.57 mm width. Additionally to the 2D spatial readout it has been possible to measure the signal of the charge traversing the stack induced on the GEMs. For this reason there are metal meshes (grids) inserted in-between the layers. They decouple the GEMs electrically from each other and reduce the total amount of charge as the GEMs have to be operated at an effective gain larger than 1 to achieve clear signatures due to the large readout capacity and weak signals. The same ASIC is used for the GEM channels, which are called the z-coordinate. As the solid boron coatings are very thin, the conversion layer identification leads to a very high time-of-flight resolution.

A high integration density and charge sensitivity is achieved by using the 64-channel CIPix ASIC[12], which was developed on the basis of the HELIX128 chip as a readout amplifier for the inner tracker of the H1 detector at HERAIt is operated at a constant frequency of 10 MHz leading to a time resolution of 100 ns. In total 5 CIPix boards are present in the system, whereas 4 are reserved for the (x,y) readout and one for the z coordinate. The digital signals are handled by a Virtex-2 FPGA board[1], which can either produce a zero-suppressed raw data output or it can analyze the data patterns by firmware algorithms in the triggered event mode. This generates and stores ready-to-use histograms, for example for stacking time-of-flight measurements. In the Spin Echo mode the Spin Echo group pattern can be followed synchronized to the coil frequency by means of a phase locked loop unit. For a MIEZE instrument this detector shows competitive performance to other NRSE setups. An efficiency of approximately 25 % for thermal neutrons and a spatial resolution of $\sigma = 1.45$ mm have been measured. At 16x sampling of the MIEZE modulated signal frequencies up to 625 kHz can be used.

3. Conversion products and their energy signatures

The thermal neutron capture leads the boron to fragment into a helium ion and the rest of the nucleus, which is referred to as a lithium ion. They either receive the full kinetic energy of 2.8 MeV with a probability of 6.4 % or an additional photon of 0.48 MeV is produced reducing the phase space accordingly. Whereas the helium ion can be detected easily, the lithium ion not only has less kinetic energy due to its higher mass but also the energy loss in the medium is higher. Therefore the range of these ions in boron is less than half compared to the 4 μm of helium. Simulations of the energy spectra of the particles after leaving the converter were done by the Monte-Carlo code URANOS (presented in [13]), which is linked to the stopping range calculations of SRIM . The results are shown in Fig. 3 for thicknesses of (100-1900) nm. Due to the stopping power the spectra extend towards lower energies for increasing coating layers. The helium ion signature can be detected far above 2000 nm, whereas the lithium ion does not contribute any more to the efficiency beyond 1500 nm. Additionally to the work presented here, analytical descriptions of the spectra have been presented in [14] and calculations focusing on the efficiency of multi-layer systems have been proposed in [15]. For our specific case we measured these spectra in a test detector with a single boron coated GEM and a large drift volume. Fig. 4 shows as an example the pulse height spectra for three different gas gains and in comparison

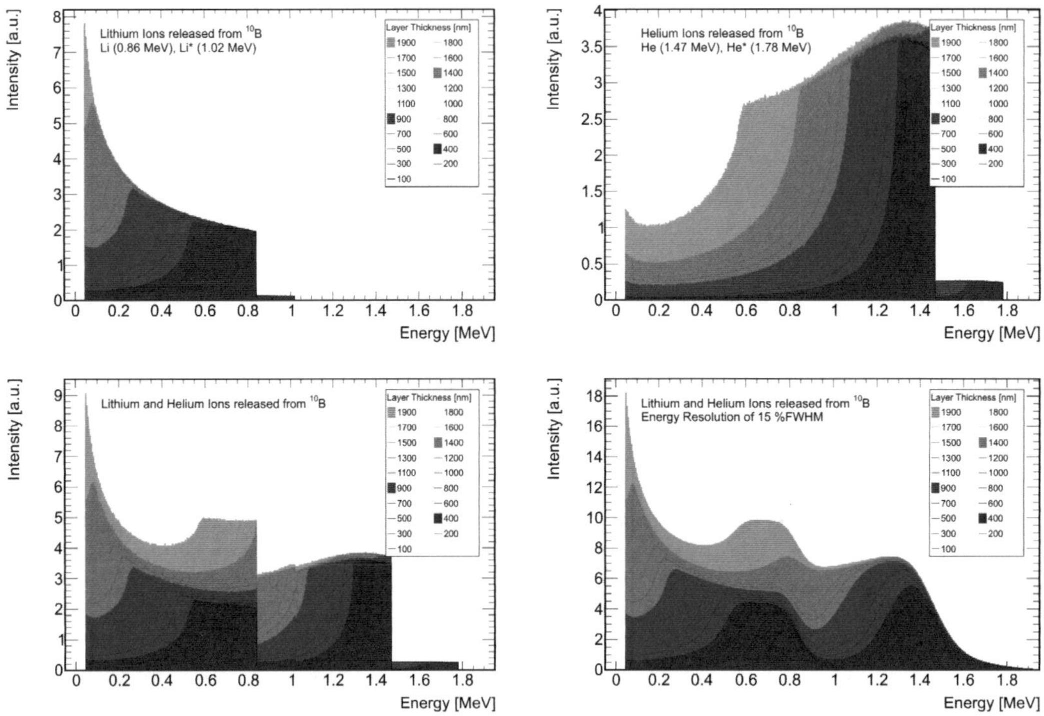

Figure 3. Simulation of the energy spectra of conversion products leaving boron-10 layers of (100-1900) nm, colored in groups of 500 nm. (top) Separated spectra by particle type: lithium (left) and helium (right). (bottom left) Combined spectra and (bottom right) combined spectra convoluted by an energy resolution of 15 % FWHM.

the simulated spectra of Fig. 3, which have been convoluted by a Gaussian resolution function of the obtained peak full width half maximum.

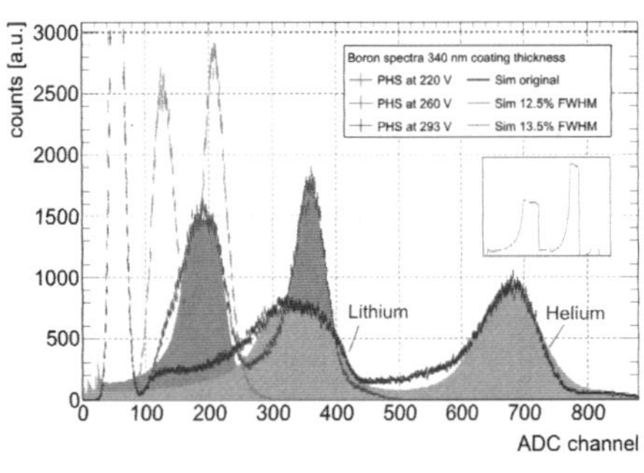

Figure 4. Pulse height spectra of the test detector, composed of a single coated GEM and the readout, at 3 different gas amplifications. The data is compared to results from the simulation of the system with the original spectrum (inlay) being convoluted by a Gaussian energy resolution function.

The gas ionization energy depositions here are slightly shifted downwards as the tracks emerging from the GEM are partly tallied by projection onto the surface instead of into a hole. In the CASCADE detector these spectra will be tallied approximately at 500 keV of energy deposited in the gas. The track length of the conversion products of about 10 mm in Argon:CO_2 is much higher than the available space between the layers of (2-4) mm. In Fig. 5 a simulation of the cumulated energy deposited in the detector at RESEDA before 2013 is shown. The tail below 200 keV comes from the thicker coatings on the front and back. For calculating the total

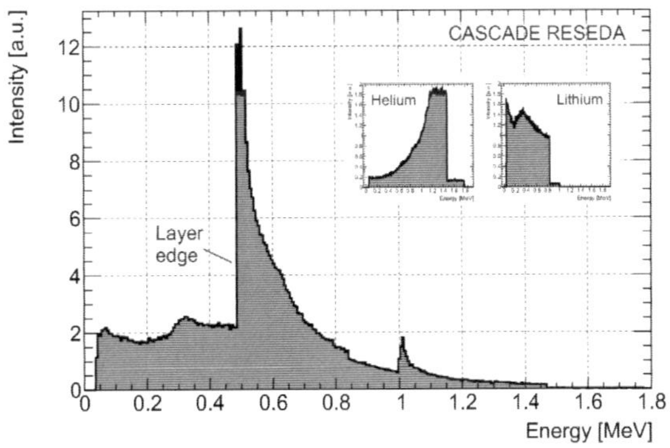

Figure 5. Simulation of the gas energy deposition of the conversion products for a CASCADE detector in the RESEDA configuration. The inlays show the energy of the particles escaping the boron coatings. As different layer thicknesses are present in the detector, the distributions are not uniform. The geometry of each layer limits the track length and therefore the available energy.

neutrons measured additionally a detection threshold of (50-100) keV has to be considered. This especially has a non negligible effect as there is a pile up in the spectra on low energies for larger converter thicknesses. The digitized signals of the CIPix ASIC do not feature a simultaneous time over threshold (TOT) or pulse height measurement, instead the charge sensitive channels require a certain time to read out a specific amount of charge. For this reason it takes several clock cycles of 100 ns to process signals present in the detector. This allows for a pseudo-TOT measurement of the charge, which is primarily important for the z-channels as the firmware algorithms can for example identify crosstalk between the layers based on this information. In normal operation the signals of the RESEDA detector as like depicted in Fig. 6. The panel shows the signal lengths of charge propagating through the stack for each of the three layers.

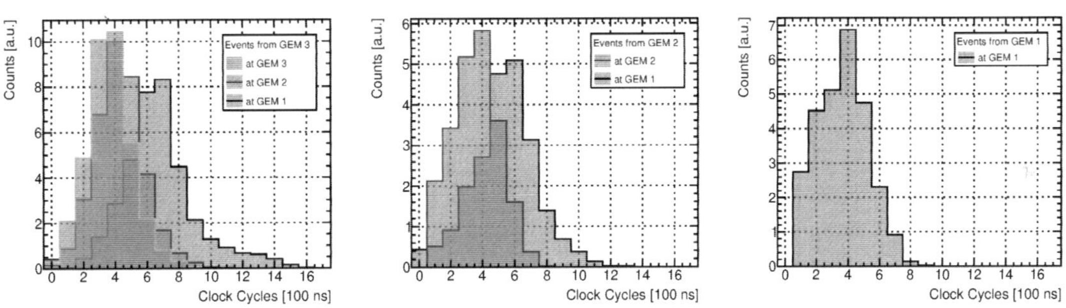

Figure 6. Raw data spectra in units of clock cycles to read out the z-channels of the charge traversing the stack at a gain voltage of 350 V per GEM of the RESEDA detector. GEM 3 denotes the layer next to the drift, the charge of a neutron event can traverse the whole stack and is measured in three layers. GEM 2 and GEM 1 are located in drift direction towards the readout.

The left graph shows neutron events originating from the boron coating of the drift cathode. As the effective gas gain is larger than 1, the amount of charge increases each layer. It has to be noted that the average pseudo-TOT is (40-50) ns and some events take up to 150 ns. A simple (x,y,z) correlation would limit the rate capability. For this reason the firmware algorithm analyzes the spatial and temporal structure of the data and dynamically masks active channels. As a result the long events in the time distribution can be separated and their effect on the rate capability marginalized. Furthermore as far as two consecutive neutrons do not convert in the same layer their origin can be effectively reconstructed well below the average pseudo-TOT.

Journal of Physics: Conference Series **746** (2016) 012003 doi:10.1088/1742-6596/746/1/012003

4. The detector in Spin Echo mode

Spin Echo measurements, especially MIEZE techniques, require a very high spatial and time resolution. The oscillating interference pattern challenges detector technologies, as the Spin Echo group, which is depicted in the example in Fig. 7, can have an extension in space in the order of only millimeters and has to be sampled in the time domain in the order of MHz. Recent user based measurements for example can be found in [16].

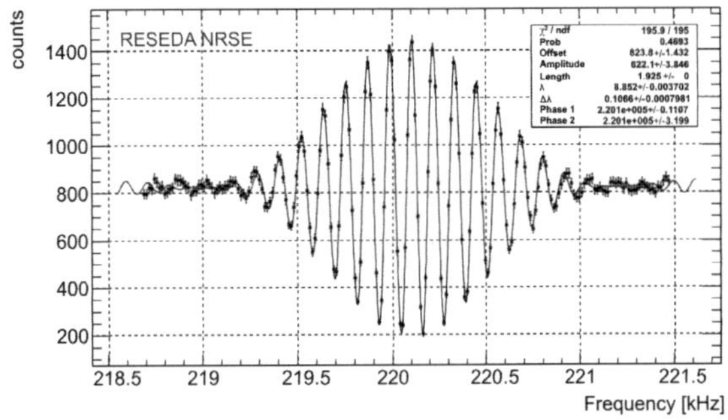

Figure 7. Example of a Neutron Resonance Spin Echo (NRSE) Group measured by the CASCADE detector at RESEDA. By changing the frequency the interference pattern is moved spatially. At 8.85 Å and $\Delta\lambda/\lambda \approx 11\,\%$ the polarization reaches 75 %. The color scale is normalized to the maximum intensity.

The multi-layer performance we want to illustrate by the capability of the detector to characterize itself in a neutron interference measurement. Whereas the data from Fig. 7 was taken in a standard NRSE setup, the following measurements were conducted in the MIEZE configuration. A thin graphite resolution sample was placed in the beam generating a homogenous illumination at the detector position. The intensity distribution integrated over the duration of the measurement is shown in Fig. 8 for each layer separately. The relative contribution of each channel depends on the boron thickness and the wavelength of the beam. As the cold part of the spectrum was used, most of the neutrons convert in the first layer.

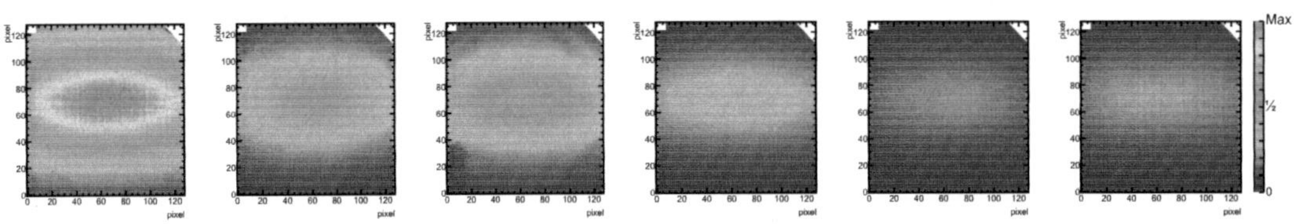

Figure 8. Intensity distribution separated by GEM channel (top layer left, bottom layer right). This MIEZE measurement at RESEDA was conducted at a frequency of 53.5 kHz and a wavelength of 8.05 Å.

In a MIEZE measurement the intensity distribution oscillates in space and time as the neutron phase travels through the detector. In the setup presented the frequency of the polarization change of 53.5 kHz is fed into the PLL and oversampled 16 times[1], which results in the detector following phase locked to the instrument at a readout frequency of 856 kHz. In every pixel on every layer of Fig. 8 the oscillating interference pattern of 9.2 mm wavelength can be measured. As an example Fig. 9 zooms into one small region and shows the time channel of each layer.

The resulting temporal sine function of the interference pattern is measured precisely. The following panel Fig. 10 shows the phase distribution across the detector at one specific point of

[1] To avoid jitter effects of the electronics, which would wash out the phase locked sine function and therefore reduce the measured polarization, the period is sampled in 16+1 time bins and afterwards recorrected. This relaxes the necessary precision to half a time bin on the interval boundaries and therefore reduces noise effects.

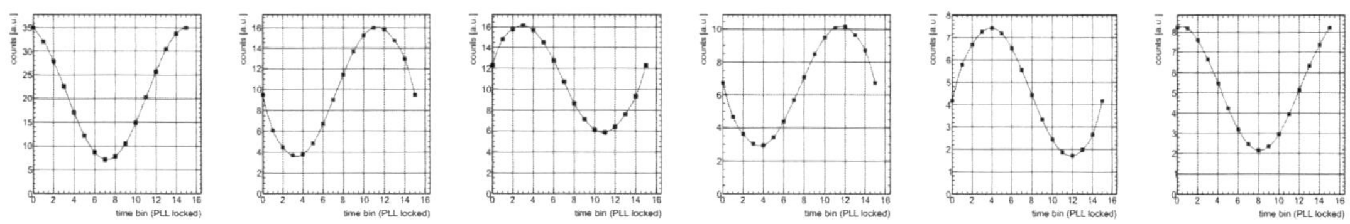

Figure 9. Measurement of the intensity distribution for each GEM channel summed over a region of 9 pixels. The MIEZE period length of 18.7 μs is oversampled by a PLL. The full cycle is divided into 16 subchannels. While the single point error is far below 1 %, the sine fit describe the data very well giving a reduced χ^2 of approximately 1.

time. The oscillation period can be followed through the layers in every pixel. This evidently shows that high spatial and time resolution is necessary to conduct this type of experiments.

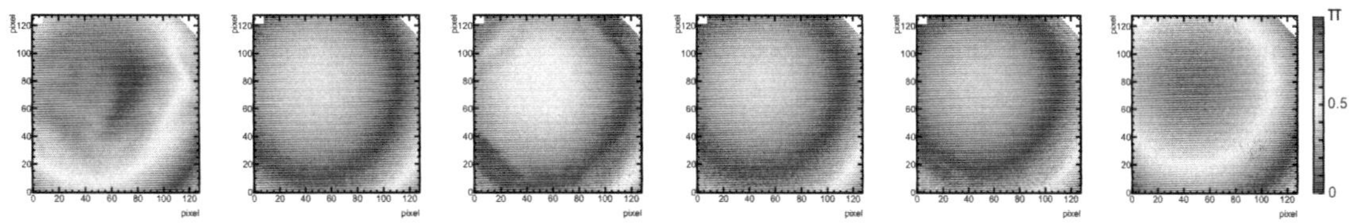

Figure 10. Phases of the neutron beam separated by GEM channel (top layer left, bottom layer right). This MIEZE measurement at RESEDA was conducted at a frequency of 53.5 kHz and a wavelength of 8.05 Å. The color code scales half a period from 0 to π, which equals 9.34 μs or 4.6 mm. The inner layers distances are approximately half a period.

We as well demonstrate how by an incidentally received bump on the front the detector can be aligned in a Spin Echo type of experiment. As the neutron speed and oscillation period are well known by the instrument setup, the phase is used to characterize the layers of the detection system spatially in respect to the beam axis and towards each other. As an example the front to back distance of the first and the last conversion layer is depicted in Fig. 11. As the bottom drift cathode is straight and therefore acts as a reference the displacement is attributed to the top layer. We can determine the bump depth of 1.7 mm with a z-axis resolution of 0.1 mm. For a shorter oscillation period length than what is available for this accidental measurement the analysis can easily be improved by a factor of 10. Using this time of flight principle it is possible to physically characterize and align the detector itself by the precision of the neutron interference pattern.

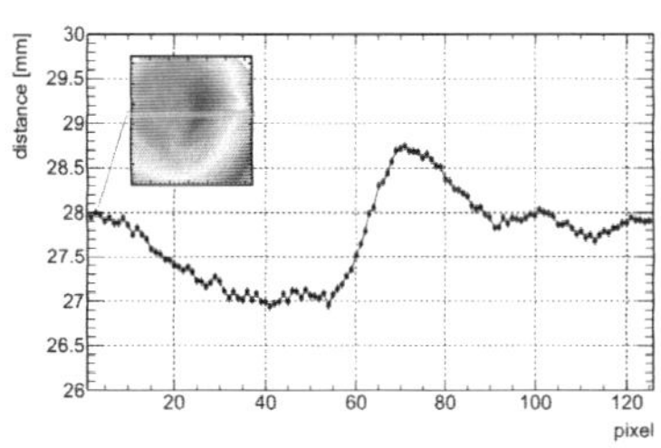

Figure 11. Profile view of the top to bottom displacement measurement by the neutron phase front. The cut view through the middle part of the detector depicts the bump on the drift cathode. The total displacement of 1.7 mm can be determined down to a precision of 0.1 mm.

5. Conclusion

The CASCADE detector has become an established device for Spin Echo measurements. Besides the high precision requirements of a NRSE instrument the detector proposes a design for a helium-3 replacement system by technology transfer from developments of particle physics. This paper presents the working multi-layer CASCADE implementation at RESEDA in the technical realization, electronics and firmware. The use of Monte-Carlo techniques to model the detector is necessary to understand the details of the underlying physics. In exemplary measurements we show the multi-layer operation in the Spin Echo mode by tracking the neutron interference pattern spatially and time resolved. The detector can be operated up to a rate capability of 2 MHz on its active surface of 20x20 cm^2. It features a time resolution of 100 ns and the spatial resolution has been determined to $\sigma = 1.45$ mm. The absolute detection efficiency of the detector at RESEDA is 21 % for thermal neutrons or 47 % at 5.5 Å. More details on the characteristics will be presented in further publications.

6. Acknowledgments

MK thanks W. Häußler for support during his active period as responsible at the RESEDA instrument at the FRM II and T. Weber as well as R. Georgii. This work has been funded by the German Federal Ministry of Education and Research (BMBF) for the project "Neutron Detectors for the MIEZE method", grant identifier 05K10VHA.

References

[1] Shea, DA and Morgan, D 2010 URL https://www.fas.org/sgp/crs/misc/R41419.pdf
[2] Feder T 2009 *Physics Today* **October** 21–23
[3] Peggs, S *et al.* 2013 *European Spallation Source Technical Design Report* Technical Design Report ESS (Lundt: ESS) ESS-2013-001 URL http://esss.se/documents/tdr/TDR_final.pdf
[4] Klein M and Schmidt C J 2011 *Nuclear Instruments and Methods in Physics Research Section A: Accelerators, Spectrometers, Detectors and Associated Equipment* **628** 9 – 18 ISSN 0168-9002 URL http://www.sciencedirect.com/science/article/pii/S0168900210014683
[5] Mezei F 1972 *Zeitschrift für Physik* **255** 146–160 ISSN 0044-3328 URL http://dx.doi.org/10.1007/BF01394523
[6] Golub R and Gähler R 1987 *Physics Letters A* **123** 43 – 48 ISSN 0375-9601 URL http://www.sciencedirect.com/science/article/pii/0375960187907602
[7] Gähler R, Golub R and Keller T 1992 *Physica B: Condensed Matter* **180** 899 – 902 ISSN 0921-4526 URL http://www.sciencedirect.com/science/article/pii/092145269290503K
[8] Franz C and Schröder T 2015 *Journal of large-scale research facilities* **1** URL http://jlsrf.org/index.php/lsf/article/view/37
[9] Schmidt C J, Groitl F, Klein M, Schmidt U and ler W H 2010 *Journal of Physics: Conference Series* **251** 012067 URL http://stacks.iop.org/1742-6596/251/i=1/a=012067
[10] Georgii R and Seemann K 2015 *Journal of large-scale research facilities* **1** URL http://jlsrf.org/index.php/lsf/article/view/21
[11] Sauli F 1997 *Nuclear Instruments and Methods in Physics Research Section A: Accelerators, Spectrometers, Detectors and Associated Equipment* **386** 531 – 534 ISSN 0168-9002 URL http://www.sciencedirect.com/science/article/pii/S0168900296011722
[12] Baumeister D 1999 *Entwicklung und Charakterisierung eines ASICs zur Kathodenauslese von MWPCs fü das H1 Experiment bei HERA* Master's thesis Physikalisches Institut, Heidelberg University
[13] Köhli M, Schrön M, Zreda M, Schmidt U, Dietrich P and Zacharias S 2015 *Water Resources Research* **51** 5772–5790 ISSN 1944-7973 URL http://dx.doi.org/10.1002/2015WR017169
[14] Klein M 2000 *Experimente zur Quantenmechanik mit ultrakalten Neutronen und Entwicklung eines neuen Detektors zum ortsaufgelösten Nachweis von thermischen Neutronen auf groen Flächen* Ph.D. thesis Physikalisches Institut, Heidelberg University
[15] Piscitelli F and Esch P V 2013 *Journal of Instrumentation* **8** P04020 URL http://stacks.iop.org/1748-0221/8/i=04/a=P04020
[16] Kindervater, J, Martin, N, Häußler, W, Krautloher, M, Fuchs, C, Mühlbauer, S, Lim, JA, Blackburn, E, Böni, P and Pfleiderer, C 2015 *EPJ Web of Conferences* **83** 03008 URL http://dx.doi.org/10.1051/epjconf/20158303008

Progress in High-resolution Neutron Imaging at the Paul Scherrer Institut – The Neutron Microscope Project

Pavel Trtik[1], **Eberhard H. Lehmann**
Neutron Imaging & Activation Group, Laboratory for Neutron Scattering & Imaging, Paul Scherrer Institut, CH-5232 Villigen PSI, Switzerland

Email: pavel.trtik@psi.ch; ptrtik@gmail.com

Abstract. Here we report the recent advances in the Neutron Microscope project at the Paul Scherrer Institut. We demonstrate the recent improvement on the capability of neutron imaging that allows us to acquire neutron images with isotropic spatial resolution of about 5 micrometres.

1. Introduction

Neutron imaging of the highest possible spatial resolution is a major demand of the neutron imaging user community with several scientific disciplines applying this technology. However, while X-ray imaging can be performed routinely with µm and even sub-µm spatial resolution even using table-top instrumentation, the spatial resolution of isotropic neutron imaging has long been limited to the deca-micrometre domain [1-4].

The sub-10 micrometre spatial resolution neutron imaging has been hitherto routinely available (in one direction only) in the case of anisotropic imaging by using a tilted detector setup (Boillat et al. [5]). There are also hybrid detectors that apparently allow for sub-10 µm spatial resolution of neutron imaging; however, their performance is currently rather limited due to their temporal resolution [6]. Another interesting development that may provide a sub-10 µm spatial resolution is based on axisymmetric focussing mirrors (so-called Wolter optics) that promise possible "dramatic

[1] Address for correspondence:
Dr Pavel Trtik, Neutron Imaging & Activation Group, Laboratory for Neutron Scattering & Imaging, Paul Scherrer Institut, CH-5232 Villigen, Switzerland
Tel: +41-56-310-5579; Fax: +41-56-310-3131
Email: pavel.trtik@psi.ch, ptrtik@gmail.com

improvements in the signal rate and (spatial) resolution" [7]. However, the current resolution of cold-neutron Wolter optics is reported to be on the scale of many deca-micrometres ([7-10]).

A recent addition to high-resolution neutron imaging systems is a microscope developed within the "Neutron Microscope Project" at PSI by Trtik et al. [11]. The first prototype of the microscope (Neutron Microscope 1.01) uses a high numerical aperture magnifying lens, which projects the image from a 4-μm thick gadolinium oxysulfide (Gd_2O_2S:Tb) scintillator screen onto a high-resolution pixelated detector magnified by a factor of 4.3. Because of this magnification factor, the pixel size of the neutron images made with the Neutron Microscope 1.01 reached 1.5 μm, which translates to a true spatial resolution of 7.6 μm [11]. Therefore, the first prototype of the Neutron Microscope delivered nearly fourfold enhancement of the spatial resolution compared to the standard neutron imaging available at PSI [4]. It should be highlighted here that, despite its name, the Neutron Microscope we are currently developing is not *per se* a neutron microscope that uses neutrons for magnification, because we do not focus neutrons using our device. Our microscope can be described more precisely as a "very high resolution, very high-numerical aperture optical microscope with the light input based on the high-performance neutron-sensitive scintillators".

In this paper, we report on a very recent enhancement of high-resolution neutron imaging capabilities within the "Neutron Microscope project".

2. Status and preliminary results

For the further enhancement of our "Neutron Microscope", we employ a tailor-made dedicated objective with an even higher (five-times) magnification factor and link it with recently developed isotopically-enriched 157-gadolinium (capture cross-section 254,000 barn) oxysulfide scintillator screens [12].

The dedicated objective is composed of thirteen individual lenses. As the objective is to be intrinsically exposed to scattered neutrons and thus possibly suffer from radiation damage, the first three lenses (i.e. the ones closest to the scintillator) are made of a radiation hard glass. A schematic of an axial cross-section is shown in Figure 1. The objective is designed to provide a spatial resolution better than 5 μm, while keeping the light transmission above 70% for the light spectral range of terbium doped gadolinium oxysulfide.

Fig. 1: Schematic of cross-sectional view of the PSI Neutron Microscope showing (from right to left): the sample holder (pink), adaptor with the scintillator screen (green and yellow), the dedicated magnifying objective composed of 13 individual lenses (light grey), and a CCD detector (red and black).

The dedicated objective is designed for a field of view of 10 mm x 10 mm and the object distance is 60 mm. The scintillator screen is based on the recently developed isotopically-enriched 157-gadolinium oxysulfide ($^{157}Gd_2O_2S$:Tb; hereafter referred to as 157-gadox). The thickness of the scintillator is approximately 3.5 μm, while the mean particle size of the 157-gadox powder was approximately 2.0 μm.

The magnifying objective linked with the isotopically–enriched 157-gadox scintillator was preliminarily tested at the ICON beamline, PSI Villigen. The test arrangement was positioned slightly downstream from measuring position No. 2 (L = 9.0 m). A full polychromatic beam with a 20 mm neutron aperture was utilized. Therefore, L/D ratio equalled 450.

The new spatial resolution test-object (a gadolinium Siemens Star) was developed in-house with the size of individual spokes as small as 4 μm. The test object was imaged using the Neutron Microscope and a sCMOS detector of a pixel size of 6.5 μm (Hamamatsu ORCA Flash 4.0). Given the magnification power of the objective, the nominal pixel size of acquired images is 1.3 μm. Figure 2 shows the neutron image of the mentioned gadolinium Siemens Star test object. The images presented in Figure 2 are based on 33 single frames each with an exposure time of 30 seconds, giving a total exposure time of 16.5 minutes.

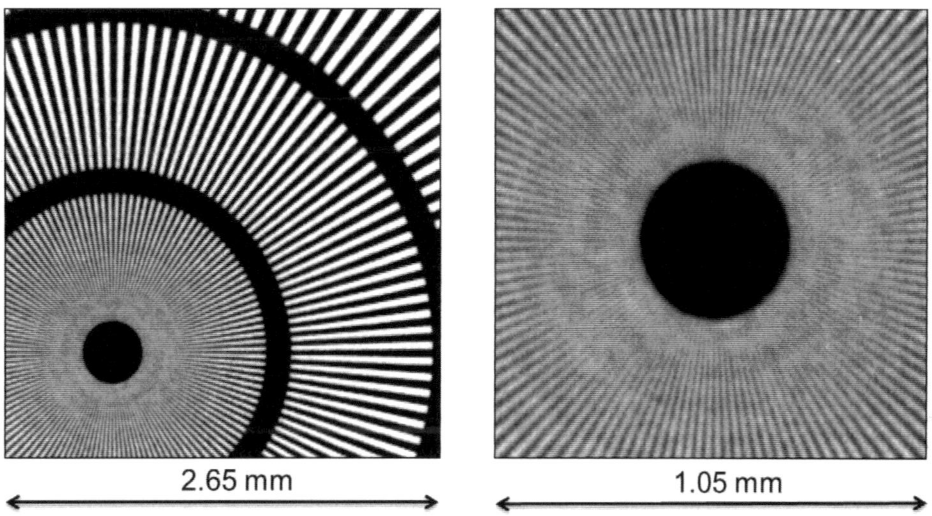

Fig. 2: Neutron radiography of the Siemens Star test object made with the PSI Neutron Microscope. The full field of view is shown on the left, and the right is zoomed-in to the high-resolution area. The Fourier ring correlation (FRC)-based spatial resolution of the image is equal to 5.4 μm.

The spatial resolution of the images was assessed using the Fourier ring correlation (FRC) technique [13, 14] and was calculated to be 5.4 μm. However, as can be seen in Figure 2, some (albeit faint) modulation was revealed at the thinnest parts of the spokes (4 μm). When compared to the spatial resolution available from the Neutron Microscope prototype 1.01, the spatial resolution improved by about 30%.

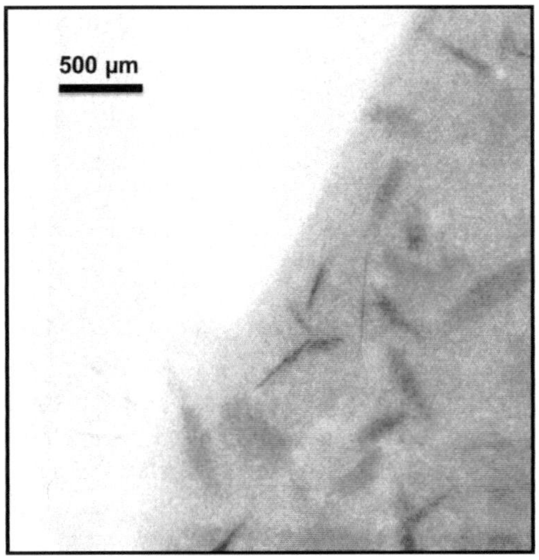

Fig. 3: Neutron radiograph of 'Wood's metal' alloy based on the Neutron Microscope test arrangement. The image clearly shows dark contrasting segments representing the cadmium containing microstructural features.

In addition to the test object, we also imaged a thin sample (about 0.3 mm) of the Wood's metal alloy. The mean distance between the sample and the scintillator was approximately 0.5 mm. The Wood's metal is a eutectic alloy composed of bismuth, lead, tin, and cadmium (about 10%) with a melting point of approximately 70°C. As bismuth, lead, and tin are rather neutron-transparent, the vast majority of the contrast in the neutron image is therefore based on cadmium (capture cross-section 2,450 barn). The neutron radiography clearly revealed needle/platelet-like structures of cadmium in the microstructure of Wood's metal alloy. Dynamic studies of relatively slow processes of solidification/segregation of such eutectic alloys is therefore one of the application fields foreseen for the Neutron Microscope instrument.

3. Outlook & discussion

The neutron capture efficiency in thin gadolinium-based scintillators has been recently improved by the provision of the isotopically-enriched version of this phosphor [12]. Next, we will attempt to produce microstructured scintillators based on this phosphor. The challenge of production of such scintillator screens lies in the limited amount of the input materials (both for the production of the isotopically-enriched phosphor itself and the manufacturing of the scintillator screens using such isotopically-enriched phosphors).

In order to incorporate the Neutron Microscope instrument into the user-programme of Paul Scherrer Institut, a new sample positioning stage with superior accuracy must be designed and implemented. The plan is to operate the Neutron Microscope device as a self-contained instrument independent from the current operating infrastructure of the PSI neutron imaging beamlines. This plan has two major consequences:

(i) it will be possible to transfer the Neutron Microscope between different beamlines with only a moderate effort,

(ii) due to the limited field of view of both the Neutron Microscope and the standard high-resolution setup (MICRO-setup) with respect to the size of the beam (e.g. at the measuring position 2 at ICON beamline), it is foreseen that both of the devices can be operated at the same time, thus utilizing the available (and costly) neutron beamtime more effectively.

It is foreseen that high-aspect ratio neutron beam collimators based on microchannel plates will be developed for both conditioning the divergence of the primary neutron beam and to suppress the signal from scattered neutrons [15]. As these collimators will lead to a decrease in the useful neutron beam intensity, the more efficient scintillators, and the tailored magnifying optics (as described above) will prove very important for the further successful development of the Neutron Microscope facility. In addition, the temporal resolution of imaging using the Neutron Microscope can be even further improved by using neutron sources with higher flux than that available at SINQ, PSI (e.g. ANTARES and potentially ESS and ILL).

4. Conclusions

In this paper, we reported on the progress of high-resolution neutron imaging at Paul Scherrer Institute under the Neutron Microscope Project. We demonstrated that the tailor-made high-numerical aperture magnifying optics linked together with a thin isotopically-enriched $^{157}Gd_2O_2S$:Tb scintillator screens can provide images with a spatial resolution of about 5 μm, which translates to approximately 30% enhancement in spatial resolution compared to the first Neutron Microscope prototype 1.01 [11].

5. Acknowledgements

The project is supported by financial contributions from SNF R'Equip, CCMX and PSI-CROSS funding bodies. The magnifying objective was designed in collaboration with Braunecker Engineering GmbH (Dr Bernhard Braunecker, Dr Massimo Biber) and produced at POG AG, Gera, Germany. Mr Heinz Richner and Mrs Gabrielle Peschke (ETH Zurich, Switzerland) are kindly thanked for the provision of the sample of Wood's metal alloy. We gratefully acknowledge manifold support for mechanical engineering by J. Hovind, M. Blumer, J. Welte and M. Meyer (all PSI). Our thanks also go to Ch. Grünzweig, A. Kaestner and P. Boillat (all PSI) for their submission of test devices and other important assistance with setting up the Neutron Microscope. Also, we kindly thank Ch. David and T. Neiger (both PSI) and the staff of Dr Gerhard Gassler GmbH for assistance with the production of scintillator screens. Last, but not the least, we would like to thank an anonymous reviewer for the kind expression of several comments that improved the quality of this paper.

[1] Kaestner, A.P, Hartmann, S., Kühne, G., Frei, G., Grünzweig, C., Josic, L., Schmid, F., Lehmann, E.H, 2011. The ICON beamline – A facility for cold neutron imaging at SINQ, Nuclear Instruments and Methods in Physics Research A. 659, 387–393

[2] Williams, S.H., Hilger, A., Kardjilov, N., Manke, I., Strobl M., Douissard, P.A, Martin, T., Riesemeier, H., Banhart, J., 2012. Detection system for microimaging with neutrons. Journal of Instrumentation 7, P02014

[3] Tremsin, A.S., Kardjilov, N., Dawson, M., Strobl, M., Manke, I., McPhate, J.B., Vallerga, J.V., Siegmund, O.H.W., Feller, W.B., 2011. Scatter rejection in quantitative thermal and cold neutron imaging. Nuclear Instruments and Methods in Physics Research Section A. 6

[4] Lehmann, E.H., Frei, G., Kuehne, G., Boillat, P., 2007. The micro-setup for neutron imaging: A major step forward to improve the spatial resolution. Nuclear Instruments and Methods in Physics Research A 576 (2-3): 389-396

[5] Boillat, P., Kramer, D., Seyfang, B., Frei, G., Lehmann, E., Scherer, G., Wokaun, A., Ichikawa, Y., Tasaki, Y., Shinohara, K. 2008. In situ observation of the water distribution across a PEFC using high resolution neutron radiography, Electrochemistry Communications, 10, 546–550, doi:10.1016/j.elecom.2008.01.018,

[6] Vavrik, D., Jakubek, J., Krejci, F., Zemlicka, J., Kaestner, A.P., 2014. Modular pixelated detector system for neutron imaging with micrometric resolution. In 10th World Conference on Neutron Radiography, Grindelwald, Switzerland

[7] Liu, D., Hussey, D., Gubarev, M. V., Ramsey, B. D., Jacobson, D., Arif, M., Moncton, D. E., and Khaykovich, B., 2013. Demonstration of achromatic cold-neutron microscope utilizing axisymmetric focusing mirrors, Applied Physics Letters, 102, doi:10.1063/1.4804178

[8] Trtik, P., Lehmann, E.H., 2013. Comment on "demonstration of achromatic cold-neutron microscope utilizing axisymmetric focusing mirrors" [Appl. Phys. Lett. 102, 183508 (2013)], Applied Physics Letters, 103, 236101

[9] Liu, D., Hussey, D., Gubarev, M. V., Ramsey, B. D., Jacobson, D., Arif, M., Moncton, D. E., Khaykovich, B., 2013. Response to "comment on 'Demonstration of achromatic cold-neutron microscope utilizing axisymmetric focusing mirrors'" [Appl. Phys. Lett. 103, 236101 (2013)] , Applied Physics Letters, 103, 236102

[10] Hussey, D., 2015. A Neutron Microscope for Energy and Materials Research, presented at VI European Conference on Neutron Scattering, ECNS2015, Zaragoza, Spain. September 2015

[11] Trtik, P., Hovind, J., Gruenzweig, Ch., Bollhalder, A., Thominet, V., David, Ch., Kaestner, A., Lehmann, E. H., 2015. Improving the spatial resolution of neutron imaging at Paul Scherrer Institut – The Neutron Microscope Project, Physics Procedia, 69, 169-176

[12] Trtik, P., Lehmann, E. H., 2015. Isotopically-enriched gadolinium-157 oxysulfide scintillator screens for the high-resolution neutron imaging, Nuclear Instruments and Methods in Physics Research Section A, 788, 67–70, doi:10.1016/j.nima.2015.03.076

[13] van Heel, M., Schatz, M., 2005. Fourier shell correlation threshold criteria. Journal of Structural Biology.151, 250–262

[14] Vila-Comamala, J., Diaz, A., Guizar-Sicairos, M., Mantion, A., Kewish, C.M., Menzel, A., Bunk, O., David, C., 2011. Characterization of high-resolution diffractive X-ray optics by ptychographic coherent diffractive imaging. Optics Express. 19, 21333-21344

[15] Tremsin, A.S., Kardjilov, N., Dawson, M., Strobl, M., Manke, I., McPhate, J.B., Vallerga, J.V., Siegmund, O.H.W., Feller, W.B., 2011. Scatter rejection in quantitative thermal and cold neutron imaging. Nuclear Instruments and Methods in Physics Research Section A. 651, 145-14

Vertical neutron beam focusing with bent mosaic crystals

P. Courtois

Institut Max von Laue - Paul Langevin, 71 Avenue des Martyrs, CS 20156, 38042 Grenoble Cedex 9, France

E-mail: courtois@ill.fr

Abstract. We report on the performance of bent mosaic crystals when used as a vertical focusing neutron monochromator. High-quality Cu(200) and Ge(335) mosaic crystals with a controlled curvature have been successfully produced at the ILL using plastic deformation at high temperature. As expected from simple geometrical considerations, they exhibit excellent properties for focusing a neutron beam vertically when examined on a high-resolution diffractometer installed on an $m = 1$ thermal neutron guide. Both Cu(200) and Ge(335) curved crystals allow a significant reduction of the focal image size at the sample position compared with a flat crystal with the same defect concentration. As a result, significant gain factors of 6 to 7 in intensity were obtained by replacing a flat crystal of 30 mm with a bent crystal.

1. Introduction

A vertical focusing monochromator provides an efficient mean of producing an intense beam of monochromatic neutrons from a white source [1-2]. To obtain a reasonable diffracted intensity, mosaic crystals are used in preference to perfect crystals, as they can match the incoming beam divergence in the scattering plane [3]. Moreover, the mosaic distribution in the direction perpendicular to the diffracting plane should be as small as possible in order to enable efficient beam compression vertically. Traditionally, a focusing monochromator consists of an assembly of flat rectangular monocrystalline plates mounted onto a mechanical support structure. The orientation of each crystal is adjusted in order to achieve the appropriate curvature, which in the case of symmetric diffraction geometry is related to the lens formula [4]:

$$\frac{1}{L_0}+\frac{1}{L_{MS}}=\frac{2\,sin\theta}{R_V} \qquad (1)$$

L_0 is the distance from the source to the monochromator, L_{MS} the distance from the monochromator to the sample position, θ is the Bragg angle and R_v the radius of curvature (figure 1). The curvature is approximated by line segments. The angular step ψ between two segments, i.e. adjacent crystals, is $\psi = h/Rv$, where h is the crystal height. The Bragg angle stays constant over the height of the monochromator. Focusing in the vertical plane leads to a large gain factor in neutron flux, of between 2 and 10, at the sample position without affecting instrument resolution [5-7]. For a fully parallel beam, it produces a focal image of the source with dimensions limited by the height of the plates used to build the monochromator. The choice of the crystal monochromator is therefore of vital importance but not straightforward. A good compromise in practice is to choose the crystal height so that it matches the characteristic height of the sample under study. It will then produce a focus size comparable to the sample dimensions, typically in the range of 20 mm to 60 mm.

Better focusing can be achieved by using curved crystals, as they can perfectly match a given curvature. In the present paper we show that high-quality bent Cu and Ge mosaic crystals can be used as efficient monochromators if properly designed. Based on simple geometrical considerations, the performance of bent crystals in terms of beam compression and intensity gain was compared directly to that of flat samples with the same characteristics. We also report on our neutron characterisation studies, during which we verified the uniformity of both the curvature and the mosaic distribution. The production of bent Cu and Ge crystals with controlled defect concentration is briefly described.

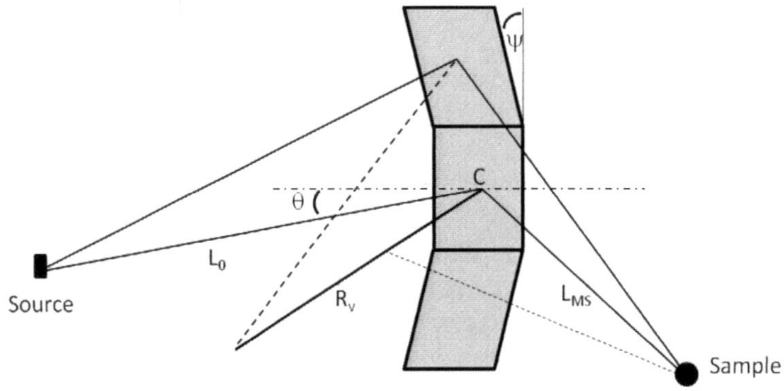

Figure 1. Vertical focusing monochromator composed of flat crystals.

2. Size of the focal image

For a flat crystal, the size of the focal image in the vertical direction is predominantly determined by the dimensions of the plate used to build the monochromator. Additional broadening due to imperfect geometry caused by crystal mosaic spread and incoming beam divergence may also affect focus size. The focus size H_f from a flat crystal of height h_c can be estimated from simple geometrical considerations (figure 2): $H_f = h_c + 2\,\gamma\,L_{MS}\,sin\theta$, where 2γ is the total angular deviation of the reflected beam. Assuming Gaussian distributions, we obtain $\gamma = \sqrt{\alpha^2 + \beta^2}$, where α is the beam divergence and β the crystal mosaic along the bending direction. The Full Width Half Maximum of the intensity distribution in the vertical direction is then:

$$H_f(FWHM) = h_c + 2\sqrt{\alpha^2 + \beta^2}\,L_{MS}\,sin\theta \qquad (2)$$

The α value depends on the neutron source (guide, beam tube or virtual point source) and the L_{MS} parameter is defined by the instrument configuration. Since the mosaic β is chosen to be as narrow as possible, it is easy to see that a significant reduction in the focal spot can only be achieved by reducing the value of h_c. However, it may be not practical to use crystals with small dimensions (less than 10 mm) due to the major difficulty involved in orienting the crystallographic planes. Indeed, in a monochromator made of a large number of pieces, the individual crystals need to be accurately aligned with respect to one another to within a fraction of a degree, i.e. 0.05°. For crystals with dimensions of 5 mm, a positioning accuracy better than 5 μm is required. This represents not only a real challenge but is also time-consuming.

In contrast to a flat crystal, a cylindrically bent mosaic crystal matches a given curvature perfectly. The Bragg planes will satisfy the vertical focusing conditions. As a direct consequence, the focus size in the vertical direction H_b is independent of the crystal height and is only affected by imperfect geometry (figure 2). We obtain:

$$H_b(FWHM) = 2\sqrt{\alpha^2 + \beta^2}\,L_{MS}\,sin\theta \qquad (3)$$

The gain factor G in beam compression for a curved crystal compared with a flat crystal can be expressed as:

$$G = \frac{H_f}{H_b} = 1 + \frac{h_c}{2\sqrt{\alpha^2 + \beta^2}\,L_{MS}\,sin\theta} \qquad (4)$$

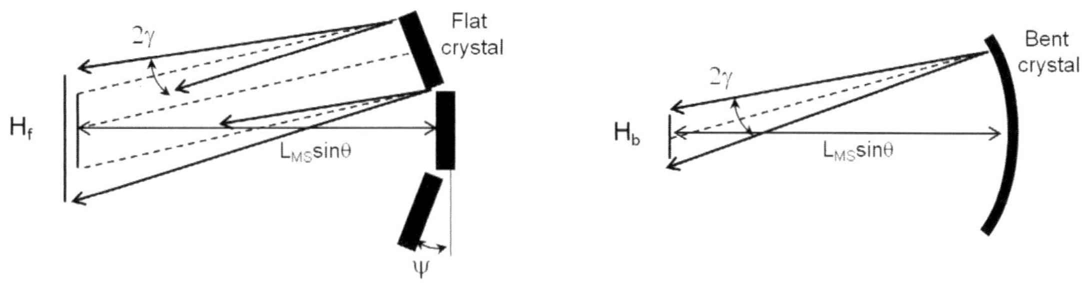

Figure 2. Geometry of focusing in the vertical plane illustrating the broadening of the focal image due to imperfect geometry. γ is the angular deviation from the perfect geometry, taking into account the beam divergence and the crystal mosaic distribution. It should be noted that length scales are not respected (H_b and $H_f << L_{MS}\sin\theta$).

As an example, some typical values are reported in table 1. The gain in beam compression has been estimated for a Bragg angle of 30°, a mosaic β of 0.1° and for crystal heights of 10 mm and 30 mm. Calculations were performed for incoming beam divergences of 0.24° and 1°. These values were chosen to be representative of the neutron beams transported by neutron guides at the ILL: α = 0.24° and α = 1° are the divergences for an m = 2 thermal guide at a wavelength of 1.2 Å and 5 Å, respectively. The contribution of the crystal height was evaluated for monochromator-sample distances L_{MS} of 2 m and 3m. The calculations show that in the case of low beam divergences the focal image size can be reduced vertically by a factor of 2 to 4 by using bent crystals. Obviously, the gain factor decreases when the beam divergence or focal length is increased. Nevertheless, the use of bent crystals remains worthwhile, especially when large single pieces are used.

	beam divergence α = 0.24°				beam divergence α = 1°			
h_c(mm)	30	10	30	10	30	10	30	10
L_{MS}(mm)	2000	2000	3000	3000	2000	2000	3000	3000
H_f(mm)	39,1	19,1	43,6	23,6	65,1	45,1	82,6	62,6
H_b(mm)	9,1	9,1	13,6	13,6	35,1	35,1	52,6	52,6
Gain factor G	4,3	2,1	3,2	1,7	1,9	1,3	1,6	1,2

Table 1. Theoretical sizes of the focal image produced by a flat crystal (H_f) and a bent crystal (H_b) as a function of crystal height and focal length. α is the incoming neutron beam divergence. The crystal mosaic β was fixed at 0.1° and the Bragg angle set to 30°. The gain factor G is defined as the ratio H_f/H_b.

3. Experimental approach

3.1 Production of bent Cu(200) and Ge(335) mosaic crystals

Plastic deformation is a well-known technique for producing copper and germanium crystals with controlled mosaic distribution [8]. By applying a suitable pressure perpendicular to the surface of a high-quality single-crystal plate at high temperature, the mosaic spread can be increased up to the desired value, in the range of 0.1° to 0.5° [8-9]. This technique was developed recently for preparing mosaic crystals with a controlled curvature. Bending was achieved without affecting the quality of the crystals by using smooth plastic deformations applied with a cylindrically shaped anvil with the desired radius of curvature. In addition, through careful positioning in the hot-pressed furnace, it was possible to obtain a mosaic distribution which was reasonably homogeneous throughout the length of the crystal. Starting from high-quality single crystals, several Cu and Ge samples were successfully prepared.

3.2 Neutron experiments

Neutron diffraction properties were studied on the double-crystal test instrument T13A at the ILL. A perfect Ge(hkl) monochromator crystal selected the desired wavelength from an $m = 1$ thermal guide tube with a cross-section of 0.3 (width) x 0.5 (height) cm^2. An adjustable vertical slit was positioned in front of the sample allowing the incoming beam height to be varied from 0.1 to 5 cm. The beam width was 25 mm for all experiments. Two detectors were used for this work: a single high-efficiency ^{3}He detector to record rocking curves and a 64 x 64 mm^2 two-dimensional area detector (32 x 32 pixels) to study focusing properties. The 2D detector was mounted on a y translation stage in order to be able to adjust the distance between the sample and the detector.

The effective neutron mosaic spread was determined as the full width at half height of the rocking curves recorded in reflection geometry. The double-crystal configuration was parallel and the arrangement almost dispersion-free (d-spacing monochromator = d-spacing sample). Thus the width of the rocking curve directly represented the mosaic distribution of the sample under study. The mosaic distribution was measured along a direction parallel ($FWHM_H$) and perpendicular ($FWHM_V$) to the bending direction. The curvature of the single-crystal plates was calculated from the peak shift $\Delta\theta$ of the angular positions along the sample. For these measurements the width of the incoming beam was limited to 0.1 cm. To investigate focusing properties, diffraction images were recorded using the 2D detector. The focal point was determined by varying the sample-detector distance L_{MS} and was defined as the smallest image size.

4. Results and discussion

4.1 Crystal quality

The present study focused on two bent Cu(200) and Ge(335) crystals of dimensions 60 x 20 x 5 mm^3. Single-crystal plates were plastically bent using a cylindrically shaped anvil with a radius of curvature of 1 m. Figure 3 shows the rocking curves from Cu(200) at a neutron wavelength of 1.8 Å provided by a Ge(220) monochromator. The mosaic distribution $FWHM_V$ in the direction perpendicular to the bending is 0.2°. The diffraction pattern indicates a symmetric single peak with a nearly Gaussian shape, demonstrating the homogeneous mosaic structure. Along the bending direction, the global width of the rocking curve is 2.8°. This result is consistent to a peak shift $\Delta\theta = l/Rv$ with $Rv \sim 1$m, where l is the length of the plate illuminated by the incoming beam ($l = 50$ mm). A narrow beam of 1 mm width was used to determine more accurately the value of the local mosaic spread $FWHM_H$. The rocking curve is wider than the local mosaic spread $FWHM_H$ because the reflection curve is convoluted with the curvature. After deconvolution, provided that distributions are Gaussian, we found that the local mosaic spread $FWHM_H$ is very close to 0.2°. The variation of the peak position as a function of the position y along the length of the crystal indicates a linear shift of 0.06° /cm, which is compatible with a uniform curvature. It corresponds to a radius of curvature $Rv = 1.047$ m. Smooth variations in peak intensity are due to inhomogeneities in the mosaic structure.

Similar measurements on a bent Ge(335) crystal at a neutron wavelength $\lambda = 1.1$ Å are shown in figure 3. The mosaic spread $FWHM_V$ is 0.25°. The mechanical properties of germanium make it difficult to introduce a spatially homogeneous microstructure in large single crystals using plastic deformation. Nevertheless, crystal quality is still acceptable for neutron applications. The peak shift along the bending direction is again linear and varies by 2° over 40 mm, which corresponds to a 1140 mm radius of curvature. The local mosaic spread $FWHM_H$ is 0.2°.

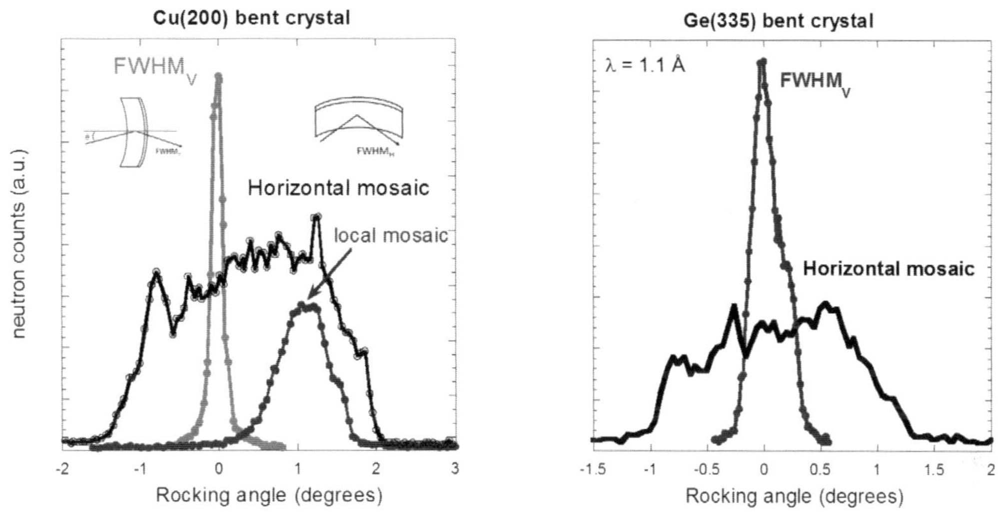

Figure 3. Experimental neutron rocking curves obtained with Cu(200) and Ge(335) mosaic crystals in reflection geometry (for more details, see text).

4.2 Focusing properties of bent crystals

The performance of bent Cu(200) and Ge(335) crystals in terms of peak intensity and beam compression was compared with that of flat crystals with the same mosaic distribution. The vertical size of the incoming beam was 30 mm. Diffraction images from Cu(200) crystals are presented in figure 4. Clearly visible is the focusing effect obtained with the bent crystal when increasing the detector-sample distance L_{MS} from 500 mm (under-focused beam) to 1010 mm (focal point). Beyond the focal point, we observed an increase in the image size resulting in over-focused geometry (figure 4c). The vertical cross-section of the diffraction pattern indicates that the spatial distribution of the intensity is perfectly Gaussian with a Full Width Half Maximum of 5.2 mm (figure 4e). Compared with a flat crystal, the reduction in the focus size is significant. It corresponds to a gain in intensity of a factor 6 resulting in a geometrical beam compression factor of 6, whereas the integrated reflectivity remains constant.

Very similar results were obtained with Ge(335) crystals (see figure 5). The focal point was found to be at a sample-detector distance of 850 mm (figures 5a and 5b). The vertical peak profile is again perfectly Gaussian with a *FWHM* of 4 mm (figure 5c). The incoming beam size has been reduced by a factor 7 at the focal point, leading to an increase in peak intensity also by a factor 7.

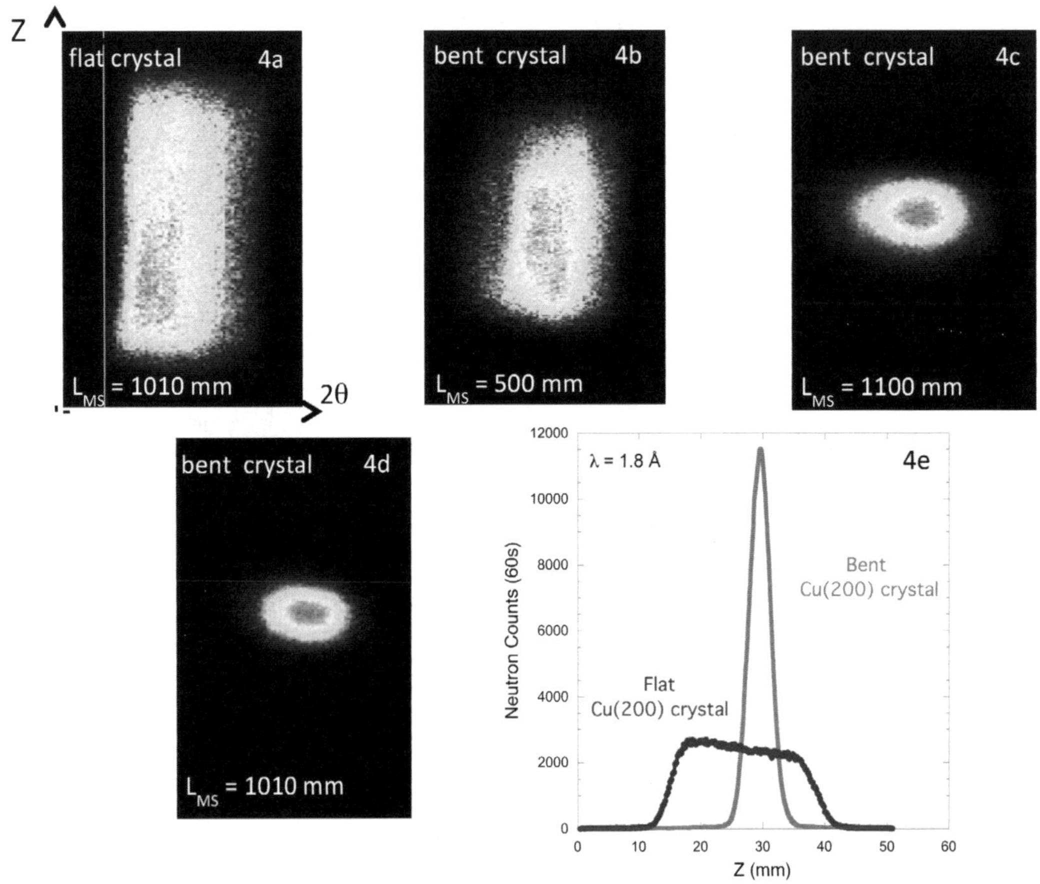

Figure 4. Diffraction images recorded at $\lambda = 1.8$Å with a flat Cu(200) crystal (Fig. 4a) and with a bent Cu(200) crystal for different detector-sample distances L_{MS}: under-focused (Fig. 4b), over-focused (Fig. 4c) and focused (Fig. 4d) geometries. Spatial distribution of the intensity at the focal point for a flat and a bent Cu(200) crystal (Fig. 4e).

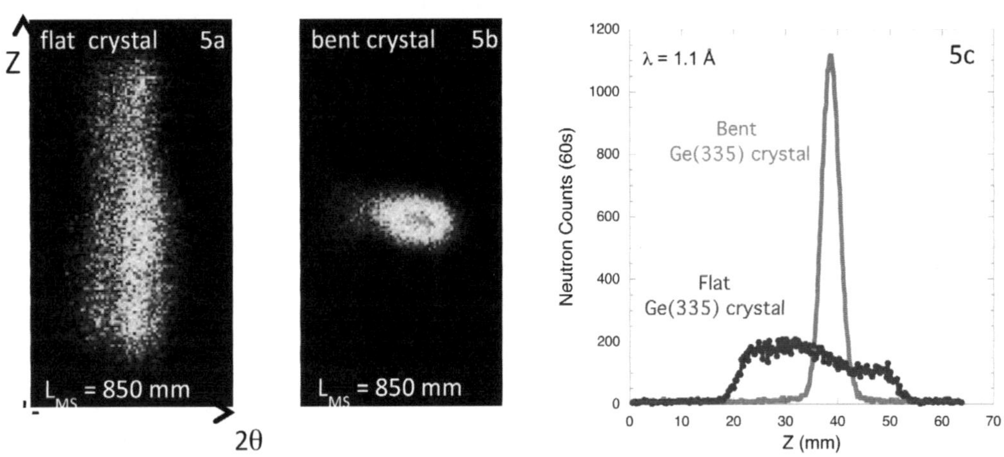

Figure 5. Diffraction images recorded at the focal point (L_{MS} = 850 mm) with a flat Ge(335) crystal (Fig. 5a) and with a curved Ge(335) crystal (Fig. 5b) and the corresponding spatial distributions of the intensity along the vertical direction at $\lambda = 1.1$ Å (Fig. 5c).

4.3 Discussion

The main results for focusing properties obtained from bent and flat crystals using neutron diffractometry on T13A are summarised in table 2. Theoretical predictions are also reported for comparison. Equations 1, 2, 3 and 4 were used for the calculations. The incoming beam divergence was set to $\alpha = 0.18°$ at $\lambda = 1.8$ Å and $\alpha = 0.11°$ at $\lambda = 1.1$ Å for Cu and Ge samples, respectively. This corresponds to the divergence values provided by an $m = 1$ neutron guide. It must be pointed out that the vertical divergence is unaffected by the reflection from the perfect Ge crystal used as a monochromator on the test instrument. The value of $\beta = FWHM_H$ was determined from the width of the neutron rocking curves, as described in section 3. It was fixed at 0.2° for all samples. Focal lengths were estimated for a 1m radius of curvature when considering a source at infinity, i.e. $L_0 \rightarrow \infty$. As can be seen in table 2, the experimental data are in good agreement with the theoretical predictions. Absolute values of both image size and focal length are very close to the observed ones for all samples. The gain in intensity is proportional to the reduction in beam size.

It must be emphasised that the focus size in the vertical direction produced by the curved Ge(335) and Cu(200) crystals is not affected by the height of the incoming beam. This result is consistent with basic optics, where the crystal acts as a lens. In accordance with the magnification equation, the crystal produces an image of the vertical source of height $h_i = h_s\ (L_{MS}/L_0)$. For a neutron beam transported by a neutron guide, the source height h_s is not well defined in practice but depends on the incoming beam divergence and crystal properties. In other words, when considering a curved crystal, each point of the crystal only reflects neutrons from the source within a vertical angular distribution γ (see section 2). The effective source height is then $h_s = 2\gamma\ L_0\ sin\theta$ and finally we obtain equation (3), as presented in section 2, i.e. $h_i = h_s\ (L_{MS}/L_0) = 2\gamma\ L_0\ sin\theta$.

	L_{MS} experimental	L_{MS} calculated	Image size (*FWHM*)	Image size (calculated)	*G* factor (experimental)	*G* factor (calculated)
Cu(200) flat	∞	∞	34	34.7		
Cu(200) R1m	1010	1003	5	4.7	6.8	7.4
Ge(335) flat	∞	∞	33	34.0		
Ge(335) R1m	850	783	4	4.0	8.3	8.5

Table 2. Experimental and theoretical results for focusing properties of bent Cu(200) and Ge(335) crystals compared with flat crystals with the same characteristics. Both focal length and image size are expressed in mm.

Conclusion

Theoretical predictions based on a simple geometrical approach suggest that a bent mosaic crystal should exhibit better performances in terms of focusing efficiency compared with the flat crystal commonly used to build neutron monochromators. To validate this assumption, Cu(200) and Ge(335) mosaic crystals with a 1 m radius of curvature were successfully produced at the ILL using plastic deformation at high temperature. The uniformity of the mosaic distribution and curvature has been determined by high-resolution neutron diffraction, demonstrating the high structural quality of the samples. The vertical size of the focal spot and the related spatial intensity distribution obtained from bent crystals were studied using a 2D detector. Spot sizes in the range 4-5 mm were then observed for all values of the incoming neutron beam dimension. As a direct consequence, compared with a flat crystal of 30 mm height, a gain factor of 6 and 7 in intensity was obtained with Cu(200) and Ge(335) curved samples, respectively.

Following this promising result, we decided to build a full-scale prototype monochromator for the single crystal diffractometer D10 at the ILL [10]. It is composed of 12 Cu(200) crystals with a 2.5m radius of curvature characterised by a mosaic distribution of 0.2°. It has been designed to replace the existing Cu(200) monochromator with the aim of improving the performance of the instrument at a neutron wavelength of 1.26 Å.

Looking further ahead, it is important to understand that future neutron scattering instruments will require neutron optical devices that allow experiments to be performed under extreme conditions. Very high pressure, magnetic field or temperature will require the use of small sample volumes in the mm^3 range. As a direct consequence, neutron optics devices such as focusing monochromators will have to be capable of efficiently focusing large neutron beams into very small volumes at the sample position. A monochromator built with bent mosaic crystals will therefore lead to significant improvements in instrument performance in terms of neutron flux and reduction of background.

Acknowledgements
The author is grateful to the ILL staff members who provided assistance during the experiments: Paolo Mutti and his team (Instrument Control Service) for data acquisition; Bruno Guérard and his team (Neutron Detector Service) for the use of the 2D detector.

References

[1] Currat R 1973 *Nuclear Instruments and Methods* **107** pp 21-28
[2] Nunes A C and Shirane G 1971 *Nuclear Instruments and Methods* **95** pp 445-452
[3] Freund A K 1979 *Lecture Notes in Physics* **112** pp 381-395
[4] Riste T 1970 *Nuclear Instruments and Methods* **90** pp 1-4
[5] Wildgruber U and Frey F 1987 *Journal of Physics E: Scientific Instruments* **20** pp 451-455
[6] Kulda J, Saroun J, Courtois P, Enderle M,Thomas M, P. Flores P 2002 *Appl.Phys.A* **74** pp s246-s248
[7] Boehm M, Steffens P, Kulda J, Klicpera M, Roux S, Courtois P, Svoboda P, Saroun J, Sechovsky V 2015 *Neutron News* 26 pp 18-21
[8] Courtois P,Hamelin B and Andersen K H 2004, *Nucl. Instr. Methods Phys. Res.* A529 pp 157-161
[9] Courtois P et al. 2006 *Physica B* 385–386, Part 2, pp 1271–1273
[10] *http://www.ill.eu/html/instruments-support/instruments-groups/instruments/d10*

ESSENSE: Ultra high resolution spectroscopy for the ESS

Stefano Pasini, Michael Monkenbusch and Tadeusz Kozielewski

Forschungszentrum Jülich, JCNS, Jülich, Germany

E-mail: m.monkenbusch@fz-juelich.de

Abstract. The instrument concept for a very high intensity neutron spin-echo spectrometer with ultimate resolution properties has been developed and submitted as an instrument proposal to ESS. Effective intensity gain factors up to 30 compared to the best current instruments are anticipated. In addition the resolution will be boosted to the technical limits by newly designed superconducting precession solenoids. The intensity gain results from the use of an optimized guide transporting the high flux from the ESS cold moderator on the one side and from the utilization of an extended wavelength frame of 8 Å yielding a multiplication of information collection rate on the other side. The instrument thus enables novel views on soft matter systems ranging from polymers, functional gels and more to to dynamics of biological molecules with relevance for MD development; the employment of new techniques for surface NSE (GINSE) may contribute to new knowledge in tribology and lubrication and other surface phenomena that currently are hampered by low intensity. New developments in "intelligent" polymers as e.g. self-healing, the properties of which depend on molecular mobility and dynamics, require observation at many 100 ns of correlation times with high intensity, which can be made with ESSENSE.

1. Introduction

In this article we present the salient features of the ESSENSE Neutron Spin-Echo(NSE)-instrument proposal for the ESS. The neutron spin-echo technique is a time-of-flight analysis that is sensitive to small changes of the neutron velocity; these changes are encoded into differences of the total Larmor-precession angle after the spin dynamics is inverted before and after the scattering. NSE measures the intermediate scattering function [1, 2, 3]. NSE is the only instrument type that enables effective **energy resolution** down to **neV** with a **dynamical range** that extends over **6 decades** [4, 5]. Its resolution surpasses all other types of neutron spectrometers by three orders of magnitude. It will predominantly provide the dynamical window for the scattering from nano- and mesoscopic structures that are abundant in soft- and biological matter. The applications range from classical polymer topics and emerging new questions like e.g. the role of entanglements in macromolecules that differ from linear polymer chains or linear chains, which differ from ideal coils, to the interaction, diffusion and internal dynamics of proteins and other bio-molecules. Many of the relevant systems exhibit low concentrations of the mobile species and high background from the matrix material. They need very high intensity and high stability and reproducibility. The high resolution of a NSE spectrometer is expressed in terms of the maximum useful Fourier-time $t_{\max} = \min(0.18\,[\mathrm{ns}/(\mathrm{Tm\,Å}^3)] \times J_{\max}\lambda^3\,,\, H\lambda^2)$. The proposed NSE has a maximum field integral

$J_{\max} \geq 1.5\,\mathrm{Tm}$ and a 2.5 times improved field integral homogeneity parameter [5] $H \simeq 2.5\,\mathrm{ns/Å}^2$. An effective gain of 30 is obtained by combining the high time-average intensity of the ESS with a larger guide cross section ($4 \cdots 6$ higher spectral neutron flux at the sample as compared to IN15 at ILL) and further gains of the order of g = 5 due to the wavelength frame width of 7 to 8 Å delivered by each pulse to the detector. The proper (secondary) spectrometer has a new magnetic design [6] with improved field integral homogeneity allowing to perform many experiments with shorter wavelength than today and with shorter counting times due to the corresponding flux advantage.

At the time of this writing the NSE-instrument has priority 17 in the proposed list for 16 reference instruments.

2. Science case

The trends in condensed matter research, both fundamental and applied, are directed towards complex matter and mesoscopic time and length scales. A large fraction of the interesting complex systems pertain to "soft matter" and increasingly to "biological matter" from single protein molecules to assemblies and interacting collectives. Increasing focus is on responsive and functional materials in the "soft matter" class and understanding of pre-requisites and key properties for function in biological system [7, 8, 9]. Here the use of scattering angles of only a few degrees (the "SANS-regime") tunes the spatial resolution to the size of macromolecules, proteins or other mesoscopic structures and aggregates. Energy resolutions far beyond 1μeV or correlation times much larger than 1 ns are required. Among the neutron spectroscopic methods these are only reached by the spin-echo method. Correspondingly slow relaxations at shorter length scales (extending beyond the SANS regime) are observed in glass-forming systems as well as in materials with slow diffusion of small molecules or even atoms or protons. In "hard-matter" increasingly complex magnetic structures, possibly with coupling to electric fields, mechanical stress etc. will become candidates for novel electronic devices or examples for fundamental physics effects. Slow or critical (paramagnetic) fluctuations can be analyzed by NSE in a time domain from a few ps to many ns. Neutron scattering and in particular NSE, in combination with computer simulations, already proved to boost the potential to obtain detailed microscopic pictures of soft-matter and biological systems [10, 11, 8]. Incoherent scattering (from protons) is also an important source of information on the transport properties of ions and protons in electrolyte and hydrogen storage materials that pertain to energy storage techniques in batteries or fuel cells, or in combination with hydrogen storage [12].

Finally we remark that the high intensity and the possibility to simultaneously cover a range of q and τ values within one wavelength frame may enable kinetic observations focused on dynamic properties. For example the loss of mobility in the course of chemical or physical cross-linking or due to external stimuli, e.g. photochemical changes upon illumination or temperature or pressure jumps.

3. Layout of the instrument

The ultimate limit of resolution can only be reached by a combination of measures: 1. highest intensity at long wavelength, 2. lowest possible intrinsic effective field integral inhomogeneity, 3. best available correction for the unavoidable inhomogeneity, 4. stability over sufficiently long counting times. At the spectrometer side intensity can only be increased by using sufficient divergence for the illumination of an as-large-as-possible sample area, low intrinsic inhomogeneity can only be achieved by an optimized effective field shape. ESSENSE relies on the latest fully optimized implementation of these criteria.

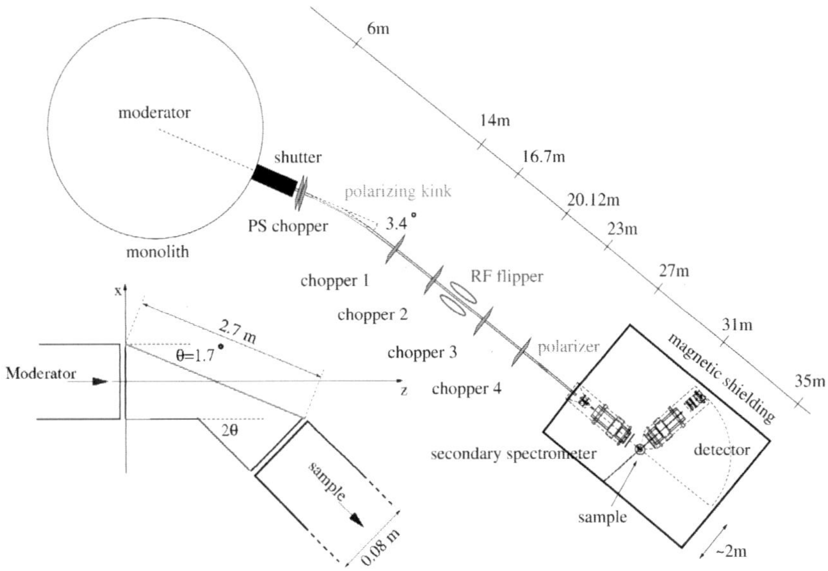

Figure 1. Overview of the whole spectrometer with the dimensions and its main components. The insert (lower left) shows the polarizing kink bender in more detail, the red line represents the polarizing mirror.

3.1. Primary spectrometer: beam transport

The neutron guide of ESSENSE for a cold flat "pancake" moderator of 3 cm height starts at a distance of 2 m from the moderator, has a 3.4^0 kink with a polarizing FeSi multilayer mirror in order to avoid direct sight into the moderator and then extends to 27 m.

The spectrometer itself consists of two main superconducting solenoid sets on the same carriers that also host the flippers, the auxiliary coils, the polarization analyzer and the detector, the latter positioned at $35 \cdots 36$ m from the moderator. This distance yields a wavelength frame-width of $\Delta\lambda = 7 \cdots 8$ Å , which is sufficient to effectively harvest the pulse gain factor. Due to the finite width of the pulse the wavelength-spread at any instant is 0.3 Å. The layout of the transport system, the positions of the chopper system and of the polarization device are shown in Figure 1.

3.2. The chopper system

Frame selection is realized by a system of 4 disc choppers in the distance range between 14 m and 23 m (e.g. C1 at 14 m, C2 at 16.7 m, C3 at 20.12 m, C4 at 23 m). With a rotation frequency of 14 Hz and sector openings of 110.9°, 138.7°, 173.8° and 203.5° respectively, the choppers make use of the whole pulse. This system is able to select any frame within (and beyond) the useful operation wavelength range between $\lambda = 3$Å and $\lambda \simeq 25$ Å without any contamination of neutrons with wavelengths shorter than 80 Å . A distance of 14 m for the first chopper is the largest distance that yields a viable frame selection with no more than 4 choppers. Optionally a pulse shaping chopper may be installed at a distance of $6 \cdots 7$ m . It consists of two closely spaced synchronously co-rotating disks each with two symmetrically placed 90^0 sectors. By phasing of the two disks any effective sector width between 0^0 and 90^0 can be set. The chopper opens twice per revolution, its frequency range is between $n\times$ 7 Hz $\rightarrow$ 7 - 168 Hz , $n = 1, 2, \cdots 24$, allowing a wavelength-spread reduction to virtually any value between $0.03 \cdots 0.3$Å.

3.3. The neutron guide system

The sample has typical/maximal dimensions of 3 cm × 3 cm and is placed at 4 m from the end of the guide. A guide cross-section of 8 cm×8 cm at the guide exit still guarantees efficient brilliance transfer for λ larger than a limit wavelength of about 4 Å within maximum illumination divergence at the sample of ±0.8°. Even if the flat moderator geometry is not optimal for

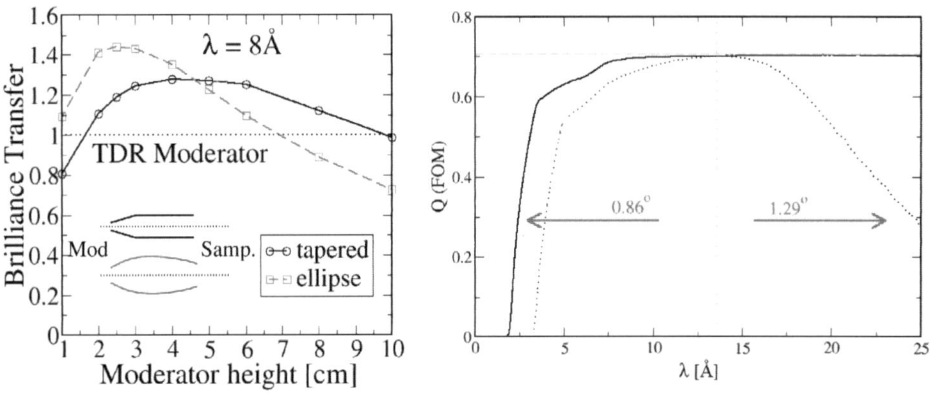

Figure 2. Left: Optimization study for a flat moderator (3 cm height) for $\lambda = 8$ Å. The picture shows two possible solutions for the vertical profile of the beam transport optics that are results of the optimization: a tapered and an elliptic. A third solution consisting of a mixed guide layout is also possible and it has been investigated by Mads Bertelsen. Such a solution would provide an average gain of 1.5 between 3 and 7 cm. The displayed neutron intensity ratio is calculated with respect to the original TDR moderator (12×12 cm^2). Right: Figure of merit (FOM) for a tapered channel analyzer with FeCoV supermirror with $m_{up} = 3$ (right). The angle of inclination is explained in the text. The FOM is the product of the polarization times the square root of the transmission ($Q = P\sqrt{T}$).

the NSE requirements, the neutron guide system can collect up to 50% of the threefold gain depending on the degree of guide adaption, see left of Figure 2. The polarizing device consists of an inclined polarizing mirror (kink). With a total length of 2.7 m, it is a sufficiently compact insert that can be placed between the 8 m after the moderator and shutter and before the first chopper at 14 m. As supermirror a coating with Fe/Si (m=4 $\cdots$ 5) multilayers is envisaged. With the inclination of $\theta = 1.7°$ and a length of the polarizing mirror of 2.7 m the kink provides a polarization of more than 95% up to a wavelength $\lambda \simeq 15$ Å dropping to 80% at 20 Å and it effectively avoids direct view from guide exit to moderator . To improve the polarisation for wavelengths beyond 15 Å an additional transmission polarizer (FeSi on silicon mirror, inclined 4.6°, preceded by a broadband adiabatic RF flipper) is inserted to improve P at wavelengths larger than 11 Å .

3.4. Analyzer

The challenge for the analyzer is to accept the divergence from the sample and to cope with a broad wavelength band $\Delta\lambda = 8$ Å simultaneously. The analyzer in front of the detector has the dimension of the detector (30 cm diameter). The design parameters are 40 cm length with 4 mm wide and 30 cm high channels with (FeCoV multilayer coating with $m_{up} = 3$ or higher and $m_{down} = 0.35$) on both vertical walls of each channel. An optimization of the angle of inclination of the channels (i.e. the orientation towards a source positioned at some centimeters off the real one) is needed in order to maximize the polarization after the analyzer (aimed > 95% polarization at 25 Å). The expected performance is shown in Figure 2.

3.5. Detector

A detector 30 cm × 30 cm and with a moderate (2D) position resolution (better than 3 cm or 1 inch), high efficiency and high maximum count rate (better than 5 KHz/cm^2) is required. A robust possible solution would consist of an array of 100-200 front-end counting tubes (^{3}He) with square cross-section (∼ 15 l ^{3}He).

3.6. Instrument layout: Secondary (proper) spectrometer

The proper spectrometer is of the generic IN11 type [4, 13], but with a new sophisticated magnetic design in order to extend the technical limit for the Fourier-time (τ) given by the residual field-integral inhomogeneity, which remains after the action of correction elements in the beam paths ("Fresnel" correction coils). The present state-of-the-art correction elements seem to be resistant to further improvement, therefore the most promising means towards the ultimate limits of resolution is a reduction of the intrinsic inhomogeneity of the main solenoids, a measure that has been first selected at the ILL to enhance the IN15 properties within a currently conducted refurbishment. The potential of the application of this strategy is about a factor between 2 and 3 inhomogeneity reduction compared to existing instruments as may be expressed by an increase of the homogeneity parameter H from about 1 to $2 \cdots 3$ ns/Å^2. Reduction of intrinsic inhomogeneity is achieved by a somewhat enhanced field in the center of the solenoids (resembling C. Zeyen's optimal field-shape solution for a narrow beam [14]) and enlargement of the solenoid diameter [6]. The latter, however, is limited by the requirement to avoid extensive fringe fields and to keep the total length of the instrument within feasible limits. Optimization consists in finding the best compromise among these contradicting requirements. By using superconducting partial solenoids, that are fringe field compensated in a similar way as the solenoids of the SNS-NSE spectrometer, these requirements can be fulfilled with a distance of 4 m from the sample to the $\pi/2$-flipper.

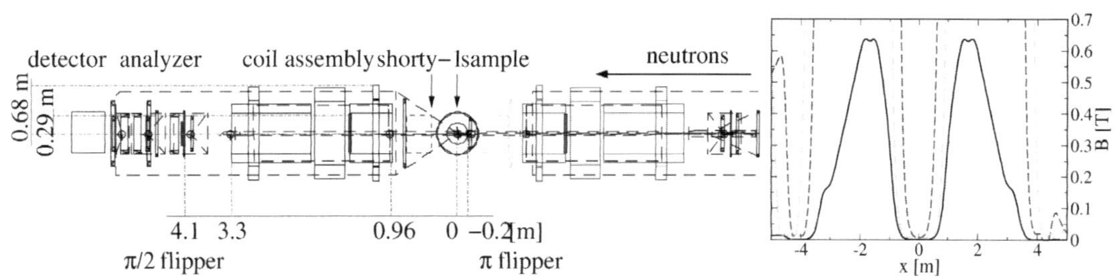

Figure 3. Left: horizontal section of the secondary spectrometer. Each pair of lines delimites the contour of a magnetic coil, see e.g. in the right arm the inner red coils ($+NI$) and the corresponding concentric, outer compensating coil ($-NI/2$) in blue. The superconducting main coils are between 0.96 and 3 m. Right: Profile of the magnetic field (solid line) along the axis for both arms of the coil. Dashed line: $B \times 40$, point-dashed line: $B \times 40$ for a configuration with an extra free sample space of 30 cm radius.

3.7. The coil assembly.

The superconducting main coil assembly is about 2.4 m long with a maximum coil radius of 0.68 m and consists of five fully compensated, superconducting parts. The magnetic field optimizations were performed with a π flipper positioned between 10 cm to 20 cm from the sample. If required by bulky sample environments or special setups more sample space can be supplied thanks to the possibility of including extra distance pieces between spectrometer arms and sample stage. The coils of Figure 3 have an intrinsic inhomogeneity $\sqrt{\langle \Delta J^2 \rangle}/J_0 = 3.4 \times 10^{-4}$ for a neutron beam with 10 cm radius at the $\pi/2$ flipper. Figure (3) also shows the magnetic field along the axis for both arms; the typical asymmetric triangle shape that is common to all optimized solutions is visible. The still necessary corrections will be performed by two correctors at optimized positions in each of the main solenoid sets. The full dipole moment compensation of the main solenoids allows easy access to large scattering angles , e.g. for 45 degrees the computed corrected inhomogeneity (by 2 quadratic ideal current distribution type correctors) is increased by magnetic cross-talk between both spectrometer arms merely from 1ppm to 4ppm. By solely shifting the correctors by less than 0.5 mm this can be reduced to 2ppm again. Further, the fringe fields are compatible with a magnetic shielding enclosure [5]. In the design optimization and assessment of the main coils it was assumed that ideal radial coils with quadratic correction effects are in place. That allows for a correction of the uniformity of the precession field to less than 1 ppm, i.e. the deviation of the main field from quadratic errors are sufficiently low and the necessary quadratic correction strength is reduced, corresponding to the reduced intrinsic inhomogeneity properties of the new coil design.

3.8. Short times option

The ultimate short time limit of the spectrometer in standard configuration is around 50 ps. In order to extend this down to 1 ps two methods may be used. Both require a few additional smaller coils and two additional flippers.

(i) A "small" version of an NSE setup within the sample space comprising two $\pi/2$-flippers and auxiliary "main" solenoids. The rest of the instruments just supplies the guide field.

(ii) Inserting extra π-flippers between the main coil ends and the "normal" $\pi/2$-flippers and adding short extra coils in the space between these flippers will enable an operation where, the effective field integral, i.e. the Fourier-time can be set to arbitrary small values. This is achieved by inverting the precession angle in the second π-flipper thus subtracting the field integral inside the main coil (at very low current) from that generated by the short extra coil. This mode has already been proven at the SNS-NSE spectrometer. The shortest time here is not given by the arbitrarily small effective field integral but rather by the condition that the field integral δJ_π variation, necessary to probe at least 1/2 echo oscillation is less than 10% to 20% of J_{eff}, which yields $\tau_{\text{min}} = (J_{\text{eff}}/\delta J_\pi)(m_n/2h)\lambda^2$, about $1 \cdots 2$ ps at $\lambda = 4$ Å.

The second option leaves more open room in the sample space.

3.9. Magnetic shielding

The coil design is compatible with a magnetic shielding around the spectrometer. If realized it would consist of a double walled mu-metal enclosure with a space of about $0.3 \cdots 0.5$ m between the layers (thickness ≈1.5-2 mm). The walls must be 2 m apart from the magnetic axis. The space in-between can host the radiation shielding around the spectrometer [5].

4. Performance

Concerning the expected performance of the instrument there are two main distinct but still entangled aspects: effective flux and resolution. The resolution issues do not depend on the nature of the neutron source and are independent from its pulse structure. However, a high neutron intensity at long wavelength is essential to reach the highest Fourier-times. The longest useful wavelength largely depends on available intensity and together with field integral homogeneity and/or the maximum available field integral, determines the largest Fourier time. The resolution of a NSE spectrometer is expressed in terms of the maximum Fourier time or better by the range where the resolution function R is still larger than 1/e. $R \simeq \exp\{-(\tau/(H\lambda^2)^2\}$, where H is a measure of the relative field integral homogeneity. Combining the new coil design with the present performance of the correction elements we expect a parameter $H = 2\cdots 2.5\text{ns}/\text{Å}^2$ leading to a maximum Fourier time of $800\cdots 1000$ ns at $\lambda = 20$ Å ($1200\cdots 1600$ ns at $\lambda = 25$ Å) and of $H\lambda^2 = 40$ ns for $\lambda = 4$ Å. The number of scattered neutrons depends on the flux at the sample and the sample size. It has huge influence on the data quality that can be obtained. The average flux (from McStas simulation [15]) at the sample for the new ESS cold source model is given in table 1. The simulated fluxes are multiplied by attenuation factors coming from the (assumed) 4 cm Al, 5 m He and 1 m air that the neutrons will inevitably find on their path in the real instrument. Finally the larger frame width yields

λ/ Å	simulated Φ 10^8 n/s/cm^2/Å	attenuation factor	Φ [**ESSENSE**] 10^8 n/s/cm^2/Å		Φ [IN15] 10^8 n/s/cm^2/Å
			TDR	"pancake"	(*)
6.3	1.3	0.77 x 0.84 x 0.98	0.82	1.1÷1.2	0.21
7	0.9	0.76 x 0.81 x 0.97	0.54	0.76÷0.81	0.18
17	0.025	0.56 x 0.73 x 0.94	0.01	0.01÷0.015	0.0035

Table 1. Fluxes at the sample position (31m) for polarized neutrons. (*) Courtesy of B. Farago.

a gain factor, g, compared to reactor based instruments which is $g = \ln(\lambda_{\max}/\lambda_{\min})/w$, where the effective relative selector width of the reactor instrument is w. Assuming $w = 0.15$ for an instrument with 15% resolution like IN15 and a wavelength window from 7 to 14 Å the gain factor is $g \simeq 5$. Thus, the new high-resolution NSE will have an effective intensity gain factor around 30 with respect to IN15. In Figure 4 these relations are illustrated by a complete experimental cycle with the scattering intensity computed for standard polymer samples from h/d mixtures of polyethylene simulated with the data of new ESS spectrometer and for the new IN15. Resolution functions were modeled by the Gaussian approximation given above, assuming H=2.5 ns/Å^2. The integrated detector area was 100 cm^2 in all cases, i.e. 1/6 of the full detector. Figure 4 shows simulated incoherent scattering from a fully protonated PE sample with 0.5 mm thickness (size 3 cm×3 cm) and it demonstrates the advantage of a large wavelength frame. The larger average intensity yields more than two times smaller error bars in the corresponding ESS curve (red points). In both cases the field-integral range was 0.001-1 Tm (ESSENSE) or 0.001-0.25 Tm (IN15, pertains omission of the last 4 tau-points). The dynamical features were assumed to be simple Rouse type polymer dynamics, in detail real polymers will show somewhat different, yet similar $S(Q,t)$ behavior. The comparisons show that beyond the larger average flux ratio of 3-4 a big advantage of the ESS spectrometer lies in the fact that between 7 and 3 effective experiments, all yielding useful data, are performed simultaneously. The significant increase in error bars at the large Fourier-time side are due to the approach of the resolution limit (residual inhomogeneity of the main coils).

5. Conclusion

An ultra high-resolution NSE is indispensable to extend the dynamic range of the ESS spectroscopic suite thus enabling measurements with 1000 times smaller energy transfer than

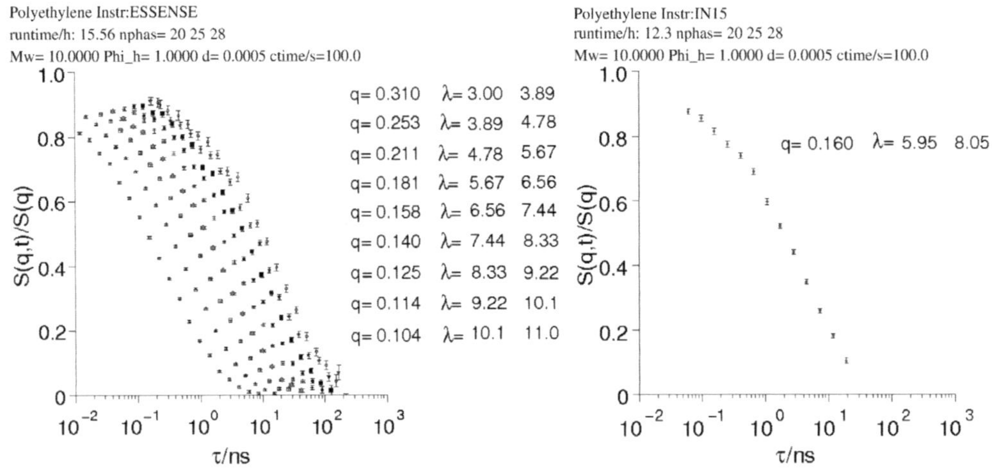

Figure 4. Simulations of an experiment with polyethylene (molecular weight 10, sample thickness d=0.5mm), for incoherent scattering (contrast ratio Phi=1) at the future ESS spectrometer (left) and at IN15 (right) with either the optimized (for ESS) or the IN15 specifications of the instrument, both cases with 15% wavelength-band resolution using data from 100 cm^2 detector area. The assumed counting times (ctimes) per field-integral setting for the ESS and the IN15 cases were equal; the total time (runtime) for the simulated experiment differs. We considered 20 phase points, 5 for spin up and 3 for spin down (npaths). In the left panel the curve at the top corresponds to the lowest q=0.104 $Å^{-1}$, the one at the bottom to the highest one. The curve corresponding to q$\sim$0.16 $Å^{-1}$ -comparable to the one in the right panel - is marked with red symbols.

any other spectrometers. ESSENSE is the high-resolution NSE (of the IN11-type) proposed to the ESS with an innovative design of the precession coils that would allow such measurements.

Acknowledgment

We are grateful to the members of the STAP as well as to the members of ESS for helping us in developing this conceptual design through many meetings and fruitful discussions. We gratefully acknowledge Melissa Sharp as the coordinator of the spin-echo work-package at ESS and Mad Bertelsen for his work on the optimization of the guide system for the flat moderator.

This work was performed with financial support by BMBF through the ESS Design-Update Project 05E10CJ1.

References

[1] Mezei F 1980 Neutron Spin Echo *Lecture Notes in Physics Vol. 128, Springer*
[2] Mezei F, Pappas C, and Gutberlet T 2003 Neutron Spin Echo Spectroscopy. *Lecture Notes in Physics, Springer*
[3] Farago B. 2009 Recent developments and applications of NSE in soft matter. *Current Opinion in Colloid & Interface Science*, **14** 391
[4] Mezei F 1979 *Nucl Instrum and Meth*, **164** 153
[5] Ohl M et al 2012 The spin-echo spectrometer at the spallation neutron source (sns). *Nucl Instrum and Meth A*, **696**
[6] Pasini S and Monkenbusch M 2015 Optimized superconducting coils for a high-resolution neutron spin-echo spectrometer at the european spallation source, ess. *Meas Sci Technol*, **26** 035501
[7] Biehl R, Monkenbusch M, and Richter D 2011 Exploring internal protein dynamics by neutron spin echo spectroscopy. *Soft Matter*, **7** 1299

[8] Smolin N, Biehl R, Kneller G R, Richter D, and Smith J C 2012 Functional domain motions in proteins on the similar to 1-100 ns timescale: Comparison of neutron spin-echo spectroscopy of phosphoglycerate kinase with molecular- dynamics simulation. *Biophys J*, **102** 1108

[9] Bu Z and Callaway D J E 2011 Dynamic propagation of long-range allosteric signals by nanoscale protein domain motion revealed by neutron spin echo spectroscopy. *Biophys J*, **100** 223

[10] Brodeck M, Alvarez F, Colmenero J, and Richter D 2012 Single chain dynamic structure factor of poly(ethylene oxide) in dynamically asymmetric blends with poly(methyl methacrylate). neutron scattering and molecular dynamics simulations. *Macromolecules*, **45** 536

[11] Arbe A, Alvarez F, and Colmenero J 2012 Neutron scattering and molecular dynamics simulations: synergetic tools to unravel structure and dynamics in polymers. *Soft Matter*, **8** 8257

[12] Karlsson M 2013 Perspectives of neutron scattering on proton conducting oxides. *Dalton Transactions*, **42** 317

[13] Farago B 1999 Recent neutron spin-echo developments at the ill (in11 and in15). *Physica B*, **267** 270

[14] Zeyen C and Rem P 1996 Optimal larmor precession magnetic field shapes: Application to neutron spin echo three-axis spectrometry. *Meas SciTechnol*, **7** 782

[15] Mcstas version 1.12c; the fluxes simulated for the 12×12 cm^2 reference moderator are consistent with the source-ess-2012 (zanini) of vitess.

Final design of the Energy-Resolved Neutron Imaging System "RADEN" at J-PARC

T. Shinohara[1], T. Kai[1], K. Oikawa[1], M. Segawa[1], M. Harada[1], T. Nakatani[1], M. Ooi[1], K. Aizawa[1], H. Sato[2], T. Kamiyama[2], H. Yokota[3], T. Sera[4], K. Mochiki[5], Y. Kiyanagi[6]

[1]J-PARC Center, Japan Atomic Energy Agency, Tokai 319-1195, Japan
[2]Faculty of Engineering, Hokkaido University, Sapporo 060-8628, Japan
[3]Center for Advanced Photonics, RIKEN, Wako 351-0198, Japan
[4]Faculty of Engineering, Kyushu University, Fukuoka 819-0395, Japan
[5]Faculty of Engineering, Tokyo City University, Tokyo 319-1195, Japan
[6]Graduate School of Engineering, Nagoya University, Nagoya 464-8603, Japan

takenao.shinohara@j-parc.jp

Abstract. A new pulsed-neutron instrument, named the Energy-Resolved Neutron Imaging System "RADEN", has been constructed at the beam line of BL22 in the Materials and Life Science Experimental Facility (MLF) of J-PARC. The primary purpose of this instrument is to perform energy-resolved neutron imaging experiments through the effective utilization of the pulsed nature of the neutron beam, making this the world's first instrument dedicated to pulsed neutron imaging experiments. RADEN was designed to cover a broad energy range: from cold neutrons with energy down to 1.05 meV (or wavelength up to 8.8 Å) with a good wavelength resolution of 0.20% to high-energy neutrons with energy of several tens keV (or wavelength of 10^{-3} Å). In addition, this instrument is intended to perform state-of-the-art neutron radiography and tomography experiments in Japan. Hence, a maximum beam size of 300 mm square and a high L/D value of up to 7500 are provided.

1. Introduction

Neutron imaging is a fundamental technique for visualizing the internal structure of objects and is regarded as an indispensable tool for non-destructive inspection, especially in the field of industry. Recently, a new neutron imaging technique, which utilizes neutron energy-dependent transmission, called "energy-resolved neutron imaging", has become a very attractive technique because of its capability to not only visualize but also to quantify physical quantities and elemental distribution or composition with spatial resolution [1]. Therefore, this technique is expected to provide a chance for neutron imaging to develop into a new scientific method, one which is not a simple inspection of internal structure but a quantitative characterization of many properties inside objects with spatial resolution. Following recent development work, several energy-resolved imaging techniques have been successfully performed, i.e. Bragg-edge imaging for crystallographic information [2], resonance absorption imaging for elemental composition and thermal information [3], and polarized neutron imaging for magnetic field information [4]. For conducting these imaging experiments, a pulsed neutron beam has several advantages, owing to its pulsed nature, as compared to a monochromatic neutron beam from a continuous source. For example, it is possible to measure energy dependent phenomena efficiently by means of the Time-of-Flight (TOF) method. Additionally, the energy or wavelength resolution is intrinsically fine in the case of the short-pulsed neutron source (better than 1 %). Thus, a wide range of neutron energies, from epithermal to cold neutrons, is available simultaneously. Until now, the low intensity of pulsed neutron beams from low-power spallation neutron sources has been a limiting factor for the application of this imaging technique. However the construction of MW-class spallation neutron sources at J-PARC and SNS has changed this situation.

In Japan, intensive technical development of pulsed neutron imaging has been done using the small accelerator-driven neutron source at Hokkaido University (HUNS) and the pulsed neutron facility (Materials and Life Science Experimental Facility, MLF) at J-PARC. Furthermore, in the Japanese pulsed neutron imaging project, a new beam line dedicated to pulsed neutron imaging, named "RADEN", has been proposed so as to bring the pulsed neutron imaging technique into the operational phase, and its construction was started from 2012 at beam line BL22 of the MLF. In addition to the main purpose of this instrument, namely, to perform energy-resolved neutron imaging experiments, it has a role as the state-of-the-art neutron imaging instrument in Japan with sufficient neutron flux, selectable L/D values up to several thousands, and monochromatic neutron imaging capability, achieving a performance comparable to the world's several-tens MW-class reactor sources. In this paper, we report the final design of the instrument and discuss the expected performance.

2. Beam line design

The requirements for the instrument, which have been discussed at the initial design stage in our previous report [5, 6], are briefly described here. The necessary performance for each energy-resolved neutron imaging technique was considered in terms of the wavelength resolution and the wavelength/energy coverage. On the other hand, for conventional neutron radiography and tomography experiments, other aspects, such as available beam flux, size and L/D value, which is directly coupled with the accessible spatial resolution, becomes important. A minimum wavelength resolution of 0.2% and high neutron energies up to several tens of keV are required for Bragg-edge and resonance absorption imaging techniques, respectively. In addition, because the availability of longer neutron wavelengths is preferable for conducting experiments using the neutron's optical properties, the wavelength range was expanded to around 9 Å within the first frame. These requirements are achieved by viewing a decoupled moderator, which possesses the best balance in the performance of the pulse width and neutron intensity from among the three types of hydrogen moderators at the J-PARC MLF, and by setting two sample positions. At the near sample position, located at a distance of 18 m from the source, a high time-averaged beam flux of less than 9.8×10^7 n/sec/cm^2 and broad wavelength coverage are expected, while a large beam size up to 300 mm square and a fine wavelength resolution of 0.2% will be available at the far sample position at 23 m from the source.

The resulting beam line layout of RADEN is shown in Figure 1 and the expected parameters are summarized in Table 1.

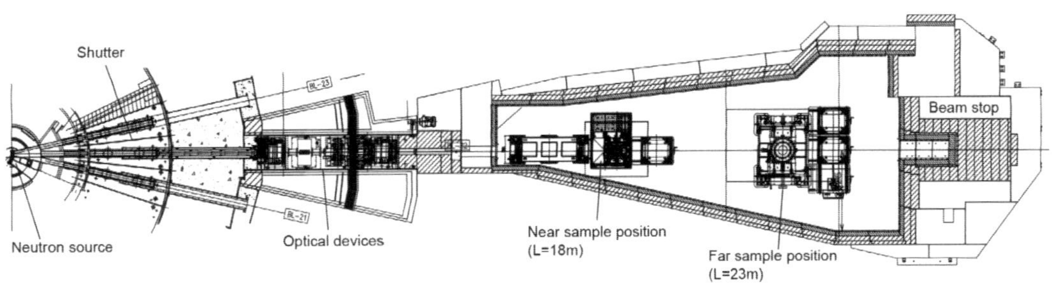

Figure 1. Layout of the RADEN instrument.

Table 1. Basic parameters of RADEN.

Beam line	BL22	
Moderator type	Supercritical hydrogen decoupled moderator	
Sample position	18 m	23 m
Beam Size	< 221 x 221 mm^2	< 300 x 300 mm^2
L/D ratio	180 ~ 5000	230 ~ 7500
Wavelength resolution (cold)	0.26%	0.20%
Energy resolution (epithermal)	1.6%	1.2%
Longest wavelength (first frame)	8.8 Å	6.9 Å

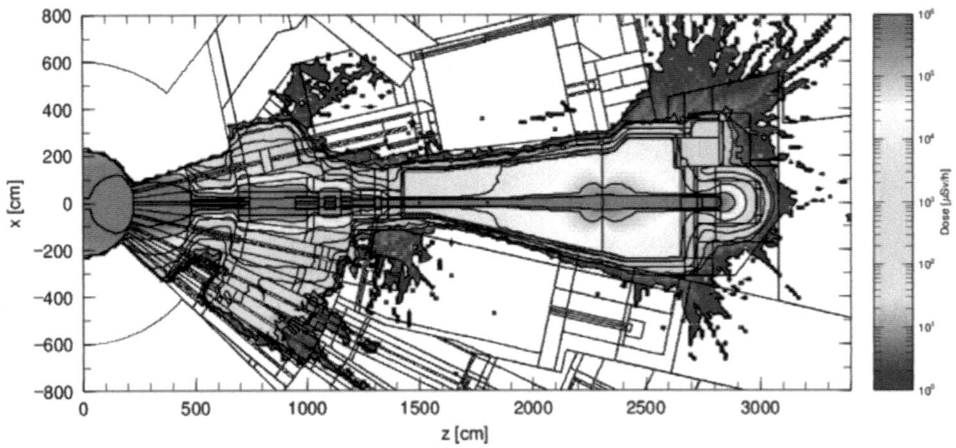

Figure 2. Typical calculation result of the dose rate using PHITS code. The sum of neutron and gamma ray doses is shown.

2.1. Shielding design

To conduct neutron imaging experiments, the capability to accommodate large objects and sufficient flexibility in the experimental arrangement are quite important. Therefore, the shielding should be designed as thin as possible to provide a wide experimental area. However, as RADEN utilizes the biggest beam and sample size in the MLF along with high-energy neutrons, massive shielding is necessary to attenuate the radiation efficiently. To design and optimize the shielding structure, we have performed dose rate calculations and evaluated the performance using PHITS code [7] (Figure 2). As a result, it was decided to use a shielding wall consisting of 4 layers, i.e., from the inside out, boron-included mortar, polyethylene, steel and concrete. Moreover, because the neutron imaging detectors are placed directly in the beam line, scattered neutrons and radiated gamma rays from the beam stop directly irradiate the detectors from the rear and cause a huge background. To suppress such background, the beam stop was placed as far as possible from the detector position. Finally, two heavy doors, which are driven electrically, are used to intercept the radiation at the gateway to the experimental area. While most of the other instruments at the MLF adopt a labyrinth structure with a light, unshielded door, the unique two-door configuration for RADEN was chosen to provide good accessibility to the experimental area.

2.2. Optical devices

Upstream of the experimental area, several optical devices are installed as shown in Figure 3. The optical devices of RADEN are divided into two groups: the beam collimation and shaping device group and the energy selection device group. The beam collimation and shaping device group is composed of a heavy shutter, rotary collimators and slits. In addition, a light shutter, referred to as a blocker, is included to suppress undesired sample activation. To enlarge the beam size at the sample position, an aperture must be placed as close as possible to the source. Thus, a heavy shutter containing three collimator insertion devices, with hole sizes 100 mm x 100 mm, 50 mm- ϕ and 26 mm-ϕ, was installed near the source [8]. Using these collimators, beam sizes ranging from 100 x 100 mm^2 to 300 x 300 mm^2 will be achieved at the far sample position (23 m from the neutron source), with the smallest L/D value producing the highest neutron flux. Inside the forward section of the shielding, two rotary collimators with four aperture slots are installed. The upstream rotary collimator uses small apertures to make higher L/D values for experiments requiring high spatial resolution, while the downstream one is used to limit the beam size as adequate for a given sample size. Because RADEN uses high-energy neutrons with energies over 1 keV, these collimators are made of thick steel and polyethylene and have an artifice on the inner wall to reduce neutron reflection.

The energy selection device group is composed of filters, a low-speed double disk chopper and a T0 chopper. Five kinds of filters, namely, Bi (25 mm, 25mm, 50 mm thicknesses), Pb (25 mm, 50 mm thicknesses), Cd (1 mm thickness), acrylic resin (5 mm thickness), and borosilicate glass (2 mm thickness), are placed upstream of the first rotary collimator and can be remotely moved in or out of the beam depending on experimental requirements. The Bi and Pb filters are used for gamma ray attenuation, especially in the case of resonance absorption imaging experiments, for which the T0 chopper is stopped to allow access to the high-energy neutron range. The Cd filter is also used in resonance absorption experiments to cut thermal and cold neutrons. The acrylic resin and borosilicate glass are used for neutron intensity attenuation. The double disk chopper, which can be operated at 25 Hz and 12.5 Hz, defines the available wavelength range in one frame. Changing the opening angle by adjusting the delay of each disk enables us to fine-tune the bandwidth at the different sample positions. Additionally, a quasi-monochromatic neutron beam with a wavelength resolution of around 10% can be provided when the opening angle is adjusted to its smallest value. The T0 chopper is installed to eliminate the prompt neutron pulse and flash gamma rays produced at the instant of neutron generation. This chopper has one heavy hammer made of Inconel alloy, whose rotation is synchronized with the accelerator, only bringing the hammer into the beam at T = 0. By means of the T0 chopper, the neutron background caused by fast neutrons is decreased effectively, and the shortest neutron wavelength accessible becomes 0.43 Å (or an energy of 0.44 eV).

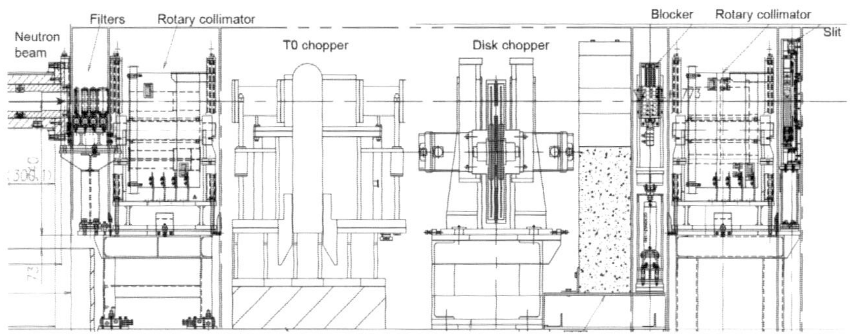

Figure 3. Arrangement of optical devices located upstream of the experimental area.

2.3. Sample environment

Inside the experimental area, several stages are installed for sample mounting and detector positioning. Also, exchangers for optional equipment are provided (Figure 4). Three sample stages are provided for mounting and positioning samples, and their topmost movable axis is rotational to easily allow for computed tomography (CT) measurements. The basic specifications of each sample stage are shown in Table 2. The stages are controlled remotely from outside of the shielding using device control software. The top table of each stage conforms to the MLF standard for sample environments so as to facilitate the mounting of various devices such as cryostats, electromagnets, furnaces, etc. A large sample stage placed at a distance of 23 m from the neutron source can be equipped with mechanical test devices developed at the engineering materials diffractometer TAKUMI [9] for use in Bragg-edge imaging experiments. A medium sample stage sits on the optical bench located at the near sample position. When the medium sample stage is removed, optical tables for polarized neutron experiments or other experiments using special devices can be slid into the beam path by an automated exchanger. Finally, a small rotation stage for CT measurements can be placed on top of the large and medium sample stages.

Flight tubes, which are filled with He gas and whose inner walls are covered with a neutron absorber, are placed in the gaps of the beam line to suppress neutron beam attenuation due to scattering in air. Additionally, a lift located upstream of the near sample position is used to exchange the flight tube with equipment for polarization analysis. With this lift, we can easily switch the setup from an unpolarized to a polarized neutron beam.

Optionally, RADEN can be equipped with an apparatus for neutron diffraction measurements consisting of ^{3}He position sensitive detector tubes and a Soller collimator (complementary to Bragg-edge imaging) and with a gamma ray detector to provide precise analysis for resonance absorption.

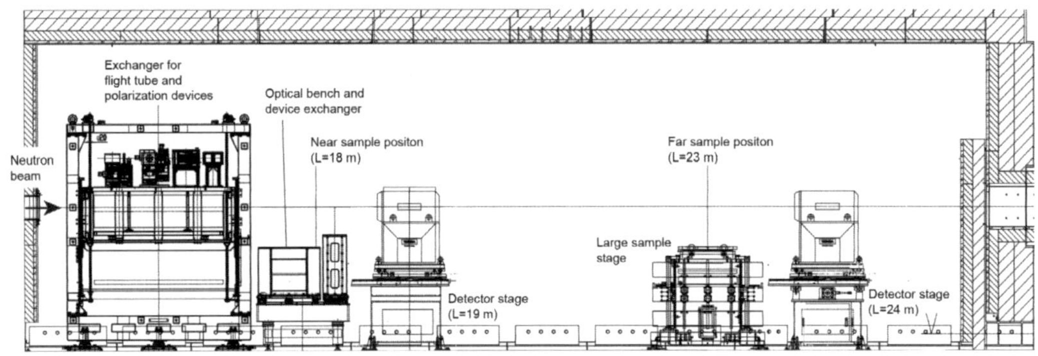

Figure 4. Arrangement of stages and exchangers in the experimental area.

Table 2. Specifications of the sample stages.

Stage	Position	Movable axis			Max. load	Table size
		θ	X, Y, Z	R_X, R_Y		
Large	L=23m	±173°	±300 mm	-	1.0 ton	700 mm-ф / 750 mm x 750 mm
Medium	L=18m	±175°	±100 mm	±5 °	600 kg	300 mm-ф / 700 mm-ф
Small	Portable	360°	-	±5 °	10 kg	150 mm-ф

2.4. Detectors

To conduct energy-resolved neutron imaging using pulsed neutrons, precise TOF information is essential, requiring neutron detectors with fine time resolution. Counting-type imaging detectors are promising, owing to their sub-microsecond time resolution, while their spatial resolution is relatively course. Of these detectors, we selected three candidates for RADEN: the μNID detector [10], the GEM detector [11] and a pixel detector composed of an array of small rectangular Li-glass scintillators coupled to a multi-anode photomultiplier. On the other hand, for conventional neutron imaging, fine spatial resolution is desirable. Hence, a cooled CCD camera system with a thin ^{6}LiF/ZnS scintillator screen and a CMOS camera combined with a neutron image intensifier are prepared in conjunction with the counting-type detectors.

3. Conclusion

Construction of the Energy-Resolved Neutron Imaging System RADEN began from 2012 at beam line BL22 of the J-PARC MLF. Because this system is designed to fully utilize the short-pulsed neutron beam of the MLF, all types of energy-resolved neutron imaging experiments can be carried out with fine energy resolution. RADEN also has an important role as the state-of-the-art neutron radiography and tomography instrument in Japan. Therefore, RADEN was designed with the flexibility to satisfy the requirements for both of these imaging techniques. After completion of the construction phase, the first neutron beam was delivered to this beam line in November 2014, and following an initial commissioning study, RADEN began to accept user programs from April 2015.

Acknowledgement

This work was partially supported by the Grant-in-Aid for Scientific Research (S) 23226018 from the Japan Society for the Promotion of Science.

References

[1] Kiyanagi Y, Sato H, Kamiyama T, Shinohara T, 2012 *J. Phys.: Conf. Ser.* **340** 012010.
[2] Sato H, Takada O, Iwase K, Kamiyama T, Kiyanagi Y, 2010 *J. Phys.: Conf. Ser.* **251** 012070.
[3] Kaneko K, Kamiyama T, Kiyanagi Y, Sakuma T, Ikeda S, 1999 *J. Phys. Chem. Solids* **60** 1499.
[4] Shinohara T, Sakai K, Ohi M, Kai T, Harada M, Oikawa K, Maekawa F, Suzuki J, Oku T, Takata S, Aizawa K, Arai M, Kiyanagi Y, 2011 *Nucl. Inst. Meth. Phys. Res.* **A 651** 121.
[5] Kiyanagi Y, Kamiyama T, Sato H, Shinohara T, Kai T, Aizawa K, Arai M, Harada M, Sakai K, Oikawa K, Ohi M, Maekawa F, Sakai T, Matsubayashi M, Segawa M, Kureta M, 2011 *Nucl. Inst. Meth. Phys. Res.* **A 651** 16.
[6] Kiyanagi Y, Shinohara T, Kai T, Kamiyama T, Sato H, Kino K, Aizawa K, Arai M, Harada M, Sakai K, Oikawa K, Ooi M, Maekawa F, Iikura H, Sakai T, Matsubayashi M, Segawa M, Kureta M, 2013 *Physics Proc.* **43** 92.
[7] Sato T, Niita K, Matsuda N, Hashimoto S, Iwamoto Y, Noda S, Ogawa T, Iwase H, Nakashima H, Fukahori T, Okumura K, Kai T, Chiba S, Furuta T, Sihver L, 2013 *J. Nucl. Sci. Technol.* **50** 913.
[8] Kenichi OIKAWA, Masahide HARADA, Takenao SHINOHARA, Tetsuya KAI, Motoki OOI, and Yoshiaki KIYANAGI, 2015 *JPS Conf. Proc.* **8**, 036002.
[9] Harjo S, Moriai A, Torii S, Suzuki H, Suzuya K, Morii Y, Arai M, Tomota Y, Akita K, Akiniwa Y 2006 *Mater. Sci. Forum* **524-525** 199.
[10] Parker J D, Hattori K, Fujioka H, Harada M, Iwaki S, Kabuki S, Kishimoto Y, Kubo H, Kurosawa S, Miuchi K, Nagae T, Nishimura H, Oku T, Sawano T, Shinohara T, Suzuki J, Takada A, Tanimori T, Ueno K, 2013 *Nucl. Inst. Meth. Phys. Res.* **A 697** 23.
[11] Shoji M, Uno S, Uchida T, Murakami T, Sekimoto M, Koike T, Nakano E, Yamamoto K, Hoshi Y, Nagashima F, 2012 *J. Inst.* **7** C05003.

Journal of Physics: Conference Series **746** (2016) 012008 doi:10.1088/1742-6596/746/1/012008

Dynamically polarized samples for neutron protein crystallography at the Spallation Neutron Source

Jinkui Zhao[1a], Josh Pierce[a], Dean Myles[b], J. L. Robertson[a], Kenneth W. Herwig[a], Bob Standaert[c], Matt Cuneo[b], Le Li[b], Flora Meilleur[b]
[a]Instrument and Source Division, Spallation Neutron Source, Oak Ridge National Laboratory, Oak Ridge, Tennessee 37831, USA
[b]Biology and Soft Matter Division, Spallation Neutron Source, Oak Ridge National Laboratory, Oak Ridge, Tennessee 37831, USA
[c]Biosciences Division, Oak Ridge National Laboratory, Oak Ridge, Tennessee 37831, USA

E-mail: zhaoj@ornl.gov

Abstract. To prepare for the next generation neutron scattering instruments for the planned second target station at the Spallation Neutron Source (SNS) and to broaden the scientific impact of neutron protein crystallography at the Oak Ridge National Laboratory, we have recently ramped up our efforts to develop a dynamically polarized target for neutron protein crystallography at the SNS. Proteins contain a large amount of hydrogen which contributes to incoherent diffraction background and limits the sensitivity of neutron protein crystallography. This incoherent background can be suppressed by using polarized neutron diffraction, which in the same time also improves the coherent diffraction signal. Our plan is to develop a custom Dynamic Nuclear Polarization (DNP) setup tailored to neutron protein diffraction instruments. Protein crystals will be polarized at a magnetic field of 5 T and temperatures of below 1 K. After the dynamic polarization process, the sample will be brought to a frozen-spin mode in a 0.5 T holding field and at temperatures below 100 mK. In a parallel effort, we are also investigating various ways of incorporating polarization agents needed for DNP, such as site specific spin labels, into protein crystals.

1. Introduction

One of the difficulties in neutron protein crystallography is the incoherent hydrogen scattering background. One way to reduce this background is sample perdeuteration [1]. However, perdeuteration is still expensive and growing deuterated crystals is not always as easy as compared to nondeuterated ones. An alternative technique is to use polarized neutron scattering from dynamically polarized proteins [2–5]. The technique exploits the high polarization-dependent neutron scattering cross-section of hydrogen. When proton spins are aligned parallel to those of the scattering neutrons, the large, 80 barn incoherent hydrogen scattering cross-section is reduced to zero. In the meantime, the 1.8 barn coherent cross-section is increased to 14.7 barns (Fig 1). The main challenge to polarized neutron protein crystallography is to achieve a high degree of hydrogen or proton polarization. For proteins and other adiabatic samples, the only viable technique today is Dynamic Nuclear Polarization (DNP). A typical DNP setup for high nuclear polarization requires a magnet field of about 2.5 T – 5 T and a sample temperature of 1 K or lower. Under these conditions, the static hydrogen nuclear polarization is only a small fraction of a percent, which is negligible for neutron scattering. In the meantime and under the same conditions, electron spins are close to being fully polarized. DNP takes advantage of the fact that electron spins are readily polarized. A small amount of paramagnetic impurities (on the order of ~10^{19} electron spins per millilitre) are introduced into the sample. Microwave radiation is then used to drive polarization transfer from unpaired electron spins of the paramagnetic centers to nuclei spins in the sample. As an experimental technique, DNP is well established many decades ago. Its

[1] To whom any correspondence should be addressed.

potential to neutron science was first demonstrated in neutron diffraction from a lanthanum magnesium nitrate (LMN) crystal [2]. Since then, it has been successfully adapted to scattering from non-crystalline samples [6,7]. However, there exists no DNP equipment today that is suitable for high performance neutron protein crystallography. On the one hand, protein crystals are comparably small and require only compact DNP setups for polarization. On the other hand, the complicated DNP equipment makes it very challenging to provide the large scattering angles required for high performance crystallography. Additionally, high dynamic nuclear polarizations in the past have only been achieved in non-crystalline samples. Further work is still needed for protein crystals, especially in finding more suitable paramagnetic centers required by DNP.

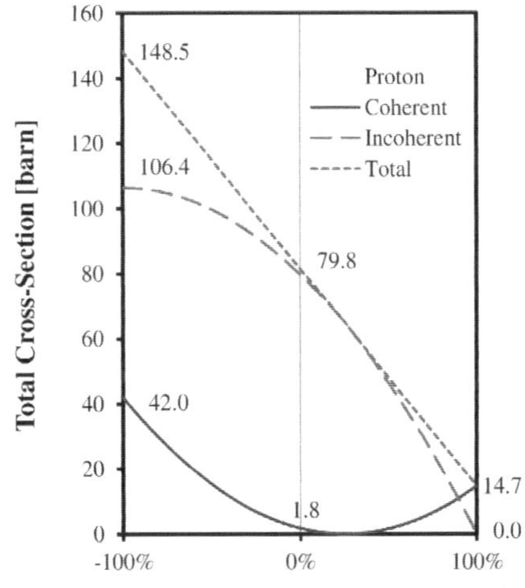

Fig. 1. Coherent and incoherent neutron scattering cross-sections of hydrogen as a function of the proton polarization. Neutron polarization is assumed to be 100%.

2. Initial polarization works at the SNS

Our first effort on polarized neutron protein crystallography from dynamically polarized samples at the Spallation Neutron Source aimed at demonstrating the feasibility of nuclear polarization in protein crystals and building a technical foundation for future applications of polarized neutron protein crystallography [5, 8-10]. The DNP system for this initial effort consisted of a 5 T magnet, a 1 K ^{4}He cryostat, and a 140 GHz microwave generator. One of our main research focuses was to achieve a good degree of hydrogen polarization in protein crystals. Selecting a right polarization agent is the first important step to achieving a high degree of nuclear polarization. The exact correlation between the type of agents and the highest achievable nuclear polarization is still under investigation. However, there are a couple of known radicals that have excellent DNP performances for various types of samples. For biological samples in solutions, for example, the Ethyl-hydroxy-butyric-acid Cr(V) complex (EHBA-Cr(V)) [3] and the Tetra-methyl-piperidinyl-oxy molecules (TEMPO) [11] have both proven to be good options. The EHBA-Cr(V) complex is extensively used in polarized neutron scattering from solution samples [3,5-7, 12,13]. Its downside is that the complex is stable in acidic solutions only. To use with protein crystals, its instability could pose a problem for some samples if it has to be soaked into the crystal for an extended period of time. Therefore, our initial DNP efforts on protein crystals focused on various forms of TEMPO, including those that can act as spin labels, such as the 4-(2-Iodoacetamido)-TEMPO. The 4-(2-Iodoacetamido)-TEMPO can bind to cysteine residues on protein surface. X-rays diffraction data from lysozyme crystals with co-crystallized 4-(2-Iodoacetamido)-TEMPO indicate a high likelihood that TEMPO molecules are bound to the protein. Figure 2 shows some of lysozyme crystal samples with and without the presence of TEMPO [5]. Polarization tests showed that proton polarizations of greater than 30% were readily achievable. While these initial results are promising, these polarization values are too low for protein diffractions and more work is thus needed. In solution scattering, it has

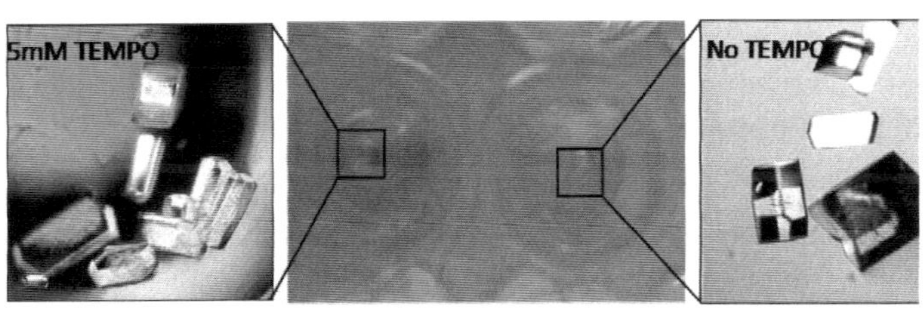

Fig. 2. Lysozyme crystals in 5% NaCl and 50mM Sodium Acetate. Left: with 5mM TEMPO. Right: No TEMPO [5].

been observed that proton polarizations in the bulk and around the paramagnetic centers can be different [7,11]. Such difference can be attributed mainly to two factors. First, the dipole moment of the electronic spin causes the local magnetic field to be different from that of the bulk. Second, the electron spins are presumed to couple with its surrounding nuclear spins only. Nuclear polarization in the bulk is achieved via polarization propagation. In any case, these inhomogeneity in nuclear polarization did not pose a serious problem in solution scattering [3,6,7] due to the low resolution nature of the scattering technique. For protein crystallography, it is unclear whether and how the crystal structure will further influence the distribution of nuclear polarization and affect the diffraction data. For our current effort, while we do not believe polarization inhomogeneity within protein crystals will be our main challenge, we have initiated various studies to investigate the nuclear polarization specific to protein crystals.

3. A dedicated DNP setup for protein crystallography

In order to prepare for next generation neutron scattering instruments for the planned second target station at the SNS, we are currently designing and constructing customized DNP sample environment equipment dedicated to neutron protein crystallography. With the huge potential of increasing the coherent signal and decreasing the incoherent signal, we aim at bring the ability to study radically small crystals, protein complexes, and membrane proteins to neutron protein diffractometers at the Oak Ridge National Laboratory.

As mentioned above, dynamic nuclear polarization for scattering typically operates in a magnetic field of 2.5 T – 5 T, which we decided to adhere in our current effort as well. For DNP, field uniformity of better than 10^{-4} over the whole sample is commonly required. This is primarily due to the fact that Bohr magnetons are more than three orders of magnitude larger than nuclear magnetons. A better than 10^{-4} field uniformity will ensure that there will be a clear nuclear Zeeman splitting in the coupled electron-nuclear spin system. Unfortunately, the high field uniformity requirement conflicts with the need of having a large scattering angle in high performance neutron crystallography. For a 2.5 T – 5 T magnet with better than 10^{-4} field uniformity, it is difficult to achieve scattering angles larger than ±15°. The detector on the current neutron protein diffractometer at Oak Ridge covers close to ±180° in azimuthal angle and larger than ±60° in vertical angle. To bridge this gap in scattering angle coverage, we plan to conduct DNP on protein crystals *ex situ*. We have already procured a 5 T, warm bore solenoid magnet and a dilution refrigerator. A customized tail piece will be made for the fridge that incorporates all the needs for the DNP setup. In addition, a small (~ 0.5 T) super conductor holding magnet will also be built into the tail piece. During the dynamic sample polarization process, the holding field is switched off and the sample-containing tail piece is inserted into the 5 T solenoid magnet. After the polarization, samples will then be cooled down to as close to the fridge base temperature (~10 mK) as possible. After that, the holding field is turned on and the solenoid field is removed. At the low temperature of a few tens of mK, a 0.5 T holding field should be strong enough for a good T_1 nuclear relaxation time and a long neutron data collection time. The solenoid magnet is highly homogeneous and can easily exceed DNP's requirement. The homogeneity of the holding field, on the other hand, can be greatly relaxed. This means that very large scattering angles can be achieved with the holding field. The detail of the holding field coil is still being designed. Our current plan is to have it in the vertical direction and allow a close to ±180° horizontal scattering angle and a ±50° vertical scattering angle.

To continue our research polarization research on protein crystals, we aimed at understanding the connection between the Electron Paramagnetic Resonant (EPR) spectra of the paramagnetic centers and the maximum achievable polarization in protein crystals. Additionally, in protein crystals, crystal orientation may impact the EPR spectra and the maximum achievable nuclear polarization. Gaining such knowledge will help us develop and optimize bulk and site specific spin labelling strategies. In collaboration with various research groups, we have established multiple efforts in this regard.

The commissioning and initial polarized neutron diffraction experiment will be carried out on the IMAGINE instrument [14]. Neutron polarization will be achieved using polarizing supermirrors, which we have tested recently. Our initial experiment will be focusing on well understood proteins such as lysozyme. In the next step, we will move on to more challenging problems such as polymerase involved in DNA repair and photosynthetic antenna complexes.

Acknowledgements
This manuscript has been authored by UT-Battelle, LLC under Contract No. DE-AC05-00OR22725 with the U.S. Department of Energy. The United States Government retains and the publisher, by accepting the article for publication, acknowledges that the United States Government retains a non-exclusive, paid-up, irrevocable, world-wide license to publish or reproduce the published form of this manuscript, or allow others to do so, for United States Government purposes. The Department of Energy will provide public access to these results of federally sponsored research in accordance with the DOE Public Access Plan (http://energy.gov/downloads/doe-public-access-plan).

References

[1] Blakeley MP, Teixeira SCM, Petit-Haertlein I, Hazemann I, Mitschler A, Haertlein M, Howard E, Podjarny AD. 2010 Neutron macromolecular crystallography with LADI-III. *Acta Cryst* D **66** 1198-1205.
[2] Hayter JB, Jenkins GT, White JW. 1974 Polarized-Neutron Diffraction From Spin-Polarized Protoons – Tools in Structure Determination. *Phys. Rev. Lett.* **33** 696-699.
[3] Stuhrmann HB, Scharpf O, Krumpolc M, Niinikoski, TO, Rieubland M, Rijlart A. 1986 Dynamic nuclear polarization of biological matter. *Eur. Biophys. J* **14** 1-6.
[4] Piegsa FM, Karlsson M, van den Brandt B, Carlile C J, Forgan EM, Hautle P, Konter JA, McIntyre GJ, Zimmer O, 2013 Polarized neutron Laue diffraction on a crystal containing dynamically polarized proton spins. *J. Appl. Cryst.* **46** 30-34.
[5] Zhao JK, Robertson L, Herwig K, Crabb D, 2013 Polarized neutron in structural biology – present and future outlook, *Physics Procedia* **42** 39-45.
[6] Knop W, Hirai M, Schink HJ, Stuhrmann HB, Wagner R, Zhao J, Schärpf O, Crichton RR, Krumpolc M, Nierhaus KH, Rijllart A, Niinikoski TO, 1992 A new polarized target for neutron scattering studies on biomolecules: first results from apoferritin and the deuterated 50S subunit of ribosomes, *J. Appl. Cryst.* **25** (2) 155-165.
[7] Zhao J, Meerwinck W, Niinikoski T, Rijllart A, Schmitt M, Willumeit R, Stuhrmann H, 1995 The polarized target station at GKSS, Nucl. Inst. Meth.A **356** 133-137.
[8] Zhao JK, Neutron Scattering Application of Polarized Solid Target in Materials Research, 2008 AIP Conference Proceedings **980**, 317.
[9] Pierce J, Crabb DG, Zhao JK, 2009 Dynamically Polarized Sample for Neutron Scattering At the Spallation Neutron Source, *SPIN PHYSICS*: 18th International Spin Physics Symposium **1149** (1) 872-875.
[10]Pierce J, Crabb DG, Tomanicek S, Demarse N, Maxwell J, Mulholland J, Zhao JK, 2010 Dynamically polarized sample at the Spallation Neutron Source, *J of Phys.* **251**, 012088.
[11]van den Brandt B, Glättli H, Hautle P, Kohlbrecher J, Konter JA, Michels A, Stuhrmann HB, Zimmer O. 2006 Creating local contrast in small-angle neutron scattering by dynamic nuclear polarization. *J. Appl. Cryst.* **40** s106-s110.
[12]Willumeit R, Burkhardt N, Diedrich G, Zhao J, Nierhaus KH, Stuhrmann HB, 1996 The in situ structure of ribosomal proteins from polarized neutron scattering *J mol struct* **383** 201-211.
[13]Zhao J, 2005 Dynamically polarized hydrogenous materials for neutron scattering, *Physica* B. **356**,168-173.
[14]Meilleur F, Munshi P, Robertson L, Stoica, AD, Crow, L, Kovalevsky A, Koritsanszky T, Chakoumakos, BC, Blessing R, Myles, DAA, 2013 The IMAGINE instrument: first neutron protein structure and new capabilities for neutron macromolecular crystallography, *J Appl Cryst.* 69, 2157-2160.

BEER – The Beamline for European Materials Engineering Research at the ESS

J Fenske[1], M Rouijaa[1], J Šaroun[2], R Kampmann[1], P Staron[1], G Nowak[1], J Pilch[2,3], P Beran[2], P Šittner[3], P Strunz[2], H-G Brokmeier[4], V Ryukhtin[2], L Kadeřávek[2,3], M Strobl[5,6], M Müller[1], P Lukáš[2] and A Schreyer[1]

[1] Helmholtz-Zentrum Geesthacht, Max-Planck-Str. 1, 21502 Geesthacht, Germany
[2] Nuclear Physics Institute, ASCR, Hlavní 130, 25068 Řež, Czech Republic
[3] Institute of Physics, ASCR, Na Slovance 2, 18221 Praha 8, Czech Republic
[4] Clausthal University of Technology, Adolph Roemer-Straße 2a, 38678 Clausthal-Zellerfeld, Germany
[5] European Spallation Source, Tunavägen 24, SE-22100 Lund, Sweden
[6] University of Copenhagen, Niels Bohr Institute, Universitetsparken 5, 2100 Copenhagen, Denmark.

E-mail: jochen.fenske@hzg.de

Abstract. The Beamline for European Materials Engineering Research (BEER) will be built at the European Spallation Source (ESS). The diffractometer utilizes the high brilliance of the long-pulse neutron source and offers high instrument flexibility. It includes a novel chopper technique that extracts several short pulses out of the long pulse, leading to substantial intensity gain of up to an order of magnitude compared to pulse shaping methods for materials with high crystal symmetry. This intensity gain is achieved without compromising resolution. Materials of lower crystal symmetry or multi-phase materials will be investigated by additional pulse shaping methods. The different chopper set-ups and advanced beam extracting techniques offer an extremely broad intensity/resolution range. Furthermore, BEER offers an option of simultaneous SANS or imaging measurements without compromising diffraction investigations. This flexibility opens up new possibilities for in-situ experiments studying materials processing and performance under operation conditions. To fulfil this task, advanced sample environments, dedicated to thermo-mechanical processing, are foreseen.

1. Introduction

Engineering materials will have paramount importance for the technological progress of mankind in the coming decades. There is presently a transition towards more complex multi-phase and/or composite materials with specially designed microstructures and tailored functional properties for particular purposes of usage. Examples of such materials are high temperature and corrosion resistant intermetallics for gas turbines, lightweight materials for transport applications or multi-functional coatings in industrial gas turbines. These novel materials, together with modern material production technologies, are urgently needed to tackle societal challenges related to sustainable development, particularly future means of transportation and mobility, energy production, distribution and storage as well as medical devices for health care and smart structures for civil engineering.

Progress in development, fabrication, optimization, and degradation monitoring of modern engineering materials is essential for the production of more efficient, more environmentally friendly and more durable engineering components. To achieve such ambitious goals, employment of science-based approaches towards material design and development as well as adoption of new methods for production, thermomechanical processing, testing and characterization of materials is required [1].

Neutron diffraction is a well-established tool for characterization of engineering materials [2,3] and components, e.g. for analysis of internal stresses and textures [4,5] or phase analysis [6]. The strength of neutron scattering, however, lies especially in in-situ investigations of advanced materials in complex sample environments [7]. Particularly for material engineering research, it will be extremely helpful to replicate real fabrication, processing and in-operando conditions in neutron beams. It will move analytical processing and performance research from post mortem analysis to yet unparalleled in-situ or in-operando analysis. Such research will lead to breakthroughs in optimization of various engineering materials processing, e.g. development of advanced methods of joining such as friction stir welding or improvements of industrial processing such as casting, hot rolling, forging and annealing. Therefore academic as well as industrial users will benefit from new in-situ and in-operando neutron diffraction experiments.

The unique concept of BEER (**B**eamline for **E**uropean Materials **E**ngineering **R**esearch) was developed to offer these possibilities. The diffractometer is based on a significant improvement in data acquisition times compared to current materials engineering flagship instruments, a flexible detector coverage, additional small-angle scattering (SANS) and imaging options and complex sample environments. The concept is mainly driven by:

(I) enabling time-resolved in-situ and in-operando investigations of structural materials during processing and exposure to simulated service environments,

(II) adopting state-of-the-art technologies for efficient and precise characterization of residual stresses, crystallographic textures and phase compositions in structural materials.

Thus, new opportunities will be offered to material engineers for studying the evolution of micro and nanostructures, phase transformations, textures and internal stresses at industrially relevant temperatures, strain rates and complex loading conditions. This will help to investigate and develop thermomechanical processing procedures as well as to study fatigue mechanisms under service conditions.

2. General Instrument Layout

BEER will be a 159 m long time-of-flight (ToF) diffractometer supplemented by a detector for optional SANS measurements. Figure 1 shows the instrument layout with the basic components: a bi-spectral extraction system inside the target monolith, a cave after the monolith with choppers for pulse shaping (PSC), frame definition (FC) and modulation (MC), followed by expanding and curved guide sections, a divergence slit, an exchangeable focusing guide and a sample stage surrounded by detectors. These components and their functions are described in the following sections.

The neutron optics design and performance parameters presented here are based on the ESS moderator concept of 2013 [8]. Since then, the neutron optics concept has been under further development to match the newly proposed low-dimensional moderators [9]. Although this modification does not affect the overall concept, some of the parameters are affected [10].

2.1. Source and bi-spectral extraction optics

The neutron optics system has to collect neutrons from both the thermal and cold moderators with an active area of 120×120 mm^2, arranged side by side with a 10 mm gap, according to the ESS moderator concept of 2013 [8]. The proposed bi-spectral extraction method follows the study by Zendler et al. [11]. It consists of an array of 7 semi-transparent mirrors on 0.5 mm thick Si substrate with a length of 0.5 m, which transmit thermal neutrons from the thermal moderator and reflect low-energy neutrons from the cold moderator into the same guide. The supermirror coating with m=4.7 (m=1 corresponds to the critical angle of Ni, 0.1°/Å) and the inclination angle of 0.95° matches the crossover wavelength of 2.0 Å between the cold and thermal spectra. The distance of the extraction system from the source of 3.7 m and the mirrors length of 0.5 m result from the beam size and source geometry. Further optimization by Monte Carlo simulations indicated an improvement of transmittance if this mirror array is slightly convergent with an angle of about ~0.75°.

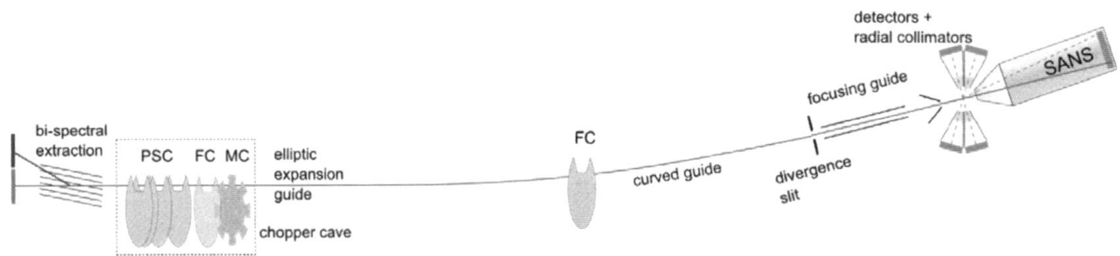

Figure 1: Schematic drawing of the BEER layout with the key components, e.g. pulse shaping (PSC), frame definition (FC) and modulation choppers (MC): for details see the text.

2.2. Neutron transport

The neutron transport optics is 157 m long and starts at 2 m behind the source with a 4 m long monolith insert including the bi-spectral extraction optics and a pair of vertically reflecting mirrors. It serves as a feeder for a narrow parallel guide (20×80 mm^2) in the chopper section, which is 3.5 m long and is interrupted by several gaps for the choppers. This section is followed by a 15 m long guide expanding elliptically from the width of 20 mm to 40 mm; the height of 80 mm is conserved. A consecutive parallel, horizontally curved guide with a length of 129 m, a cross section of 40×80 mm^2 and a radius of curvature of 20 km blocks the direct line of sight at about 85 m from the source, which helps to reduce radiation background in the experimental area. An adjustable slit for tuning the beam divergence marks the end of the long guide. It is followed by an optional 5.5 m long focusing guide that is elliptically tapered in vertical direction and parallel in horizontal direction. This focusing guide ends 1 m before the sample leaving enough space for large sample environments. The guide is divided in two sub-sections with lengths of 3.5 m and 2 m, which are removable and can be optionally replaced by slits or a horizontally focusing multichannel segment. These options allow for adjustment of the beam divergence in a wide range and, together with variable chopper settings, to efficiently trade resolution for intensity as required by the variety of experiments to be carried out at BEER. The neutron optics concept is described in more details in [10].

2.3. Pulse-shaping choppers

Pulse-shaping choppers (PSC) close to the source are required to tune the $\Delta\lambda/\lambda$ resolution in a range suitable for diffraction. Four such choppers will be installed in the cavity after the monolith, with the first chopper at a minimum practically feasible distance of 6.45 m (the basic chopper parameters are listed in table 1). This minimum distance together with the ESS pulse length of τ=2.86 ms and the repetition rate of 14 Hz basically restricts the bandwidth to $\Delta\lambda$<1.73 Å. The path length which optimally fills the interval between subsequent ESS pulses at the detector (T=71.4 ms) is L_D=163 m. Two PSC's are operated at the same time in optically blind mode, i.e. the second disk opens when the first disk closes. This technique has been introduced by van Well [12] and is nowadays a common technique used at various neutron instruments. The main advantage of this concept is that the wavelength resolution ($\Delta\lambda/\lambda$) does not depend on the wavelength, which permits more efficient use of the source flux by using longer opening times for longer wavelengths. The different distances between the choppers (150 – 1200 mm) and combinations of the four choppers allow for a $\Delta\lambda/\lambda$ resolution between 0.1 and 0.8 %.

Table 1: Basic parameters of the choppers for pulse shaping (PSC), beam modulation (MC) and frame definition (FC).

ID	distance [m]	frequency [Hz]	beam width/height [mm]	window width [deg]
			Pulse Shaping	
PSC1	6.45	168	20/80	144
PSC2	6.6	168	20/80	144
PSC3	6.9	168	20/80	144
PSC4	7.65	168	20/80	144
			Modulation	
MCa	8.95	42 … 280	20/80	16 × 4°, distance 22.5°
MCb	9.00	42 … 280	20/80	4 × 4°, distance 90°
MCc	9.50	42 … 70	20/80	1 × 180°, followed by 7 × 4°, distance 22.5°
			Frame Definition	
FC1a	8.28	14/7	20/80	70
FC1b	8.32	63/70	20/80	180
FC2a	79.55	14	40/80	180
FC2b	79.59	7	40/80	90

2.4. Frame definition choppers

There are two double disc choppers available for wavelength frame definition, FC1 and FC2, located at 8 m and 80 m from the source, respectively. Further details are listed in table 1. They permit to define either the standard bandwidth of $\Delta\lambda\approx1.7$ Å, as given by the source pulse length of 2.86 ms and the PSC distance, or an extended bandwidth. The latter can be achieved by two methods. The first one consists in the suppression of one or more subsequent pulses and tuning FC2 so that it selects a quasi-continuous wavelength band of the width $\Delta\lambda\approx n\times1.7$ Å. This mode can only be used in combination with the pulse multiplexing technique described in the following section, since the pulse shaping system does not allow for sufficient suppression of frame overlaps. The second method is the so-called alternating wavelength frame mode. In this case, the FC choppers are set in such a way that alternately either thermal or cold neutrons arrive at the detectors at subsequent time frames. In this case, the two wavelength frames are separated by the gap $\Delta\lambda_s$=1.77 Å.

2.5. Modulation choppers

For the modulation techniques three different choppers are alternatively available MCa, MCb and MCc located at a distance between 8.95 and 9.5 m from the source (the basic parameters of the MCs are listed in table 1). The choppers are operated together with the frame definition choppers FC1 and FC2 for setting the wavelength frame while the pulse shaping choppers PSC are stopped. The pulse modulation technique works as follows: Seen from one moment in time from the detector the ESS source pulse of 2.86 ms creates at the modulation chopper position at about 9 m from the moderator a virtual source with a pulse length of 2.70 ms. The MC's modulate this virtual source into a chain of sub-pulses with a pulse-to-pulse interval Δt_{ptp} depending on the frequency and the slit distance of the MC's. The number of sub-pulses is given by M_d=2.70ms/Δt_{ptp}. M_d is called hereafter as multiplexing degree. At the highest frequency of 280 Hz, up to 12 sub-pulses are formed by the MCa chopper. Thus, several neutron pulses arrive at the detectors at the same time, which are well distinguished in wavelength. At a lower frequency less sub-pulses are formed (e.g. M_d=3.5 for f=70 Hz). The pulse modulation results in a multiplexing of Bragg reflections as shown in figure 2 for a frequency of 210 Hz. Thus, the pulse modulation technique leads to a gain in intensity of 2–12 (corresponding to the multiplexing degree M_d) without relaxing the Q-resolution. This operation mode gives a huge advantage, e.g. for increasing the time resolution of in-situ experiments or reducing counting times for mapping of residual strains. The distance between split sub-peaks can be increased while keeping the resolution, if for example the peak tails shall be analyzed, by using the MCb chopper instead of MCa. Here, the larger distance between the slits leads to a larger distance between the sub-peaks.

If the pulse distance is larger than the length of the virtual source, M_d becomes <1 which means that at any moment in time neutrons of only one wavelength arrive at the detector, referred to in the following as "one-λ-mode". This is the case if MCb runs with a frequency below ≤98 Hz.

The pulse modulation technique and the one-λ-mode can be combined with the pulse suppression technique to extend the bandwidth, e.g. for capturing texture changes in in-situ experiments where the sample cannot be rotated. The unique design of MCc in combination with the alternating wavelength frame mode allows for the modulation of thermal neutrons while neutrons from the cold wavelength range pass the chopper unhindered giving an opportunity to combine diffraction and SANS measurements.

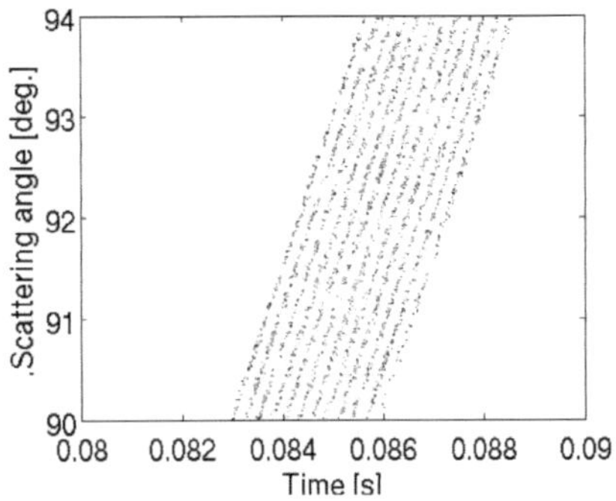

Figure 2: Examples of a multiplexed Bragg reflection (Al powder, (220)) as simulated by means of McStas [13] with a MCa frequency of 210 Hz.

2.6. Detectors

The variety of complex sample environments foreseen at the BEER instrument requires a flexible system of detectors allowing for both angular and time analysis of diffraction data. The arrangement of detectors and their distances have to allow for large in-situ devices such as the GLEEBLE® system [14] to be installed at the sample. Therefore the detectors are placed at a distance of 2 m from the sample position. The proposed detector arrangement leaves the necessary space along the axis of deformation devices. Four main detectors (D1–D4), each covering about 30×30 deg^2 (area 1 m^2) are positioned as shown in figure 3. The detectors are mounted together with attached radial collimators on rails, which permit precise radial movement and opening of the experimental space during installation of a large sample environment. A smaller detector at a shorter distance (~1.5 m) with medium spatial resolution (~5×5 mm^2) placed at one side of the incident beam allows for measurements in backscattering geometry, offering high resolution data especially for structural refinement. In the transmitted beam, a small (~40×40 mm^2) high-resolution (app. 100 μm) position-sensitive detector serves for imaging studies, with the possibility of energy analysis, i.e. Bragg edge imaging. Alternatively a 1×1 m^2 detector for SANS measurements is available in a vacuum tank at up to 6.5 m behind the sample. This maximum distance is determined by the distance of the divergence slit in front of the sample, which will be used for both diffraction and SANS.

An additional arc detector bank in the plane perpendicular to the incident beam enables sufficient angular coverage for texture and strain analyses. This detector arc will have a shorter distance (≈1 m) from the sample to optimise its active area for required angular coverage (e.g. 3 segments, 0.6×0.5 m^2 each, covering together an arc of about 100°×30°).

2.7. Sample environments

BEER will be equipped with dedicated sample environments for in-situ measurements of different kinds, covering both fast and slow processes. This includes devices for in-situ physical simulation of materials processing such as the GLEEBLE® simulator [14] or a quenching and deformation dilatometer [15]. A midsize deformation rig is foreseen for in-situ material testing as well as a small-size deformation rig positioned with a goniometer or a robot. Additionally, deformation rigs for long-term material testing and shared standard sample environments e.g. cryostats, magnets, furnaces will be used. The beamline layout allows for horizontal access to the sample area and for installation of user supplied sample environments. The main positioning table with a load capacity of 3 tons, rotation range ±360°, and vertical movement is required for precise positioning of sample environments. Several other standard positioning systems (hexapod, precise xyz stage, goniometer, robot, etc.) can be mounted on the main positioning table and will ensure a secondary precision movement of the sample or smaller sample environments.

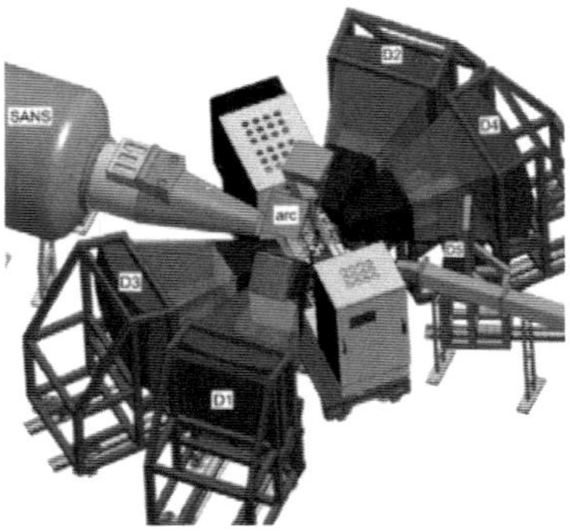

Figure 3: View of the sample area with proposed detector banks. Detectors can be retracted for bringing in large sample environment, e.g. as shown the GLEEBLE® simulator [14]. In the horizontal plane, the detectors are centred at 2θ=90° (D1), -90° (D2), 50° (D3), -130° (D4) and -160° (D5). In addition, an arc of detectors spans about 100° in the plane normal to the incident beam, filing partly the gap between the detectors D1 and D2. This arc is required mainly for texture analysis.

3. Performance

BEER offers a broad range of operation modes, e.g. pulse shaping or modulation. The neutron fluxes and resolutions for the various operation modes were obtained by Monte Carlo simulations assuming the moderator brightness baseline of 2013 [8]. The results are summarized in figure 3 together with similar data available for world leading ToF engineering diffractometers. For the ESS data, resolutions were evaluated from the peak widths of the Fe (211) and Fe (110) diffraction lines simulated for the detector bank at 2θ=75°...105°. For the other instruments, simulated data is shown when available (ENGIN-X [16], VULCAN at cold λ-range) or the data from the instruments web pages (TAKUMI [17], IMAT [18], VULCAN at thermal λ-range [19–21]). It should be underlined that these integrated intensities do not allow for direct calculation of count rates at a given diffraction line and the simulations do not include some aspects of physical reality such as absorption in Al windows or air scattering. Moreover, the flux values for the other instruments correspond to different wavelength intervals because of the different lengths and source frequencies. In spite of this, figure 3 clearly shows the excellent performance of BEER in terms of flexibility in trading resolution for high flux. The combination of variable $\Delta\lambda/\lambda$ resolutions and focusing optics allows for a significantly wider range of choices concerning tailoring flux and resolution to the needs of specific measurements as compared to the existing instruments at short-pulse sources. In addition, the pulse modulation method can preserve high flux at high resolutions, provided that the crystal structure allows for evaluation of the multiplexed diffraction lines.

The simulations show that the expected high flux enables very fast measurements. Counting times shorter than 1 s can be reached in the high flux (HF) and pulse modulation (MLR) modes with focusing optics. Even single pulse measurements (sampling at 14 Hz) seem to be possible for specific samples and studies, e.g. in-situ hot-compression tests on 2Mn-0.2C steel rods with a diameter of 7 mm using the medium resolution mode (MR).

The performance of BEER allows for residual stress investigations and for texture measurements. The detector arrangement provides the opportunity to measure a complete pole figure by the rotation of the sample around just the tension axis.

Furthermore, BEER offers the opportunity to combine diffraction measurements with SANS or imaging without a penalty for the diffraction performance. The slit at a distance of L=6.5 m from the sample for tuning the beam divergence for diffraction can also be used to define an aperture for imaging. Its size, D is continuously adjustable, allowing thus for imaging measurements with the L/D collimation ratio ranging from 115 to more than 1000. Although the imaging characteristics are in general worse when compared to ODIN [22,23] the proposed dedicated imaging instrument at ESS, they become complementary for high energy resolution. For $\Delta\lambda/\lambda$ ~0.3% (at λ=2.0...3.7 Å) and an L/D ratio of 504 the simulated flux is about $3.4\cdot10^6$ n/s/cm^2. The corresponding flux per wavelength interval (flux density) is about a factor 2.6 higher than at ODIN and allows for sub-second exposure times. The tuneable wavelength resolution is suitable for energy resolved imaging, i.e. Bragg edge analysis, and hence mapping of various microstructural characteristics, e.g. phase composition, texture and strains [24,25].

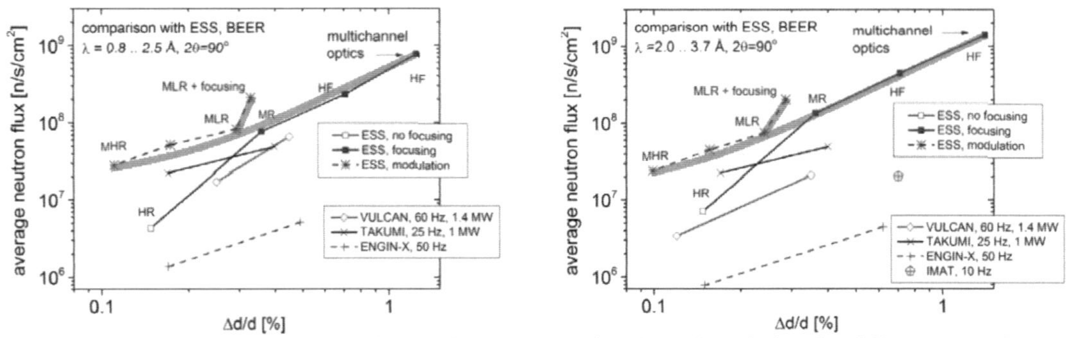

Figure 3: Simulated time-averaged flux at the sample as function of resolution for different operation modes (Pulse modulation with high resolution (MHR) ($\Delta\lambda/\lambda$=0.07%, divergence slit: 10×80 mm²), pulse modulation with low resolution (MLR) (0.27%, 40×80 mm²), high resolution (HR) (0.1%, 15×80 mm²), medium resolution (MR) (0.3%, 40×80 mm²) and high flux (HF) (0.8%, 40×80 mm²)) of BEER at thermal (left side) and cold (right side) wavelength ranges. The values for the modulation mode were simulated without focusing, except of the last point as indicated. Data available for VULCAN (SNS), TAKUMI (MLF), ENGIN-X and IMAT (both ISIS) are shown for comparison.

Like for imaging, the divergence aperture can be used to increase the resolution for simultaneous diffraction and SANS measurements. Its position determines the optimum SANS detector distance to be about 6.5 m. The alternating frame technique has to be used to define thermal and cold neutron wavelength bands in subsequent pulses in order to carry out simultaneous SANS and diffraction measurements with pulse shaping choppers. Positions of these bands can be varied synchronously by changing the chopper phases in order to extend the Q-range. Table 2 gives an example of conditions under which the simultaneous SANS and diffraction measurement are feasible. The calculation assumes a medium resolution diffraction mode with the divergence slit size of 40×40 mm^2 and $\Delta\lambda/\lambda$=0.3%. Although the SANS characteristics are worse when compared to proposed dedicated instruments at the ESS, e.g. SKADI [26], they are still significantly better than e.g. for the well established V4 instrument at HZB [27] (flux ~4.6·10^6 n/s/cm^2 for λ=5 Å, $\Delta\lambda/\lambda$=10%, collimation and detector distances L=2 m, divergence slit 30×50 mm^2). It can thus be concluded that extremely important information about kinetics of formation of pores and precipitates can be obtained by the combination of diffraction measurements with SANS without compromising diffraction investigations.

Table 2: Examples of estimated SANS and diffraction characteristics in the case of simultaneous SANS and diffraction measurements.

	flux	wavelength	resolution	d-range
diffraction	9.6·10^6 [n/s/cm²]	1.2 ... 2.9 Å	Δd/d ~ 0.4 %	0.7 ... 2.3 Å
SANS	6.2·10^6 [n/s/cm²]	4.7 ... 6.3 Å	δQ ~ 0.003 Å^{-1}	20 ... 350 Å

4. Summary

BEER will be a unique instrument that offers a broad range of operation modes, spanning over one decade in resolution. The simulations predict a neutron flux approaching 10^9 n/cm^2/s which unlocks an unparalleled potential for fast kinetic experiments with sub-second time resolution. Under favourable circumstances, e.g. strong scattering and suitable sample volume, measurements with a single pulse, i.e. a 14 Hz sampling rate, should be feasible. In addition, the novel pulse modulation technique permits to preserve high flux also at high resolution in the case of materials with high symmetry, i.e. with well-separated diffraction lines. The possibility of simultaneous diffraction and SANS or imaging measurements without penalty for the diffraction performance strengthens the ability of the instrument to address growing interest in microstructure characterization during material processing. BEER will also provide a large experimental area at the sample position, allowing for unique in-situ experiments with large high-power sample environment devices required for such studies.

Acknowledgments
This work has been supported by the BMBF under the contract number 05E11G1 within the German contribution to the Design-Update project of the European Spallation Source. Preparation of the instrument project was further supported by the Czech Ministry of Education, Youth and Sports under the project LM 2010011.

References

[1] Jarvis D 2012 *Metallurgy Europe - A Rennaissance Programme for 2012-2022* (Strasbourg) <http://www.esf.org/fileadmin/Public_documents/Publications/metallurgy_europe.pdf>
[2] Wang X L 2006 *JOM* **58** 52
[3] Reimers W, Pyzalla A, Schreyer A and Clemens H 2008 *Neutron and Synchrotron Radiation in Engineering Materials Science* (Weinheim: Wiley-VCH)
[4] Krawitz A D 2001 *Introduction to diffraction in material science and engineering* (New York: Wiley)
[5] Brokmeier H-G and Yi S B 2008 *Neutron and Synchrotron Radiation in Engineering Materials Science,* ed W Reimers, A Pyzalla, A Schreyer and H Clemens (Weinheim: Wiley-VCH) pp 57-77
[6] Billinge S J L 2006 *JOM* **58** 47
[7] Kannengiesser T, Babu S S, Komizo Y and Ramirez A J 2010 *In situ Studies with Photons, Neutrons and Electrons Scattering* (Berlin Heidelberg: Springer)
[8] Peggs S 2013 *ESS Technical Design Report* (Lund) <http://eval.esss.lu.se/cgi-bin/public/DocDB/ShowDocument?docid=274> 2015-10-07
[9] Ferenc Mezei F, Zanini L, Takibayev A, Batkov K, Klinkby E B, Pitcher E and Schönfeldt T 2014 *Journal of Neutron Research* **17** 101
[10] Šaroun J, Fenske J, Rouijaa M, Beran P, Navrátil J, Lukáš P, Schreyer A and Strobl M *Neutron optics concept for the materials engineering diffractometer at the ESS,* this issue
[11] Zendler C, Lieutnant K, Nekrassov D, Cussen L D and Strobl M 2013 *Nucl. Inst. Meth.* A **704** 68
[12] van Well A A 1992 *Physica* B **180–181** 959
[13] Willendrup P, Farhi E and Lefmann K 2004 *Physica* B **350** 735
[14] <gleeble.com> 2015-10-07
[15] Staron P et al. 2011 *Advanced Engineering Materials* **13** 658
[16] private communication with Shu Yan Zhang 2015-10-01
[17] <http://j-parc.jp/researcher/MatLife/en/instrumentation/ns_spec.html#bl19> 2015-10-07
[18] Kockelman W, Zhang S Y, Kelleher J F, Nightingale J B, Burca G and James J A 2013 *Physics Procedia* **43**, 100
[19] <http://neutrons.ornl.gov/sites/default/files/Instrument_7.pdf> 2015-10-07
[20] An K, Skorpenske H D, Stoica A D, Ma D, Wang X L and Cakmak E 2011 *Metallurgical and Materials Transactions* A **42** 95
[21] Wang X L, Holden T M, Stoica A D, An K, Skorpenske H D, Jones A B, Rennich G Q and Iverson E B 2010 *Mater. Sci. Forum* **652** 105
[22] <https://europeanspallationsource.se/sites/default/files/odin_imaging_instrument_construction_proposal.pdf> 2015-10-07
[23] Strobl M 2015 *Physics Procedia* **69** 18-26
[24] Santisteban J R, Edwards L, Steuwer A and Withers P J 2001 *J. Appl. Cryst.* **34** 289
[25] Strobl M 2009 *Nucl. Instr. and Meth.* A **604** 646–652
[26] <https://europeanspallationsource.se/sites/default/files/skadi_proposal.pdf> 2015-10-07
[27] <https://www.helmholtz-berlin.de/pubbin/igama_output?modus=einzel&sprache=en&gid=1706&typoid=39942> 2015-10-07

The HB-2D Polarized Neutron Development Beamline at the High Flux Isotope Reactor

Lowell Crow, WA Hamilton, JK Zhao, and JL Robertson

Instrument and Source Division
Oak Ridge National Laboratory, Oak Ridge, TN 37831, USA

E-mail: crowmljr@ornl.gov

Abstract. The Polarized Neutron Development beamline, recently commissioned at the HB-2D position on the High Flux Isotope Reactor (HFIR) at Oak Ridge National Laboratory, provides a tool for development and testing of polarizers, polarized neutron devices, and prototyping of polarized neutron techniques. With available monochromators including pyrolytic graphite and polarizing enriched Fe-57 (Si), the instrument has operated at 4.25 and 2.6 Å wavelengths, using crystal, supermirror, or He-3 polarizers and analyzers in various configurations. The Neutron Optics and Development Team has used the beamline for testing of He-3 polarizers for use at other HFIR and Spallation Neutron Source (SNS) instruments, as well as a variety of flipper devices. Recently, we have acquired new supermirror polarizers which have improved the instrument performance. The team and collaborators also have continuing demonstration experiments of spin-echo focusing techniques, and plans to conduct polarized diffraction measurements. The beamline is also used to support a growing use of polarization techniques at present and future instruments at SNS and HFIR.

1. Introduction

Oak Ridge National Laboratory (ORNL) operates two major neutron facilities, the Spallation Neutron Source and the High Flux Isotope Reactor (HFIR). The HFIR has been operational at ORNL since 1965 [1]. It is a light water cooled and moderated, beryllium reflected reactor with a cylindrical "flux trap" core structure. Thermal neutron flux up to 2.5 x 10^{15} n/cm^{2}s is available for isotope irradiation, and the source thermal neutron flux for the beam tubes is up to 1.2 x 10^{15} n/cm^{2}s.

The HB-2 beam tube has a radial view of the HFIR core, so that both thermal and fast neutron flux, in addition to core gamma rays, are present. A cooled sapphire crystal assembly serves as a fast neutron filter. The shield assembly at the end of the HB-2 beam tube allows for installation of crystal monochromators for 4 beamlines. Neutron user instruments presently occupy three of the positions: (1) the HB-2A Powder Diffractometer [2], (2) the HB-2B Neutron Residual Stress Facility, and (3) the HB-2C Wide Angle Neutron Diffractometer. The HB-2D position has a variable angle monochromator drum, and has previously hosted the MIRROR neutron reflectometer, and the SNS detector test station. Recently, we have recommissioned HB-2D as the Polarized Neutron Development Beamline.

The monochromator drum, originally intended for a triple-axis spectrometer, is operated at a fixed angle for purposes of the present instrument. The monochromator position is 10.9 m from the end of the beam tube. The 78.6° monochromator angle yields a wavelength of 4.25 Å for a graphite monochromator, far from the approximately 1 Å spectral peak of the thermal beam tube. With a 100 mm thick Be filter at room temperature, the flux in this configuration is a modest 2 x 10^5 n/cm^{2}s. Reconfiguration for other monochromators at the same angle is straightforward. For example, we have tried, but not yet optimized for good polarization, an ^{57}Fe(Si) polarizing monochromator, which yields a primary wavelength of 2.6 Å and a flux of about 10^6 n/cm^{2}s.

Figure 1. This figure shows the HB-2D monochromator drum (upper left), and the variable height lift tables between the exit port and the beamstop.

The beamline (Figure 1) has about 4 m between the exit port and the beam stop. There are 4 variable height hydraulic lift breadboard tables, enabling rapid reconfiguration of beam components. The instrument has necessary components to operate as a polarimeter, a reflectometer, or a diffractometer, depending on requirements of the measurement.

Very recently, the instrument team has acquired two modern supermirror polarizers, which enable much better polarization performance. One is a Neutron Optics Berlin S-bender, optimized for cold neutron wavelengths. The second is a Swiss Neutronics V-cavity, which operates well at wavelengths as short as 2.4 Å.

2. Examples of component measurements.

One function of HB-2D is testing of polarization components for development or for use on other ORNL neutron instruments. It has hosted measurements of ^{3}He polarizers, including SEOP in-situ units [3], "drop-in" cells, and an experimental cell to measure the internal ^{3}He gas temperature. It has been used in a continuation of flipper development started on CG1 [4,5] with various flipper prototypes including cryoflippers, Drabkin coils, rf flippers, and dc flippers.

As an example, we show the results of an attempt to create a cryoflipper by tiling YBCO films on sapphire, using the configuration shown in Figure 2 [6]. The YBCO films were arranged in three different patterns as shown. In each case the YBCO was cooled to 90 K and translated across a 1 mm wide slit in the polarized beam. A single slab shows a uniform flipping performance across its width. With gaps between the films, the flipping effect is nearly absent within the film gaps, as expected. When the films are overlapped, there is clearly enough field leakage to compromise the flipping performance. This indicates that edge effects are a potential issue in any application using YBCO films as neutron Meissner screens.

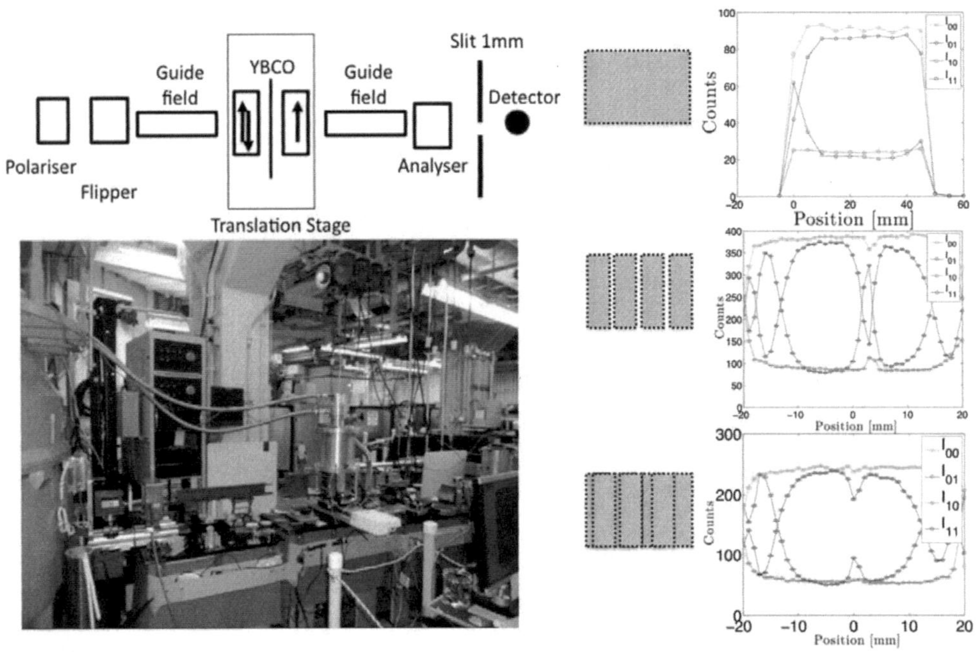

Figure 2. The schematic in the upper left illustrates the experimental arrangement for measuring the effect of gaps in YBCO assemblies used as cryoflippers. The solid slab (upper right) shows a uniform response as the film is translated across the beam. The assembly with gaps (middle right) shows that the flipping effect nearly disappears within the gaps. The assembly with overlaps (lower right) shows that even with an overlap, there is a large change due to the use of multiple slabs.

3. Tests of Spin Polarization Instrument prototypes.

HB-2D has hosted tests of spin manipulation assemblies. The polarized neutron group from Indiana University has developed a "Cryo-Cup" Spherical Neutron Polarimeter [7]. In the SNP apparatus, cooled to 20 K, the incoming and outgoing neutron polarization are controlled separately using fields generated by high-Tc coils. YBCO films serve as Meissner shields for non-adiabatic neutron spin transitions. Measurements of a 10 micron thick slab of permalloy with the easy axis in-plane showed a three-dimensional polarization matrix consistent with magnetization measurements [8].

Another device demonstrated at HB-2D is the prototype Modulated Intensity by Combined Effort (MICE) [9]. In the MICE technique, the neutron spin is encoded by rf flippers. The neutrons travel through an adjustable-current solenoid between the rf flippers; since it does not rely on a zero-field region, the performance should be robust in moderate magnetic field environments. The initial demonstration experiment, using the apparatus shown in Figure 3, clearly demonstrated time focusing, as shown in Figure 4.

Figure 3. The Modulated Intensity by Combined Effort system is shown in its initial configuration. The square coils before and after the cylindrical solenoid are rf flippers. The solenoid current is varied to conduct a scan. A neutron detector downstream detects neutrons in time-of-flight, with the time base synchronized to the coils.

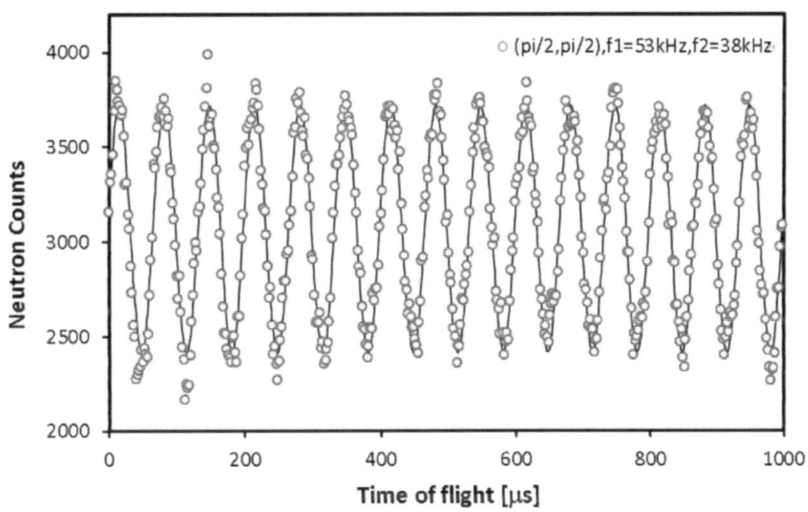

Figure 4. A typical MICE pattern is shown, with an initial coil frequency of 53 kHz and a final coil frequency of 38 kHz. The 15 fringes in 1000 μs is as predicted for this configuration.

4. Summary and future plans

Activities underway at HB-2D include continued development of the MICE technique and tests of a Wollaston prism prototype from Indiana University [10]. Measurements of components such as flippers and ^{3}He polarizers in support of other ORNL instruments will continue. The instrument now has a working diffractometer arm, which will be used in demonstration polarized diffraction measurements in the near future. The work reported here was conducted before improved polarizers were available, so the improved polarization performance will lead to better measurements.

Acknowledgments

The authors thank SR Parnell and T Wang for participating as users at HB-2D and forwarding preliminary reports of their work. This manuscript has been authored by UT-Battelle, LLC under Contract No. DE-AC05-00OR22725 with the U.S. Department of Energy. The United States Government retains and the publisher, by accepting the article for publication, acknowledges that the United States Government retains a non-exclusive, paid-up, irrevocable, world-wide license to publish or reproduce the published form of this manuscript, or allow others to do so, for United States Government purposes. The Department of Energy will provide public access to these results of federally sponsored research in accordance with the DOE Public Access Plan (http://energy.gov/downloads/doe-public-access-plan).

References

[1] Weinberg A 1967 *Phys. Today* **20**(6) 23
[2] Garlea VO, Chakoumakos BC, Moore SA, Taylor GB, Chae T, Maples RG, Riedel RA, Lynn GW, Selby DL 2010 *Appl. Phys. A – Mat. Sci. and Processing* **99**(3) *Special Issue* 531
[3] Jiang CY, Tong X, Brown DR, Chi S, Christianson AD, Kadron BJ, Robertson JL, Winn BL 2014 *Rev. Sci. Instr.* **85** 075112
[4] Parnell SR, Washington AL, Kaiser H, Li F, Wang T, Hamilton WA, Baxter DV, and Pynn R 2013 *Nucl. Instr. and Meth. A* **722** 20
[5] Crow L, Robertson L, Bilheux H, Fleenor M, Iverson E, Tong X, Stoica D, Lee WT 2011 *Nucl. Instr. and Meth. A* **634** S71
[6] Parnell SR and Hamilton WA 2015 Private Communication
[7] Wang T, Li F, Parnell SR, Hamilton WA, Kaiser H, Washington AL, Baxter DV, and Pynn R 2014 *J. Phys.: Conf. Ser.* **528** 012024
[8] Wang T 2015 Private Communication
[9] Zhao J, Hamilton WA, Lee S-W, Robertson JL, Crow L and Kang YW 2015 *Appl. Phys. Lett.* **107**, 113508
[10] Li FK and Pynn R 2014 *J. Appl. Cryst.* **47**, 1849

Neutron optics concept for the materials engineering diffractometer at the ESS

J Šaroun[1], J Fenske[2], M Rouijaa[2], P Beran[1], J Navrátil[1,3], P Lukáš[1], A Schreyer[2] and M Strobl[4]

[1] Nuclear Physical Institute, ASCR, Hlavní 130, 25068 Řež ,Czech Republic
[2] Helmholtz-Zentrum Geesthacht, Max-Planck-Str.1, 21502 Geesthacht, Germany
[3] Czech Technical University, Faculty of Nuclear Sciences and Physical Engineering, Břehová 7, 11519 Praha 1, Czech Republic.
[4] European Spallation Source, Tunavägen 24, 22363 Lund, Sweden

E-mail: saroun@ujf.cas.cz

Abstract. The Beamline for European Materials Engineering Research (BEER) has been recently proposed to be built at the European Spallation Source (ESS). The presented concept of neutron delivery optics for this instrument addresses the problems of bi-spectral beam extraction from a small moderator, optimization of neutron guides profile for long-range neutron transport and focusing at the sample under various constraints. They include free space before and after the guides, a narrow guide section with gaps for choppers, closing of direct line of sight and cost reduction by optimization of the guides cross-section and coating. A system of slits and exchangeable focusing optics is proposed in order to match various wavelength resolution options provided by the pulse shaping and modulation choppers, which permits to efficiently trade resolution for intensity in a wide range. Simulated performance characteristics such as brilliance transfer ratio are complemented by the analysis of the histories of "useful" neutrons obtained by back tracing neutrons hitting the sample, which helps to optimize some of the neutron guide parameters such as supermirror coating.

1. Introduction

The long pulse concept of the European Spallation Source (ESS) brings about new challenges for the design of neutron scattering instruments. While other spallation sources provide sufficiently short pulses which can be directly used for wavelength definition, the 2.86 ms long pulse of the ESS needs to be further tailored in most cases in order to achieve sufficient resolution. It is particularly the case of diffraction experiments, where the relative wavelength resolution $\Delta\lambda/\lambda$ below 1% must be achieved at short wavelengths. This leads to an increased complexity and cost of such instruments due to the installation of choppers for pulse shaping. On the other hand, instruments at ESS can profit from the pulse tailoring as it permits to adjust instrument resolution to a particular problem. This feature can bring significant advantage over the short pulse sources, where the pulse width and its dependence on the wavelength is determined by the moderator and the possibility to trade resolution for intensity is limited to adjusting beam divergence [1-3]. However, the possibility of choosing wavelength resolution does not provide more flexibility by itself, it has to be accompanied by properly designed neutron optics which allows to match the beam properties, especially the divergence, to the wavelength resolution.

This paper introduces the concept of neutron delivery optics for the diffractometer BEER (Beamline for European materials Engineering Research) at ESS, which allows for such a flexibility. It is a 159 m long time-of-flight (ToF) powder diffractometer that has been proposed to address the needs of materials engineers from both academic and industrial areas. It will provide new capacity for in-situ and in-operando studies of materials under external conditions, which are relevant for materials production, treatment and use, but also in the traditional domain of residual strain mapping. In addition, optional measurements of small-angle scattering (SANS) and direct imaging are envisaged [4]. This broad scale of techniques calls particularly for the above mentioned adaptability of neutron optics. The required quantitative characteristics of the neutron beam in the main operation modes are derived in Section 2, which is followed by the description of the main components of the neutron transport system (Section 3). Section 4 then summarizes the performance of the system in terms of brilliance transfer efficiency.

2. Required neutron beam characteristics

Regarding the neutron beam properties, the main criteria for an engineering diffractometer are the wavelength resolution and bandwidth, gauge volume size and angular divergence. The neutron transport system can be well characterized by the quantity called brilliance transfer (BT) ratio defined as the neutron intensity integrated over a pre-defined beam area, solid angle and wavelength interval, relative to the source brightness. Due to the Liouville's theorem, BT ratio ≤ 1. It describes the quality of the neutron delivery optics alone, regardless of the source spectrum. It is however important to define the integration intervals adequately with respect to the assumed mode of operation. In addition, qualitative criteria such as uniformity of angular distribution should be considered.

2.1. Wavelength

While it is relatively easy to transport cold neutrons, the short wavelength limit is critical for the design of neutron guides. In contrast to general purpose powder diffractometers, experiments at an engineering diffractometer are usually carried out on materials with known crystallographic structure of main constituting phases. This partly relieves the requirements on wavelength bandwidth since a smaller set of reflections is required for Rietveld analysis to yield data on phase composition, lattice strain or texture compared to full structure refinement. From this point of view, the BT ratio close to the lower wavelength limit of $\lambda \geq 1$ Å is a suitable figure of merit for optimizations. Although the instrument can operate below 1 Å, its performance there can be traded for other important characteristics without significant impact on the scientific output. On the other hand, delivery of cold neutrons up to $\lambda \cong 6$ Å or more is also needed in order to be able to measure diffraction in transmission mode (Bragg edge), to increase contrast for imaging experiments and to reach suitable momentum transfer range for SANS. Therefore, a bi-spectral extraction optics has to be incorporated into the neutron guide feeder.

2.2. Gauge volume

The required beam area varies significantly according to the experiment type. It may be very small (below 1 mm^2) in the case of residual strain mapping, but also rather large in the case of in-situ experiments (typically 5 x 10 mm^2). The latter value has been chosen as the integration interval for BT evaluation since this is the typical operation mode assumed at BEER.

2.3. Trading resolution for intensity

Tuning of the beam divergence with respect to the required resolution in lattice parameter, $\Delta d/d$ is assumed for the reasons explained in Section 1. Relaxing resolution in order to boost the neutron flux is meaningful only if it leads to an increase, or at least conservation, of the overall figure of merit. Whereas various proposals for such a figure of merit in the case of a powder diffractometer can be found in literature, the traditional definition as the *integral intensity over the square of the peak width*,

$$FoM = I\left(\frac{\Delta d}{d}\right)^{-2} \tag{1}$$

is well justified for an engineering diffractometer [5]. The intensity, I is proportional to the phase space volume transported by the neutron guides and choppers from the source, that is to the product of the widths in wavelength, $\Delta\lambda$, horizontal divergence, $\Delta\alpha$, and vertical divergence, $\Delta\beta$,

$$I \propto \Delta\lambda \cdot \Delta\alpha \cdot \Delta\beta \,. \tag{2}$$

These three components also contribute to the instrument resolution, which can be formally expressed as

$$\left(\frac{\Delta d}{d}\right)^2 \cong \left(\frac{\Delta\lambda}{\lambda}\right)^2 + \left(\frac{\Delta\alpha}{2\tan\theta}\right)^2 + \left(\frac{h\,\Delta\beta}{8\,L_D\tan\theta}\right)^2 \tag{3}$$

where h and L_D are the detector height and distance, respectively and 2θ is the scattering angle. This equation holds only approximately for 2θ near $\pm$ 90°, it neglects other contributions (sample size, detector resolution) and second order effects (aberration), but it still provides a useful guide for the estimation of optimum beam divergences. It is important to note that the widths $\Delta\lambda$, $\Delta\alpha$ and $\Delta\beta$ in the above equations should take into account different distribution functions of the respective variables. They are therefore calculated from the corresponding variances as $\Delta x^2 = 8\ln 2\ [\langle x^2\rangle - \langle x\rangle^2]$.

The highest value of the figure of merit is achieved if all the three terms in equation (3) keep approximately the same value. Neutron intensity then increases with the 3rd power of resolution as desired. This condition permits to estimate optimum beam divergences for various settings of the pulse shaping choppers. Examples for three different resolution modes are given in Table 1, where $\Delta\lambda/\lambda$ was calculated from pulse profiles simulated at λ=2 Å.

Table 1. Wavelength resolutions provided by various chopper settings and corresponding optimum horizontal and vertical divergences.

resolution	$\Delta\lambda/\lambda$ [%]	$\Delta\alpha$ [mrad]	$\Delta\beta$ [mrad]
high	0.08	1.5	12
medium	0.20	4.0	32
low	0.52	10	83

2.4. Constraints

Apart of the beam characteristics given above, the concept of neutron optics must comply with several restrictions. One of them is the free space of 2 m between the source and the guide system entry required due to radiation heating and source design. This is a major obstacle for efficient neutron transport from low dimensional moderators, such as the 30 mm high pancake moderator [6] considered in this work. A similar problem occurs on the opposite side, where the minimum 1 m distance between the guide exit and sample is required in order to provide space for the large sample environment devices planned for BEER. Another limitation arises from the need to install a cascade of choppers in the cavity just after the exit from the target monolith, between 6 and 9.5 m from the source. There, several gaps in the neutron guide have to be provided for the choppers and the beam width is reduced to about 20 mm in order to achieve short transition times. Closing the direct line of sight in order to reduce background arising mainly from fast neutrons is relatively easily achieved by bending of the long neutron guide. However, the high power of the ESS source requires that this closure takes place before one half of the instrument length. This obviously affects the brilliance transport efficiency at short wavelengths. Last, but not least, cost is an important factor for such a long neutron transport system. Significant cost reduction can be achieved by reducing the cross-section of neutron guides and the critical angle of their multilayer coating. Therefore, a trade-off between these parameters and instrument performance has to be an integral part of the conceptual design of neutron guides. The following section shows how all these requirements were incorporated into the neutron optics concept of BEER.

3. Components of the neutron optics system

The initial concept of the instrument has been naturally created on the basis of existing knowledge and experience regarding efficient neutron transport over long distances. Recent studies show that very high BT ratios can be achieved with elliptic guides or ballistic guides with elliptical ends [7-9]. Although an elliptic guide is theoretically an ideal tool for point-to-point focusing, aberration effects strongly influence its real performance already with rather small source and target sizes of few millimetres. Several strategies for reducing the aberration effects have been suggested in literature, such as the double ellipse [10-11] or hybrid elliptic-parabolic guide [12] configurations. However, the extreme eccentricity of the elliptic profile required for neutron transport makes the analytical approach based on a single or double reflection model unusable in many practical situations, especially in the case of long transport distances. Quantitative evaluation of the guide system performance and its optimization therefore requires numerical simulations. The advantage of simulations is not only in the high level of physical reality they can describe, but also in the ability to yield information that can't be obtained by other means, including measurements. Namely, recording of neutron histories during ray-tracing permits to back-trace neutrons that hit the sample and that are thus relevant for an experiment. Analysis of these "useful" neutrons can reveal where they come from, where they get absorbed or reflected etc. We have used this approach to more deeply analyse properties of the proposed neutron optics system and to optimize some of its parameters, such as supermirror coating.

Neutron optics concept for the BEER instrument is therefore based on the ballistic guide geometry with elliptical ends, which was further modified to match the requirements described in Section 2 and optimized by ray-tracing simulations with the program SIMRES [13]. For numerical optimizations by the particle swarm algorithm [14], we have defined the figure of merit as the BT ratio calculated for the beam area 5 x 10 mm^2 (width x height) and solid angle 5 x 30 mrad^2. This choice corresponds to the assumed most usual operation mode with in-situ experiments at the medium resolution (see Table 1). The following sections describe the three main parts of the neutron transport system: (i) beam extraction, (ii) long-range transport and (iii) focusing at the sample. Their configuration was optimized for the ESS pancake source concept [6], assuming 10 cm wide, 3 cm high thermal and cold moderators placed side by side, separated by a 2 cm gap. This concept has been recently replaced by a new "butterfly" configuration, therefore the parameters given in this paper are probably not the final ones to be used in the instrument construction. However, the two moderator geometries are rather similar as for the moderators height and separation and the neutron optics can thus be rather easily adapted to the new geometry without major modification of the presented concept.

3.1. Neutron beam extraction

The beam extraction starts at 2 m from the source by the 4 m long monolith insert, including the bi-spectral multi-mirror optics and neutron guides. It is followed by a 3.5 m long chopper cave where the guide is interrupted by several gaps and the width is reduced to 20 mm. The guide than expands to reach the final cross-section of 40 x 80 mm^2 at 15.5 m from the source (figure 1a,b). Vertically, the guide profile forms part of a semi-ellipse with the focal distance optimized for the source height of 30 mm.

In the horizontal plane, the feeder is less important as the moderators are sufficiently wide to over-illuminate the 2 cm wide beam port at the exit from the monolith. The acceptance area of this setup fits well the moderator geometry (see figure 1c).

The bi-spectral extraction system proposed for BEER has been studied earlier by C. Zendler et al. [15]. It consists of an array of 8 semi-transparent mirrors on 0.5 mm thick single crystal Si substrate, which transmit thermal neutrons from the thermal moderator, but reflect low-energy neutrons from the cold moderator into the same guide. The length of 0.5 m, supermirror coating with m=4 (m=1 corresponding to the critical angle of Ni, 0.1°/Å) and inclination angle 0.78° were calculated to match the beam size and the crossover of the cold and thermal spectra at λ_c=1.95 Å. Given the separation of the moderators, the mirrors have to start at about 4.15 m from the source. Good performance of such a system was predicted by ray-tracing simulations, showing the efficiency between 70 % and 80 % for both thermal and cold neutrons. Note that the "useful" neutrons are extracted from a smaller area on the cold source (figure 1c) due to the focusing effect of the multi-mirror setup which acts as a Laue lens. This effect would be enhanced if the device can be placed in the middle between the source and the 20 mm beam port.

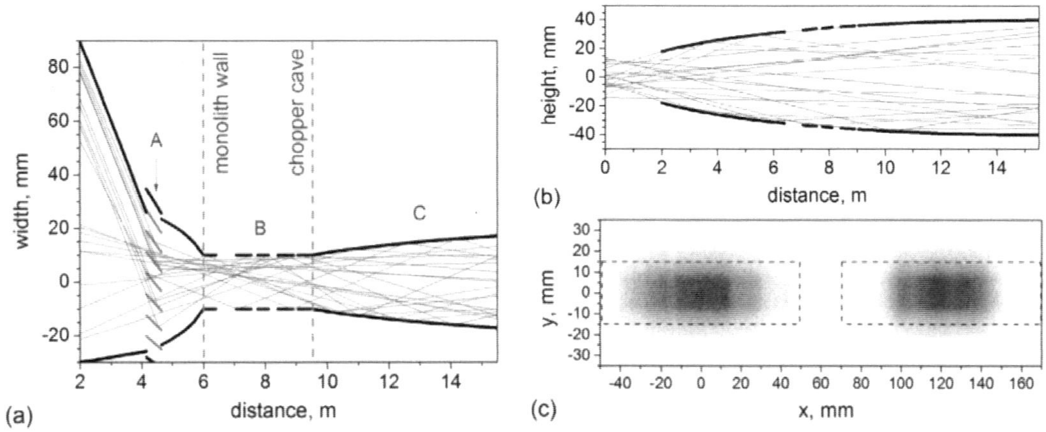

Figure 1. (a) Horizontal profile of the beam extraction including the bi-spectral optics (A), the narrow chopper section (B) and the expanding section (C). (b) Corresponding vertical profile. The thin lines show a sample of neutron trajectories. (c) Acceptance diagram of the guide system represented by the intensity of "useful" neutrons emitted at the source position (the rectangles show the active moderator areas).

3.2. Long-range neutron transport

The long guide spanning the distance between the extraction and focusing sections has to tackle two main problems: (i) transport of a large vertical divergence over the long distance and (ii) closing the direct line of sight for fast particles. Optimization of the vertical profile was carried out in order to minimize the guide height (and therefore cost), while keeping a high BT ratio. It has been modeled by a ballistic guide of the total length L=154 m spanning the whole length of the guide system. The focal length, f of the elliptic ends (practically equal to the major semi-axis length) was taken together with the entry and exit heights as a free parameter for numerical optimization. Resulting BT values as a function of the guide height (minor axis of the ellipse) are plotted in figure 2.

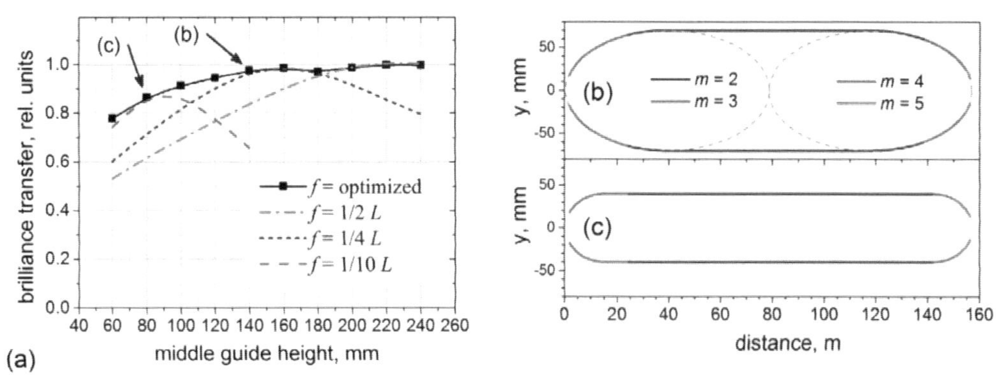

Figure 2. (a) Simulated brilliance transfer ratio as a function of the maximum guide height with optimized (points) and fixed (dashed lines) lengths of the elliptic sections. The arrows point to the two solutions on the right panel. (b) The optimum profile corresponding to a double ellipse. (c) The cost-efficient solution with reduced height.

This result shows that the guide height can be reduced from a full ellipse configuration (f=L/2) down do a ballistic profile enveloping a double ellipse, f=L/4 (figure 2b) without any loss of transport efficiency. The guide height can be further reduced if a small performance reduction is accepted, leading to the proposed setup with the height of 80 mm (figure 2c) and BT reduction to about 85 % of the maximum. Closing the direct line of sight as close as possible to the source requires a compromise to be made between the guide curvature and transmission of short wavelength neutrons. Figure 3a shows the intensity of illumination of the guide walls by straight rays emitted from the source as simulated by the ray-tracing method. The curvature $1/R > 0.06$ km^{-1} is necessary in order to suppress the direct line of sight before one half of the instrument length. Figure 3b then illustrates that this value is also near the limit of what is acceptable regarding transport efficiency, costing about 8% of brilliance transfer at λ =1 Å. The 126 m long parallel guide with cross-section 40 x 80 mm^2 and curvature 0.06 km^{-1} is therefore proposed for BEER as a cost-efficient solution.

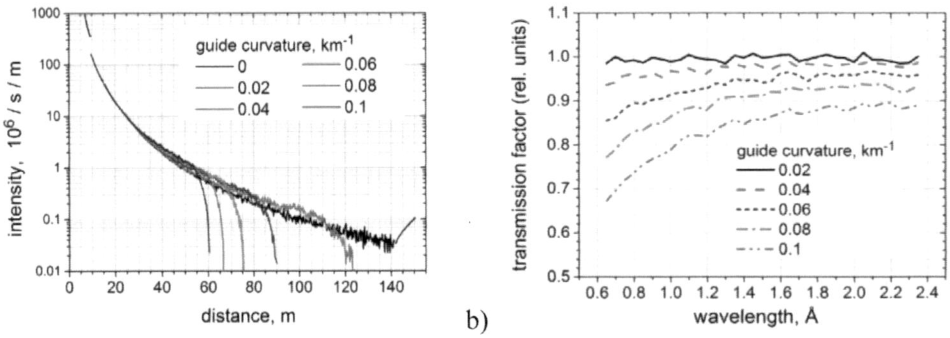

Figure 3. (a) Illumination of the guide walls by direct rays from the source. (b) Decrease of the brilliance transfer ratio due to the guide bending as a function of wavelength.

3.3. *Focusing at the sample*

After the bent parallel guide, neutrons are focused vertically by a 14.5 m long elliptic guide at the sample, which is placed 1 m after the guide exit leaving thus enough space for large sample environment. The guide width remains constant at 40 mm. This guide is actually divided in 3 sub-sections with the lengths 9 m, 3.5 m and 2 m, respectively (figure 4a).

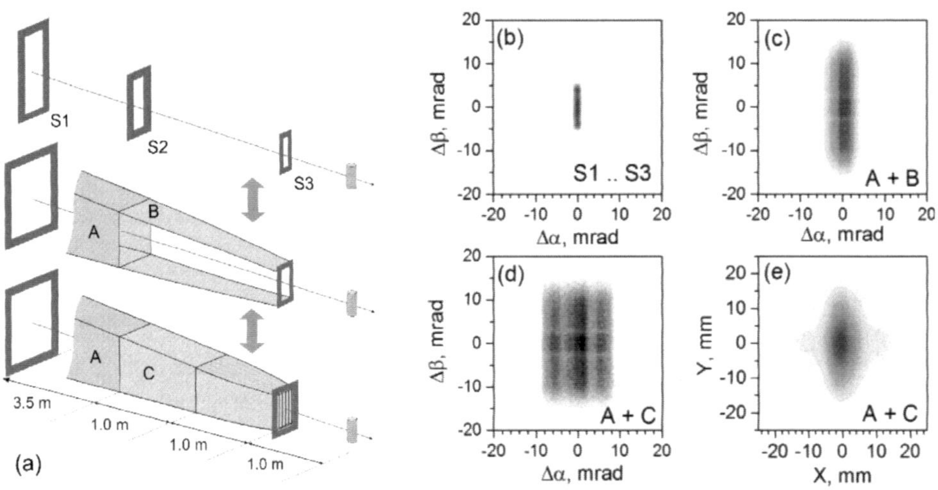

Figure 4. (a) Focusing section with the exchangeable guide segments. (b-d) Divergence distribution at the sample for three resolution modes. (e) Flux distribution at the sample with the horizontally focusing segment.

The last two sections (A,B) are placed in between a pair of adjustable slits (S1 and S3). These guides are removable and can be optionally replaced by another slit (S2) or by another vertically tapered guide, which includes a horizontally focusing multichannel segment (C). This system of three slits (S1-S3) and three guides (A-C) allows to adjust beam divergence in a wide range (figure 4b-e) as required, although the extreme vertical divergence of 83 mrad suitable for low resolution measurements (Table 1) could not be reached.

3.4. Coating optimization

Cost of neutron guides strongly depends on the number of coating layers and hence the critical angle. Recording of neutron histories during ray-tracing permits to calculate statistical distribution of bouncing angles along the guides and thus to find the maximum *m* value required for the guide coating at given positions. Fig. 5 shows such a distribution recorded for the proposed neutron guide system. In this simulation, *m*=5 was considered on all reflecting walls in order to allow for higher reflecting angles. However, the figure shows that *m*=2 is sufficient for most of the neutron guides length except the initial part of the beam extraction, the chopper cave, and the tip of the focusing guide.

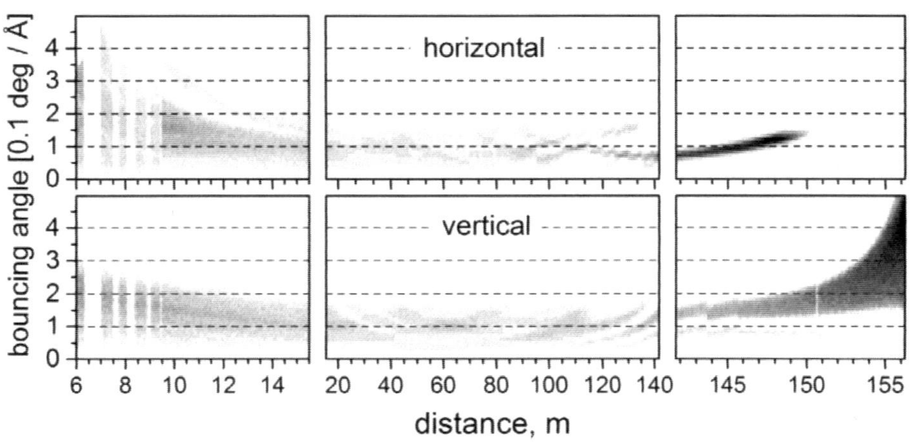

Figure 5. Distribution of bouncing angles along the guides for the left/right (a) and top/bottom (b) mirrors. Only "useful" neutrons that hit the sample are recorded.

4. Summary characteristics

The performance of the neutron guide system described in the preceding section has been assessed by the simulation of the BT ratio as a function of wavelength over the assumed operation range (see figure 6) for the medium resolution mode (see figure 4c) with the target area of 5 x 10 mm^2 and the acceptance solid angle of 5 x 30 $mrad^2$. Physical reality which was taken into account included gravity, Gaussian mirror waviness (σ=0.2 mrad), the sequence of gaps for choppers (figure 1a) and an approximation of the elliptic walls by flat segments with lengths varying from 1 m in the middle section to 0.1 m at the ends. Without the bi-spectral optics, the BT ratio of 50% was achieved for cold neutrons, which then decreased down to 30% at λ=1Å. The bi-spectral optics add another factor of 70 to 80 % for most of the wavelength range of interest. This is less than one may achieve under ideal conditions [8-9], but closer to the real situation when various constraints imposed by other technologies (e.g. choppers, sample environment) or cost limits have to be taken into account. The loss of BT ratio is mainly due to the non-uniform filling of the phase space (apparent in figure 4), which results from the combination of two factors: (i) small moderator height and the safety distance of 2 m, which do not permit to over-illuminate the entrance port of the neutron guide system, and (ii) choosing less than optimal guide height (see Section 3.2) for cost efficiency.

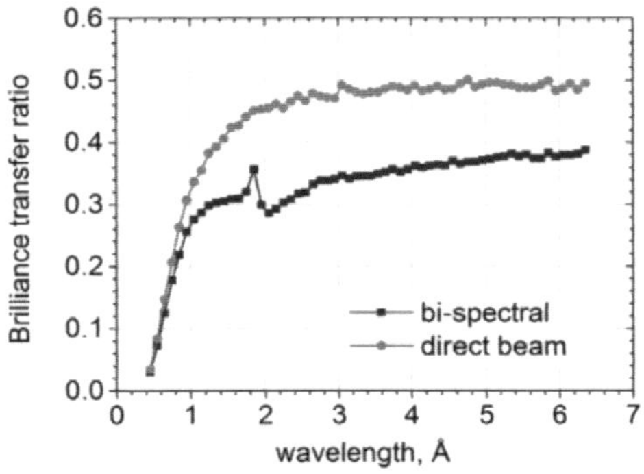

Figure 6. Brilliance transfer ratio for the whole guide system (solid line) and for the direct beam without bi-spectral extraction (guides only, dashed line).

5. Conclusions

The proposed neutron optical system for the BEER instrument addresses successfully most of the instrument requirements and constraints. In particular, it permits to efficiently trade resolution for intensity in a broad range facilitated by the pulse shaping and modulation chopper techniques. The reduced brilliance transfer ratio due to the bi-spectral extraction, small moderator height and other spatial and cost constraints is compensated by the high brightness of the low-dimensional moderator and the outlined instrument flexibility. The neutron optics system is well adapted to other key components of the instrument, such as the chopper system, detectors and specialized sample environment for materials engineering. As the final design of the neutron delivery system for BEER is yet under development, this concept is presented with the prospect of serving as a reference for further modifications improving the instrument performance.

Acknowledgements

This work has been supported by the Czech Ministry of Education, Youth and Sports under the project LM 2010011, and by the BMBF (contract number 05E11G1) within the German contribution to the Design-Update project of the European Spallation Source.

References

[1] Wang X L, Holden T M, Rennich G Q, Stoica A D, Liaw P K, Choo H and Hubbard C R 2006 *Physica* B **385–386** 673
[2] Santisteban J R, Daymond M R, James J A and Edwards L 2006 *J. Appl. Cryst.* **39** 812
[3] Harjo S, Ito T, Aizawa K, Arima H, Abe J, Moriai A, Iwahashi T, Kamiyama T 2011 *Materials Science Forum* **681** 443
[4] Fenske J et al., *J Phys Conf Series*, this issue
[5] Johnson M W & Daymond M R 2002 *J. Appl. Cryst.* **35** 49
[6] Ferenc Mezei F, Zanini L, Takibayev A, Batkov K, Klinkby E B, Pitcher E and Schönfeldt T 2014 *Journal of Neutron Research* **17** 101
[7] Schanzer C, Böni P, Filges U and Hils T, 2004 *Nucl. Instr. Meth.* A **529** 63
[8] Klenø K H, Lieutenant K, Andersen K and Lefmann K, 2012 *Nucl. Instr. Meth.* A **696** 75
[9] Bertelsen M, Jacobsen H, Hansen U B, Carlsen H H, Lefmann K 2013 *Nucl. Instr. Meth.* A **729** 387
[10] Zendler C, Nekrassov D and Lieutenant K 2014 *Nucl. Instr. Meth.* A **746** 39
[11] Cussen L D, Krist T and Lieutenant K 2015 *Nucl. Instr. Meth.* A **777** 6
[12] Bentley P M, Kennedy S J, Andersen K H, Rodríguez D M and Mildner D F R 2012 *Nucl. Instr. Meth.* A **693** 268
[13] Šaroun J, RESTRAX, http://neutron.ujf.cas.cz/restrax
[14] Šaroun J and Beran P 2014 *Journal of Neutron Research* **17** 53
[15] Zendler C, Lieutenant K, Nekrassov D, Cussen L D and Strobl M 2013 *Nucl. Instr. Meth.* A **704** 68

MeV Neutron Production from Thermal Neutron Capture in ^{6}Li Simulated With Geant4

Valentina Santoro[1], **Douglas D. DiJulio**[1,2], **Phillip M. Bentley**[1,3]

[1] European Spallation Source ERIC, SE-221 00 Lund, Sweden
[2] Department of Physics, Lund University, SE-221 00 Lund, Sweden
[3] Department of Physics and Astronomy, Uppsala University, 751 05 Uppsala, Sweden

E-mail: valentina.santoro@esss.se

Abstract. Various Li compounds are commonly used at neutron facilities as neutron absorbers. These compounds provide one of the highest ratios of neutron attenuation to γ-ray production. Unfortunately, the usage of these compounds can also give rise to fast neutron emission with energies up to almost 16 MeV. Historically, some details in this fast neutron production mechanism can be absent from some modeling packages under some optimization scenarios. In this work, we tested Geant4 to assess the performance of this simulation toolkit for the fast neutron generation mechanism. We compare the results of simulations performed with Geant4 to available measurements. The outcome of our study shows that results of the Geant4 simulations are in good agreement with the available measurements for ^{6}Li fast neutron production, and suitable for neutron instrument background evaluation at spallation neutron sources.

1. Introduction

The long (3ms) proton pulse of the European Spallation Source (ESS) [1] gives rise to unique background challenges for the instrument suite [2, 3, 4]. In such a long pulse spallation source, one of the main instrument performance limits is that defined by the instrument background, feeding into the Signal to Noise (S/N) ratio. The instrument requirements for ESS signal to noise ratio is $> 10^6$ [5, 6] for many instruments. Such a signal to noise ratio is achievable, but absolutely cutting edge at spallation neutron sources, and in some cases is approaching electronic noise levels.

To achieve the required performance for each instrument, full beam line simulations must be carried out while taking into account all beam line components, choppers, shutters, collimator blocks and the beam dump, with different shielding configurations.

Several different materials can be used in the beam line to shield or to attenuate the neutron background. Amongst the different possibilities, the most used are boron or lithium compounds. The main disadvantage of boron compounds is the fact that the attenuation of the neutron beam is provided through neutron capture in ^{10}B which releases an alpha particle, a ^{7}Li ion and a 0.48 MeV photon. Whilst this gamma photon is considerably easier to manage compared to those from other absorbers, such as gadolinium or cadmium, in some cases it is an unwanted background contributor.

Li compounds on the other hand provide very good attenuation for thermal neutrons with almost no γ-ray production. Unfortunately, Li compounds can also produce fast neutrons of energies up to almost 16 MeV. The physics process behind this is that, after being absorbed by ^{6}Li, a neutron can induce a triton and an alpha particle with a cross section of 955 barns at 0.0253 eV [7]. The triton emitted with energy of 2.73 MeV slows down in the target and interacts with Li and other constituents of the compound to produce fast neutrons and γ-rays. Table 1 shows a summary of the physics processes involved.

This fast neutron production mechanism may be missing from some Monte-Carlo modeling packages often used for shielding design [8], depending on the optimization being used. At ESS, one of the radiation transport codes used for instrument background and shielding optimisation is Geant4 [9]. For the reasons explained above, it is important to assess that this toolkit is able to describe the fast neutron production process. To check the capabilities of Geant4, we simulated the fast neutron production from Li compounds and compared the results to available measurements from the following paper [10]. In that work, fast neutron yields from several Li and B compounds were measured and a summary of the results can be found in Table 2.

Table 1: Fast neutron and γ ray producing reactions

Primary Thermal neutron induced reactions		
Target	Branch +Q(MeV), E_{max}(MeV)	Ref
^{6}Li	$\rightarrow {}^4He + t + 4.785 \cdot E_t = 2.73$	[11, 12]
Secondary triton induced reactions at $E_t \leq 2.8$ MeV		
Target	Branch +Q(MeV),	Ref
^{6}Li	$\rightarrow {}^9Be + \gamma + 17.7$	[11, 12, 13, 14, 15, 16]
	$\rightarrow {}^8Be + n + 16.02$	
	$\rightarrow 2\,{}^4He + n + 16.115$	
	$\rightarrow {}^5He + {}^4He + 15.15$	

Table 2: Fast neutron yields from targets of ^{6}Li compounds from thermal neutron capture from the reference paper [10] and from Geant4 simulation.

Target	^{6}Li abundance (%)	measured yield per 10^6 neutrons from reference paper [1]	simulated yield per 10^6 neutrons with Geant4 [2]
^{6}Li	96	84 ± 26	172 ± 13
LiH	7.5	201 ± 61	82 ± 9
LiF	7.5	178 ±54	36 ± 6
Li_2CO_3	7.5	107 ± 33	17 ± 4

2. Geant4 Simulation toolkit

Geant4 is a Monte-Carlo toolkit for the simulation of the passage of particles through matter. The toolkit is developed by a worldwide collaboration of physicists, implemented in C++ and has an open source license. The code is used across a number of different scientific fields,

[1] The uncertainty in the yields includes the uncertainty due to the incident flux measurement, detector efficiency and any contribution of γ background for more details see [10]. These uncertainty do not include any possible contributions from sample impurity.

[2] This yield number is obtained using the QGSP_BERT_HP physics list

including high-energy physics, accelerator physics and medical and space science. The physics processes offered cover a comprehensive range, from meV and extending to TeV energies, and include electromagnetic, hadronic and optical processes with a large set of long-lived particles, materials and elements. The Geant4 set of physics processes to model the behavior of particles are decoupled from each another (for the most part) and the user selects these components in custom-designed physics lists that depend on the particles, the energy range and the process which are to be simulated.

In this work, the Geant4 calculations were carried out using Geant4 version 10.0 patch 3. The physics lists used for the calculations performed in this work are the QGSP_BERT_HP and QGSP_INCLXX_HP list. These lists are recommended for shielding applications [17]. QGS stands for quark-gluon string, which describes the high-energy interactions of protons, neutrons, pions, and kaons and nuclei. The high-energy interaction creates an exited nucleus, which is passed to the precompound model (P) which models the nuclear de-excitation. The label BERT refers to the Bertini intra-nuclear cascade model [18, 19], INCL to the Liege Intranuclear Cascade model [20], and HP for the high-precision neutron package [21]. The HP package uses evaluated neutron data, named G4NDL4.4, for interactions of neutrons below 20 MeV. This data largely comes from the ENDF/B-VII.0 libraries. Additionally included in this package is thermal scattering data for neutrons below 4 eV.

The Geant4 simulations presented in this work were performed with the ESS Detector Group framework [22]. The framework is a Geant4-based Python/C++ simulation environment which contains developments useful for neutron shielding and detector calculations.

3. Simulation of fast neutron production from ^{6}Li

To compare the Geant4 results we used a simulation setup that matches as closely as possible the geometry described in the reference paper [10], to avoid that our results could be affected by any geometrical differences. As a neutron source, we used 10^6 mono-energetic neutrons, with a circular beam profile of diameter 2.5 cm, and with an energy of 0.025 eV. The ^{6}Li target intercepted the whole of the thermal neutron beam and had a thickness of 1 mm. This is considered enough for a complete attenuation of the thermal neutron beam [10]. The lithium sample was surrounded by a sensitive detector fully covering 4π steradians, to measure the energy of the neutrons coming out from the lithium sample and to avoid any neutron losses.

The observed neutron spectrum for the two different physics lists described above is shown in Fig. 1. This spectrum has the following interesting features:

- there are several fast neutrons with energies above 14 MeV that are consistent with the production mechanism from triton secondary interactions as described above.
- the shape of the fast neutron spectrum is similar to the spectrum from available measurements. (See Fig.5. of [10])
- the neutron yields obtained in the simulation (172) for the QGSP_BERT_HP and (151) for the QGSP_INCLXX_HP are on the same order of magnitude, 10^2, as the measurements from the reference paper (see Table 2).

As an additional check of the physics processes, we also studied the energy spectrum of the mother particle from which the fast neutrons are produced. The study shows that all the fast neutrons were produced by triton induced reactions. The energy distribution of the tritons that generated the fast neutrons is shown in Fig. 2. The energy spectrum is consistent with tritons being produced from interactions of a thermal neutrons with ^{6}Li and the spectrum is almost monoenergetic due to conservation of energy and momentum. In summary, all the features illustrated above show that the primary and secondary reactions are described reasonably well in the Geant4 simulation package.

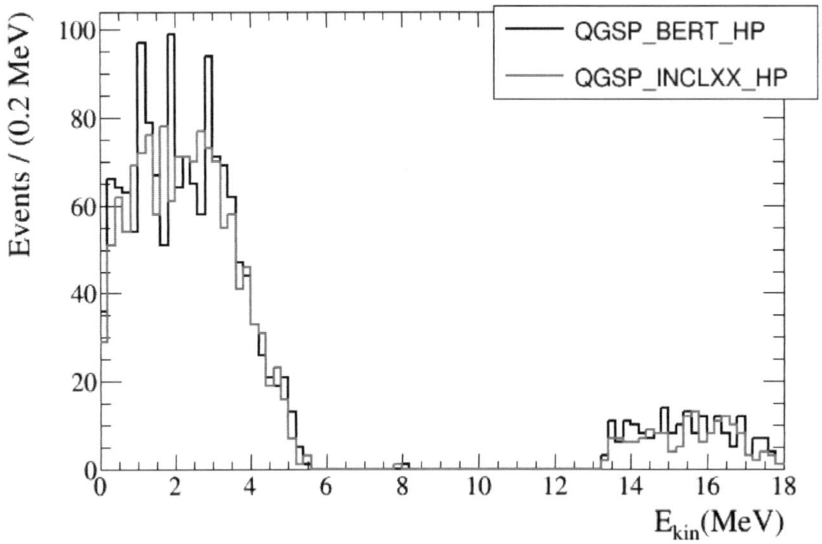

Figure 1: Energy spectrum of the fast neutrons produced by thermal neutron capture in ^{6}Li for two different physics lists.

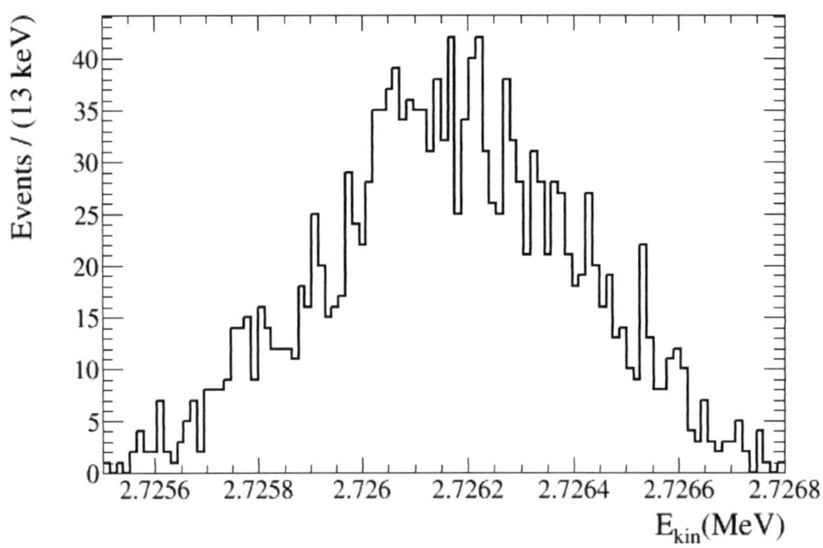

Figure 2: Energy spectrum of tritons produced by the reaction ^{6}Li(n,t)^{4}He.

4. Lithium compound simulations

For completeness, we also tried to simulate several different Li compounds to compare with the expected yield from Table 2. The resulting spectra are shown in Fig. 3. The shape of the spectra for LiH, LiF and Li_2CO_3 seems consistent with what is shown in the reference paper, but the yields are somewhat lower than expected (see Table 2). The differences between the measured and simulated spectra may be due to a number of factors. These include the lack of precise thermal neutron scattering data for some of the materials of interest; geometric and material composition differences that are not fully documented; and the reliance on nuclear models for the charged particle interactions. This last point may be addressed with an up and coming package, titled G4ParticleHP [23], for Geant4. This package will introduce evaluated data for simulating the interactions of charged particles in Geant4. In order to further understand the origins of the differences between the measured and calculated spectra, a detailed investigation of the multi-step process would be necessary for each material.

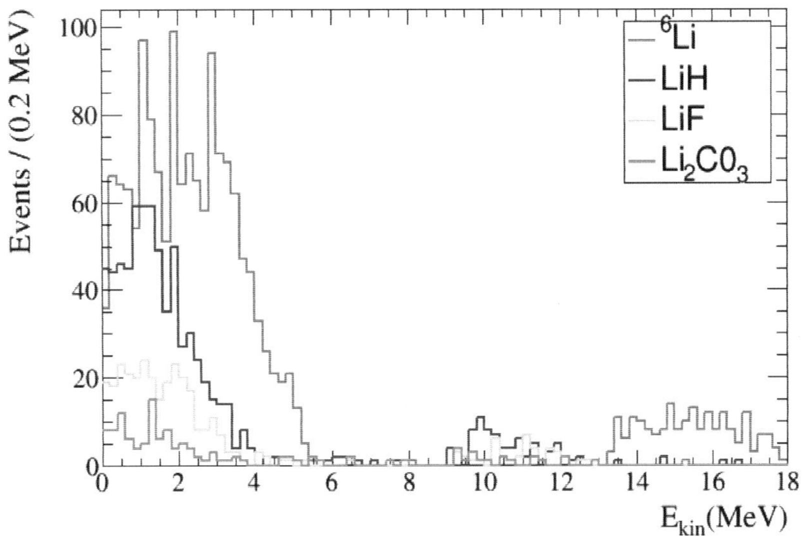

Figure 3: Energy spectrum of the fast neutrons produced by thermal neutron capture in different Lithium compounds

5. Conclusions

We have presented simulations of the fast neutron production channel of lithium absorption of thermal neutrons using Geant4. The simulations compare favourably with available measurements for ^{6}Li fast neutron production in the literature. As such, we conclude that sufficient sensitivity and reliability exists in Geant4 for full beam line models of different instruments that are planned at ESS, where ^{6}Li absorbers are the primary means of achieving efficient neutron shielding.

References

[1] S. Peggs *et. al.*, ESS Technical Design Report, Lund, Sweden, 2013
[2] N. Cherkashyna *et. al.*, J. Phys. Conf. Ser. **528** 012013
[3] N. Cherkashyna *et. al.*, arXiv:1501.02364
[4] N. Cherkashyna *et. al.*, Phys. Rev. ST Accel. Beams 18, 083501
[5] I. Ioffe *et. al*, ESS Instrument Construction Proposal Heritage, 2015
[6] X. Fabrèges *et. al.*, ESS Instrument Construction Proposal Magnetic Single Crystal Diffractometer Magic, 2015
[7] G. F. Knoll, Radiation Detection and Measurement (Wiley, New York, 1979).
[8] G.-b. Wang *et. al.*, Nucl. Instrum. Methods Phys. Res. Sect. B: Beam Interact. Mater. At., 287 (2012), p. 19
[9] S. Agostinelli *et. al.*, Nucl. Instrum. Methods Phys. Res., Sect. A. 506, 250 (2003)
[10] M.A. Lone D.C. Santry and W.M. Inglis., Nuclear Instruments and Methods 174 (1980) 521-529
[11] C. Maples, G.W. Goth and J. Cerny, Nucl. Data A2 (1966) 429

[12] F.K. McGowan and W.T. Milner, Atomic and nuclear data reprints, vol.2. (Ed. Katherine Way; Academic Press, New York, 1973).
[13] V.I. Serov and B.Ya. Guzhovskii, At. Energiya 12 (1962) 11.
[14] N. Jarmie and B.C. Driven, Nucl. Sci. Eng. 17 (1963) 433.
[15] R.W. Crews, Phys. Rev. 82 (1951) 100
[16] A.K. Valmmer, P.I. Vatset, L.Ya. Kolensnikov, S.G. Tonapetyan, K.K. Chernyavskii and A.I. Shpetnyi, At. Energiya 10 (1961) 577.
[17] Geant4 Physics Reference Manual http://geant4.web.cern.ch/geant4/UserDocumentation/UsersGuides/PhysicsReferenceManual/fo/PhysicsReferenceManual.pdf
[18] H.W. Bertini, Phys. Rev. 131 (1963) 1801.
[19] H.W. Bertini, Phys. Rev. 188 (1969) 1711.
[20] S. Leray, *et. al*, J. Phys. Conf. Series 420 (2013) 012065.
[21] P. Speckmayer. Impact of the choice of physics list on geant4 simulations of hadronic showers in tungsten. *LCD-Note-2010-002*, 2010.
[22] T Kittelmann *et. al* J. Phys. Conf. Ser. **513** 022017 2014
[23] E. Mendoza et al., 2011 IEEE Nuclear Science Symposium Conference record, N53-5, pg. 2242

DREAM – a versatile powder diffractometer at the ESS

W Schweika[1,2], N Violini[1], K Lieutenant[3], C Zendler[2], D Nekrassov[3], A Houben[4], P Jacobs[4] and P F Henry[2]

[1] Jülich Centre for Neutron Science, JCNS, Forschungszentrum Jülich, Germany
[2] European Spallation Source ESS ERIC, Lund, Sweden
[3] Helmholtz-Centre Berlin, Germany
[4] Institute of Inorganic Chemistry, RWTH Aachen University, Germany

E-mail: werner.schweika@esss.se

Abstract. The instrument DREAM, in construction at the long pulse European Spallation Source (ESS), is a new type of neutron time-of-flight powder diffractometer, which utilizes additional choppers to meet the typical high resolution requests. Pulses will be of symmetric shape and their width can be varied from 10 μs to 1 ms, providing an unprecedented flexibility from highest to low resolution with optimized intensities at the superior brightness of the 5 MW source. The design is driven particularly by the needs and challenges for small and complex samples, large unit cell materials, thermoelectric cage structures or metal-organic framework structures, multiphase battery materials and complex magnetic structures. Therefore, the chosen wavelength bandwidth of 3.7 Å may cover well the peak intensities of the thermal and cold moderator used simultaneously and provides a sufficient Q (and d) range for obtaining diffraction patterns in a single setting. VITESS simulations show a performance that is about two orders of magnitude higher than current best instruments.

1. Introduction

Time-of-flight (TOF) neutron powder diffractometers benefit from the brightness of pulsed neutron sources and have gained increasing importance over the past decades with respect to monochromatic instruments [1] at continuous sources. Short-pulse spallation sources (SPSS) are well suited for powder diffraction offering the brilliance from the sharp neutron pulses extending to shortest wavelengths, with new applications in measuring the pair distribution functions from total scattering with high spatial resolution[2]. In contrast to monochromatic instruments at reactors, which measure angle dispersive like x-ray diffractometers, in neutron TOF diffraction another dimension is offered by measuring additionally energy-dispersive, that has immediately found new applications for special sample environments, by confining the gauge volume by diffraction angles near 90 degree [3], and for high resolution studies in backscattering essentially determined by the ratio τ/T of pulse width to neutron TOF to the detector. The time structure of SPSS moderators yields a wavelength dependent pulse length and keeps the relative time and wavelength resolution approximately constant over a broad wavelength range. By adapting further the detector geometry, it is possible to achieve a constant relative resolution $\Delta d/d$, as exemplified by the design of the POWGEN instrument [4] at the Spallation Neutron Source (SNS) in Oak Ridge.

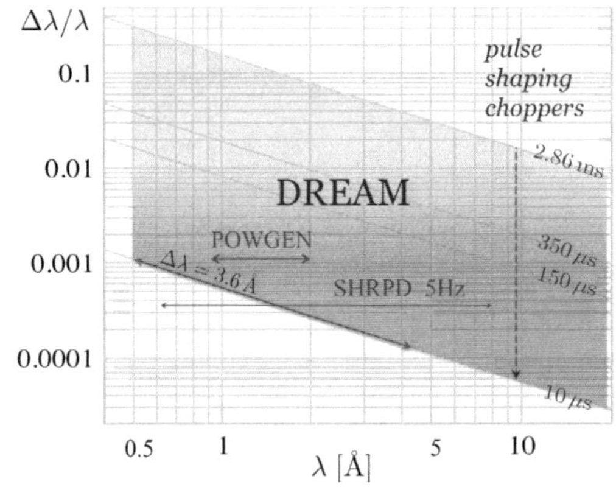

Figure 1. Backscattering resolution of DREAM *versus* wavelength for varying pulse shaping. All TOF powder diffractometers can shift the bandwidth $\Delta\lambda$ with chopper phase. In addition, the time resolution of DREAM is flexible from $10\mu s$ to 2.86 ms.

In contrast, the new European Spallation Source (ESS) will have a long neutron pulse essentially determined by the proton pulse duration on the target. The pulse length of 2.86 ms is apparently not of practical use for typical powder diffraction tasks, since already medium time resolution $\tau/T \approx \Delta d/d = 0.005$ requires an extreme flight path $L = v(\lambda)\tau(\Delta d/d)^{-1}$, *e. g.* more than 2 km for $\lambda = 1$ Å neutrons. However, the neutron pulse of the 5 MW ESS will also be the most intense pulse with a superior peak brightness. Therefore, generating shorter pulses will still be gainful. It will create new opportunities and particularly a valuable new versatility in neutron powder diffraction.

The optimization strategies and instrument design at long pulse sources need to be novel and have led to the here presented new type of a neutron TOF powder diffractometer DREAM, which is highly flexible in resolution and versatile for applications. The obvious tools to produce shorter pulses of thermal and cold neutrons are fast, counter-rotating disc choppers producing a symmetric pulse shape, which is preferable in view of data refinement. Pulses can be shaped with high flexibility by varying the choppers speed, changing the time resolution over three orders of magnitude from $10\,\mu s$ to ms-range, and trading resolution for intensity easily. This virtue of flexible resolution is illustrated in Figure 1. It shows a large accessible region, which will even extend current limitations of high-resolution neutron powder diffraction.

The pulse τ is simply chosen independent of the neutron wavelength and fixes the minimal absolute resolution Δd in backscattering. With acceptable increase in chopper complexity [5] it would be possible to change to a mode of constant relative resolution $\Delta\lambda/\lambda$, which however implies a limitation on the achievable best resolution. Instead of aiming at a uniform relative $\Delta d/d$ resolution function for integrating the Debye-Scherrer lines to single peaks, the present concept rather intends to make best simultaneous use of high resolution backscattering and high intensity diffraction at lower angles. Therefore, we will imply an wavelength- and angle-dispersive 2D Rietveld refinement method [6], which has been developed for the TOF powder diffractometer POWTEX [8, 9] under construction at the research reactor FRM II in Munich. As soon as such constraints for a uniform resolution function can be overcome by an adapted 2D data refinement method, the instrument design and particularly, the detector and its geometry can be better optimized with respect to cost and efficiency and less complexity. Since this instrument is specifically designed to benefit from measuring and analyzing data versus wavelength and diffraction angle, we have given it the name DREAM, which stands for *Diffraction Resolved by Energy and Angle Measurements*. Beyond the instrument's dedication for powder diffraction,

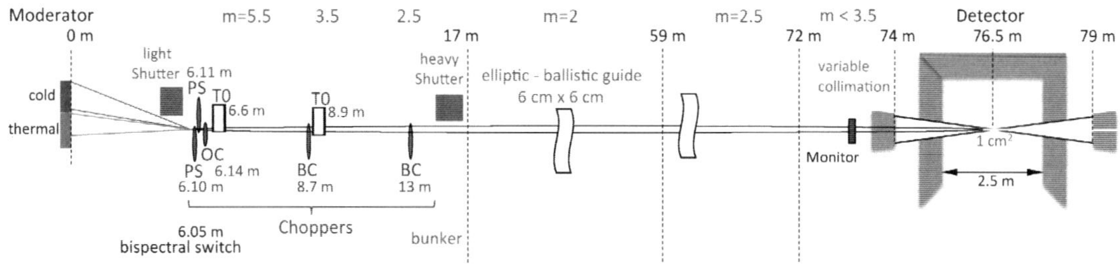

Figure 2. Scheme of the instrument.

the name implies further capabilities in resolving the out-of plane dependence of the scattering to study preferred orientation, texture and finally single crystal diffraction in neutron TOF Laue technique. Here, the instrument benefits from the covered large solid angle of detection including 2D position sensitivity, the flexibility in time-resolution, and the optimization for small samples.

Below, we present further design considerations of the instrument, the proposed instrument layout, and finally VITESS [11, 12] simulations of its performance, which is compared to existing world class neutron TOF powder diffraction instruments.

2. Design considerations and layout of the instrument

An overview on the scheme of the instrument and its components is given in Fig. 2.

2.1. Source parameters and bandwidth

The ESS parameters have been defined in the Technical Design Report [7] with a source frequency of 14 Hz, which in reverse gives a TOF frame of $T_s = 1/14\text{Hz} = 71.4\,\text{ms}$. The duration of the proton pulse on the target is $\tau_s = T_s/25 = 2.86\,\text{ms}$ and large compared to moderation time. Target and moderator are surrounded by the monolith shielding and shutter extending 6 m from the center. A Pulse Shaping Chopper (PSC) to be placed outside the monolith at $L_{PSC} > 6\,\text{m}$ defines the instrument length L from the moderator to the detector position, $L = L_{PSC}(T_s/n\tau_s + 1)$ and the wavelength frame $\Delta\lambda \leq 1.88\,\text{Å}$ by L_{PSC}/τ. Here n is the integer number of wavelength frames $\Delta\lambda$ used in a single TOF frame of the source, T_s, and $n\Delta\lambda$ gives the total bandwidth.

Looking at choices for current instrumentation, a bandwidth of 1 Å as chosen for POWGEN is usually not sufficient to complete diffraction studies and most users request two or more settings for data refinements. The SHRPD [13] at JPARC operates most of its time at 5 Hz, *id est* with 80% pulse suppression to achieve a desirable larger bandwidth of 8 Å instead of 2 Å in normal 25 Hz source mode. The instrument WISH [14] at ISIS target station 2 with 10 Hz has a bandwidth of 9.4 Å. In principle, it is possible to cover a larger wavelength band in sequential steps with the same efficiency, while a narrow bandwidth might advantageously focus to special diffraction peaks of interest. On the other hand a sufficiently large bandwidth is important for real time experiments on irreversible processes, such as studies of chemical reactions.

DREAM has been designed as a bi-spectral instrument using two wavelength frames, $2\Delta\lambda \approx 3.7\,\text{Å}$, to cover simultaneously the peak spectra of both thermal and cold fluxes with a dynamic range that perfectly suits for diffraction experiments in a single setting. While pulse suppression like for SHRPD gives similar opportunities for a longer instruments, the choice to set the instrument length to approximately 75 m is motivated by the reduced cost for a shorter instrument and better neutron transport by the neutron optics when taking gravity into account.

2.2. Neutron optic elements and guide system

The original design was based on the assumption that two moderators, thermal and cold, each with 12x12 cm^2 surface, were placed side by side. Shielding of the moderator monolith and shutters extends to 6 m from the source. This geometry comfortably allows for extracting a suitable divergence through a pinhole at 6 m, without neutron guides in the monolith. The desirable phase space density at the pinhole can be efficiently transported to the sample position. The simplest geometry would be two consecutive ellipses with a common focal point and pinhole and sample placed in the outer foci. The reason for having two but one ellipse follows from the phase space transformation by an ellipse, which is restored by the second ellipse [9]. The optimization supported by simulations led to two modifications. First, the central part of this guide system could have simpler straight shape, with even small gain in transport efficiency. Second, including gravity effects a larger vertical opening at the pinhole accepting a larger phase space is essentially able to retrieve the phase space density for a sample smaller than the pinhole. This design with a straight view from sample to moderator is superior particularly for short wavelength neutrons. The pinhole concept combined with the defocussing and focusing optics accept essentially the useful neutrons, which largely reduces the background.

Recent developments to the current concept of a so-called "butterfly" moderator design [10] have increased the extractable brightness by reducing the surface area to 3 cm height. Our investigations have shown that this increased brightness can be fully exploited by shifting the vertical ellipse focus from 6 m to the source using guide inserts starting 2 m from the source. With respect to the horizontal guide layout, the pinhole near 6 m can be kept, only an additional mirror inside the monolith compensates for the slightly reduced source size. The instrument axis points straight to the center of the thermal moderator. Neutrons from the cold source placed next to the thermal moderator are mirrored into the beam by a bi-spectral device [15, 16] that is inclined about 0.5 degree with respect to the thermal beam axis. We considered various options to inflect the cold beam into the neutron guide system, (i) a straight mirror 0.5 mm thin glass coated with a NiTi supermirror (ii) the same materials assembled to a shorter 25 cm long mirror stack with 1 mm air gaps, (iii) a solid-state bender [17] made of a 60 mm short stack of 150 μm thin Si-wafers with total reflection at the Si/NiTi interface, (iv) the same assembled to a straight mirror stack of only 16 mm length. Our idea of using the compact Si-wafers has the advantage that it allows to place the device for better access outside the monolith near to the horizontal focal point at 6 m, and it can reduce actually the path through solid materials compared to the cases (i) und (ii). In VITESS simulations the bender (iii) provides best performance for cold neutrons, the transmission of the thermal neutrons however, is affected by Laue extinction. A straight Si-stack (iv) will be about 15% less efficient in inflecting neutrons from the cold moderator, however, it largely avoids Laue extinction. Since such a device is only of 16 mm length, it is of least impact to the transmission of the straight thermal beam.

The neutron guide is straight for optimal transport of short wavelengths neutrons. In combination with T0 choppers and a heavy shutter, we expect a reduction of background, of instrument shielding and of costs in comparison to a curved guide option only. The guide cross section opens up to $6\,cm \times 6\,cm$ as it results from the optimization with a figure of merit of intensity for a small sample cross section of 1 cm^2 within a divergence of $\pm0.25°$(FWHM). Near the sample the focusing section can be extended by an exchangeable part, thus approaching 50 cm distance to the sample focusing with $\pm0.8°$ (base to base) can be achieved, which will be valuable for small samples, *e.g.*, under high pressure. A set of slits in final 4 m guide part allows further control of the divergence (see WISH [14]).

2.3. Chopper system

The chopper system and chopper positions are shown in Fig. 1. The system consists of two fast, counter-rotating disc choppers to shape the pulse (PSC), a 14 Hz overlap chopper (OC)

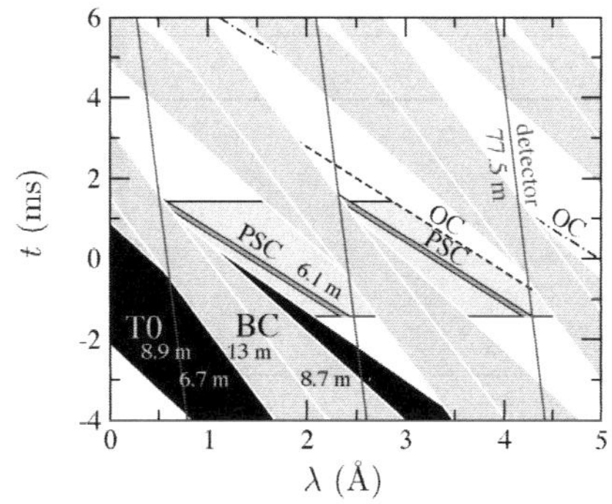

Figure 3. Chopper acceptance diagram $t \propto s\lambda$ choosing $t = 0$ at the moderator position. T0-choppers at 6.7 and 8.9 m block the prompt fast neutron pulse. Pulse shaping choppers (PSC) define two pulses of variable width (here 150 μs) separated in time by the source pulse width of 2.86 ms for two wavelength frames of 1.8 Å each. Note the band control chopper (BC) at 8.7 m suppresses also the "afterglow" passing the PSC. The overlap chopper OC provides a coarse 14 Hz selection for only the depicted TOF frame up to $\lambda <$ 46 Å.

to eliminate all higher harmonics from the PSC, two band control choppers (BC), and 28 Hz T0-choppers to suppress the prompt pulse. The interplay of these choppers is depicted in the acceptance diagram, see Fig. 3, showing neutron time versus wavelength, $t \propto s\lambda$. In the given coordinates the slope is related to the distance s from the source with $t = 0$ at $s = 0$. [18] The figure shows a typical setting for 0.6 Å $< \lambda <$ 4.2 Å at a medium time resolution of 150 μs. The PSC have a 75 cm diameter disc and may operate in multiples of the 14 Hz source frequency up to 294 Hz for highest resolution. There are different pairs of openings for each frequency to achieve the pulse shaping for the two wavelength bands. The BC provide a better definition of the wavelength band and further suppress the exponentially decaying "after-glow" background from the source for 6 ms. PSC, BC, and OC may shift in phase to choose the wavelength range. The T0-chopper concept is still under development. Current simulations of high energy neutrons favor a solution using essentially several W-absorbers, each approximately 5 cm thick.

2.4. *Diffractometer and detector*

The diffractometer has a detector with 6 sr solid angle coverage, including backward, forward and a 2 sr section to one side, within a cylindrical geometry along the beam axis. Sample to detector distance is $\geq$ 1.25 m, the cylinder radius. The geometry is favorably adapted to the Debye-Scherrer cones. The spatial resolution of 5 mm (FWHM) at $2\theta = 90°$ in two dimensions provides a better angular resolution definition, $\Delta\theta = 0.13°$ (FWHM), than given by the incoming divergence, and is similar to contribution from the sample size. The concept includes two further detectors at 2.5 m distance in order to access higher angles in backscattering and smaller angles in forward direction side to extend the Q-range to the scale of nano-particles covering in total three orders of magnitude in length scale with $0.01\,\text{Å}^{-1} < Q < 25\,\text{Å}^{-1}$.

The detector itself is novel and has been developed for the POWTEX instrument. It is a "Jalousie" type detector with ^{10}B absorbing cathode layers in a multi-wire gas detection chamber [19]. In order to achieve sufficient absorption efficiency, several layers covered with 1 μm ^{10}B are composed in an inclined geometry creating a volume detector with a predicted and measured detection efficiency of 54% at 1 Å wavelength. Noteworthy features of such a volume detector are the higher count rate capability due to a depth of approximately 20 anode layers, the particular insensitivity to gamma background by coincidence measurement, intrinsic collimation properties

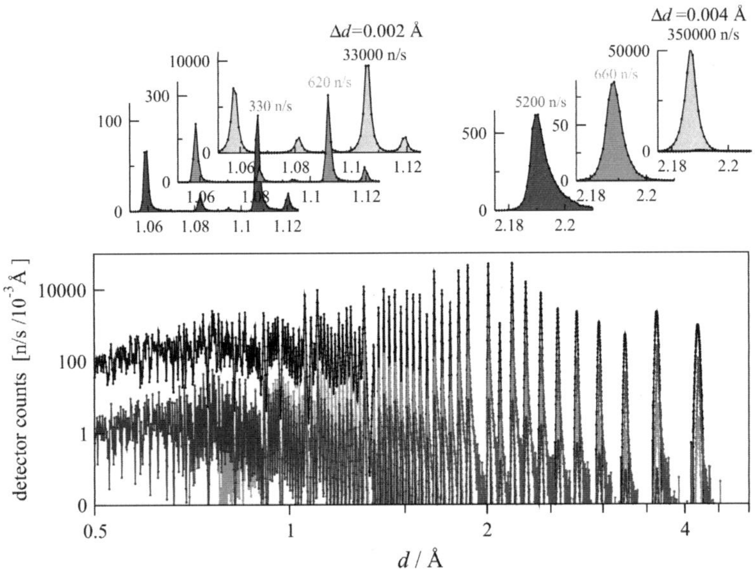

Figure 4. Simulated diffraction patterns for a reference sample $Na_2Ca_3Al_2F_{14}$ show for a similar and high resolution option that the performance of DREAM (black-grey: $0.8\,Å < \lambda < 4.6\,Å$) is almost two orders of magnitude higher compared to POWGEN at SNS (light brown: $1.5\,Å < \lambda < 2.5\,Å$, dark brown: $0.8\,Å < \lambda < 1.8\,Å$, and WISH at ISIS (blue: $1.4\,Å < \lambda < 10.8\,Å$). *Upper panel*: Peak shape, line width and integrated peak intensity are exemplified for two peaks near $d \approx 1.1$ Å and 2.2 Å (WISH, POWGEN, DREAM data displayed from bottom to top).

by tracing diffraction signals back to the sample source, and finally to mention is the intrinsic energy sensitivity that can be obtained by analyzing the measured absorption depth profiles.

The diffractometer has top and side access to the sample position and will offer an open platform for various sample environments for experiments at low to high temperatures, applying magnetic fields and high pressure, for using chemical reaction cells, and with capabilities to scan larger objects.

3. Instrument performance

The instrument has been simulated by including all components in detail verifying a high brilliance transfer along the neutron guide system and a clean and flexible time structure achieved by the chopper system. Finally, we studied the performance of DREAM for different resolution options by VITESS simulating the diffraction pattern of a common reference sample of $Na_2Ca_3Al_2F_{14}$ cubic ($I2_13$) a=10.257(1) Å [20]. Here, we chose a small volume 0.4 cm^3, which will be a typical sample size for the instrument. We have not included many details of the detector but account properly for distance and efficiency. These simulations were based on the 2013 ESS thermal and cold moderators. The new "butterfly" moderator yields gains of 2.57 and 3.4 for the thermal and cold spectra respectively, which will increase the DREAM performance accordingly and which have been used to scale the results shown in Fig. 4. We have also modeled other diffractometers POWGEN, WISH, GEM, D20 in VITESS in order to benchmark the performance. Here, we chose a typical but comparatively larger sample size of 0.8 cm^3. The flux of WISH for the high resolution setting was smaller in our model than given in Ref. [14], probably due to actual changes in the moderator compared to the available model. This has been taken into account by scaling appropriately the data by a factor 2 to

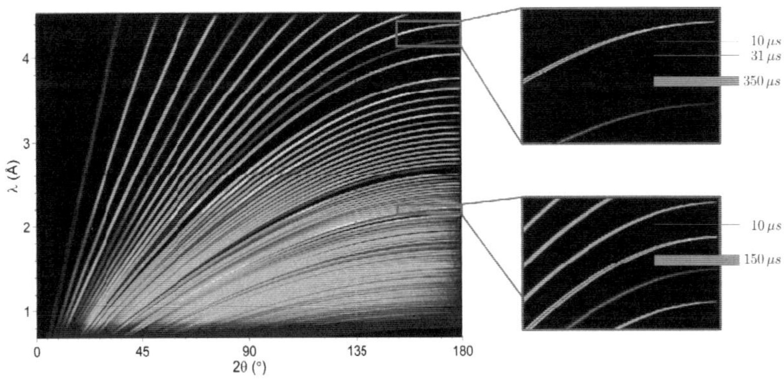

Figure 5. Simulated diffraction pattern *versus* wavelength and diffraction angle. The resolution varies along the Bragg diffraction lines. The insets show clear separation of diffraction lines. Resolution is best for backscattering, where shortest pulses of $10\,\mu$s yield a resolution of $\Delta d = 2.8 \times 10^{-4}$Å. Instead of integrating data to $I(d_{hkl})$, see Fig. 4, a two-dimensional data analysis will optimally exploit intensity and resolution. Relaxing time-resolution may yield very high intensities with a flux of 10^9 n/(cm^{2}s) at the sample and count rates larger than 10^6 n/s for strongest peaks with better than 1% resolution.

the experimental flux at the sample of 1.08×10^7 n/(cm^{2}s) [14]. For better comparison, we tried to match the resolution of POWGEN and WISH approximately also in the simulation of DREAM, using 40 μs pulse width and $\pm 0.167°$, $\pm 0.33°$ divergence horizontal and vertical. For better clarity two peaks are selected and presented in the upper panel of Fig. 4. The comparison reveals the strength of DREAM, symmetric peak shapes and about two orders of magnitude higher intensities at comparable resolution. The diffraction pattern also reveals that the 3.7 Å bandwidth is rather suitable choice. With respect to the POWGEN instrument we considered two settings of wavelength frame for better comparability.

The comparison made in Fig. 4 is based on the conventional integration along the Bragg lines for each d-spacing. This is particularly fine for the instrument POWGEN, which by design has essentially a constant resolution, however, it is less appropriate for DREAM and also for WISH. Fig. 5 shows the λ and θ dependence of the diffraction pattern for the DREAM simulation. Note that the graphical resolution is misleading, the instrument resolution can be better seen by the zooming inserts to the right, showing that the peaks can in fact be nicely separated with a better resolution of $\Delta d = 0.001$. The line widths change with θ by the typical divergence contribution proportional to $\cot\theta$. While it is certainly possible to refine the diffraction pattern shown in Fig. 4, the new approach of a two dimensional Rietveld refinement [6] will make much better use of the data quality. Furthermore, the data inspection and analysis in its λ and θ dependence will give better control on background and also better reveal systematic deviations, such as structured background from sample environment or problems with detector banks.

4. Final remarks

The DREAM powder diffractometer will offer truly new opportunities with its flexibility that builds on the ESS source parameters. The example shown in the simulation does not show the best resolution, which will be close to best synchrotron beam lines. The case of medium resolution with 10^6 n/s for the strongest peak will enable real-time studies on ms-time scale for irreversible processes. The intense peak flux of the cold coupled moderator combined with shortest wavelength neutrons available, yields a wavelength band that is most valuable for meeting the challenges of the science case contributions to the DREAM proposal, which were highly appreciated by the Scientific Advisory Committee. The cases discussed for magnetism, functional materials, metal-organic frame work structures [22], thermo-electrics [23], batteries [24] have in common complex structures with relative large unit cells. Cases particularly appropriate for neutrons in view of the sensitivity to even small magnetic moments as well as for light elements on heavier atom background, however, very demanding in diffraction quality with respect to resolution and the ability to distinguish small near to strong peaks [21]. Challenges from material and applied science point towards studies of devices in operandi, for example super-capacitors, material assemblies at short-time scales.

There are many fascinating capabilities and only few can be spotted here, for example, structure determination of magnetic nano-particles by total scattering over multiple length scales, and by future possibilities to combine this with neutron polarization and relatively simple polarization analysis in forward direction. Studies at highest pressure are limited by the flux on small samples and limits can be pushed significantly with the DREAM instrument. However, it further requires to improve signal to background and we expect that this can be largely improved by the proposed two and three dimensional data analysis. Resolving diffraction data with position- and, as we propose, even volume-sensitive detectors with intrinsic sensitivity to direction and energy transfer is as appealing as challenging from the point of data analysis.

As a final point, scientific advisors addressed the need for single crystal studies for signals too weak for powder diffraction. The instrument will fulfill such needs by neutron time-of-flight Laue diffraction and even with best gains from the ESS long pulse structure.

Acknowledgment
We gratefully acknowledge the valuable support from the scientific community which defined the science case of the DREAM proposal and the funding from the German Federal Ministry of Education and Research (BMBF) in the frame of the ESS Design and Update Programme.

References

[1] Hewat A W 2006 *Physica B* **385-386** 979
[2] Egami T and Billinge S J L *Underneath the Bragg Peaks* 2003 Pergamon Materials Series **16**
[3] Jorgensen J D 1990 *High Press. Res.* **4** 441-3
[4] Huq A, Hodges J P, Gourdon O, Heroux I 2011 *Z. Kristallogr. Proc.* **1** 127-35
[5] Cubitt R, Fragneto G 2002 *Appl. Phys. A* **74** S329
[6] Jacobs P, Houben A, Schweika W, Tschougréeff A, Dronskowski R, 2015 *J. Appl. Cryst.* **48**
[7] Carlile C J *et al* 2013 *European Spallation Source Technical Design Report* ISBN: 978-91-980173-2-8
[8] Conrad H, Brückel T, Schäfer W, Voigt J 2008 *J. Appl. Cryst.* **41** 836-845
[9] Houben A, Schweika W, Brückel T, Dronskowski R 2012 *Nucl. Instr. and Meth. A* **680** 124-33
[10] Zanini L, Batkov K, Klinkby E, Mezei F, Schoenfeldt T, Takibayev A, 2015 *ibid.*
Mezei F, Zanini L *et al* 2014 *Journal of Neutron Research* **17** 101-105
[11] Zsigmond G, Lieutenant K, Mezei F 2002 *Neutron News* **13** 11-14
[12] Lieutenant K, Zendler C, Manoshin S, Fromme M, Houben A, Nekrassov D 2014 *J. Neutron Res.* **17** 45-51
[13] Torii S, Yonemura M, Surya Panca Putra T Y *et al* 2011 *J. Phys. Soc. Jpn.* **80** SB020
[14] Chapon L C , Manuel P, Radaelli *et al* 2011 *Neutron News* **22** 22-5
[15] Russina M and Mezei F 2002 *Proc. SPIE (Int.Soc. f. Opt. Eng.)* **24** 4785
[16] Lieutenant K and Mezei F 2006 *J. Neutron Res.* **14** (2) 177
[17] Krist Th 2004 *Nucl. Instr. and Meth. A* **529** 50
[18] Copley J R D 2003 *Nucl. Instr. and Meth. A* **510** 318-24
[19] Modzel G *et al Nucl. Instr. and Meth. A* **743** 90-5
[20] Courbion G and Ferey G 1988 *J. Solid State Chem.* **76** 426
[21] Zhao L *et al* 2012 *Advanced Materials* **24** 2469-73
[22] Wu H *et al* 2013 *J. Am. Chem. Soc.* **135** 10525.
[23] Zevalkink A *et al* 2014 *Dalton Trans.* **43** 4720
[24] Dolotko O, Senyshyn A, Muehlbauer M, *et. al.* 2014 *Journal of Power Sources* **255** 197-203

Probing the dynamics of high-viscosity entangled polymers under shear using Neutron Spin Echo spectroscopy

M Kawecki [1], P Gutfreund [2], F A Adlmann [1], E Lindholm [3], S Longeville [4], A Lapp [4] and M Wolff [1]

[1] Department of Physics and Astronomy, Uppsala University, Box 516, Uppsala, SE
[2] Institut Laue Langevin, EPN Science Campus, Grenoble, FR
[3] Angstrom Workshop, Uppsala University, Box 516, Uppsala, SE
[4] Laboratoire Leon Brillouin, CEA Saclay, Gif-sur-Yvette, FR

E-mail: maciej.kawecki@physics.uu.se, max.wolff@physics.uu.se

Abstract. Neutron Spin Echo spectroscopy provides unique insight into molecular and sub-molecular dynamics as well as intra- and inter-molecular interactions in soft matter. These dynamics may change drastically under shear flow. In particular in polymer physics a stress plateau is observed, which might be explained by an entanglement-disentanglement transition. However, such a transition is difficult to identify directly by experiments. Neutron Spin Echo has been proven to provide information about entanglement length and degree by probing the local dynamics of the polymer chains. Combining shear experiments and neutron spin echo is challenging since, first the beam polarisation has to be preserved during scattering and second, Doppler scattered neutrons may cause inelastic scattering. In this paper we present a new shear device adapted for these needs. We demonstrate that a high beam polarisation can be preserved and present first data on an entangled polymer solution under shear. To complement the experiments on the dynamics we present novel SANS data revealing shear-induced conformational changes in highly entangled polymers.

1. Introduction

Neutron Spin Echo spectroscopy (NSE) provides exceptional energy resolution corresponding to long Fourier times in the intermediate scattering function and allows to observe the diffusive modes on molecular scales. The predictions of the reptation model proposed by de Gennes have been verified experimentally through the means of NSE, through quantifying the topological interactions in polymeric fluids and revealing the tube-like confinement [1–3]. In particular NSE probes the intermediate scattering function. The reptation model predicts a Q dependent plateau in this function for longer Fourier times reflecting the fact that the motion of the individual chains is constrained by the entanglement from the other chains. Only the outstanding energy resolution or long Fourier times accessible at modern NSE instruments allowed to unambiguously prove the deviations of polymer chain dynamics from the Rouse dynamics at long Fourier times. Even more it was shown that the scaling regarding chain and entanglement length are correctly predicted by the reptation model. However, apart from very few exceptions only static samples have been probed. Actually, the only study known to the authors describes elongational flow of 0.3 wt% 4 kg/mol polyethylenoxide in D_2O [4] investigated in a flow cell.

Studies of liquids under shear using NSE are challenging since the beam polarisation needs to be preserved and Doppler scattering should be avoided. Still such measurements are extremely important to explain changes in the viscosity of highly entangled polymers subjected to increasing deformations or shear rates. It is known that the macroscopic viscosity, probed by rheology, is connected to the internal relaxation times of polymers. The dramatic changes in viscosity under shear, e.g. shear thinning or stress plateaus for increasing strain, should be reflected in the internal relaxation processes. In this paper we present a new shear cell overcoming the challenges of combined NSE and shear experiments. We demonstrate that Doppler scattering and depolarisation of the beam can be sufficiently minimised and present first experimental results on an entangled polymer solution.

2. Doppler scattering

Neutron spin echo spectroscopy makes use of the Larmor precession to resolve the energy transferred during inelastic and quasi-elastic scattering processes [5, 6]. The total phase angle φ accumulated by a single neutron passing through the spectrometer (equation 1) is a priori independent from the initial neutron wavelength and is a function of the wavelength change $\delta\lambda$ resulting from the scattering as well as the magnetic field B generated by the coils of the spectrometer[1].

$$\varphi = \gamma \left(\frac{m\delta\lambda}{h} \right) Bl \tag{1}$$

Here $\gamma = -1.83247188 \cdot 10^8\ \mathrm{T}^{-1}\mathrm{s}^{-1}$ is the gyromagnetic ratio of the neutron and both coils of the spectrometer are assumed to be of the same length l and generating a fully homogeneous magnetic field of strength B.

The echo measured can be described by equation (2) and corresponds to the number of neutrons $n(i)$ passing the analyser and reaching the detector as the magnetic field of the second coil is detuned with respect to the first one. The detuning allows to extract the beam polarisation P_e by normalizing to the difference $(n(i)_{max} - n(i)_{min})$ between $n(i)_{max}$ and $n(i)_{min}$, the maximum and minimum number of neutrons counted at the detector, respectively.

$$n(i) = \frac{N}{2} \left(1 + P_e \cdot \cos(C \cdot (i - i_e)\lambda_0 + \varphi_p) \cdot e^{-\frac{C^2(i-i_e)^2\sigma^2}{2}} \right) \tag{2}$$

Above, i is the current in the tuning coil and i_e the current in the tuning coil that provides identical path integrals over the magnetic field in the two coils. λ_0 is the mean neutron wavelength of the beam, N the number of neutrons reaching the analyser and C an instrument-dependent constant. To extract the echo several measurements are required, varying the current i around the value i_e. P_e, the polarisation of the beam at equilibrium, can be shown to correspond to a single point in the intermediate scattering function $S(Q, t)$ as long as $\delta\lambda \ll \lambda_0$ holds [5]. The Fourier time t in $S(Q, t)$ is defined through the magnetic field in the coils and the Fourier time range that can be covered is proportional to the mean neutron wavelength λ_0 to the power of three. From equation (1) it can be seen that Doppler scattering results in a shift of the phase angle. For a single constant velocity, the Doppler scattering results in a certain single phase shift φ_p in equation (2). As the shape of the echo remains unchanged the polarisation P_e can be recovered. Even though not required, the coils could be tuned out of resonance to extract P_e and compensate for the energy transfer due to Doppler scattering. However, for a distribution of velocities the situation changes. A spread of phase shifts results in a net depolarisation of the beam. Thus, Doppler scattering resulting from a distribution of velocities has a damping effect on P_e. If not correctly accounted for, this can result in an apparent faster decay of the intermediate scattering function.

[1] A deeper elaboration can be found in [7] and an alternative derivation of this property is depicted in [8]

3. A shear device for NSE

Plate-plate or cone-plate shear devices provide a linear flow profile as compared to the parabolic one in flow cells. Further, typically samples for neutron scattering experiments are rather large but thin, which implies that high pressures are required to pump highly viscous polymers through a flow cell. As a result the walls of a flow cell encounter large forces which can become an engineering challenge but even more important the high pressures may alter the structure or dynamics of the sample [9]. To overcome these difficulties with flow cells we have decided to construct a cone-plate shear device. A vertical cone to plate geometry (figures 1 and 2) allows to minimise Doppler scattering and reduces the sample amount compared to a Couette geometry and provides a constant shear rate over a large sample area.

3.1. Cell architecture

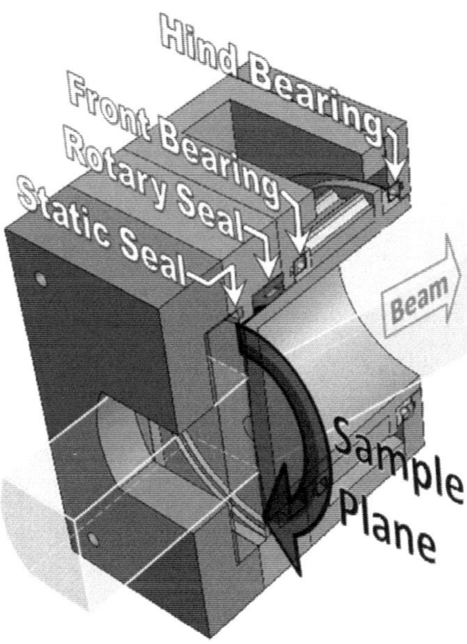

Figure 1. Sketch of the vertical cone-to-plate shear cell constructed for NSE spectroscopy. The black arrow indicates the curvature of the sample flow, the yellow arrow the beam path.

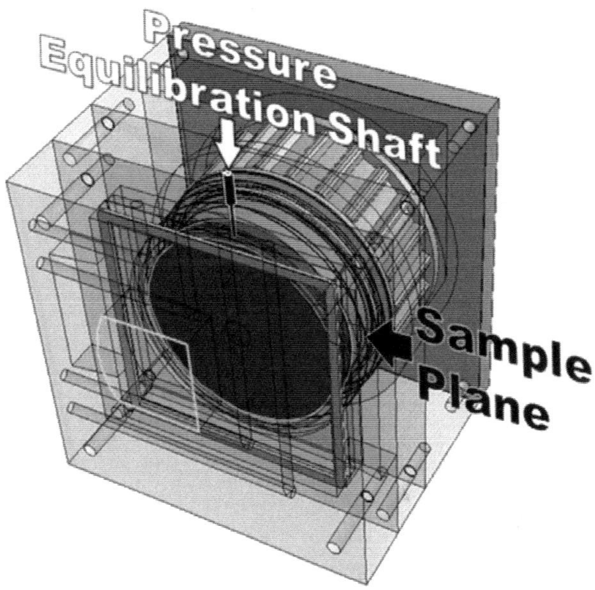

Figure 2. Engineering design of the cone-to-plate shear cell designed for NSE (front view). The front casing contains the cooling ducts and is exchangeable. The modular structure of the cell allows grazing incidence scattering as well.

The casing of the shear device consists of four modules facilitating the exchange of sealings or bearings. The shear counter surface consists of a 70 × 70 × 10 mm polished silicon wafer replaceable by a quartz window. Nonmagnetic A4 steel screws are used throughout the shear cell while shaft, cone and casing are made from aluminium 6082.

The hollow shaft (see figure 1) allows both dynamical sealing and bearings outside of the beam path while keeping the cone radius small (30 mm). This is a compromise between allowing for a large neutron beam as well as for low surface velocities. The latter is limited by the dynamical sealing. Non-magnetic bearings of 60 mm inner diameter are commercially scarce. We decided to use interchangeable PTFE (polytetrafluoroethylene) solid bearings and custom POM (polyoxomethylene, a.k.a. acetal) ball bearings with a brass frame in the shear device. Both of the mutually fully interchangeable bearing solutions provide axial as well as radial support while minimising vibrations. When mounted, the front bearing is locked on the shaft against the pulley and the rear bearing against a ledge fixing the pulley position. Figure 1 indicates the bearing position inside the shear device and figures 3 and 4 depict both bearing solutions schematically.

In the solid bearings two disks of PTFE slide directly over each other. PTFE acts self-lubricating and the friction coefficient between two moving PTFE surfaces is 0.04 [10]. Pure PTFE was used for the manufacturing of the solid bearings, as fibre reinforcement would increase the friction coefficient even though improving the surface wear properties. The custom ball bearings consist each of two brass rings, one on the shaft and one embedded in the casing. The rings are separated by 63 POM precision balls of 3 mm diameter. The precision balls (Klasse G200, Sorte 0) were manufactured and purchased through KGM Kugelfabrik Gmbh & Co. KG[2].

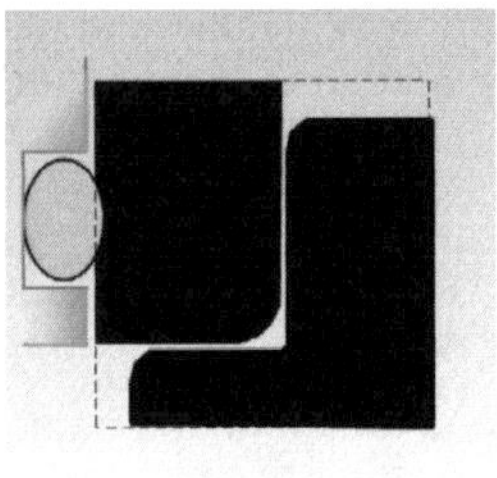

Figure 3. Vertical cut through the solid bearing. Black: cut through the PTFE discs. Turquoise: tensioning O-ring pressing the discs together.

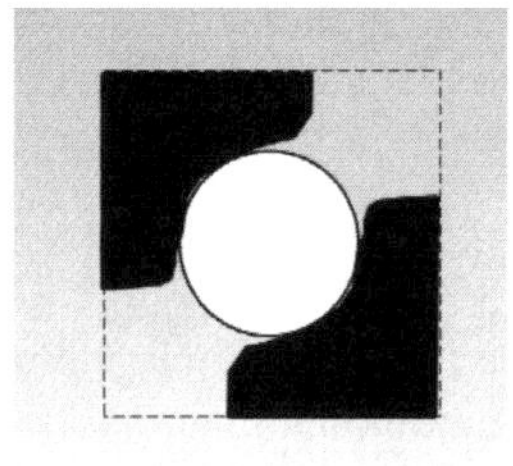

Figure 4. Vertical cut through the custom made ball bearing. Black: main structure made of brass. White: 3 mm diameter POM precision balls.

The relatively long measurement times required for NSE and the large inner diameter of the sealing result in considerable shaft wear at the dynamical sealing. Mitigation is provided through a 0.254 mm thick sleeve of hardened stainless steel (hardness: 95 HRB) slid over the frontal part of the shaft acting as counter surface for the sealing (see figure 5). The 59.99 mm inner diameter sleeve is heat-expanded when mounting and provides a tight grip to the shaft surface once it is cooled to room temperature. Assuring no evaporation or creep of the sample material at the shaft/sleeve interface a 1.78×56.87 mm FFKM (perfluoroelastomer) O-ring is incorporated in a groove in the shaft and sealing against the sleeve.

The dynamical sealing is a FKM (fluoroelastomer) lip-sealing providing excellent tightness and wear resistance. The sealing contains a ferromagnetic skeleton but does not depolarise the beam. Tests on IN15 at a neutron wavelength of 16 Å have revealed a depolarisation of less than 2 %. The lip sealing is mounted with the cavity facing away from the sample in order to minimise the sample material. An FKM O-ring provides the static sealing against the silicon wafer or quartz window.

The filling of the shear cell is performed by removing the frontal lid. Sample material is directly distributed over the cone inside the static sealing. A pressure equilibration duct of 1 mm diameter in the casing module containing the sealings allows to enclose the sample without air inclusion (figure 2). Once the lid is fastened this duct can be sealed with a corresponding screw.

3.2. Drive and velocity control

In order to minimise magnetic stray fields and a resulting depolarisation of the beam a pneumatic drive (PMO-0450 pneumatic motor purchased through P.T.M. Production[3]) is used. The motor is mounted above the shear device. The distances between the motor and beam path can be

[2] Company web page: *http://kgm-kugeln.de*

[3] Company web page: *www.ptmgmbh.com*

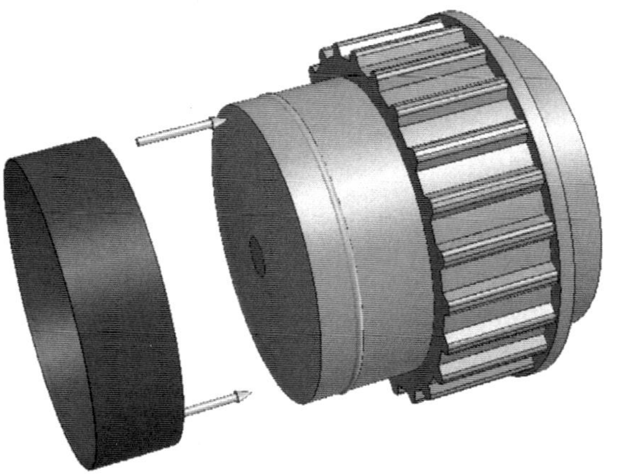

Figure 5. Cone and hollow rotor. Red: steel sleeve acting as counter surface for dynamical sealing. Turquoise: FFKM O-ring preventing creep underneath the sleeve.

adapted to the instrumental needs by the use of timing belts. This solution is very flexible and even allows to decouple the motor from the shear device and operate it at distances of meters by transmitting the torque via the timing belt.

To account for possible load fluctuations a feedback control system regulating the pressure and keeping the rotational velocity of the cone constant was implemented. The PMO-0450-5-0-020 motor has a pre-installed Hall sensor measuring the rotational speed and a PMO-DRZ-030300 speed controller provides the necessary electronics for feedback. An EVD-3900-008GAP solenoid valve (CKD[4] purchased through P.T.M. Production) is used as electro pneumatic regulator, controlled by the 0-10 V output signal of the speed controller. This allows to regulate the airflow to the motor according to the desired shear rate. The valve generates magnetic stray fields and needs to be placed far enough from the neutron beam path.

3.3. Doppler scattering

A velocity component $v_{\|}$, parallel to the transfer of momentum experienced by the neutrons, persists due to the curved flow and large beam footprint. The distribution of $v_{\|}$ is limited to a value $\pm v_{\|,max}$, which is a function of shear rate $\dot{\gamma}$, cone angle θ and vertical width of the sample window d_w (equation 3). A small cone angle allows to reach high shear rates at low rotational velocities. A vertical narrowing of the sample window using B_4C allows to reduce the horizontal velocity component in the beam path.

$$v_{\|,max} \approx \dot{\gamma} \cdot 0.03\tan(\theta) \cdot \frac{d_w}{0.06} \tag{3}$$

For the sample window vertically narrowed down to 10 mm, a cone angle leading to $0.03\tan(\theta) = 8 * 10^{-4}$ and a shear rate of 100 s^{-1}, $v_{\|}$ becomes 1.3 cm/s. $Q \cdot v_{\|,max} t \leq 1$ holds for Fourier times up to 150 ns at $Q = 0.5\ \mathrm{nm}^{-1}$. The accessible Fourier times are inversely proportional to the shear rate.

The energy transferred to the neutron due to Doppler scattering can be expressed by equation (4) [6, p.154f]. Figure 6 depicts the transferred energy due to Doppler scattering for a range of Q and angles between Q and velocity.

$$\Delta E = h(\boldsymbol{Q} \cdot \boldsymbol{v}) = hQv_{\|} \tag{4}$$

[4] Company web page: *http://www.ckd.co.jp/english/index.htm*

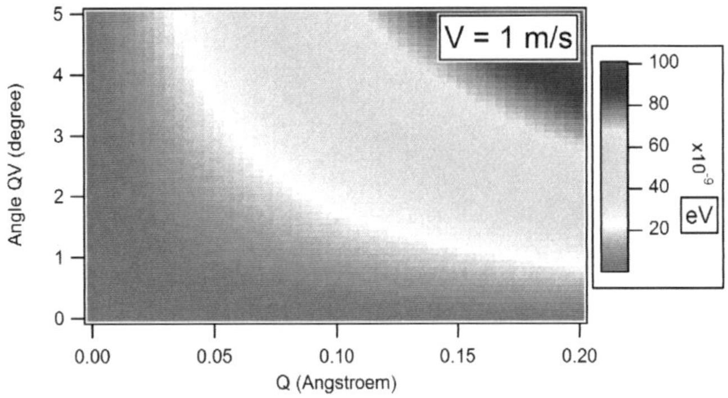

Figure 6. The shift in energy due to Doppler scattering as a function of momentum transfer and angle between velocity and momentum transfer.

4. Experimental results

The shear cell was used in measurements to probe a possible disentanglement transition of polymers under shear as suggested e.g. by Tapadia and Wang [11, 12]. Small Angle Neutron Scattering (SANS) and NSE measurements were conducted at Laboratoire Léon Brillouin, France. The sample was a 20 wt% of fully deuterated 575 kg/mol polystyrene (PS) and 10 wt% of protonated 520 kg/mol PS dissolved in deuterated toluene. Measurements were conducted both without shear and at a shear rate of 300 s^{-1}. The SANS measurements at PAXY using a wavelength of 12 Å and a detector distance of 6.75 m revealed a flattening of the signal with applied shear indicating a swelling of the coils in flow direction. Perpendicular to flow, no conformational change was observed. Figure 7 depicts the azimuthal integration of the two-dimensional detector image separated into its component parallel and perpendicular to the flow direction for the sample under shear. The slope of -2 for the intensity plotted versus Q on a log-log scale confirms a Gaussian shape of the coils in the static sample. In the direction perpendicular to the flow the distribution remains Gaussian under shear, whereas in flow-direction the slope slightly decreases to -1.85. In literature, detailed studies on the shear rate dependence of the structure factor at various temperatures in dilute and semidilute PS solutions can be found [13]. These studies were performed for shear rates up to 300 s^{-1} and show a butterfly like scattering pattern at low Q-values, revealing concentration fluctuations. Studies using labelling that allows to observe the coil conformation confirm an elliptical anisotropy in the high-shear flow regime in dilute and semidilute PS solutions [14, 15]. The studies above conclude that for the single chain conformation the radius of gyration parallel to the flow direction increases while it remains unchanged perpendicular to the flow. Both observations correspond very well to our observations on the highly entangled polystyrene solution.

The NSE measurements were conducted on the instrument MUSES (Laboratoire Léon Brillouin, France). Figure 8 depicts two spectra obtained at a neutron wavelength of 5 Å and $Q = 0.25$ Å^{-1}. The red squares correspond to the signal without shear, whereas the blue circles were obtained with a shear rate of 300 s^{-1}. Extensive NSE studies on polymer dynamics without shear can be found in literature [16, 17]. Even using less rigid polymers than Polystyrene and when measured in the melt, entanglement constraints become only visible at Fourier times of several nanoseconds [18]. Modern NSE instruments such as IN15 cover the range from the picosecond regime to several hundred nanoseconds [19]. The NSE measurements conducted on MUSES did not reach high enough Fourier times to probe entanglement effects. However, the measurements clearly show that the shear device does not influence the NSE experiments. Future measurements with the shear device are planned and will probe Fourier times up to 100 ns. Preparatory polarisation measurements with the empty shear device on IN15 have shown a depolarisation of the beam below 2% at 16 Å.

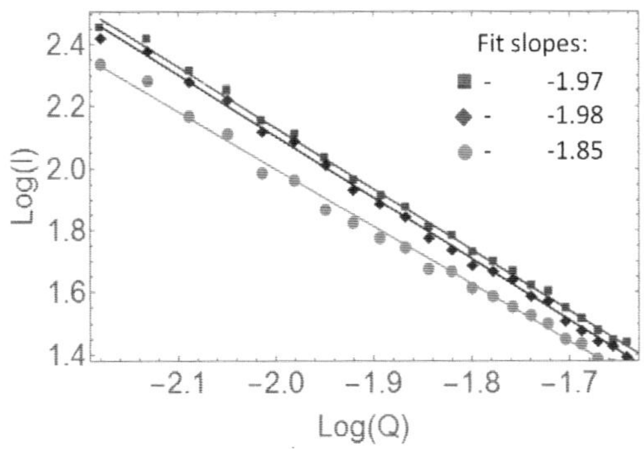

Figure 7. Red squares: the azimuthally integrated isotropic signal obtained when no shear was present. Blue diamonds and Green circles: the horizontal respectively the vertical integration of the scattering image under shear.

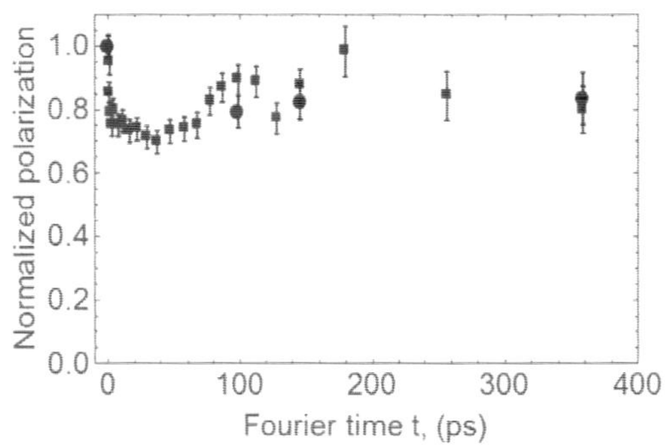

Figure 8. NSE measurements of the sample at a wavelength of 5 Å, $Q = 0.25$ Å^{-1} and measuring times up to 1 ns. Red diamonds: signal for measurements without shear. Blue circles: signal for measurements with a shear rate of 300 s^{-1}.

5. Conclusion and outlook

We have constructed a cone-plate shear device meeting the requirements of NSE measurements. Polarisation tests at wavelength up to 16 Å show negligable beam depolarisation. From calculations it is clear that Doppler scattering can be sufficiently reduced if the geometry of the experiment is chosen correctly. On the other hand if needed Doppler scattering could also be used to get complementary information on the macroscopic flow [20, 21]. As first experimental results using the cell on a SANS machine we extend the knowledge on polymer chain conformation under shear in dilute and semi-dilute solutions to highly entangled (moderately dense) solutions. We report a swelling of the coils in flow direction while no change in the direction perpendicular can be observed.

Acknowledgments

We thank Laboratoire Léon Brillouin for beamtimes on PAXY and MUSES. We acknowledge Airidas Korolkovas for help at the beam times at LLB and fruitful discussions. Further, we are very grateful for the polarisation tests of the shear device at large wavelengths, performed by Bela Farago and Peter Falus at the Institute Laue-Langevin. Financial support from STINT under contract number contract No. IG-2011-2067 as well as the Swedish research council under contract No. C0511501 and within the Swedish infrastructure project Super ADAM is acknowledged as well.

References

[1] Richter D, Ewen B, Farago B and Wagner T 1989 *Phys. Rev. Lett.* **62** 2140
[2] Schleger P, Farago B, Lartigue C, Kollmar A and Richter D 1998 *Phys. Rev. Lett.* **81** 124
[3] Zamponi M, Monkenbusch M, Willner L, Wischnewski A, Farago B and Richter D 2005 *Europhys. Lett.* **72** 1039
[4] Rheinstädter M C, Sattler R, Häussler W and Wagner C 2010 *Physica B* **405** 3690
[5] Mezei F *et al.* 1980 *Neutron Spin Echo - Proc. of a Laue-Langevin Institut Workshop* (*Lecture Notes in Physics* vol 128) ed Mezei F (Berlin Heidelberg: Springer Verlag)
[6] Mezei F *et al.* 2003 *Neutron Spin Echo Spectroscopy - Basics, Trends and Applications* (*Lecture Notes in Physics* vol 601) ed Mezei F, Pappas C and Gutberlet T (Berlin Heidelberg: Springer Verlag)
[7] Kawecki M 2015 *Uppsala University Publications* (*Preprint* `urn:nbn:se:uu:diva-262370`)
[8] Lechner R E and Longeville S 2006 Quasielastic neutron scattering in biology, part 1: Methods *Neutron*

Journal of Physics: Conference Series **746** (2016) 012014 doi:10.1088/1742-6596/746/1/012014

Scattering in Biology - Techniques and Applications ed Fitter J, Gurberlet T and Katsaras J (Berlin Heidelberg: Springer Verlag) pp 309–354
[9] Mortensen K, Schwan D and Janssen S 1993 *Phys. Rev. Lett.* **71** 1728
[10] Walker J S 2010 *Physics* (San Francisco: Addison-Wesley) p 182 4th ed
[11] Tapadia P and Wang S Q 2003 *Phys. Rev. Lett.* **91** 198301
[12] Tapadia P and Wang S Q 2004 *Macromolecules* **37** 9083
[13] Morfin I, Lindner P and Boué F 1999 *Macromolecules* **32** 7208
[14] Lindner P and Oberthür R C 1988 *Colloid Polym. Sci.* **226** 886
[15] Nakatani A I, Douglas J F, Ban Y B and Han C C 1994 *The Journal of Chemical Physics* **100** 3224–3232
[16] Ewen B and Richter D 1997 Neutron spin echo investigations on the segmental dynamics of polymers in melts, networks and solutions (*Advances in Polymer Science* vol 134) (Berlin Heidelberg: Springer Verlag) pp 1–132
[17] Richter D, Monkenbusch M, Arbe A and Colmenero J 2005 Neutron spin echo in polymer systems (*Advances in Polymer Science* vol 174) (Berlin Heidelberg: Springer Verlag) pp 1–221
[18] Richter D, Butera R, Fetters L J, Huang J S, Farago B and Ewen B 1992 *Macromolecules* **25** 6156
[19] Schleger P *et al.* 1998 *Physica B* **241-243** 164
[20] Wolff M, Magerl A, Frick B and Zabel H 2003 *Materialwiss. Werkst.* **34** 568
[21] Wolff M, Magerl A, Frick B and Zabel H 2003 *J. Phys. Condens. Mat.* **15** 337

How to polarise all neutrons in one beam : a high performance polariser and neutron transport system

D. Martin Rodriguez[a], P. M. Bentley[a,c], C. Pappas[b,*]

[a]*European Spallation Source ESS ERIC, Box 176, 221 00 Lund, Sweden*
[b]*Faculty of Applied Sciences, Delft University of Technology, Netherlands*
[c]*Department of Physics and Astronomy, University of Uppsala, Uppsala, 751 20, Sweden*

Abstract

Polarised neutron beams are used in disciplines as diverse as magnetism,soft matter or biology. However, most of these applications often suffer from low flux also because the existing neutron polarising methods imply the filtering of one of the spin states, with a transmission of 50% at maximum. With the purpose of using all neutrons that are usually discarded, we propose a system that splits them according to their polarisation, flips them to match the spin direction, and then focuses them at the sample. Monte Carlo (MC) simulations show that this is achievable over a wide wavelength range and with an outstanding performance at the price of a more divergent neutron beam at the sample position.

Keywords: neutron guides, neutron polarisers, Monte Carlo simulations

1. Introduction

Polarised neutron beams fully exploit the strength of neutron in investigating the structure and motion of atoms and molecules in the bulk. This leads eventually to a broad range of applications and a high demand for the highest possible intensities. The usual polarisation methods involve polarising supermirrors [1, 2, 3], ^{3}He polarisers [4, 5, 6] or Heussler monochromatos [7]. All these methods, however, act as spin filters, i. e. they transmit the part of the neutron beam that has the *correct* spin state while discard all neutrons with the *wrong* spin state. Therefore and in order to maximise the neutron intensity, it would be highly beneficial to design a neutron polarising system that uses the full beam and after some manipulation brings all neutrons in the same spin state and recombines them at the sample position.

*Correspondig author
Email address: damian.rodriguez@esss.se (D. Martin Rodriguez)

Such a device has been proposed by Badurek et al. [8], who showed that by applying the dynamic polarisation technique it is possible to polarise the full neutron beam and reach a high polarisation. However, this solution requires either a highly monochromatic beam, or a time-of-flight polychromatic beam with high time resolution. In both cases the neutron intensity is severely reduced, which is a major drawback. Therefore, for general purpose applications a different, more versatile, solution is required with an approach, which overcomes the limitations of conventional neutron polarisers.

2. Principle and description of the system

The principle of the proposed system consists in separating the neutron beams according to their spin states using a supermirror polarisering cavity. The neutrons are then converted in to the same spin state by adequate use of spin flippers and finally focussed and recombined at the sample position of an instrument. A schematic drawing of the setup is given in fig. 1. The starting point and key component is a

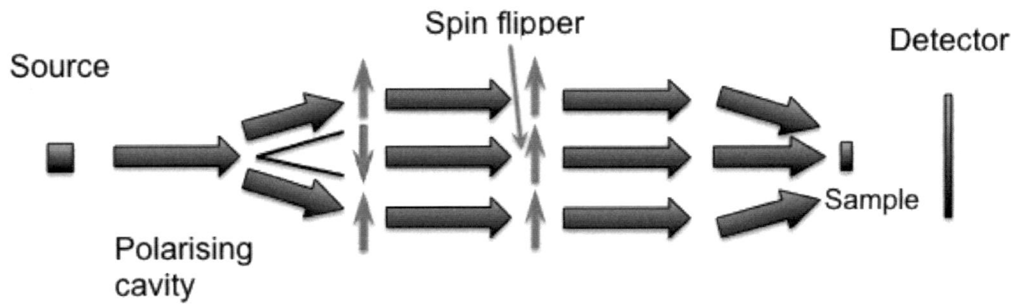

Figure 1: Principle of the beam transport system proposed in this work, illustrating the separation of the incoming depolarised neutron beam in three beams with different polarisations. Using spin flipper(s) all beams are converted into the same polarisation and finally recombined at the sample position.

V-shaped supermirror polarising cavity, which separates the neutron beams according to their spin states. Such a device has been realised at the Helmholtz Zentrum Berlin (previously Hahn-Meitner Institute) and provided polarised neutron beams to two instruments: the transmitted beam with a polarisation $|-\rangle$ was serving the reflectometer V6 while the reflected beam with a polarisation $|+\rangle$ was directed to the neutron-spin-echo spectrometer V5 [9]. The solution adopted here is a variant of the Berlin setup and is schematically shown shown in fig. 1-3. It includes a cavity consisting of m=2.5 (supermirrors) and m=1 (Ni) coated Si wafers inserted in the neutron guide as specified in 2(b). The cavity splits the original neutron beam in three: a transmitted central beam with a polarisation $|-\rangle$, and two reflected lateral beams with a polarisation $|+\rangle$. After the cavity a guide system transports the central and lateral beams to the sample. Broadband RF flipper(s) as schematically shown in fig. 2(a) and in the artist's view of fig. 3 bring the three beams to the same polarisation state, e.g. in $|+\rangle$. Figures 2(a)-(b) and 3 illustrate the parameters on

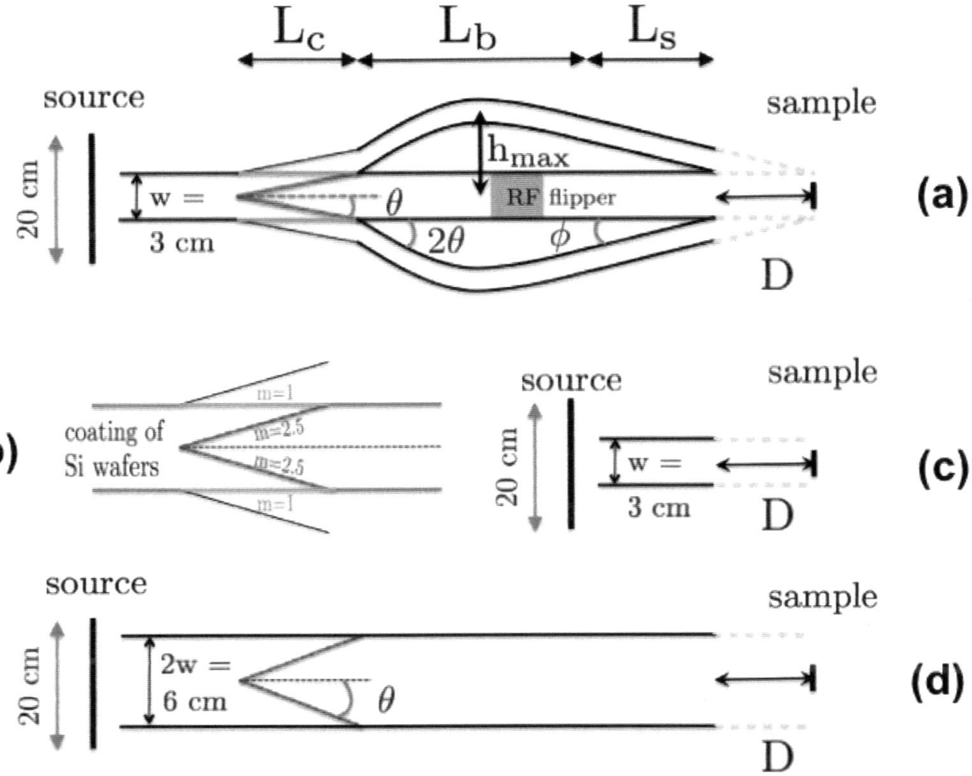

Figure 2: (a) Schematic representation of the system illustrating the fixed dimensions and the free parameters (angles θ and ϕ, lengths h_{max}, L_b and L_s) used for the Monte Carlo simulations. The elements in orange and red designate the parts that are reflecting and also transmitting neutrons. In a real setup these would be consisting of Si wafers coated with m=2.5 suppermirrors (red lines) and m=1, i.e. Ni (orange lines), as illustrated by the closeup view of (b). The reference setup used to normalise the neutron intensities at the sample is shown in (c). The performance of the system (a) has been compared with that of a double width polarising cavity shown in (d). All drawings are not in scale.

the basis of which the system was optimised using Monte Carlo simulations. If θ is the angle between the polarising supermirrors of the V-cavity and the central beam propagation axis, the reflected beams will be at an angle of 2θ, and for this reason at the sides of the cavity two curved guides take over with a respective inclination of

2θ. The curved guide sections have a linear length of L_b and separate the beams by h_{max} by following an arc with an angular opening of $2\theta+\phi$, where ϕ is determined by :

$$\phi \approx \tan\phi = \frac{3W}{4D} \tag{1}$$

with W the width of the initial beam and D the distance between the end of the guide and the sample position. Finally, a straight guide with an inclination of ϕ with respect of the central guide direction, transports the lateral beams to the sample. Obviously the width of the lateral guides is half that of the central beam and the successful implementation of the concept relies on the optimum design of the lateral beam sections. The system potentially uses all incoming neutrons, but at the price of a more divergent beam. The successful implementation of the device requires the optimum superposition of the three beams at the sample with a smooth beam profile and a homogeneous angular divergence. We will show that this is feasible. Furthermore, the system provides the flexibility to block the lateral beams and use only the less divergent central beam if required.

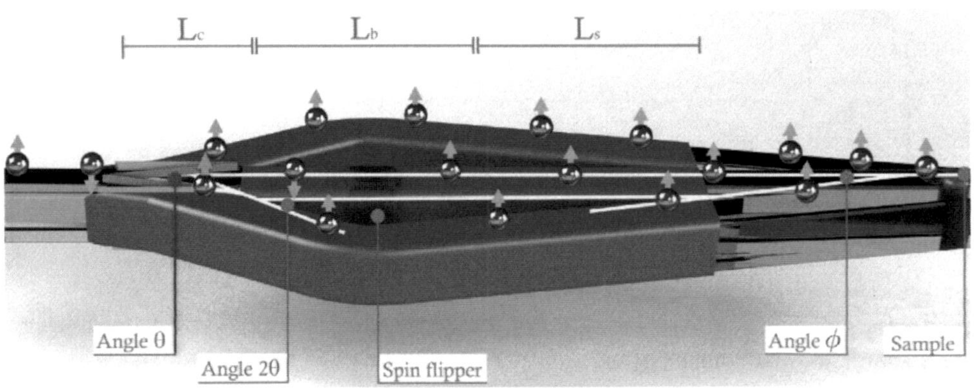

Figure 3: Artist's view of the setup of fig. 2 illustrating the geometry and an RF flipper positioned at the central beam. The drawing is not in scale.

3. Simulation procedure and results

The simulations were performed with the VITESS program [10], assuming the ILL_cold_2011 neutron source. The source had a diameter of 20 cm, large enough to fully illuminate with full divergence and at all wavelengths the neutron guides considered in this work. At 2 m from the source a 10 m long guide was placed with a width of W=3 cm, a height of 6 cm and all walls coated with m=1 (Ni). After this straight guide section the V-shaped polarising cavity was placed as shown in fig. 2(b). The coating of the polarising supermirrors corresponded to Fe/Si multilayers with m=2.5, while the rest of the mirrors was coated with m=1. An important element for the successful implementation of the concept are the m=1 coated Si wafers, sketched by the orange lines in fig. 2(a)-(b) and 3, which were also included in Berlin setup [9]. The curved lateral parts had a non-polarising coating of m=2 for the outer walls and m=1 for the rest. The central guide was also coated with m=1 over the whole length. The sample had a cross-section of 2 x 2 cm^2 and was placed at D = 4 m from the end of the guide assembly.

The results of the simulations will be discussed in two sections, in which the performance of the proposed system (from now on named *the system*) is discussed as a function of its total length and the geometry of the polarising cavity respectively. The neutron intensities at the sample were normalising the corresponding intensity of the reference setup shown in fig. 2(c), which corresponds to the arrangement of fig. 2(a) without the polarising system: i.e. at 2 m from the source a 10 m long guide is placed with a width of W=3 cm, a height of 6 cm and all walls coated with m=1 (Ni). This normalised neutron intensity together with the polarisation, horizontal profile and divergence of the neutron beam at the sample position were used to assess the performance of the system.

3.1. Length of the device

The length of the system depends on the length of the cavity ($L_c = W/2\tan\theta \approx W/2\theta$) and the requirements to redirect the lateral beams to the sample. With the angles defined as in fig. 2 the angular length of the arc is fixed to $2\theta+\phi$. However, the length depends on the radius of curvature ρ, which can be obtained following a well established geometrical approach [11]. For the assessment of the performance three extreme cases were considered by comparing the position of the curved section exit window with respect to the line-of-sight (LOS) of the entrance window : (a) half in the LOS (b) exactly out of LOS and (c) twice out of LOS. The parameters used for the simulations are summarised in table 1.

The results are given in fig. 4 and show that in all cases the polarisation is high and the normalised neutron intensity, which in this case can be associated with the transmission of the system, higher than 50 %, reaching even 80 %. The

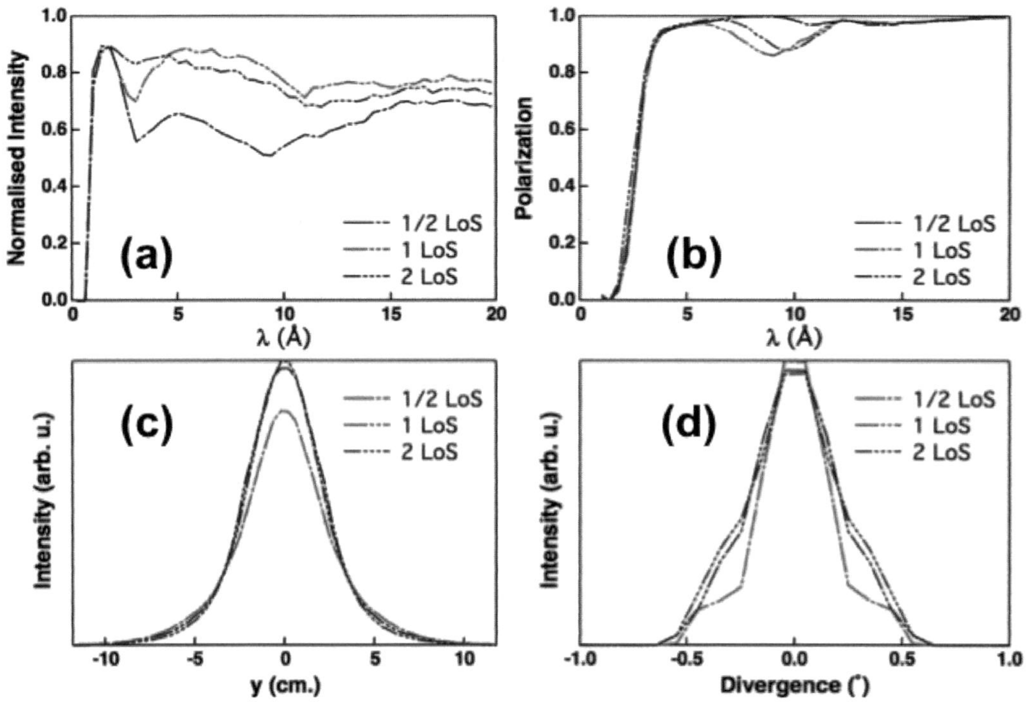

Figure 4: Normalised neutron intensity (a), polarisation (b), horizontal beam profile (c) and divergence (d) of the system for three positions of the curved section exit window with respect to the line-of-sight (LOS) of the entrance window. The characteristic lengths used in the simulations are given in Table 1.

Table 1: Parameters of the systems simulated in 3.1.

θ [°]	ρ [m]	$L_c[m]$	L_b [m]	L_s [m]	Remarks
0.6	44.34	2.86	1.16	1.84	1/2 LOS
0.6	177.38	2.86	4.61	7.25	1 LOS
0.6	709.51	2.86	19.65	20.78	2 LOS

cases "1 LOS" and "2 LOS" give very similar results. It is also noticeable that the polarisation of the lateral beams goes through a minimum at about 9 Å for the "1 LO" and "2 LOS" cases but not for "1/2 LOS". This effect should therefore be brought in relation with ρ and the reflectivity of the outer walls, which are m=2 for both spin states and thus not polarising. The higher values of ρ lead to lower incident angles of the neutrons on the outer walls, which increases the transmission for all neutrons including those with the wrong spin state that are not completely filtered out by the polarising cavity. It is indeed known that such a "leakage" exists for the reflected beams [12], which is closely related with the geometry of the cavity and will be discussed in the following sub-section. The "leakage" could be eliminated by a polarising supermirror coating of the outer walls, as it was suggested in [12] and realised in Berlin [9].

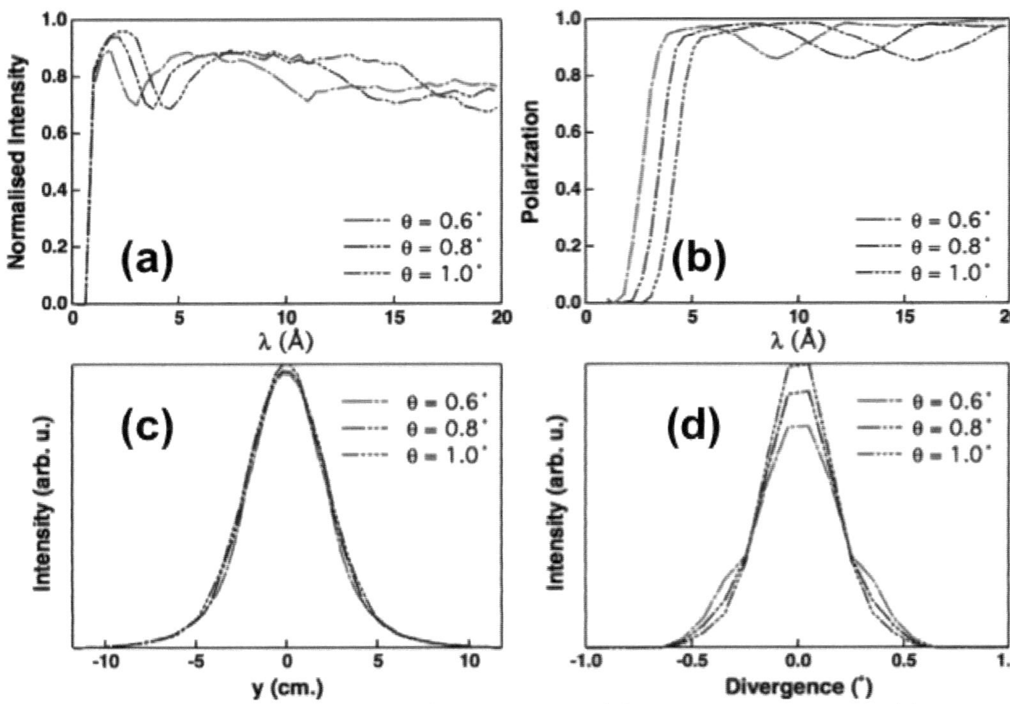

Figure 5: Normalised neutron intensity (a), polarisation (b), horizontal beam profile (c) and divergence (d) of the system for three values of the angle θ for the polarising cavity. The simulations correspond to the case "1 LOS" assuming the characteristic lengths given in Table 2.

3.2. Geometry of the polarising cavity

The performance of polarising cavities depends on the combination of the angle θ and the supermirror coating [12, 3]. In this work we fixed the supermirror coating to m=2.5 as shown in fig. 2b and changed θ to 0.6°, 0.8° and 1.0°, while keeping the "1 LOS" configuration for the lateral beams with the lengths given in Table 2. The deduced performance is given in figure 5, where similar trends and similar shapes are found for these three cases. As expected, the higher values of θ shift the optimum performance towards longer wavelengths and the "leakage" seen in the polarisation occurs at longer wavelengths as well.

Table 2: Parameters of the systems simulated in 3.2.

θ [°]	ρ [m]	L_c [m]	L_b [m]	L_s [m]	Remarks
0.6	177.38	2.86	4.61	7.25	1 LOS
0.8	111.47	2.15	3.66	8.75	1 LOS
1.0	76.48	1.72	3.03	10.00	1 LOS

4. Discussion

The system can efficiently split the incoming neutron beam in three polarised beams, and after proper spin manipulation with spin flippers, it has the potential to redirect them to the sample position. In this way the intensity is practically doubled with respect to what would have been delivered only by the supermirror cavity in transmission, i.e. by the central beam of the system shown in fig. 2a. In the following we will show that this result is in many aspects similar to what would have been obtained by doubling the dimensions of the central beam as shown in fig. 2d. The performance of the different setups (always at the sample position) is given in fig. 6 for the $\theta = 0.6°$ and "1 LOS" case. All setups lead to a highly polarised beam. On the other hand, the intensity at the sample is the lowest for the central beam alone and is practically doubled either by doubling the width of the guide or by using the proposed system. A similar trend is found on the divergence of the beam, which is the lowest for the central beam alone and broader but with a similar distribution both for the double width and the proposed system cases. The beam profile in the horizontal direction is different for each of the cases: it is the broadest for the double width beam whereas it shows a sharp profile for the proposed system illustrating its focusing effect.

A pre-requisite for the optimum performance of the system is the possibility

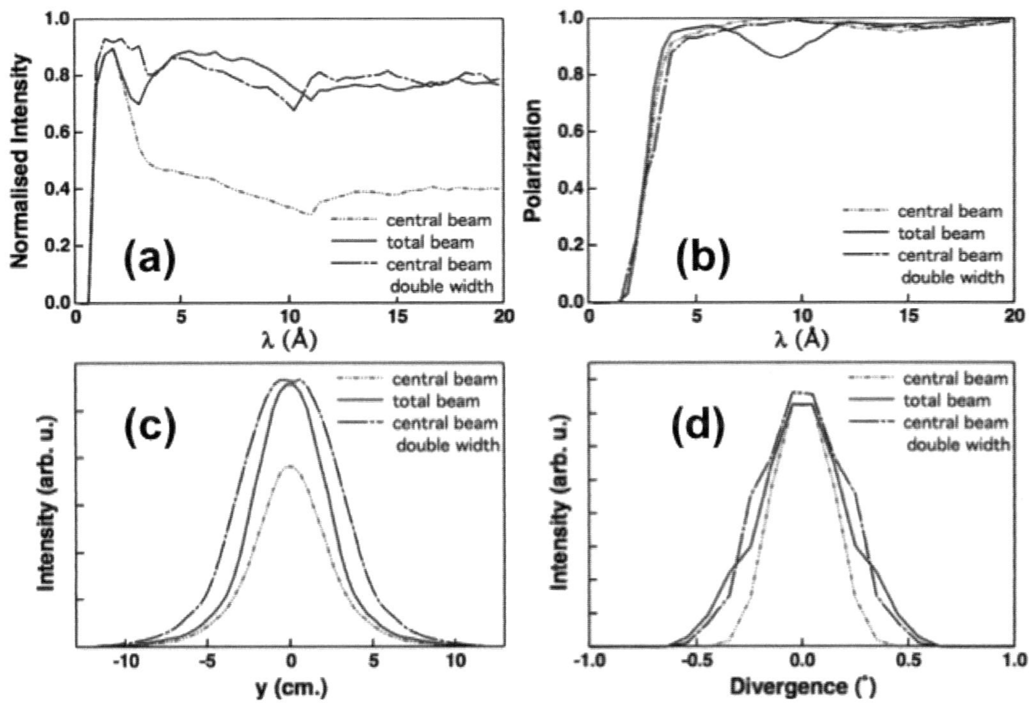

Figure 6: Comparison of (a) the normalised intensity, (b) the polarisation, (c) divergence integrated over the full spectrum and (d) beam profile in the horizontal direction (y) at the sample position for three cases: the central beam of the system shown in fig. 2a, the combined central and lateral beams of fig. 2a, and for the double width central beam with the polarising cavity in transmission shown in fig.2d. All results refer to the sample position, and to the case of "1 LOS" with $\theta = 0.6°$.

to apply magnetic fields over the whole length of the device. The magnetic fields should be high enough to polarise the beam at the cavity and then sufficient to guide and maintain the beam polarisation. In fact several realisations of polarised cavities have addressed this point [14, 13] and a variety of solutions have been successfully adopted involving either coils or the strong permanent magnets available nowadays.

The design of the device implies that the strong magnetic fields needed to saturate the supermirrors and polarise the neutron beam are applied only at the cavity and thus far away from the sample position. This is an advantage for applications, like in Neutron Spin Echo spectroscopy, where a high magnetic field homogeneity is required. and any cross-talk with other magnetic fields must be minimised.

Last but not least the performance of the device depends on the capability to flip the beam polarisation of the central and/or the lateral beams with a high efficiency over a broad wavelength range. For this purpose RF adiabatic flippers [15] can be used, which combine a compact design with broadband high flipping efficiency [13, 16]. These flippers consist of a longitudinal RF coil, which can practically have the dimensions of the neutron guide and can even slide along the guide [16], and a vertical magnetic field, with a longitudinal gradient, which can be realised by placing permanent magnets below and above the device as a whole.

To conclude, from todays perspective there are no technological risks in adopting the proposed design for future polarised neutron instruments, in particular neutron spin echo spectrometers. In this way it is possible to improve the performance of beam extraction systems by polarising also those neutrons that are usually discarded.

Acknowledgments

M.R. would like to thank Bela Farago for enlightening discussions. Funding is acknowledged from the European Union Seventh Framework Programme [FP7/2007-2013] under grant agreement Nr. 283883.

References

[1] Mezei F., 1976 *Comm. Phys.* **1** 81

[2] Mezei F., Dagleish P.A., 1977 *Commun. Phys.* **2** 41

[3] P. Böni et al., 2009 *Physica B* **404** 2620

[4] Petoukhov A. et al., 2006 *NIMA* **560** 480

[5] McIver J. W. et al., 2009 *Rev. Sci. Instrum.* **80** 063905

[6] Ye Q. et al. 2013 *Phys. Procedia* **42** 206

[7] Courtois P. 1999 *Physica B* **267-268** 363-366

[8] Badurek G., Rauch H. and Zeilinger A. 1980 *Z Physik B* **38** 303-311

[9] Krist Th. et al., 1995 *Physica B* **213-214** 941

[10] Lieutenant K., Zsigmond G., Manoshin S., Fromme M., Bordallo H., Champion D., Peters J. and Mezei F. 2004 *Proc. SPIE* 5536 134

[11] Schirmer A. and Mildner D.F.R. 1991 *Meas. Sci. Technol* **2** 1059-1063

[12] Mezei F. 1989 in: Neutron Optical Devices, ed. C. Majkrzak, Proceedings of the SPIE p. 10-17 (doi:10.1117/12.948741)

[13] Keller Th. et al., 2000 *NIMA* **451** 474

[14] Krist Th. et al., 2010 *Journal of Physics: Conference Series* **251** 012081

[15] Bazhenov A.N. et al., 1993 *NIMA* **332** 534

[16] Plomp J. 2009 Spin-echo development for a time-of-flight neutron reflectometer, Ph.D. Thesis, Delft, ISBN:978-90-9024161-6

Recent advancements of wide-angle polarization analysis with ^{3}He neutron spin filters

W.C. Chen[1,2], T.R. Gentile[1], Q. Ye[1,2], A. Kirchhoff[1], S.M. Watson[1], J.A. Rodriguez-Rivera[1,2], Y. Qiu[1], C. Broholm[1,3]

[1] National Institute of Standards and Technology, Gaithersburg, Maryland 20899, USA
[2] University of Maryland, College Park, Maryland 20742, USA
[3] John Hopkins University, Baltimore, Maryland 21218, USA

E-mail: wcchen@nist.gov

Abstract.
Wide-angle polarization analysis with polarized ^{3}He based neutron spin filters (NSFs) has recently been employed on the Multi-Axis Crystal Spectrometer (MACS) at the National Institute of Standards and Technology Center for Neutron Research (NCNR). Over the past several years, the apparatus has undergone many upgrades to address the fundamental requirements for wide angle polarization analysis using spin exchange optical pumping based ^{3}He NSFs. In this paper, we report substantial improvements in the on-beam-line performance of the apparatus and progress toward routine user capability. We discuss new standard samples used for ^{3}He NSF characterization and the flipping ratio measurement on MACS. We further discuss the management of stray magnetic fields produced by operation of superconducting magnets on the MACS instrument, which can significantly reduce the ^{3}He polarization relaxation time. Finally, we present the results of recent development of horseshoe-shaped wide angle cells.
Key words: wide angle polarization analysis, wide angle ^{3}He cell, horseshoe-shaped ^{3}He cell, polarized ^{3}He SEOP, ^{3}He NSF

1. Introduction

Polarized neutron scattering is one of the most challenging neutron instrumentation techniques due to significant intensity reduction from polarized beam production and analysis, necessity of precise spin manipulation, and complicated data reduction and analysis. Supermirrors and Heusler crystals have been successfully employed to polarize the incident beam and analyze the scattered beam for decades. Recently, many modern neutron instrumentation techniques call for larger beam angular divergences and wavelength bandwidths, and a larger area of detector banks in the interest of large on-sample neutron fluxes and improved detection efficiency. This has made it difficult to apply the standard neutron polarization techniques using supermirrors or Heusler crystals. ^{3}He neutron spin filters (NSF) are advantageous in (1) polarizing a broad wavelength band of neutrons effectively; (2) polarizing large area and widely divergent neutron beams; and (3) efficiently flipping the neutron polarization by reversing the ^{3}He nuclear polarization using the adiabatic fast passage (AFP) nuclear magnetic resonance (NMR) technique[1]. The development of ^{3}He NSFs has been advanced significantly during the last decade and these NSF devices have been applied or planned to be applied to many polarized neutron instrumentation techniques in most neutron facilities worldwide[2, 3, 4, 5, 6, 7]. Recently, there has been much increased interest in wide angle polarization analysis using ^{3}He NSFs[6, 8, 9, 10, 11, 12].

The Multi-Axis Crystal Spectrometer (MACS)[13] at the National Institute of Standards and Technology (NIST) Center for Neutron Research (NCNR) is one of such modernized spectrometers where polarized beam instrumentation cannot be implemented with conventional neutron polarizing devices. Two existing and operating instruments with wide angle polarization analysis, D7[14] at the Institut Laue-Langevin and DNS[15] at the Jülich Centre for Neutron Science (JCNS), have employed supermirrors with a total cross sectional area of several hundred of square meters as the wide angle spin analyzer. A ^{3}He NSF-based wide angle polarization analysis apparatus has recently been developed on MACS[10, 11]. Combined with high incident beam flux and multi-detector banks, MACS provides an improvement in polarized neutron detection efficiency by at least two orders of magnitude compared to the conventional cold triple-axis spectrometer SPINS at the NCNR. Very recently, we have primarily focused on improving the on-beam-line performance of the apparatus for routine user experiments. We describe substantial improvements of the wide angle polarization analysis capability, including standard samples for ^{3}He NSF characterization, reduced magnetic field gradients, and fabrication of horseshoe cells. These advancements, combined with experience from several user experiments done recently, will make routine polarized MACS capability available soon.

2. Description of wide angle polarization analysis setup on MACS

MACS has recently been moved to the BT-9 beam line at the NCNR with a dedicated cold hydrogen source, and is operational now for routine user experiments[13]. MACS covers an incident energy range from 2.3 meV to 17 meV and typical final energies between 2.5 meV and 5 meV. MACS has a 34 cm by 42 cm monochromator consisting of 357 Pyrolytic Graphite (PG) crystals. This allows for doubly focusing the incident beam at the sample, providing an incident neutron beam flux up to 5×10^8 cm^{-2}s^{-1}[13]. The detector system on MACS consists of 20 detector banks, each separated by 8 °. Each detector channel has a conventional 90′ Söller collimator. The scattering angle coverage depends on the incident energy and is different between negative and positive sides of the scattering angle. Following the neutron beam trajectory, positive scattering angles are located on the right side of the beam and negative scattering angles are located on the left side of the beam. The scattering angle reaches the maximum, about 128 ° on the negative scattering angle side at the highest incident energy and about 115 ° on the positive scattering angle side at the lowest incident energy. MACS allows for high sensitivity to access physics in small samples and is ideal for probing slowly propagating excitations in hard condensed matter.

Wide-angle polarization analysis on MACS has been developed using ^{3}He NSFs and such an apparatus on MACS has been described in detail in previous papers[10, 11]. Here we outline the polarized beam setup. The polarized beam apparatus consists of a cylindrical ^{3}He spin filter cell in a radio frequency(RF)-shielded square solenoid to polarize the incident neutron beam and two three-sectioned wide-angle ^{3}He cells located near the sample and on opposite sides of the main beam to spin-analyze the scattered beam. Both the cells and the sample are contained within a uniform magnetic field of less than 4 mT provided by a vertical, neutron compatible solenoid. Therefore, the wide angle polarization analysis capability on MACS refers to uniaxial polarization analysis with the neutron polarization axis perpendicular to the scattering vector. Neutron spin flipping of the polarized beam is accomplished by inverting the ^{3}He polarization using the AFP NMR technique. When flipping the neutron spin of the polarizer, the polarization loss to the analyzer is ensured to be negligible. The loss for the polarizer was 0.10 % per flip and the associated loss in the analyzer cell was 0.005 % per flip[11]. During the experiment, FID is used to monitor the ^{3}He polarization in the spin filter cells. The ^{3}He gas is polarized in sealed cells off the neutron beam line by spin-exchange optical pumping (SEOP) and the cells are transported to the beam line in a portable solenoid. Recently, we reported several upgrades to this apparatus[11], in particular the development of GE180[16] wide-angle cells with relaxation times between 100 h and 400 h, and ^{3}He polarization values, P_{He}, between 0.65 and 0.80. Using conventional cylindrical cells with a small angular coverage[17], the apparatus was employed for a study of superconducting $Fe_{1+y}Te_{0.62}Se_{0.38}$[18]. After substantial improvements of polarized MACS capability reported in reference[11] and discussed in this paper, several user experiments involving wide angle polarization analysis have been successfully completed and data analysis is in progress.

3. Standardized samples for ^{3}He NSF characterization

A sapphire powder sample was used at the very beginning of the polarized MACS development. Sapphire powder has several Bragg peaks. However, only the (012) peak with a d-spacing of 3.482 Å can be observed for both positive and negative scattering angles on the MACS instrument at 5 meV. One of the disadvantages of the sapphire sample is that the Bragg peak intensity is not very strong, requiring a fully open incident beam for quick and accurate measurements of the ^{3}He polarization. This makes it inconvenient to perform transmission measurements of a ^{3}He NSF[19] for a routine user experiment. Peak scans on a strong nuclear Bragg peak are typically used for transmission measurements so that the background from the sample and the ^{3}He cell glass material can be removed. In order to observe Bragg peaks on both sides of the neutron beam at energies less than 5 meV, a large d-spacing crystal is necessary. Naturally occurring materials with a d-spacing on the order of 7-10 Å are not generally available. We are aware of two synthetic materials, mica[20] and intercalated graphite[21]. We chose to employ mica due to availability at the NCNR. The lattice spacing of the synthetic mica, fluorophlogopite, has been measured to be $d = 9.963$ Å. So the scattering angles 2θ are 33.4 °, 27.3 °, and 23.4 ° at neutron energies of 2.5 meV, 3.7 meV and 5 meV, respectively. At 5 meV, we also use a PG crystal, which yields a scattering angle of 74.6 °. The synthetic mica and PG crystals are a great resource for accessing all final energies on MACS for ^{3}He polarization and flipping ratio measurements. In addition, since the intensities on the Bragg peaks from mica and PG are much stronger than that from the sapphire powder sample, we can significantly reduce the incident beam flux with a strong attenuator. This allows for significantly reducing the radiation level near the sample, but still optimizing the ^{3}He polarization and flipping ratio measurements.

In order to map the angular dependence of the opacity for a typical wide-angle cell within the full scattering angle range on MACS, we use a plastic sample so the scattered neutrons uniformly illuminate the cell. We discuss the angular dependent opacity of the wide angle cells in Sec. 5.

4. Reduction of magnetic field gradients on MACS

One of the main goals for experiments on MACS is to address scientific applications at high magnetic fields. More than 50 % of user experiments may utilize superconducting magnets with fields up to 11 T. Due to heavy usage of the superconducting magnet, MACS has been designed to be as non-magnetic as possible within a radius of 135 cm near the sample. Although all of the structure materials were carefully chosen when the instrument was built, some small parts near the sample area are not truly non-magnetic. These small parts are magnetized after use of the superconducting magnet. In addition, structure elements below the sample region are not non-magnetic. Although the biological shielding is mostly fabricated from non-magnetic stainless steel, we found that two locations inside the biological shielding on both sides of the neutron beam presented some field gradients. These two locations are symmetric relative to the main beam. All of these components become magnetized after use of the superconducting magnet, leading to field gradients. We found that these field gradients have significantly reduced the relaxation times of the polarizer and two analyzers.

The magnetic field gradient ($|\vec{\nabla} B_{\perp}/B|$) induced ^{3}He polarization relaxation rate is proportional to the square of the gradient and inversely proportional to the partial ^{3}He pressure[22]. On the beam line, the space is often limited for implementation of ^{3}He NSFs. However it is still desirable to have a field gradient of 5×10^{-4} cm^{-1} or better so that the field gradient contribution to the relaxation time of the ^{3}He polarization is 600 h or longer at a ^{3}He pressure of 1 bar. We note that the larger field gradients can be tolerated for an in-situ ^{3}He SEOP polarizer or analyzer.

We have established three diagnostic tools to evaluate the field gradients near the polarizer or the analyzer on MACS: (1) mapping the field directly, (2) determining the gradients from the relaxation time T_1[23], (3) determining the gradient from the NMR frequency difference among the polarizer and analyzers. Using a fluxgate magnetometer with a measurement sensitivity of 10^{-6} mT, we mapped the residual fields along three orthogonal X, Y, Z directions. Here X denotes along the neutron beam direction; Y denotes the direction perpendicular to the beam in the scattering plane, Z denotes the direction normal to the scattering plane. T_1 is a signature of the volume averaged gradient in a ^{3}He cell. The frequency difference provides information about the gradient between two cell locations.

After use of the superconducting magnet, the residual field was between 0.3 mT and 0.5 mT mainly in the Z direction, while the residual field in the scattering plane was negligible. The relaxation times were about 25 h and 110 h for the polarizer cell Mourvedre and the analyzer cell Giant at a holding field of 3 mT. Given that we obtained T_1's of 330 h for the polarizer Mourvedre and 400 h for the analyzer Giant in a homogenous field with a gradient better than 2×10^{-4} cm^{-1}, this indicated a field gradient of 2.5×10^{-3} cm^{-1} for the polarizer cell Mourvedre and 1.1×10^{-3} cm^{-1} for the analyzer cell Giant after use of the magnet. The difference in FID NMR frequency between the polarizer and one of the analyzers was about 2.3 kHz out of 99 kHz (about 3 mT), which yielded a field gradient of 1.2×10^{-3} cm^{-1} for a distance of 20 cm between the polarizer and analyzer pickup coils. Therefore the resulting field gradients for the polarizer and analyzer after use of the superconducting magnet are consistent among the direct field measurement, T_1 data and the difference in FID frequency. The field gradient at the analyzer was better than that at the polarizer. This is mainly due to the presence of weakly magnetized pre-sample optics and strongly magnetized magnetic cables used for controlling the beam size slit.

The short relaxation times after use of the magnet made it difficult to perform a polarized user experiment on MACS. Hence we found an appropriate demagnetization procedure to improve the relaxation times of the cells in the MACS solenoid. We started with the same maximum field applied during the magnet operation but in a reversed direction, then gradually decreased the applied field and switched the direction between two consecutive fields. This method reliably decreased the field gradient near the ^{3}He cells. It was also important to scan the entire MACS energy range at each applied field to ensure that demagnetization occurred at each possible cell location. We found a possible field gradient dependence on cell location if demagnetization was done at one incident energy. So the relaxation times might be different at different incident energies, which would complicate time-dependent polarization efficiency corrections. An example of the demagnetization sequence after running the magnet at 9 T is (-9 T $\rightarrow$ 8 T $\rightarrow$ -7 T $\rightarrow$ 6 T $\rightarrow$ -5 T $\rightarrow$ 4 T $\rightarrow$ -3 T $\rightarrow$ 2 T $\rightarrow$ -1 T $\rightarrow$ 0.5 T $\rightarrow$ -0.2 T $\rightarrow$ 0.1 T $\rightarrow$ -0.05 T $\rightarrow$ 0.02 T $\rightarrow$ -0.01 T $\rightarrow$ 0 T). It typically takes about 3 hours to perform a complete demagnetization procedure. After an adequate demagnetization procedure, there are three notable features of the residual field and more importantly the gradient: (1) the field is lower than 0.07 mT, mainly along the Z direction. The X and Y components in the scattering plane are negligible; (2) the fields at two locations near the biological shielding as mentioned earlier are significantly smaller than those measured before demagnetization; (3) the fields from the magnetized actuator cables near the polarizer are substantially lower than those measured before demagnetization. Consequently the relaxation times of the cells are significantly improved as shown in Fig. 1a. The relaxation times of the polarizer cell Mourvedre and the analyzer cell Giant are up to 257 h and 350 h, respectively, at a holding field of 3 mT. This implies a field gradient of 3.7×10^{-4} cm^{-1} and 3×10^{-4} cm^{-1} for the cells Mourvedre and Giant, respectively. These indicated improvement in field gradients by a factor of 6.7 and 3.7 for the polarizer and analyzer. The shorter relaxation time of the analyzer cell Reliance compared to the analyzer cell Giant was due to a shorter relaxation time even in a very homogenous field. Most importantly, the relaxation times for the polarizer and two analyzer cells did not change between 3.7 meV and 14 meV, although they changed slightly between 2.5 meV and 3.7 meV. This allows us to perform polarized inelastic experiments for nearly the entire energy range of the MACS instrument. Furthermore the improvement in gradient was confirmed by checking the FID NMR frequency difference, which is shown in Fig. 1b. The frequency difference between the polarizer and analyzer is significantly reduced from 2.3 kHz to less than 0.4 kHz at 2.5 meV and even less than 0.2 kHz between 3.7 meV and 14 meV, which corresponds to field gradients of 2×10^{-4} cm^{-1} and 1×10^{-4} cm^{-1}. Since the relaxation times for the polarizer and analyzer cells we measured on MACS were close to the intrinsic ones that were measured in a homogenous field with a gradient better than 2×10^{-4} cm^{-1}, we do not plan to replace the weakly magnetic parts.

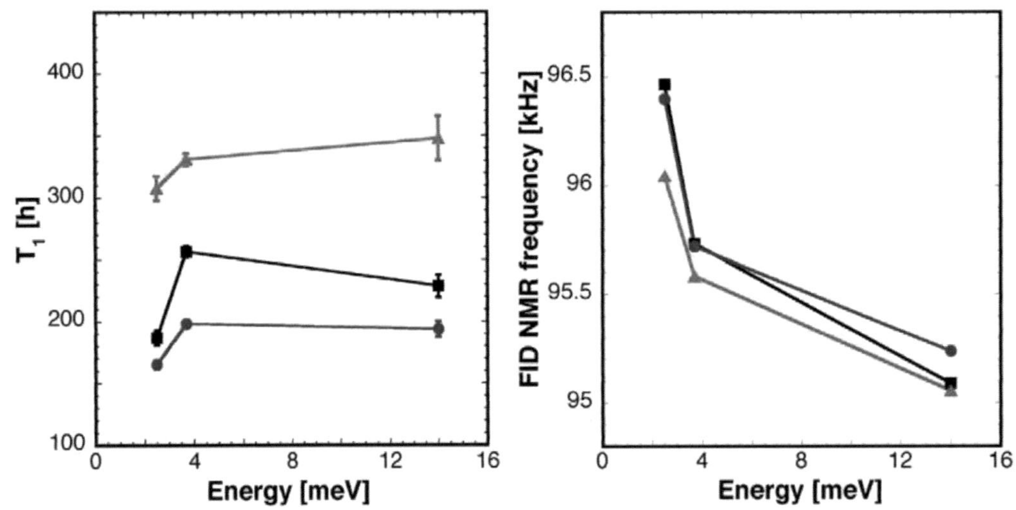

Figure 1. (Color online) (a) Measured relaxation times of the polarizer cell Mourvedre (solid box), the analyzer cell Giant (solid triangle) and Reliance (solid circle) at different scattered energies. (b) Fit FID NMR frequency of the polarizer cell Mourvedre (solid box), the analyzer cell Giant (solid triangle) and Reliance (solid circle) at different scattered energies. The uncertainty of the fit FID frequencies are much smaller than the size of the figure marker. All measurements were performed after performing a complete demagnetization on MACS. The demagnetization procedure is described in the text. For both measurements, the cell Mourvedre was used as polarizer and cells Giant and Reliance were used for analyzers at the positive and negative scattering angle sides, respectively. The solid lines are to guide the eye.

5. Wide-angle cell development

Wide-angle cell fabrication has been one of the most challenging and key elements in wide-angle polarization analysis using ^{3}He spin filters. Substantial progress has already been made for MEOP-based, valved, wide-angle quartz cells[9], but quartz has proved to be unsuitable for SEOP due to temperature-dependent relaxation[11, 17]. During the last decade, we have had great success in achieving long relaxation times and high ^{3}He polarizations using sealed SEOP cells [17, 25]. These cells have mainly been fabricated from boron-free aluminosilicate glass GE180[16] due to the low ^{3}He permeability, good alkali resistance, and great reliability to achieve long ^{3}He polarization relaxation times.

5.1. Three-sectioned cell development

We have successfully made sealed, two or three-sectioned, reblown GE180 cells[11]. Fig. 2 shows an example of a sealed, three-sectioned GE180 cell named Reliance. These three-sectioned GE180 cells employed Rb-K mixtures instead of pure Rb to improve the optical pumping efficiency[26, 27]. These cells showed relaxation times up to 400 h with achievable polarization up to 80 % for the cell volume up to 1290 cm^3. Using one of the NCNR SEOP systems equipped with three 100 W laser diode bars, we were able to polarize the 1290 cm^3 cell to 80 % with a time constant of 9 hours for optical pumping. Similar to the flat window cells in Ref. [17], it was not as reliable to obtain long relaxation times in these three-sectioned GE180 cells as compared to fully reblown GE180 cells, however the improvement did present a significant step toward success of wide angle polarization analysis using the SEOP method.

The beam size at the sample is tunable and up to 2 cm wide by 4 cm tall[13]. The beam sizes at both the polarizer and analyzer locations are calculated from the beam size at the sample, the distance from the sample, and divergences for both incident and scattered beams. When the polarizer and analyzer ^{3}He cells were designed, all cells that are used for routine user experiments were ensured to accept the maximum beam size at each corresponding configuration[10, 11]. The maximum beam height is expected to be less than 6 cm and 5.5 cm for the three-sectioned and horseshoe-shaped cells (see Sec. 5.2), respectively. The three-sectioned cells typically have a diameter slightly larger than 8 cm, and the horseshoe-shaped cell Mozart has a height of 7.3 cm, which are large enough to cover the full beam height. A cross-sectional view of both three-sectioned and horseshoe-shaped cells is also shown in Fig. 2. The cross section is circular for the three-sectioned cell and rectangular for the horseshoe-shaped cell.

One drawback to the use of these cells as analyzers for wide angle polarization analysis is the angular dependence of the path length along the scattered beam direction. To measure the scattering angle dependence of the opacity of the analyzer cell, a plastic sample (as mentioned in Sec. 3) was used to uniformly scatter neutrons through two analyzer cells situated in experiment conditions. The opacity is a characteristic parameter of a ^{3}He NSF, and is linearly proportional to the ^{3}He gas density, the wavelength, and the length of the cell. The measurement was done at 3.7 meV. After correcting the detector efficiency, a map of the scattering angle dependent opacity is shown in Fig. 3. The cell Beethoven was placed on the negative scattering angle side and Reliance was placed on the positive side. Since both cells have the same geometry and similar dimensions, their angle dependent opacity showed an anti-symmetric trend, as expected. The shape of the angular dependent opacity is expected and can be modeled from the geometric shape of three-sectioned cylinders. The opacity differences between the lowest and highest values for their usable angular range are 6.2 % and 8.9 % for Reliance and Beethoven, respectively. The resulting transmission differences for an unpolarized beam would be 5.3 % and 6.9 % at a ^{3}He polarization of 75 % for Reliance and Beethoven, respectively. The resulting neutron polarization differences would be less than 1 % for both cells. The angle dependent opacity, transmission and neutron polarization can be corrected when performing the polarization efficiency correction although this presents additional complexity due to a multi-detector bank configuration on MACS.

5.2. Horseshoe-shaped cell development

With three-sectioned GE180 cells we have begun polarized beam experiments on MACS, but have not yet reached a routine capability. We have also continued to develop more suitable wide angle cells. Recently, Salhi and Babcock at the JCNS developed a very large donut-shaped cell with an outside diameter of 22 cm[28]. Very recently, we have developed a method to produce a horseshoe-shaped cell by pinching off a section from a donut-shaped cell. The donut-shaped cell is produced using the similar procedure as described in the Ref. [28]. The first such cell named Mozart is shown in Fig. 2. The cell Mozart has an outside diameter of 23 cm and inner diameter of 9.2 cm with an overall volume of 1675 cm^3, covering an angular range of at least 250 °. The inner diameter accommodates a 7.6 cm outside diameter cryostat tail piece. A horseshoe-shaped cell has four advantages over a three-sectioned cell: 1) since the cell is made from fully blown glass, long relaxation times can be obtained with greater reliability, 2) the circular shape yields a uniform gas path length for all scattering angles, 3) the cross section of the cell is close to rectangular, yielding a more uniform gas path length compared to a round cross section, and 4) only one ^{3}He spin analyzer is necessary, reducing the demand on our optical pumping system and FID NMR monitoring and simplifying polarization corrections from ^{3}He polarization time-dependence.

Before the cell Mozart was filled with gases of ^{3}He and nitrogen and alkali-metals, the transmission measurement was carried out to determine the variation of glass thickness with scattering angle. The measurements were done at both 2.5 meV and 5 meV using a plastic sample with and without the cell Mozart on the cold triple axis spectrometer SPINS at the NCNR. The result is shown in Fig. 4. At 2.5 meV, the glass transmission is constant for the scattering angle range between 20 ° and 120 °. At 5 meV, the transmission is nearly independent of the scattering angle. This indicated the glass thickness is uniform for the entire scattering angular range, an important result given that the cell is fully reblown and the cell fabrication process is complicated. In the future, we plan to measure this transmission on MACS. We filled the cell Mozart to 1.07 bar ^{3}He gas and distilled an appropriate ratio of rubidium and potassium metals[17]. We were successful in obtaining a relaxation time of 350 h in the first attempt. The K/Rb vapor mixture ratio at an operating temperature of 215 °C was measured to be 5, optimized for optical pumping efficiency and ^{3}He polarization. For comparison, a map of the opacity for the cell Mozart as a function of the scattering angle is also shown in Fig. 3. The cell Mozart is optimized at 3.7 meV. Unlike both cells Reliance and Beethoven, the cell Mozart showed a scattering-angle independent opacity on both sides of the main beam. The opacity is slightly lower at positive angles for the cell Mozart than that at negative ones, which is due to imperfect coincidence of the centers of the inner hole and the outer circle when forming a donut-shaped cell. More horseshoe cells will be made in the future for a routine capability of wide angle polarization analysis on MACS. The cell Mozart was recently polarized using one of the NCNR SEOP systems[2] equipped with three 100 W laser diode bars. We obtained a ^{3}He polarization of 78 % based on a neutron transmission measurement.

The opacity can change with the beam height as this has been modelled for other instrument geometries[30]. For polarized MACS experiments we have done so far, we have measured opacities for each cell to accommodate a different sample size. A typical sample height was about 2.5 cm. We took particular care to ensure a three-sectioned analyzer cell being reliably located on the beam line in the same way for every experiment. To accomplish this, we designed an aluminum structure (referred to as the thimble[10]) to (1) support the cryostat and (2) to isolate the analyzer cells from the cryostat so that the ^{3}He glass cells are protected. Since there is only a few millimeter gap between the thimble and the analyzer cells due to space constraint, aligning the analyzer cells is straightforward by placing the cells close to the thimble simulator with an additional alignment tool between the two cell ends on either side of the main beam.

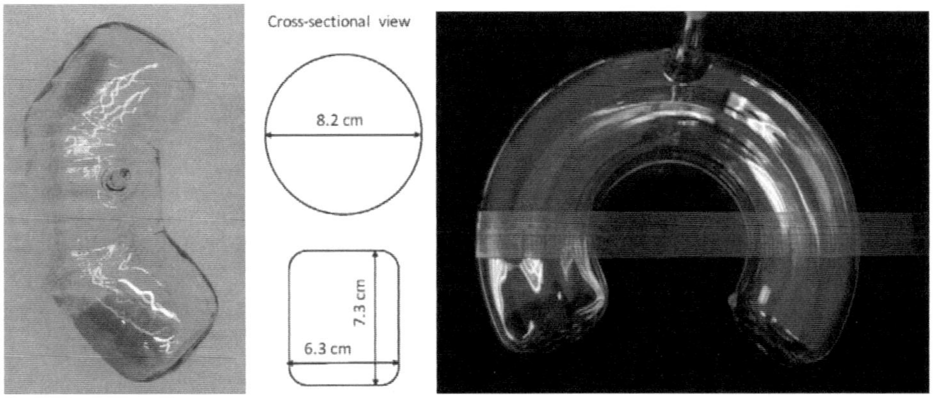

Figure 2. (Color online) Three-sectioned wide-angle cell as developed[11] (left photo). Each cell covers a scattering angle range between 90 ° and 110 °; An example of the recently developed horseshoe shaped wide angle cells (right photo). A cross-sectional view of both cells is shown in the middle of the photo. A circular cross section indicates the three-sectioned cell and a rectangular cross section indicates the horseshoe-shaped cell. A single horseshoe cell covers scattering angle of up to 260 °. The cell Mozart has an outside diameter of 23 cm, an inner diameter of 9.2 cm and a height of 7.5 cm.

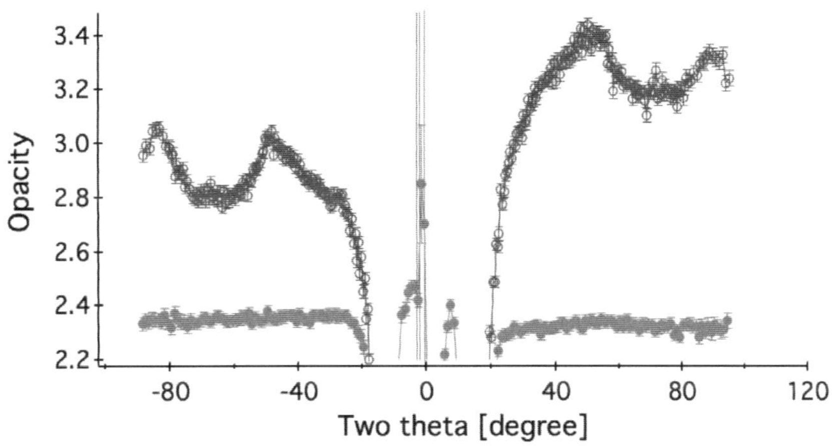

Figure 3. (Color online) Measured opacity as a function of the scattering angle for the three-sectioned cells (open circles) and the horseshoe cell (solid circles). The opacity of a ^{3}He cell is linearly proportional to the ^{3}He gas density, the wavelength, and the length of the cell. The measurements were done at 3.7 meV on MACS. For three-sectioned cell measurements, the cell Beethoven was located on the negative scattering angle side and the cell Reliance was located on the positive scattering angle side.

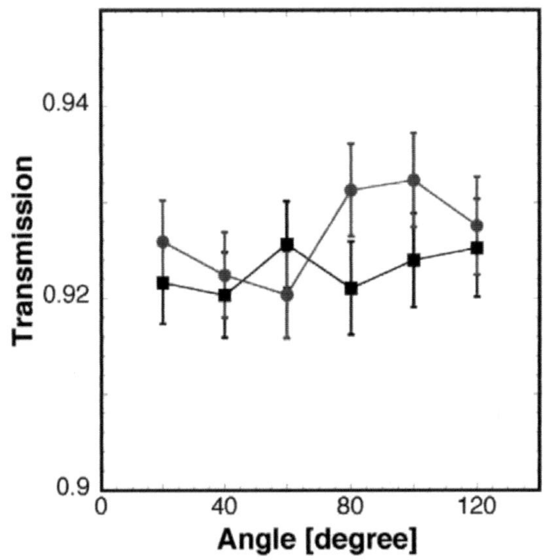

Figure 4. (Color online) Measured glass transmission of the unfilled Mozart at different scattering angles at energies of 2.5 meV (solid box) and 5 meV (solid circle). The measurements were done on the cold triple axis spectrometer SPINS at the NCNR. Nearly scattering-independent transmission implies that the glass thickness is uniform for the entire scattering angle range. Given that the cell is fully reblown, it is remarkable. The solid lines are to guide the eye.

6. Conclusions

We have established a new procedure for the ^{3}He spin filter characterization using different standard samples, either synthetic mica or PG crystals depending on the energy. This allows for more efficient and convenient measurements of transmission and flipping ratio measurements. We have found an appropriate demagnetization procedure to significantly reduce the residual field gradient present on MACS, which allows us to reliably achieve relaxation times of 200 h or higher for the polarizer and 300 h or higher for the analyzer. Most importantly, these relaxation times are independent of the incident energy, a characteristic feature necessary for polarization analysis of inelastic scattering. We have developed horseshoe shaped cells that present advantages over three-sectioned wide angle cells in achieving longer relaxation times with greater reliability, scattering-angle independent cell opacity, and more efficient operation in polarized beam preparation and polarization efficiency correction. We obtained a relaxation time of 350 h and achieved ^{3}He polarization of 78 % for the first filled horseshoe cell. With this substantial progress, we will soon have routine capability in wide angle polarization analysis on MACS.

Acknowledgments

We thank Jeff Anderson and Jack Fuller of the NIST glassblowing and optical shops. In addition we thank J.W. Lynn and C.F. Majkrzak for their help in loaning synthetic mica crystals and a PG crystal. We acknowledge Leland Harriger for his help in the glass transmission measurement on SPINS. The development and application of neutron spin filters was supported in part by the U.S. Dept. of Energy, Basic Energy Sciences. The work utilized facilities supported in part by the National Science Foundation under Agreement No. DMR-1508249.

References

[1] Abragam A 1961 *The Principles of Nuclear Magnetism* Oxford University Press, Oxford, England
[2] Chen WC *et al.* 2014 *Journal of Physics: Conference Series* **528** 012014
[3] Andersen KH *et al.* 2005 *Physica B* **356** 103
[4] Babcock E, Ioffe, A 2011 *Journal of Physics: Conference Series* **294** 012005
[5] Jiang CY *et al.* 2013 *Physics Procedia* **42** 191
[6] Beecham CJ *et al.* 2011 *Physica B* **406** 2429
[7] Kira H *et al.* 2011 *Journal of Physics: Conference Series* **294** 012014
[8] Heil W *et al.* 2002 *Nucl. Instrum. Meth. A* **485** 551
[9] Andersen KH *et al.* 2009 *Physica B* **404** 2652
[10] Fu CB *et al* 2011 *Physica B* **406** 2419
[11] Ye Q *et al* 2013 *Physics Procedia* **42** 206
[12] Babcock E *et al.* 2014 *Journal of Physics: Conference Series* **528** 012018
[13] Rodriguez-Rivera A *et al* 2008 *Meas. Sci. Technol.* **19**, 034023
[14] http://www.ill.eu/instruments-support/instruments-groups/instruments/d7/description
[15] http://www.mlz-garching.de/dns
[16] GE Lighting Component Sales, Bldg. 315D, 1975 Noble Rd., Cleveland, OH 44117. The largest standard GE180 tubing diameter is 16 mm; the 25 mm tubing was procured through a special glass run. Certain trade names and company products are mentioned in the text or identified in an illustration in order to adequately specify the experimental procedure and equipment used. In no case does such identification imply recommendation or endorsement by the National Institute of Standards and Technology, nor does it imply that the products are necessarily the best available for the purpose.
[17] Chen WC *et al.* 2011 *Journal of Physics: Conference Series* **294** 012003, and references therein
[18] Thampy V *et al.* 2012 *Phys. Rev. Lett.* **108**, 107002
[19] Jones GL*et al.* 2000 *Nucl. Instrum. Meth. A* **440** 772
[20] McCauley JW, Newnham RE, and Gibbs GV 1973 *American Mineralogist* **58** 249
[21] Boeuf A *et al.* 1983 *Synthetic Metals* **8** 307
[22] Cates GD, Schaefer SR and Happer W 1988 *Phys. Rev. A* **37** 2877
[23] McIver JW, Erwin R, Chen WC and Gentile TR 2009 *Rev. Sci. Instrum.* **80** 063905.
[24] Babcock E *et al.* 2006 *Phys. Rev. Lett.* **96** 083003
[25] Chen WC *et al.* 2014 *J. Appl. Phys.* **116** 014903
[26] Babcock E *et al.* 2003 *Phys. Rev. Lett.* **91** 123003
[27] Chen WC, Gentile TR, Walker TG, and Babcock E 2007 *Phys. Rev. A* **75** 013416
[28] Salhi Z, Babcock E, Pistel, P, and Ioffe A 2014 *Journal of Physics: Conference Series* **528** 012015
[29] Goland AN, Sondericker Jr. JH and Antal JJ 1959 *Rev. Sci. Instrum.* **30** 269.
[30] Chen WC, Erwin R, and Watson SM, 2013 *Physics Procedia* **42** 163-170.

Neutron imaging data processing using the Mantid framework

Federico M. Pouzols[1], Nicholas Draper[2], Sri Nagella[3], Erica Yang[3], Ahmed Sajid[3], Derek Ross[3], Brian Ritchie[3], John Hill[1], Genoveva Burca[1], Triestino Minniti[1,4], Christopher Moreton-Smith[1], Winfried Kockelmann[1]

[1] ISIS Pulsed Neutron and Muon Source, Rutherford Appleton Laboratory, Science & Technology Facilities Council STFC, Harwell Oxford, OX11 0QX, UK
[2] Tessella Ltd, Abingdon, Oxfordshire, OX14 3YS, UK
[3] Scientific Computing Department, Science & Technology Facilities Council - STFC, Harwell Oxford, OX11 0QX, UK
[4] CNR-IPCF, Viale Ferdinando Stagno d´Alcontres, n. 37 98158, Messina, Italy

E-mail: `federico.montesino-pouzols@stfc.ac.uk`, `nicholas.draper@tessella.com`, `winfried.kockelmann@stfc.ac.uk`

Abstract. Several imaging instruments are currently being constructed at neutron sources around the world. The Mantid software project provides an extensible framework that supports high-performance computing for data manipulation, analysis and visualisation of scientific data. At ISIS, IMAT (Imaging and Materials Science & Engineering) will offer unique time-of-flight neutron imaging techniques which impose several software requirements to control the data reduction and analysis. Here we outline the extensions currently being added to Mantid to provide specific support for neutron imaging requirements.

1. Introduction

Several imaging instruments are currently being constructed at neutron sources around the world, including IMAT [1] at the ISIS facility, VENUS at the SNS and ODIN at the ESS. IMAT (Imaging and Materials Science & Engineering) is beginning its scientific commissioning phase in Autumn 2015. It offers unique time-of-flight diffraction techniques by capitalising on latest image reconstruction procedures and event mode data collection schemes which impose several software requirements to control the data reduction and analysis of its imaging data.

IMAT will provide neutron radiography (2D), neutron tomography (3D), energy-selective and energy-dispersive neutron imaging [1, 2]. Flexible neutron wavelength discrimination along this *fourth imaging dimension* via time-of-flight measurements is available at a pulsed-source facility like ISIS which opens up new opportunities for contrast-enhanced and microstructure imaging. For energy-selective analyses, wider and narrower energy bands can be selected at will, and time-resolving cameras are able to discriminate quasi-monochromatic neutron channels. For *white-beam* applications that do not need energy discrimination neutron counts in a wide energy band can be summed up to provide higher counting statistics. Standard data processing and reconstruction algorithms can be used for wide and narrow energy bands, with the limitation only being the reduced counting statistical errors for the latter. In this sense, energy-selective neutron imaging is a straightforward extension of the two-dimensional radiography and three-dimensional tomography methods. It is obvious that for best exploiting the information content of such 4D data sets the use of high performance computer (HPC) clusters comes into consideration, especially for time evolving experiments (5D) or on-the-fly image reconstructions during data acquisitions. Furthermore, in addition to this direct imaging approach (wavelength band selection), the transmission spectra of the imaging camera pixels may be analysed using model-based fitting methods providing physical, elemental or structural parameters that can then be reconstructed and displayed in two and three dimensions.

2. Frameworks

The Mantid software project [3, 4] provides an extensible framework that supports high-performance computing for data manipulation, analysis and visualisation of scientific data [5]. The Mantid framework is mature and widely used across many facilities and techniques. It is being extended to provide specific support for neutron imaging requirements. Being part of the Mantid framework, the software introduced here effectively integrates with a vast array of capabilities and methods readily available in Mantid. Mantid is open source, supported on multiple platforms (Windows, Linux, Mac OS X), and packages are available for easy deployment.

Here we introduce the imaging functionality that is being integrated in Mantid. A new custom graphical user interface integrates reconstruction and visualisation capabilities for neutron imaging. This interface (Figure 1) gives access to new features of Mantid to load images and stacks of images into standard Mantid data structures and process imaging data. It also provides access to third party software such as packages for tomography reconstruction that support different 3D image reconstruction approaches for neutron tomography. Mantid has the capability to run image reconstruction jobs either locally or in compute clusters and supercomputing facilities thereby enabling quick evaluation and refinement of the reconstruction pipeline.

Different facilities frequently choose different job scheduling tools and hardware for their computing facilities. For example, the computer cluster SCARF (Scientific Computing Application Resource for Facilities), which is available on-site for ISIS users, uses Platform Load Sharing Facility (Platform LSF [6]). On the other hand the clusters at the Oak Ridge National Laboratory and the SNS use the MOAB workload manager [7]. The Mantid framework provides the middleware required to control jobs in different computing facilities handling the different job schedulers and/or web service interfaces. The Mantid framework itself can be used through its scripting interface via a web service API. This addresses challenges that arise at different facilities, such as different authentication methods, job schedulers, and location of compute resources.

Together with reconstruction functionality, Mantid also integrates multidimensional visualisation tools that leverage on the VSI (Visualisation simple interface of the Mantid project) [5], and the ParaView data analysis and visualisation application [8]. This provides powerful tools to visualise the ≥ 4 dimensional data that can result from imaging at a time of flight spallation source (three spatial dimensions, wavelength, and conditions in time evolving experiments). The VSI is a specialized application for neutron scattering data while ParaView is a more generic application with which users can quickly build visualisations to analyze their data, using quantitative and qualitative techniques. With both tools, the data can be explored interactively in 3D visualisations, or programmatically via a command line interface, scripting, and batch processing capabilities.

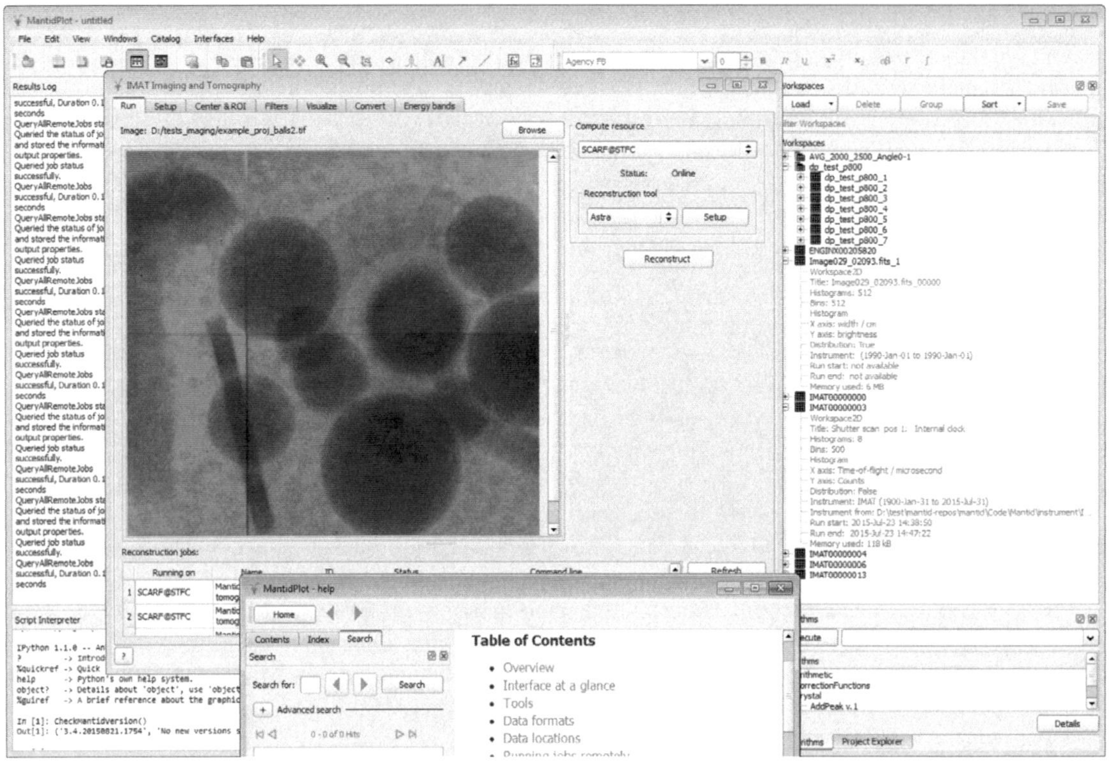

Figure 1. Mantid imaging interface integrated in the general Mantid graphical user interface with functionality for general manipulation and analysis of neutron data.

3. Compute resources at ISIS, STFC

Within ISIS, the SCARF and Emerald computer clusters, administered by the Scientific Computing Department at STFC (Science & Technology Facilities Council), will allow rapid refinements during the imaging experiments. A key hurdle is the ability to take full advantage of available computing resources. Different facilities frequently choose different job scheduling tools, interfaces and hardware for their computing resources.

At ISIS, users of the IMAT instrument will benefit from the high performance computing services provided by the ULTRA platform (Figure 2). This project, led by the Scientific Computing Department at STFC, aims at providing secure and efficient high performance computing and data services to support operations at ISIS and the Diamond Light Source.

The IMAT instrument requires advanced tomographic image reconstruction software that is able to deal with the energy dependencies in neutron imaging experiments. As a result, rapid 3D volume data reconstruction and in-experiment data rendering and visualisation with HPC based infrastructure are critical to IMAT experiments. In-experiment visual feedbackto inform and guide experiments will optimise the use of beam time. Mantid is capable of using ULTRA2 high performance computing services in a way that is transparent to the user, via a standardised web service.

4. Tomography: reconstruction tools

The imaging interface being integrated in Mantid supports third party software for 3D image tomographic reconstruction. A number of tools are currently supported and/or being trialed: TomoPy [9], Astra Tomography Toolbox [10–12], and Octopus [13], and in the future the

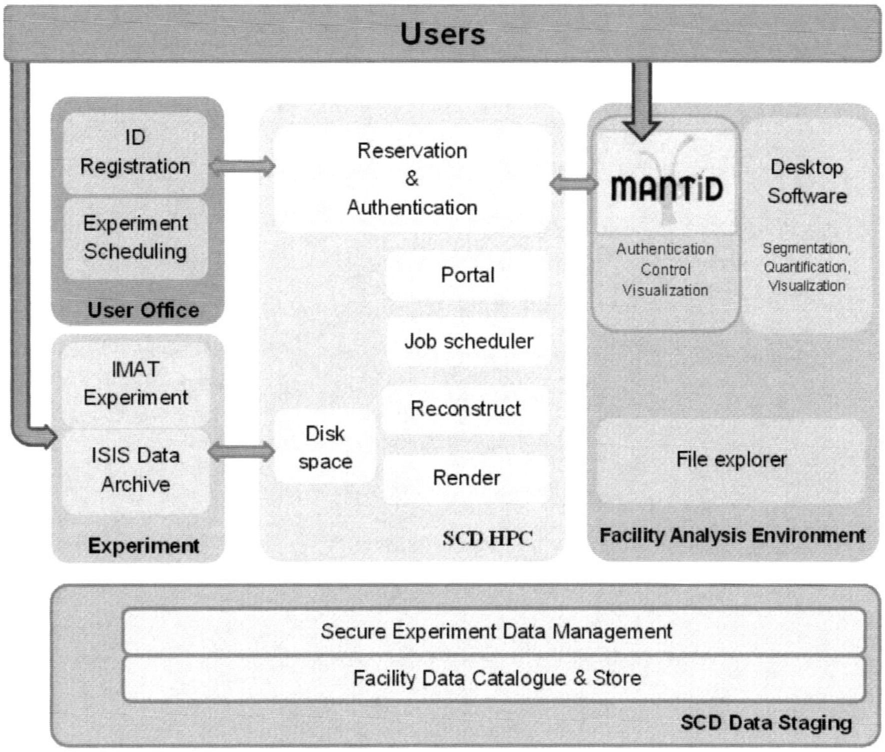

Figure 2. Simplified system architecture of the project ULTRA2 (Ultra High-throughput HPC Platform for Scientific Applications).

reconstruction pipeline software Savu [14, 15] developed by the Diamond Light Source. Different reconstruction approaches are thus supported, namely filtered back projection and iterative methods such as simultaneous iterative reconstruction technique, but also variants of these developed in different tools.

5. Pre-/post-processing and visualisation

The new imaging features of Mantid are being tested with simulated and sample images. Mantid provides a harmonised and simple interface to different, heterogeneous tomographic reconstruction methods and tools which provide a varying degree of support for image pre-processing tasks.

Mantid is being extended with support for several input/output formats and additional parameterisation and pre-/post-processing functionality. This includes for example a graphical interface to select different areas, such as the region of interest for reconstruction, and the area for open beam normalisation (Figures 3 and 4). Support is being added for different data formats (FITS [17], NeXus NXtomo [16], TIFF [18], PNG [19], etc.) and variants of them. Additional capabilities are provided to process multi-dimensional stacks of images from energy-selective analyses, see Figure 5 for an illustration of the differences between energy neutron radiographies for different energy bands.

Image format extensions currently being defined for more specific processing of imaging data from IMAT also require integration in the framework. Different tomographic reconstruction tools support pre-processing and filtering tasks to a largely varying degree and in heterogeneous ways. Several pre-processing tasks that are essential for effective tomographic reconstruction are

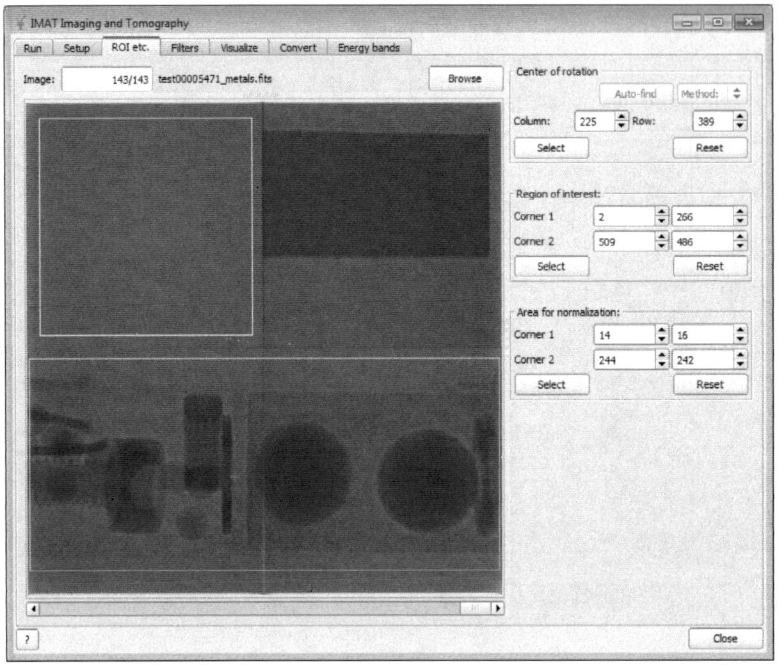

Figure 3. Section of the imaging interface for interactive selection of different image parameters. The parameters, including center of rotation, region of interest, and open beam normalisation (or air region), are selected interactively in an interface that visualises a stack of images.

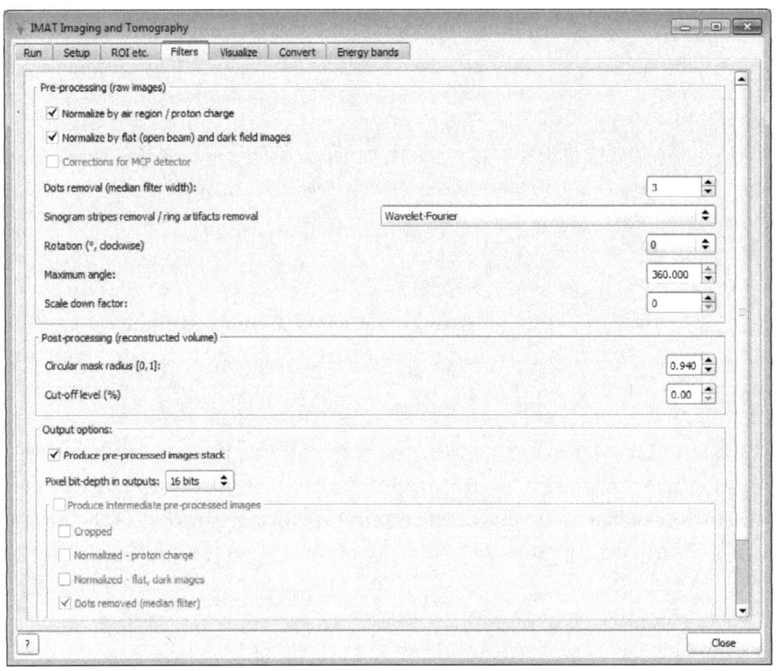

Figure 4. Section of the imaging interface where different pre-/post-processing filters and options can be defined. This includes for example enabling different types of normalization, detector specific corrections, and several pre-processing techniques for improving tomographic reconstruction results.

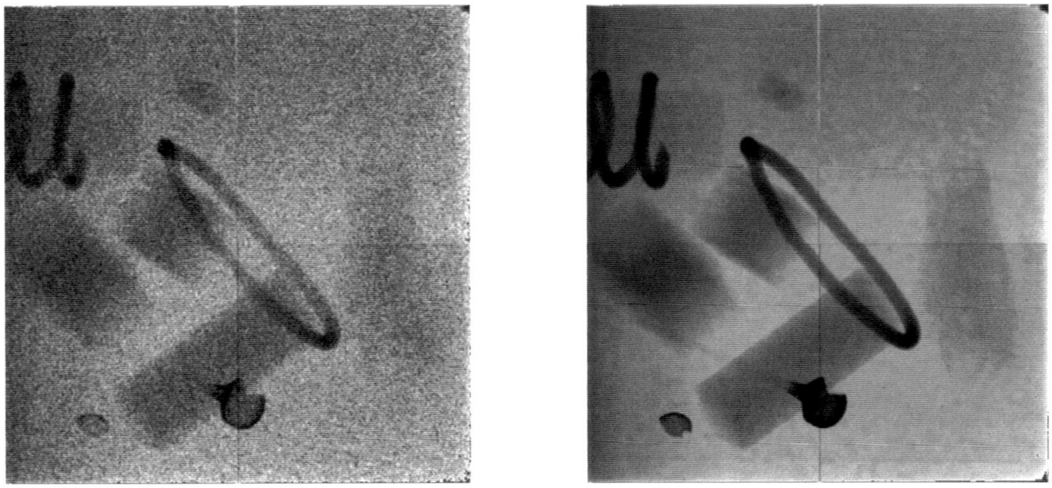

(a) Single energy band. (b) Average for a wide energy band.

Figure 5. Example images taken for testing purposes with the IMAT MCP detector, on a test beamline at ISIS.

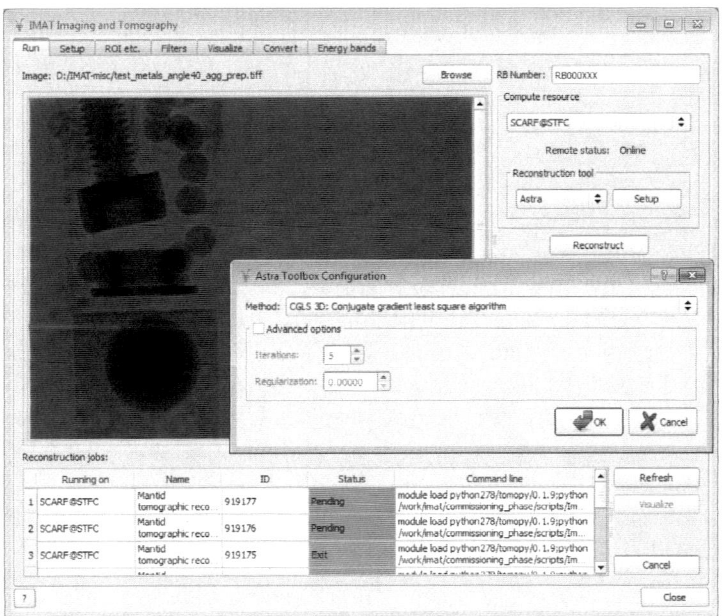

Figure 6. Interface for the submission and control of remote reconstruction jobs. Multiple jobs can be submitted simultaneously, using common and specific parameters, and taking full advantage of the compute resources available on site for rapid in-experiment analysis and also off-site data analysis.

being integrated in a harmonised manner, considering different forms of image normalisation, filtering xand selection of regions of interest. All these components are integrated in a common graphical interface with which users can make an effective use of the tomographic reconstruction tools and HPC resources available (Figures 6 and 7).

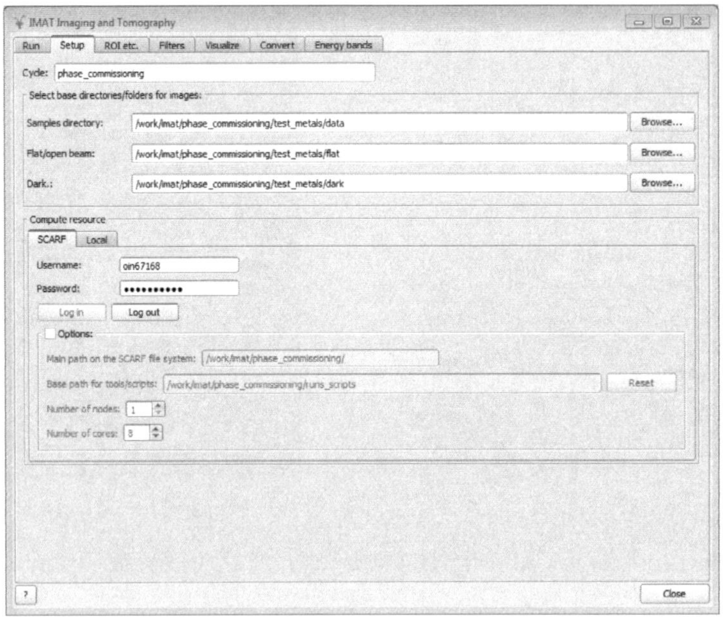

Figure 7. Remote compute resource settings in the imaging interface of Mantid, including parameters, dataset selection, and user credentials for remote execution of reconstruction jobs on a HPC platform.

6. Future

Mantid and its imaging features are being actively developed. A custom graphical user interface, aiming at simplifying and harmonizing pre-processing and visualisation tasks, enables IMAT users to perform on-the-fly analysis of tomography data using high-performance computing resources. Long-term objectives include dealing with the multiple challenges posed by the complex datasets with three or more dimensions produced by energy-dependent tomography experiments: data volumes, better tailored pre-/post-processing, ease of analysis, etc. Equally important is aiding our users through the process of building their analysis workflows using the new features, and extending the tools on both the visualisation and manipulation and analysis sides as needed to support and optimise operations.

7. Acknowledgment

We acknowledge the computing resources provided by STFC Scientific Computing Department's SCARF cluster (http://www.scarf.rl.ac.uk).

References

[1] Kockelmann W, Zhang S Y, Kelleher J F, Nightingale J B, Burca G and James J A 2013 IMAT - A New Imaging and Diffraction Instrument at ISIS *Physics Procedia* **43** 100–110

[2] Minniti T, Kockelmann W, Burca G, Kelleher J F, Kabra S, Zhang S Y, Pooley D E, Schooneveld E M, Mutamba Q, Sykora J, Rhodes N J, Pouzols F M, Nightingale J B, Aliottaa F, Bonaccorsi L M, Ponterio R, Salvato G, Trusso S, Vasi C, Tremsin A S and Gorini G 2013 Materials analysis opportunities on the new neutron imaging facility IMAT@ISIS *Journal of Instrumentation* **11** 1–16

[3] Mantid Project URL `http://www.mantidproject.org`

[4] Mantid 2015 Manipulation and Analysis Toolkit for Instrument Data. Mantid Project URL http://dx.doi.org/10.5286/SOFTWARE/MANTID3.5

[5] Arnold O *et al.* 2014 Mantid - Data analysis and visualization package for neutron scattering and μSR experiments *Nuclear Instruments and Methods in Physics Research Section A* **764** 156–166

[6] IBM Corporation 2016 IBM Platform LSF URL http://www.ibm.com/support/knowledgecenter/SSETD4/product_welcome_lsf.html

[7] Adaptive Computing Enterprises Inc 2016 MOAB HPC Suite URL http://docs.adaptivecomputing.com

[8] ParaView URL http://www.paraview.org/

[9] Gürsoy D, Carlo F D, Xiao X and Jacobsen C 2014 Tomopy: a framework for the analysis of synchrotron tomographic data *Journal of Synchrotron Radiation* **21** 1188–1193

[10] van Aarle W, Palenstijn W J, De Beenhouwer J, Altantzis T, Bals S, Batenburg K J and Sijbers J 2015 The ASTRA Toolbox, a platform for advanced algorithm development in electron tomography *Ultramicroscopy* **157** 35–47

[11] Palenstijn W J, Batenburg K J and Sijbers J 2011 Performance improvements for iterative electron tomography reconstruction using graphics processing units (GPUs) *Journal of Structural Biology* **176** 250–253

[12] The ASTRA Toolbox URL https://github.com/astra-toolbox/astra-toolbox/

[13] Octopus Imaging Software — Inside Matters URL http://octopusimaging.eu

[14] Atwood R C, Bodey A J, Price S W T, Basham M and Drakopoulos M 2015 A high-throughput system for high-quality tomographic reconstruction of large datasets at diamond light source *Philosophical Transactions A* **373** 20140398 https://github.com/DiamondLightSource/Savu

[15] Savu: Tomography Reconstructon Pipeline URL https://github.com/DiamondLightSource/Savu

[16] NeXus Application Definition NXtomo URL http://download.nexusformat.org/sphinx/classes/applications/NXtomo.html

[17] FITS Working Group, International Astronomical Union 2008 Definition of the Flexible Image Transport System (FITS). FITS Standard Version 3.0 URL http://fits.gsfc.nasa.gov/iaufwg/

[18] Adobe Developers Association 1992 TIFF^{TM}. Revision 6 URL http://partners.adobe.com/public/developer/en/tiff/TIFF6.pdf

[19] W3C 2003 Portable Network Graphics (PNG) Specification (Second Edition). Information technology Computer graphics and image processing Portable Network Graphics (PNG): Functional specification. ISO/IEC 15948:2003 (E) URL http://www.w3.org/TR/PNG/

Compact chopper spectrometers for pulsed sources

J. Voigt[1], **N. Violini**[1], **W. Schweika**[1,2],

JCNS and PGI, JARA-FIT, Forschungszentrum Jülich GmbH, 52425 Jülich, Germany
European Spallation Source ESS ERIC, Lund, Sweden

Abstract. We report on the opportunities for direct geometry chopper spectrometers (DGCS) by polychromatic illumination of the sample. At pulsed sources the use of multiple initial neutron energies appears naturally, if the repetition rate of chopper in front of the sample is larger than the repetition rate of the source. As a consequence, a large part of the spectrum is measured redundantly with variable energy and momentum transfer resolution. This can be used to optimize a chopper instrument for deep inelastic scattering, relaxing the requirements on the pulse length, by which the sample is illuminated, and on the secondary flight path, while the width of the spectral distribution must be narrowed down. This can open the path to new types of compact direct geometry chopper spectrometers, which need comparably small areas of detector coverage and allow very high repetition rates to provide a high intensity even if sample size and divergence distributions are limited.

1. Introduction

The high peak flux of MW spallation sources allows the preparation of neutron beams with a very well defined spectral width via sophisticated chopper systems. For the short pulse spallation source the very short pulses in particular of thermal and epithermal neutrons even limits the instrument length when balancing energy resolution and intensity. For cold neutrons and at the long pulse spallation source, **P**ulse shaping by the P-chopper can be employed to provide instruments with unprecedented energy resolution. For the long instruments proposed for the ESS, the effective pulse length to realize high energy resolution varies between a few 100 μs for (epi-)thermal neutrons up to the full pulse length of 2.86 ms for cold neutrons.

The mismatch between the period of the neutron source 17 ms $< T_{\rm src} <$ 100 ms and the typical time needed to record the essential part of the scattered neutron spectrum 1 ms$< T_{\rm frame} <$ 10 ms can be overcome by using several spectral lines passing through the M-chopper, i.e. the **M**onochromatizing chopper placed as close as possible in front of the sample, which allows the illumination of the sample with a short pulse to resolve the energy of the scattered neutrons by time-of-flight analysis. This has been termed repetition rate multiplication (RRM) [1] or multiple E_i method [2] and happens naturally, when the M chopper spins at a multiple frequency of the source without special precautions.

The instruments at the long pulse spallation sources add now more flexibility to illuminate the samples by discrete polychromatic lines. While 4SEASONS today is using four different lines today, the long pulse of the ESS and a compact secondary spectrometer will allow the use of up to 48 lines. The spacing between subsequent lines $\delta\lambda$ and the instrument bandwidth $\Delta\lambda$ can be chosen, as the instrument length is restricted by the finite pulse length of the source only for very long wavelength λ and very long instruments:

$$\delta\lambda = \frac{h}{m_n}(f_{\mathrm{M}}L_{\mathrm{M}})^{-1} \quad (1)$$

$$\Delta\lambda = \frac{h}{m_n}(f_{\mathrm{src}}L_{\mathrm{Det}})^{-1}. \quad (2)$$

$L_{\mathrm{Det}} \approx L_{\mathrm{M}}$ denote the distances of the detector and the M chopper, respectively, from the neutron source. In contrast to a monochromatic DGCS, which probes the entire range of energy transfers with the same initial wave vector, one may now illuminate the sample with different wave vectors to cover specific ranges of energy and momentum transfer with ideally matched definitions of the initial and final wave vectors. We will emphasize how this can be used to design a compact spectrometer for high energy transfer, which enables a high illumination rate and an efficient solid angle coverage. The idea has been taken from our developments for the inelastic option of the ESS instrument MAGIC, where it is favorably combined with polarization analysis.

Here we discuss general aspects of compact polychromatic spectrometers, which can achieve very good energy and momentum transfer resolution for deep inelastic scattering.

2. Resolution requirements

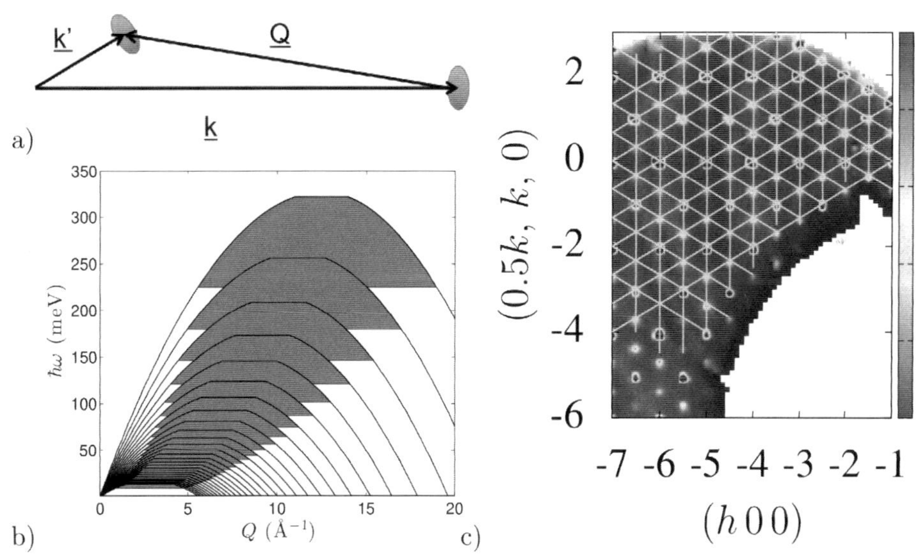

Figure 1. a), b) Kinematics of deep inelastic scattering: (a) Schematic scattering triangle for deep inelastic scattering ($\frac{\hbar\omega}{E_i} = 0.75$). (b) Dynamic range for pulses with 0.5 Å$< \lambda <$ 2.2 Å. The red regions indicate more than 75% energy loss. c) Elastic energy cut through the 4D $S(\vec{Q}, \hbar\omega)$ of the layered Kagome system $CaBaCo_2Fe_2O_7$. The lines indicate Brillouin zone boundaries in the selected plane. The large area, which is probed in reciprocal space, requires a very fine $\vec{Q}$ and hence $\vec{k}$ resolution to probe dispersive features within one Brillouin zone. Courtesy of Johannes Reim [3].

Fig. 1 a) illustrates the scattering triangle and the resolution ellipsoids for the initial and final wave vectors $\vec{k}, \vec{k}'$ for deep inelastic scattering. Note that we use the convention $\hbar\omega > 0$ for neutron energy loss throughout the paper. In particular the initial wave vector must be defined very precisely in that case. The example in Fig. 1c) shows the huge reciprocal space area that is covered by $k \approx 6\text{Å}^{-1}$. To resolve dispersive features in a Brillouin zone, it is necessary to control the length of the wave vector by the appropriate chopper settings and the direction of the wave vector by a narrow collimation. The new feature of the concept proposed here is clarified in Fig. 1b), which shows the dynamic range covered by the individual discrete pulses: Thanks to the narrowly spaced discrete lines, by which the sample is illuminated, one can explore a large range in energy and momentum transfer by combining the data from the region $\hbar\omega > 0.75E_i$, as indicated by the red areas.

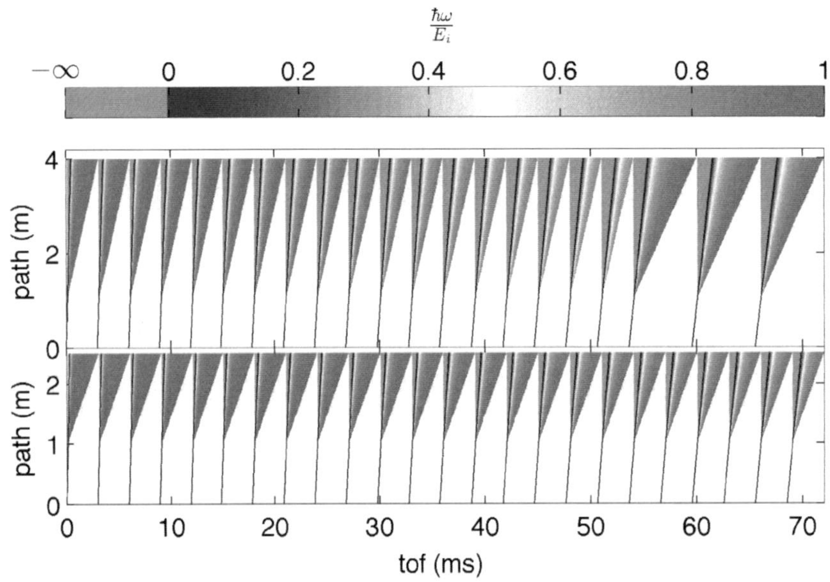

Figure 2. Flight path diagram for a secondary spectrometer with $L_{\text{SD}} = 1.5$ m (bottom) and $L_{\text{SD}} = 3$ m (top) during one ESS pulse. The colors code the relative energy transfer of neutrons arriving at the detector at a certain time.

In Fig. 2 we compare the relative energy transfer, which is recorded at the detector after the illumination through the M-chopper for all sub-pulses within one ESS pulse using a band 0.5 Å$< \lambda <$ 2.2 Åfor different sample-to-detector distances L_{SD}. The reddish colors indicates, that the neutron, which are recorded at the detector have experienced an energy loss of more than 50 %, while blue represents a modest energy loss and green refers to energy gain scattering. Comparing the red and blue fractions for different sample-to-detector distances it is obvious, that the detector closer to the sample records deep inelastic scattering for most of the time. To measure clean spectra up to very high energy transfers, the frame overlap with the up-scattering events from the subsequent pulse must be excluded within the time frame, which is defined by the highest required energy transfer:

$$T_{rec} = \frac{m_n}{h} L_{SD} \lambda \frac{1}{\sqrt{1 - \frac{\hbar\omega_{max}}{E_i}}}, \quad (3)$$

depending on the energy transfer range of interest. In Fig. 2 the criterion $\hbar\omega_{max} = 0.8E_i$ was chosen and therefore three pulses must be suppressed, e.g. by the action of a Fan chopper as suggested by one of the authors in [4].

The energy resolution of a chopper spectrometer can be expressed according to Lechner [5][6] as:

$$\Delta\hbar\omega = \frac{h^3}{m_n^2}\frac{\sqrt{A^2+B^2+C^2}}{\lambda'^3 L_{\mathrm{SD}} L_{\mathrm{PM}}}, \tag{4}$$

with m_n being the mass of neutron, h the Planck's constant,

$$A = \tau_{\mathrm{M}}(L_{\mathrm{PM}} + L_{\mathrm{MS}} + (\lambda'/\lambda)^3 L_{\mathrm{SD}}), \tag{5}$$

$$B = \tau_{\mathrm{P}}(L_{\mathrm{MS}} + (\lambda'/\lambda)^3 L_{\mathrm{SD}}), \tag{6}$$

$$C = \frac{m_n}{h} L_{\mathrm{PM}} \cdot \lambda' \cdot \Delta L, \tag{7}$$

to separate the resolution contributions of the pulse length $\tau_{\mathrm{M}}, \tau_{\mathrm{P}}$ A and B of the initial velocity defining M and P choppers and the flight path uncertainties C, respectively. Here λ, λ' denote the initial and the scattered neutron wavelength and $L_{\mathrm{PM}}, L_{\mathrm{MS}}, L_{\mathrm{SD}}$ the distance between the resolution defining P and M choppers (typically pairs of counter rotating discs), between final M chopper and sample and between sample and detector. A more general treatment of the 4d resolution function of a DGCS can be found in [7], but for the present discussion we assume that the total flight path uncertainties sum up to ΔL.

While shorter $\tau_{\mathrm{M}}, \tau_{\mathrm{P}}$ improve the energy resolution, the intensity passing through the temporal 'slit system' is proportional to the product:

$$F \propto \frac{\tau_{\mathrm{P}} + \tau_{\mathrm{M}}}{L_{\mathrm{PM}}} \times \tau_{\mathrm{M}}. \tag{8}$$

The first term on the right hand side describes the spectral width of the incoming neutron beam and the second term takes into account that the illumination time of the sample must be restricted to analyze the energy transfer by measurement of the secondary time-of-flight. Optimizing the energy resolution under the additional condition of a maximized flux using the method of Lagrange multipliers yields

$$\tau_{\mathrm{P}} = \frac{\tau_{\mathrm{M}}}{2} \times \left(\sqrt{4\left(\frac{L_{\mathrm{PM}} + L_{\mathrm{MS}} + (\lambda'/\lambda)^3 L_{\mathrm{SD}}}{L_{\mathrm{MS}} + (\lambda'/\lambda)^3 L_{\mathrm{SD}}}\right)^2 + 1} - 1\right). \tag{9}$$

This result is different from Lechner [5], as the flux in eq. (8) depends quadratically on τ_{M}, but linearly on τ_{P}. In the case of $L_{\mathrm{PM}} \gg L_{\mathrm{MS}} + (\lambda'/\lambda)^3 L_{\mathrm{SD}}$ eq. (9) simplifies to

$$\tau_{\mathrm{P}} = \tau_{\mathrm{M}} \times \left(\frac{L_{\mathrm{PM}} + L_{\mathrm{MS}} + (\lambda'/\lambda)^3 L_{\mathrm{SD}}}{L_{\mathrm{MS}} + (\lambda'/\lambda)^3 L_{\mathrm{SD}}}\right) \tag{10}$$

as in [5]. According to eq. (9) the ideal pulse shaping time varies strongly depending on the energy transfer, as shown in Fig. 4, which compares the ideal ratio of τ_{P} and τ_{M} as a function of relative energy transfer for two different sample-to-detector distances and with chopper positions as suggested for the long ESS chopper spectrometers CSPEC and T-REX. With increasing relative energy transfer a shorter pulse length τ_{P} is needed to define the initial wave vector and hence energy more precise. This is evident from Fig. 2, too: For the M-chopper pulses that arrive later within one ESS pulse the fraction of deep inelastic scattering is reduced and therefore the initial wavelength definition can be relaxed, i.e. the pulse length τ_{P} can be increased for later pulses with longer wavelength.

This is exactly the property of a optical blind chopper configuration proposed originally for reflectometers by [8]. Here a pair of disc choppers rotate in the same sense at a certain distance

Z. The phase is set such, that the second chopper only starts opening the beam, when the first chopper finishes closing the beam. As a result, the chopper is closed for infinitely fast neutrons, but for neutrons of finite velocity the choppers can be passed with a pulse length, that increases linearly according to the distance between the choppers:

$$\tau(\lambda) = \tau_{max} - |\tau_{max} - \frac{m_n}{h}\lambda Z| \quad (11)$$

$$\tau_{max} = \frac{w}{2\pi R f} \quad (12)$$

with w denoting the width of the neutron beam, R the radius and f the chopper pair frequency. If the right hand expression of eq. (11) would evaluate negativ, the chopper does not transmit. Recently this concept has been extended by realizing the first and second chopper by pairs of counter rotating choppers to double the speed, by which the beam cross section can be opened and closed [9]. This design not only provides a wavelength dependent pulse length, but decouples the pulse length to a certain extend from the chopper frequency: As the first pair opens the beam and the second pair closes the beam, the pulse length can be adjusted shifting the relative phase ϕ of the two pairs yielding

$$\tau(\lambda) = \tau_{max} - |\tau_{max} - \frac{m_n}{h}(\lambda - \lambda_1)Z| \quad (13)$$

with $\lambda_1 = \frac{h}{m_n}\frac{\phi}{2\pi f Z}$. As seen in Fig. 3 the pulse length passing through a quartet of choppers for a given wavelength varies with the phase between the first pair and the second pair without changing the chopper frequency.

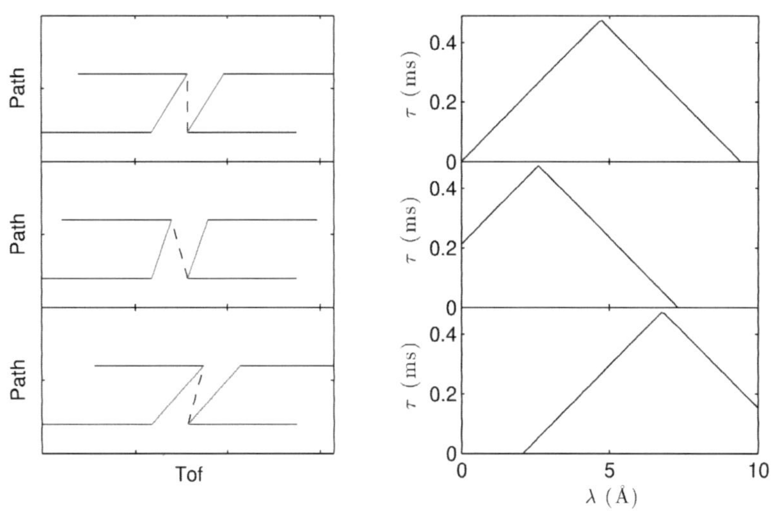

Figure 3. Schematic flight path diagrams (left) for chopper pairs at the distance $Z = 0.4$ m for three different phases and the resulting $\tau(\lambda)$ graphs (right). (top) $\lambda_1 = 0$Å, (center) $\lambda_1 = -2.1$Å, (bottom) $\lambda_1 = 2.1$Å,

For the optimization of the M-chopper pulse length we consider the secondary time-of-flight of the slowest final neutrons, which can be detected without frame overlap:

$$\lambda'_{max} = \frac{h}{m_n}\frac{1 + \frac{L_{\mathrm{MS}}}{L_{\mathrm{PM}}}}{L_{\mathrm{SD}} f_{\mathrm{M}}}. \quad (14)$$

This does not change, if the time frame is not changed, i.e. no pulses are skipped. Hence we keep τ_{M} constant for all pulses.

We evaluate now the resulting energy resolution for a constant $\tau_{\mathrm{M}} = 30\mu$s and varying τ_{P} according to the ratio $\frac{\tau_{\mathrm{P}}}{\tau_{\mathrm{M}}}$ shown in Fig. 4a) in the range from 0.8 $\frac{-\hbar\omega}{E_i}$ to 0.5 for the different sample-to-detector distances $L_{\mathrm{SD}} = 1.5, 3$ m. This yields τ_{P} varying from 0.105 ms to 0.285 ms and from 0.06 to 0.166 ms within a band $\Delta\lambda = 1.77$ Å for the assumed instrument length. The pulse length gradient is realized by a distance of 0.4 m and 0.24 m between both P chopper pairs. The slits on each of the four chopper discs must be chosen significantly wider than the beam cross section to enable the variation of the pulse length. This requirement seems technically achievable: First, the pulse length is rather long ($\tau_{\mathrm{P}} > 100\mu$s) and second, the requirement of a narrow divergent beam to achieve a good $\vec{Q}$-resolution will limit the guide dimensions also for a ballistic guide concept (note that the guide cross section of the T-REX spectrometer at the position of the P-chopper, which is 60 mm × 90 mm, can be closed sufficiently fast).

The relaxed resolution for the shorter secondary flight path is fully compensated by a flux gain, as the flux is proportional to τ_{P}. It should be noted, that even the short instrument realizes a resolution of 11 meV for 320 meV excitation energy and 0.4 meV for 18 meV excitation energy. For this sample-to-detector distance a large solid angle coverage is possible with detector areas well below 10 m^2. Concerning line shape investigations the narrowing of the resolution curves matches with usually longer lifetimes of lower energy excitations in contrast to monochromatic DGCS, where the resolution gets broader for lower energy transfer.

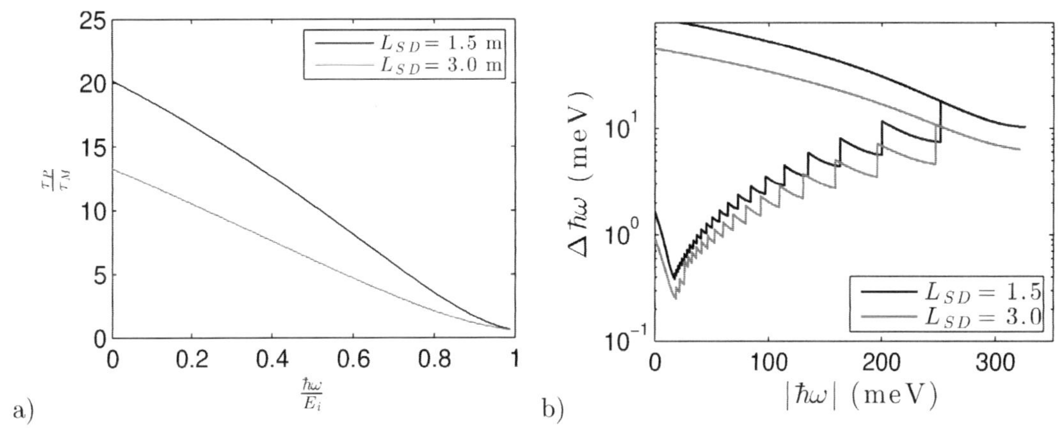

Figure 4. left: Ratio $\frac{\tau_{\mathrm{P}}}{\tau_{\mathrm{M}}}$ as a function of rel. energy transfer. right: Energy resolution as a function of energy transfer for the following parameters: $f_M = 350$ Hz, $L_{\mathrm{M}} = 165$ m, $L_{\mathrm{P}} = 110$ m, $L_{\mathrm{MS}} = 1$ m, $\Delta L = 10$ mm. Except for the highest and lowest energy pulses, only the best resolution for a given energy transfer is displayed.

3. Conclusion

We propose here a new concept for a compact chopper spectrometer optimized for deep inelastic scattering and good momentum transfer resolution. It makes use of the polychromatic illumination of a sample at a pulsed neutron source, where each pulse is optimized to a narrow energy transfer region. The illumination by multiple narrow spectral lines, which is natural at pulsed sources is crucial to explore a large overal dynamical range. The definition of the initial neutron wave vector can be tuned to achieve excellent energy transfer and momentum transfer resolution even by relaxing the timing conditions on the chopper directly in front of the sample, which usually sets the technical limit for the energy resolution. Furthermore the secondary flight path can be reduced to allow an efficient solid angle coverage at reasonable cost.

References

[1] Russina M and Mezei F. 2009 *Nucl. Instr. Meth. A* **604(3)**:624–631.

[2] Nakamura M, Kajimoto R, Inamura Y, Mizuno F, Fujita M, Yokoo T, and Arai M 2009 *Jour. Phys. Soc. Jap.*, **78(9)** 093002.

[3] Reim J 2015 Ph. D. thesis, RWTH Aachen.

[4] Voigt J, Violini N, and Brueckel T 2014. *Nucl. Instr. Meth. A*, **741** 26–32.

[5] Lechner R E 1991 In M. Misawa, M. Furusaka, H. Ikeda, and N. Watanabe, editors, *Proceedings ICANS-IX* KEK.

[6] Lechner R E 1984 In R. Scherm and H. Stiller, editors, *Proceedings "Workshop on neutron scattering instrumentation for SNQ"*, volume Jül-1954. Forschungszentrum Jülich, 1984.

[7] Violini N, Voigt J, Pasini S, and Brueckel T 2014 *Nucl. Instr. Meth. A*, **736** 31–39.

[8] van Well A A 1992 *Physica B: Condensed Matter* **180–181**, Part 2(0):959–961.

[9] Vickery A and Deen P P 2014 *Rev. Sci. Instr.*, **85(11)** 115103.

Superconducting thermal neutron detectors

V. Merlo[1], A. Pietropaolo[2,3,*], G. Celentano[2], M. Cirillo[1], M. Lucci[1], I. Ottaviani[1], M. Salvato[1], A. Scherillo[4], E.M. Schooneveld[4], A. Vannozzi[2]

1. Università degli Studi di Roma Tor Vergata, Physics Department and MINAS Lab, via della Ricerca Scientifica 1, 00133 Roma, Italy
2. **ENEA**, Fusion and Nuclear Safety Technologies Dept. Via E. Fermi 45, 00044 Frascati (Roma), Italy
3. **MIFP**-Mediterranean Institute of Fundamental Physics, Via Appia Nuova 31, 00040 Marino (Rome), Italy
4. Science and Technology Facility Council, ISIS Facility Chilton, Didcot, Oxfordshire OX11 0QX, United Kingdom

* **Corresponding author**: antonino.pietropaolo@enea.it

Abstract. A neutron detection concept is presented that is based on superconductive niobium nitride (NbN) strips coated by a boron (B) layer. The working principle is well described by a hot spot mechanism: upon the occurrence of the nuclear reactions $n + ^{10}B \rightarrow \alpha + ^{7}Li + 2.8\ \text{MeV}$, the energy released by the secondary particles into the strip induces a superconducting-normal state transition. The latter is recognized as a voltage signal which is the evidence of the incident neutron. The above described detection principle has been experimentally assessed and verified by irradiating the samples with a pulsed neutron beam at the ISIS spallation neutron source (UK). It is found that the boron coated superconducting strips, kept at a temperature T below 11K and current-biased below the critical current I_C, are driven into the normal state upon thermal neutron irradiation. Measurements on the counting rate of the device are presented and the basic physical features of the detector are discussed and compared to those of a borated Nb superconducting strip.

1. Introduction

The capability of superconducting materials to act as radiation detectors for charged particle (minimum and/or minimum ionizing) as well as x and gamma rays is well established in the literature [1-3].

The main mechanism is explained in terms of the so-called "hot-spot" model (see figure 1) [4]: the ionizing radiation releases its energy in the bulk giving rise to a local temperature enhancement. The heat then propagates over the material. In the case of a charged particle, where ionization occurs along the particle track the hot spot is in a simplified model cylindrical and spread over an area provided by the relation [5]:

$$S = \frac{dE}{dx}\frac{1}{\Delta h} \quad (1)$$

With:

$\frac{dE}{dx}$: energy release per unit length into the material

Δh : enthalpy difference per unit volume for the superconducting transition

In this model, the transport (or bias) current I_B circulating into the strip featuring a resistivity ρ, is forced to pass through the reduced superconducting region, left by the definite hot spot (normal) region. When the density of transport current J_B exceeds the critical density current, J_C, the whole strip undergoes a transition from the superconducting to normal region that gives rise to the onset of an ohmic current and thus to a potential drop:

$$V = 2\rho J_C \left(\frac{dE}{dx} \frac{1}{\pi \Delta h} \right) \tag{2}$$

tagging the presence of the traversing particle.
Figure 1 illustrates schematically the mechanism of the hot spot.

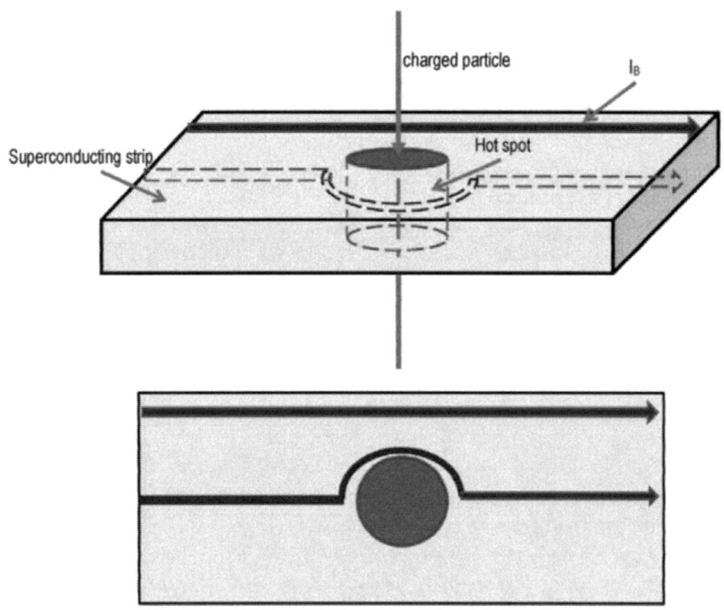

Figure 1: Schematic mechanism for the hot spot generation upon particle energy release

The tests performed on superconducting radiation detectors and presented in different experimental papers (see Refs. [1-3] for example) have shown that, depending on the material and at the proper thermodynamic conditions voltage pulses related to charged particles (minimum and/or non-minimum

ionizing) up to 50-60 mV. Recently, a Nb superconducting strip coated with boron was tested for neutron detection at the ISIS spallation neutron source [6].
In this contribution, the characterization of NbN superconducting strips coated with thin film of boron is presented. These devices were tested at the ISIS spallation neutron source and a brief overview of results are discussed. Eventually a comparison between NbN and Nb strips is highlighted in terms of counting rate. NbN offers, when compared to Nb, the advantage of a higher transition temperature (about 16K, opposed to the 9K of Nb) and a superior mechanical hardness. Future perspectives are briefly outlined in the concluding section.

2. Device characterization

In this section the electrical characterization of the NbN strip and the boron coating procedure and analysis are briefly described.

2.1. Superconducting strips

Niobium Nitride (NbN) was grown on a 2 inches silicon wafer with a 1 μm SiO_2 layer on the top. The geometrical patterns of the strips were defined by a lift-off lithographic technique: a "negative" pattern of resist was first impressed on the substrate and subsequently the superconductive film was deposited by reactive magnetron sputtering in DC mode. The removal of the residual resist left only the patterned strips on the sample. During the sputtering process we had a gas mixture composition of 9:1 Ar/N_2 flux and the total pressure in the deposition chamber was 10^{-1} Pa. In these conditions a 150 nm (thickness measured by an ex-situ profilometer) NbN film thickness was obtained. The liftoff process could be employed just due to the fact that the substrate was not heated during deposition and its temperature could reasonably be kept below 370K during deposition.
The properties and the quality of the NbN films were checked also characterized by XRD and XPS spectroscopy according to a procedure described in [7] in order to verify their properties.
In figure 2, a typical R(T) curve measured on a NbN strip by means of a 4-probe setup is shown.

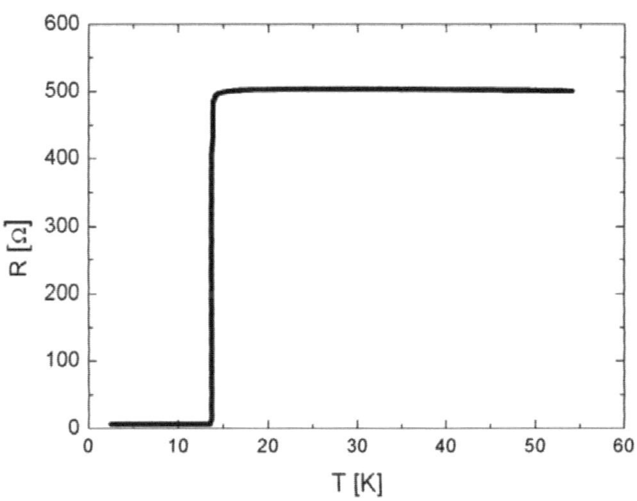

Figure 2: Trend of a NbN thin film resistance as a function of the bath temperature

2.2. Boron deposition

B coating was obtained by means of electron beam evaporation technique with a multi-crucible Thermionics 3 kW electron beam system with 4 kV of electron acceleration voltage and 750 mA of maximum current. As source for evaporation Alfa Aesar crystalline natural B pieces, 10 – 30 mm in size and purity 99.5 at% were used. Film growth rate was monitored by a quartz crystal controller and fixed at 0.2–0.3 nm/s by tuning the electron beam current, I_b. The distance between evaporation source and sample holder was $d_{T\text{-}S}$ = 12.5 cm. Boron film thickness was 450 nm equivalent to a 90 nm thick film of ^{10}B considering the natural ^{10}B isotope abundance. A sketch of the deposition set-up is shown in Figure 3. NbN strips were protected by a stainless steel mask in order to allow film growth on a defined circular area 1 mm in diameter. In Figure 4, the x-ray diffraction θ-2θ pattern is plotted for a typical Boron film deposited on Al_2O_3 substrate. As can be seen, peaks related to the β-Boron phase are observed, in addition to other weak reflections indicating the presence of minor films contaminations such as Boron oxides and carbides. A more quantitative analysis carried out by XPS spectroscopy techniques has revealed that the sample contains only B, O and C [8]. The total Boron content can be as high as 80 at.%.

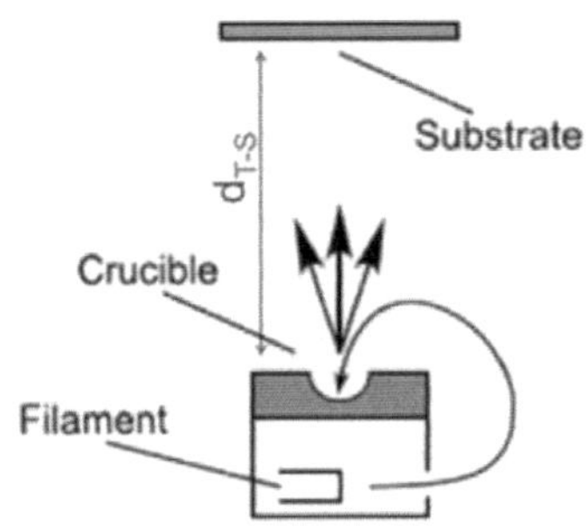

Figure 3: Schematic layout of the e-beam deposition system used for NbN boron coating.

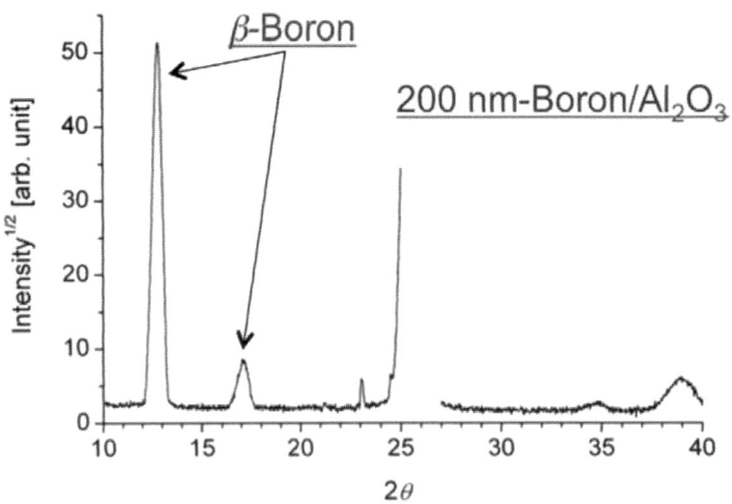

Figure 4: X-ray diffraction pattern from a test boron film deposited onto a Al_2O_3 substrate

3. Tests on beam

The NbN strip with 450 nm boron deposition was tested at the INES diffractometer [9] at the ISIS spallation neutron source [10]. The instrument is equipped with B4C mechanical jaws that are used to tail the thermal neutron beam. Figure 5 shows the I/V characteristic of the NbN strip in two different cases: i) without neutron irradiation (jaws closed) and ii) during irradiation (jaws open). In the former case the transition to the normal state occurs when I_B exceeds I_C (~ 32 mA as noted in figure 5). During neutron irradiation the transition occurs at a lower value of I_B (~ 27 mA). During detector test, the jaws were opened simultaneously to I_B supply. This is explained in terms of the hot spot model: indeed, the hot spot reduces the effective volume of the superconducting region of the strip. Thus a lower I_B can provide a $J_B > J_C$.

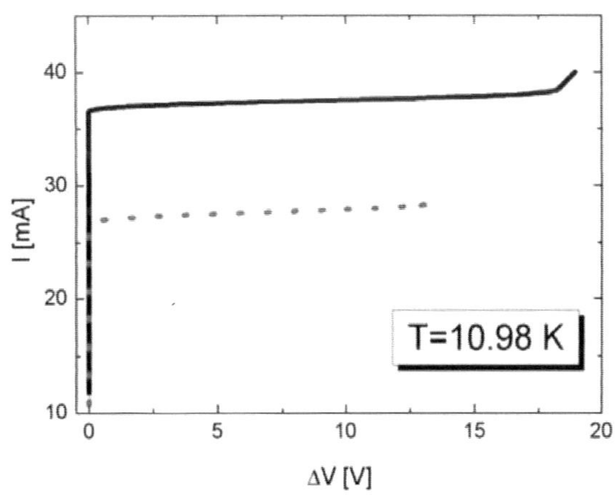

Figure 5: I/V characteristic of the NbN strip without (continuous black line) and with (dashed red line) neutron irradiation

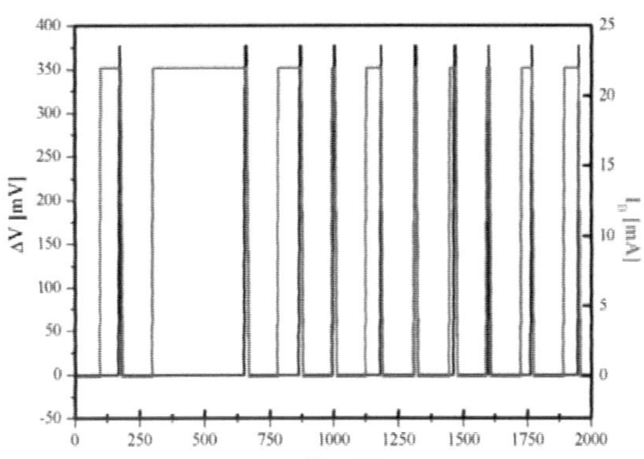

Figure 6: V-t measurement for a NbN strip detector with 450 nm deposition of natural boron, in the continuous operation mode. The red line (scale on the right) illustrates the on-off bias current which switches simultaneously to the open-close jaws status.
The black line (scale on the left) is the measured voltage strip across the strip.

The count rate has been extracted from the experimental data by counting the number of leading edges of the voltage time traces (see Figure6). This value was then corrected taking into account the detector dead time using a non-paralyzable counter model [11]. In the case here discussed, dead time was fixed at 120 s, suggested (although a little bit overestimated) by the time needed to allow the strip to bring back to the superconducting state after the hot spot transient. Considering the particular geometry and the finite thickness, the escape probability cannot be taken as unity (i.e. particles emitted at wide angles are stopped in boron or are out the active volume of the detector. The estimate was obtained following Ref. [9], where the detailed calculation for a somewhat similar geometry was carried out. This approach provided a calculated efficiency of about 0.1% and 0.5% at the thermal peak (25 meV), for the NbN and NB strips, respectively. The count rate increases as the I_B is raised [2,6] and figure 7 shows the comparison between the count rates measured for the Nb and NbN strips at T=11 and 8 K, respectively.

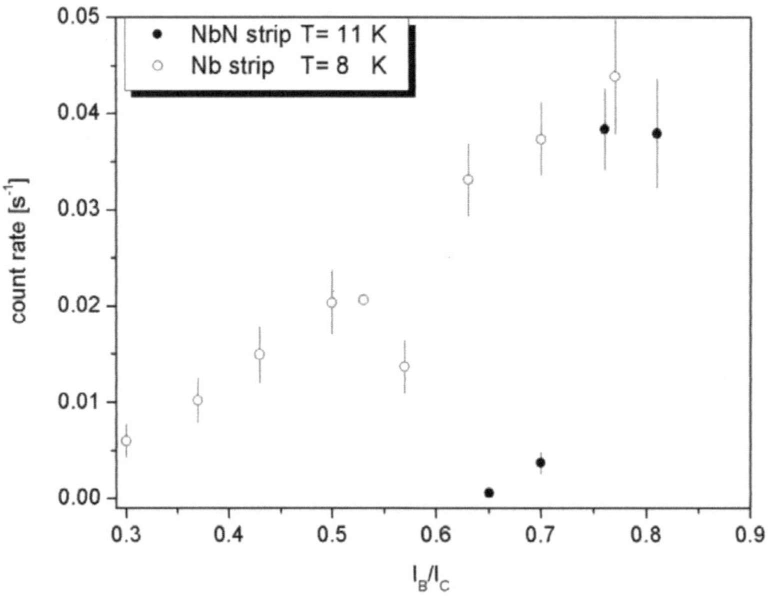

Figure 7: Comparison between the NbN (full dots) and Nb (empty dots, form Ref. [6]) SCNDs.

In the case of the NbN detector (full dots in Figure 7) the behavior in the range 0.65-0.8 seems not linear. This trend was already evidenced in Refs. [2,3] and explained as assort of "threshold" where for a given material and geometry, the transition to normal state, in absence of irradiation, occurs beyond a certain value of I_B. Thus upon irradiation the trend is conserved although the threshold value might differ (lower) due to the effect of the hot spot.

4. Conclusions

In this paper, we presented a detection device relying on the use of NbN superconducting strips coated with natural boron, thus making the material sensitive to thermal neutrons. The secondary alpha and lithium particles emerging from the n-^{10}B nuclear reactions release energy in the bulk of the strip, owing to a transition of the overall strip to the normal state whose mechanism is well described within the framework of the hot spot model.

The detector was characterized on beam at the ISIS spallation neutron source using the INES beam line. The count rate, in the best case, is in the order of 0.04 s^{-1}, corresponding to an efficiency of 10^{-3}

Acknowledgments

The authors warmly acknowledge the professional support of the ISIS detector and computing groups. The cooperation Agreement no. 06/20018 between CNR and STFC, concerning collaboration in scientific research at the spallation neutron source ISIS (UK), is gratefully acknowledged. One of the authors (MS) acknowledges the CNR-SPIN support.

References

[1] D. H. Andrews et al. Phys. Rev. **76**, 154 (1949)
[2] E. C. Crittenden and D. E. Spiel. J. App. Phys. **42**, 3182 (1971)
[3] A. Gabutti et al. Nucl. Instrum. Meth. A **289**, 274 (1990)
[4] A. Engel and A. Schilling, J. Appl. Phys. **114**, 214501 (2013)
[5] J. Kitaygorsky, "Photon and Dark Counts in NbN Superconducting Single-Photon Detectors and Nanostripes", PhD Thesis Univ. Rochester, New York (US) 2008
[6] V. Merlo, M. Salvato, M. Cirillo, M. Lucci, I. Ottaviani, A. Scherillo, G. Celentano, and A. Pietropaolo, Appl. Phys. Lett. **106**, 113502 (2015)
[7] M. Lucci, S. Sanna, G. Contini, N. Zema , V. Merlo, M. Salvato, H.N. Thanh, I. Davoli, Surf. Sci. **601,** 2647 (2007)
[8] G. Celentano, A. Vannozzi, A.Mancini, A. Santoni, A. Pietropaolo, G. Claps, E.Bemporad, M. Renzelli, F. Murtas, L. Quintieri, Surf. & Coat. Tech. **265**, 160 (2015).
[9] F. Grazzi, M. Celli, S. Siano, and M. Zoppi, Nuovo Cim. C **30**, 59 (2007).
[10] ISIS website: http://www.isis.stfc.ac.uk/
[11] G. F. Knoll, in Radiation detection and measurements, John Wiliey & Sons Inc. (2000)

7C2, the new neutron diffractometer for liquids and disordered materials at LLB

G. J. Cuello[1], J. Darpentigny[2], L. Hennet[2,3], L. Cormier[4], J. Dupont[2], B. Homatter[2], B. Beuneu[2]

[1] Institut Laue Langevin, 71, av des Martyrs, 38042 Grenoble, France
[2] Laboratoire Léon Brillouin, CEA-CNRS, CEA Saclay, 91191 Gif sur Yvette Cedex, France
[3] CEMHTI and University of Orléans, 45071 Orléans Cedex 02, France
[4] Institut de Minéralogie, de Physique des Matériaux et de Cosmochimie (IMPMC), Sorbonne Universités, UPMC Univ Paris 06, CNRS UMR 7590, Museum National d'Histoire Naturelle, IRD UMR 206, 4 place Jussieu, 75005 Paris, France

E-mail: `cuello@ill.eu`

Abstract.
The disordered-materials diffractometer 7C2 at the Laboratoire Léon Brillouin (LLB), Saclay, France has been upgraded through the replacement of its detector. The old one, a banana like BF_3 detector, has been replaced by a new ensemble of ^{3}He tubes arranged on a cylindrical surface concentric with the sample position, covering an angular range of 133°, with an angular step of 0.52°. At the shortest wavelength, 0.57 Å, this represents a maximum momentum transfer of 20 Å^{-1}. The gas pressure in the ^{3}He tubes is 30 bars, which increases the efficiency to 80% for 0.72 Å neutrons, compared to 17% for the previous detector. A larger solid angle of detection and an improved efficiency have increased the total counting rate by a factor of 22, thereby reducing random error in the diffraction measurements. The banana-like multidetector gives to 7C2 the possibility of registering a complete diffractogram over the whole angular range in one shot, which is a comparative advantage with respect to other hot-neutrons two-axis diffractometers.

1. Introduction

The pair distribution function (PDF) for liquids and amorphous systems (or more generally, systems with no long-range order: nanoparticles, poorly crystallised materials, etc.) is the Fourier transform of the structure factor determined from diffraction experiments, giving detailed information on the short to intermediate length scales. This Fourier transform is an integral over the momentum transfer Q, which should extend up to infinity [1]. This is the reason why the accurate determination of structure factors for disordered materials requires the precise determination of the scattered intensity over a wide range of momentum transfer. This momentum transfer is usually approximated by that corresponding to elastic scattering and given by the Bragg's law ($Q = 4\pi/\lambda\ \sin\theta$, where λ is the neutron wavelength and 2θ is the scattering angle).

In the case of X-ray diffraction, because the interaction happens with the electronic cloud and can not be considered as punctual, the atomic form factor (strongly decreasing with Q) produces structural information only in a limited Q-range. On the other hand, neutron diffraction is a very well suited technique because the scattering power of a given nucleus is independent of the momentum transfer, as consequence of the short range interaction between the neutron and the nucleus [2]. This scattering power is proportional to the square of the tabulated scattering lengths [3, 4].

Considering again the Bragg's law, for a given scattering angle, the shorter is the wavelength the larger is the corresponding Q-value. Thermal neutrons (with energies around 25 meV) will produce diffractograms with a maximum momentum transfer of about 7 to 10 $Å^{-1}$, which is rather small for a good Fourier transformation. But using neutrons of wavelengths shorter than 1 Å, i.e., with some hundreds of meV, the momentum transfer range can be increased up to 25 or even 30 $Å^{-1}$. In reactor sources, like Orphée reactor at Laboratoire Léon Brillouin (LLB), this is achieved using a hot source kept at high temperature (1400 K), where neutrons are thermalised at about 120 meV. The High Flux Reactor at ILL (Institut Laue Langevin) has a hot source at 2400 K, where neutrons are thermalised at about 170 meV. In spallation sources, like ISIS in UK or SNS in USA, the attainable Q-range extents even up to 50 $Å^{-1}$ [5, 6, 7].

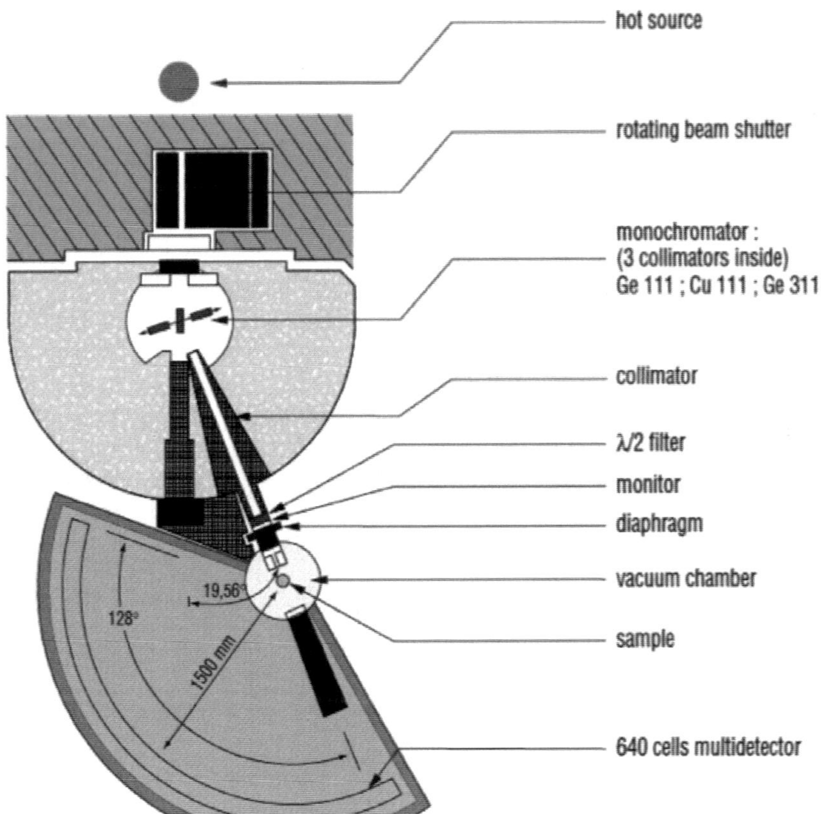

Figure 1. Sketch of the 7C2 instrument before changing the detector. A white beam of hot neutrons comes from the reactor's hot source and is monochromatised by one of the three available monochromators. Then the beam pass the monochromator shielding through a collimation tube, before being transmitted by a filter which removes the harmonic wavelengths. An almost transparent monitor register the beam intensity for normalisation purposes. A diaphragm defines the beam geometry and size on the sample, which is located in the centre of a cylindrical vacuum chamber. Finally, the diffracted beam is registered by the banana-like multidetector.

The main advantages of the reactor-based instruments (when compared to spallation-based ones) are the simplicity of the corrections and the counting stability (extremely important for isotopic substitution experiments). The two-axis diffractometers dedicated to liquids and amorphous systems in Europe are D4 at ILL [13] and 7C2 (Fig. 1) at LLB [8], and the former is considered as a reference in the field of disordered matter. In spite of its lower intensity (25 times smaller that D4), 7C2 had (and still has) the advantage of its banana-like multidetector and its sample changer. The upgrade of the instrument was then justified by the need of keeping it competitive with other available instruments.

2. The instrument

The 7C2 instrument is located on one of the two beam-tubes looking at the hot source of the Orphée reactor at Saclay [8]. A system of three monochromators (Ge 111, Cu 111 and Ge 311) on a rotating table allows changing the wavelength among three standard values (1.1, 0.72 and 0.57 Å), which are determined by the availability of suitable $\lambda/2$ filters.

After the monochromator, the beam is filtered in order to remove the harmonics of the main wavelength λ ($\lambda/2$, $\lambda/3$, etc.), then it passes through an almost transparent detector serving as monitor for normalisation purposes. The final flux on the sample is 10^6-10^7 neutrons/cm^2/s.

Finally, a diaphragm defines the size and geometry of the beam at the sample position, which is at the centre of a cylindrical vacuum chamber (Fig. 1). In addition to the standard room temperature experiments, the big vacuum chamber allows several sample environments, like cryostats, furnaces or pressure cells. The instrument has the possibility of using a 5-positions sample changer (Fig. 2), but only for room temperature experiments.

Thanks to the multidetector, a diffractogram can be registered in one shot allowing in situ experiments, like thermodiffractometry. This is a possibility that other dedicated reactor instruments for liquids and amorphous systems can not offer.

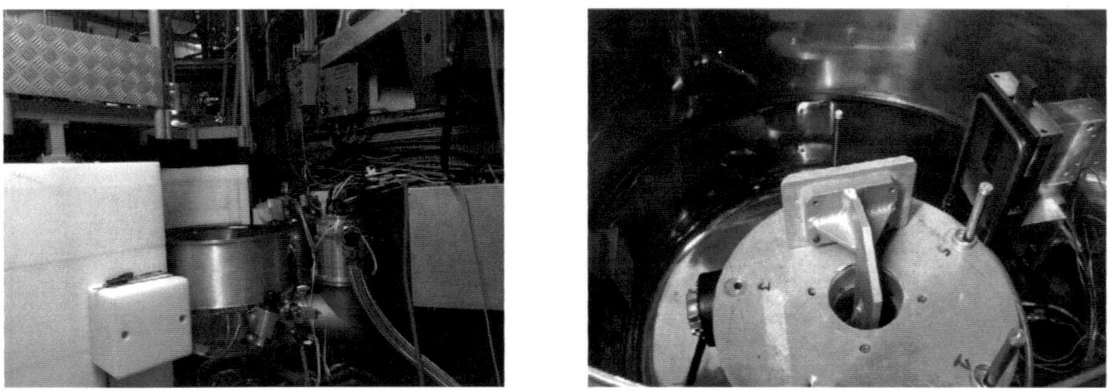

Figure 2. Left: a view of the cylindrical vacuum chamber; the sample is located at the center of this cylinder.
Right: the 5-positions sample changer, with a shielding to reduce the contribution from the other 4 samples; on the right we see the diaphragm defining the size and shape of the beam.

3. The detector

The previous detector was a banana like multidetector, i.e., a one dimensional position sensitive detector, consisting of 32 blocks of 20 cells (Fig. 1). Each block was composed of a cathode with anode wires of 40 μm in diameter, at working voltages of -2700 V and $+2900$ V, respectively.

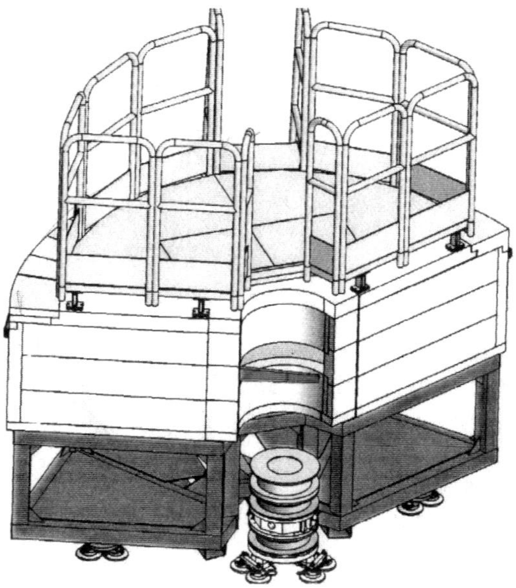

Figure 3. General view of the new detector. In white we see the polyethylene serving as shielding for the detector tubes. The ensemble is mounted on air pads allowing the movement of the whole detector array around the monochromator position.

The filling gas was $^{10}BF_3$ at a pressure of 1 bar, giving efficiencies of 30%, 17% and 12% for wavelengths of 1.1 Å, 0.72 Å and 0.57 Å, respectively. The spacing between two cells was 5.2 mm, which combined with a sample-detector distance of 1.5 m gave an angular step of 0.2° and an angular range of 128°. The main drawback of this detector was the low efficiency, as compared to the efficiency of ^{3}He gas detectors, which motivated a project for replacing this $^{10}BF_3$ detector.

At present, 256 ^{3}He gas tubes replace the $^{10}BF_3$ detector. These tubes are 0.5 inches diameter, 30 bars ^{3}He filled, position sensitive detectors (PSD) from Reuter-Stokes (GE Energy company), working at high voltage (1700 V). They are also located in a banana like configuration, at 1.5 m from the sample position, as it is shown in Figures 3 and 4(top). The distance between tubes is 13.7 mm, giving an angular step of 0.52° and covering an angular range of 133°. The high pressure in the tubes, 30 bar, gives an efficiency of 80% for neutrons of 0.72 Å. This efficiency must be compared to the previous efficiency at the same wavelength, i.e., only 17%.

The tubes have a length of 500 mm and are equipped with PSD electronics from Mesytec company. This charge division electronics makes the detectors also sensitive to the vertical position, then we divide tubes in 64 vertical pixels. In this way, the detector becomes a 2D-detector. Figure 4 (bottom) shows a typical diffractogram for the Ni powder sample, which is used to calibrate the wavelength and zero-angle correction. We observe the intersection of the Debye-Scherrer cones with the detector surface, which are clear at low angles but become less apparent at higher angles.

With the new detector the angular definition passed from 0.2° to 0.52°, but this has no important consequences to the average resolution of the instrument in Q-scale. The low-to medium-resolution in Q-scale is the consequence of using short-wavelength neutrons, and depends on the scattering angle 2θ. The best value is found at around 20°, where it is $\Delta Q/Q = 0.02$, corresponding to $\Delta 2\theta \approx 0.4°$, and it is worst elsewhere. Then the angular definition of 0.52° does not affect the resolution of the structure factor. This low resolution could be a problem for polycrystalline samples, but not for disordered systems.

The new detector array represents a factor 22 in gain when compared with the previous detector: a factor 5.5 thanks to the bigger solid angle and a factor 4 because the better efficiency. Now the new 7C2 has a global efficiency comparable to the reference instrument in the field. Nowadays, the main advantage of D4 is the fact of being located in a reactor 5 times more intense than Orphée reactor.

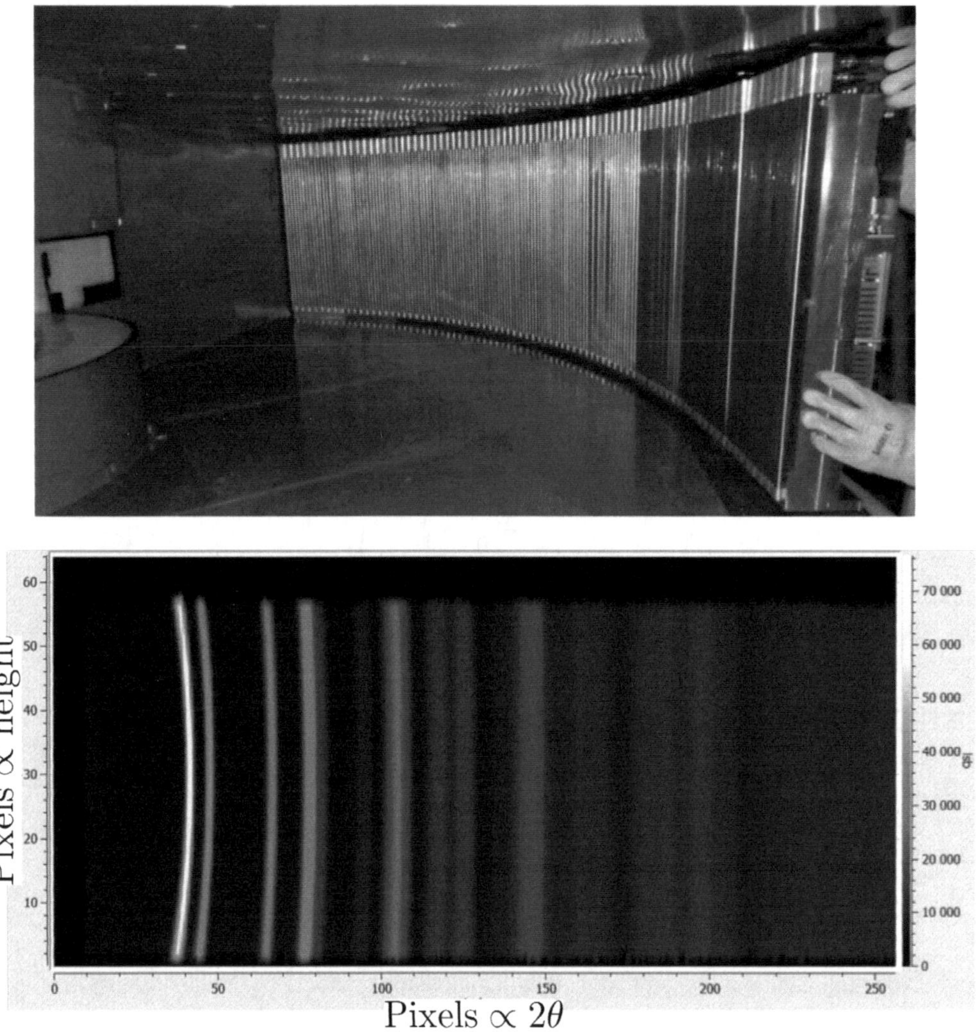

Figure 4. Top: the 16 moduli of detectors, each one composed of 16 detectors, placed at 1.5 m from the sample in a circular section centred at the sample position.
Bottom: A 2D diffractogram corresponding to a polycrystalline sample of Ni powder. The coordinates are given in pixels (256 × 64), but the abscissa represents the scattering angle and the ordinate is the vertical position along the tubes. Several peaks are clearly distinguished at low angles, but then the other peaks are less apparent due to the low resolution of the instrument.

4. Data reduction

The raw data, as those shown in Fig. 4(bottom) must be corrected for the relative efficiency of the detectors and radially integrated in order to obtain a 1D-diffractogram. The resulting diffractograms are corrected for inelasticity [9], multiple scattering [10] and attenuation [11] using standard programs like CORRECT [12] or others, which finally leads to the static structure factor (see Ref. [1]). Then, this structure factor is Fourier transformed to determine the final pair distribution functions in real space.

5. Discussion

The new 7C2 instrument with its factor of gain about 22 with respect to the previous version, is almost as performant as its main competitor D4 [13], also a reactor based instrument at the Institut Laue Langevin. The flux on the sample is still 5 times lower than that at D4, but this is only due to the different operation power of both reactors. In fact, 7C2 has two advantages when compared with D4. The first one is the banana-like multidetector covering 132° and the possibility of registering a whole diffractogram in one shot, which allows in situ experiments as thermodiffractometry. The second advantage is the 5-positions sample changer, which is an extremely useful tool for room-temperature experiments.

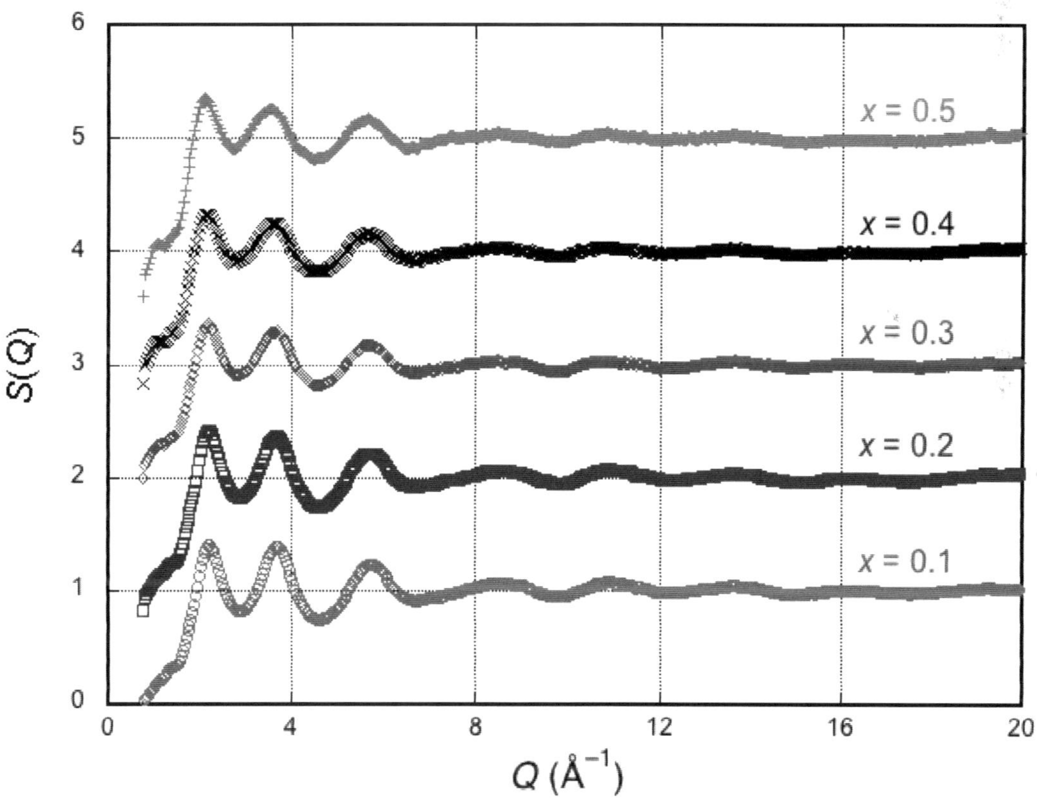

Figure 5. The neutron static structure factor for a series of samples of $(HgI_2)_x$-$(As_2Se_3)_{1-x}$ where $x = 0.1, 0.2, 0.3, 0.4, 0.5$, as determined from 7C2 data after about 8 h of counting. For the sake of clarity, the curves are shifted by $10x - 1$.

Comparing the instrument with those in pulsed sources, the main disadvantages are the limited momentum transfer range (which could be as high as 50 Å^{-1} in pulsed sources [5, 6, 7]) and the low resolution in reciprocal space. But the latter requirement is less critical for high temperature measurements, for which the oscillations do not extend over a large Q-range. The widely used vanadium furnace, allowing experiments up to 1400 K [8], lead to an accurate determination of structural features in important silicate liquids, such as the window glass/melt [14].

6. Preliminary results

As an example of data recently collected at 7C2, we chose the structure factor determination of a series of chalcogenide glasses.

Chalcohalide glasses have optical applications that need a wide transparency window especially in the far-IR region. Increasing the atomic number of a glass former causes a significant red shift in the low-frequency spectral domain from 11 μm in sulphide, to 15 μm in selenide and finally to 19-22 μm in telluride glassy binaries. Moreover, the average atomic number in the glass can be further increased by heavy-metal and heavy halogen doping [15]. Promising chalcogenide and chalcohalide glass systems include thus heavy-metal iodides: SbI_3, PbI2 and HgI2. While structural features of antimony and lead iodide and chalcogenide glasses have been studied in the past using diffraction and spectroscopic techniques, the atomic structure of mercury containing chalcogenide glasses is hardly known. The high absorption cross section of Hg makes the neutron diffraction experiments difficult, but the use of the hot-neutron spectrum at Orphée reactor reduces that absorption to 1/3 of the absorption for thermal neutrons.

The Figure 5 shows preliminary results on the static structure factor for the series $(HgI_2)_x$-$(As_2Se_3)_{1-x}$ where $x = 0.1, 0.2, 0.3, 0.4, 0.5$. For the sake of clarity, the data were shifted in the vertical axis by 0, 1, 2, 3 and 4, respectively. Each diffractogram corresponds to about 8 hours of measurement on a sample of 3 g. The working wavelength of 0.576 Å allows a Q-range extending to almost 20 Å^{-1}.

7. Conclusions

The two-axis instrument 7C2 is a new diffractometer dedicated to the structure of liquid, amorphous and disordered materials. It has been recently commissioned (March 2012) at the reactor Orphée in Saclay, France, substituting the previous version [8]. This is a state-of-the-art instrument comparable with their main competitors in the field [5, 6, 7, 13]. Its banana-like multidetector allows in situ experiments, such as thermodiffractometry.

Thanks to the hot neutron source, the momentum transfer range is long enough to perform proper Fourier transforms of the static structure factor, leading to the PDF with good resolution in real space. The instrument allows also experiments on nano-particles, for which the PDF analysis has demonstrated to be specially useful [16].

In addition, the short-wavelength neutrons reduce the absorption cross section, increasing the feasibility of neutron diffraction experiments on highly-absorbent samples like those containing mercury or rare earths (Sm, Eu, etc.)

Acknowledgments

We gratefully thank M. Kassem and E. Bychkov for the use of their data taken at 7C2 in June 2015.

References

[1] G. J. Cuello, J. Phys.: Cond. Matter **20**, 244109 (2008).
[2] G. L. Squires, **Introduction to the Theory of Thermal Neutron Scattering**, ISBN 0-486-69447-X, Dover Publications Inc, Mineola, New York (1978).
[3] V. F. Sears, Adv. Phys. **24**, 1 (1975).

[4] J. Dawidowski, J. R. Granada, J. R. Santisteban, F. Cantargi, L. A. Rodríguez Palomino, Experimental Methods in the Physical Sciences, **Neutron Scattering-Fundamentals**, ISBN 978-0-12-398374-9 **44**, Elsevier Inc (2013).
[5] D. T. Bowron, A. K. Soper, K. Jones, S. Ansell, S. Birch, J. Norris, L. Perrott, D. Riedel, N. J. Rhodes, A. Botti, M.-A. Ricci, F. Grazzi, M. Zoppi, Rev. Sci. Instr. **81**, 033905 (2010).
[6] C. J. Benmore, A. K. Soper, RAL Technical Reports, RAL-TR-1998-006 (1998).
[7] J. Neuefeind, M. Feygenson, J. Carruth, R. Hoffmann, K. Chipley, Nucl. Instr. Meth. Phys. Res. B **287**, 68 (2012).
[8] J. P. Ambroise, M.-C. Bellissent-Funel, R. Bellissent, Rev. Phys. Appl. (Paris) **19**, 731 (1984).
[9] G. Placzek, Phys. Rev. **86**, 377 (1952).
[10] I. A. Blech, B. L. Averbach, Phys. Rev. **137**, A1113 (1965).
[11] H. H. Paalman, C. J. Pings, J. Appl. Phys. **33**, 2635 (1962).
[12] M. A. Howe, R. L. McGreevy, P. Zetterström, Computer Code CORRECT: Correction Program for Neutron Diffraction Data, NFL Studsvik (1996).
[13] H. E. Fischer, G. J. Cuello, P. Palleau, D. Feltin, A. C. Barnes, Y. S. Badyal, J. M. Simonson Appl. Phys. A: Material Sci. Processing **74**, S160 (2002).
[14] L. Cormier, G. Calas, B. Beuneu, J. Non-Cryst. Solids **357**, 926 (2011).
[15] A. Masuno, S. Kohara, A. C. Hannon, E. Bychkov, H. Inoue, Chem. Mater. **25**, 3056 (2013).
[16] S. J. L. Billinge, Z. Kristallogr. Suppl. **26**, 17-26 (2007).

Time-of-flight Fourier Spectrometry with UCN

G.V. Kulin[1][1], A.I. Frank[1], S.V. Goryunov[1], P. Geltenbort[2], M. Jentschel[2], V.A. Bushuev[3], B. Lauss[4], Ph. Schmidt-Wellenburg[4], A. Panzarella[5], Y. Fuchs[5]

[1]Joint Institute for Nuclear Research, Dubna, Russia
[2]Institut Lauer-Langevin, Grenoble, France
[3]Moscow State University, Russia
[4]Paul Scherrer Institut, Switzerland
[5]ESRF, The European Synchrotron, Grenoble, France

E mail: kulin@nf.jinr.ru

Abstract. The report presents the first experience of using a time-of-flight Fourier spectrometer of ultracold neutrons (UCN). The description of the spectrometer design and first results of its testing are presented. The results of the first experiments show that the spectrometer may be used for obtaining UCN energy spectra in the energy range of 60÷200 neV with a resolution of about 5 neV. The application of TOF Fourier spectrometry technique allowed us to obtain the energy spectra from the diffraction of monochromatic ultracold neutrons on a moving grating. Lines of 0, ±1 and ±2 diffraction orders were simultaneously recorded, which had previously been impossible to be done by other methods. These results have made it possible to make a comparison with the recent theoretical calculations based on the dynamical theory of neutron diffraction on a moving phase grating.

1. Introduction

The UCN spectrometry has been successfully developing for a number of decades. Over the past fifteen years a number of experiments with the gravity UCN spectrometer with interference filters have been performed [1-5]. In this method the range of energies to be measured is limited by mgH, where m is the neutron mass, g is the free fall acceleration and H is the maximum possible distance between both filters. In practice, the energy range did not exceed 20-30 neV. In particular, the effect of neutron energy quantization in UCN diffraction by a moving grating was observed for the first time using such a spectrometer [2]. Later this effect was used to test the weak equivalence principle for the neutron [5].

The continuation of this research [6] and recent theoretical results [7] have brought us to the understanding of the need for more detailed investigation of UCN spectra in diffraction by a moving grating. The problem is that the energy range of the resulting spectrum is of the order of 100 neV, which is comparable with the UCN initial energy. At the same time, it is desirable that the energy of the spectral lines could be measured with an accuracy of 1÷1.5 neV. The gravity UCN spectrometry with interference filters appears to be unsuitable for these purposes.

As it follows from simple estimations, the classical TOF method is not suited for this task either, because the ratio of the initial time pulse duration to the pulse repetition period cannot exceed the ratio of the necessary resolution to the width of the time-of-flight spectrum. In our case this ratio is of the order of 10^{-2}. Such losses in luminosity are absolutely unacceptable.

It is likely that the solution of the problem is the use of any correlation time-of-flight methods. It may be the correlation spectroscopy with pseudo-random flux modulation [8-16] or Fourier spectroscopy with periodic or quasiperiodic flux modulation [17-25]. Having compared the two approaches, we have chosen the Fourier TOF spectroscopy [26]. The point is that we had a UCN

[1] To whom any correspondence should be addressed

spectrometer with periodic flux modulation [6], which could be relatively easily transformed into a TOF Fourier-spectrometer. On the other hand, the use of pseudo-random flux modulation was bound to lead to a dramatic loss of intensity, as its realization would severely limit the beam cross-section, which in our case had a shape of a ring. Such a UCN Fourier spectrometer was built and further on we report the first experience of its using.

2. Principle of the Time-of-flight Fourier spectroscopy

Let the aim of the measurement be the distribution of times $I(t)$ that neutrons spend for the flight along the known path length. This time-of-flight spectrum is evidently related to the velocity distribution. The spectrum might be directly measured, if neutrons would be generated in an ideally narrow δ-like time pulse. In the general case, the detector response is

$$Z(t) = \int_0^\infty I(t')\theta\left(t-t'\right)dt', \qquad (1)$$

where $\theta(t)$ describes the time dependence of the initial flux. If the initial flux is modulated harmonically, then

$$Z(t) = \int_0^\infty I(t')\sin\left[\omega\left(t-t'\right)\right]dt'. \qquad (2)$$

Each small part of the time spectrum contributes its harmonic to the detector count rate and the phase shift of this harmonic relative to the modulation phase is defined by the time of flight and modulation frequency. Since the time spectrum may be represented by a Fourier expansion

$$I(t) = \int_0^\infty R(\omega)\,sin\left[\omega t - \varphi(\omega)\right]d\omega\ . \qquad (3)$$

$Z(t)$ in equation (2) can be interpreted as a Fourier-harmonic of the initial spectrum. For the reconstruction of the initial spectrum in an ideal case, it is necessary to find functions $R(\omega)$ and $\varphi(\omega)$ in an infinitely large range of frequencies ω.

Since there was a device at our disposal with a fairly stable low-frequency beam modulator of periodic operation, we decided that at the initial stage of the work the measurement with a discrete set of frequencies ω_k would be acceptable. Then the results of the measurement are the amplitudes R_k and phases φ_k of the count rate oscillation for all frequencies ω_k. Having a rather large amount of such data, it is possible to reconstruct the initial time-of-flight spectrum in any time interval

$$I_{\exp}(t) = \frac{\pi}{2}\sum_k R_k \sin\left(\omega_k t + \varphi_k\right)\ . \qquad (4)$$

3. Experimental setup and measurement procedure

The experiment was performed at the PF2 source of the Institute Laue-Langevin (Grenoble, France). The spectrometer is a slightly modified version of the device described in [6]. It is shown in Figure 1. Ultracold neutrons are fed to the entrance chamber through the UCN neutron guide and, after a number of reflections, fall down the annular channel with the lower section closed by a monochromator, which is a five-layer Ni-Ti interference filter [27, 28]. To suppress the background of neutrons with energies higher than the effective potential of nickel, the filter-monochromator is combined with a multilayer "superwindow" filter [27]. In the experiments with a moving grating (see [6] for details) the latter was placed directly below the monochromator. It could be rotated by a motor. The neutron flux with the spectrum formed by the combination of two filters and transformed by the moving diffraction grating (if the latter is used) enters the spectrometric part of the device. It comprises a Fourier-chopper, vertical neutron guide, and detector. The Fourier chopper consists of a rotating rotor and stator. The rotor is a 2-mm-thick titanium disc of about 40 cm in diameter with twelve radial slits (see Figure 2). The stator is a titanium diaphragm with only one slit. It is placed at the inlet section of the vertical neutron guide.

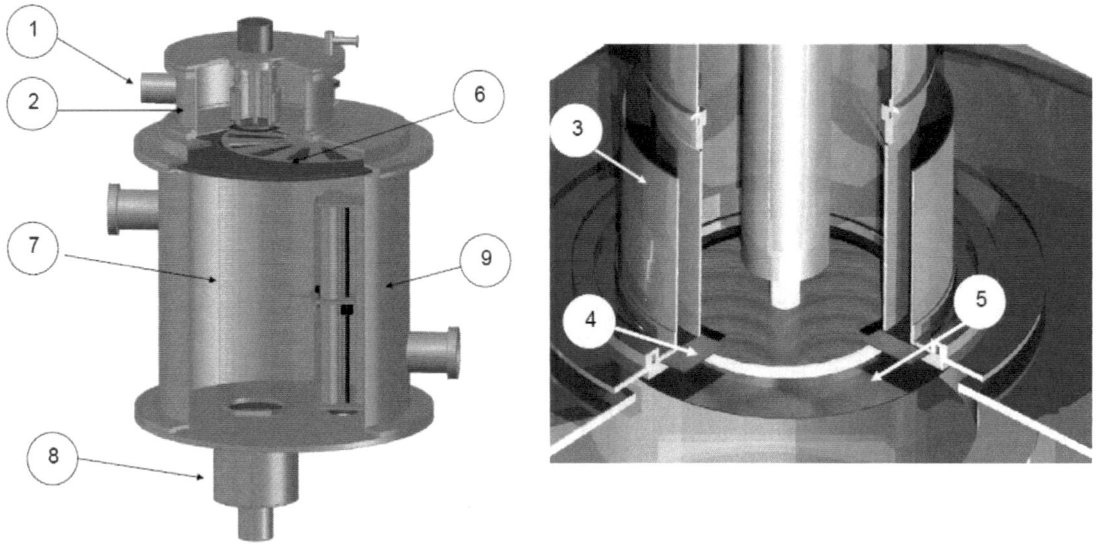

Figure 1. Time-of-flight Fourier spectrometer: general view (on the left) and its upper part (on the right): 1 - feeding guide, 2 - entrance chamber, 3 - annular channel, 4 - filter-monochromator, 5 - grating, 6 - rotor of the Fourier modulator, 7 - vertical glass guide, 8 - detector, 9 - vacuum vessel.

The rotor is driven by a stepper motor located outside the vacuum volume and connected to the rotor via a toothed belt and magnetic coupling. The rotation frequency of the rotor may reach 1800 rpm, which corresponds to a modulation frequency of 360 Hz. For the use of the electronic control system an infrared sensor and a small slot at the periphery of the rotor are used. Stability of the rotor rotation frequency is of the order of 10^{-4}. UCNs passing through the modulator come to the vertical neutron guide formed by six 95x680 mm float glass plates and reach the scintillation detector.

Figure 2. Rotors for the Fourier chopper.

The measurement system registered the time of arrival of pulses from the sensor of the chopper and from the detector and these data were recorded sequentially to a file. The modulation frequency f_k ranged from 6 to 360 Hz with a step of 6 Hz and upon reaching the maximum value decreased again.

Due to the vertical orientation of the spectrometer and the effect of the Earth's gravity the neutron time of flight does not linearly depend on the initial velocity. Sometimes for the correct interpretation of the results it was necessary to recalculate TOF spectrum to the energy scale. In addition to the relation between TOF and energy given by the equation

$$E = \frac{M}{2}\left(\frac{H^2}{t^2} - gH + \frac{g^2t^2}{4}\right), \qquad (5)$$

where M is the neutron mass, g is the free fall gravity acceleration and H is the difference in height between the Fourier modulator and detector. The latter was 72.5 cm. It was also necessary to take into account the nonlinear relation between the widths of energy and time channels on the abscissa axis. The relation between N_t and N_E values, which were proportional to the number of counts in time and energy channels, is given by

$$N_E = N_t\left[M\left(\frac{H^2}{t^3} - \frac{g^2t}{4}\right)\right]^{-1}. \qquad (6)$$

4. Test measurements with neutron interference filters

To test the spectrometer, three neutron interference filters (NIFs) [27, 28] were used. Filter 1 is a five-layer filter with alternating layers of NiMo and TiZr. It has a narrow transmission line. Filter 2 is a nine-layer filter with alternating layers of the same materials. It has a wide transmission line. Filter 3 is a five-layer filter with alternating layers of Ni(N) and TiZr. It was characterized by the splitting of the transmission line (see details in [27, 28]). The time-of-flight spectra measured by the spectrometer are shown in Figure 3.

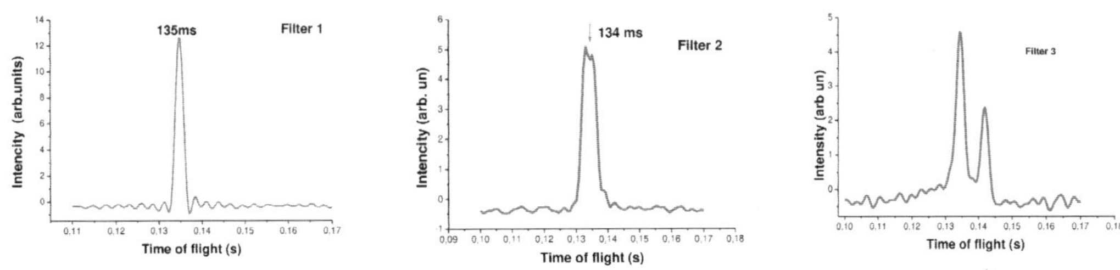

Figure 3. Time-of-flight spectra measured by the Fourier spectrometer. Energy spectra formed by filters 1-3 in combination with filter "superwindow".

These spectra demonstrate that TOF UCN Fourier spectrometer have a resonable resolution and may be used for physical measurements.

5. Energy spectra of UCN diffracted by a moving grating

As in our previous works [2, 4], the diffraction grating was prepared on the surface of a silicon disc of 150 mm in diameter and 0.6 mm in thickness. Radial grooves (see Figure 4) were made in the peripheral region of the disc, which is a ring with an average diameter of 12 cm and a width of about 2 cm. The widths of the grooves are proportional to the radius and this proportionality ensures a constant angular distance between the grooves equal to a half period. The angular period of the structure is exactly known to be $\alpha = 2\pi/N$ with $N = 94500$. The design depth of the grooves of 0.144 mkm was chosen to ensure a phase difference $\Delta\varphi = \pi$ between the neutron waves passing through the neighboring elements of the grating. The grating was manufactured by Qudos Technology Ltd.

The grating was examined using an atomic force microscope. Two fragments of the grating positioned at different distances from the center were scanned and the image of one of these is shown in Figure 5. Subsequent analysis has shown that the real grating parameters are quite satisfactory agreement with the designed values.

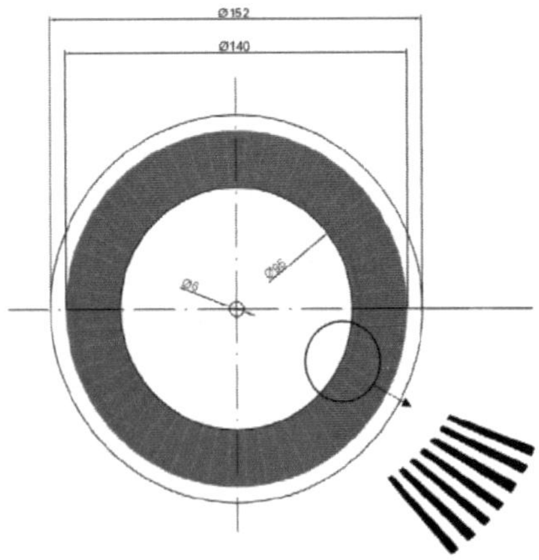

Figure 4. Diffraction grating: dimensions and orientation of grooves.

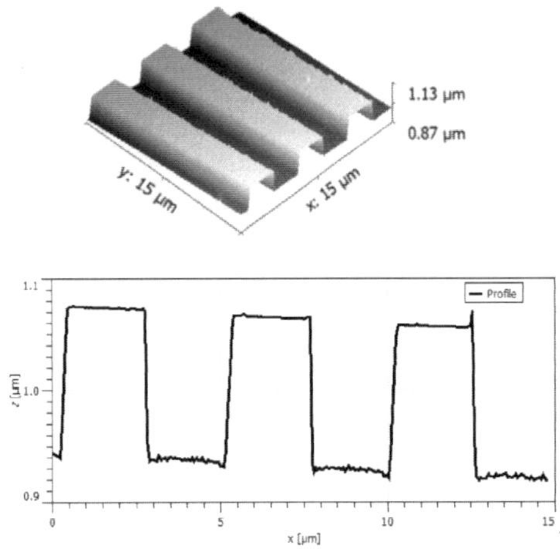

Figure 5. 3D image of the grating fragment obtained with an atomic force microscope.

Energy spectra of UCN passing through the grating were measured both for the grating at rest and for the grating rotating at frequencies of 1500, 3600 and 4800 rpm. With the fixed/specified number of grooves N the expected value of the energy splitting is $\Delta E = 2\pi\hbar Nf$, where f is the rotation frequency. Time-of-flight spectra reconstructed in accordance with (4) were then recalculated into the energy scale using eqs. (5), (6).

These spectra were compared with the results of calculation based on the dynamical theory of diffraction [7]. In these calculations the normal component of the neutron velocity at the surface of the grating was taken to be V_z = 4.67 m/s (energy of 114 neV), which corresponds to the position of the maximum of the monochromator transmission spectrum measured earlier by the time-of-flight method. The distribution of the velocity V_z was assumed to be a Gaussian with the degree of monochromatization $\Delta V_z / V_z = 0.0175$, where ΔV_z is FWHM of the distribution. The dependence of the resolution on the time of flight was neglected. The poorly known spectrum of horizontal velocities was also assumed to have a Gaussian distribution with a zero average value and FWHM equal to 3 m/s.

Figure 6 shows the TOF spectrum of UCN passing through the stationary grating and its corresponding energy spectrum. In addition to the main peak one can also clearly see a small peak at an energy of 255 neV. Apparently, this is a peak at the energy slightly higher than the value of the potential significantly suppressed by the transmission of the "superwindow".

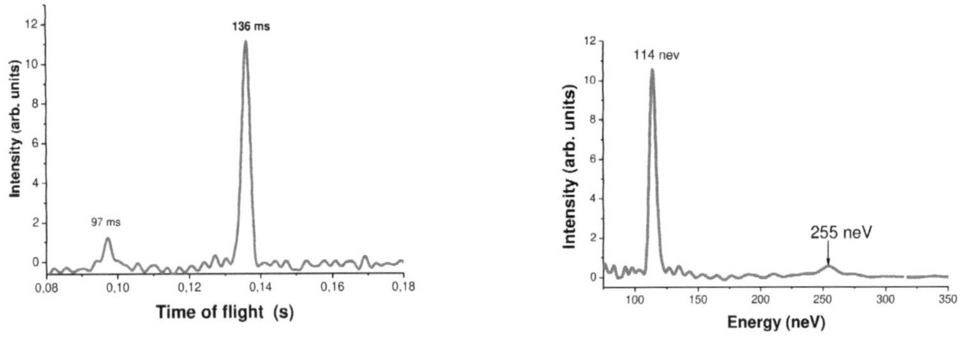

Figure 6. TOF spectrum (on the left) and energy spectra of UCNs passing through the stationary grating (on the right).

Figure 7 illustrates the main results of the experiment. The obtained time and energy spectra measured at three rotation velocities of the grating are displayed together with the results of the calculation. The absolute normalization of spectra was done for better visualization purposes.

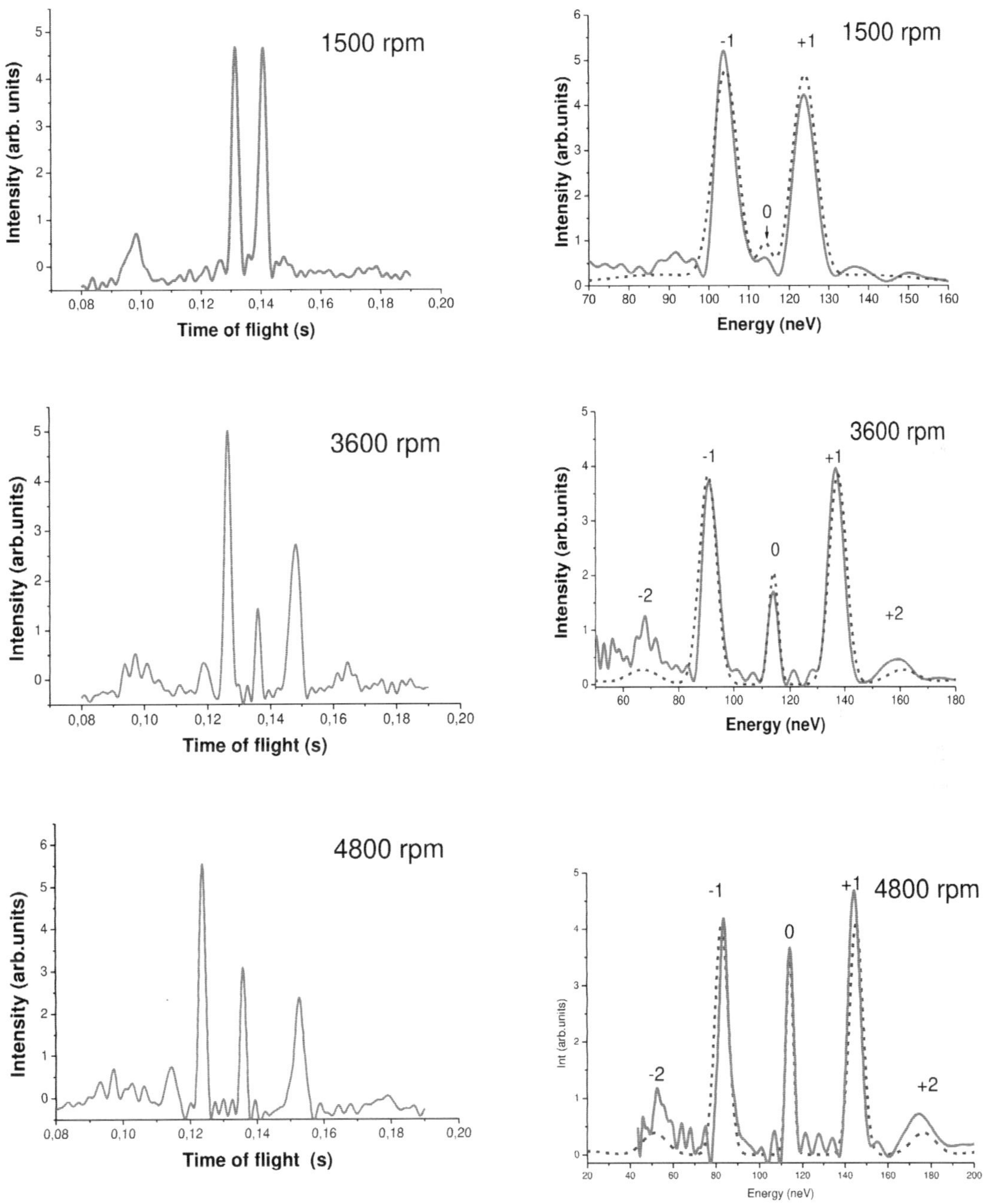

Figure 7. Time-of-flight (on the left) and energy (on the right) spectra of neutrons diffracted by a moving grating measured at different rotation frequencies. Lines of 0, ±1 and ±2 diffraction orders are clearly seen. Experimental (red solid line) and theoretical (blue dotted line) spectra are shown.

The obtained results demonstrate that using this method it is possible to detect spectra with the energy range comparable with the UCN initial energy. That permit us to compare the intensities of lines of five diffraction orders with predictions of the dynamic theory of neutron diffraction [7]. The experimental data were found in a good agreement with this theory.

6. Conclusion

We reported on the development of the time-of-flight UCN Fourier spectrometer designed mainly for studying the non-stationary phenomenon of neutron diffraction by a moving grating. This device is a modification of the gravity spectrometer [6] and therefore the first tests of the new device were carried out in the regime of the discrete spectrum of beam modulation and off-line data processing. This regime is not optimal, but it made it possible to obtain the first results in a relatively short time. The results of the test have shown that this device allows UCN energy spectra to be measured in the energy range of 60 ÷ 200 neV with the resolution of the order of 5 neV.

UCN spectra appearing in the experiments on neutron diffraction by a moving grating were measured using the time-of-flight Fourier spectrometer. In spite of the fact that the first measurements were made with not very high statistics, the results allow for a reliable comparison with the theoretical prediction. The obtained data testify that the experiment is in rather satisfactory agreement with the theoretical predictions based on the multiwave dynamical theory of neutron diffraction in the approximation of slowly changing amplitudes [7].

The activities on further improvement of the device will be aimed at increasing its luminosity. In addition, the spectrometer will be modified for using a continuous spectrum of modulation frequencies, which will improve the parameters of the device.

7. Acknowledgements

The authors are very grateful to Tomas Brenner for his outstanding technical support. This work was partly supported by the Russian Foundation for Basic Research (RFBR grants 15-02-02367 and 15-02-02509).

References

[1] Frank A I, Bodnarchuk V I, P. Geltenbort P et al. 2003 *Phys. Atom. Nuclei* **66** 1831
[2] Frank A I, Balashov S N, Bondarenko I V et al. 2003 *Phys. Lett.A*, **311** 6
[3] Frank A I, Geltenbort P, Kulin G V and Strepetov A N 2003 *JETP Letters*, **78** 188
[4] Frank A I, Geltenbort P, Kulin G V et al. 2005 *JETP Letters* **81** 427
[5] Frank A I, Geltenbort P, Jentschel M et al., 2007 *JETP Letters*, **86** 225
[6] Kulin G V, Frank A I, Goryunov S V et al. 2015 *Nucl. Instr. Meth.* A **792** 38
[7] Bushuev V A, Frank A I and Kulin G V arXiv:1502.04751v1 [physics.optics]
Bushuev V A, Frank A I and Kulin G V 2015 *JETP* **148** (in press)
[8] Cook-Yahrborough H E 1964 in: *Instrumentation Techniques in Nuclear Pulse Analysis, Washington DC* 207
[9] Mogilner A I, Shalnikov O A and Timochin L A 1966 *Sov.J.Techn. Phys.* **2** 22 (in Russian)
[10] Sköld K 1968 *Nucl. Inst. Meth.* **63** 114
[11] Gordon J, Kroo N, Orban G, et al. 1968 *Phys. Lett.* A **26** 122
[12] Kroo N, Cher L 1977 *PEPAN* **8** 1412
[13] Cser L, Ferenczy F, Kroó N, Rubin Gy et al. 1981 *Nucl. Insr. Meth.*,**184** 431
[14] Freudenberg U, Glaser W 1986 *Nucl. Instr. Meth.* A **243** 429
[15] Gutsmiedl E, Golub R, Butterworth J 1991 *Physica B: Condensed Matter* **169** 503
[16] Novopoltsev M I and Pokotilovski Yu N 2010 *Instr. and Exp. Tech.* **53** 635
[17] Hiismäki P, Trunov VA, Antson O et al. *1985 in Proc. Conf. on Neutron Scatt. in the Nineties,*

Julich, Germany,. (Vienna: IAEA, 1985) 453
[18] Virjo A 1969 *Nucl. Instr. and Meth.* A, **73** 189
[19] Colwell J F, Lehinan S R, Miller Jr. P H and Whittemore W L 1969 *Nucl. Inst.Meth.*A **76** 135
[20] Nunes A C, Natans R and Schoenborn B P 1971 *Acta Cryst.* A**27** 284
[21] Pöyry H, Hiismäki P, Virjo A 1975 *Nucl. Inst.Meth* **126** 421
[22] Pöyry H 1978 *Nucl. Instrum. Methods* **156** 499
[23] Schröder J, Kudryashev V A, Keuter J M et al. 1994 *Neutron Res.* **2** 129
[24] Aksenov V L, Balagurov A M 1996 *Phys.Uspechi*, **39** 897
[25] Maayouf R M A, Abdel-Latif I, El-Kady A, El-Shafey A, Khalil M, El-Shaer Y 1997 *Nucl. Inst.Meth A* **398** 295
[26] Kulin G V, Kustov D V, Frank A I et al. 2014 *JINR Communication P3-2014 -72* (in Russian).
[27] Bondarenko I V, Bodnarchuk V I, Balashov S N 1999 *Phys. Atom. Nuclei.* **62** 721
[28] Bondarenko I V, Balashov S N, Cimmino A et al. 2000 *Nucl. Instr. Meth.* A **440** 591

Journal of Physics: Conference Series **746** (2016) 012022 doi:10.1088/1742-6596/746/1/012022

Advancing data analysis for reflectivity measurements of holographic nanocomposite gratings

Jürgen Klepp[1,*], Christian Pruner[2], Yasuo Tomita[3], Peter Geltenbort[4], Joachim Kohlbrecher[5] and Martin Fally[1]

[1]Faculty of Physics, University of Vienna, Boltzmanngasse 5, A-1090 Vienna, Austria
[2]Department of Materials Science and Physics, University of Salzburg, Hellbrunnerstrasse 34, A-5020 Salzburg, Austria
[3]Department of Engineering Science, University of Electro-Communications, 1-5-1 Chofugaoka, Chofu, Tokyo 182-8585, Japan
[4]Institut Laue-Langevin, 71 avenue des Martyrs, 38000 Grenoble, France
[5]Laboratory for neutron scattering, Paul Scherrer Institut, 5232 Villigen PSI, Switzerland

E-mail: *juergen.klepp@univie.ac.at

Abstract. We discuss the analysis of neutron rocking curves measured with holographic nanoparticle-polymer composite gratings with SANS instruments and at very-cold neutron beamlines. In particular, the effects of broad incident wavelength-distributions and of nonuniformity of the refractive-index profile along the grating depth are assessed.

1. Introduction

Photosensitive materials in combination with holographic techniques can be used to produce diffraction gratings for neutron-optics. One-dimensional sinusoidal holograms recorded in such materials exhibit a periodic light- and neutron-refractive-index pattern arising from a light-induced density modulation. They work as diffraction gratings. The recorded refractive-index pattern is described by $n(x) = n_0 + \Delta n_1 \cos(2\pi x/\Lambda)$, where n_0, Δn_1 and Λ are the mean refractive index, the refractive-index modulation amplitude and the grating period, respectively. $\Delta n_1 = \lambda^2 b_c \Delta\rho/2\pi$ for neutrons, with the scattering-length density modulation amplitude $b_c\Delta\rho$. Examples for such materials already tested with neutrons are deuterated (poly)methylmethacrylate [1, 2, 3, 4, 5], holographic polymer-dispersed liquid crystals [6, 7] and nanoparticle-polymer composites (NPCs, see [8, 9, 10, 11] and Fig. 1).

A decisive advantage of holographic NPC gratings is their versatility due to the range of possible choices of nanoparticle composition, thereby selecting the most suitable value of the coherent scattering length/absorption cross section for a specific application. For instance, magnetic or absorptive terms could be added to the periodic optical potential by incorporating ferromagnetic/superparamagnetic [12] nanoparticles or Gd-based nanoparticles, respectively. Magnetic grating-structures could be used as polarizing beam splitters for retrofitting small-angle neutron scattering (SANS) instruments. Absorption gratings with small Λ could find applications in very-cold neutron (VCN) interferometry [13]. Recently investigated material types [14, 15] exhibited an increase in reflectivity of about 40% as compared to earlier experiments.

Figure 1. A transmission electron microscope image of the cross section of a plane-wave grating recorded in a ZrO_2 NPC sample at grating spacing of 0.6 μm. Note that the darker portions in the image correspond to ZrO_2 nanoparticle rich regions, indicating holographic assembly of nanoparticles in the formed polymer [9, 10].

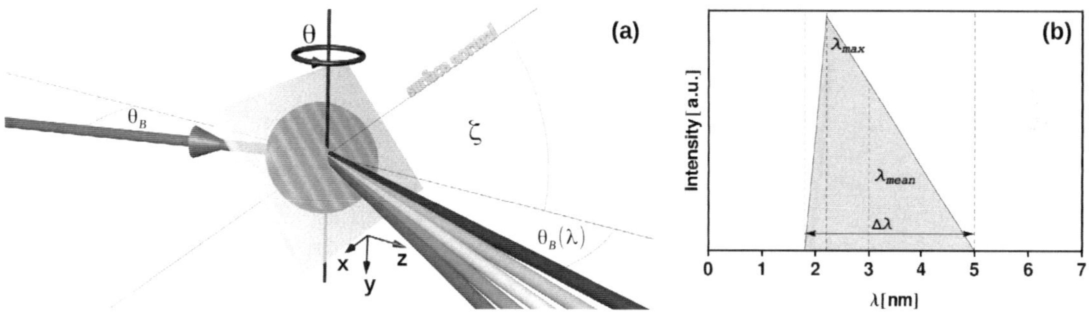

Figure 2. (a) Sketch of the experimental setup. (b) Our model of the asymmetric wavelength distribution of PF2VCN.

Measured reflectivity curves ('rocking curves', i.e., neutron reflectivity vs. angle of incidence) of NPC gratings are usually fitted using the theoretical models by Kogelnik [16] or the equivalent dynamical diffraction theory (DDT) [17, 18]. For gratings with grating-strength decaying exponentially along the direction of the grating thickness, Uchida's approach [19] is applied. For similarities and differences between the approaches, see Refs. [20, 21, 22, 23], for instance.

In comparison to crystal neutron optics it should be noted that NPC gratings are usually designed to work as relatively thin (in an optical sense [24]) diffraction gratings: In some cases, more than two diffraction spots can be visible for a given angle of incidence, so that the two-beam approximation is not applicable. In that case, the so-called rigorous coupled-wave analysis (RCWA) [20, 25, 26], that accounts for an arbitrary number of, even overlapping, peaks must be adopted for data analysis. Furthermore, the Pendellösungs length of NPC gratings can be of the order of 1 mm, so that the oscillatory structure of rocking curves is observable with realistic beam collimation. As a consequence, and most importantly, it is possible to beat the usual limit of about 50% that is imposed on the reflectivity of thick crystals in Laue (transmission) geometry by the mere fact that measured rocking curves are usually smeared out over a vast number of reflectivity oscillations [17]. In contrast, NPC gratings can in principle reach mirror-like reflectivity and even exhibit full Pendellösung oscillations as a function of angle of incidence θ, neutron wavelength λ, scattering length density modulation amplitude $b_c\Delta\rho$ or grating thickness d [27, 28, 29, 30, 31]. DDT predicts that—due to the particular superposition state within a periodic structure—the neutron intensity oscillates between reflected and transmitted beams

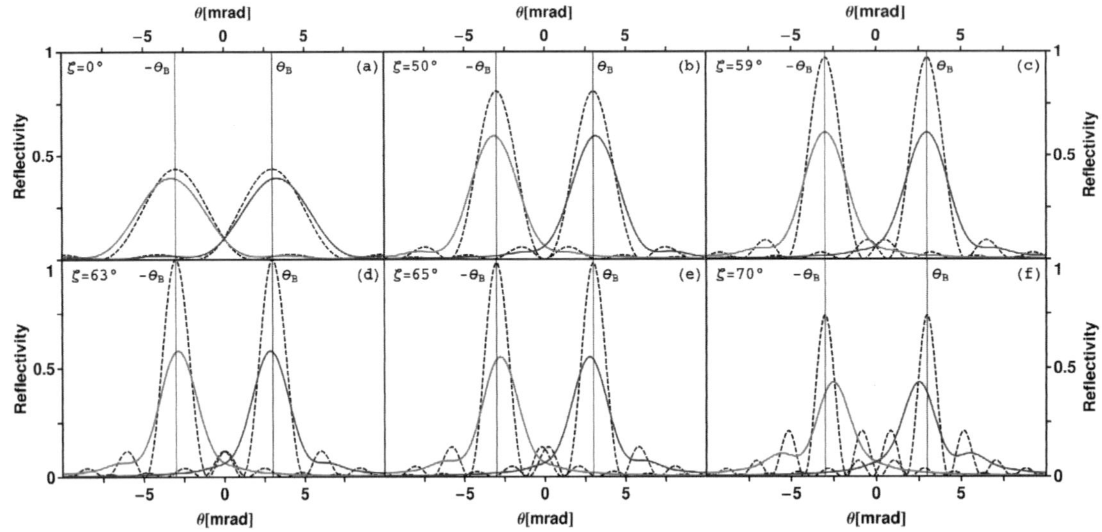

Figure 3. Theoretical rocking curves calculated using DDT (dashed lines) and convolutions of DDT with a slit function (determined by the collimation) and the wavelength distribution shown in Fig. 2, for various tilt angles (a) – (f).

(Pendellösung interference). It has been shown that NPC gratings can be used as 50:50 beam splitters [32, 33], mirrors [34] or three-port beam splitters (30:30:30) [35] for cold- and very-cold neutrons (VCN). Furthermore, in the course of the experiments described in the above references, it has been asserted that the grating properties and quality have not changed substantially for many years, with the NPC gratings stored under ambient conditions.

In the present paper, some issues arising in the data analysis of reflectivity measurements with NPC gratings, related to the above facts, are discussed.

2. A typical experiment

In Fig. 2 (a), a schematic of the experimental setup used at the VCN beam position of the ILL (PF2VCN) is shown: We use a collimated beam with divergence limited both horizontally and vertically to about 10^{-3} rad. The collimation allows to observe the side maxima of the rocking curves at reasonable flux. The wavelength spectrum of the incident beam is very broad and asymmetric at PF2VCN, as shown in Fig. 2 (b) (for a measured VCN spectrum see, for instance, [13]). The cut-off at shorter wavelengths is due to reflection from a neutron (super)mirror used to redirect the incident beam, while the large width is due to the spectrum originating from the vertical and curved guide that feeds the ultra-cold neutron turbine [36]. Of course, further shaping of the incident spectrum is possible, but making use of the entire incident flux is preferable. Since, according to the Bragg equation $\lambda = 2\Lambda \sin\theta_B$ and due to gravity, longer λ means larger Bragg angles $\theta_B(\lambda)$ and lower impact positions on the detector (for given sample-detector distance), respectively. The longer λ, the further away from the 0th diffraction order spot (both horizontally and vertically) do diffracted neutrons hit a subsequent 2D detector at a particular distance from the grating. Consequently, we can discriminate the (mean) wavelengths on the 2D detector (pixel-size $\approx 2 \times 2\,\text{mm}^2$) at about 2 m sample-detector distance by analyzing each pixel line in the diffraction spots on the detector separately to exploit the full wavelength-distribution of the VCN beam at PF2 at once (wavelength-multiplexing). Note that the sample-detector distance is usually insufficient to obtain narrower wavelength spectra for

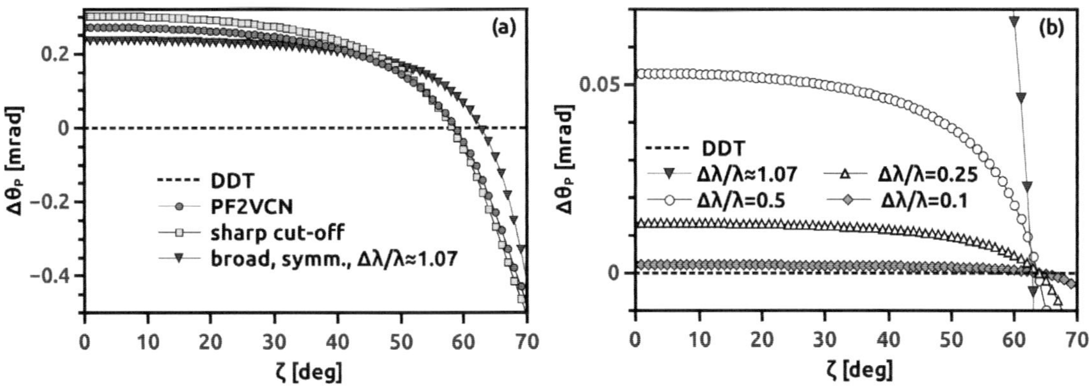

Figure 4. $\Delta\theta_P$ vs. the grating tilt angle ζ for incident spectra of (a) varying asymmetry (see text), (b) various widths $\Delta\lambda/\lambda$.

individual pixel lines, as compared to the width of the incident spectrum. However, the *mean* wavelength is increasing from the top to the bottom pixel line. We found that the intensity loss to absorption/incoherent scattering by an NPC grating is about 20%. To compare this with time-of-flight for PF2VCN in a similar configuration (2 m distance and spectrum ranging from 3 nm to 6 nm with good collimation, say), one suffers at least 80% loss due to the chopper alone. Of course, the wavelength resolution is much better for time-of-flight.

In Fig. 2 (a), one notes that the grating is not mounted in straightforward transmission geometry (also called Laue geometry), but is tilted about the grating vector by the angle ζ. We use such tilt to overcome thickness limitations of NPC gratings or to fine-tune the effective thickness in order to reach a particular reflectivity [33].

In the following section, peculiarities in data analysis that arise due to the VCN incident spectrum are discussed.

3. Appearance of shifted peaks

$\pm1^{\mathrm{st}}$ diffraction order peaks of rocking curves calculated using DDT for a particular NPC grating (parameters: $d = 100\,\mu$m, $b_c\Delta\rho = 4.8\,\mu\mathrm{m}^{-2}$, $\Lambda = 500$ nm) at different tilt angles ζ are shown by the dashed lines in Figs. 3 (a)–(f). The solid, vertical lines are the respective positions of $\pm\theta_B$, which do not depend on ζ. The calculations are carried out for $\lambda = 3$ nm, which is also the mean wavelength of the spectrum shown in Fig. 2 (b). The solid curves are the $\pm1^{\mathrm{st}}$ diffraction order peaks of rocking curves calculated numerically by a convolution of the DDT curves with a slit function (according to the divergence of $\approx$1 mrad in our experiments) and the assumed wavelength distribution as shown in Fig. 2 (b). Taking a closer look at the peak positions $\pm\theta_P$ of the solid curves, one observes slight deviations from $\pm\theta_B$ depending on ζ. While for $\zeta = 0°$ [Fig. 3 (a)] the peak positions $\pm\theta_P$ are larger than $\pm\theta_B$, the former decrease with increasing ζ until $\pm\theta_P = \pm\theta_B$ at $\zeta \approx 59°$ [Fig. 3 (c)], only to be shifted further towards $\theta = 0°$ for higher ζ. For $\zeta = 70°$ [Fig. 3 (f)], one can see that the solid curves become also quite asymmetric.

Though surprising at first sight, it is clear that the shift of θ_P stems from incoherent superposition of the rocking curves measured 'at once' with a wide and asymmetric incident wavelength spectrum: Δn_1 is proportional to λ^2, so the reflectivity is much higher at longer wavelengths. Furthermore, curves for longer λ exhibit larger diffraction angles. Fig. 4 (a) shows deviations $\Delta\theta_P = \theta_P - \theta_B$ (calculated) from the DDT peak-position θ_B as a function of ζ for several different incident wavelength spectra that all exhibit the same mean wavelength of $\lambda_{\mathrm{mean}} = 3$ nm. The horizontal, dashed line is the DDT prediction for a perfectly collimated, monochromatic incident beam. The filled red circles represent the results for the PF2VCN model-spectrum of Fig. 2 (b) with a width of $\Delta\lambda = 3.2\,\mathrm{nm}$. The filled green squares represent data computed assuming a very sharp wavelength cut-off, for which λ_{max} (the wavelength of highest flux) is at the left border of the distribution. The filled, blue upside-down triangles assume a distribution symmetrical around λ_{mean} with the same large $\Delta\lambda$ as above, which—due to its symmetry—can be characterized by $\Delta\lambda/\lambda \approx 1.07$. The latter data set is also shown in Fig. 4 (b), together with data extracted from simulations assuming symmetrical distributions with decreasing values of $\Delta\lambda/\lambda$. In particular, further data are shown for $\Delta\lambda/\lambda = 0.5$ (filled yellow circles), $\Delta\lambda/\lambda = 0.25$ (empty triangles) and $\Delta\lambda/\lambda = 0.1$ (filled orange squares). Inspecting the difference in scaling of the y−axes in Figs. 4 (a) and (b), one can infer that the exact form (the asymmetry, in this case) plays a minor role in the shifts of θ_P as compared to the actual width of the incident wavelength distribution.

If the incident mean wavelength is to be estimated only from peak positions and the Bragg condition, the shifts account for errors of up to 10%. With respect to our experimental method (wavelength-multiplexing by line-by-line analysis of diffraction spots) proposed above, it is clear that one cannot obtain a consistent estimation of the mean wavelength incident at a particular pixel line or other parameters aimed at, without taking the above findings into consideration.

Note that even for $\Delta\lambda/\lambda = 0.1$, as is the case of widths of typical SANS incident wavelength-spectra, there is some deviation in θ_P to be expected [Fig. 4 (b)]. However, it would be only of the order of 10^{-3} mrad or—translated to wavelengths—10^{-2} Å. Note also that the actual maximum tilt angle we have reached in experiments so far is about 80°. The behavior of θ_P in the range $70° \leq \zeta < 90°$ exhibits interesting resonance-like features, which will be discussed elsewhere.

4. Rocking curve shape-imperfection due to attenuation of the grating-strength along the sample depth

As already mentioned, in Ref. [35] an NPC grating was used as three-port beam splitter for neutrons at a wavelength $\lambda = 1.7\,\mathrm{nm}$. Grating parameters deduced were mainly due to an estimation at the Bragg angles, applying a two-beam coupling model [16]. This model, however, is only a rough approximation for two reasons: *(i)* standard two beam-coupling theories must fail as at least the consideration of three beams is vital to assess the situation, *(ii)* the side minima of the rocking curve are obviously lifted from the zero line.

To tackle problem *(i)* it would be sufficient to apply the RCWA [25, 26] by fitting it to the data. As discussed in the previous section, *(ii)* might be caused by a convolution of the wavelength distribution with the rocking curve. While this is important for diffraction measurements using the broad spectrum of PF2VCN, the situation is different for the rather narrow wavelength distribution at the SANS I instrument of PSI ($\Delta\lambda/\lambda \approx 0.1$), where the measurements shown in [35] were performed.

Another well known effect that might lead to *(ii)*, is a nonuniformity of the grating strength along the sample depth [19] like, e.g., attenuation of the grating modulation-amplitude along the grating depth. The tedious task of including such a behavior into the RCWA was performed in Ref. [37] by dividing the hologram in a large number of layers along the sample depth, each of them treated by the RCWA and considering the boundary conditions between subsequent layers. We choose a more pragmatic approach. By inspecting the data shown in Fig. 5 we note that only five diffraction orders ($\pm 2, \pm 1, 0$) are observed. Therefore, we set up a five-wave coupling theory allowing for a profile of the coupling parameters along the sample depth and solve a set of coupled first-order ordinary differential equations. Second-order derivatives are disregarded (slowly-varying amplitude approximation) as the refractive-index modulation is moderate and reflection of neutrons at the surfaces can be neglected ($n_0 \approx 1$). This procedure gives rise to an ambiguity for the choice of boundary conditions [23, 38, 39, 40, 41]. Here, we opt for the

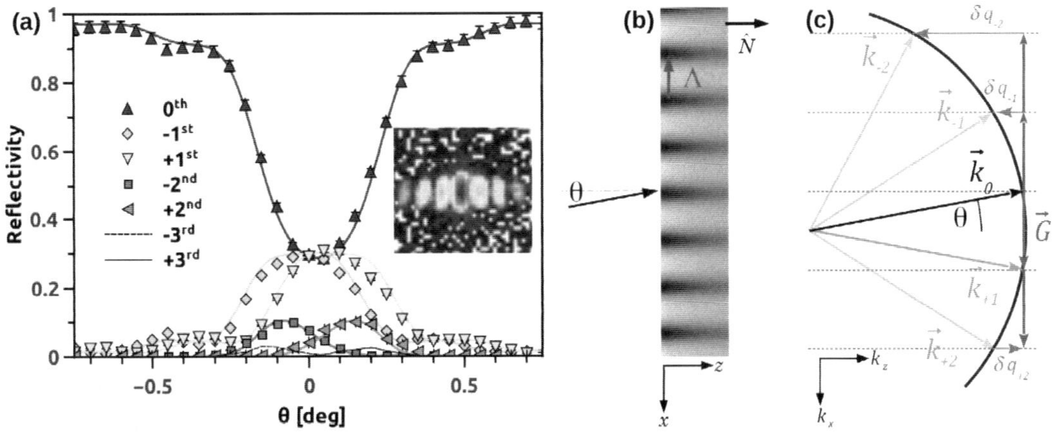

Figure 5. (a) Rocking curve for an NPC three-port beam splitter ($\zeta \approx 56°$). The inset shows a sum of the detector counts for all measured angular positions. (b) Geometrical configuration of a $x-z$-planar diffraction grating. $\vec{G} = G(1,0,0) = 2\pi/\Lambda(1,0,0)$, $\hat{N} = (0,0,1)$. (c) Ewald sphere at Bragg incidence for $+1^{\rm st}$ order (not to scale), Floquet-condition.

beta-value method (see e.g. [38]), i.e., $|\vec{k}_0| = |\vec{k}_{\pm 1}| = |\vec{k}_{\pm 2}| = 2\pi/\lambda := \beta$. The system of ODEs is solved numerically to fit it to the experimentally determined reflectivity. We refine the description of the refractive index modulation in our gratings (see Sec. 1) by allowing for an arbitrary profile along the sample depth z [see Fig. 5 (b)]:

$$n(x,z) = 1 + \Delta n_1(z)\cos(Gx) + \Delta n_2(z)\cos(2Gx) + \ldots \quad (1)$$

Here, $\Delta n_i(z)$ are the depth-dependent Fourier-components of the refractive index, $G = 2\pi/\Lambda$ is the modulus of the grating vector (spatial frequency). Neutrons are assumed to travel only in the $x-z$ plane. The wave equation (deduced from the time-independent Schrödinger equation) reads:

$$\left(\nabla^2 + k^2\right)\psi(x,z) = 0, \quad (2)$$

where $k = \beta n(x,z)$ and $\psi(x,z)$ is the wave function. For the latter we use a coupled-wave ansatz $\psi(x,z) = \sum_j S_j(z)e^{-\imath\vec{k}_j\cdot\vec{x}}$, where $S_j(z)$ are the diffraction amplitudes to be determined. The Floquet condition requires that

$$\vec{k}_j = \vec{k}_0 + j\vec{G} - \delta q_j\hat{N}, \quad (3)$$

with the (scalar) mismatch parameter

$$\delta q_j = \beta\left[\cos\theta - \sqrt{1 - (\sin\theta + jG/\beta)^2}\right]. \quad (4)$$

Here, θ denotes the angle between the sample surface normal $\hat{N}$ and the wavevector $\vec{k}_0$ of the incident beam. Inserting Eqs. (1), (3) and (4) into the wave equation (2), disregarding the second order derivatives and sorting the terms according to their propagation directions yields the coupled differential equations to be solved for the $S_j(z)$ with the boundary conditions $S_0(z=0) = 1$ and $S_j(z=0) = 0\ \forall j \neq 0$:

$$\begin{aligned} 2\imath k_{j,z}S_j(z)' &= \beta^2\left[\Delta n_1(z)\left(S_{j+1}(z)e^{\imath(\delta q_{j+1}-\delta q_j)z} + S_{j-1}(z)e^{\imath(\delta q_{j-1}-\delta q_j)z}\right)\right. \\ &+ \left.\Delta n_2(z)\left(S_{j+2}(z)e^{\imath(\delta q_{j+2}-\delta q_j)z} + S_{j-2}(z)e^{\imath(\delta q_{j-2}-\delta q_j)z}\right)\right], \end{aligned} \quad (5)$$

Table 1. Summary of fitting results.

	$\zeta \approx 55°$	$\zeta \approx 0°$
λ	(1.72 ± 0.03) nm	(1.65 ± 0.05) nm
Δn_1^0	$(5.05 \pm 0.10) \times 10^{-6}$	$(5.18 \pm 0.15) \times 10^{-6}$
Δn_2^0	$-(9.4 \pm 0.6) \times 10^{-7}$	$-(1.12 \pm 0.05) \times 10^{-6}$
d	$(186 \pm 1)\ \mu$m	$(99 \pm 1)\ \mu$m
L	$(109 \pm 2)\ \mu$m	$(65 \pm 1)\ \mu$m
$\langle b_c \Delta\rho_1 \rangle$	$(5.1 \pm 0.2)/\mu\text{m}^2$	$(6.2 \pm 0.4)/\mu\text{m}^2$
$\langle b_c \Delta\rho_2 \rangle$	$-(0.96 \pm 0.07)/\mu\text{m}^2$	$-(1.3 \pm 0.1)/\mu\text{m}^2$

where we use the abbreviation $S_j(z)' = \partial S_j(z)/\partial z$. The j-th order reflectivity R_j is given by $R_j = k_{j,z}/k_{0,z}|S_j(z = d)|^2$.

4.1. Experimental data and fitting

Rocking curves were measured for two tilt angles ($\zeta \approx 0°, 55°$). As explained in Sec. 2, longer wavelength neutrons hit the detector at lower positions and larger Bragg angles. However, it can be seen from Fig. 5 (inset) that the rather narrow wavelength distribution of SANS I is negligible: the longish diffraction spots are almost vertical and parallel. Therefore, we evaluated the angular dependence of the reflectivity by summing up all the counts of each diffraction spot in contrast to the line-by-line analysis. Next, we assume the functional dependence of the refractive-index depth-profile to be exponential with a decay length L such that $\Delta n_j(z) = \Delta n_j^0 e^{-z/L}$. This is a quite obvious choice as usually light absorption and—as in our case—linear light scattering processes lead to the exponential attenuation of the interference-fringe amplitude along the z-direction during recording [42, 43]. However, our model is not restricted to this particular choice. The profile can be of arbitrary form since the *mean* Fourier components (proportional to the mean values $\langle b_c \Delta\rho_j \rangle$) relate to the Fourier components Δn_j^0 simply by averaging over d (integration along sample depth, Eq. (5) in Ref. [33]).

Using the numerical solutions to Eqs. (5) for five waves ($\pm 2, \pm 1, 0$ diffraction orders), a fit to the rocking curves at two tilt angles ($\zeta \approx 0°, 55°$) yielded the parameters in Tab. 1. Note that the modulation parameters Δn_i^0 and the wavelengths for the two tilt angles are perfectly consistent and the thicknesses are simply related by $d = d(\zeta = 0)/\cos\zeta$. Similar reasoning applies to the decay lengths L [33]. Further, for VCN measurements we have to consider that the coherent scattering-length density modulation amplitudes $b_c \Delta\rho_j$ for the j-th Fourier coefficient are related to the refractive-index modulation amplitudes Δn_j via $b_c \Delta\rho_j = 2\pi/(\lambda^2)\Delta n_j$ and are, thus, wavelength dependent.

The mean wavelength for the measurements at PF2VCN was about 3.6 nm. A sum of counts in detector images at all angles of incidence is shown in the inset of Fig. 6. The skewed diffraction spots in the inset indicate the rather wide wavelength distribution, similar to our model distribution in Fig. 2 (b). In this case, it is mandatory to apply the line-by-line analysis. We employ the same model as in the previous (SANS I) case, using the same values for d and L as obtained from the previous evaluation (Tab. 1). The results obtained for two different horizontal detector lines, i.e., two different mean wavelengths, are shown in Fig. 6.

As expected, the fit to the data becomes worse as the wavelength increases. We attribute this observation to the fact that we did not consider the wavelength spectrum and the beam divergence. A more detailed study on these features is under way.

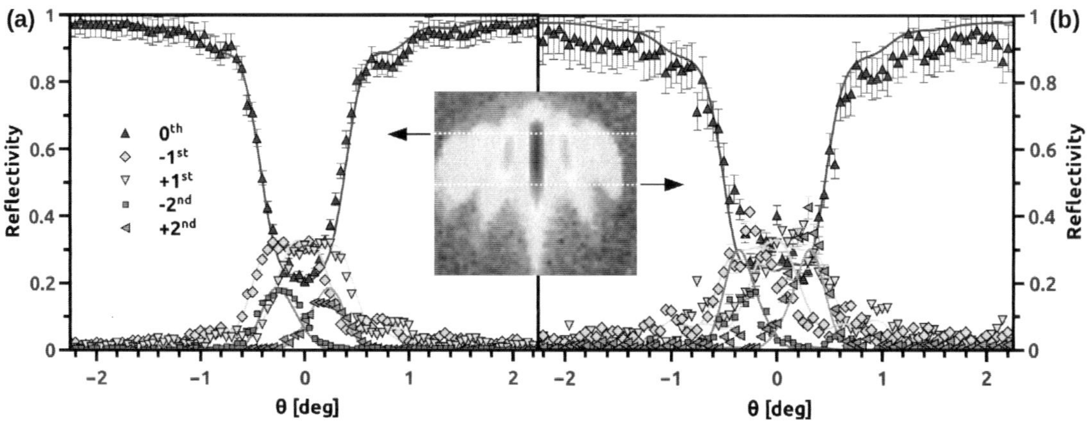

Figure 6. Rocking curves for two different pixel-lines. (a) $\lambda = (3.73 \pm 0.07)$ nm (fit result) and (b) $\lambda = (5.0 \pm 0.1)$ nm (fit result). The inset shows the sum of 2D-detector counts for all measured angles of incidence.

References

[1] Rupp R A, Hehmann J, Matull R and Ibel K 1990 *Phys. Rev. Lett.* **64** 301–302
URL http://link.aps.org/doi/10.1103/PhysRevLett.64.301

[2] Havermeyer F, Lyuksyutov S F, Rupp R A, Eckerlebe H, Staron P and Vollbrandt J 1998 *Phys. Rev. Lett.* **80**(15) 3272–3275
URL http://link.aps.org/doi/10.1103/PhysRevLett.80.3272

[3] Schellhorn U, Rupp R A, Breer S and May R P 1997 *Physica B: Condensed Matter* **234236** 1068 – 1070 proceedings of the First European Conference on Neutron Scattering
URL http://www.sciencedirect.com/science/article/pii/S092145269700015X

[4] Pruner C, Fally M, Rupp R A, May R P and Vollbrandt J 2006 *Nucl. Instrum. Methods A* **560** 598 – 605
URL http://www.sciencedirect.com/science/article/pii/S0168900205026811

[5] Fally M 2002 *Appl. Phys. B* **75** 405
URL http://dx.doi.org/10.1007/s00340-002-1035-0

[6] Fally M, Drevenšek-Olenik I, Ellabban M A, Pranzas K P and Vollbrandt J 2006 *Phys. Rev. Lett.* **97** 167803
URL http://link.aps.org/doi/10.1103/PhysRevLett.97.167803

[7] Fally M, Bichler M, Ellabban M A, Drevenšek-Olenik I, Pruner C, Eckerlebe H and Pranzas K P 2009 *J. Opt. A: Pure Appl. Opt.* **11** 024019
URL http://stacks.iop.org/1464-4258/11/i=2/a=024019

[8] Suzuki N, Tomita Y and Kojima T 2002 *Appl. Phys. Lett.* **81** 4121
URL http://scitation.aip.org/content/aip/journal/apl/81/22/10.1063/1.1525391

[9] Tomita Y, Suzuki N and Chikama K 2005 *Opt. Lett.* **30** 839
URL http://ol.osa.org/abstract.cfm?URI=ol-30-8-839

[10] Chikama K, Mastubara K, Oyama S and Tomita Y 2008 *Journal of Applied Physics* **103**
URL http://scitation.aip.org/content/aip/journal/jap/103/11/10.1063/1.2938849

[11] Tomita Y, Hata E, Momose K, Takayama S, Liu X, Chikama K, Klepp J, Pruner C and Fally M 2016 *Journal of Modern Optics* **63** S1–S31
URL http://dx.doi.org/10.1080/09500340.2016.1143534

[12] Klepp J, Drevenšek-Olenik I, Gyergyek S, Pruner C, Rupp R A and Fally M 2012 *J. Phys.: Conf. Ser.* **340** 012031
URL http://stacks.iop.org/1742-6596/340/i=1/a=012031

[13] Klepp J, Pruner C, Tomita Y, Geltenbort P, Drevenšek-Olenik I, Gyergyek S, Kohlbrecher J and Fally M 2012 *Materials* **5** 2788–2815
URL http://www.mdpi.com/1996-1944/5/12/2788

[14] Fujii R, Guo J, Klepp J, Pruner C, Fally M and Tomita Y 2014 *Opt. Lett.* **39** 3453–3456
URL http://ol.osa.org/abstract.cfm?URI=ol-39-12-3453

[15] Guo J, Fujii R, Ono T, Klepp J, Pruner C, Fally M and Tomita Y 2014 *Opt. Lett.* **39** 6743–6746
URL http://ol.osa.org/abstract.cfm?URI=ol-39-23-6743

[16] Kogelnik H 1969 *Bell System Technical Journal* **48** 2909–2947 ISSN 1538-7305
URL `http://dx.doi.org/10.1002/j.1538-7305.1969.tb01198.x`
[17] Sears V F 1989 *Neutron Optics* (Oxford, UK: Oxford University Press)
[18] Rauch H and Werner S A 2015 *Neutron Interferometry* (Oxford, UK: Oxford University Press)
[19] Uchida N 1973 *J. Opt. Soc. Am.* **63** 280–287
URL `http://www.osapublishing.org/abstract.cfm?URI=josa-63-3-280`
[20] Gaylord T and Moharam M 1982 *Applied Physics B* **28** 1–14 ISSN 0946-2171
URL `http://dx.doi.org/10.1007/BF00693885`
[21] Klepp J, Pruner C, Ellabban M A, Tomita Y, Lemmel H, Rauch H and Fally M 2011 *Nucl. Instrum. Methods A* **634** S59–S62
URL `http://www.sciencedirect.com/science/article/pii/S0168900210015779`
[22] Fally M, Klepp J and Tomita Y 2012 *Appl. Phys. B* **108** 89–96 ISSN 0946-2171
URL `http://dx.doi.org/10.1007/s00340-012-5090-x`
[23] Prijatelj M, Klepp J, Tomita Y and Fally M 2013 *Phys. Rev. A* **87**(6) 063810
URL `http://link.aps.org/doi/10.1103/PhysRevA.87.063810`
[24] Gaylord T K and Moharam M G 1981 *Appl. Opt.* **20** 3271–3273
URL `http://ao.osa.org/abstract.cfm?URI=ao-20-19-3271`
[25] Moharam M G and Gaylord T K 1981 *J. Opt. Soc. Am.* **71** 811–818
URL `http://www.osapublishing.org/abstract.cfm?URI=josa-71-7-811`
[26] Moharam M G, Gaylord T K, Grann E B and Pommet D A 1995 *J. Opt. Soc. Am. A* **12** 1068–1076
URL `http://josaa.osa.org/abstract.cfm?URI=josaa-12-5-1068`
[27] Sippel D, Kleinstueck K and Schulze G 1965 *Physics Letters* **14** 174 – 175 ISSN 0031-9163
URL `http://www.sciencedirect.com/science/article/pii/0031916365905706`
[28] Shull C G 1968 *Phys. Rev. Lett.* **21**(23) 1585–1589
URL `http://link.aps.org/doi/10.1103/PhysRevLett.21.1585`
[29] Shull C G 1973 *Journal of Applied Crystallography* **6** 257–266 ISSN 1600-5767
URL `http://dx.doi.org/10.1107/S0021889873008654`
[30] Somenkov V A, Shilstein S S, Belova N E and Utemisov K 1978 *Solid State Commun.* **25** 593–595
URL `http://www.sciencedirect.com/science/article/pii/0038109878914977`
[31] Klepp J, Pruner C, Tomita Y, Geltenbort P and Fally M 2016 In preperation
[32] Fally M, Klepp J, Tomita Y, Nakamura T, Pruner C, Ellabban M A, Rupp R A, Bichler M, Drevenšek-Olenik I, Kohlbrecher J, Eckerlebe H, Lemmel H and Rauch H 2010 *Phys. Rev. Lett.* **105** 123904
URL `http://link.aps.org/doi/10.1103/PhysRevLett.105.123904`
[33] Klepp J, Pruner C, Tomita Y, Plonka-Spehr C, Geltenbort P, Ivanov S, Manzin G, Andersen K H, Kohlbrecher J, Ellabban M A and Fally M 2011 *Phys. Rev. A* **84** 013621
URL `http://link.aps.org/doi/10.1103/PhysRevA.84.013621`
[34] Klepp J, Pruner C, Tomita Y, Mitsube K, Geltenbort P and Fally M 2012 *Appl. Phys. Lett.* **100** 214104
URL `http://scitation.aip.org/content/aip/journal/apl/100/21/10.1063/1.4720511`
[35] Klepp J, Tomita Y, Pruner C, Kohlbrecher J and Fally M 2012 *Appl. Phys. Lett.* **101** 154104
URL `http://scitation.aip.org/content/aip/journal/apl/101/15/10.1063/1.4758686`
[36] Steyerl A, Nagel H, Schreiber F X, Steinhauser K A, Gähler R, Gläser W, Ageron P, Astruc J, Drexel W, Gervais G and Mampe W 1986 *Phys. Lett. A* **116** 347 – 352
URL `http://www.sciencedirect.com/science/article/pii/0375960186905876`
[37] Neipp C, Sheridan J, Gallego S, Ortuño M, Márquez A, Pascual I and Beléndez A 2004 *Optics Communications* **233** 311 – 322 ISSN 0030-4018
URL `http://www.sciencedirect.com/science/article/pii/S0030401804001075`
[38] Sheridan J 1992 *Journal of Modern Optics* **39** 1709–1718
URL `http://www.tandfonline.com/doi/abs/10.1080/713823578`
[39] Kong J A 1977 *J. Opt. Soc. Am.* **67** 825–829
URL `http://www.osapublishing.org/abstract.cfm?URI=josa-67-6-825`
[40] Yeh P 1993 *Introduction to Photorefractive Nonlinear Optics* Wiley Series in Pure and Applied Optics (New York: Wiley)
[41] Russell P 1981 *Physics Reports* **71** 209 – 312
URL `http://www.sciencedirect.com/science/article/pii/0370157381901964`
[42] Suzuki N and Tomita Y 2004 *Appl. Opt.* **43** 2125
URL `http://ao.osa.org/abstract.cfm?URI=ao-43-10-2125`
[43] Suzuki N and Tomita Y 2007 *Appl. Opt.* **46** 6809–6814
URL `http://ao.osa.org/abstract.cfm?URI=ao-46-27-6809`

Neutron Scattering Collimation Wheel Instrument for Imaging Research

E Van Every[1] and A Deyhim[2]
[1]Director of Operations, *Advanced Design Consulting USA, Inc.*, 126 Ridge Road, Lansing, NY 14882, USA
[2]President, *Advanced Design Consulting USA, Inc.*, 126 Ridge Road, Lansing, NY 14882, USA

E-mail: adc@adc9001.com

Abstract. The design of a state-of-the-art selector wheel instrument to support the area of neutron imaging research (neutron radiography/ tomography) is discussed. The selector wheel is installed on the DINGO Radiography instrument at the Bragg Institute HB2 beamline at ANSTO in Sidney Aus. The selector wheel consists of a single axis drum filled with a wax/steel shielding mixture and six square cutouts for neutron optics and a larger solid shielding sector to act as a shutter. This paper focuses on the details of design and shielding of the selector wheel.

1. Introduction

The DINGO Radiography instrument is the first at the Bragg Institute to use the HB2 thermal neutron beam with an estimated flux of up to 4.7 x 107 [n/cm2s] at the sample [2]. The main components of the DINGO instrument consist of a primary collimator, collimation wheel, Tertiary & Fast Shutter, sample stage; detector camera and beam stop (Figure 2). The collimation wheel is a single axis drum filled with a wax/steel shielding mixture and six square cutouts for neutron optics and a larger solid shielding sector to act as a shutter.

The DINGO instrument covers a large area of scientific research such as medical applications, biology and environmental science, geology and engineering science as well industrial application, which are key areas for future technology and industrial developments in Australia. The selector wheel shutter unit, shown in figure 1 below, combines two different functions in one item. The first function is to work as an instrument shutter and the second one is to work as a selector wheel. The selector wheel assembly component provides selective aperture options for high resolution or phase contrast imaging and separates the two beams coming from the in-pile collimator. One beam is blocked and the other passes the selector wheel insert and is used for an imaging experiment. A positioning accuracy of 0.01° step width is achieved. The selector wheel was designed with a stepped housing to mate with the inserts and prevent direct shine from the beam.

The main component of the selector wheel is a 2.5 ton aperture wheel filled with a shielding mixture. It provides six openings, which are equipped with various neutron beam optics plus a solid 'shutter' section to block the beam. Utilizing the solid section as a shutter requires complex interaction between the safety [1] interlock and motion control systems. A standardized Galil based motion system controls the movement of the wheel while a Pilz safety PLC specifies the desired position and handles other safety aspects of the instrument. A shielded absolute SSI encoder is employed to give high accuracy feedback on the position in conjunction with a number or limit switches.

Figure 1. ANSTO Selector Wheel Shutter Unit

The selector wheel is located in the DINGO beamline just after the neutron guides but before the window slits and collimator leading up to the sample stage. Please see figure 2 below showing the beamline layout as well as a photo of the actual sample stage area.

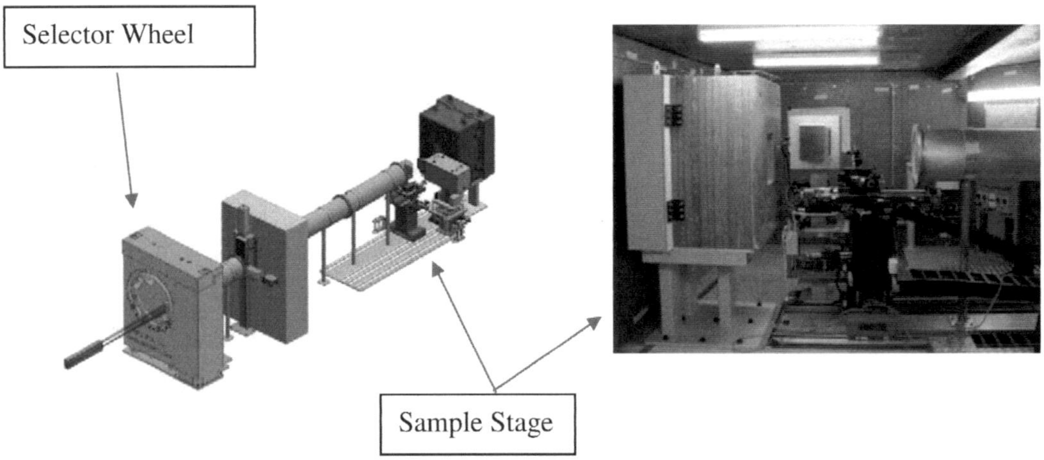

Figure 2. DINGO instrument layout

2. Shutter Wheel Design Parameters

The design criteria for shutter wheel instrument includes: an indexing wheel with six insert locations; inserts that are adjustable in all 6 degrees of freedom; a wheel that rotates at a speed of 6º per sec; end of travel limit switches and safety switch; neutron shielding; and finally no direct neutron path was allowed. Major components of the selector wheel include; slew bearing driving the rotational motion; radial bearing constraining the upstream end of the indexing wheel; precision machining on large diameter components; stepped chambers filled with neutron shielding to eliminate direct neutron paths; inserts that were designed to adjust in all 6 degrees of freedom and can be flipped. The engineering, FEA modeling, design, and detail of the system from concept to final design took 5 months.

The range of rotation was 360 degrees to allow access to all apertures. The radial resolution of position was 0.1^{O} with repeatability of 0.2^{O} and absolute accuracy of 0.01^{O}. These specifications were chosen to ensure the beam passed consistently through the center of the optics. The drive motor is located in radiation and must be radiation hard. The rotation speed is 6°/min. All structural components are non-magnetic. These parameters are summarized in Table 1 below.

Table 1. Collimation Wheel Instrument Design Requirements

Range	360°
Resolution	0.1°
Repeatability	0.2°
Absolute Accuracy	0.01°
Travel Speed	6°/min
Radiation Hardness	Yes
Vacuum Rating	not required
Non-Magnetic Material	Yes

A heavy duty rotation bearing from a German manufactured supplier was used. Upon arrival at ADC, inspection showed errors in manufacturing of the bearing. The bearing was returned to the manufacturer for corrective action. A radiation hard motor was used but also failed initial QA testing and was returned.. Each of these issues was properly document as part of ADC ISO 9001 quality control and corrective actions were taken immediately to meet project requirements.

3. Neutron Shielding Mix Prototyping

A prototype of the neutron shielding for the shutter wheel assembly was performed at ADC. ANSTO had provided requirements for the shielding mix including materials and density Specifications include: mild steel shot, boron carbide, and refined paraffin. ADC was required to develop a process to implement the shielding mix.

ADC first constructed a steel chamber with a right triangular cross section, 197 mm by 203 mm, and with a height of 404 mm, for a volume of $8.08x10^6$ mm^3, or approximately 8 liters. The shape was chosen to test sharp angles due to the angles among braces inside the main housing assembly, and the depth is comparable to the internal depth of the main housing piece, at 550 mm. The structure was welded together and all joints were sealed with a silicon based caulk. Aluminum plates where attached to the chamber with cable ties, with approximately 80 mm extending above the top of the steel chamber to act as a reservoir for molten wax. The mold was filled with the steel shot, boron carbide, and paraffin wax. Figure 3 below shows the prototype vessel.

Figure 3. Triangular cross section steel test chamber with aluminum plate reservoir and heating tape.

A thermocouple was inserted as deep into the mix as possible. The steel mold was wrapped in heating tape and then covered with aluminum foil in half width strips, covering the top third, middle third, then lower third and base. The mold was wrapped in foam insulation as shown in figure 4 below, and then slowly heated.

Figure 4. The steel mold filled with proper mixtures while heating with tape heaters

The internal temperature was recorded, close to the center of the steel and boron carbide mixture, and on the outside of the aluminum foil sheathing. After melting, the temperature of the paraffin was monitored, to ensure that it reached the required specification. The lower portion was removed of the aluminum foil, exposing the bottom third of the mold to air. The mold was supported on three feet allowing air to pass under the chamber to facilitate bottom up cooling and solidification.

Results

The top couple centimeters of the shielding mix are not as hard or regular as the rest, and were not fully wetted by the wax. This is a result of a leak which did not permit us to have an adequate reservoir of molten wax. The surface of the shielding block looked smooth over half its area. In figure 5 below, one can clearly see shot which has stuck to the walls of the mold when it was split open. The entire surface was solid, although grit could be chiseled or cut away. This bock had a volume of approximately 7780 cm^3, and a mass of 40.5 kg for a density of 5.2 g/cm^3. The shielding density specification was greater than 5.0 g/cm^3, thereby exceeding requirements.

The block of shielding material was removed from the mold and then split open, approximately on one third and two thirds up, along what was the horizontal axis while pouring. Figure 6 below shows sections of shielding material after removal from the block.

The final result was quite satisfactory, other than near the pouring surface which was partially depleted of wax. When a portion of the block was exposed it was solid, *i.e.* all the shot were well held in the matrix and only became loose directly from cutting or chiseling. The block was both solid and homogeneous.

Figure 5. The mold with the finished shielding block, opened for testing

Figure 6. Faces of the shielding block after being broken open

4. Factory Acceptance Testing

A factory acceptance test (FAT) document describing the tests and methods was developed by ADC to the satisfaction of ANSTO. At the FAT, the selector wheel was verified to meet all specifications. Open loop and closed loop positioning tests were performed and the wheel was rotated more times than was expected in its lifetime. Rotational position accuracy, resolution, and repeatability, were verified using an electronic tilt meter mounted to the face of the wheel which compared the actual position with the measured position. The results of the FAT are listed below in Table 2. Please note that each of the items in this list were verified with independent instruments and were found to exceed specifications.

Other than density measurements, the effectiveness of the shielding can only be tested in a neutron beam. For this reason, we relied on ANSTO's experience regarding the density of the mix for adequacy of neutron shielding.

Table 2. Summary of FAT

No	Areas of testing	FAT Conclusion
1	Dimensional requirements	Comply*
2	Accuracy	Comply*
3	Repeatability	Comply*
4	Adjustable inserts	Comply*
5	Multiple revolutions	Comply*
6	Total time for one complete revolution at various speeds	Comply*
7	Motion Control Cam and limit switches activated	Comply*
8	All cams have sufficient adjustability	Comply*
9	All limit switches have sufficient adjustability	Comply*
	Safety home cam activates safety limit switch	Comply*
	Backlash is within a requirement range	Comply*
	Stopping time at various speeds	Comply*
	Acceleration/deceleration times sufficient such that the cams do not overshoot limit switches	Comply*
	Check each position	Comply*
	The wheel is able to be moved from a position where the motion control cam is activating motion control limit switch 1 through 355 degrees where it contacts motion control limit switch 2.	Comply*
	Provisions for cabling	Comply*
	Use a probe for repeatability so something comes down and touches the probe, goes away and then comes back.	Comply*
	Does the wheel lock?	Comply*

*During the Factory Acceptance Test (FAT) extensive test were conducted and the results showed the Neutron Scattering Collimation Wheel Instrument was in compliant or exceeded the design requirements.

Additional data supporting the FAT results is provided in figure 7 below. This is an actual acceptance sheet required by ANSTO.

Item	Type of Control	Design Parameter	Range of provided motion (This is the range of this motion that will be provided in the adjustment system) - i.e. Adjustability	Absolute Accuracy (This is the level to which you will be able to repeatadly get to the same spot as last time)	Importance Level
1	Selector Wheel & Base plate alignment	Height	± 1mm		B
		Length	± 1mm		B
		Width	± 1mm		B
		Tilt (Roll)	± 1º	± 0.1º	B
		Tilt (Pitch)	± 1º	± 0.1º	B
		Yaw	± 1º	± 0.1º	B
2	Selector Wheel Housing Proximity Location with existing Walls.	Clearance between existing infrastructure: Range	± 10mm		A
3	Selector Wheel	Range	360º		A
		Resolution	0.1º		B
		Precision	0.01º		B
		Repeatability	0.2º		B
		Travel Speed	6º / min.		C
4	Shutter Inserts (*) *Refer to Figure Below:*	X, Y and Z	± 2mm	± 0.2mm	B
		Tilt (Roll)	± 1º	± 0.1º	B
		Tilt (Pitch)	± 1º	± 0.1º	B
		Yaw	± 1º	± 0.1º	B
5	Distance Between Reactor Face and Inserts	Length	<50mm (close as possible)		A

* the absolute accuracy required for the shutter inserts with respect to the position of the neutron beam may be found using a fixed insert solution.

Definitions:

Resolution - *This is the minimum change on the axes that can be detected on the encoder*

Precision - *This is the minimum change that you will receive when one step is sent to the motor*

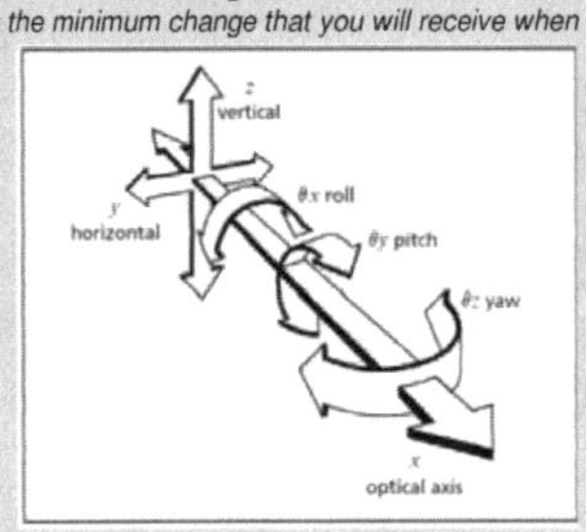

Figure 7.Selector Wheel Housing Unit Accuracies

5. Conclusions

The DINGO instrument was successfully commissioned during 2013 and saw its first neutron radiography image during August of that year. The ADC selector wheel has been shown to be a robust and reliable component of this beamline.

Journal of Physics: Conference Series **746** (2016) 012023 doi:10.1088/1742-6596/746/1/012023

5. References

[1] The Interaction Between Safety Interlock And Motion Control Systems On The Dingo Radiography Instrument At The Opal Research Reactor; Proceedings of ICALEPCS2013, San Francisco, CA, USA

[2] The Neutron Beam Expansion Program at the Bragg Institute; International Workshop on Neutron Optics and Detectors (NOP&D 2013) Journal of Physics: Conference Series 528 (2014) 012026

[3] E. Van Every[1] Director of Operations, *Advanced Design Consulting USA, Inc.*, 126 Ridge Road, Lansing, NY 14882, USA

[4] A. Deyhim[2] President, *Advanced Design Consulting USA, Inc.*, 126 Ridge Road, Lansing, NY 14882, USA E-mail: adc@adc9001.com

Journal of Physics: Conference Series **746** (2016) 012024 doi:10.1088/1742-6596/746/1/012024

Neutron Optics: Towards Applications for Hot Neutrons

C Schanzer, M Schneider, and P Böni

SwissNeutronics AG, Brühlstrasse 28, CH-5313 Klingnau, Switzerland

E-mail: christian.schanzer@swissneutronics.ch

Abstract. Supermirrors with large critical angles of reflection, i.e. large index m are an essential ingredient to transport, focus and polarise neutrons over a wide range of energy. Here we summarise the recent developments of supermirror with very large critical angles of reflection and high reflectivity that were conducted at SwissNeutronics as well as their implementation in devices. Approaching critical angles $m = 8$ times the critical angle of natural nickel makes new applications possible and extends the use of reflection optics towards the regime of hot and epithermal neutrons. Based on comparisons of simulations with experiment we demonstrate future possibilities of applications of large-m supermirrors towards devices for neutrons with short wavelength.

1. Introduction

The introduction of the concept of supermirrors in 1967 [1] and its realisation [2,3] laid the foundation to increase the efficiency for the transport of neutrons by internal reflection in guides and the manufacturing of polarising mirrors by accepting reflection angles larger than the total reflection angle of Ni. In 1989, Hayter and Mook published a formalism to compute an adequate layer sequence for supermirror [4]. Using this algorithm, the supermirror technology was applied exclusively for the first time for the complete guide system at the Swiss Spallation Source SINQ at the Paul Scherrer Institute (Switzerland) using mirrors with $m = 2$ [5]. In passing we note that the critical angle for supermirror is given by $\theta_c = 0.099m\lambda$. Here the maximum angle of reflection θ_c is given in degrees and the wavelength λ in Å. The following years witnessed enormous developments for neutron guides combining new guide concepts, e.g. ballistic guides [6] and non-linearly tapered parabolic and elliptic guides [7] with supermirrors whose performance has been improved continuously [8]. Moreover supermirrors with high reflectivity on metallic substrates [9] have enabled new applications namely extending neutron guides close to the moderator and for beam focusing based on Montel optics thus improving the performance of beam lines for neutron scattering and imaging further. As a result of the continuous developments of the deposition techniques for supermirrors and the super-polishing of substrates on large areas, non-magnetic and polarising supermirrors are available with large m-values, excellent reflectivity and high polarization. Here we report about the state-of-the-art of these supermirrors and their possible use towards applications for hot neutrons.

2. Performance of non-magnetic supermirrors

2.1. Preparation of supermirrors

Atomically smooth substrates and layers are required to obtain a high reflectivity of neutron supermirrors. Dedicated processes are developed for the super-polishing and coating of glass, silicon, and metallic substrates on large areas of the order of 350 × 500 mm². We probe the substrate surface by atomic force microscopy (AFM) using a desktop instrument easyScan 2 AFM from Nanosurf. Figure 1 shows the AFM image of a typical surface scanned on an area of 10 μm × 10 μm. The RMS roughness is $\sigma < 1$ Å. This quality is regularly obtained on glass, silicon and metallic substrates, and is required for the excellent reflectivities presented below.

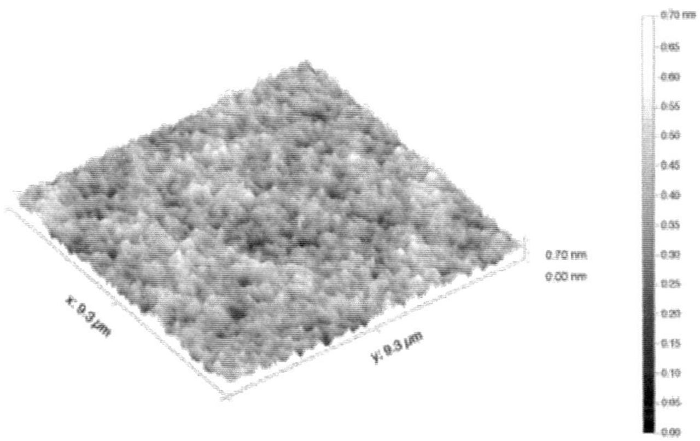

Figure 1. Surface image of a super-polished substrate measured with AFM. The RMS roughness over the scan area of 10 × 10 μm² is $\sigma < 1$ Å.

The supermirrors are coated using DC magnetron sputtering. The process is highly sophisticated in order to obtain layers with extremely low roughness and low internal stress. The thicknesses of the single layers are computed using the formalism by Hayter and Mook [4]. Non-magnetic supermirrors comprise a combination of materials of nickel-molybdenum alloy and titanium.

2.2. Reflectivity of supermirrors

The reflectivity of all supermirrors is measured at the reflectometer NARZISS at SINQ (PSI), which is operated in an angular dispersive mode at a wavelength of $\lambda = 5$ Å [10]. Figure 2 shows a compilation of measurements of supermirrors with $2 \leq m \leq 8$. The slope of the reflectivity is similar for all *m*-values, e.g. supermirrors with $m = 2$ and $m = 8$ have essentially the same reflectivity at $m = 2$. The observation of an almost constant slope is remarkable because when fabricating a supermirror with $m = 8$, more than 16500 thin layers for $2 < m \leq 8$ are deposited first before the layers for the reflection at $m = 2$ are coated. The high reflectivity at $m = 2$ demonstrates that the growth mode of the layers does not change, i.e. the interface roughness remains essentially constant. As a result, supermirrors with large *m*-values do not compromise the performance of the supermirror for intermediate and small *m*-values. Therefore, the increase of the *m*-values leads to a true gain of intensity due to the increased divergence of the neutron beam.

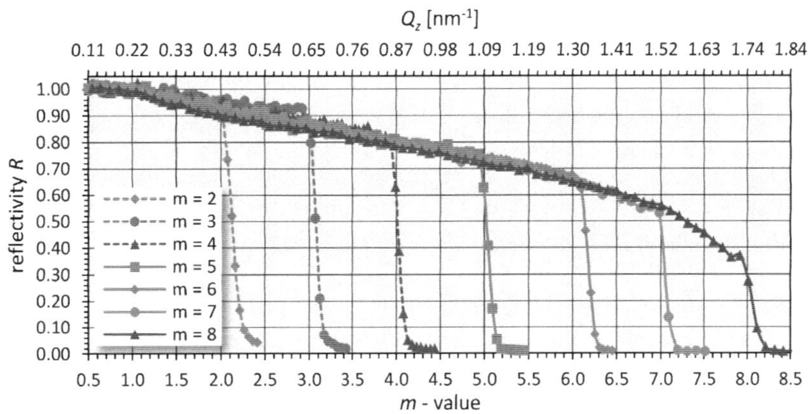

Figure 2. Reflectivity profiles of supermirrors with *m*-values $2 \leq m \leq 8$. The slope of *R* is largely independent of the *m*-value indicating a stable interface roughness, which is essentially constant for all layers.

The supermirror with $m = 8$ shows a pronounced decrease of the reflectivity in the range $7 \leq m \leq 8$. For a proper understanding we have conducted computations of the reflectivity. The good agreement between the model calculation of the supermirror sequence with $m = 8$ and the reflectivity data (Figure 3) indicates that the physics of the deposition process is well understood. As we do not want to disclose the details of the coating process, we simply summarise the essential features of the model as follows: The reflectivity was calculated using the matrix formalism from [11]. The real and imaginary (absorption, incoherent scattering) parts of the scattering length densities for Ni and Ti were taken from the literature [12]. The roughness at the interfaces was assumed to attain the value as measured at the surface of the substrate by x-ray reflectivity and was taken into account in the simulations by a static Debye-Waller factor. In addition, the non-constant yield of the sputter targets due to the consumption of the materials was taken into account.

The results in Figure 3 show that absorption (compare black-dotted and red lines) causes a significant reduction of the reflectivity of about 10% at large m-values. The deviation of the sputter yield of the target from its nominal value during the very long sputter runs leads to significant deviations of the thickness of the layers from the sequence calculated by Hayter and Mook [4]. The change in deposition rate is clearly visible at very large m leading to the observed drop in reflectivity. In future runs this effect can be compensated by adjusting the parameters of the sputter process. Note also the small dip in reflectivity around $m = 7.9$. It is caused by an incomplete single layer due to a software error during the run. The dip is reproduced in the simulation by eliminating layer #15701 from the layer sequence.

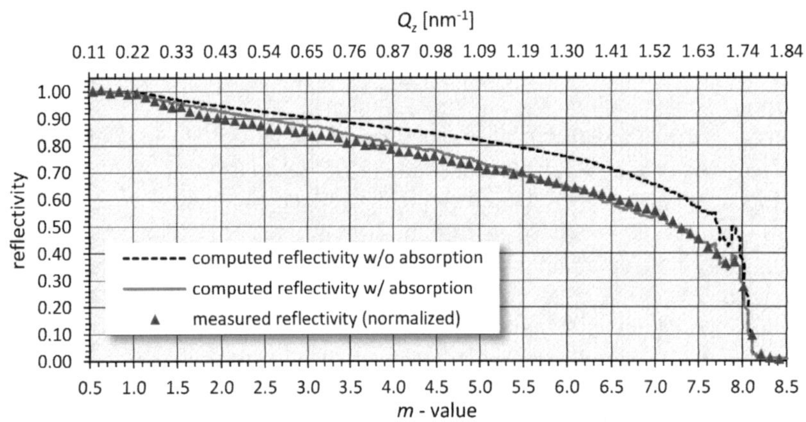

Figure 3. Measured (triangles) and computed reflectivity of a supermirror with $m = 8$. The red and black-dotted lines show the reflectivity with and without taking absorption by the Ni and Ti layers into account. Roughness, non-constant target yield and the missing layer #15701 have been taken into account (see text).

2.3. Long-term stability of supermirror

As shown above, the performance of present-day supermirror [13] is excellent up to $m = 8$. This raises the question about the long-term stability of the supermirrors. The comparison of the reflectivity of the very first supermirror with $m = 6$ as produced by SwissNeutronics in 2006 with measurements conducted in the following years as shown in Figure 4 demonstrates the perfect stability of the mirror sequence. Obviously, the measured reflectivity does not change. Visual inspections show neither a degradation nor a delamination of the supermirror.

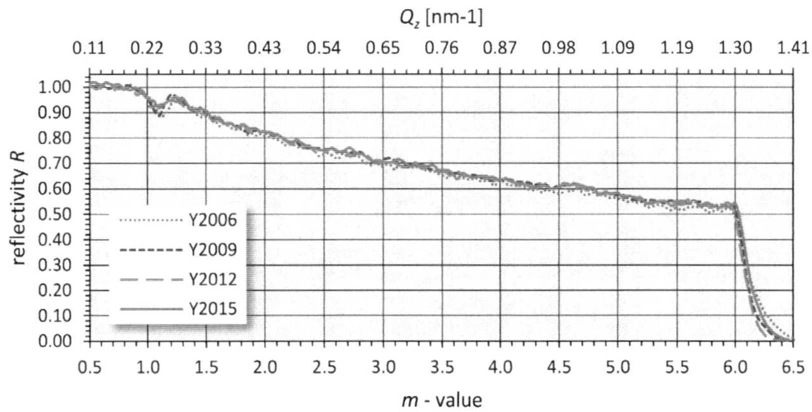

Figure 4. Reflectivity of an $m = 6$ supermirror measured directly after production in 2006 and the following years. No aging is observed. The dip near $m = 1$ is due to a missing of the last Ti layer.

2.4. Applications of supermirror with large m-value

Supermirrors with large m allow reflecting neutrons at large momentum transfers Q, i.e. at large angles of reflection even for beams with short wavelength. The large-m supermirrors are typically required for the focusing of neutrons and for their efficient transport by means of non-linearly tapered guides. Using various optical concepts such as parabolic or elliptic focusing guides or even Montel geometries [14,15], significant gains in intensity can be realised because the brilliance transfer [16,17] can be improved. Indeed, the installation of an elliptic neutron guide for the powder diffractometer HRPD at ISIS has witnessed an intensity gain of up to two orders of magnitude [18] when compared with the previous conventional guide system. Increasing the m-value of its coating from $m = 3$ to $m = 8$ would extend the wavelength range towards the hot regime, i.e. $\lambda \cong 0.3$ Å.

In addition, the use of focusing mirrors allows concentrating neutron beams at the sample position [19]. Therefore, small samples can be investigated even under extreme conditions because the scattering from the sample environment can be strongly reduced leading to an improved signal-to-noise ratio. Various experiments with cold neutrons demonstrate the benefit of focusing optics [20,21,22]. Supermirrors with large m will be essential for an efficient operation of the planned beam lines at the new European Spallation Source ESS as they allow an efficient extraction of the neutrons from the moderator and the possibility to focus the neutrons while maintaining a large brilliance transfer [17]. In all applications the large m-values enable also the handling of neutrons with higher energies, e.g. hot and even epithermal neutrons.

3. Performance of polarising Fe/Si supermirrors

3.1. Reflectivity, transmission and polarisation of magnetised Fe/Si supermirrors

Polarising Fe/Si supermirrors are prepared using the same sputtering facility as for the non-magnetic supermirrors (section 2.1). Spin-dependent neutron reflectivity and transmission measurements are performed at the reflectometer NARZISS at SINQ (PSI). Figure 5 shows the spin-dependent reflectivity R and the polarisation P of a polarising supermirror Fe/Si with $m = 5.5$ coated on borosilicate glass with a thickness $t = 0.3$ mm. The reflectivity for the spin-up state is $R > 70\%$ at the critical edge and $P \cong 1$ across the entire supermirror regime. Spin-down neutrons are reflected below $m < \approx 0.7$ because of the mismatch between the nuclear and magnetic scattering length density of Fe.

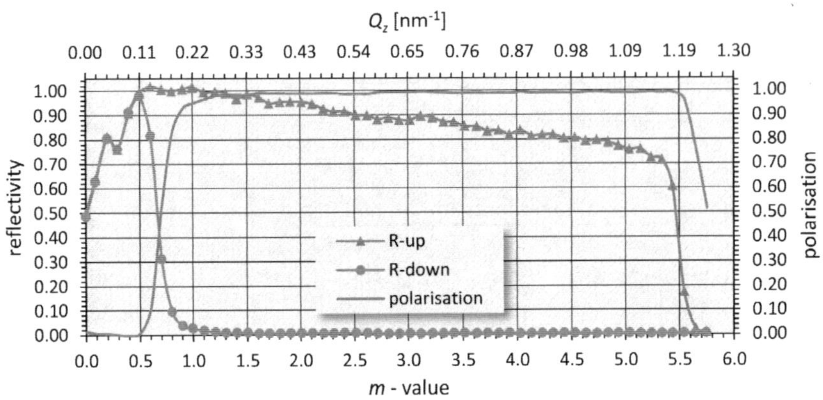

Figure 5. Spin-dependent reflectivity R and polarisation P of a supermirror Fe/Si with m = 5.5. R > 70% at the critical edge and $P \cong 1$ for the entire regime of the supermirror. For m < 0.5 the sample is over-illuminated.

Figure 6 shows the spin-dependent transmission and the polarisation from a Si wafer that is coated on both sides with polarising supermirror Fe/Si with m = 5. The thickness of the Si wafer is 0.3 mm. Below $m \approx 0.5$ the effects of over- illumination at grazing incidence are visible. For m < 0.7 both spin-states are totally reflected leading to P = 0. Within the supermirror regime the transmission for the spin-down neutrons increases while it is essentially suppressed for the spin-up neutrons. The polarisation is close to unity for 1 < m < 2.5 and decreases to 0.91 at m = 5. The decrease of P is a result of the finite reflectivity of the supermirror, which leads to the slight increase of the transmission for the spin-up neutrons near m = 5. The average polarisation within the supermirror regime is $P \cong$ 0.96.

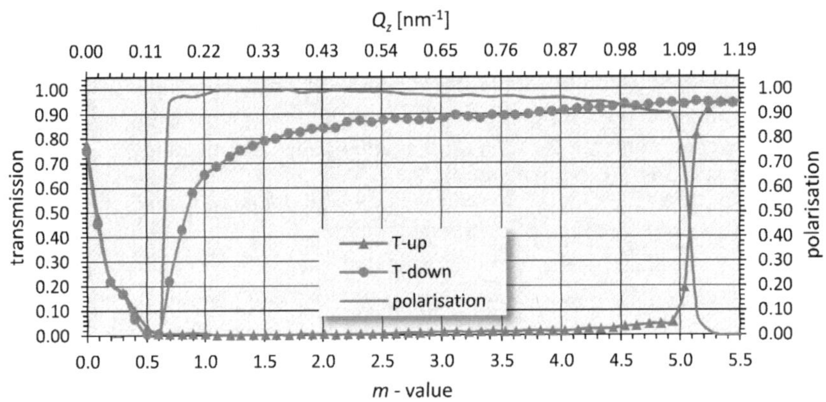

Figure 6. Spin-dependent transmission and polarisation of a polarising supermirror Fe/Si with m = 5 coated on both sides of a Si wafer with a thickness t = 0.3 mm. The average polarisation is $P \cong 0.96$.

3.2. Applications of polarising supermirror with large m-value

Polarising supermirrors are an essential ingredient for a large number of polarising devices such as V-cavities, single and S-shape curved benders, reflectors, and wide angle analysers. In particular V-cavities became a very convenient concept for polarising neutron beams. V-cavities can be arranged parallel and/or in series to shorten the device and/or increase the polarisation, respectively [23]. Figure 7 shows spin-dependent neutron transmission measurements of a V-cavity comprising 11 channels.

Each channel is equipped with 2 Vs in series coated with polarizing supermirror $m = 5$ allowing to polarise neutrons with a lower critical wavelength $\lambda_c = 2.1$ Å. The transmission is measured as a function of the lateral position across the entrance of the cavity ($W = 74$ mm), i.e. the cavity is translated across a well-collimated neutron beam with a wavelength of $\lambda = 5$ Å. Pronounced features of the dividing walls and the overlaps of the Si wafers at the tips of the Vs are observed because the beam is blocked there. The average transmission is 26% (100 unpolarised neutrons in – 26 polarised neutrons out). The average polarisation is $P = 0.99$. To simulate the performance of the V-cavity we have written a dedicated component for the Monte-Carlo simulation tool McStas [24,25] that takes into account various features of the cavity such as the measured spin-dependent reflectivity of the double-sided coated Si wafers, absorption by the coating and the Si wafers, and the precise arrangement of the Si wafers namely the overlap at the V-tip and the intrusions at the sides of the channels. Figure 8 shows the transmission T of the spin-down neutrons. All features are precisely reproduced. The simulated value $T = 29\%$ is slightly higher than measured. The reason for the difference is because the model does not yet consider refraction of the neutron beam, which increases the path length of the neutrons in the material and thus the absorption. An analytic estimation of the effect of refraction increases the absorption by approximately 2.5%. The excellent agreement between measurement and simulation proves that polarising devices can be easily designed for any anticipated application.

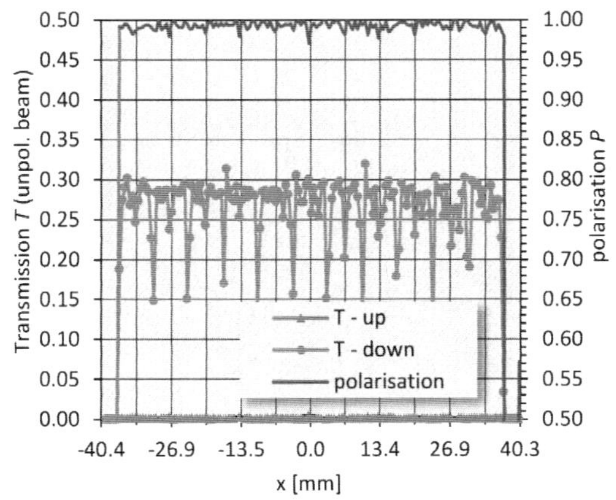

Figure 7. Spin-dependent transmission and polarisation of an 11-channel double-V-cavity as measured across the width of the cavity.

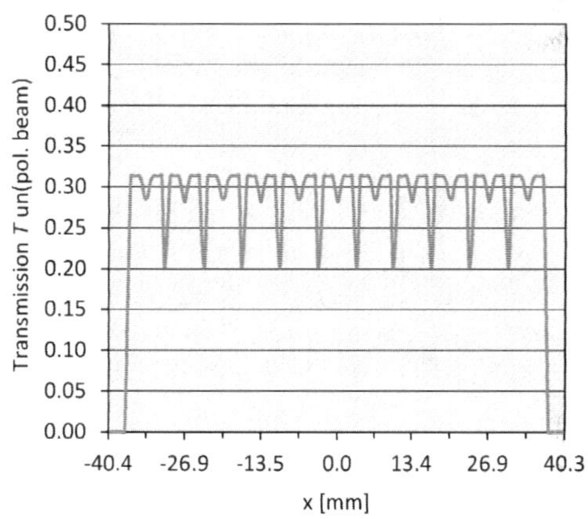

Figure 8. Simulated transmission of an 11-channel double-V-cavity as function of the lateral position across the width of the cavity.

The presently available polarising mirrors with $m = 5.5$ lead to large critical angles of reflection even for neutrons with short wavelengths. These mirrors are ideally suited for analysing the polarisation of thermal neutrons over large angular ranges. Recently, such polarising analysers were realised for the time-of-flight spectrometers HYSPEC at SNS and POLANO at J-PARC using remanent supermirror FeCoV/TiN_x [26] and Fe/Si, respectively. They allow analysing neutrons with energies as large as approximately 30 meV. The optimisation was performed using McStas.

4. Conclusions

We have shown that with modern sputtering technologies it is possible to control roughness and interdiffusion of multilayers on an atomic scale for heterostructures with the number of layers exceeding $N = 16000$. The thinnest layers for supermirror with $m = 8$ have a thickness $t \cong 18$ Å which is comparable with the lattice constant of novel materials in condensed matter physics. Moreover, the mechanical properties, i.e. the stress can be essentially reduced to zero thus preventing the coatings from delamination. Last but not least the coatings are radiation resistant. As the critical angle of reflection θ_c for Ni/Ti supermirror is directly proportional to m and the number of required layers is approximately given by $N = 4m^4$ [27], N increases dramatically with increasing θ_c. Our model calculations indicate that it is the absorption that essentially limits the number of layers. As the physical properties of Ni and Ti are given, a reasonable upper limit may be around $m = 10$. Clearly, it depends on the specified reflectivity to be obtained. Using a combination of ^{58}Ni and ^{62}Ni [1] will lead to a significant decrease of N due to the very large scattering contrast between the two isotopes thus allowing the production of supermirror with m exceeding 10. The upper limit of m for polarising Fe/Si supermirror may also be significantly increased because monocrystalline Fe develops the full moment already for a layer thickness $t < 10$ Å. It has to be evaluated at what minimal t the Fe-layers develop monocrystalline grains in a sputtering process. Anyway, the limit will be above $m = 6$.

Obviously, the availability of supermirror $m \cong 8$ facilitates the transport and focusing of hot neutrons because reasonably divergent beams can be handled. For example, for $\lambda \cong 0.3$ Å (900 meV) neutrons with a divergence $\alpha \cong 0.5°$ can be processed. Using non-linear tapering, α can be increased further. It is indeed the availability of large-m supermirror in combination with elliptic geometries [7] and metallic substrates [9] that will make the efficient use of hot and epithermal neutrons at spallation sources feasible.

References

[1] Turchin V F 1967 *Atomic Energy* **22** (Issue 2) 124
[2] Mezei F 1976 *Commun. Phys.* **1** 81
[3] Mezei F and Dagliesh P A 1977 *Commun. Phys.* **2** 41
[4] Hayter J B and Mook H A 1989 *J. Appl. Cryst.* **22** 35
[5] Wagner W, Bauer G S, Duppich J, Janssen S, Lehmann E, Luthy M and Spitzer H 1998 *J. Neutron Res.* **6** 249
[6] Mezei F and Russina B 2000 *Physica B* **283** 318
[7] Schanzer C, Böni P, Filges U and Hils T 2004 *Nucl. Instr. and Meth. A* **529** 63
[8] Hino M, Sunohara H, Yoshimura Y, Maruyama R, Tasaki S, Yoshino H and Kawabata Y 2004 *Nucl. Instr. and Meth. A* **529** 54
[9] Schanzer C, Böni P and Schneider M 2010 *J. Phys.: Conf. Series* **251** 012082
[10] http://www.swissneutronics.ch/index.php?id=21
[11] Yamada S, Ebisawa T, Achiwa N, Akiyoshi T and Okamoto S 1978 *Annu. Rep. Res. Reactor Inst. Kyoto Univ.* **11** 8
[12] Sears V F 1992 *Neutron News* **23** 26
[13] http://www.swissneutronics.ch/index.php?id=24
[14] Montel M 1957 *X-Ray Microscopy with Catamegonic Roof Mirrors, in X-ray Microscopy and Microradiography* (Academic Press, New York) pp 177-185
[15] Stahn J, Panzner T, Filges U, Marcelot C and Böni P 2011 *Nucl. Instr. and Meth. A* **634** 12
[16] Böni P 2014 *J. Phys.: Conf. Series* **502** 012047
[17] Klenø K H , Lieutenant K, Andersen K H and Lefmann K 2012 *Nucl. Instr. and Meth. A* **696** 75
[18] Ibberson R M 2009 *Nucl. Instr. and Meth. A* **600** 47
[19] Hils T, Böni P and Stahn J 2004 *Physica B* **350** 166
[20] Böni P, Roessli B and Hradil K 2011 *J. Phys.: Condens. Matter* **23** 254209
[21] Adams T, Brandl G, Chacon A, Wagner J N, Rahn M, Mühlbauer S, Georgii R, Pfleiderer C and Böni P 2014 *Appl. Phys. Lett.* **105** 123505
[22] Brandl G, Georgii R, Dunsiger S R, Tsurkan V, Loidl V, Adams T, Pfleiderer C and Böni P 2015 *Appl. Phys. Lett.* **107** 253505
[23] Böni P, Münzer W and Ostermann A 2009 *Physica B* **404** 2620
[24] Nielsen K and Lefmann K 2000 *Physica B* **283** 426
[25] http://neutron.risoe.dk/mcstas
[26] Semadeni F, Roessli B and Böni P 2001 *Physica B* **297** 152
[27] Böni P 1997 *Physica B* **234-236** 1038

New Very Small Angle Neutron Scattering (VSANS) Instrument

E Van Every[1] A Deyhim[2] and J Kulesza[3]
[1] Director of Operations, *Advanced Design Consulting USA, Inc.*, 126 Ridge Road, Lansing NY 14882, USA
[2] President, *Advanced Design Consulting USA, Inc.*, 126 Ridge Road, Lansing NY 14882, USA
[3] Instrumentation and Control, *Advanced Design Consulting USA, Inc.*, 126 Ridge Road, Lansing NY 14882, USA

E-mail: adc@adc9001.com

Abstract. The design of a new Very Small Angle Neutron Scattering (VSANS) Instrument for use in National Institute of Standards And Technology (NIST) will be discussed. This instrument is similar to a shorter instrument we designed and delivered to ANSTO in Australia called the Bilby SANS instrument. The NIST VSANS and the ANSTO Bilby SANS instruments have very similar dimensions for length and diameter and have similar requirements for internal detector motion, top access port, walkway supports, and ports; however, the Bilby SANS instrument vacuum requirement was lower (7.5x10-5 Torr) and the entire (60,000 pound) vessel was required to move 1.5 meters on external rails with a repeatability of 100 um, which ADC achieved. The NIST VSANS length is 24 meter, internal diameter 2.3 meter with three internal carriages.

The NIST VSANS instrument, which covers the usual SANS range will also allow configuration to cover the range between $q \approx 10^{-4}$ Å^{-1} to 10^{-3} Å^{-1} with a sample beam current of (10^4 neutrons/s). The key requirements are a second position-sensitive detector system having a 1 mm pixel size and a longer sample-detector flight path of 20 m (i.e., a 40 m instrument).

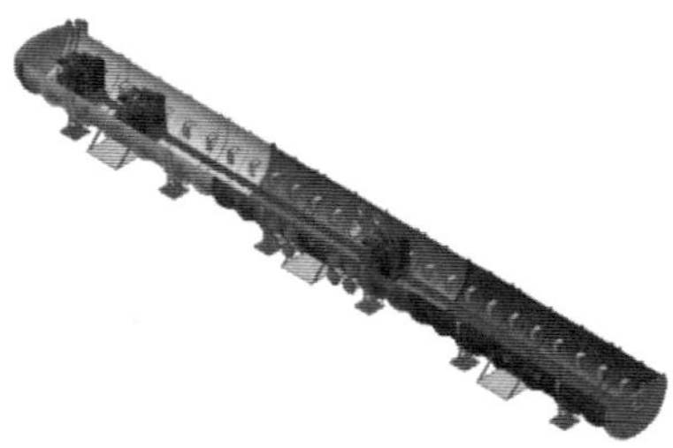

Figure 1. NIST VSANS

Figure 2. A Pair of Dector Units

1. Introduction

The VSANS instrument, which will cover the usual SANS range, will also allow configuration to cover the range between $q \approx 10^{-4}$ Å^{-1} to 10^{-3} Å^{-1} with a sample beam current of (10^4 neutrons/s). The key requirements are a second position-sensitive detector system having a 1 mm pixel size and a longer sample-detector flight path of 20 m (i.e., a 40 m instrument).

The NIST NCNR VSANS vessel consists of a steel vacuum vessel capable of 1 x 10^{-3} torr. The vessel length is 24000 mm and is built in 3 sections roughly 8000 mm each. The vessel material is 516 steel. Each section is supported on 4 adjustable feet that rest on the floor. Each section must have external ribs designed to withstand the compressive load of the vacuum. Each section is joined to the other at a flange with a captured O-ring on one side and a flat surface on the other. The vessel leak rate is .01 torr/hr. Alignment features accurately join the sections. The front section has a flat aluminum cover that supports a vacuum gate valve and 4" thick high-density polyethylene (HDPE). The rear section has a domed hatch on an articulated hinge. The first and last sections have manway hatches on the top. Ports are provided for viewing and electrical/air connections to the interior. Two supports for a cable troughs are provided on either side of the vessel. Each section has welded lifting tabs at the top.

The interior of the 3 assembled sections have two welded steps in the bottom that support profile rails. The rails support and guide 3 carriages that travel in-line. A detector cassette mounts to each carriage. The carriages are driven by a pinion gear and motor on a common rack. Each carriage has an absolute encoder. Hard stops and limits are provided for each carriage as well as an anti-collision sensor on each that limits the closest approach between carriages to 2 meters. Cables in cable chains supported by trays are provided for each carriage and detector. Design was based on ASME BPV Section VIII.

2. Key specifications

Key specifications are identified in the chart below which also identifies the ANSTO Bilby SANS vessel requirements and the achievement of each key specification by ADC.

Table 1. NIST VSANS and ANSTO Bilby Spec Comparison

Specification	**NIST VSANS**	**ANSTO Bilby**	**Achieved on Bilby**
Length	24,000 mm	20,020 mm	20025 mm
Diameter	2300 mm	2300 mm	2300 mm
Vacuum	1 x 10^{-3} Torr	7.5 x 10^{-5} Torr	7.5 x 10^{-5} Torr
Vessel Straightness	+/- 12 mm	+/- 10 mm	+3, -7.5 mm
Leak Rate	.01 Torr/hr	.000075 Torr/hr	.00002 Torr/hr
Carriage Load	800 Kg	2000 Kg	2000 Kg
Carriage Travel	17,400 mm Max (Mid)	18,800 mm	18,800 mm
Carriage XY Deviation	+/- 2 mm	+/- 1 mm	+/- .7 mm
Adjustment Vertical	+/- 25 mm	+/- 10 mm	+/- 10 mm
Adjustment Lateral	+/- 25 mm	+/- 20 mm	+/- 20 mm
Adjustment Axial	+/- 25 mm	1.5 meters	1.5 meters
Codes	ASME BPV Section VII	AS 1210	ASME BPV Section VII, AS 1210
Material	Steel 516	Stainless Steel 304	Stainless Steel 304
Sections	3	3	3
Rear Hinge	Yes	Yes	Articulated
O-Rings	Buna- N	Buna- N	Buna- N
Walkway Support	2000 lbs./tab	250 kg/m2	7300 lbs. Centrally located on Section
Manway Hatch	# TBD, 66 cm 3 sections	x1, 120 cm Diameter	x1 Central Section

3. Straightness of Travel

The straightness of travel for each detector is the most important spec and possibly the hardest to achieve. Vessel straightness, roundness, compression deflection, are not as important as the straightness requirement of the internal steps and the adjustability of the rails to make the travel of the center of the detectors parallel to the centerline of the vessel. ADC has devised a method for adjusting and supporting the rails on the internal step for the much heavier Bilby detectors for ANSTO. In addition, the ***entire*** 64,000 pound Bilby chamber moved on external rails 1.5 meters. The Bilby vessel movement feature placed constrains on the support of each section that can be greatly improved on the NIST VSANS vessel (which does not move).

Figure 3. ANSTO Bilby SANS Vessel

ADC designed the step for maximum stiffness by using thick solid 304 SS material to prevent deflection of the rails as the detector is moved. The steps ran the full length of each section and were machined in one piece as can be seen in Figure 4 below. They were stitch welded to the shell to limit warp. For the NIST VSANS, access is provided in the vertical step for paint access under the step.

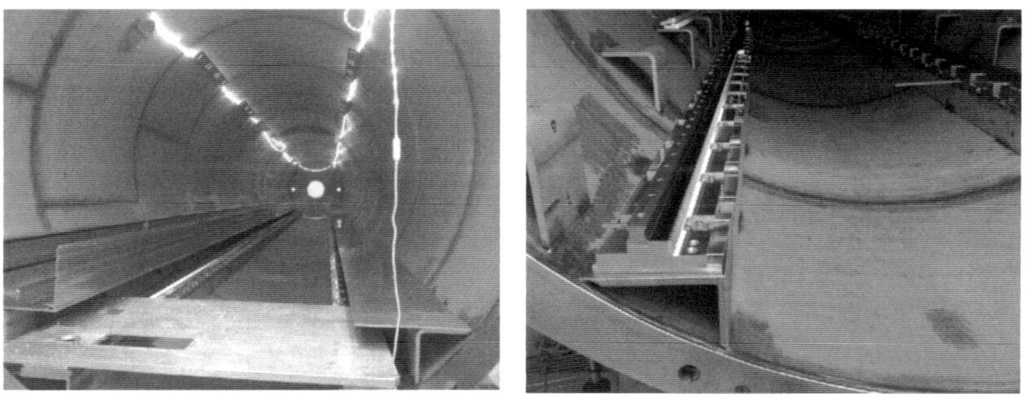

Figure 4. ADC's Rail Shimming Design

ADC used a laser to measure the error in the steps (down to 10 um) and thereby calculate the thickness of the shim plates. The vertical error was corrected to +/- 0.35 mm as shown in the chart below in Figure 5.

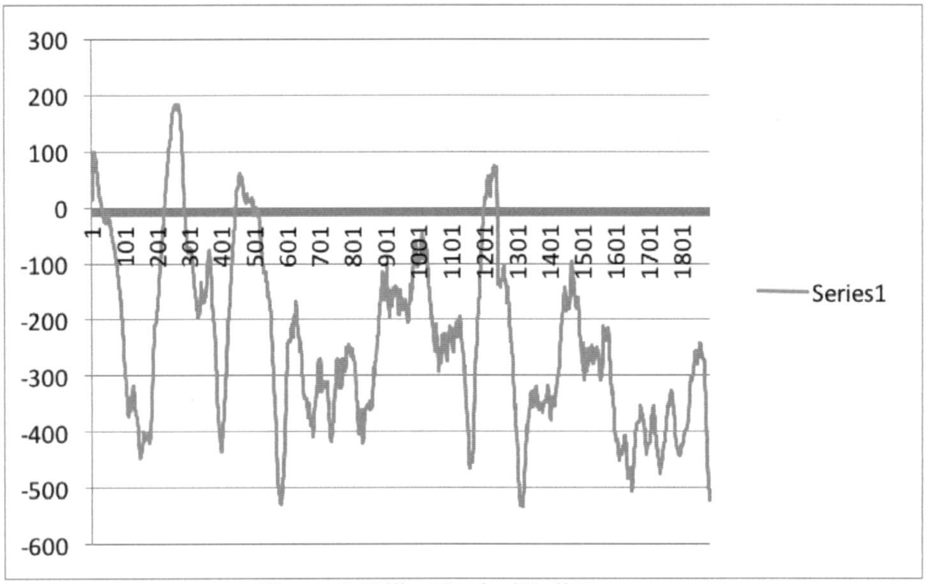

Figure 5. Bilby Vertical Rail Error

4. Vessel Material

The Bilby SANS vessel was constructed of 304 stainless steel and was unpainted inside to achieve 7.5 x 10^{-5} Torr. The NIST VSANS vessel is constructed of carbon steel (516) and is painted inside and out; however, the vacuum requirement is 1 x 10^{-3} Torr. Carbon steel is stronger than stainless and a thinner shell may be possible which reduces the weight. The Bilby vessel had a thick shell and was reinforced with a double shell at key locations such as at the lifting tabs. Bilby was designed so that lifting would not permanently deform the shape of the vessel (primarily the internal steps) and so that a 250 pound man on the catwalk would not affect the experiment if one was in progress.

5. Detector Motion System

The motion system for the detectors consists of 3 carriages that are driven on the same rack and guidance rails. Each carriage is driven by a pinion gear attached to a gear box and stepper motor. An absolute rotary encoder is mounted to a separate pinion gear on each carriage to encode the position. Two profile rails are provided on steps welded to the bottom of the vessel shell. The motor and gearbox are sized to provide .1 m/s^2 accel/decel on 800 Kg payload while resisting the IGUS cable chain load and bearing truck friction. Top speed of 200 mm/sec (7.9 inches/sec) can be achieved.

IGUS chains and trays are provided for each carriage and convey the cables and hoses necessary for the detector as well as the carriage motor and encoder.

6. References

[1] The 30 m Small-Angle Neutron Scattering Instruments at the National Institute of Standards and Technology; *J. Appl. Cryst.* (1998). **31**, 430-445

[2] D33—a third small-angle neutron scattering instrument at the Institut Laue Langevin; IOP PUBLISHING *Meas. Sci. Technol.* **19** (2008) 034007 (8pp)

[3] Novel multiple-beam very small angle neutron scattering (VSANS) using a conventional SANS instrument; *Journal of Applied Crystallography*; Volume **47**, Part 4 (August 2014)

[4] The Neutron Beam Expansion Program at the Bragg Institute; International Workshop on Neutron Optics and Detectors (NOP&D 2013) *Journal of Physics*: Conference Series 528 (2014) 012026

A new fast detection system at the KWS-2 high-intensity SANS diffractometer of the JCNS at MLZ – prototype test

A Radulescu[1]*, N Arend[1], M Drochner[2], A Ioffe[1], G Kemmerling [1], V Ossovyi[1], S Staringer[1], G Vehres[3], K McKinny[4], B Olechnowicz[4], D Yen[4]
[1]Jülich Centre for Neutron Science (JCNS), Outstation at MLZ, Forschungszentrum Jülich GmbH, 85747 Garching, Germany
[2]Central Institute for Engineering, Electronics and Analytics (ZEA-2), Forschungszentrum Jülich GmbH, 25425 Jülich, Germany
[3]Jülich Centre for Neutron Science (JCNS-1), Forschungszentrum Jülich GmbH, 25425 Jülich, Germany
[4]Reuter Stokes Products, GE Measurement & Control, General Electric Company, Twinsburg, Ohio USA 44087

a.radulescu@fz-juelich.de

Abstract. A new detection system based on an array of ^{3}He tubes and innovative fast detection electronics was designed and produced by GE Reuter Stokes for the high-intensity small-angle neutron scattering diffractometer KWS-2, operated by the Jülich Centre for Neutron Science (JCNS) at the Heinz Meier-Leibnitz Zentrum (MLZ). The new detector consists of a panel array of 144 ^{3}He tubes and a new fast read-out electronics. The electronics is mounted in a closed case in the backside of the ^{3}He tubes panel array and will operate at ambient atmosphere under cooling air stream. The new detection system is composed of eighteen 8-pack modules of ^{3}He-tubes that work independently of one another (each unit has its own processor and electronics). Knowing beforehand the performance of one detector unit and of one single tube detector is prerequisite for tuning and maximizing the performance of the complete detection system. In this paper we present the results of the tests of the prototyped 8-pack of ^{3}He-tubes and corresponding electronics, which have been carried out at the JCNS instruments KWS-2 (in high flux conditions) and TREFF.

1. Introduction

The small-angle neutron diffractometer KWS-2, operated by the JCNS at MLZ, is dedicated to the investigation of mesoscopic structures and structural changes due to rapid kinetic processes in soft condensed matter and biophysical systems. The high neutron flux (Fig. 1), comparable with that of the world leading SANS instruments, which is supplied by the neutron delivery system [1, 2] – cold neutron source, neutron guide, velocity selector, and the possibility to use large sample area, using lenses (for the same resolution as for conventional pinhole mode), enable high-intensity measurements and time resolved studies at tens of milliseconds time resolution [3]. On demand, the instrument resolution can be tuned by varying the wavelength spread between 2% and 20% by using a double-disc chopper with adjustable slit opening [4]. By means of lenses and a secondary high-resolution

* To whom any correspondence should be addressed.

detector (resolution about 1mm) a Q_{min} down to 1 x 10^{-4}Å^{-1} can be achieved, which, in combination with the pinhole mode, permits the exploration of a continuous length scale from 1nm to one micron [3].

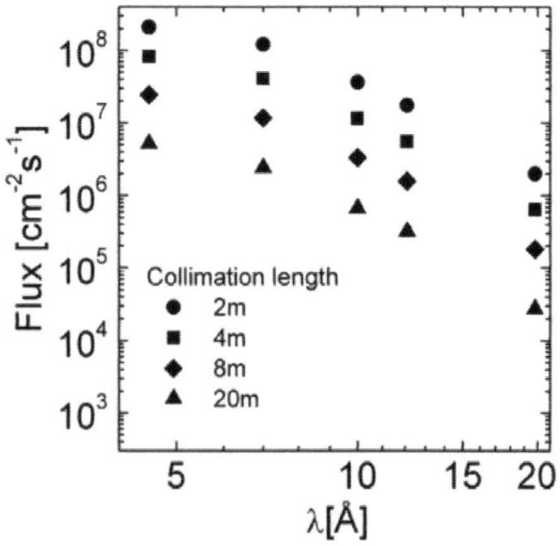

Figure 1. Absolute neutron flux at the sample position as a function of wavelength λ for different collimation lengths L_C and full beam-size (50mm x 50mm), as provided by the neutron guide.

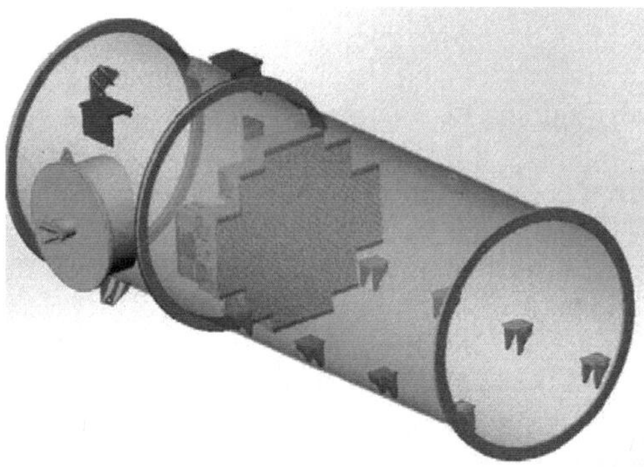

Figure 2. Detector upgrade proposal for KWS-2: the diameter of the vacuum tank is 1.40m.

The main detector of the instrument, which was in use until the summer of 2015, was a Anger camera based on an array of 4x4 ^{6}Li glass scintillator and 8x8 photomultipliers [5] and provided a 60x60cm^2 active detection area with a space resolution of about 7mm. The limited count rate at 10% dead-time (ca. 150kHz), the gamma sensitivity and the inhomogeneous dead-time over the detection area for the cases of samples exhibiting strong scattering in the forward direction, were a drawback, especially for investigations on small soft-matter and biological molecules that typically deliver at high Q very weak scattering signal above the buffer or solvent level, and thus demand very high neutron intensity and a high stability in time of the pixel response. In the end of 2013 it was decided to replace this detector with a new detection system with a performance that enables the optimal use of the high flux of the instrument (up to 2x10^8 n cm^{-2} s^{-1} are available on the sample). A collaborative work between JCNS and GE Reuter Stokes Inc. was started, with the main goal to install at KWS-2 a new detection system consisting of an array of position sensitive ^{3}He tubes with a larger active area (towards 1m^2) and an innovative rapid electronics, capable for measuring a count rate of 1MHz, or even higher, at 10% dead-time. The new detector should have a modular structure consisting of several 8-pack units of ^{3}He

tubes (8mm) that should be combined together at appropriate number and lengths, so as to maximize the active area by filling completely the cross-section of the KWS-2 tank (Fig. 2).
The high-voltage and read-out electronics should be mounted in a closed case at the backside of the ^{3}He tubes panel array and should operate at ambient atmosphere under cooling air stream. As the new detection system will be composed of many 8-pack units that will work independently of one another (each 8-pack unit has its own processor and electronics), the performance of one detector unit and of one single tube are needed to be known beforehand. This helps for tuning and optimizing the performance of the full system by knowing the expectations regarding the maximal performance of several similar independent units that will work in parallel while assembled together into the complete detection system.
In this paper we present the results of the tests of one prototyped 8-pack unit of ^{3}He-tubes and corresponding electronics, which have been carried out at the JCNS instruments KWS-2 (in high flux and vacuum conditions) and TREFF.

2. Technical and experimental details

The prototype 8-pack detector module consisted of eight ^{3}He tubes of a diameter of 8mm and an active length of 822mm, two preamplifiers, two high-voltage (HV) modules and a digital processing board (platform processor). The ^{3}He pressure is 12.6 bar, which corresponds to an efficiency of 85% for λ=5Å [6]. An air box containing an Ethernet router (Hirschman) and the mechanical (cooling lines) and electrical (wired Ethernet cabling for signal and power) interfaces between the 8-pack and the air box and router were attached to the prototyped detector module.
At KWS-2 (Radulescu et al., 2012) the detector module was placed in the vacuum tank (2x10^{-2} mbar), in an out-of-direct beam position. Two airlines (one supply, one exhaust) were connecting the air box with the outside of the vacuum tank and cooling air at a flow rate of 800 l/min and humidity less than 5% were provided to the air box and further to the 8-pack electronics. One optical Ethernet cable (for signal) and one power cable were routed through the exhaust airline from the air box to outside the vacuum tank and further to an external router (Hirschman) and power supply (Toellner). The two Hirschmann routers (one in the air box, one outside the vacuum tank) were capable of supporting high speed data (1Gbit), power-over-Ethernet (PoE) and precision time protocol (PTP).
At TREFF the detector module was placed in air, having further the cooling lines, the Ethernet and power cables and the external router connected to it.
All data have been collected in 1024 channels (pixels) over the length of each tube, which is the maximum resolution supported by the detector electronics.
Measurement of count rate and stability of the pixel response were carried out at KWS-2 using a incoherent scattering Plexiglas (PMMA) standard sample that provided an flat homogeneous scattering pattern on the detector. The detector was placed at 1m and 2m after the sample and various collimation lengths (8m, 4m and 2m with a 50x50mm^2 entrance aperture) and beam sizes (sample aperture between 8x8mm^2 and 25x25mm^2) were used for varying the neutron intensity on the sample, and thus the count rate on the detector. For determining the dead time and the count rate behavior an almost dead time free fission chamber (LND) was mounted after the sample in an out of detector view position [3], in order to collect in parallel the incoherently scattered neutrons on the sample. The check of the stability of pixel response involved conducting of repeated measurements (30 minutes each) at constant intensity conditions and over a long time period during day and night, in order to have variable EMV and neutron background environment dictated by the regular activities taking place around the instrument. Different count rates between 4 and 16cps per pixel have been tested. Measurements of linearity, resolution and stability of resolution were carried out in direct beam at TREFF using a narrow vertical slit (1mm x 10cm) and moving the detector in beam along its long axis with constant step (2cm) that allowed the local illumination of all eight tubes at once and the scanning along the active length of tubes.

3. Results and discussion

3.1. Flood test

Fig. 3 shows the intensity profile on each tube for homogeneous scattering on the 8-pack provided by the incoherently scattering Plexiglas sample at a count rate of about 4000cps on each tube. A narrow Cd stripe (5mm) with a thickness of 0.5mm was mounted in the middle of the 8-pack, across all tubes, with the purpose of adjusting the pixels across the 8-pack by using the sharp drop in the intensity profile provided by the Cd-mask. The weak waviness observed for some tubes is due to small imperfections along the anode wire and turned out to be stable in time and over a broad range of count rates, thus correctable by applying a sensitivity calibration.

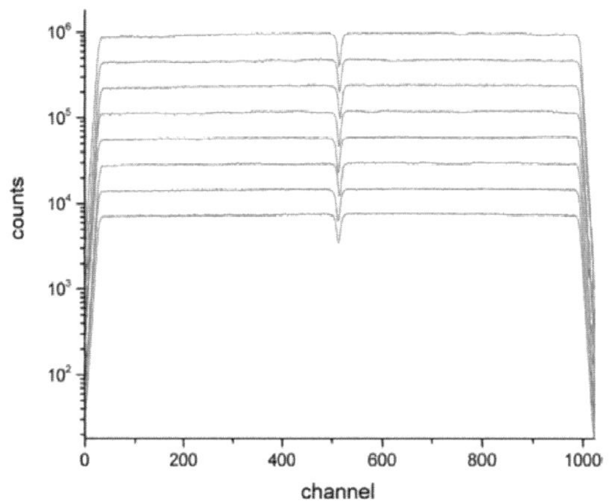

Figure 3. Intensity profile of the detector flood (homogeneous illumination) for each ^{3}He tube of the 8-pack; the results are scaled up with a factor of two, for clarity.

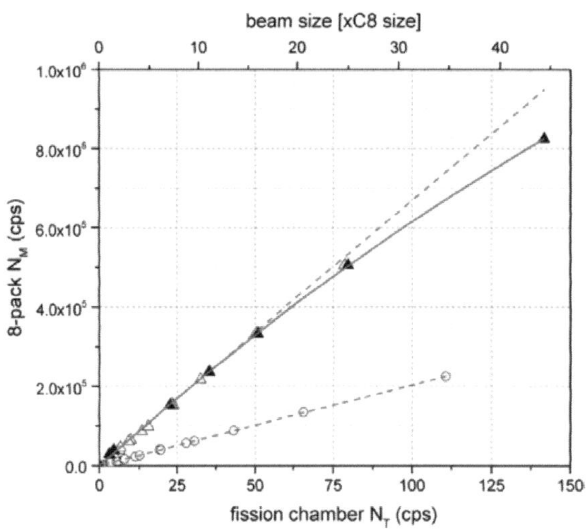

Figure 4. Dead time measurement on the ^{3}He 8-pack positioned at 1m (triangles) and 2m (circles) after the sample; the results are presented versus the intensity measured in parallel with a fission chamber (open symbols) or as a function of beam size (filled symbols).

3.2. Dead time and count rate

Fig. 4 shows the measured count rate N_M on the detector at two sample-to-detector distances (L_D=1m and L_D=2m) versus the count rate N_T given by the fission chamber [3], which is proportional to the flux of incoherently scattered neutrons. Additionally, the measured intensities at L_D=1m are shown versus the beam size [7, 8], which was varied by changing the collimation length and size of source (entrance) and sample apertures. Therefore, the beam size is expressed in multiples of standard beam-size, which, for the old scintillation detector at KWS-2, it was considered for the collimation length L_C=8m and the aperture sizes of 30mm x 30mm (source) and 8mm x 8mm (sample). By considering a homogeneous beam, the intensity is proportional to the aperture sizes and squared ratio between collimation lengths. The full line in Fig. 4 are the data fitting with the paralyzable model [9], and the dotted lines are linear functions. No deviation from the linearity is observed for count rates as high as $3x10^5$cps for the whole 8-pack. At higher count rates deviations from the ideal linear behavior due to dead time are observed. The dead time τ is found 70ns, and corresponds to a 10% loss at around N_M=700kHz. Given that each 8-pack module has its own platform processor, it is expected that in an ideal case, a larger 2D detector consisting of many 8-packs, like that shown in Fig.2, will have a count rate that will scale with the number of 8-packs. Thus, with the optimized data transfer and storage, several MHz at 10% dead time are expected in this case.

3.3. Stability of the pixel response

In order to check the stability of the pixel response over a long time, long series of similar Plexiglas measurements were carried out at constant count rate. The stability was checked by dividing results obtained on two similar measurements performed at different time and calculating the standard deviation of the ratios. Typically, 24 measurements each of 30 minutes were performed during the day and during the night time, when the EMV and background conditions around the instrument were different. The procedure was repeated for low and high count rates. Fig. 5 presents the ratio between two similar measurements on tube No. 6 carried out at two count rates, 4cps per channel and 15.5cps per channel. In both cases the results it is around 1 along the entire tube length (end channels were excluded due to the high uncertainty) and the standard deviation of the ratios lies within statistical uncertainty of the measurements.

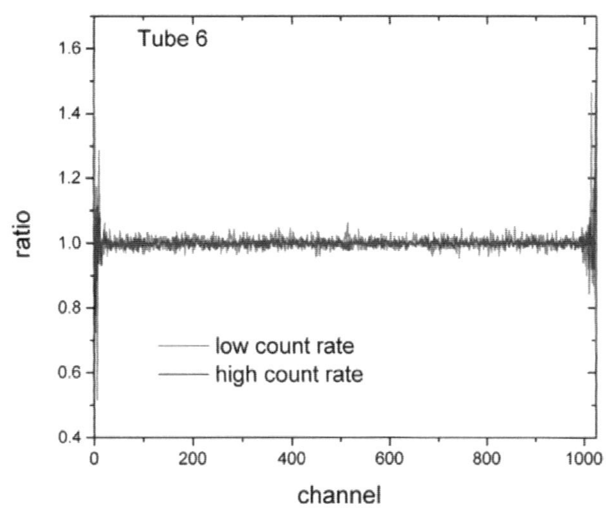

Figure 5. The ratio between two similar measurements with constant count rate performed at different times; two count rates were tested.

3.4. Position calibration and linearity

The position resolution was tested along the anode wires (with a 2cm step) by using a Cd-mask with 1mm slit which stretches across the entire 8-pack (Fig. 6a). Thus, all eight ^{3}He tubes were illuminated at once. The position linearity was tested (Fig. 7b) and the fit function of the measured positions to the actual positions was y=(1±0.000471)x+(0.525±0.1088). The linear correlation coefficient was 1.024, which showed a good linearity.

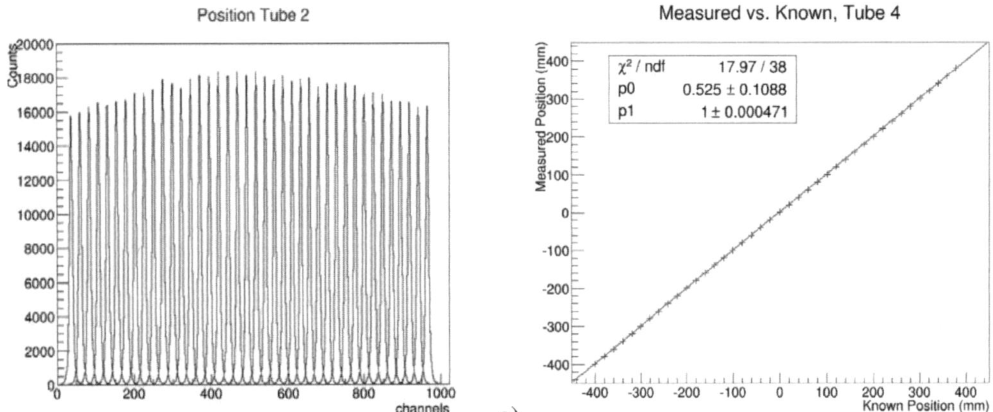

a) b)

Figure 7. Example of results of the scan measurement along the anode wire (a) and of the position reconstruction (b).

3.5. Position Resolution

Gaussian fit to peaks measured along the tubes by using the 1mm slit (Fig. 7a) delivered the position resolution (Fig. 8), which is better than 8mm for all eight tubes. By repeating the measurement at different times it was found that the position is stable within 0.06mm RMS.

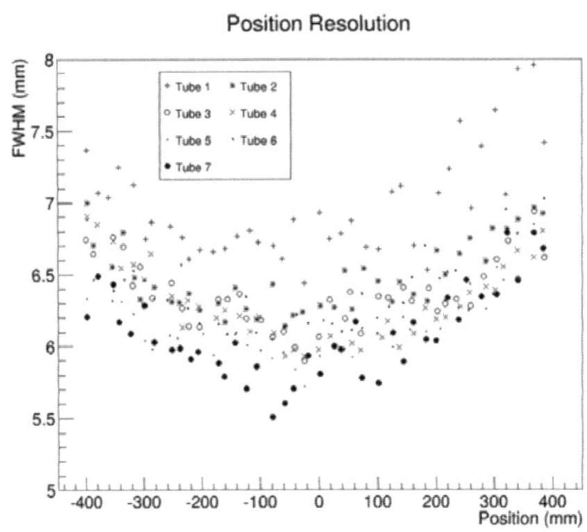

Figure 8. Position resolution along the anode wire for all eight ^{3}He tubes in the 8-pack

4. Summary

A new fast detection system is planned for the high intensity SANS diffractometer KWS-2, which is operated by the JCNS at MLZ. The new detector should consist of 144 ^{3}He tubes, which will be grouped in 8-packs that will work in parallel, each 8-pack having its own microprocessor. An 8-pack prototype was produced and was tested at KWS-2 (in high flux conditions) and TREFF instruments of JCNS, in order to learn about the capabilities of the new detector and how tuning and maximizing the performance of the complete system can be achieved.

A summary of the test results on the 8-pack prototype is presented in Table 1. It is thus expected that the new detection system will show a high stability of the pixel response and position, a position resolution better than 8mm and a high count rate (several MHz) at 10% dead time.

Table 1. Parameters (targeted and measured) of the ^{3}He fast detector prototype for KWS-2

Parameter	**Target**	**Measured**
count rate (tube)	28000 cps	~90000 cps
count rate (8-pack)	224000cps	~720000cps
position resolution	<8mm	<8mm
position stability (RMS)	≤ 1.5mm	< 0.06mm
position accuracy / linearity	–	1mm
pixel rate stability	<< 10%	<0.4% within statistical uncertainty

References

[1] A. Radulescu and A. Ioffe, Nucl. Instrum. Methods Phys. Res. Sect. A 586, 55-58, 2008.
[2] A. Radulescu, V. Pipich and A. Ioffe, Nucl. Instrum. Methods Phys. Res. Sect. A 689. 1-6, 2012.
[3] A. Radulescu et al., J. Phys. Conf. Ser. 351, 012026, 2012.
[4] A. Radulescu et al., J. Appl. Cryst. (in press).
[5] G. Kemmerling et al., IEEE Trans. Nucl. Sci. 51, 1098-1152, 2004.
[6] N. Johnson, GE Reuter-Stokes, private communication, 2013.
[7] I. Grillo, in Soft Matter Characterization, Ed. R. Borsali and R. Pecora, Springer, 2008.
[8] P. Strunz, K. Mortensen and S. Janssen, Physica B 350, e783–e786, 2004.
[9] G.F. Knoll and J. Wihley, in P. Convert and J.B. Forsyth (ed.), Radiation Detection and Measurement, Academic Press, New York, 1989.

Journal of Physics: Conference Series **746** (2016) 012027 doi:10.1088/1742-6596/746/1/012027

Neutron guide shielding for the BIFROST spectrometer at ESS

K. Mantulnikovs[1], M. Bertelsen[1], C.P. Cooper-Jensen[2,3], K. Lefmann[1] and E.B. Klinkby[3,4]

[1] Nano-Science Center, Niels Bohr Institute, University of Copenhagen, Denmark
[2] Department of Physics and Astronomy, Uppsala University, Uppsala, Sweden
[3] European Spallation Source ESS AB, Lund, Sweden
[4] Center for Nuclear Technologies, Technical University of Denmark, Denmark

E-mail: esbe@dtu.dk

Abstract. We report on the study of fast-neutron background for the BIFROST spectrometer at ESS. We investigate the effect of background radiation induced by the interaction of fast neutrons from the source with the material of the neutron guide and devise a reasonable fast, thermal/cold neutron shielding solution for the current guide geometry using McStas and MCNPX. We investigate the effectiveness of the steel shielding around the guide by running simulations with three different steel thicknesses. The same approach is used to study the efficiencies of the steel wall a flat cylinder pierced by the guide in the middle and the polyethylene layer. The final model presented here has a 3 cm thick steel shielding around the guide, 30 cm of polyethylene around the shielding, two 5 mm thick B_4C layers and a steel wall at position Z = 38 m, being 1 m thick and 10 m in radius. The final model finally proves that it is sufficient to bring the background level below the cosmic neutron rate, which defines an order of magnitude of the lowest obtainable background in the instruments.

1. Introduction

The worlds strongest neutron source for the study of materials and biosystems will be the European Spallation Source (ESS) [1], which is presently under construction in Lund, Sweden. At ESS, neutrons will be produced by a 2 GeV protons impinging on a rotating tungsten target. The beam power will reach 5 MW resulting in unprecedented cold and thermal neutron brightness, but will also give rise to experimental backgrounds to a level beyond what is observed at existing neutron facilities [2].

Prompt neutrons escaping the target monolith have energies reaching up to the energy of the initial proton beam, and thus the task of instrument shielding is completely different compared to the case of reactor sources, based on which most shielding experience relies. At neutron energies exceeding 10 MeV, the scattering cross section of most commonly used shielding materials drops dramatically meaning that the task of instrument shielding at the ESS is even more complex than what the proton beam power dictates.

Having in mind also that ultimately the performance of most instruments mostly depends on the signal-to-noise ratio (S/N), and that shielding is expected to be a significant cost driver of the facility, the shielding design at ESS is as important as it is complex.

The present study focusses on evaluating the shielding options of the BIFROST spectrometer, which will in 2016 enter Phase 1 of its technical design at ESS. The beam optics of BIFROST is designed and optimised using the ray-tracing code McStas [3, 4, 5, 6, 7]. While McStas is well-recognised for its capability to precisely describe neutron scattering instruments in terms of signal distributions, it does lack in the description of backgrounds. To remedy this challenge an exact one-to-one implementation of the instrument is developed MCNPX [8, 9] - which is the standard Monte Carlo tool used for shielding calculations. Below we study the performance of various shielding design options considered for the BIFROST instrument. The aim of such study is to outline the shielding design prior to the instrument construction, hereby allowing to iterate the instrument and shielding design to a common optimum.

2. Methods

MCNP and MCNPX are general purpose Monte Carlo radiation transport codes that have the ability to track many particle types over a broad energy range. These capabilities include but are not limited to tracking protons and electrons. MCNP and MCNPX stands for Monte Carlo N-Particle and Monte Carlo N-Particle Extended. The extended version has been developed to simulate 34 different particle types and more than 2000 heavy ions in a broad range of energies.

McStas is a ray tracing software package for simulating neutron scattering experiments from the moderator to the detector. Such experiments can be described by a series of components, where the ray is propagated through each one without the ability to move back, which enables the code to be relatively simple, fast and modular. For this reason, McStas is well suited to exploration of possibilities, as development times are manageable, and the resulting simulations are fast enough to be used with numerical optimisers.

When designing a neutron guide system, there is a overwhelming amount of possibilities, even when only considering the geometry of the guide system. Writing McStas code for every possibility is unfeasible for a single person designing a neutron instrument, meaning that a small number of possible solutions can be investigated. The program guide_bot [10] is meant to reduce the time spent coding in these initial stages of project design, as it will write McStas guide optimisations tailored to the specific requirements of the instrument from a very limited amount of user input. The optimised guides are then automatically compared with comprehensive performance analysis for each case, making the decision making in neutron guide design more informed.

This project also makes use of the McStas-MCNPX coupling interface [11, 12], which takes advantage of the specific areas of expertise of each software package. More precisely, this coupling is used due to the fact that there are no models in MCNPX that account for reflectivity of the neutron guide – this is the area where one would need to use McStas, which is made to perform ray-tracing simulation of neutron transport in the neutron guides.

Since MCNPX and McStas are both Monte Carlo based software packages one would need significant computer power to run simulations on. In this project the cluster of the ESS Science division based in Data Management and Software Centre (DMSC) in Copenhagen is used. The cluster consists of the following components [13]:

- 42 compute nodes, each consisting of 2 processors (Intel Xeon 2.66 GHz) with 6 cores each and 48 Gb memory;
- 50 Tb of storage;
- Management network, used for maintenance;
- InfiniBand network, connecting the nodes in between each other as well as connecting them with the storage system;
- A batch-system for handling jobs.

3. Model description

The BIFROST spectrometer [14, 15] was accepted for construction in 2014 and is designed to achieve high detection efficiency in the horizontal scattering plane. Apart from the spectrometer

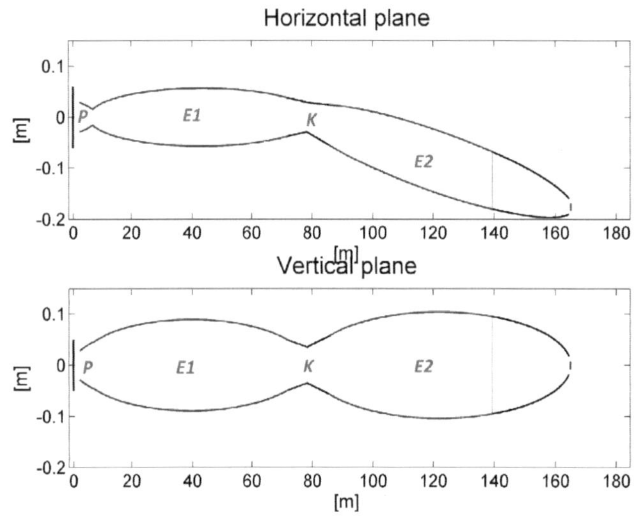

Figure 1. ZX (top) and ZY (bottom) cross sections of the BIFROST neutron guide obtained from McStas. The direction of the guide (the direction of the guide's optical axis) is along the Z axis, the X axis is represented in top image, Y axis – in the bottom one. The guide is composed out of four main sections: P – parabolic feeder, E1 and E2 – two elliptic sections that are connected by the kink – K. Parabolic and elliptic shapes are used to improve neutron intake and transport to the sample, while at the kink the guide's second elliptic section E2 is rotated out of line-of-sight (line-of-sight is broken at the magenta line) to avoid direct beam from the moderator.

itself BIFROST has a neutron guide – a tube that is used to transport cold neutrons to the sample position. The direction of the guide is along the Z axis with the Y axis pointing upwards and X to the right, forming a left-handed coordinate system. The guide, seen in figure 1 utilizes a parabolic feeder to improve neutron intake as well as to decrease the parasitic background [16]. The following neutron guide sections are shaped as a double ellipses to improve neutron transport to the sample position and the optical axis has a kink between the two ellipses to avoid line-of-sight. To take advantage of the time-of-flight measurement technique BIFROST's neutron guide is a long guide with length of 162.24 m and the sample position is situated 164.24 m away from the moderator.

The MCNPX model is based on MIRROTRON's metal-glass sandwich technology [17] with a slight alteration. The world outside the guide is represented by a box, filled with air ($\rho = 1.296 \cdot 10^{-3}$ g/cm^3), spanning ±100 m in the X and Y directions and 164.41 m along Z.

The model used in this study uses the combination of the conventional shielding measures such as steel and polyethylene along with the boron carbide (B_4C) as well as has a steel wall, which mimics the steel end wall of the planned shielding bunker. The modelled wall is basically a flat disk pierced by the guide in the middle. The reasoning for placing a steel wall is that high energy neutrons can get past the initial monolith and guide shielding due to presence of windows (minima) in the cross section. Fast neutrons having energies within windows would then travel far away from the source and increase the background locally [18]. The steel wall is expected to stop these stray neutrons close to the monolith shielding and prevent their propagation closer to the sample position. The material stack-up in the guide in XY cross section is as follows and can be seen in figure 2:

- 2 μm thick Ni/Ti coating ($\rho = 6.45$ g/cm^3), which serves as a supermirror;
- 1 cm thick substrate of variable material, serving as a structure material;
- $1, 2$ or 3 cm thick stainless steel shielding ($\rho = 8.03$ g/cm^3) around the guide;
- $20, 30$ or 40 cm thick polyethylene (PE) layer ($\rho = 0.97$ g/cm^3) around the guide to moderate and scatter fast neutrons;
- Two 5 mm thick B_4C layers: one, around the steel shielding (only last 10 m of the guide) and another around the polyethylene to absorb neutrons moderated by the PE layer.

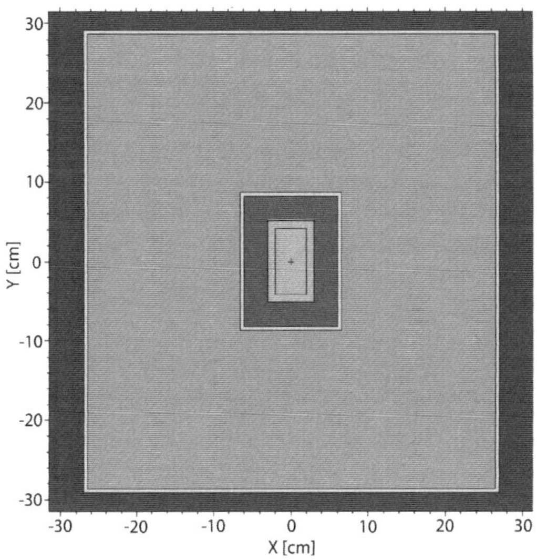

Figure 2. XY cross sections of the final model for BIFROST spectrometer's neutron guide shielding. The inner green is the guide vacuum (low pressure helium atmosphere) inside the Ni/Ti supermirror coating, which is supported by the substrate (yellow frame around the vacuum). The substrate is covered with stainless steel (cyan), serving as initial shielding. The thin yellow frame around steel is the boron carbide layer, which is present only in the last 10 m of the guide. Next layer is the PE layer – a conventional material to moderate fast neutrons and finally the same thin yellow frame of boron carbide to absorb the moderated neutrons. The outside (dark blue) is air.

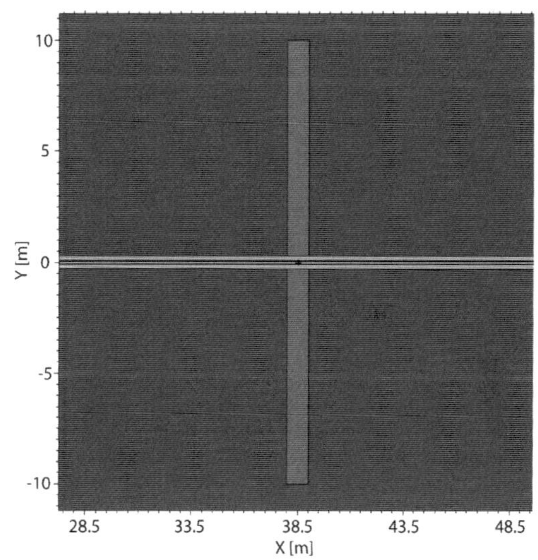

Figure 3. ZX cross sections of the final model for BIFROST spectrometer's neutron guide shielding explicitly showing the steel wall. The outside of the guide (dark blue) is air and the shield is depicted in cyan. It has a 10 m radius and 1 m thickness, being placed at position Z = 38 m.

The variable material of the guide's substrate changes depending on the position along the guide. First 20 m there is a copper substrate ($\rho = 8.96$ g/cm^3), the next 58 m the material is aluminium ($\rho = 2.73$ g/cm^3) and then until the end of the guide – BORKRON glass ($\rho = 2.39$ g/cm^3). Note that parallel to the work described in this paper it has been shown that the sections made of copper and aluminium are going to be much shorter and the distance previously covered by copper or aluminium sections will be made out of glass. Apart from the lateral stack described above there is a stainless steel wall, seen in figure 3, with radius of 10 m and 0.5, 1 or 2 m thickness for further shielding of fast neutrons placed at position Z = 38 m. This is just before the place where the first elliptic section of the guide is the thickest, thus placing the shield here we expect neutrons to see less material with which they can interact and consequently bring the background level up.

The monolith shielding is made as a steel cylinder with wall thickness of approximately 3.3 m and is centred around the centre of the moderator, which is at Z = 15.54 cm. The choice of the wall thickness is mostly arbitrary but has the idea that even when the construction of the monolith shielding is unknown, we want to study the effect of neutrons entering the guide rather then focusing on the monolith shielding. Thus the walls have been chosen to be thick enough to stop most of the neutrons that come from the source but do not end up in the guide opening.

The MCNPX detectors are placed in two positions. First, just before the kink (Z = 71.8 m), before the guide have turned and the model is symmetric with respect to the origin, so that the data acquired at this position are easy to interpret. This detector is referred to as symmetric plane detector or SPD. Second detector which should give real information about the background, at the sample position just after the end of the guide (Z = 164.4 m). This detector is referred to as endplane detector or EPD. Both of the detectors are planes spanning ± 100 m in X and Y directions and utilize the Surface Source Write (SSW) [11, 21] functionality of MCNPX, which allows to obtain energy and position, among other parameters, of every neutron crossing the detector plane.

Note that apart from the statistical error present in the simulation, there is an expected 15 % systematic error associated with nuclear interaction models and nuclear data libraries when running simulations in MCNPX, which has been shown in [1].

4. The source

The neutron source for the simulations in this work has been modelled from running a full MCNPX simulation of the ESS target and moderator [19]. Further details on model of the source in this MCNPX model can be found in [20]. Neutrons that are produced in the process of spallation in the target arrive to the moderator and are slowed down. "The focusing" effect in MCNPX that attracts statistics has been used to boost statistics while sacrificing each particle's weight. To do this a circular area with radius of 12 cm has been specified to attract statistics. The choice of such an area is dictated by the approximate size of a neutron guide entrance, thus giving a boost to statistics of neutrons that enter the guide rather than focusing on monolith shielding, modelling which lies beyond the scope of this paper as mentioned earlier. Neutrons are then tracked back to the plane of the moderator surface and data on each neutron is written into a file for further use.

The ESS source covers a broad range of energies up to 3 GeV, which would mean that to simulate all the energy ranges to the same extent a long simulation is needed. To avoid having very long and complicated simulations we decided to take out some degree of complexity by simplifying the source. The position and angular distributions are left the same, while the energy distribution is changed to a fixed energy of 1 GeV. Check runs have been performed for energies of 1 MeV, 10 MeV, 100 MeV and 1 GeV. Their comparison showed that the simplified simulations preserve the all relevant physical effects and the only difference is the intensity. The reasoning is that the fastest neutrons, although few in number, are hardest to shield. Hence shielding efficiency against this radiation is likely to be even better for the lower energy neutrons. This is in greater detail explained in [20].

Choosing one energy for the emitted neutrons means that the detectors described in the previous section would detect neutrons that have either the energy of the source – 1 GeV or lower energies. The first case can occur when the neutron does not interact with any material and does not loose energy, thus being detected as emitted. In the other case, neutrons interact with matter, for example, via a spallation, neutron capture process or generating a shower [22], which leads to energy loss of the neutron and possibly also re-emission of secondary, lower energy neutrons.

Here we need to make one very important note: only neutrons coming from the moderator are considered. We cannot and are not making any statements considering the overall background, which most definitely needs to be explored in the near future, given the fact that during the time of this study the bunker design has been developing.

5. Runs with different thicknesses of steel shielding

This section describes our study of the thickness of the guide steel shielding. The steel shielding around the guide's substrate should act as shielding and this section presents the study of variations of the thickness of the shielding. Three thicknesses have been tested: 1 cm, 2 cm and 3 cm.

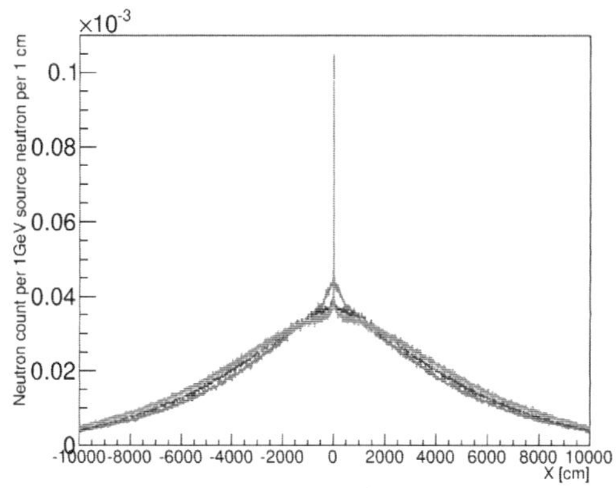

Figure 4. Neutron intensity projections on the X axis at the symmetric detector plane for three different steel shielding thicknesses around the guide. Red curve represents the run without steel shielding, blue – 1 cm of steel around the guide, green – 2 cm, magenta – 3 cm. The runs are made with 1 GeV neutron source.

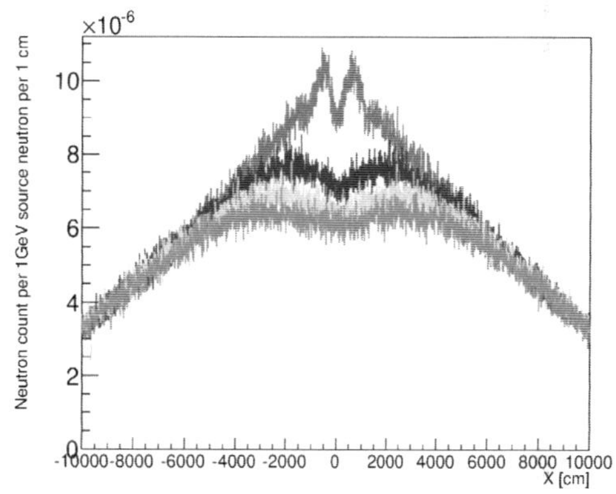

Figure 5. Neutron intensity projections on the X axis at the endplane detector for three different steel shielding thicknesses around the guide. Red curve represents the run without steel shielding, blue – 1 cm of steel around the guide, green – 2 cm, magenta – 3 cm. The runs are made with 1 GeV neutron source.

The results of these runs are presented in the form of intensity projections on the X axis and can be seen in figures 4 and 5. From the SPD plot in figure 4 one can see that the background level actually grows slightly in the region around the guide ($X < \pm 20$ m) with increasing steel thickness and gets even higher than the background solely from the substrate and coating, depicted in red. This can happen due to the presence of spallation in the model – having more material for the fast neutrons from the source to interact with also brings more secondary neutrons coming from spallation. Yet far away from the guide ($X > \pm 20$ m) the guide itself and it's close vicinity the background level falls with increasing thickness of the steel.

The background levels at the sample position (EPD) falls off with increased steel thickness. Two other very important observations are that, firstly, the background spreads in a wide range and, secondly, the peak from the direct view of the moderator present in the SPD plot is gone at the sample position – EPD plot, which is the effect of going out of line-of-sight.

To get a better quantitative grasp of the situation at the sample position we resort to looking at the integrals under the intensity projections shown in figure 5. Since the integral values are a measure of how much the background level has dropped, this will help to compare the change in the different steel shielding thicknesses efficiencies via calculating a relative change with respect to the run with only substrate and coating, i.e. in the absence of the steel shielding whatsoever. The values are shown in table 1.

Table 1. Intensity integrals under the projection curves and relative change in integral value, calculated with respect to the model with only substrate and coating, for models with different steel thicknesses at the sample position (EPD).

Model	Integral	Relative change
Substrate+coating	0.01303±0.00004	n/a
1 cm steel	0.01213±0.00003	6.9±0.4 %
2 cm steel	0.01147±0.00003	12.0±0.4 %
3 cm steel	0.01095±0.00003	15.9±0.4 %

Judging from table 1 the increase in thickness from 1 cm to 2 cm gives almost a doubled increase in efficiency as compared to the change of solely adding a 1 cm steel shielding around the guide, going from 6.9±0.4 % to 12.0±0.4 %. Further increase of the thickness to 3 cm a slightly smaller increase of 3.9±0.6 %, suggesting that at 2 or 3 cm the effect is starting to decay slowly.

6. Runs with different steel wall thicknesses

Apart from adding a steel wall the rest of the model for these runs is unchanged and a 3 cm thick stainless steel shielding around the guide is used. Once again a simplified run with the source emitting only neutrons with energy of 1 GeV is used.

The results of the effect of the thickness of the stainless steel wall are presented as intensity projection on the X axis in figures 6 and 7. Compared to the plots in figures 4 and 5 we can clearly see that the background levels fall off in the region around the guide when the wall has been introduced into the model. Moreover there is a direct correlation with the walls thickness – the bigger the thickness, the lower the background drop.

A more quantitative representation of the date in the plots are the integrals under the curves, which are presented in table 2. From this, as well as from the plots, we can see that addition of even of a 0.5 m wall introduces a noticeable drop in background, which corresponds to 13.5±0.4 % relative change in background at sample position. Doubling the thickness to 1 m increases the relative change even more but to a smaller extent giving 19.5±0.4 %. Doubling the thickness once more to 2 m gives an increase but only by 1.3±0.6 % increasing the relative change to 20.8±0.4 %. From this we can conclude that a 1 m wall is the one that performs the most efficiently as compared to the additional price for a thicker wall.

7. Runs with different thicknesses of the polyethylene

We cover the guide with a layer of polyethylene (PE). This is a conventional material that is used for shielding purposes to moderate and scatter high energy neutrons, due to its high hydrogen content. The material that absorbs neutrons in this case is boron carbide (B_4C) in form of two 5 mm layers as mentioned above. The runs were performed with three different thicknesses of

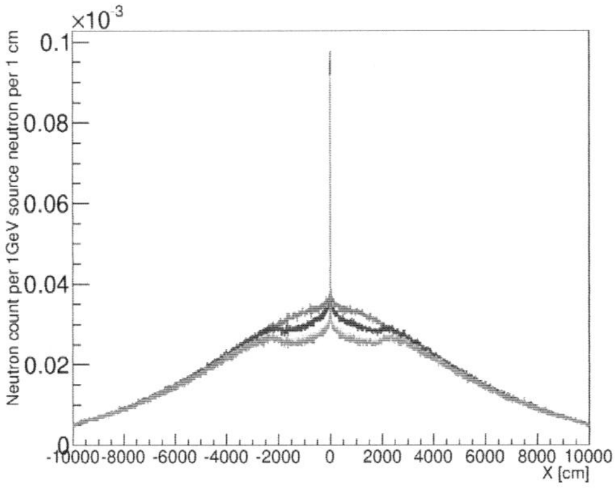

Figure 6. Neutron intensity projections on the X axis at the symmetric detector plane for three different steel wall thicknesses. 3 cm thick stainless steel shielding is used in the model. Red curve represents the run without the steel wall, blue – 0.5 m thick wall, green – 1 m thick, magenta – 2 m thick. The runs are made with 1 GeV neutron source.

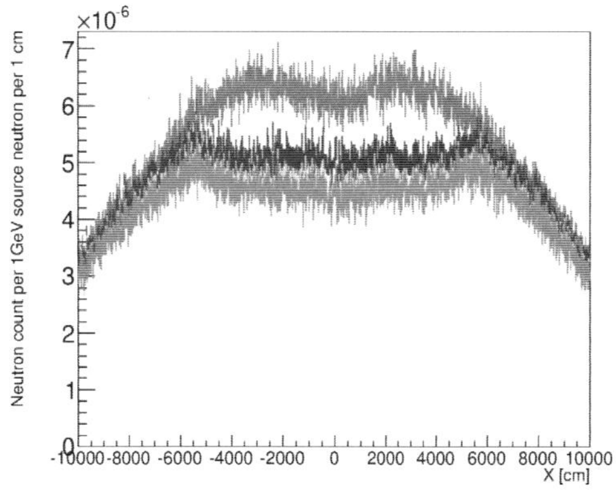

Figure 7. Neutron intensity projections on the X axis at the endplane detector for three different steel wall thicknesses. 3 cm thick stainless steel shielding is used in the model. Red curve represents the run without the steel wall, blue – 0.5 m thick wal, green – 1 m thick, magenta – 2 m thick. The runs are made with 1 GeV neutron source.

Table 2. Intensity integrals under the projection curves and relative change in integral value, calculated in respect to the model with 3 cm of steel shielding around the guide, for models with different wall thicknesses at the sample position (EPD).

Model	Integral	Relative change
3 cm steel	0.01095±0.00003	n/a
0.5 m wall	0.00948±0.00003	13.5±0.4 %
1 m wall	0.00882±0.00003	19.5±0.4 %
2 m wall	0.00867±0.00003	20.8±0.4 %

20 cm, 30 cm and 40 cm. The rest of the model is the same as described above, combining a 3 cm stainless steel shielding around and a 1 m thick, 10 m radius steel wall at position Z = 38 m. The runs are made with a simplified 1 GeV neutron source.

The results of such runs are presented in figures 8 and 9. As compared to the run of the model without any shielding, i.e. only substrate and coating, the background level has dropped significantly. Once again we see that there is a direct correlation between the thickness of the PE layer and the efficiency of the layer – the thicker the layer, the greater the efficiency at both SPD and EPD. From the plots we can also see that a 20 cm thick layer of borated PE (5 wt% B) works almost as efficient as a 20 cm thick PE layer combined with two 5 mm B_4C layers.

A more quantitative picture can be seen by looking at the integral values and relative changes of the integral values underneath the curves and this data is presented in table 3. We can see that the relative change by adding PE layer, 3 cm of shielding around the guide and a 1 m thick stainless steel wall is quite big – 75.9±0.4 %. Increasing the thickness of the PE layer by another 10 cm we get an increase of about 6.2±0.6 %. Another step in increasing the thickness by yet another 10 cm, making the PE layer 40 cm thick, gives only a minor 4.2±0.6 % increase.

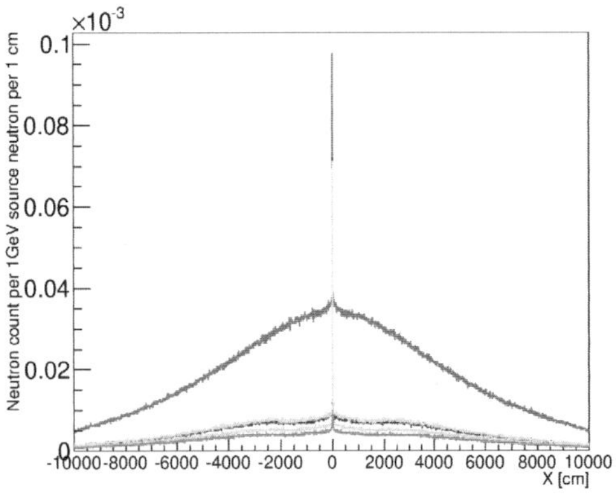

Figure 8. Neutron intensity projections on the X axis at the symmetric detector plane for three different polyethylene layer thicknesses. Red curve represents the run only with 3 cm of steel shielding around the guide, blue – with 3 cm of steel shielding around the guide, 1 m thick wall, 20 cm thick layer of polyethylene and two 5 mm thick layers of B_4C, green – the same as blue but with 30 cm thick layer of polyethylene, magenta – the same as blue but with 40 cm thick layer of polyethylene, cyan – the model which only has 20 cm thick layer of borated polyethylene (5 wt% Boron) around the guide without boron carbide layers. The runs are made with 1 GeV neutron source.

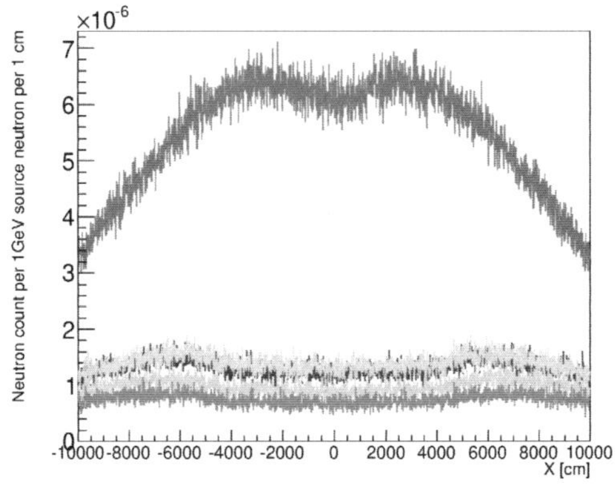

Figure 9. Neutron intensity projections on the X axis at the endplane detector for three different polyethylene layer thicknesses. Red curve represents the run only with 3 cm of steel shielding around the guide, blue – with 3 cm of steel shielding around the guide, 1 m thick wall, 20 cm thick layer of polyethylene and two 5 mm thick layers of B_4C, green – the same as blue but with 30 cm thick layer of polyethylene, magenta – the same as blue but with 40 cm thick layer of polyethylene, cyan – the model which only has 20 cm thick layer of borated polyethylene (5 wt% Boron) around the guide without boron carbide layers. The runs are made with 1 GeV neutron source.

Table 3. Intensity integrals under the projection curves and relative change in integral value, calculated in respect to the model with 3 cm of steel shielding around the guide, for models with 3 cm of steel shielding, 1 m thick steel wall, two B_4C layers and with different PE layer thicknesses at the sample position (EPD).

Model	Integral	Relative change
3 cm steel	0.01095±0.00003	n/a
20 cm PE layer	0.00264±0.00002	75.9±0.4 %
30 cm PE layer	0.00196±0.00001	82.1±0.4 %
40 cm PE layer	0.00150±0.00001	86.3±0.4 %
20 cm borated PE layer	0.00279±0.00002	74.5±0.4 %

From this we can conclude that the most reasonable thicknesses to use range between 20 cm and 30 cm. Using an expensive 20 cm borated polyethylene layer gives the same efficiency as the use of PE layer combined with two boron carbide layers, suggesting it is not the best material of choice.

For the model, described in this paper we decided to use a 20 cm, which reduces the total background level in a 1 GeV run from $(1.3\pm0.1)\cdot10^{-3}$ neutrons per cm^2 per 1 GeV source neutron in model without the shielding to $(2.6\pm0.5)\cdot10^{-4}$ neutrons per cm^2 per 1 GeV source neutron for the model with full shielding setup as described in this section.

8. Comparison of neutron spectra at sample position.

After having investigated the effectiveness of various shielding parts and their configurations in the simplified 1 GeV runs it is very important to perform the run with a full ESS spectrum. Having obtained the data we want to compare the neutron spectra at the sample position for various setups, namely: the model with only supermirror coating and substrate, model with steel shielding around the guide, the final model with shielding around the guide, PE, B_4C and the wall and some kind of benchmark. The benchmark chosen is the cosmic neutron rate at sea level. The idea behind comparing neutron spectra at the sample position to the spectrum of the cosmic neutrons at sea level is the cosmic neutron background defines an order of magnitude of the lowest obtainable background in the instruments. This cosmic neutron spectra is roughly sketched in figure 10 with a black line and is taken from [23].

As mentioned earlier in the text, only neutrons coming from the moderator are considered and we cannot and are not making any statements considering the overall background, which most definitely needs to be explored in the near future, given the fact that during the time of this study the bunker design has been developing.

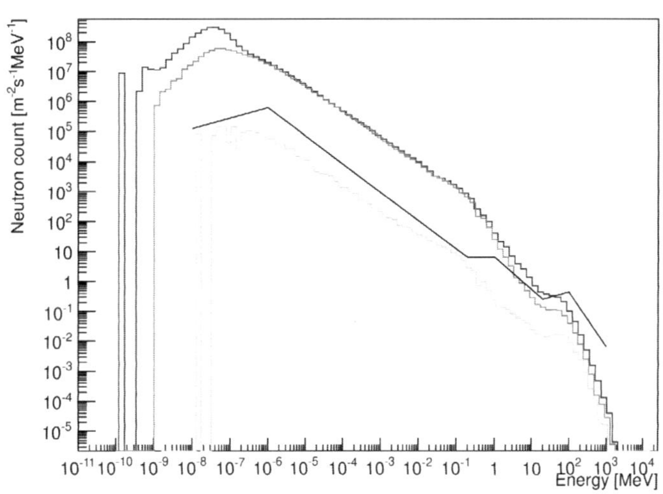

Figure 10. Neutron spectra comparison at the sample position, simulated with the full ESS neutron spectrum. Blue represents the model with only supermirror coating and substrate, red – the model with 3 cm of steel shielding around the guide, green – the final model with 3 cm of steel shielding and 20 cm PE layer around the guide, two 5 mm layers of boron carbide and the steel wall. The black line is a sketch of the cosmic neutron spectra at sea level, taken from [23].

From the plot in figure 10 we can see that the background without any shielding, just with supermirror coating and substrate (blue) is two orders of magnitude above the cosmic neutron rate, which does not satisfy requirements for detector operation. Having added 3 cm of steel shielding around the guide (red) we observe that some part of fast neutron intensities in energy range 10 – 100 MeV are cut down, which also leads to decrease in the amount of neutrons in thermal range $\approx$10 meV, presumably, due to the fact that there are less fast neutrons to induce secondary neutron emission.

Adding a 1 m thick, 10 m radius steel wall at position Z = 38 m to shield fast neutrons that get through the initial shielding, 20 cm thick layer of polyethylene to moderate fast neutrons and two 5 mm thick layers of boron carbide to absorb slowed down neutrons we move to the green curve. This curve goes below the black one in a broad energy range. Most important being the cold and thermal range – these are the neutrons that get scattered and detected in the experiments. More quantitatively the total background level has dropped from $(4.4\pm0.3)\cdot10^5$ $cm^{-2}s^{-1}$ to $(2.7\pm0.2)\cdot10^3$ $cm^{-2}s^{-1}$, which is a significant two orders of magnitude drop.

9. Conclusion

In this paper a shielding model against fast and thermal/cold neutrons for the 162 m long guide system of the BIFROST spectrometer at ESS has been proposed and tested using MCNPX. Only neutrons coming from the moderator have been considered. The BIFROST neutron guide is modelled using the metal-substrate sandwich technology, which has its roots in MIRROTRON's metal-glass sandwich neutron guide construction technology.

Different parts of the shielding model have been tested for their efficiency in a simplified mode with source neutrons having a fixed energy of 1 GeV, namely: three different thicknesses of steel shielding around the guide, three different thicknesses of the steel wall and three different thicknesses of polyethylene layers surrounding the steel shielding of the guide.

After the simplified runs the model with chosen thicknesses of the above mentioned parts was tested with the full ESS spectrum and the results have shown that the background level drops below the cosmic neutron rate at sea level in a broad range of energies, most importantly in the cold and thermal ranges, since neutrons of these energies are the ones being scattered by the sample and detected by the detectors.

Acknowledgements

We than Stuart Ansell for many illuminating discussions and Jonas Okkels Birk for providing the initial model of the BIFROST guide system.

References

[1] Peggs S *et al* 2013 *ESS Technical Design Report* ESS 2013-001 **10** 20
[2] Cherkashyna N, DiJulio D D, Panzner T, Rantsiou E, Filges U, Ehlers G and Bentley P M 2015 *Phys. Rev.* ST Accel. Beams **18**
[3] Webpage www.mcstas.org
[4] Lefmann K and Nielsen N 1999 *Neutron News* **10** 20
[5] Willendrup P, Farhi E and Lefmann K 2004 *Physica* B **350** 735
[6] Willendrup P K, Knudsen E B, Klinkby E B, Nielsen T, Farhi E, Filges U and Lefmann K 2014 *J. Phys.: Conf. Series* **528**
[7] Willendrup P, Farhi E, Knudsen E B, Filges U and Lefmann K 2014 *Journal of Neutron Research* **17** issue 1 pp 35 - 43
[8] Waters L S *et al* 2007 *The MCNPX Monte Carlo radiation transport code* AIP Conf.Proc. **896** pp 81-90
[9] X-5 Monte Carlo Team 1987 MCNP – A General Monte Carlo N-Particle Transport Code, Version 5, *LA-UR-03-1987*
[10] Bertelsen M 2014 *Optimizing neutron guides using the minimalist principle and guide_bot* University of Copenhagen
[11] Klinkby E B *et al* 2013 Interfacing MCNPX and McStas for simulations of neutron transport *Nuclear Instruments & Methods in Physics Research* A **700**
[12] Klinkby E B, Knudsen E B, Willendrup P K, Lauritzen B, Nobœl E, Bentley P, Filges U 2014 Application of the MCNPX-McStas interface for shielding calculations and guide design at ESS *J. Phys.: Conf. Series* **528**
[13] Webpage http://europeanspallationsource.se/cluster *DMSC*
[14] Freeman P G *et al* 2015 CAMEA ESS - The continuous angle multi-energy analysis indirect geometry spectrometer for the european spallation source *Phys. Web of Conf.* **83**
[15] Birk J O, Marko M, Freeman P G, Jacobsen J, Hansen R L, Christensen N B, Niedermayer C, Månsson M, Rønnow H M, Lefmann K 2014 Prismatic analyser concept for neutron spectrometers *Rev. Sci. Instr.* **85**

[16] Bertelsen M, Jacobsen H, Hansen U B, Carlsen H H, Lefmann K 2013 Prismatic analyzer concept for neutron spectrometers *Nuclear Instruments & Methods in Physics Research* A **85**
[17] Webpage 2006 http://www.mirrotron.hu/media/Brosh/Brosh10.pdf *MIRROTRON*
[18] Cherkashyna N *et al* 2015 *Overcoming High Energy Backgrounds at Pulsed Spallation Sources* Proceedings of ICANS XXI
[19] Batkov K, Takibayev A, Zanini L and Mezei F 2013 Unperturbed moderator brightness in pulsed neutron sources *Nuclear Instr. Meth.* A **729**
[20] Mantulnikovs K 2015 *Neutron guide shielding for the BIFROST spectrometer at ESS* University of Copenhagen
[21] Knudsen E B, Klinkby E B and Willendrup P K 2014 McStas event logger: Definition and applications *Nuclear Instruments and Methods in Physics Research* Section A: Accelerators, Spectrometers, Detectors and Associated Equipment **738** pp 20 - 24
[22] Cherkashyna N *et al* 2014 High energy particle background at neutron spallation sources and possible solutions *J. Phys.: Conf. Series* **528**
[23] Hagmann C, Lange D and Wright D 2012 Monte carlo simulation of proton-induced cosmic-ray cascades in the atmosphere *Lawrence Livermore National Laboratory*

On the Edge Refraction Contrast Imaging on a Conventional Neutron Diffractometer Employing Dispersive Double-Crystal Monochromator

P Mikula[1], M Vrána[1], V Em[2], J Čapek[3] and D Korytár[4]

[1]Nuclear Physics Institute ASCR, v.v.i., 25068 Řež, Czech Republic
[2]National Research Centre "Kurchatov Institute", Moscow, 123182, Russia
[3]Faculty of Math. and Physics, Charles University, Prague, Czech Republic
[4]Institue of Physics SAS, Dúbravská cesta 9, 845 11 Bratislava, Slovak Republic

E-mail: mikula@ujf.cas.cz

Abstract. Conventional radiography is based on absorption effects, which depend only on the imaginary part of the refractive index. However, slits and macroscopic objects such as edges give rise to refraction and this phenomenon can be effectively used in the so called phase-contrast imaging. In addition to the absorption contrast in the conventional radiography, it exploits also contributions of phase shifts induced by the propagation of a coherent neutron (or X-ray) beam through an investigated sample. The difference in real part of the refrative index between a studied detail and its vicinity results in a phase shift between the wave transmitted through and outside the detail. Due to this phase shift, the two waves interfere. However, the used radiation has to possess a sufficiently high spatial coherence permitting one to detect resulting interference pattern. In this contribution we present two simple neutron diffractometer configurations based on two bent perfect Si crystals which provide a high spatial coherence beam of sufficient intensity which could be potentially used in some phase contrast radiography experiments.

1. Introduction

It is well known that depending on geometry or performance of diffraction by single crystals one can easily manipulate e.g. with resolution, cross-section and focusing properties of the diffracted beam in the scattering plane. In the case of manipulation of the cross-section of the beam, the simplest way is to use asymmetric diffraction geometry, which has been studied many times and described in many text books related to neutron or X-ray diffraction. Namely, it is often used for special monochromator systems on scattering instruments using synchrotron radiation. The extreme case of the asymmetric diffraction geometry is so called fully asymmetric diffraction geometry (FAD geometry). Such a unique diffraction geometry employing crystal of large dimensions, namely the length, is practically applicable only in neutron diffraction. Neutron has a high penetration ability and bent perfect crystals (BPC) of Si are very suitable for such experiments and studies. For observation and imaging of edge

[1] To whom any correspondence should be addressed.

refraction effects one needs to have not only a beam of a sufficiently large cross-section but also the beam has to have a required spatial coherence [1-7]. High coherence of the beam, however, can be easily achieved by using a dispersive or quasi-dispersive double-crystal diffraction setting employing optimally bent perfect crystals (see figures 1 and 2). In the case of using the FAD geometry, two alternatives can be distinguished: the FAD geometry with the output beam compression and the opposite one - the FAD geometry with the output beam expansion. In the former case e.g. a wide incident polychromatic beam is impinging the crystal and a narrow (compressed) beam is diffracted when passing the crystal perpendicularly to its thickness e.g. along its longest edge. Of course, such alternative is not suitable for studies of edge refraction effects. This alternative was already tested in the eighties for exploitation in the designing of new types of neutron monochromators [9-10]. In the latter case a narrow beam enters the bent-crystal slab through the end face and, after passing along the longest edge of the slab, the diffracted monochromated beam is expanded to a large cross-section (see figure 2). Such FAD geometry has been then succesfully employed in the high resolution double bent crystal (1,-1) SANS diffractometer with the FAD analyzer as a second crystal [11-17] which provided

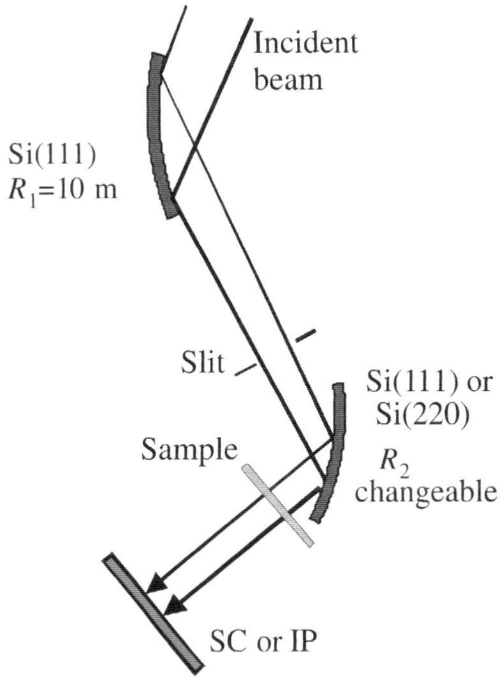

Figure 1. Scheme of the quasi-dispersive diffractometer performance. SC - scintillation camera, IP - imaging plate.

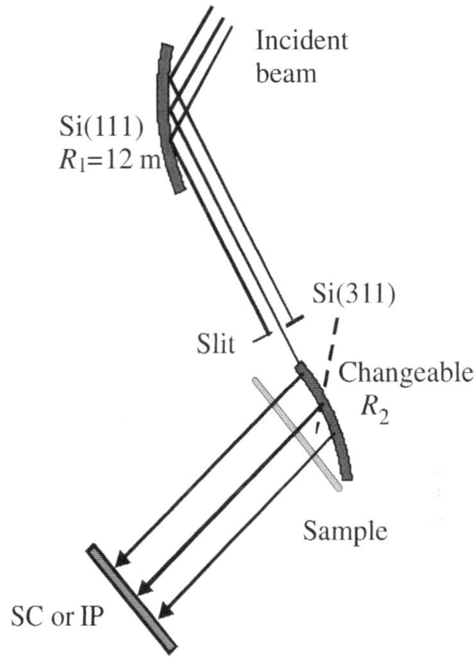

Figure 2. Scheme of the quasi-dispersive diffractometer performance with the BPC Si(311) crystal in FAD geometry.

a substantial enlargement of the range of analyzed angles that could be transformed to a linear spatial dimension. It opened the possibility to use a one dimensional position sensitive detector for the analysis of the SANS signal instead of the step-by-step analysis with the conventional symmetric diffraction geometry of the crystal analyzer. Recently, double bent crystal dispersive (n,-m) arrangement with the second FAD crystal in the the output beam expansion diffraction geometry has been studied for the imaging of edge refraction effects [10,11] by using rather complicated three axis diffractometer setting. In this case, before the double-crystal system a bent perfect premonochromator was used. In the present paper we introduce simpler versions providing monochromatic beams of acceptable spacial coherence permitting to observe edge refraction effects.

2. Experimental details

The experiments were carried out on the neutron optics diffractometer operating at λ=0.162 nm which is installed at the medium power research reactor LWR-15 in Řež. Figures 1 and 2 show schematically

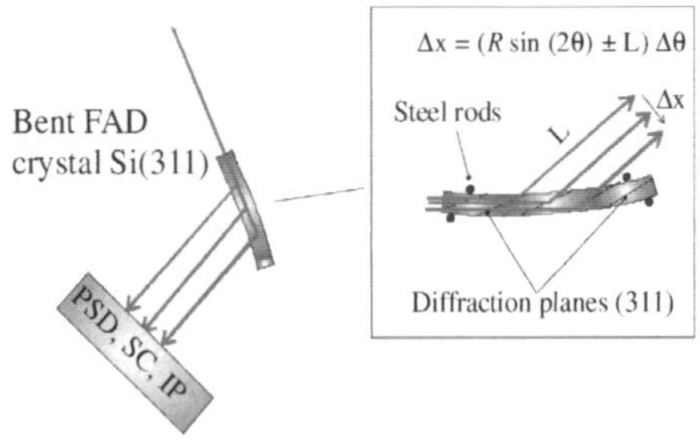

Figure 3. The detail of the FAD geometry transforming an angular deviation $\Delta\theta$ from the mean Bragg position.

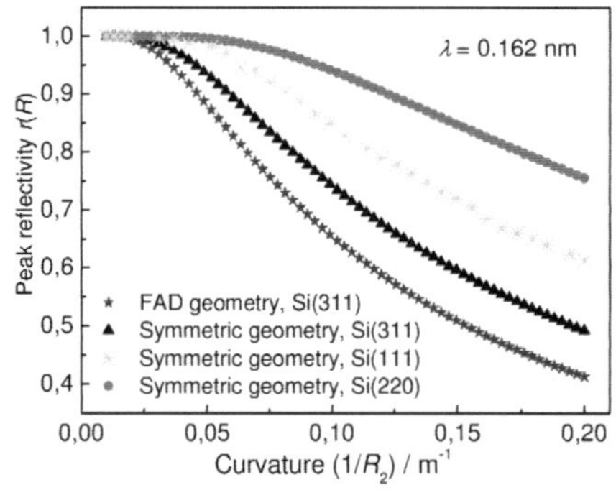

Figure 4. Peak reflectivity vs crystal curvature for several crystal slabs.

the double bent crystal arrangements. Figure 3 shows the detail of the FAD diffraction transforming an angular deviation $\Delta\theta$ from the mean Bragg position, e.g. in the middle of the FAD-crystal, to the linear spatial shift Δx [12]. In combination with the dispersive setting with the bent Si(111) crystal such double crystal configuration represents a self-collimation system providing a high spatial coherence without necessity to use any Soller collimator. For bending the BPC slabs we used four-point bending devices [16] (see the steel rods in figure 3). The reflecting properties of the BPC slabs will not be discussed here and can be found in our second paper of these proceedings or in detail in [8]. Here we introduce only the reflection probability (called peak reflectivity) of the neutron when meeting the Bragg condition in the BPC Si-crystal as a function of the crystal curvature. The dependences for Si(311) reflection with the crystal in symmetric and FAD diffraction geometries are shown in figure 4 for the sake of comparison. All Si crystal slabs used in the experiment were of the length of 200 mm and the width (height) of 40 mm. The thickness of the first Si(111) slab working as a monochromator was 4 mm. It had a fixed curvature of 0.08 nm^{-1} (R_I=12 m). As a second crystal we used successively Si(111), Si(220) and FAD-Si(311) of the thickness of 2 mm, 3 mm and 4 mm, respectively. For the imaging of the beam profiles from the second crystal (without a sample) we used scintillation camera (SC) of the spatial resolution of 0.1 mm and for imaging of the edge profiles of the beam after passing through the sample we used the Imaging plate (IP) having 50 μm spatial resolution.

3. (1,-1) diffraction setting with the 2 mm thick Si(111) second slab (see figure 1)

First, as a second crystal we used the bent Si(111) crystal in the (1,-1) setting. However, due to the opposite sense of the curvatures of the individual slabs, the setting is slightly dispersive. As

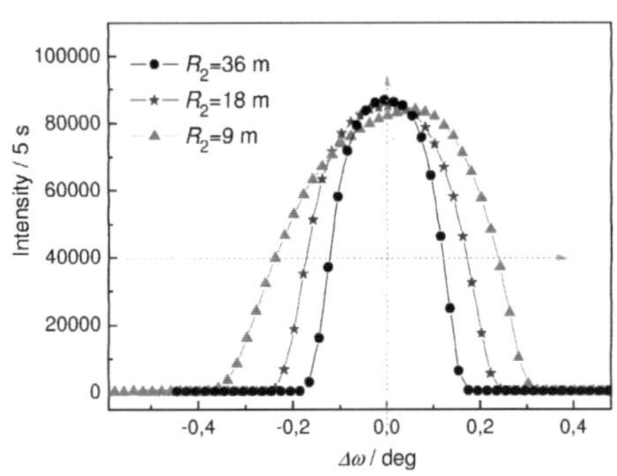

Figure 5. Rocking curves (RC) of the second Si(111) bent perfect crystal for different R_2.

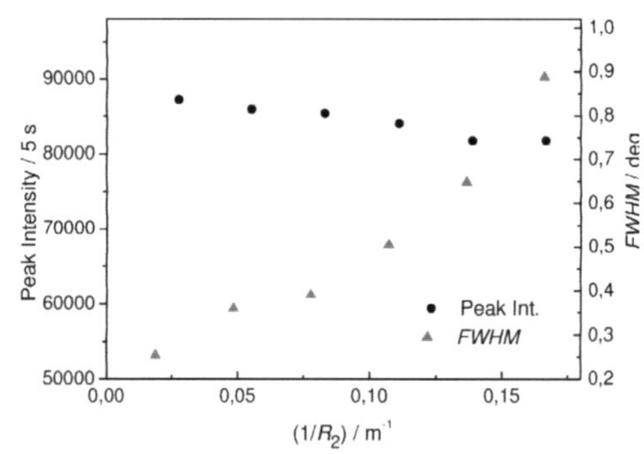

Figure 6. Parameters of the rocking curves vs the curvature of the second Si(111) slab.

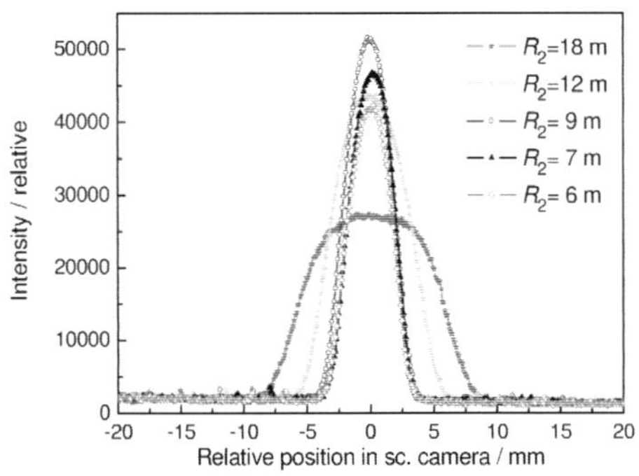

Figure 7. Diffraction profiles of the output beam taken at the peak of the rocking curve and imaged by SC at the distance of 65 cm.

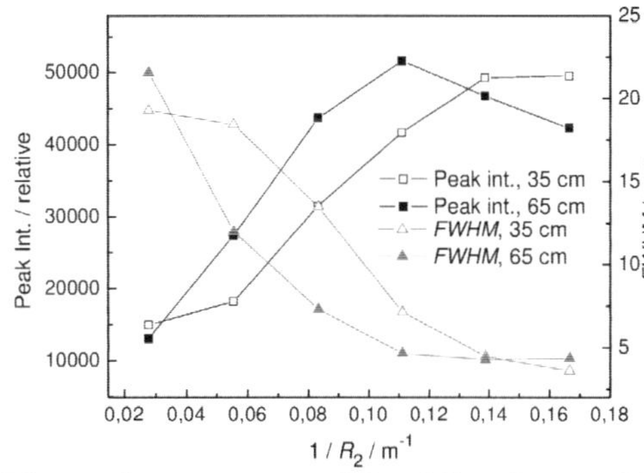

Figure 8. Parameters of the diffraction profiles registered by SC at 35 cm and 65 cm.

documented in the figures from 5 to 8, the dispersity becomes stronger with the increase of the crystal curvature. On the other hand, when the crystals have the curvatures of the same sense, the setting becomes nearly nondispersive depending on the mutual distance of the bent crystal slabs (for details we refer to [8], [9] and [12]). The figures from 9 to 11 show that a slight dispersity of the setting provides a sufficient spatial coherence required for observation of the edge refraction effects.

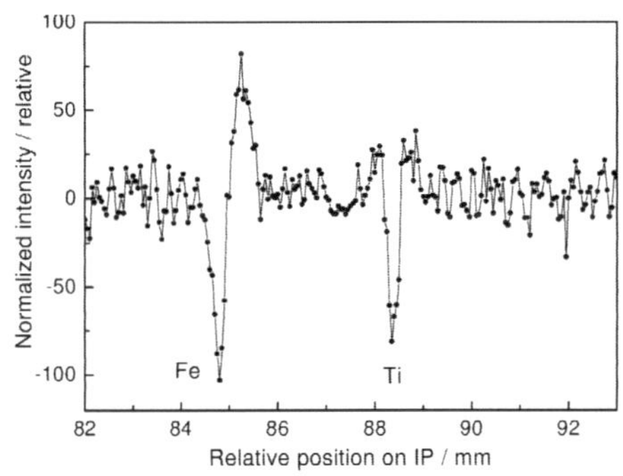

Figure 9. Refraction effects obtained on Fe (2 mm) and Ti (2 mm) edges for IP at 40 cm from the sample and R_2=36 m.

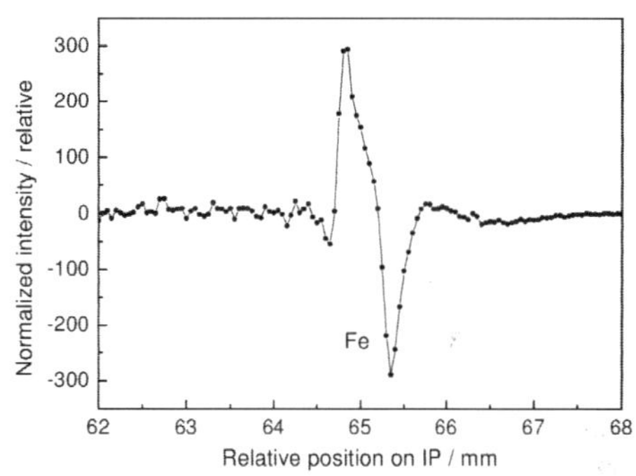

Figure 10. Refraction efffect on Fe (9 mm) edge for IP at 35 cm from the sample and R_2=36 m.

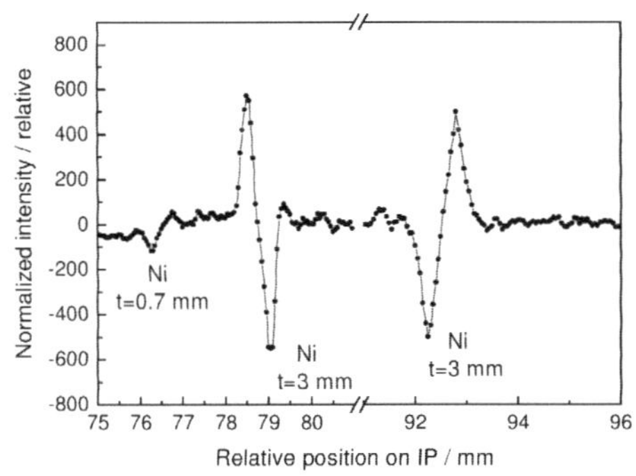

Figure 11. Refraction effects obtained on edges of Ni plates of different thickness t for IP at 40 cm and R_2=36 m.

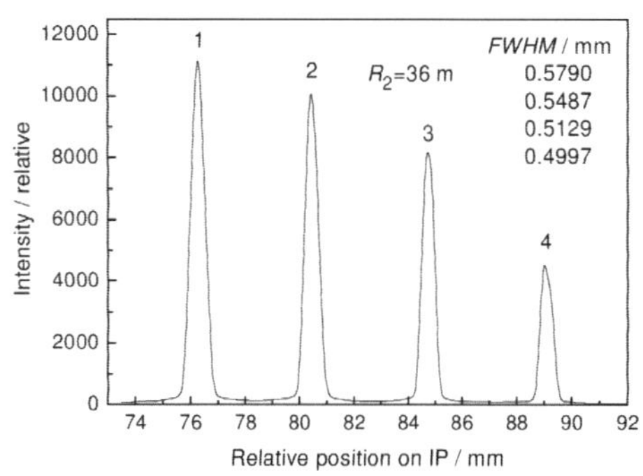

Figure 12. *FWHMs* of the profiles of the beam passing through Gd slits of different widths (see text) at the distance of 40 cm.

Figure 12 shows the collimation of the double diffracted beam when using a system of Gd slits of the width of 0.47 mm, 0.36 mm, 0.24 mm and 0.11 mm. The beam divergence was of about 1x10^{-3} rad.

4. (n,-m) diffraction setting with the 3 mm thick Si(220) second slab (see figure 1)

Similarly to the previous setting, in this case we used a Si(220 slab as a second crystal which makes together with the Si(111) monochromator quasi-dispersive perfomance. The inspection of the figures

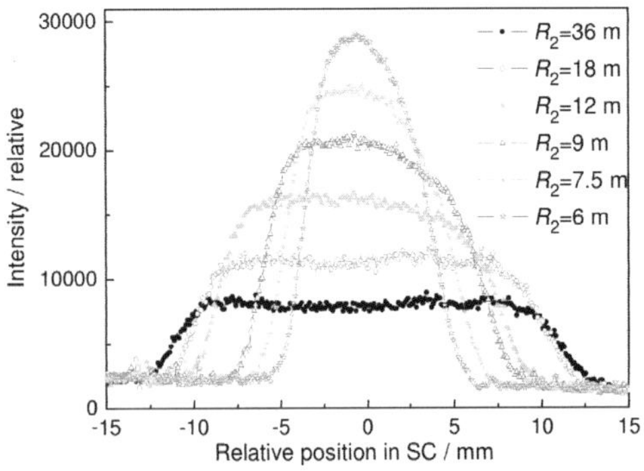

Figure 13. Diffraction profiles of the output beam taken at the peak of the rocking curve and imaged by SC at the distance of 35 cm.

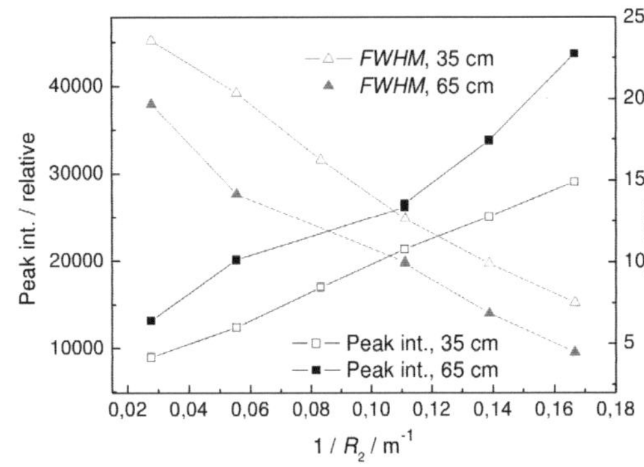

Figure 14. Parameters of the diffraction profiles registered by SC at 35 cm and 65 cm.

13 and 14 reveals that with the increase of the curvature $1/R_2$, the dispersity of the setting becomes more pronounced and thus the *FWHM* of the beam profile decreases. However, the peak intensity of

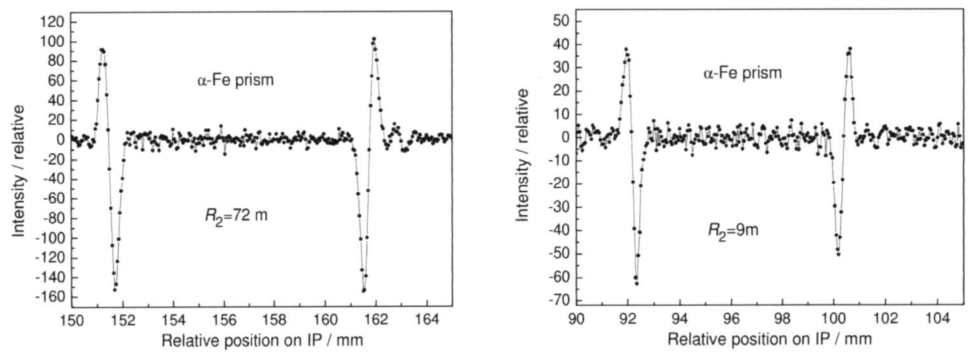

Figure 15. Refraction effects on both sides of squared 9 mm α-Fe prism imaged by IP at 40 cm from the sample and for R_2=72 m and R_2=9 m.

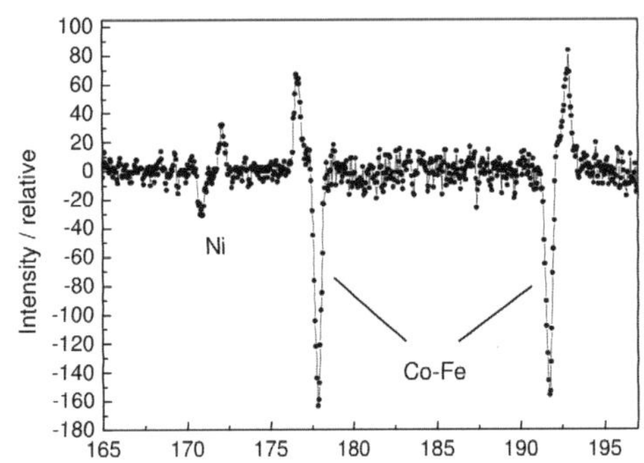

Figure 16. Refraction effects on both sides of Co-Fe plate and a Ni of 3 mm imaged by IP at 32 cm from the sample and for R_2=12.5 m.

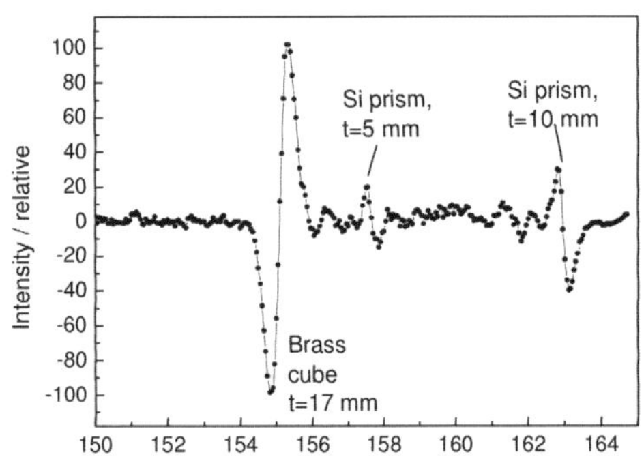

Figure 17. Refraction effects on three samples of different widths t of edges imaged by IP at 40 cm from the sample and for R_2=9 m.

the profile increases due to the increase of the effective mosaicity of the BPC slab. In the whole range of curvatures the focusing is rather weak (see the difference in *FWHM* for the distance of 35 cm and 65 cm). Figures from 15 to 17 show several examples of the imaging of the edge refraction effects and document the quality of the output double diffracted beam.

5. (n,-m) diffractometer setting with the BPC Si(311) crystal in the FAD geometry

As a most promisig configuration, the (n,-m) diffractometer setting with the BPC Si(311) crystal in the FAD geometry has appeared. Of course, depending on the crystal cut of the slab one can use the FAD diffraction geometry with another system of lattice planes. By using the Si(311) reflection the dispersity of the (n,-m) is more pronounced than in the previous case. In addition, depending on the curvature of the FAD crystal the width of the double diffracted beam could be substantially enlarged as demostrated in the next figure 18. Such a large beam was then used for the imaging of the teeths of a 8 cm wide steel comb. One part of the result can be seen in figure 19. The next figures from 20 to 23 provide further examples of refraction effects on the edges of different materials.

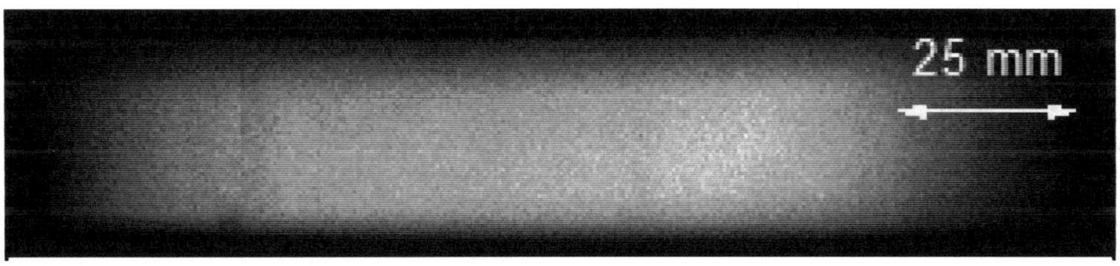

Figure 18. An example of the image of the diffracted beam by the FAD crystal for $R_{2,FAD}$=12 m.

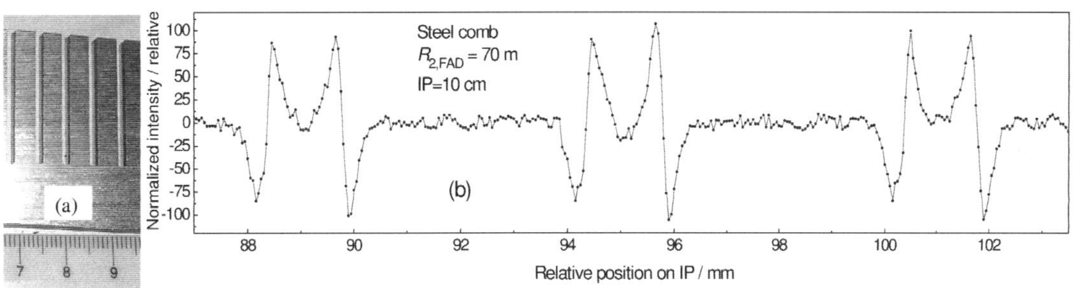

Figure 19. Imaging of the refraction effects (b) on the steel comb (a) with 4.5x4.5 mm^2 teeth.

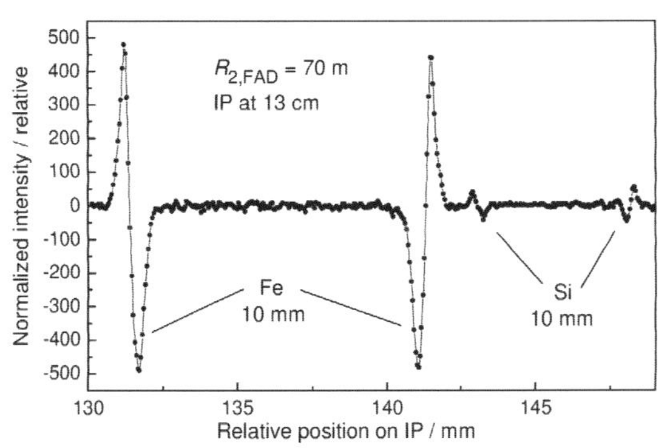

Figure 20. Refraction effects on the edges of squared α-Fe (10 mm) and Si (10 mm) samples.

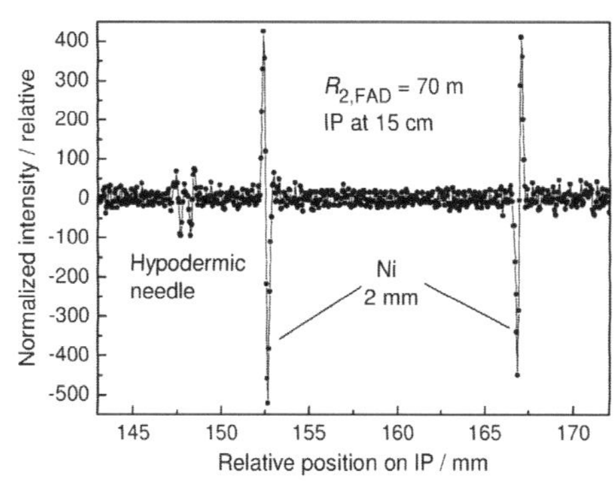

Figure 21. Refraction effects on a hypodermic needle and Ni (2 mm) edges.

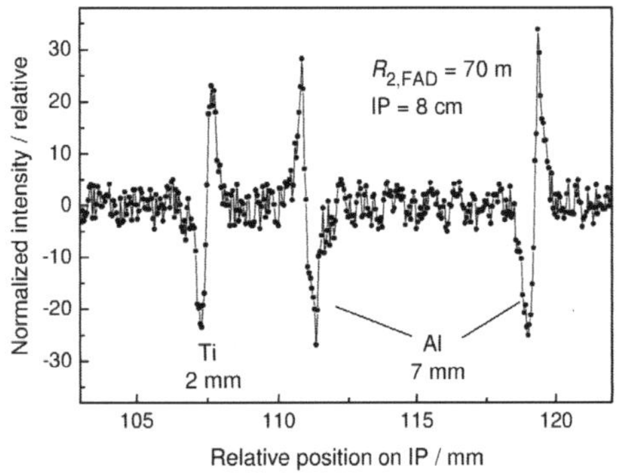

Figure 22. Refraction effects on the edges of Ti (2 mm) and Al (7 mm) samples.

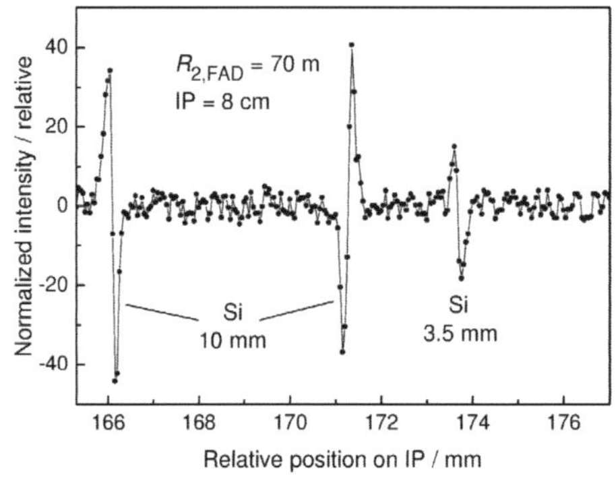

Figure 23. Refraction effects on the edges of Si (10 mm) and Si (3.5 mm) samples.

6. Summary

The quasi-dispersive double bent crystal arrangements provide a highly parallel monochromatic beam having a cross-section of several square centimetres which can be also suitable, besides other things, for conventional radiography experiments. Moreover, the beam possesses sufficiently high spatial coherence and thus, the phase shift contrast phenomenon can be also exploited. As can be seen from the presented figures, the refraction edge profiles can be observed even in the (1,-1) setting of the bent Si(111) crystals with the mutual curvatures far from the nondispersive diffraction geometry. However, it should be pointed out that the quasi-dispersive experimental performances have such excellent properties only in one dimension, i.e. in the scattering plane. As demonstrated in the FAD case, the spatial coherence and the magnifying of the beam cross-section can be (in some range) easily manipulated by changing the curvature of the FAD crystal.

Acknowledgements

The related measurements were carried out at the CANAM infrastructure of the NPI ASCR Rez. Bragg diffraction optics investigations are in the Czech Republic supported by GACR project no. 14-36566G, by the ESS project LM2010011: ‘Contribution to Partnership in Large Research Infrastructure of Pan-European Importance’ as well as by the project of EU-FP7-NMI3 II: “Integrated Infrastructure Initiative for Neutron Scattering and Muon Spectroscopy”, 2009-2015.

References

[1] Podurets K M, Somenkov E A, Chistyakov R R and Shilstein S Sh 1989 *Physica B: Condensed Matter*, **156-157** 694-697.
[2] Treimer W and Schaper J 1996 *BENSC Experimental Report*, p 321.
[3] Treimer W, Feye-Treimer U and Herzig C 1998 *Physica B* **241-243** 1197-1203.
[4] Kardjilov N, Lehmann E, Steichele E and Vontobel P 2004 *Nucl. Instrum. Methods in Phys. Res.* **A 527** 519-530.
[5] Strobl M, Treimer W, Kardjilov N and Hilger A 2005 *Nucl. Instrum. Methods in Phys. Res.* **A 542** Issues 1-3, 383-386.
[6] Strobl M, Treimer W, Kardjilov N, Hilger A and Zabler S 2008 *Nucl. Instrum. Methods in Phys. Res.* **B 266** 181-186.
[7] Strobl M, Kardjilov N, Hilger A, Kune G, Frei G and Manke I 2009 *Nucl. Instrum. Methods in Phys. Res.* **A 604** 640-645.
[8] Mikula P, Kulda J, Vrána M and Chalupa B 1984 *J. Appl. Cryst.* **17** 189-195.
[9] Mikula P, Chalupa B, Kulda J, Vrána M, Sedláková L and Michalec R 1985 *J. Appl. Cryst.* **18** 135-140.
[10] Mikula P, Kulda J, Horalík L, Chalupa B and Lukáš P 1986 *J. Appl. Cryst.* **91** 324-330.

[11] Lukáš P, Mikula P, Šaroun J and Strunz P 1994 *Nucl. Instrum. Methods in Phys. Research*, **A 338** 111-115.
[12] Šaroun J, Lukáš P, Mikula P and Alefeld B 1994 *J. Appl. Cryst.* **27** 80-88.
[13] Hempel A, Eichhorn F, Reichel P and Boede W 1996 *Nucl. Instrum. Methods in Phys. Res.* **A 381** 466-471.
[14] Mikula P, Vrána M, Seong B S, Woo W, Em V and Korytár D 2014 *IOP Publishing Journal of Physics: Conference Series*, **528** 012004.
[15] Mikula P, Vrána M, Šaroun J, Pilch J, Seong B S, Woo W and Em V 2014 *J. Appl. Cryst.* **47** Part 2, 599-605.
[16] Mikula P, Kulda J, Lukáš P, Vrána M, Ono M and Sawano J 2000 *Physica B*, **276-278** 174-176.

Journal of Physics: Conference Series **746** (2016) 012029 doi:10.1088/1742-6596/746/1/012029

Comparison of double crystal (+n,-m) and (+n,+m) settings containing a fully asymmetric diffraction geometry of a bent perfect crystal with the output beam expansion

P Mikula[1], M Vrána[1], J Šaroun[1], W Woo[2], V Em[3], J Čapek[4] and D Korytár[5]

[1]Nuclear Physics Institute ASCR, v.v.i., 25068 Rez, Czech Republic
[2]Neutron Science Division, Korea Atomic Energy Research Institute, 1045 Daedok daero, Yuseong-gu, Daejeon, 305-353, Republic of Korea
[3]National Research Centre "Kurchatov Institute", Moscow, 123182, Russia
[4]Faculty of Math. and Physics, Charles University, Prague, Czech Republic
[5]Institue of Physics SAS, Dúbravská cesta 9, 845 11 Bratislava, Slovak Republic

E-mail: mikula@ujf.cas.cz

Abstract. In this paper neutron diffraction properties of the dispersive double crystal (+n,-m) and (+n,+m) settings containing the bent perfect crystal Si(220) in symmetric diffraction geometry and the bent perfect crystal Si(311) in the fully asymmetric diffraction geometry with the output beam expansion are studied and compared. It has been found that the properties, namely, the width of the output double diffracted beam, of both settings strongly depends on the relative combination of curvatures of Si(311) crystal slab with respect to the curvature of the Si(220) one. Moreover, after the beam expansion the fully asymmetric diffraction geometry provides in both cases a highly collimated and highly monochromatic beam of a rather large cross-section of several square centimetres. Application possibilities, of such a beam, namely for neutron phase contrast radiography will be discussed.

1. Introduction

Several papers concerning studies of diffraction properties of bent perfect crystal (BPC) slabs in the fully asymmetric diffraction (FAD) geometry due to a possible use for highly efficient monochromators and analyzers were already published in the past [1-10]. Recently, we have decided to study another beneficial properties with the goal of obtaining an unconventional high-resolution neutron optics components. While in the past nondispersive (1,-1) double-bent-crystal settings were also studied [6-10], recent experimental investigations have been focused on studies of the properties of the monochromatic beam produced by a dispersive (or quasi-dispersive) double-bent-crystal settings with the second FAD crystal as an analyzer. It is clear that in practice, the FAD geometry can be effectively studied in the case of Si-BPC slabs, because the attenuation factor for thermal neutrons in Si is very low [11]. The first recent experimental investigations [12] have shown that thanks to the possibility of the variation of the curvature of both BPC crystals, one can easily manipulate with the width of the obtained monochromatic beam in the range from few to tens of millimetres. As the double crystal setting Si(111)+Si(311) is dispersive, the monochromatic beam is naturally highly collimated

[1] To whom any correspondence should be addressed.

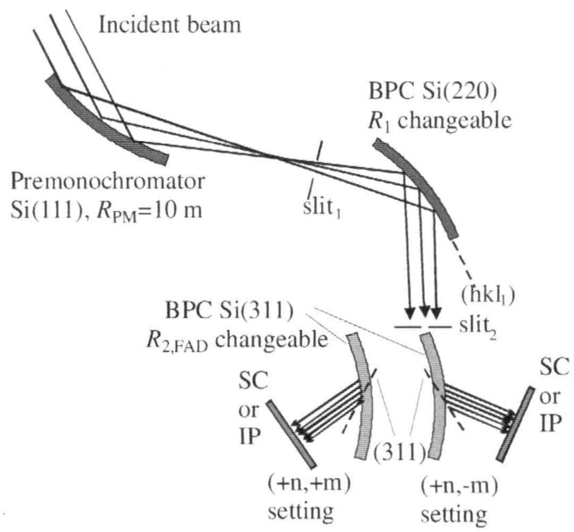

Figure 1. Schematic sketch of the experimental arrangement used for the studies of the FAD-OBE properties of the BPC Si (311) crystal in the (+n,-m) or (+n,+m) double-crystal setting with respect to the Si(220) one. SC - scintillation camera, IP - imaging plate.

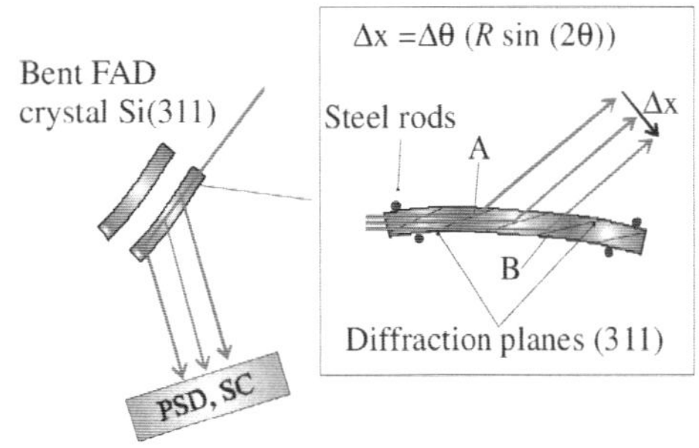

Figure 2. A detail of the sketch showing the FAD geometry of the Si crystal slab as transforming the Bragg angle deviation $\Delta\theta$ between A and B into the linear spatial one.

without the use of any collimators and the FAD crystal can work as one dimensional magnifying element. It should be mentioned that two possibilities of the FAD diffraction geometry are possible: with the output beam compression (FAD-OBC) or with the output beam expansion (FAD-OBE). In the case of the FAD-OBC geometryě a wide incident beam (several centimetres) is incident on the FAD crystal and a condensed output monochromatized beam of the width of a few millimetres is diffracted. In the present case of the opposite FAD-OBE geometry (see figures 1 and 2) a narrow beam enters the bent FAD crystal slab (of the width of several millimetres) through its end face and after passing along the longest edge of the slab, depending on its curvature, the diffracted monochromatized beam is expanded to a rather large cross-section. In this paper the comparison of these two (+n,-m) and (+n,+m) settings is carried out. Dimensions of all Si crystal slabs used in the diffractometer performances were: 200x40x4 mm^3 (length x width x thickness).

2. Reflecting properties of the bent FAD perfect crystal

Bent perfect crystals (BPC) are usually employed in symmetric or slightly asymmetric diffraction geometry. They are well known, namely, for their focusing properties in real as well as momentum space with respect to a sample. Generally, the BPC elements are attractive for an employment from the following reasons: predictability and reproducibility of the effective mosaicity, predictability and reproducibility of a rather high peak reflectivity and its uniformity over large areas of the crystal, a rather high peak reflectivity for asymmetric or transmission geometry and predictability of the integrated reflectivity. In the transmission geometry one can also benefit from natural wavelength focusing. In the limiting case of the FAD-OBE geometry which was used in the present case, some more properties can be studied which could be possibly used for the application in neutron diffractometry. As can be seen from figure 1, depending on the curvature of the crystal slab a narrow incident beam can be spatially enlarged. In such a way, it is possible to manipulate not only with focusing in real and momentum space but also with a spacial distribution (width) of the diffracted beam. Even though one can expect some focusing in real space in the diffracted beam, due to the output beam expansion, it has not to be so pronounced. On the other hand, the expansion of the diffracted beam directly influences the separation of the wavelengths of individual neutron rays within the Δx distribution (see figure 2) which is, of course, strongly correlated. It is well known that in the case of BPC symmetric and asymmetric diffraction, Bragg law introduces a correlation between the direction and the wave-vector of neutrons, but in the case of FAD geometry there is a new correlation between the position within the Δx range and the wave-vector of neutrons. In some experiments this can be successfully exploited in combination with a position sensitive detector [6-10].

The general formula for the peak reflectivity (reflection probability) of neutron passing through a homogeneously deformed perfect crystal (see figure 3) has been derived in the form [1,13]

$$r(R) = \{1 - \exp[-Q_{hkl}(\partial\Delta\theta/\partial\Delta s)^{-1}]\}, \tag{1}$$

where $(\partial\Delta\theta/\partial\Delta s)$ is the rate of the change of the Bragg angle θ on the flight path Δs in the crystal in the incident beam direction and $Q_{hkl} = (F_{hkl})^2\cdot\lambda^3/(\Omega^2\cdot\sin 2\theta)$ is the kinematical reflectivity of the crystal unit volume. F_{hkl}, λ and Ω are the structure factor, neutron wavelength and the unit-cell volume, respectively [14]. The effective mosaicity $\delta\theta = \delta\theta_1 - \delta\theta_2$ (see figure 3) and the integrated reflectivity $\rho^\theta(R)$ can be for homogeneous bending defined as $\delta\theta(R) = s\cdot(\partial\Delta\theta/\partial\Delta s)$ and $\rho^\theta(R) = \delta\theta(R)\cdot r(R)\cdot A(\mu)$, respectively [1]. $s = t/\sin(\theta+\psi)$ is the distance between the points A and B (see figure 3), t is the thickness of the crystal slab, ψ is the angle of asymmery and $A(\mu)$ is the attenuation factor. Then, for FAD geometry we can derive simple expressions for the effective mosaicity and the peak reflectivity [1]

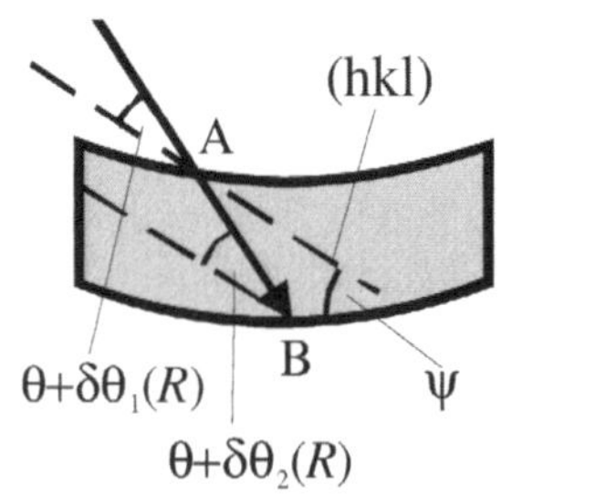

Figure 3. Detail of the incident beam passing through the BPC slab (general diffraction geometry).

$$\delta\theta(R) = s/R\,, \tag{2}$$

$$r(R) = 1 - \exp(-Q_{hkl}\cdot R)\,. \tag{3}$$

Due to the difference in the path length s in the crystal, the value of the effective mosaicity is different in the incident and diffracted beam direction. In the present FAD case, for the incident beam $\delta\theta(R) = L/R$, where L is the length of the crystal. Figure 4 shows the calculated dependences of the peak reflectivities $r(R)$ for Si(311) FAD diffraction geometry and also for Si(220) slab set in symmetric diffraction geometry (for the sake of comparison) on the crystal curvature ($1/R$). It can be seen from figure 4 that the peak reflectivity corresponding to Si(311) FAD geometry is smaller than the one corresponding to the symmetric Si(220) diffraction geometry, namely, due to the difference in $(F_{hkl})^2$ and thus in Q_{hkl}. As for the attenuation factor $A(\mu)$, then in our case of low attenuating Si crystal ($\mu = 1.9\cdot 10^{-2}$ cm^{-1}) and a rather short width of the diffracted beam (less than 3 cm, see figures 5 and 8), we can simply use for calculations the approximation expression $A(\mu) = \exp(-\mu\cdot L/2)$.

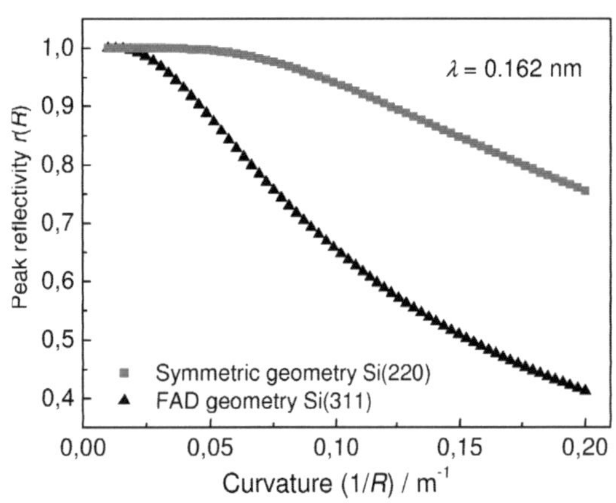

Figure 4. Peak reflectivity (reflection probability) vs crystal curvature calculated according to the formulae (1) and (3) for symmetric Si(220) as well as FAD Si(311) diffraction geometry.

3. Experimental results

The experimental studies of the (+n,-m) and (+n,+m) double-crystal settings with the FAD-OBE geometry of the second crystal-analyzer were carried out on the dedicated neutron optics diffractometer installed at the reactor LWR-15 in Řež. It operates at the fixed neutron wavelength of 0.162 nm provided by a bent Si(111) premonochromator (the radius of curvature - 10 m). No Soller collimators are used on the beam path from the reactor to the detector. Due to a small Bragg angle for Si(111) crystal (15°), the premonochromatic beam of a rather large divergence and $\Delta\lambda/\lambda$ spread of about 4.4×10^{-2} [14] can be considered from the point of view of the resolution properties of (+n,-m) and (+n,+m) settings as quasi-polychromatic. Thus, it was used for the following studies. The distance between the Si(111) and Si(220) crystals was 1.7 m and the distance between the Si(220) and Si(311) FAD crystal was 0.5 m. As the curvature of both BPC slabs Si(220) and Si(311) were changeable, the properties of the beam profiles as registered by the scintillation camera (SC) or imaging plate (IP) could be studied for different combination of the individual curvatures. The width of the incident beam entering the FAD crystal was limited by a 3 mm slit.

3.1. Parallel (+n,-m) double-crystal arrangement

Figure 5 shows the experimental results of the width (represented by *FWHM*) of the output expanded beam as a function of curvature of FAD crystal for different curvatures of the Si(220) crystal slab. It can be seen from figure 5 that for small curvatures $1/R_{2,\mathrm{FAD}}$, the length ΔL of the area inside the FAD

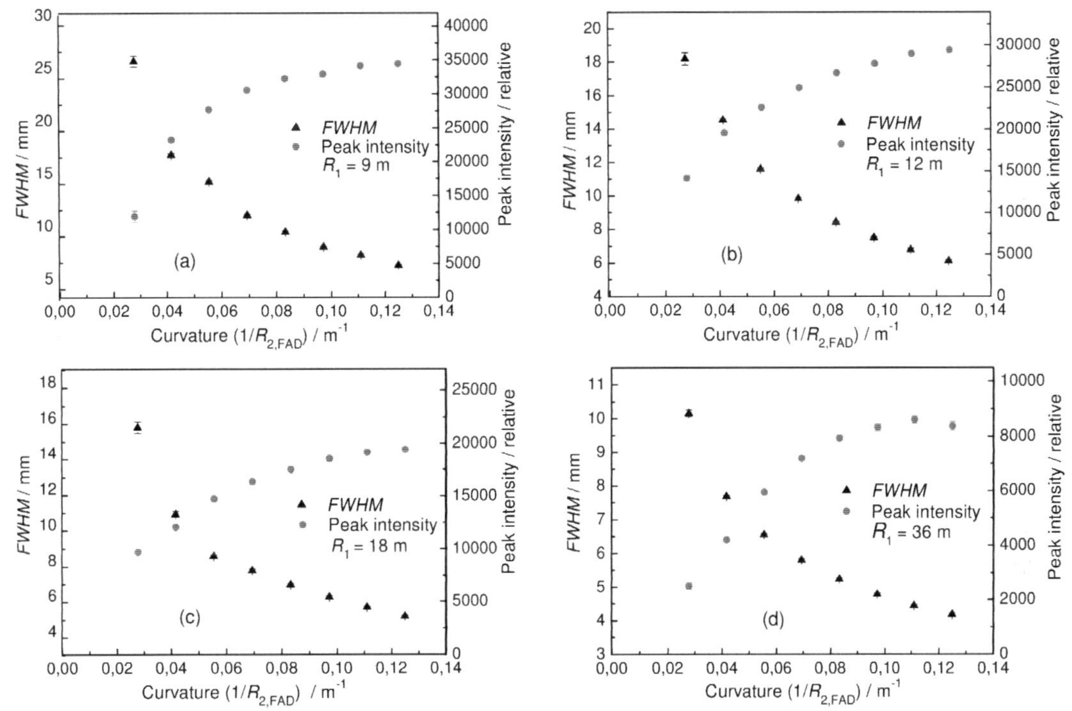

Figure 5. The dependences of the peak intensity and *FWHM* of the diffraction profile on the FAD crystal curvature as registered by the scintillation camera (at 56 cm from the FAD crystal) for fixed curvatures of the Si(220) one: (a) - 0.11 m^{-1}, (b) - 0.083 m^{-1}, (c) - 0.056 m^{-1} and (d) - 0.028 m^{-1}.

crystal where the Bragg condition for neutrons coming from the bent Si(220) crystal takes place is less than 30 mm (ΔL = *FWHM*/sin 2θ). In all cases of the fixed curvatures of the BPC Si(220) crystal, *FWHM* of the double diffracted beam decreases when the curvature of the FAD crystal increases. It achieves values of several millimeters at the place of the camera. As the corresponding peak intensity of the profiles does not change considerably for large curvatures of the FAD crystal, it means that the change of *FWHM* is not given directly by focusing in real space but by overlapping the phase space elements of individual crystals. It means that the length ΔL of the volume element where the diffraction process in the FAD crystal takes place and which then transforms to Δx, decreases with the curvature. It also means that the dispersity between the Si(220) and FAD-Si(311) crystals becomes more pronounced. Therefore, one can easily manipulate with the width of the double diffracted beam. When considering cylindrical bending of the FAD crystal, it could be expected that the deviation from the mean Bragg angle linearly changes along its longest edge. Figure 6 experimentally proves this

Journal of Physics: Conference Series **746** (2016) 012029 doi:10.1088/1742-6596/746/1/012029

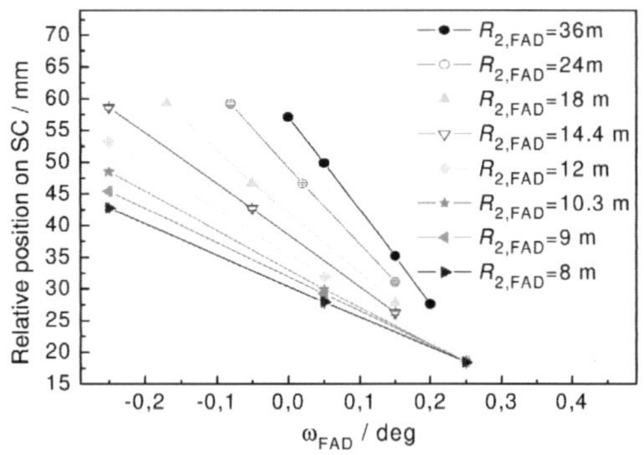

Figure 6. Calibration of the rocking angle of the FAD crystal versus the peak position of the diffraction profile on the scintillation camera.

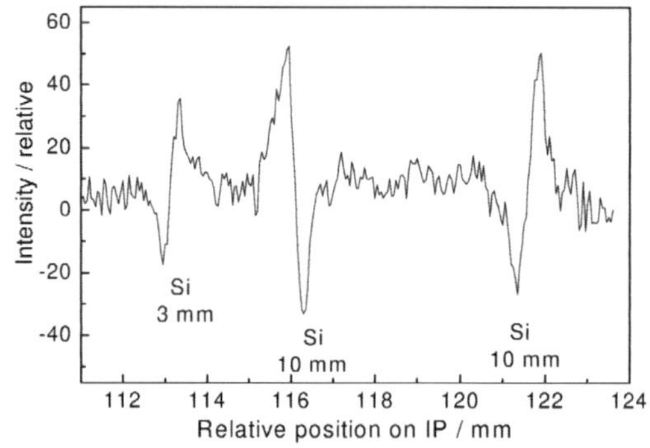

Figure 7. Edge profiles obtained on the two sides of a Si prism of the 10 mm thickness and on the one side of a 3 mm thick Si slab.

property of the linear change of the Bragg angle for different values of the bending radii when rocking the FAD crystal with respect to the bent Si(220) one. The shift of the diffraction profile on the scintillation camera can be expressed by a formula $\Delta x = \Delta\theta \cdot (R_{FAD}\sin(2\theta))$ (see the insert in figure 2). Similarly to the exploitation of FAD crystal in the nondispersive (+n,-n) setting [6-8], this property can be successfully used for analyzing some effects related to scattering to small angles. The imaged refraction edge profiles obtained on Si samples (IP was at the distance of 35 cm from the sample) which are shown in figure 7 document that the obtained output monochromatic beam is of a high spatial coherence. For imaging a similar experimental arrangement as shown in figure 9 was used.

3.2. Antiparallel (+n,+m) double-crystal arrangement

Similarly to the quasi-parallel (+n,-m) setting, the properties of the output expanded beam in the

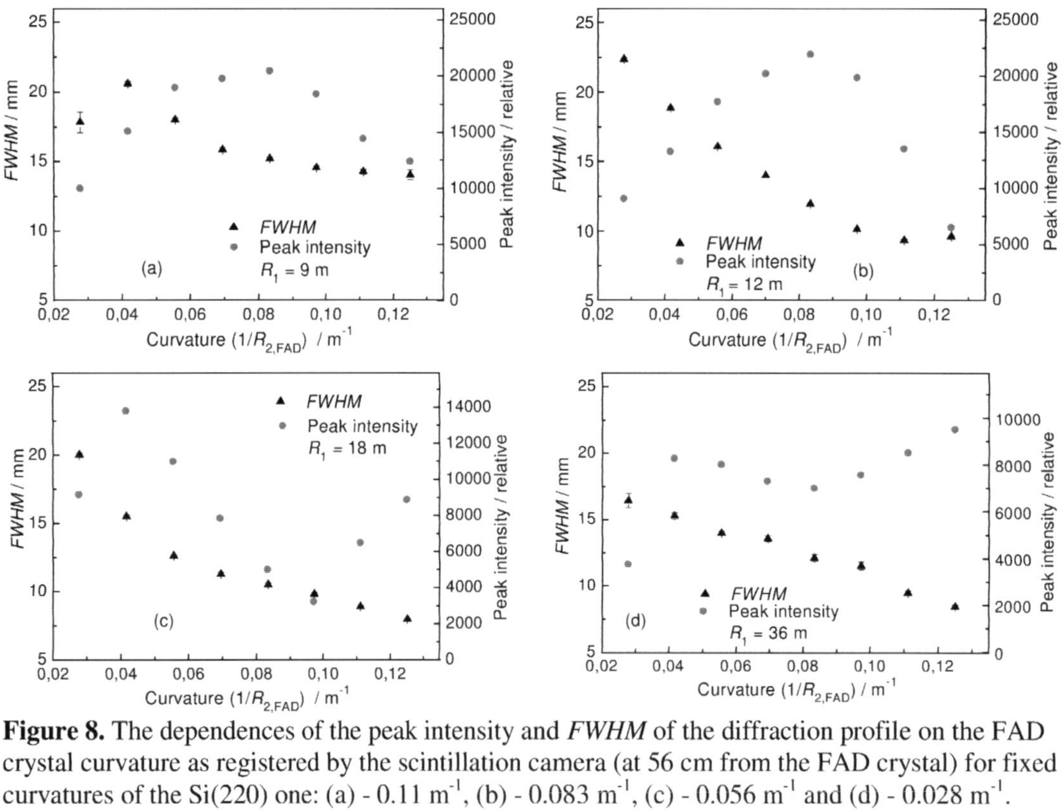

Figure 8. The dependences of the peak intensity and *FWHM* of the diffraction profile on the FAD crystal curvature as registered by the scintillation camera (at 56 cm from the FAD crystal) for fixed curvatures of the Si(220) one: (a) - 0.11 m^{-1}, (b) - 0.083 m^{-1}, (c) - 0.056 m^{-1} and (d) - 0.028 m^{-1}.

(+n,+m) setting as a function of curvature of FAD crystal for different curvatures of the Si(220) crystal slab were investigated. The obtained results are summarized in figure 8. The comparison of the results from figures 5 and 8 reveals that in the (+n,-m) setting the *FWHMs* of the output beam diffraction profiles registered by the scintillation camera are slightly smaller than in the (+n,+m) one. However, in both cases *FWHMs* have a decreasing tendency with the increase of the curvature of the FAD crystal slab. It means that also in this setting the natural dispersity between the Si(220) and FAD-Si(311) crystals strengthens with the curvature of the FAD crystal slab. Similarly, the peak intensities of the profiles registered by SC are higher in the case of (+n,-m) setting. The different variations of the peak intensities vs crystal curvature (see figures 8c and 8d) can be explained by different focusing conditions of the BPC-FAD Si(311) crystal with respect to the BPC Si(220) one, namely, in momentum space. It should be pointed out that the peak reflectivity $r(R)$ is not directly related to the peak intensity of the diffraction profile which is dependent e.g. on focusing in real and momentum space. The theoretical dependence of the peak reflectivity $r(R)$ on the crystal curvature was experimentally verified in the past [1,2]. However, the peak intensities of the individual diffraction profiles can be theoretically hard to describe. Moreover, the diffraction profiles registered by SC were taken in one angular position (peak position) of the Si(311) crystal rocking curve with respect to the Si(220) one. Thus, the values of *FWHM* presented in figures 5 and 8 are not related to the rocking curve profiles but just to the profiles of the diffracted beam at the peak position of the rocking curves. The peak position of the rocking curve of the FAD crystal also means that the volume element where neutrons are diffracted is situated in the middle ($L/2$) of the crystal slab. Even though the FAD crystal is illuminated along the whole length, due to a large homogeneous effective mosaicity ($\delta\theta(R) = L/R$), monochromatic neutrons coming from the bent Si(220) crystal are diffracted in a rather short volume element inside the FAD Si-slab. However, by rocking it, the diffraction volume element moves along the longest edge of the crystal slab and then the diffracted neutrons step by step fall outside the acceptance window of SC. The resolution demonstrating a high spatial coherence of the (+n,+m) setting was also tested on the imaging of the edge refraction effects on three different samples (two Si plates of the width of the edges of 5 mm and 10 mm, respectively and a rectangular Fe prism of the width of the edge of 9 mm) The results are shown in figure 9. As expected, the calibration of the rocking angle of the FAD crystal versus the peak position of the diffraction profile imaged by SC was found the same as in the case of the (+n,-m) setting.

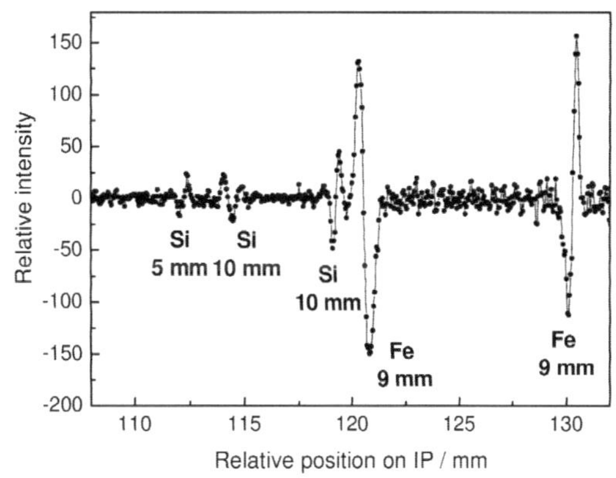

Figure 9. Photo of the irradiated samples of different edge widths with the marked window for the beam and experimental edge profiles obtained for $R_{2,FAD}$ =36 m. Imaging plate was at the distance of 35 cm from the samples.

4. Summary

Diffraction properties of the FAD geometry of the BPC Si(311) slab as a second crystal in the (+n,-m) and (+n,+m) double-crystal settings with respect to the BPC Si(220) one were investigated and mutually compared. In both cases, the output monochromatic beam provided by such a flexible double-crystal setting is highly collimated and *FWHM* of the profile could be easily manipulated by the curvature of the FAD crystal. Contrary to common opinion, though, it is the double diffraction process, the output monochromatic beam current can be sufficiently high and useful for employment in some special diffraction experiments. Thanks to a high peak reflectivity of the FAD Si(311) and Si(220) crystals (see figure 4), the obtained monochromatic neutron current after double diffraction process corresponds to the resulting phase-space element given by the overlapping of phase-space elements of the individual BPC Si(220) and FAD BPC Si(311) crystals in the (+n,-m) or (+n,+m) settings. Of course, the common peak reflectivity $r^2=r(R_{220})\cdot r(R_{311})$ should be considered The estimation of the reflectivity of the double crystal setting expressed by the product of the peak intensity and the corresponding value of *FWHM* (see figures 5 and 8) shows that both settings are from this point of view roughly equivalent. It should be also pointed out that due to the fact that the two step diffraction is carried out on crystals with different lattice spacings, the double diffracted beam has a narrow $\Delta\lambda/\lambda$ spread ($\Delta\lambda/\lambda = \Delta\theta \cot\theta$) of the order 10^{-3} or less. It is dependent on the curvatures of the individual crystals (see figure 12 and the related text in another paper of ours in this proceedings where the beam divergence for less dispersive setting was of about $1\text{x}10^{-3}$ rad). It should be said that both (+n,-m) and (+n,+m) settings as presented in this paper can be used, namely, at high-flux neutron sources for some experiments of neutron imaging when using the output beam of a large cross-section [15] or for high resolution (in scattering angle as well as $\Delta\lambda/\lambda$) diffraction experiments when using the output beam of small cross-section.

Acknowledgements

The related measurements were carried out at the CANAM infrastructure of the NPI ASCR Rez. Bragg diffraction optics investigations are in the Czech Republic supported by GACR project no. 14-36566G, by the ESS project LM2010011: ‘Contribution to Partnership in Large Research Infrastructure of Pan-European Importance’ as well as by the project of EU-FP7-NMI3 II: “Integrated Infrastructure Initiative for Neutron Scattering and Muon Spectroscopy”, 2009-2015.

References

[1] Mikula P, Kulda J, Vrána M and Chalupa B 1984 *J. Appl. Cryst.* **17** 189-195.
[2] Mikula P, Chalupa B, Kulda J, Vrána M, Sedláková L and Michalec R 1985 *J. Appl. Cryst.* **18** 135-140.
[3] Mikula P, Kulda J, Horalík L, Chalupa B and Lukáš P 1986 *J. Appl. Cryst.* **91** 324-330.
[4] Mikula P, Lukáš P and Michalec R 1987 *J. Appl. Cryst.* **20** 428-430.
[5] Mikula P, Krüger E, Scherm R and Wagner V 1990 *J. Appl. Cryst.* **23** 105-110.
[6] Mikula P, Lukáš P and Eichhorn F 1988 *J. Appl. Cryst.* **21** 33-37.
[7] Šaroun J, Lukáš P, Mikula P and Strunz P 1993 *Journal de Physique*, Coll. C7, **3** 439-442.
[8] Šaroun J, Lukáš P, Mikula P and Alefeld B 1994 *J. Appl. Cryst.* **27** 80-88.
[9] Hempel A, Eichhorn F, Reichel P and Boede W 1996 *Nucl. Instrum. Methods in Phys. Res.* A **381** 466-471.
[10] Lukáš P, Mikula P, Šaroun J and Strunz P 1994 *Nucl. Instrum. Methods in Phys. Research*, A **338** 111-115.
[11] Freund A 1983 *Nucl. Instrum. Methods,* **213** 495-511.
[12] Mikula P, Vrána M, Šaroun J, Pilch J, Seong B S, Woo W and Em V, 2014 *J. Appl. Cryst.* **47** Part 2, 599-605.
[13] Kulda J, 1984 *Acta Cryst.* **A40** 120-126; 1988 *Acta Cryst.* **A44** 283-285.
[14] G.E. Bacon, *Neutron Diffraction*, Clarendon Press, Oxford, 1975, p. 67 and p. 95.
[16] Kardjilov N, Lee S V, Lehmann E, Lim I C, Sim C M and Vontobel P 2005 *Nucl. Instrum. Methods Phys. Res.* A **542** 100–105.

A Proposal for a Next Generation European Neutron Source

K.H. Andersen[1] & C.J. Carlile[2]

[1] European Spallation Source, Box 176, 221 00 Lund, Sweden
[2] Department of Physics and Astronomy, Uppsala University, Box 516, 751 20 Uppsala, Sweden

E-mail: ken.andersen@esss.se

Abstract. We argue that it is not too early to begin the planning process for a next generation neutron source for Europe, even as the European Spallation Source is being constructed. We put forward three main arguments. Firstly, nowadays the period between the first scientific concept of a new facility being proposed and its actual realisation is approaching half a century. We show evidence for this. Secondly, there is a straightforward development of the short pulse/long pulse spallation concepts that will deliver gains in neutron brightness of more than a factor 30 over what the ESS will soon deliver and provide the optimum balance between resolution and intensity. We describe our concept, which is a spallation source where the proton pulse length is matched to the moderating time of slow neutrons. Thirdly, when we look at our colleagues in astronomy and high energy physics, we see that they have a totally different, more global and more ambitious approach to the coming generations of large facilities. We argue that it is time for the neutron community not simply to rest upon its laurels and take what is given but to be proactive..

1. Introduction

Peter Egelstaff gave talks (unpublished) in the 1980s demonstrating that the period of gestation of large scientific facilities had been increasing by one year for every year that had elapsed since the end of the second world war. He supported this assertion by comparing the gestation period of the Dido and Pluto reactors at Harwell (~5 years) with the ISIS pulsed source (~15 years), amongst other examples. Nowadays to see the truth of his statement we need look no further than the elapsed time between the original proposal for the ESS at Simonskall in Germany in 1991 [1] and its projected attainment of full specification with a complete instrument suite in 2028, a period which will approach 40 years – a scientific lifetime. We indicate this in Fig. 1 where a clear linear relationship can be discerned as a function of calendar year, whose slope falls somewhat short of Egelstaff's prediction in that the slope is nearer to 8 months per calendar year elapsed rather than 12 months. Nevertheless this would still mean that a next generation neutron source, which we refer to as NGENS, would be ~47 years in gestation if conceived and embarked upon now and would only arrive at full specification in 2062. If, on the other hand, we waited until ESS was fully functional in 2028 before approaching the matter it would then, according to Egelstaff's Law, take 55 years to complete, and would not be at full specification until 2083! Egelstaff's Law can be seen as being equivalent to Moore's Law, but in reverse gear, and it is a clear warning to the community, recalling that ESS is slated to close down in 2065 under current plans

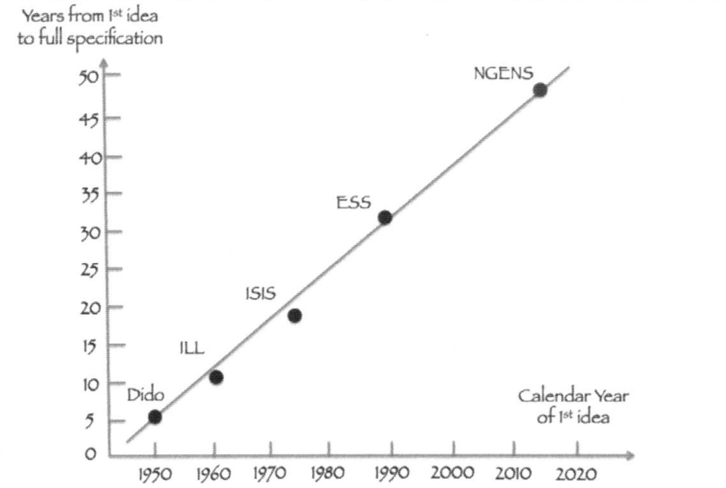

Figure 1. A graphical representation of Egelstaff's Law showing the time to realise a given neutron source as a function of the calendar year when it was first proposed. The numbers are necessarily approximate. Assuming this trend continues we have plotted the position of a next generation neutron source NGENS were it to be proposed now, as we suggest. It would take almost 50 years to reach full operations with a meaningful user programme. Data taken from Jacrot 2012 [2] and Hance 2006 [3]

Therefore it is timely, if not in fact urgent, to contemplate the idea of a next generation neutron source for Europe if neutron scattering is to progress beyond the middle of the century. We propose, as a viable candidate, a high power pulsed spallation source where the incoming proton pulse is matched to the moderating time of slow neutrons.

2. Matching proton and neutron times

The incoming proton pulse widths of the third generation MW-class spallation sources around the world (SNS, J-PARC and ESS) are all mismatched to the neutron moderating times by significant factors, either too short or too long. The moderator response for cold neutrons to a delta-function burst of protons to a spallation target is shown in Fig 2 [4]. Effectively SNS, JPARC and ISIS deliver delta function proton pulses (actually ~600ns) to their targets. The mismatch at this wavelength for a short pulse facility is very obvious. At shorter wavelengths the mismatch is less, but it is still one order of magnitude. When contemplating higher brightness sources employing narrow pulses we are limited by the instantaneous energy deposited into the target; no target is likely to survive the stresses imposed beyond a proton power of ~1MW. The forte of short pulse sources is that they are intrinsically high resolution facilities which is ideal for some applications but in many cases such as neutron spin echo, reflectometry ,small angle scattering and cold-neutron chopper and crystal-analyser spectroscopy, the resolution can be too high and hence instrument performance is below the optimum.

At the other extreme we have the long pulse concept, which ESS employs. In this situation a proton pulse of length ~3ms is used to generate the slow neutron beams and the moderating time of the neutron plays little part in the final pulse width. The moderating time of the neutrons can therefore be allowed to be much longer than in a short pulse source. In this situation increased intensity is the driver behind the source design and high resolution is achieved by chopping the pulse in time at each individual instrument. This however is at the expense of intensity. In many ways a long pulse source behaves like a quasi-continuous source.

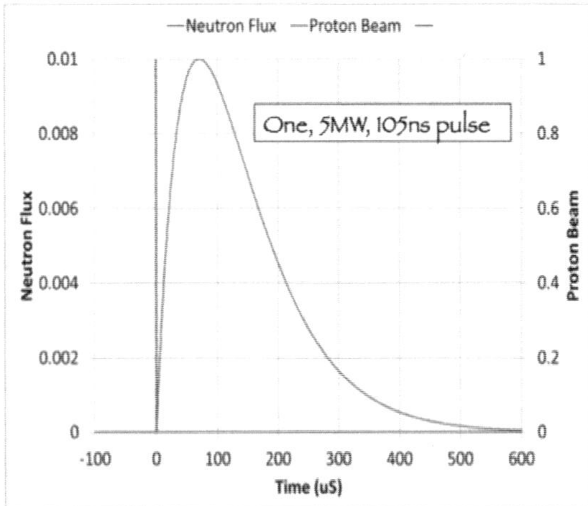

Figure 2. A monochromatic slice taken from the neutron pulse (blue) resulting from a quasi-delta-function (105ns) proton pulse (red) with the fast rise time and slow decay time clear in the neutron pulse. A useful rule of thumb for such sources is that the FWHM of the neutron pulse is approximately 7λ(Å)μs in the epithermal region and 22λ(Å)μs in the thermalised region.

In the case of a continuous source there is full flexibility to trade intensity for resolution and to be able to build a diverse set of instrumentation. Equally well continuous sources can be pulsed, with the advantage that the pulse length can be varied and the pulse repetition rate can be chosen as required, rather than being dictated by the parameters of the source itself. The design parameters of a pulsed source have a far greater consequence for the performance of the instrument suite than do the design parameters of a continuous source where spectral range and intensity are the only relevant factors.

If we consider the more or less accepted practical limits in pulsed source power today we see that short pulse sources such as SNS or J-PARC do not have ambitions beyond 1.4MW and the only long pulse source, ESS, is targeting 5MW. It is therefore appropriate to ask what would be the maximum power achievable were the pulse length to be varied away from the two extremes. Current expertise [5] suggests that, provided a solid target were used, then the full 5MW power of the ESS could be employed. The Goldilocks solution – just right - would therefore be to have the proton and neutron time constants as nearly equal as possible over the desired range of neutron wavelengths. In such a case both the peak and time-average slow neutron brightness can be simultaneously maximised for a given accelerator power. Gain factors of between one and two orders of magnitude in terms of intensity for a given resolution would accrue at the neutron instruments and would represent a very significant increase in sensitivity for neutron scattering investigations - a technique which benefits from a range of unique advantages for studies of condensed-matter science, but nevertheless facing fierce competition from the inexorable rise in intensity of photon sources. It is often stated that neutron and synchrotron sources are complementary but such complementarity begins to wane if the comparative source intensities diverge too much. Whilst pulse matching cannot be achieved at all wavelengths one could envisage a number of complementary regional sources which are optimised for different spectral ranges and hence for different ranges of scientific investigations rather in the way that different designs of telescope are diverse, being focused upon sky surveys or pin-point observations and operating in a specific wavelength band such as the UV, optical, IR or radio ranges, thus optimally serving science.

Our proposal is for an H- linear accelerator feeding into a compressor ring that generates proton pulses of some tens to hundreds of μs width [4] feeding into a rotating tungsten target similar to that of the ESS.

3. Evolution of neutron source strength

The rise in source intensity since the discovery of the neutron by Chadwick in 1932 is frequently represented by variations of the plot shown in Fig 3 [6] which takes its data from the quoted source fluxes which are often not actually useable by the instrument suite.

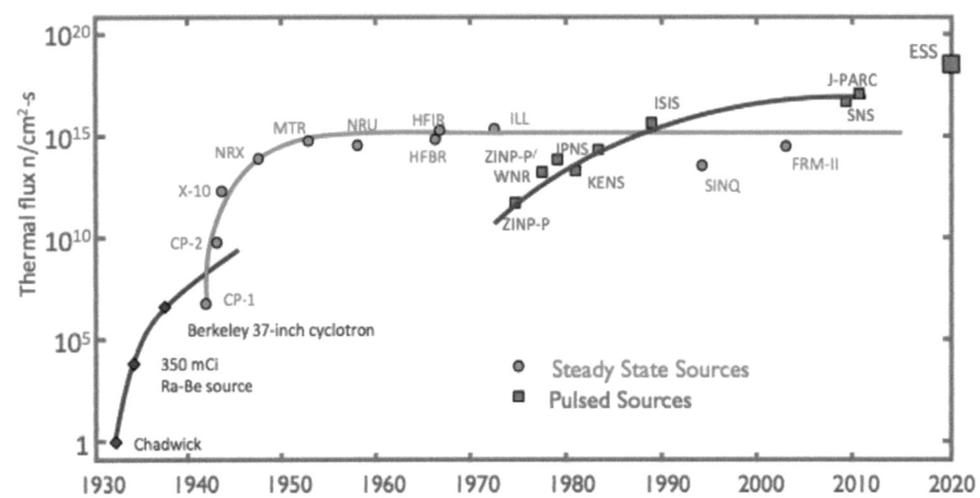

Figure 3. A representation of the effective thermal flux of various different neutron sources since the discovery of the neutron in 1932 by James Chadwick, adapted from [6]. Early sources are shown as red diamonds, reactor sources as green circles and spallation sources as red squares. An estimated flux for ESS is indicated by the larger red square.

Instead we have re-examined the data and express the same information in terms of neutron intensity available to the instruments. The results are shown as a scatter plot in Fig 4. Here we have made the distinction between reactor sources and spallation sources only and have fitted the data to straight lines in this logarithmic representation. This allows an average rise in useable effective source brightness over the past 5 decades to be derived and compared for the two kinds of source. In addition we have put upper and lower trend lines in the diagram. We find that whilst reactor facilities have risen in brightness by a modest 20% per decade on average, spallation sources have risen in brightness by a factor of 4 per decade on average. From this we can conclude that, of the two kinds of source, spallation has the most potential to deliver further gains in brightness when we come to consider possible source options for a next generation neutron source.

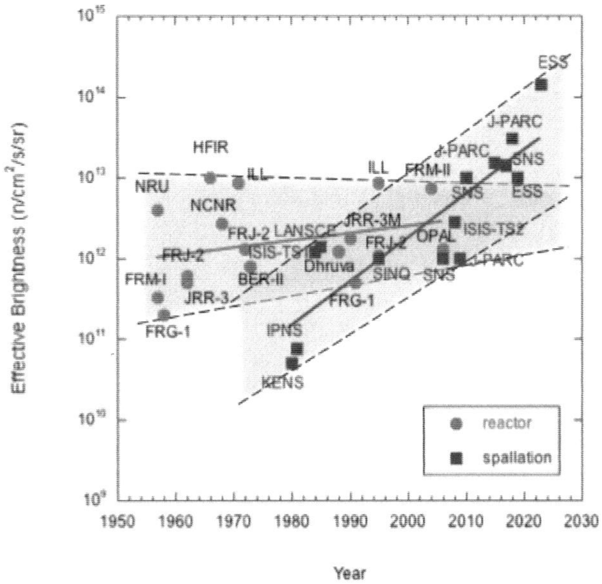

Figure 4. A scatter plot of source brightnesses where reactor sources are plotted in blue and spallation sources in red. The sources are those referred to in Tables 1 and 2.

The data shown in Fig. 4 are based on a literature search, resulting in the neutron facility list shown in Table 1.

Table 1. Neutron sources which are part of the present study

Source	Dates	In-pile thermal flux ($n/cm^2/s$)	B_{eff} ($n/cm^2/s/sr/Å$)
BER-II	1973 10MW	1.2e14 [8][9]	8e11 [a]
Berlin	2019 shutdown		
FRJ-2	1962 10MW		
Jülich	1967 15MW		
	1972 23MW		
	1990 shutdown		
	1995 20MW	1.7e14 [8]	1.1e12 [a]
	2006 shutdown		
FRG-1	1958 5MW		
Geesthacht	1990 CS added		
	1991 core size reduced	8e13 [8]	5e11 [a]
	2010 shutdown		
ILL	1971 57MW		
Grenoble	1992 shutdown		
	1995 57MW	1.3e15 [9]	8.7e12 [10]
FRM-I	1957 4MW	5e13 [8]	3.3e11 [a]
Munich	2000 shutdown		
FRM-II	2004 20MW	8e14 [8][9]	7.4e12 [11]
Munich			
Dhruva	1985 start		
Trombay	1988 100MW	1.8e14 [7][9]	1.2e12 [a]
NCNR	1967 10MW		

NIST	1985 20MW 1987 CS	4e14 [9]	2.7e12 [a]
HFIR Oak Ridge	1966 100MW 1986 shutdown 1990 85MW	1.5e9 [9]	1.0e13 [a]
NRU Chalk River	1957 200MW ^{nat}U 1964 60MW HEU 1991 135MW LEU 2018 shutdown	3e14 [12]	4e12 [a]
JRR-3(M) Tokai	1962 10MW 1983 shutdown 1990 20MW 2011 shutdown	2.7e14 [9]	1.8e12 [a]
OPAL Lucas Heights	2006 20MW	2.4e14 [9]	1.3e12 [a]
SINQ PSI	1995 1.5mA 600MeV	1.5e14 [9]	1.0e12 [a]
Lujan LANSCE	1985 120uA 800MeV 2015 shutdown		1.4e12 [b]
IPNS Argonne	1981 1.5uA 450MeV 2008 shutdown		7.7e10 [b]
KENS KEK	1980 7uA 500MeV 2006 shutdown		5.1e10 [b]
ISIS-TS1 RAL	1984 start 1990 160uA 800MeV		1.2e12 [b]
ISIS-TS2 RAL	2008 40uA 800MeV		2.8e12 [b]
SNS Oak Ridge	2006 start 2010 1MW 2017 1.4MW		1.0e13 [b]
J-PARC Tokai	2009 start 2015 500kW 2018 1MW		3.0e13 [b]
ESS Lund	2019 start 2023 5MW		4.1e14 [b]

[a] Rescaled from the value of the ILL effective brightness B_{eff}, assuming same ratio between B_{eff} and in-pile thermal flux. This will occasionally be an overestimate, as it assumes good beam-tube access to the region of peak thermal in-pile flux and an efficient cold source. Hot sources are not taken into account.

[b] Calculated from the spectral brightness values shown in Table 2, using equation (2).

Such a cross-facility comparison is fraught with difficulty. The first is conceptual; there is no simple correlation between the flux on the instruments and their scientific productivity, which ultimately is the metric against which large-scale facilities are measured [13][14]. Even if we somehow reach agreement on displaying what will inevitably be interpreted as "scientific productivity" or "usefulness to society" in terms of the neutron flux available on the instruments, there are a number of technical issues which stand in our way: the correlation between source brightness and instrument performance depends entirely on the efficiency of the neutron optics and other critical instrument design features. The availability of sample environment and data reduction and analysis software is similarly critical. These considerations, however, are somewhat beyond the scope of the present paper and we will instead focus our efforts on making the best possible comparison between individual steady-state and pulsed neutron sources. It is important that a comparison is made in order to arrive at an informed evaluation of the current and possible future developments in neutron facilities.

The time-average source brightness of pulsed sources is typically orders of magnitude below that of the steady-state sources. A claim frequently made however is that the performance of instruments on a pulsed source scales with the peak brightness, thereby redressing the balance in favour of pulsed sources.

A study [15] has been carried out of the performance of a reference instrument suite for the ESS as a function of the source time structure. At the time, the purpose of the study was to determine the optimal duty cycle of the ESS, resulting in the choice of a pulse repetition period of 14 Hz and a proton pulse length of 2.857 ms. However, one of the key findings of the paper is that, averaged over the full instrument suite, the instrument performance (expressed by flux on the sample) scales very closely to the duty cycle (pulse length divided by repetition period) to the power of 0.30. This leads us to propose a figure of merit (FoM) for all sources which is proportional to the product of the time-average brightness (B_{av}) to the power of 1/3 and the peak brightness (B_{pk}) to the power of 2/3. This corresponds to a geometric average of B_{av} and B_{pk} where B_{pk} is given twice the weight of B_{av}:

$$\mathrm{FoM} = B_{av}^{1/3} B_{pk}^{2/3} \tag{1}$$

For a continuous source, the time-average and peak brightnesses are the same, resulting in an effective FoM equal to the time-average brightness. In order to take into account the neutron wavelength-dependence of the source brightness, we propose to take an equal-weight geometric average of the FoMs evaluated using the highest available source spectral brightness at wavelengths of 1 Å and 5 Å, resulting in the “Effective Brightness” given below:

$$B_{eff} = \left(\mathrm{FoM}(1\text{Å}) \times \mathrm{FoM}(5\text{Å})\right)^{1/2} \tag{2}$$

The data needed for the evaluation of the Effective Brightness are not easily available for all neutron sources. We have therefore made some simplifying assumptions which are stated in the captions of Tables 1 and 2.

Table 2. Moderator spectral brightnesses in units of n/cm²/s/sr/Å for the pulsed sources in Table 1 at wavelengths of 1 Å and 5 Å. For each facility, the moderator with the highest brightness for that wavelength has been chosen and stated in the table. The corresponding numbers for ILL and FRM-II are shown for comparison.

	Bav (1Å)	Bpk(1Å)	FoM(1Å)	Bav(5Å)	Bpk(5Å)	FoM(5Å)
ESS 5MW	4.4e13	1.1e15	3.6e14	3.8e15	1.5e14	4.6r14
2.5mA 2GeV 14Hz	3cm tall coupled water			3cm tall coupled liquid para-H_2		
J-PARC 1MW	4.5e11	1.2e15	7.6e13	4.4e11	6.8e14	1.2e13
333uA 3GeV 25Hz	coupled liquid para-hydrogen			coupled liquid para-hydrogen		
ISIS-TS1 128kW	9e10	1.5e14	1.1e13	3.5e9	9e11	1.2e12
160uA 800MeV 50Hz	decoupled poisoned water			decoupled liquid hydrogen		
ISIS-TS2 32kW	4.5e10	3.5e14	1.5e13	7e9	5.5e12	5.3e11
40uA 800MeV 10Hz	decoupled solid methane			decoupled solid methane		
SNS 1MW	5e11	5e14	4.5e13	1.2e11	1.1e13	2.3e12
1mA 1GeV 60Hz	decoupled poisoned water			coupled liquid hydrogen		
KENS 3.5kW	2.5e9 [a]	8.2e12 [a]	4.8e11 [a]	9.6e7 [a]	4.9e10 [a]	5.5e9 [a]
7uA 500MeV 20Hz						

IPNS 7kW 15uA 450MeV 30Hz	4.7e9 [a]	1.1e13 [a]	7.3e11 [a]	1.8e8 [a]	6.3e10 [a]	8.1e9 [a]
LANSCE 96kW 120uA 800MeV 20Hz	6.8e10 [a]	2.3e14 [a]	1.3e13 [a]	2.6e9 [a]	1.4e12 [a]	1.5e11 [a]
ILL 57MW [10]	2.8e13	2.8e13	2.8e13	2.7e12	2.7e12	2.7e12
		H2 thermal beamtube			horizontal/vertical cold source	
FRM-II 20MW [11]	3.2e13	3.2e13	3.2e13	1.7e12	1.7e12	1.7e12
		thermal beamtube			cold source	

[a] extrapolated from ISIS-TS1, using the source powers and time structures, assuming similar moderator-reflector performance

4. Conclusions

We conclude that a concerted consideration of the design of a next generation neutron source for Europe should begin in the immediate future. Even so we predict that such a source would not be operational, under the most optimistic scenario, until well into the second half of this century. We put forward as a viable option for such a source a matched neutron-proton pulsed source driven by a 5MW proton accelerator with a proton pulse length of between 50 and 100 μs . In such a case we believe that peak brightnesses a factor of 35 over what ESS is calculated to achieve in the mid 2020s will accrue. This implies brightnesses over what is available now to researchers at SNS and ILL will be almost three orders of magnitude higher. This would be a considerable step forward in terms of scientific investigative power.

References

[1] A.D. Taylor & G.H. Lander (1992) Neutron News Volume 3, Issue 2, Page 4
[2] B. Jacrot (2012), Des neutrons pour la science: Histoire de l'Institut Laue-Langevin, EDP Sciences
[3] N. Hance (2006), Harwell: The Enigma Revealed
[4] D. McGinnis, M. Lindroos, R. Miyamoto, (2013), Proceedings of IPAC2013, Shanghai
[5] E. Pitcher & J. Haines (2015), Private communications
[6] K. Sköld and D. L. Price, (1986) eds. *Neutron Scattering*, Academic Press.
[7] Multipurpose research reactors, IAEA, Vienna 1988 STI/PUB/762, ISBN 92-0-050688-7
[8] IAEA-SM-360/43 The new compact core design of the FRG-1, P. Schreiner, W. Krull
[9] ESS Technical Design Report, S. Peggs (ed.) (2013)
[10] K.H. Andersen & E. Farhi, in preparation
[11] "Blue Book" Experimental facilities, Forschungs-Neutronenquelle Heinz Maier-Leibnitz (FRM II), page 11 (2012)
[12] T.M. Holden, B.M. Powell & G Dolling (1994), MRS Proceedings, 376, 7 doi:10.1557/PROC-376-7
[13] O. Hallonsten, Scientometrics, 96: 497-513 (2013)
[14] O. Hallonsten, Scientometrics 100: 483-496 (2014)
[15] K. Lefmann et al., Rev. Sci. Instr. 84, 055106 (2013)
[16] L. Zanini et al., private communication (2015)

Pelletized cold moderator of the IBR-2 reactor: current status and future development

V Ananiev[1], A Beliakov[1], M Bulavin[1], A Verkhogliadov[1], E Kulagin[1], S Kulikov[1], K Mukhin[1], E Shabalin[1] and K Loktaev[2]

[1] Joint Institute for Nuclear Research

[2]Obninsk Institute of Nuclear Power

E-mail: bulavin85@inbox.ru

Abstract. Current status and future development of the pelletized cold moderator of the IBR-2 reactor in Neutron Physics Laboratory of JINR are represented. Nowadays cold moderator works for physical experiments and allows conducting experiments in the region of wavelengths more than 4 Å up to 10-13 times faster in comparison with the warm water moderator. Future development of the pelletized cold moderator is aimed at increasing the time of its operation for experiments and is based on three components: creation of a system of continuous charging and discharging of beads, supplementation of various additives, and use of new materials, such as triphenylmethane.

1. Current status of the pelletized cold neutron moderator of the IBR-2

The IBR-2 is a fast pulsed research reactor intended for investigation of condensed matter at external neutron beams. Fast neutrons generated in the reactor lose their energy in the moderators located around the reactor core. Combined bi-spectral neutron moderator is a part of the equipment of the IBR-2. It provides with neutrons of broad spectrum (cold, thermal or mixed) for spectrometers at four neutron beams: number 7, 8, 10 and 11.

The combined moderator consists of a pelletized cold neutron moderator [1] based on aromatic hydrocarbons, and of a set of thermal moderators and pre-moderators, which are based on room-temperature water (Fig. 1).

Figure 1. Combined moderator of the IBR-2 reactor

Chamber of the pelletized cold moderator is of rectangular shape. The area of its surface facing the neutron beams is of 314 cm^2. The chamber is charged with beads of 3.5 mm in diameter each (Fig. 2), made of mixture of aromatic hydrocarbons: mesitylene and m-xylene. Charging volume for the beads is 1 L. Operation temperature of beads is 30K. The temperature is maintained by cold gaseous helium circulating through the chamber filled with beads. The moderator can operate without recharge of beads during one reactor cycle of 9-10 days. After the reactor cycle is over, cooling of the moderator is stopped, the frozen beads are melted and the irradiated liquid is removed from the moderator chamber. The chamber of the moderator is charged with frozen beads again before the new reactor cycle starts.

Figure 2. Beads of the mixture of mesitylene and m-xylene in liquid nitrogen

Cold neutron flux at the surface of the pelletized moderator in the range of long wavelengths (more than 4 Å) is 10-13 times greater than that from the surface of the water moderator (Fig. 3) [2].

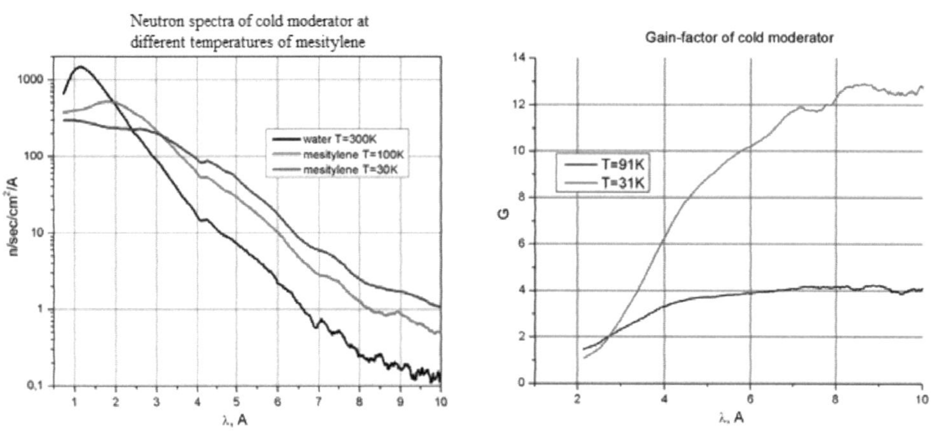

a b

Figure 3. (color online) Neutron spectra (experimental data via TOF) of the combined moderator (a) and gain factor of the flux of cold neutrons to thermal neutrons (b) for beam №8 (spectrometer «REMUR») of the IBR-2.

Since 2013 many physical experiments have been successfully implemented using the combined bi-spectral moderator of the IBR-2 reactor. They included the study of spatial beam-splitting by reflection from a magnetically non-collinear film, producing microbeams by a layered waveguide (REMUR reflectometer of the IBR-2 reactor, Dubna, Russia), neutron diffraction texture analysis of the sample of slate composed of five minerals (SKAT diffractometer, IBR-2 reactor, Dubna, Russia), etc. [3].

2. Future development of the pelletized cold moderator

Currently, the main problem of the cold moderator based on a mixture of mesitylene and m-xylene is duration of its operation cycle. At the moment, the operation time is 9 days, while a standard reactor cycle of thermal water moderator lasts up to 11 days. The operation time limit of the cold moderator is determined mainly by an increase of viscosity of the mixture of mesitylene and m-xylene because of irradiation. After 10 days of irradiation (510 MW·h or 156 MGy, essentially, due to fast neutrons, about 90%, the rest - γ radiation), the viscosity increases by 23 times (Fig. 4.). Kinematic viscosity of irradiated liquid was measured by capillar viscosimeter based on standard technics ASTM D 445. Therefore, the future development and improvement of the pelletized moderator is aimed primarily at solving the problem of increasing the time of its continuous work in a physical experiment.

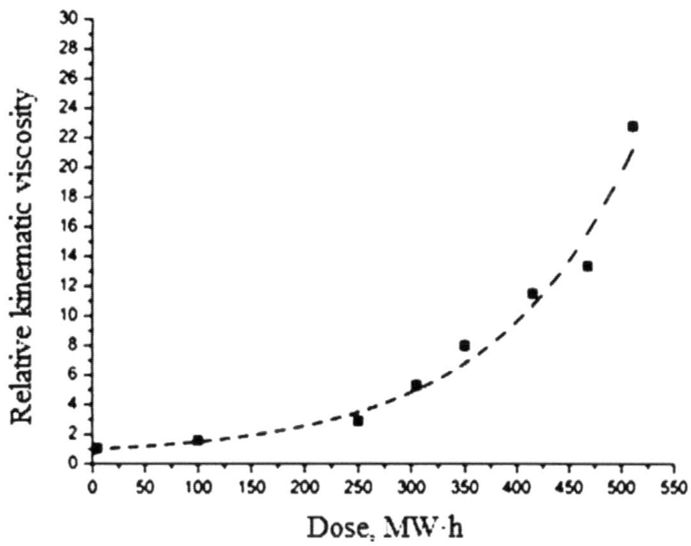

Figure 4. Relative kinematic viscosity for different doses of mesitylene and m-xylene irradiation (100 MWh=30.6 MGy). Dots correspond to experimental data of viscosity. Dashed line shows approximation.

There are several ways to solve this problem. The first one is to create a system of continuous charging and discharging of beads. This system will remove frozen beads that have already received high radiation dose, and charge the new, i.e. non-irradiated ones. Thus, we can control viscosity of the mixture during irradiation. Creation of such a system for the cold moderator at the IBR-2 involves a complex of engineering challenges. Up to date, the works on development of a prototype of such a system for a special test stand have been started.

The second way is the supplementation of various additives to the original mixture in order to minimize the viscosity induced by radiation. There are at least two possible mechanisms to reduce viscosity using additives. The first one is based on substitution of a part of mesitylene molecules with more radiation-resistant aromatic hydrocarbons. These can be molecules of naphthalene or anthracene consisting of benzene rings. Using these additives, radiation resistance of the mixture may be increased up to 1.5 times with a slight decrease in cold neutron flux. The second mechanism is based on supplementation of the mixture with a non-hydrogen substance, on which a certain part of recoil protons received from neutron moderation can be inhibited. Small volumetric replacement of molecules of mesitylene with uniformly distributed microparticles will not be essential for neutron moderation. Such additives, which should not absorb slow neutrons, may be nanopowders of diamond, zirconium or lead. Presently, the work to study the efficiency of the additives has been initiated.

The third way of increasing the operation time of the cold moderator is to replace the working mixture of mesitylene and m-xylene with another material with a better radiation resistance and proper neutron moderation. This substance might be triphenylmethane. In the literature there are indications that triphenylmethane is not inferior to mesitylene in terms of its neutron-physical properties (flux of cold neutrons, etc.) [4]. Radiation resistance of triphenylmethane is much higher than that of mesitylene [5]. In the end of 2015, the first experiment to study the radiation resistance of triphenylmethane is planned at the irradiation facility of the IBR-2 reactor [6, 7].

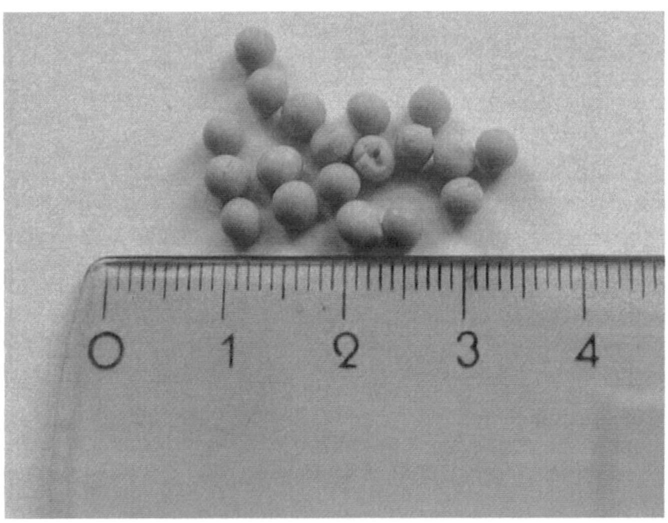

Figure 5. Triphenylmethane beads

The main advantage of triphenylmethane over mesitylene, from the point of view of manufacturing of beads, is that triphenylmethane is solid at room temperature, whereas mesitylene is liquid. The melting point of triphenylmethane is 92°C, while that of mesitylene is -50°C. Consequently, triphenylmethane beads can be kept at room temperature, while mesitylene beads – in liquid nitrogen only. So that, it is possible to manufacture beads of regular shape and size, good hardness and durability in advance for their further transportation to the cold chamber of the moderator. At the same time, the main disadvantage of triphenylmethane within the existing technology of charging/discharging of beads is the need for additional heating of the moderator system up to 92 °C to melt the solid beads and remove them as liquid, whereas mesitylene beads melt and can be removed at room temperature.

In the summer of 2015, in FLNP the works were performed proving the possibility to manufacture solid beads of triphenylmethane [Fig. 5].The first experiments on charging the triphenylmethane beads into the chamber of the test stand of the cold moderator are planned for 2016.

3. Conclusion

Since 2013 the pelletized cold moderator as a part of the first combined (bi-spectral) moderator at the IBR-2 is working successfully for physical experiments at external neutron beams. Neutron flux of the cold moderator allows conducting experiments in the region of wavelengths (of more than 4 Å) up to 10-13 times faster in comparison with the warm water moderator; at the same time, the thermal neutrons flux does not decrease dramatically. The future development of the pelletized cold moderator is aimed at increasing the time of its operation in experiments and is based on three components:

- creation of a system of continuous charging and discharging of beads;
- supplementation of various additives;
- use of new materials, such as triphenylmethane.

References

[1] Cold neutron moderator on an upgraded IBR-2 reactor: the first set of results / V. Ananiev [et al.] // Technical physics – 2014. Vol. 59, №2. pp. 283-286.

[2] The world's first pelletized cold neutron moderator at a neutron scattering facility / V. Ananiev [et al.] // Nuclear instruments and methods in physics research B. 2014. Vol. 320. pp. 70-74.

[3] Application of a cryogenic moderator in the REMUR neutron reflectometer / Kozhevnikov [et al.] // Journal of surface investigation. X-ray, Synchrotron and Neutron Techniques, 2016, Vol. 10, No. 1, pp. 1–9.

[4] Triphenylmethane, a possible moderator material / Th. Hugle [et al.] // Nuclear instruments and methods in physics research A. 2014. Vol. 738. pp. 1-5.

[5] The mechanism and kinetics of radiation-chemical reactions / S. J. Pshezhetsky. — Second edition revised and enlarged – Moscow: Chemistry, 1968. P. 368.

[6] Irradiation facility at the IBR-2 reactor for investigation of material radiation hardness / A. Verkhogliadov [et al.] // Nuclear instruments and methods in physics research B. 2015. Vol. 343. – pp. 26-29.

[7] Spectrum and density of neutron flux in the irradiation beam line №3 of the IBR-2 reactor / E. Shabalin [et al.] // Physics of particles and nuclei letters. 2015. Vol. 12, №2. pp. 336-343.

Journal of Physics: Conference Series **746** (2016) 012032 doi:10.1088/1742-6596/746/1/012032

Benchmarking Geant4 for spallation neutron source calculations

Douglas D. DiJulio[1,2], Konstantin Batkov[1], John Stenander[1,2], Nataliia Cherkashyna[1], Phillip M. Bentley[1,3]

[1] European Spallation Source ERIC, SE-221 00 Lund, Sweden
[2] Department of Physics, Lund University, SE-221 00 Lund, Sweden
[3] Department of Physics and Astronomy, Uppsala University, 751 05 Uppsala, Sweden

E-mail: douglas.dijulio@esss.se

Abstract. Geant4 is becoming increasingly used for radiation transport simulations of spallation neutron sources and related components. Historically, the code has seen little usage in this field and it is of general interest to investigate the suitability of Geant4 for such applications. For this purpose, we carried out Geant4 calculations based on simple spallation source geometries and also with the the European Spallation Source Technical Design Report target and moderator configuration. The results are compared to calculations performed with the Monte Carlo N-Particle extended code. The comparisons are carried out over the full spallation neutron source energy spectrum, from sub-eV energies up to thousands of MeV. Our preliminary results reveal that there is generally good agreement between the simulations using both codes. Additionally, we have also implemented a general weight-window generator for Geant4 based applications and present some results of the method applied to the ESS target model.

1. Introduction

Spallation neutron sources are powerful tools for materials research. These sources generate low-energy neutrons for scientific investigations through the bombardment of a heavy metal target with a high-energy proton beam. A popular set of software, used in the design of modern-day spallation sources, is the Monte Carlo N-Particle Transport extended code (MCNPX) [1, 2]. For example, the baseline design of the upcoming European Spallation Source (ESS) [3] and the subsequent optimization of the moderators were performed with MCNPX [4]. The reliance on MCNPX is partly based on it's history of benchmarking and validation, use of data driven libraries, and implementation of high-energy cascade models. Of specific interest are the pre-packaged libraries for the interactions of low-energy neutrons in materials, which compose the primary signal for materials science investigations. These are critical for calculations of moderator performance.

Geant4 [5, 6], which is a popular simulation toolkit particularity in the high-energy physics realm, has seen little application in the field of spallation neutron sources. There are however a number of reasons why it is of interest to investigate the suitability of Geant4 for spallation neutron source calculations. For example, in the recent years there has been an influx of new ideas, especially from high-energy physics labs such as CERN [7], directly into the neutron scattering community. Geant4 is also open source, freely available online, and supported by a large user community spanning a broad range of scientific fields. Lastly, modern day spallation sources

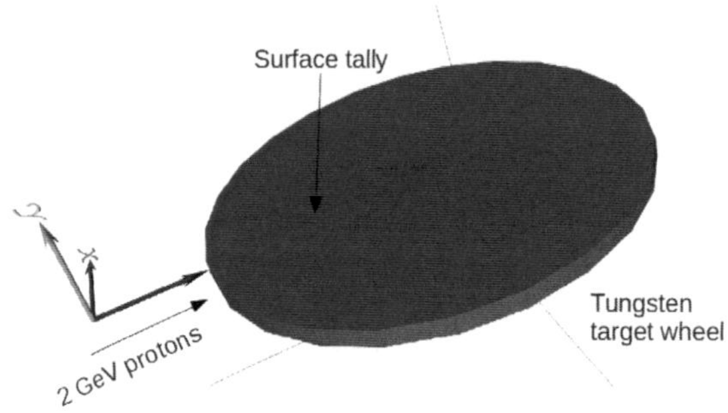

Figure 1: Rendering of the simple spallation source target model used in the Geant4 calculations.

are reaching energy ranges which are commonplace in the high-energy physics community. The Materials and Life Science Experimental Facility [8] at the Japan Proton Accelerator Research Complex (J-PARC) [9] employs a 3 GeV proton beam and the ESS will use a 2 GeV proton beam.

At ESS, there is a significant push towards building on-top of the experiences from the high-energy physics community and developing Geant4 to meet the requirements for neutron instrument detector design [10, 11]. Additionally, Geant4 is employed at ESS for calculations of the neutronic response of the shielding components of neutron scattering instruments [12, 13, 14, 15]. These two points make it possible to simulate in detail the areas where the cross-talk between radiation background levels and neutron detectors is critical for the performance of an instrument. A key question which could be asked is then how well does Geant4 simulate the spallation neutron source environment, compared to the often used MCNPX code?

The aim of the present work is to investigate the suitability of Geant4 for spallation neutron source calculations and benchmark it against MCNPX. Three test problems were selected for comparison of the two codes. The first is a simple spallation target model and the second is the spallation target model combined with a single para-hydrogen moderator. We selected the ESS baseline design highlighted in the Technical Design Report [3] for the third comparison. In the following, we present the general features of the two codes relevant for this work and the results of the three test scenarios. We focus primarily on the production of neutrons which are important for both the signal used for materials science investigations and also the background noise. Lastly, we also present an implementation of a general weight-window generator for Geant4 based applications, in particular to be used for deep-shielding calculations.

2. Details of the Monte-Carlo software

The MCNPX calculations were performed using version 2.7.0. The default nuclear interaction Bertini model [16, 17] coupled with the ENDF/B-VII.0 [18] neutron libraries was used. At thermal energies and lower, the neutron transport was handled by the prepackaged neutron scattering kernels. These were activated for hydrogen in para-hydrogen and light water, beryllium, aluminum-27, and iron-56 in the iron and steel.

The Geant4 calculations were carried out using Geant4 version 10.0 patch 3. The physics list QGSP_BERT_HP was used for the calculations [19]. This list is one of the recommended lists for shielding applications [20]. QGS stands for the quark-gluon string model, P for precompound, BERT for the Bertini intra-nuclear cascade model, and HP for the high-precision neutron package [19]. The HP package uses evaluated neutron data, named G4NDL4.4, for interactions of neutrons below 20 MeV. This data largely comes from the ENDF/B-VII.0 libraries. Included

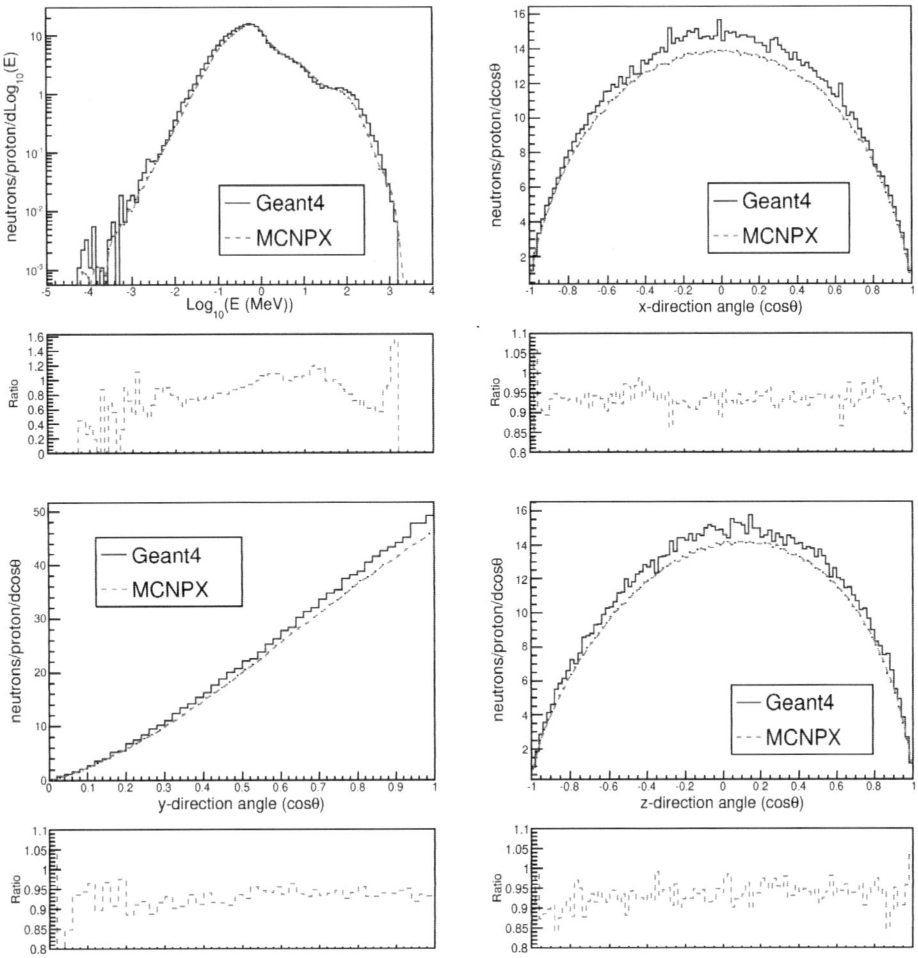

Figure 2: Results of the MCNPX and Geant4 calculations for the simple spallation source target model shown in Fig. 1. The energy spectrum of the neutrons leaving the target are shown in the top left panel, the x-directions of the neutrons leaving the target in the top right panel, the y-directions in the bottom left panel, and the z-directions in the bottom right panel.

also in this package is data for the thermal scattering of neutrons below 4 eV.

Finally, we mention that the Geant4 simulations were carried out using the ESS Detector Group framework [10]. The framework provides a Geant4-based Python/C++ simulation environment which offers a number of developments to aid neutron detection and shielding calculations. The results presented here were carried out within this framework, however no additional features were implemented beyond what is contained within the standard Geant4 release.

3. Simple spallation source models

The first test of the two codes was carried out on a simple tungsten target which was 2 m is diameter. The Geant4 rendered geometry is shown in Fig. 1. The energy of the primary proton beam was 2 GeV. The energies and directions of neutrons leaving the top surface of the target

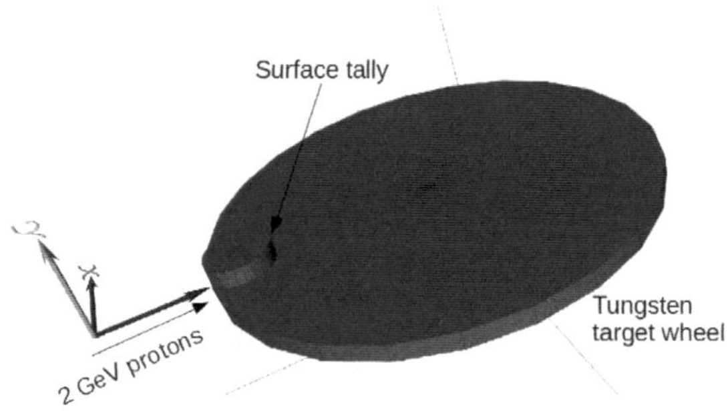

Figure 3: Rendering of the simple spallation source target model and moderator used in the Geant4 calculations.

were saved for the comparison and the results are shown in Fig. 2. The four spectra in the figure show the energies of neutrons leaving the top surface of the target and their directions. The z-direction is along the proton beam axis, the y-direction is along the height of the target, and the x-direction along the width of the target. Overall there is excellent agreement between the results of the simulations using the two codes.

The second test model was the simple tungsten target combined with a liquid para-hydrogen moderator and is shown in Fig. 3. Information regarding the neutrons leaving the side surface of the moderator was saved for comparison of the results of the two codes. Generally, there is again good agreement between the results of the two codes. One can note in particular the excellent reproduction of the moderator peak in the low-energy spectrum, which is of particular interest for neutron scattering studies. The neutron directions are also reproduced well by Geant4.

4. ESS TDR model

The third model tested in this work was the ESS TDR design of the target and moderators, described in Ref. [3]. The neutronic design of the model was carried out using MCNPX as the primary simulation code. The MCNPX model of the target zone is shown in Fig. 5. The proton beam from the accelerator, parabolic in shape with a width of 16 cm and a height of 6 cm, impinges on an 8 cm tall helium-cooled rotating tungsten target. The diameter of the wheel is 2.5 m. Above and below the target are positioned volumetric moderators of cylindrical shape with a diameter of 16 cm and a height of 13 cm. The moderators contain super-critical para-hydrogen and are partly covered by ambient light water premoderators with wings which have thicknesses of 4 cm. These make it possible to extract both cold and thermal neutrons for a single beamline. Extending from the moderators are four extraction areas of 60° for the neutron beamlines. A beryllium inner reflector, a stainless steel outer reflector, and iron shielding surround the moderator region.

The ESS TDR geometry was also implemented into a Geant4 model and the preliminary results of the Geant4 simulations compared to the MCNPX simulations are shown in Fig. 6. The energy spectra are the sum over all the neutrons leaving the four 60° extraction areas at a distance of 2 m from the moderators. While there is good agreement between the spectra, work is still underway in refining the geometry of the Geant4 model to more closely match the MCNPX model. At this moment in time, there is no tool that the authors are aware of that makes the transition from one geometrical model in MCNPX to Geant4 a simple procedure. The preliminary results presented here are however very encouraging.

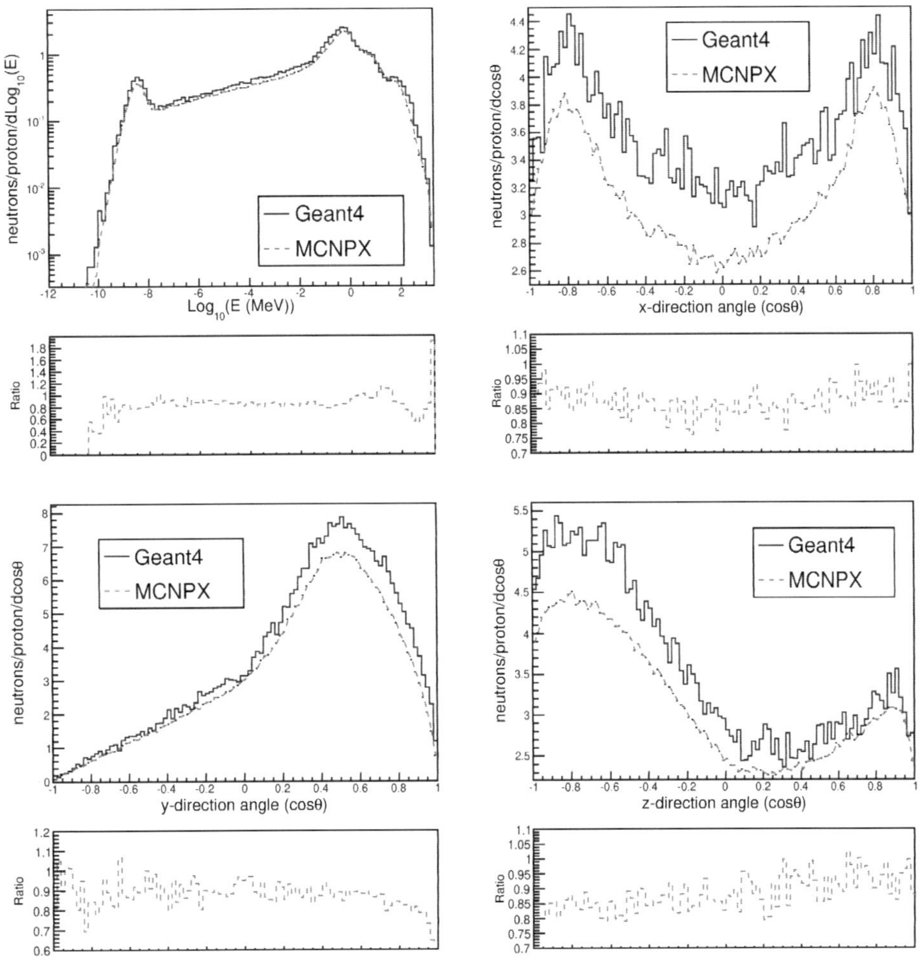

Figure 4: Results of the MCNPX and Geant4 calculations for the simple spallation source target and moderator model shown in Fig. 3. The energy spectrum of the neutrons leaving the target are shown in the top left panel, the x-directions of the neutrons leaving the target in the top right panel, the y-directions in the bottom left panel, and the z-directions in the bottom right panel.

5. Automated variance reduction methods

In addition to the benchmarking tests described above, we have also investigated the suitability of Geant4 for weight-window based calculations. While Geant4 supports the usage of the weight-window technique, it however does not come equipped with a weight-window generator. Therefore, we have implemented an automated variance reduction method for Geant4 based applications. These methods are of particular interest for deep shielding calculations, which are required for both safety reasons and estimates of instrument backgrounds due to leakage of high-energy particles from the shielding components. The implemented approach is described in detail in Ref. [15, 21] and will be briefly summarized here. The approach uses a Global Variance Reduction (GVR) method to encourage the uniform population of particles throughout a model geometry and is based on the weight-window technique. The steps for the procedure are

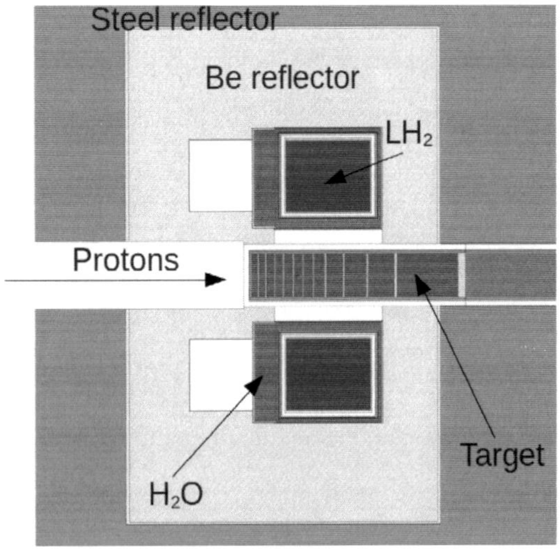

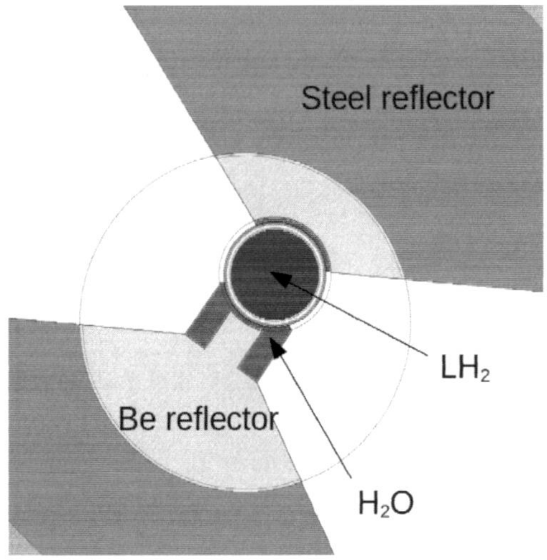

Figure 5: The left panel shows a vertical cut through the axis of the moderators in the MCNPX model and the right panel shows a horizontal cut.

described below:

1. A weight-window and scoring mesh are defined and overlaid across the mass geometry of the model.

2. An analog run (i.e a run with the weight-window technique disabled) is performed and a log file is stored which contains the relative errors and fluxes of each cell in the mesh.

3. The stored log file is then used by the application to calculate the weight-window limits using either of the following equations

$$W_L = (\frac{C_U + 1}{2})^{-1} \frac{\text{Min}(\vec{Re})}{Re_i} \tag{1}$$

or

$$W_L = (\frac{C_U + 1}{2})^{-1} \frac{\phi_i}{\text{Max}(\vec{\phi})}, \tag{2}$$

where the first is known as the flux-based method and the second as the relative-error based method [22]. The parameter W_L is the weight-window lower bound, C_U is a parameter defined by the user, and $\vec{Re}$ and $\vec{\phi}$ are vectors with the relative error or flux information of the defined cells in mesh, respectively. The first method encourages a uniform population of Monte-Carlo particles throughout the geometry while the latter encourages a flat distribution of errors throughout. The user can choose either method in the code. After the weight-window run is carried out, a new log file is written which can be used for more iterations of the procedure.

Figure 7 shows a selection of flux maps and relative error maps for an analog simulation compared to a GVR simulation using the flux-based approach and applied to the Geant4 ESS TDR target and moderator model for simulations of a similar time. The maps represent a plane above the target center. The black circular regions mark the ends of the monolith inner iron and outer concrete shielding. In the given simulation time, the GVR simulations lead to a nearly

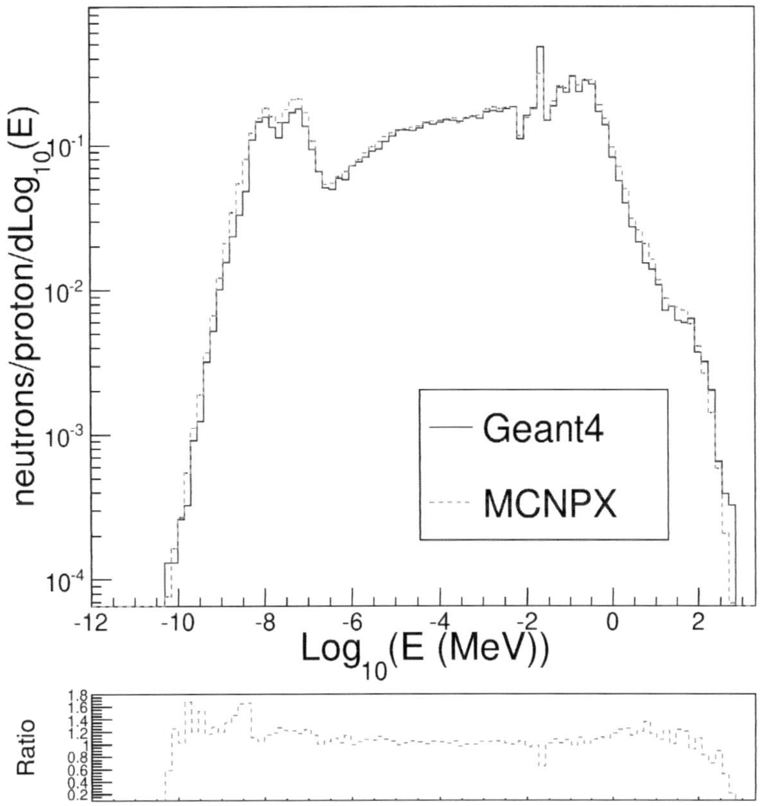

Figure 6: Preliminary comparison between MCNPX and Geant4 for the ESS TDR target and moderator design. The energy spectra represent all the neutrons detected at 2 m from the moderators and summed over all four beamline extraction areas.

complete population of the cells of the mesh in the plane and a more uniform distribution of the relative errors throughout, when compared to the analog case.

6. Conclusions

In the current work, we presented results of simulations of simple spallation targets and the ESS TDR design using the Monte-Carlo codes Geant4 and MCNPX. Overall, there is in general good agreement between the results shown in this work. Additionally, we presented an implementation of an automated weight-window generator for shielding applications and showed the benefit of using the approach compared to analog simulations. The above mentioned developments suggest that Geant4 is well suited for spallation neutron source related calculations.

References

[1] Pelowitz D *et al.* 2011 MCNPX User's Manual Version 2.7.0 LA-CP-11-00438
[2] MCNPX Home Page https://mcnpx.lanl.gov/
[3] Peggs S *et al.* 2013 ESS Technical Design Report Lund, Sweden

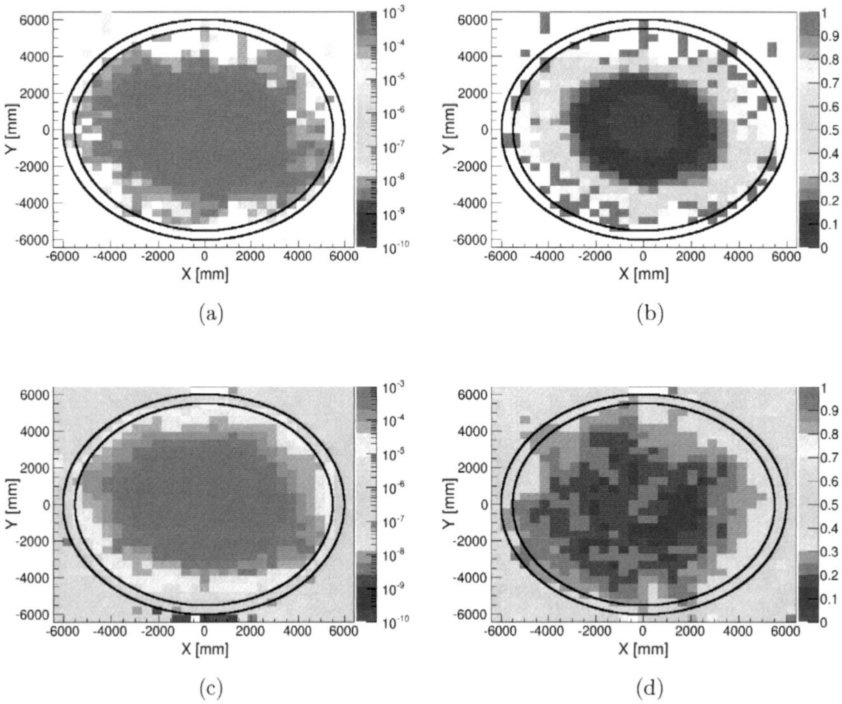

Figure 7: Flux and relative errors maps for a plane in the Geant4 ESS TDR model using an analog simulation and a simulation using the GVR approach. Panel a) shows the flux map for the analog simulation, panel b) the relative error map for the analog simulation, panel c) the flux map for the GVR method, and panel d) the relative error map for the GVR method.

[4] Batkov K *et al.* 2013 *Nucl. Instrum. and Meth. in Phys. Res. A* **729** 500–S05
[5] Agostinelli S *et al.* 2003 *Nucl. Instrum. and Meth. in Phys. Res. A* **506** 250
[6] Geant4 http://http://www.geant4.org
[7] CERN `http://home.web.cern.ch/`
[8] Materials and Life Science Experimental Facility - J-PARC `http://www.j-parc.jp/MatLife/en/`
[9] J-PARC — Japan Proton Accelerator Research Complex `http://j-parc.jp/index-e.html`
[10] Kittelmann T *et al.* 2014 *J. Phys. Conf. Ser.* **513** 022017
[11] Kittelmann T and Boin M 2015 *Comp. Phys. Comm.* **189** 114
[12] Cherkashyna N *et al.* 2014 *Journal of Physics: Conference Series* **528** 012013
[13] Cherkashyna N *et al.* 2015 Submitted to the Proceedings of ICANS XXI (International Collaboration on Advanced Neutron Sources) Mito, Japan
[14] Cherkashyna N *et al.* 2015 *Physical Review Special Topics: Accelerators and Beams* **18** 083501
[15] Stenander J and DiJulio D 2015 *Nucl. Instrum. and Meth. in Phys. Res. A* **798** 167
[16] Bertini H 1963 *Phys. Rev.* **131** 1801
[17] Bertini H 1969 *Phys. Rev.* **188** 1711
[18] Chadwick M *et al.* 2006 *Nuclear Data Sheets* **107** 2931
[19] Geant4: User documentation http://geant4.cern.ch/support/userdocuments.shtml
[20] Geant4: Reference physics lists http://geant4.cern.ch/support/proc_mod_catalog/physics_lists/useCases.shtml
[21] Stenander J Geant4 simulations of neutron flux using automated weight windows with applications to ESS Master's Thesis, Lund University (2015)
[22] van Wijk A *et al.* 2011 *Annals of Nuclear Energy* **38** 2496

Characterization of the radiation background at the Spallation Neutron Source

Douglas D. DiJulio[1,2], Nataliia Cherkashyna[1], Julius Scherzinger[1,2], Anton Khaplanov[1], Dorothea Pfeiffer[1,3], Carsten P. Cooper-Jensen[1,4], Kevin G. Fissum[1,2], Kalliopi Kanaki[1], Oliver Kirstein[1,5], Georg Ehlers[6], Franz X. Gallmeier[7], Donald E. Hornbach[7], Erik B. Iverson[7], Robert J. Newby[7], Richard J. Hall-Wilton[1,8], Phillip M. Bentley[1,4]

[1] European Spallation Source ERIC, SE-221 00 Lund, Sweden
[2] Division of Nuclear Physics, Lund University, SE-221 00 Lund, Sweden
[3] CERN, CH-1211 Geneva 23, Switzerland
[4] Department of Physics and Astronomy, Uppsala University, 751 05 Uppsala, Sweden
[5] University of Newcastle, Callaghan, NSW 2308, Australia
[6] Quantum Condensed Matter Division, ORNL, Oak Ridge TN-37831, USA
[7] Instrument and Source Division, ORNL, Oak Ridge, TN-37831, USA
[8] Mid-Sweden University, SE-851 70 Sundsvall, Sweden

E-mail: douglas.dijulio@esss.se

Abstract. We present a survey of the radiation background at the Spallation Neutron Source (SNS) at Oak Ridge National Laboratory, TN, USA during routine daily operation. A broad range of detectors was used to characterize primarily the neutron and photon fields throughout the facility. These include a WENDI-2 extended range dosimeter, a thermoscientific NRD, an Arktis ^{4}He detector, and a standard NaI photon detector. The information gathered from the detectors was used to map out the neutron dose rates throughout the facility and also the neutron dose rate and flux profiles of several different beamlines. The survey provides detailed information useful for developing future shielding concepts at spallation neutron sources, such as the European Spallation Source (ESS), currently under construction in Lund, Sweden.

1. Introduction

Spallation neutron sources provide scientists with intense fluxes of cold and thermal neutrons for condensed matter investigations. The usable signal for such experiments is typically on the order of a few meVs. However, at a spallation neutron source, the experiments may also be contaminated by the presence of a significant number of high-energy neutrons, which can extend up to the energy of the primary proton beam (several GeVs). These neutrons are generated in the primary spallation target and can reach the experimental stations via a number of mechanisms, which may include streaming down the neutron guides, penetration through the target biological shielding, or leakage from a nearby beamline. The presence of the these neutrons causes significant shielding concerns at spallation sources which are not otherwise present at reactor-based facilities [1, 2, 3]. Additionally, at a pulsed spallation source the high-energy neutrons are correlated with the proton impact on the spallation target and may show up as a prompt-pulse of particles arriving at the sample position of a particular instrument [4, 5]. This is particularly important for inelastic scattering instruments, where signals may be on the order of 10^{-6}-10^{-4} compared to elastic scattering [6]. Two such instruments relevant to the current study are the hybrid spectrometer (HYSPEC) and the cold neutron spectrometer (CNCS) instruments [7, 8], which both exhibit noticeable prompt-pulse signals [4, 9].

Understanding the mechanisms which give rise to the high-energy prompt-pulse backgrounds at modern spallation sources can provide valuable information for upcoming sources, such as the the European Spallation Source (ESS) [10], currently under construction in Lund, Sweden. Therefore, we have carried out a survey of the radiation background at the Spallation Neutron Source (SNS) at Oak Ridge National Laboratory, TN, USA during routine daily operation [11]. The measurements were carried out around the target monolith, close to the shielding of beamlines, and on cracks between shielding blocks. A broad range of detectors, including some which are not traditionally used at spallation neutron sources for background characterization, were employed. The advantages of these types of detectors, particularly for high-energy neutron detection, will be discussed below.

In the following, we present some relevant information on the shielding configurations at SNS, the methods and results of our measurements, and lastly some discussion and conclusions with the focus in particular on backgrounds primarily related to neutron and photon fields measured throughout the facility.

2. The SNS facility

The SNS is an accelerator-based facility with a proton beam of 1 GeV that is incident on a mercury target. The neutrons are moderated by three cryogenic hydrogen moderators and one ambient water moderator. The moderators are surrounded by a water cooled beryllium and steel reflector region. The biological shielding consists of approximately 5 m of iron and 1 m of high-density concrete. High-density concretes contain aggregates, such as magnetite, which are more efficient for shielding against high-energy neutrons above 1 MeV and γ-rays compared to standard concrete. In total there are 18 beamlines extending from the shielding. A detailed overview of the target station design can be found in Ref. [12].

At SNS, the neutron beamlines can be straight or curved, where the latter geometry is used to reduce the amount of high-energy particles reaching the sample position of an instrument. Each beamline has a primary shutter inside the monolith, in order to stop the incident radiation and allow access into the instrument caves during daily operation. The shutters are designed to comprise of a 1 meter long block of tungsten surrounded by steel [13]. Many beamlines also have secondary shutters, which makes it possible for multiple beamlines to share a single primary shutter [14]. The standard shielding materials for beamlines include regular concrete, steel, and high-density concrete. Depending on the particular geometrical layout of the beamline, shielding thicknesses can reach up to 2 m of regular concrete [15]. Shielding challenges at SNS have been highlighted in a number of previous publications [13, 15] and specific beamline calculations can also be found in [14, 16].

3. Equipment

The measurements were carried out at SNS using a wide range of detectors. Each detector will be briefly presented and described below. Detailed descriptions will be referenced for further information.

3.1. Neutron dosimetry

The neutron dose rates at SNS were mapped with a WENDI-2 (<5 GeV) [17] dosimeter and a Thermoscientific NRD (<10 MeV) [18]. The WENDI-2 is a cylindrical dosimeter containing a ^{3}He tube surrounded by a tungsten loaded polyethylene shell. The tungsten powder layer is 1.5 cm thick and provides a scattering layer with a large interaction probability above ~10 MeV, ultimately maintaining the sensitivity of the device at higher energies. On the other hand, the NRD is a spherical dosimeter loaded with cadmium and has a decreasing sensitivity above ~10 MeV. Thus measurements with the two dosimeters at the same locations may reveal areas where the the fast neutron dose above ~10 MeV is significant. The NRD is limited to dose rates above 0.2 μSv/hr [18] while the WENDI-2 is limited to dose rates above 0.001 μSv/hr [17]. The calibration of the two dosimeters was checked against a ^{252}Cf source, which emits neutrons predominately in the energy region where the response of the dosimeters is nearly identical. A maximum difference of ~20% in the measured dose rates with WENDI-2 and NRD was observed.

3.2. Fast-neutron detection

The Neutron Diagnostic Tool (NDT) from Arktis Radiation Detectors [19, 20, 21, 22] was used to measure the fast-neutron rates throughout the facility. The noble gas scintillation counter NDT was developed for fast neutron counting in high γ-ray background environments and is optimized for neutron energies between 0.5 MeV and 20 MeV. The active detector volume is 1.05 liters of highly pressurized (179 bar) ^{4}He gas in a 2 inch inner diameter steel tube. The ultraviolet scintillation light is shifted to longer wavelengths by a wavelength shifting coating on the reflecting sides of the tube. The transmitted light is read out using two R580 Hamamatsu [23] photo-multiplier-tubes (PMTs). The PMTs are operated in photon-counting mode and the pulses are processed by a 10 bit WaveDREAM-B16 [21] digitizer with a sampling rate of 1 GS/s. This set-up allows for combined neutron/γ-ray count rates of up to 1 MHz, identification of fast neutrons by both pulse-shape and pulse-height discrimination, and the determination of the interaction point along the active volume of the detector [22]. Rejection of the γ-ray events is made possible by the fact that high-pressure ^{4}He scintillation light exhibits fast and slow components, with decay times of a few nanoseconds and about 1 ms respectively [20]. These decay constants are attributed to the singlet and triplet states of the excited dimers formed by the excitation of atomic ^{4}He. The relative strength of the fast component to the total light yields enables the identification of incident particles by discriminating recoiling electrons correlated to γ-ray events and recoiling alpha particles correlated to neutrons.

The count rates collected by the NDT were converted to fast-neutron count rates for further analysis. For this, a correction for the non-uniform light transport to the PMTs was applied. The non-uniformity is due to light loss in the defuse reflector. The reflector is a thin teflon (PTFE) coating which has normally a high reflectivity (~90%) for visible light. On top of this is also a phosphor coating to shift the wavelength from 80 nm (emission from He) to 430 nm (PMT sensitivity). This means that for the long distance (50 cm) from the center of the detector to the PMTs, a significant fraction of light will be lost by transmission/absorption in the reflector material. The effect of the reflector is the main light-loss mechanism in the detector. The interaction position was calculated as the relative difference between the numbers of photons counted in the right PMT and in the left PMT divided by the total photon number. The maximum photon count was a function strongly dependent on the interaction point and the detection efficiency of the detector is therefore also dependent on the interaction point. In order to calculate the neutron rate, the neutron interaction efficiency was estimated to be 0.03 for the energy range between 0.5 MeV and 20 MeV and zero for all other energies. For 0.5 MeV to 6 MeV, the values were corrected for gas pressure from Ref. [20] and were used in the analysis. An average efficiency of 0.02 was estimated from Evaluated Nuclear Data Files (ENDF) cross

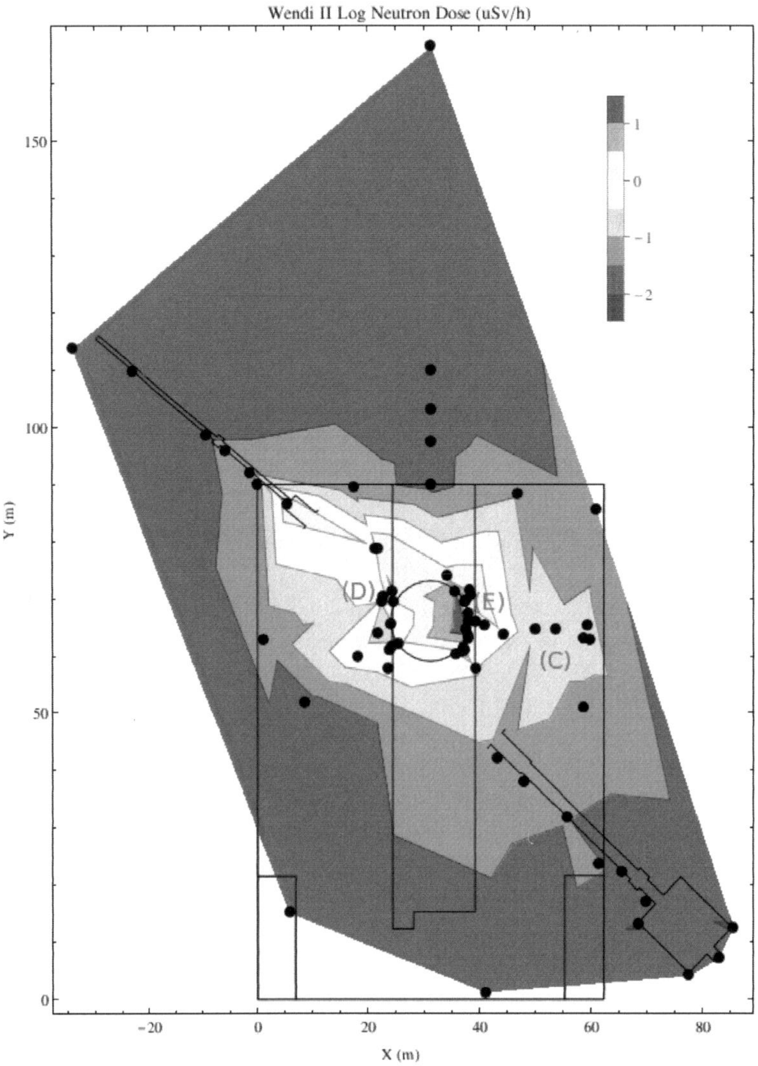

Figure 1: Contour plot of the neutron dose rate measurements carried out with the WENDI-2 dosimeter. The black dots represent the measurement points and the marked letter regions highlight several areas of interest, as described in the text.

sections [24] and the calculations from Ref. [20] for energies above 6 MeV.

3.3. γ-ray spectroscopy

A large (5x10x40cm^3) NaI detector was used for γ-ray measurements throughout the facility. The resolution of the NaI detector was not sufficient to separate the 511 keV annihilation peak from the often-present 478 keV peak, due to neutron capture on ^{10}B, which is common at neutron scattering facilities. However, the advantage of the NaI detector was its considerably large efficiency at high energies.

Table 1: Comparison between the measured and calculated neutron dose rates from the WENDI-2, NRD, and the Arktis NDT at several measurement locations throughout the area shown in Fig. 1. The letters indicate the regions where the measurements were carried out. The measured Arktis NDT count rates are also given for completeness.

Position (Region Fig. 1)	WENDI-2 (μSv hr^{-1})	NRD (μSv hr^{-1})	Arktis NDT (μSv hr^{-1})	Arktis NDT (n s^{-1}cm^{-2})
1 (C)	0.11(2)	0.44(14)	0.28(1)	0.45(2)
2 (D)	0.78(14)	0.87(29)	0.93(7)	1.48(10)
3 (D)	3.27(19)	4.37(144)	x	x
4 (D)	0.94(14)	x	0.97(12)	1.56(19)
5 (D)	2.95(21)	2.62(86)	5.82(28)	9.28(45)
6 (E)	21.85(122)	9.61(317)	28.68(63)	45.78(101)
7 (E)	10.08(79)	8.67(108)	14.71(45)	23.49(72)
8 (E)	0.92(12)	0.57(16)	1.36(14)	2.17(22)
9 (E)	0.95(12)	0.89(38)	1.43(14)	2.29(23)

4. Results and Discussion

Fig. 1 shows a neutron dose rate contour map collected with the WENDI-2 dosimeter. The indicated points represent the measurement locations and the dose rates were calculated by averaging the collected data over periods when the proton beam was stable. The central circular and rectangular regions in the middle of the map mark the edges of the biological shielding of the target monolith. The letter markers on the map indicate several regions of interest. Point (A) highlights the backscattering spectrometer BASIS [25], point (B) the powder diffractometer POWGEN [26], point (C) is an area down on the instrument floor between the shielding of the HYSPEC and FNPB [27] beamlines, and points (D) and (E) are regions where the instrument beamline shielding meets the target monolith shielding. The highest dose rate measured with the WENDI-2 was $\sim$25 μSv/hr, indicated by the red color in the map, which was in region (E) on top of a gap between an instrument shielding block and the monolith wall.

Table 1 presents a comparative study of the WENDI-2, NRD, and Arktis NDT dose rates from the measurements throughout the facility. The regions where the specified measurements were taken are indicated in the table. The Arktis NDT dose rates were calculated from the collected count rates and using a dose conversion factor of 0.000174 μSv cm^2 [28]. This is the dose conversion factor for 1 MeV neutrons, which corresponds to the maximum in the cross-section for ^{4}He neutron elastic scattering [20]. The broad agreement between the measured and calculated dose rates may support the conclusion that the dose rates arise from neutrons around 1 MeV at the selected positions. However, it should be mentioned that further comparison of the detector results requires more details regarding the response functions of the detectors, systematic uncertainties and information about the local radiation fields. Detailed simulations would also be of great value for the analysis of the current results.

In areas where beamlines are present, geometrical effects may play an important role in how the dose rate scales as a distance along the beamline. To investigate this further, we also performed measurements along specific beamlines. Fig. 2 presents the dose rates measured along the two beamlines POWGEN and BASIS with the WENDI-2 dosimeter. The dose rate measured along the straight beamline (POWGEN) was fit by a straight line and the dose rate measured along the curved beamline (BASIS) was fit with a $1/r$ function.

Similar measurements were also carried out along the HYSPEC and TOPAZ [29] beamlines

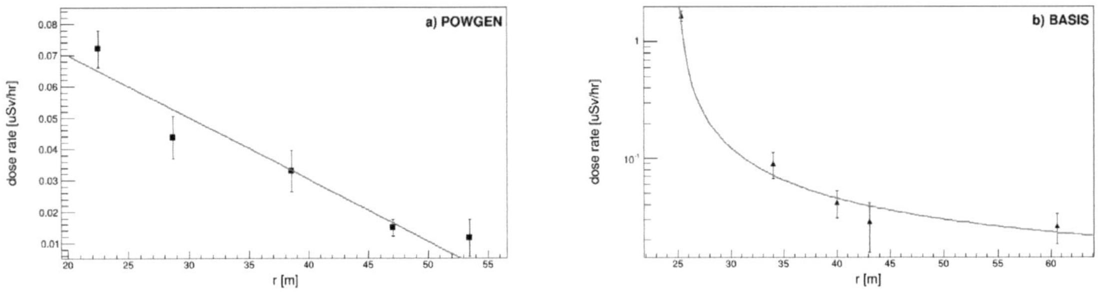

Figure 2: Neutron dose rates measured with the WENDI-2 dosimeter along the a) POWGEN and b) BASIS instruments as a function of distance from the monolith biological shielding.

using the Arktis NDT and the results are shown in Fig. 3. The two beamlines are in the same vicinity and both curves were fit with $1/r^2$ functions.

Fig. 4 presents the data collected based on traces taken with a scope and the NaI detector. The measurements were taken on the instrument beamline shielding above a gap in the shielding and close to the biological shield of the monolith. Each trace was 20 ms long and has one million samples at 20 ns intervals. The data contains 400 traces. The scope triggered randomly, but this usually happened in the beginning of an SNS pulse. Traces where this was not the case were manually removed and amounted to approximately 10% of the total. Each event contained an amplitude and a time signal. Overall, there were 72k events in the data set.

The left panel of Fig. 4 shows the number of counts collected as a function of time. The prompt-pulse is clearly visible with the repetition rate of 60 hz. The right panel shows a two-dimensional scatter plot of the amplitude vs the time. The spectra have three noticeable components. The activation γ-rays are always present and occur in the lowest amplitudes. Capture γ-rays occur up to about 0.01 V, which corresponds to $\sim$10 MeV, and this contribution dies off after a few ms. Lastly, there are the prompt γ-ray events which occur only in the beginning of the pulse, which is visible with the repetition rate of 60 hz. These have much higher energies than 10 MeV. However, it can be noted that pulse-pile up effects may contribute to a fraction of the large observed amplitudes due to the high instantaneous rate. This last energy group of photons represents an interesting possible contribution to the prompt-pulse background unique to pulsed spallation sources. High-energy photons coincident with the proton beam can lead to additional neutrons generated by photonuclear reactions through the giant dipole resonance. These neutrons have higher energies, into the MeV range, and can travel into and induce additional secondaries in the surrounding instrument shielding. The importance of this contribution to background noise at spallation neutron facilities requires more detailed investigations.

5. Conclusions and Outlook

In summary, we carried out measurements at the SNS during routine user operation in order to understand the radiation leaking from the shielding structures. We presented maps of the neutron dose rate throughout the facility and also along various neutron beamlines. These highlight potential future enhancements in shielding solutions and also differences in the leakage properties of the the individual beamlines. A comparative study between the neutron detectors suggested that the dose rates around the facility at selected positions may be contributed to neutrons on the order of 1 MeV. Lastly, our results indicated the presence of high-energy photons

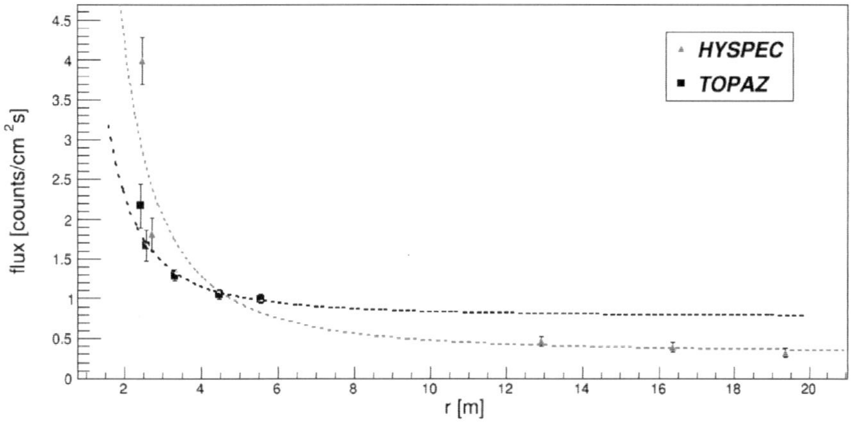

Figure 3: Neutron fluxes measured with the Arktis NDT as a function of distance from the monolith biological shielding for the HYSPEC and TOPAZ beamlines.

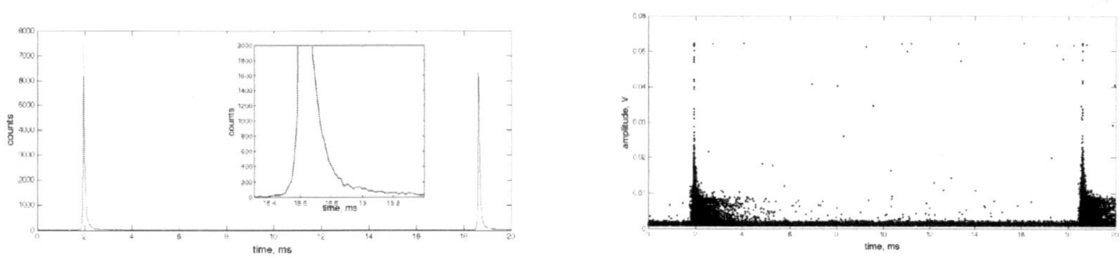

(a) (b)

Figure 4: Panel a) shows the number of counts as a function of time collected with the NaI detector for a position on top of the instrument shielding close to the monolith wall. The prompt-pulse is clearly visible every 60 hz. Panel b) shows a scatter plot of the amplitude as a function of time collected with the NaI detector. Prompt γ-rays, capture γ-rays, and γ-rays from continuous activation make up the three major components.

correlated with the primary beam impact on target on top of the instrument shielding.

The results presented in this paper highlight the need for an improved understanding of backgrounds at modern spallation neutron source facilities. Future studies could for example include an analysis of correlations between the measurements and the status of the beamline shutters throughout the facility, detailed modeling of the response functions of the detectors and local radiation fields, and also the implementation of detectors which can provide more precise determination of the energy of the neutrons leaking from the shielding components.

6. Acknowledgments

The authors would like to thank the Nuclear and Radiological Support Services staff at SNS for assisting in the measurements at the facility. We would also like to thank Professor Lawrence Weinstein and the Experimental Nuclear and Particle Physics Group of Old Dominion University for providing detectors and electronics for our measurements at short notice. This research at ORNL's Spallation Neutron Source was supported by the Scientific User Facilities Division,

Office of Basic Energy Sciences, U.S. Department of Energy.

References

[1] Russel G *et al.* 1988 Proceedings of the ICANS-X Conference Los Alamos, NM
[2] Koprivnikar I and Schachinger E 2002 *Nucl. Instrum. and Meth. in Phys. Res. A* **487** 571
[3] Cherkashyna N *et al.* 2015 *Physical Review Special Topics: Accelerators and Beams* **18** 083501
[4] Cherkashyna N *et al.* 2014 *Journal of Physics: Conference Series* **528** 012013
[5] Cherkashyna N *et al.* 2015 Submitted to the Proceedings of ICANS XXI (International Collaboration on Advanced Neutron Sources) Mito, Japan
[6] Kajimoto R *et al.* 2013 *J. Phys. Soc. Jpn.* **82** SA032
[7] Winn, Barry, Filges, Uwe, Garlea, V Ovidiu, Graves-Brook, Melissa, Hagen, Mark, Jiang, Chenyang, Kenzelmann, Michel, Passell, Larry, Shapiro, Stephen M, Tong, Xin and Zaliznyak, Igor 2015 *EPJ Web of Conferences* **83** 03017 URL `http://dx.doi.org/10.1051/epjconf/20158303017`
[8] Ehlers G, Podlesnyak A A, Niedziela J L, Iverson E B and Sokol P E 2011 *Review of Scientific Instruments* **82** 085108 ISSN 00346748 URL `http://dx.doi.org/doi/10.1063/1.3626935`
[9] Winn B and Hagen M 2012 Inelastic spurion from the prompt pulse, HYSPEC memo hyspec_memo01272012_01 http://neutrons.phy.bnl.gov/HYSPEC/memos_technical_reports.shtm
[10] Peggs S *et al.* 2013 ESS Technical Design Report Lund, Sweden
[11] Mason T E, Abernathy D, Anderson I, Ankner J, Egami T, Ehlers G, Ekkebus A, Granroth G, Hagen M, Herwig K, Hodges J, Hoffmann C, Horak C, Horton L, Klose F, Larese J, Mesecar A, Myles D, Neuefeind J, Ohl M, Tulk C, Wang X L and Zhao J 2006 *Phys. B (Amsterdam, Neth.)* **385-386** 955–960
[12] Haines J *et al.* 2014 *Nucl. Instrum. and Meth. in Phys. Res. A* **764** 94
[13] Ferguson P *et al.* 2005 *Rad. Prot. Dos.* **115** 170
[14] Popiva I *et al.* 2014 *Prog. Nucl. Sci. and Tech.* **4** 753
[15] Gallmeier F *et al.* 2005 *Rad. Prot. Dos.* **115** 23
[16] Popiva I and Gallmeier F 2014 *Prog. Nucl. Sci. and Tech.* **4** 178
[17] Olsher R *et al.* 2000 *Health Physics* **79 (2)** 170–181
[18] ASP-2e/NRD Portable Neutron REM Meter. `http://www.thermoscientific.com`
[19] Arktis Radaition Detectors `http://www.arktis-detectors.com`
[20] Chandra R *et al.* 2012 *Journal of Instrumentation* **7** C03035
[21] Friederich H *et al.* 2012 *Real Time Conference (RT), 2012 18th IEEE-NPSS* pp 1–7
[22] Chandra R *et al.* 2010 *2010 IEEE International Conference on Technologies for Homeland Security (HST)* 410 – 413
[23] Hamamatsu Photonics K.K. `http://www.hamamatsu.com`
[24] ENDF: Evaluated Nuclear Data File `https://www-nds.iaea.org/exfor/endf.htm`
[25] Mamontov E and Herwig K W 2011 *Review of Scientific Instruments* **82** 085109 URL `http://scitation.aip.org/content/aip/journal/rsi/82/8/10.1063/1.3626214`
[26] The Powder Diffractometer at SNS http://neutrons.ornl.gov/powgen/
[27] Fundamental Neutron Physics Beamline at SNS http://neutrons.ornl.gov/fnpb/
[28] ICRP Publication 116 2010, Annals of the ICRP Vol. 40 Nos 2-5
[29] TOPAZ: The Single-Crystal Diffractometer at SNS http://neutrons.ornl.gov/topaz/

Neutron imaging on the VR-1 reactor

J Crha[1], L Sklenka[1], J Soltes[1]

[1] Faculty of Nuclear Sciences and Physical Engineering, Czech Technical University in Prague

E-mail: honzacrha@seznam.cz

Abstract. Training reactor VR-1 is a low power research reactor with maximal thermal power of 1 kW. The reactor is operated by the Faculty of Nuclear Science and Physical Engineering of the Czech Technical University in Prague. Due to its low power it suits as a tool for education of university students and training of professionals. In 2015, as part of student research project, neutron imaging was introduced as another type of reactor utilization. The low available neutron flux and the limiting spatial and construction capabilities of the reactor's radial channel led to the development of a special filter/collimator insertion inside the channel and choosing a non-standard approach by placing a neutron imaging plate inside the channel. The paper describes preliminary experiments carried out on the VR-1 reactor which led to first radiographic images. It seems, that due to the reactor construction and low reactor power, the neutron imaging technique on the VR-1 reactor is feasible mainly for demonstration or educational and training purposes.

1. Introduction

Neutron radiography (also known as neutron imaging) is a non-destructive analytical method based on two-dimensional detection of neutrons that penetrate an object being imaged. Any material that forms an object has its own unique attenuation properties that are also incident neutron energy dependent. The decrease of neutron beam intensity caused by an object can be expressed in a simple form of the exponential attenuation law as follows

$$I(x) = I_0 e^{-\Sigma_{tot} x}, \tag{1}$$

where $I(x)$ is the neutron beam intensity in the detector plane, I_0 is the initial neutron beam intensity, Σ_{tot} is the macroscopic cross section summed over all isotopes occurring in the object and x is the thickness of an object.

Neutron radiography is in its principle same as X-ray radiography well known from medical diagnostics. As X-rays interact by electromagnetic interaction with electrons in the electron cloud, the probability of interaction (microscopic cross section) increases with increasing atomic number. This does not apply for neutrons. The probability of neutron interaction with matter does not exhibit regular dependence on the atomic number. Neutrons, which are not electrically charged, interact mostly by strong interaction with the atomic nucleus, furthermore neutron interaction probability with a nucleus depends on the incident neutron energy.[2] Neutron radiography and X-ray radiography are imaging methods that are complementary due to different ways of interaction of thermal and cold neutrons (which are used for neutron imaging) and X-rays with matter.[4]

Nowadays, for reactors above 250 kW, neutron radiography is a well-established imaging method ([1], [3]). The main purpose of this work was to analyze the possibility of performing neutron radiography on very low power reactors, such as the VR-1 reactor.

2. Experimental setup

A basic setup of every neutron imaging facility consists of a neutron source, a collimator, an object being imaged and a neutron detector. The VR-1 reactor is low power pool-type light water training reactor with maximal power level of 1 kW. The fuel type is IRT-4M enriched to 19.7% with negligible burnup. Light water is used as coolant, moderator, reflector and biological shielding. Due to its low power, heat removal is provided by natural convection. Its experimental equipment includes specific educational and experimental instrumentation (such as instrumentation for fast reactivity changes, instrumentation for study of void coefficients) as well as standard R&D instrumentation (such as rabbit, vertical and horizontal beam tubes). Its maximum neutron flux is at the level of 10^9 $cm^{-2}s^{-1}$.[5]

As a reference for neutron imaging the LVR-15 reactor was used. The LVR-15 reactor is tank-type light water research reactor operated by the Research Centre Rez Ltd with maximum power of 10 MW. It uses the same type of fuel as the VR-1 reactor (IRT-4M). Its maximum neutron flux is at the level of 10^{14} $cm^{-2}s^{-1}$. The LVR-15 reactor is a multipurpose reactor mainly used for radioisotope production, silicon doping and beam experiments.[8]

Neutron imaging on the VR-1 reactor was carried out in its only radial channel. The channel is a tool for extracting neutrons from the core through the reactor vessel and biological shielding to the irradiation position. The reactor instrumentation does not include a large irradiation box neither a thermal column, therefore samples were irradiated inside the radial beam port (channel). The beam port was partially closed with a plug and sealed with polyethylene bricks. For better moderation, a plastic bottle filled with water was added to the part of channel adjacent to the core. Simple neutron beam collimation was supposed to be provided by a cadmium plate rolled in cylindrical shape. As a neutron detection system, a Fujifilm neutron imaging plate BAS-ND 2025 [7] was used. Besides neutrons the imaging plate utilized for measurements is sensitive to electrons, X-rays, gamma radiation, alpha radiation. The imaging plate consist of two main parts, a phosphor material whose crystalline lattice stores the energy of incident radiation and of a neutron converter (Gd_2O_3 - Gd producing gammas that are internally converted into electrons).[2] The basic scheme of the experimental setup is shown in Figure 1. Figure 2 shows the samples being placed on the imaging plate. The samples were subsequently

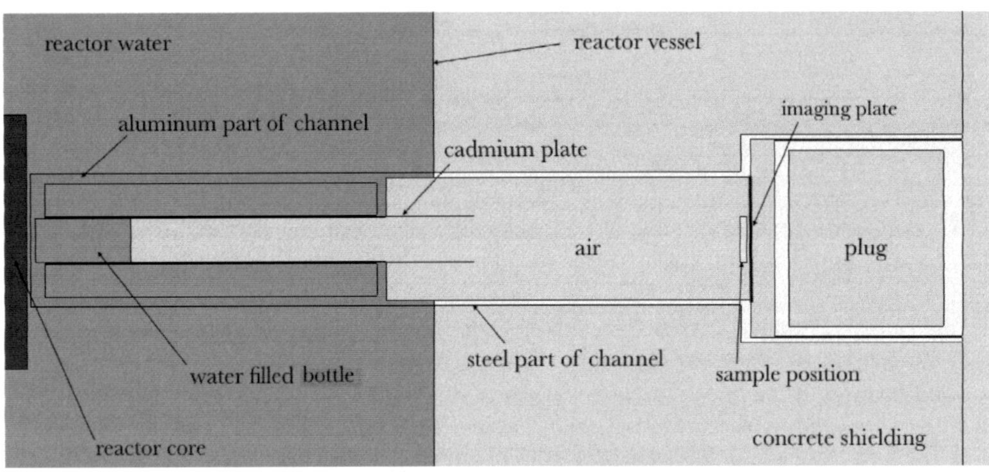

Figure 1. Experimental setup in the radial channel of the VR-1 reactor.

inserted inside the radial channel. The beam port exit was sealed by the plug and polyethylene bricks.(see Figure 3).

Figure 2. Fixing the samples on the imaging plate before irradiation.

Figure 3. Beam port exit sealed by the plug and polyethylene bricks.

3. Results

Three preliminary experiments (on the VR-1 reactor) and one reference experiment (on the LVR-15 reactor) were carried out. All radiographs from the VR-1 reactor were taken with the same exact power level. Only irradiation times were different (30 mins, 60 mins, 90 mins). During the 60 minute irradiation, a 10 cm neutron filter made of mono-crystalline silicon was added into the channel for epithermal and fast neutron beam content suppression. This filter also reduces the gamma radiation content originating from fission reactions in the core.

Radiographs of specific objects were taken during experiments on the VR-1 reactor. The set of samples for neutron imaging consisted of roller skates bearings, a computer fan, polyethylene plugs and a cadmium plate. Bearings are mostly made of stainless steel (balls and two rings). Balls, which enable revolving motion, are held by plastic in between the two rings. Inside inner diameter of two bearings out of three, a small polyethylene cylinders with a central hole (one borated polyethylene, one non-borated) were placed.

On the radiographs from the VR-1 reactor, stainless steel exhibits better attenuation than plastic parts (plastic holding balls, plastic in computer fan and polyethylene plugs) which is in contradiction with expectations and with the radiograph obtained on the LVR-15 where the neutron beam characteristics are well-know (fast neutrons and gamma is suppressed over thermal neutrons). It seems that the neutron beam from the VR-1 is probably not enough thermalized and due to the proximity to the core neutron beam is contaminated with a significant portion of gamma which also affects the images (see Figures 4, 5, 6 and 7). Also epithermal neutrons which are another possible contaminant of the beam could be thermalized in polyethylene parts of samples resulting in what could appear as a higher intensity detected. Therefore neutron spectrum in the radial channel is yet to be determined. Since this paper was focused on the feasibility of neutron imaging on a low power reactor, the spatial resolution is not yet available. The L/D ratio of the setup used on the VR-1 reactor is approximately 8.

Journal of Physics: Conference Series **746** (2016) 012034 doi:10.1088/1742-6596/746/1/012034

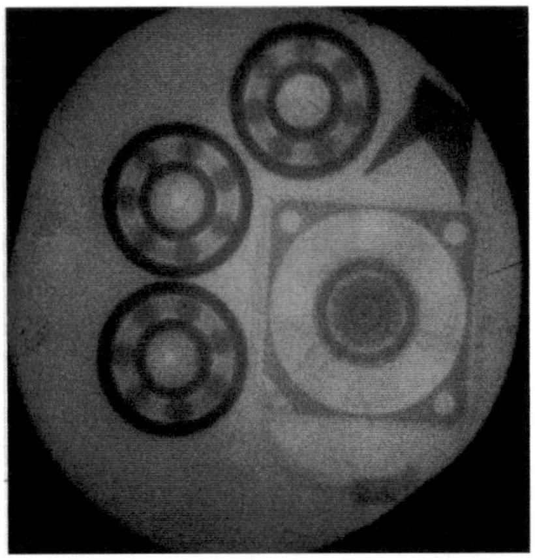

Figure 4. Neutron image at the VR-1 reactor (30 mins, without neutron filter).

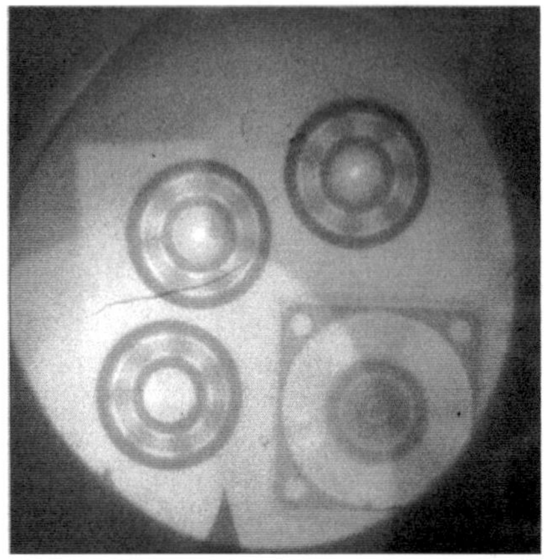

Figure 5. Neutron image at the VR-1 reactor (60 mins, with 10 cm neutron filter).

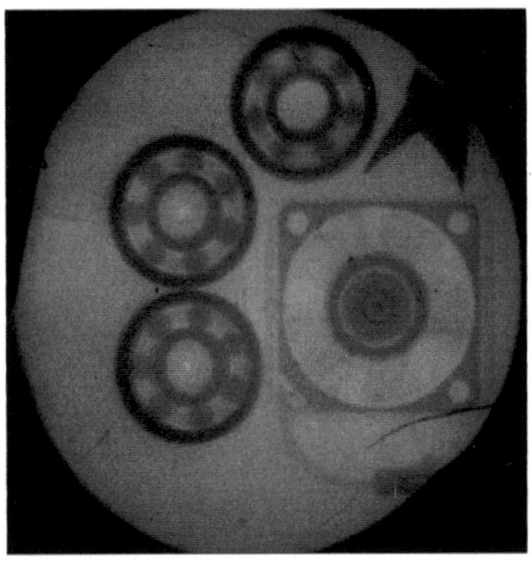

Figure 6. Neutron image at the VR-1 reactor (90 mins, with neutron filter).

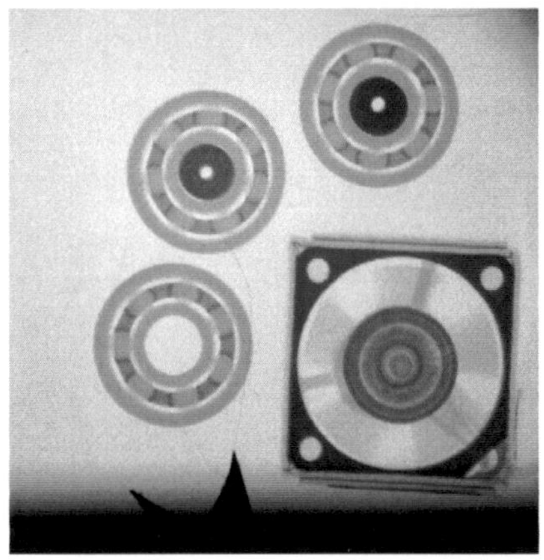

Figure 7. Neutron image at the LVR-15 reactor (5 secs, with 1 m neutron filter).

4. Conclusion and future works

Preliminary information about the neutron beam intensity at the sample position were obtained by a Monte Carlo code Serpent [6]. Although future calculations are planned to be performed with the MCNP code as it comprises a designated neutron radiography tally and also up-to-date model of the VR-1 reactor is available. Experiments carried out on the VR-1 reactor clearly showed that it is possible to perform neutron radiography on a low power reactor, at least for demonstrational and educational purposes. Future works should continue with neutron radiography investigations and with development of suitable instrumentation such as neutron filters, neutron collimators and neutron detectors and with setting the most suitable irradiation times.

Acknowledgments

This work was supported by projects SGS14/157/OHK4/2T/14 *Study of Neutron Filter Parameters on the VR-1 Trainning Reactor* and LM2011031 *VR-1 - Support of reactor operation for research activities.* We would like to thank Mr. Ladislav Viererbl from Research Centre Rez Ltd. for helping with neutron imaging experiments carried out on the LVR-15 reactor, Mrs. Bozena Michalcova from Nuclear Physics Institute, Academy of Sciences of the Czech Republic for borrowing imaging plates and for helping with analysis of radiographs. Also we would like to thank the VR-1 reactor and the LVR-15 reactor staff.

References

[1] Anderson I S, McGreevy R L and Bilheux H Z. *Neutron Imaging and Applications: A Reference for the Imaging Community. Neutron Scattering Applications and Techniques.* New York: Springer, 2009.

[2] Craft A E, Wachs D M, Okuniewski M A, Chichester D L, Williams W J, Papaioannou G C and Smolinski A T 2015 *Neutron Radiography of Irradiated Nuclear Fuel at Idaho National Laboratory. Phys Procedia* **69** 483-490

[3] FUJIFILM Imaging Plate. `http://www.buero-analytik-winden.de/app/download/13602912/Imaging+plates+ausf%C3%BChrliche+Beschreibung.pdf`.

[4] Leppanen J. *Serpent - a Continuous-energy Monte Carlo Reactor Physics Burnup Calculation Code.* User's Manual. 2015. `http://montecarlo.vtt.fi/download/Serpent_manual.pdf`

[5] Mannes D, Schmid F, Frey J, Schmidt-Ott K and Lehmann E 2015 *Combined Neutron and X-ray imaging for non-invasive investigations of cultural heritage objects. Phys Procedia* **69** 653-660

[6] Rataj J. *Extensive Utilization of VR-1 Reactor for Nuclear Education and Training.* Presented at RRFM 2010. `https://www.euronuclear.org/meetings/rrfm2010/transactions/RRFM2010-transactions-s3.pdf`.

[7] Viererbl L et al. *Horizontal Channel for Radiography and Tomography in LVR-15 Research Reactor.* Presented at RRFM/IGORR 2012. `https://euronuclear.org/meetings/rrfm2012/transactions/rrfm2012-transactions.pdf`.

[8] Zawiski M, Bastürk M, Rehacek J and Hradil Z 2004 *Neutron tomographic investigations of boron-alloyed steels. J. Nucl. Mater* **327** 188-193

Journal of Physics: Conference Series **746** (2016) 012035 doi:10.1088/1742-6596/746/1/012035

Upgrades to the SINQ Cold Neutron Source

R M Bergmann, U Filges, D Kiselev, T Reiss, V Talanov and M Wohlmuther

Paul Scherrer Institut, 5232 Villigen, Switzerland

E-mail: ryan.bergmann@psi.ch

Abstract. Changing the configuration of the cold neutron source during an extended shutdown of the Swiss Spallation Neutron Source (SINQ) at the Paul Scherrer Institut is being considered for improving performance of the instruments that use the cold source. Proposed plans include making the REH external (ensuring that it is not filled with liquid D_2) and using a REH that has been optimized to provide maximum gains between 3 and 6 Å. The optimization study was done using the MCNP Monte Carlo particle transport code, but was made tenable by developing a quick flux reconstruction technique that allows the neutron guide reflectivity to be approximated and accounted for in the optimization figure of merit. Ultimately, a wedge-shaped REH that penetrates the D_2 volume to about 4 cm from its center was determined to be optimal, and should provide an average gain of 24% from 3 and 6 Å and a peak gain of 29% at 5.5 Å.

1. Introduction

The Swiss Spallation Neutron Source (SINQ) is a spallation neutron source driven by a continuous 590 MeV proton beam at the Paul Scherrer Institut (PSI) in Villigen, Switzerland. The incoming protons impinge on a lead target cooled with heavy water, producing high energy neutrons. These neutrons are moderated by the tank of D_2O surrounding the target. The cold source is a 20 liter volume of liquid D_2 at approximately 25 K located within the D_2O moderator tank with its innermost face approximately 35 cm away from the center of the target. It serves to further reduce the energies of the "thermal" neutrons coming from the D_2O tank into a regime that is more useful to instruments which receive neutrons from its surface. Neutrons are transported from the cold source to the instruments via neutron guides, which are internally coated with neutron "supermirrors" reflective to neutrons at low energies and low incident angles [1].

There is an extended shutdown of SINQ ambitiously planned for 2018 which will provide an opportunity for changes to be made to the cold neutron source [2]. In the planned upgrade strategies, no changes to the surrounding aluminum and zirconium safety hulls will be made, and no changes to the insert tube structure will be made. In other words, all changes must fit into the existing multi-walled insert geometry. This paper presents a parametric study on changes to the physical dimensions of cold source and the predicted effects these changes have on the integral neutron flux in the neutron guides.

2. Re-entrant Hole

The goal of a re-entrant hole (REH) is to remove scattering and absorbing material between the spot with the maximum cold neutron flux and the direction of interest. In this case, this spot is at the center of the cold source and the direction of interest is towards the SINQ guide hall. The REH is presently a five-sided box with the open face at the bottom. It sits inside the D_2 volume directly next to the wall facing the guide hall. Voiding this kind of REH relies on radiative heating to boil D_2 which in turn fills the REH cavity with vapor, pushing the liquid D_2 out. Proposed plans include making the REH external, ensuring that it is not filled with liquid D_2.

Calculations done prior to these changes predicted cold neutron fluence gains from 1.2 to 1.9, with gains increasing with wavelength [3]. These increases have not been observed, and it is suspected that the REH is not fully voided [4]. All future designs will have an external REH, which will decouple the fill level from radiation heating and ensure that the REH is never filled with liquid D_2.

3. Simulation Geometry

Simulations for the optimization study were done using the Monte Carlo N-Particle code (MCNP) developed as Los Alamos National Laboratory (LANL). Figure 1 shows a zoomed-in view of the important parts of the MCNP model geometry. A CAD model of the neutron guide bundles was converted to MCNP geometry by the Monte Carlo Modeling Interface Program (MCAM) package and integrated into the model [5]. The target is made of lead rods with zircaloy cladding and is cooled by D_2O, the moderator tank is filled with D_2O, the cold source is liquid D_2, the nozzles are filled with a low-pressure helium mixture, and the shielding and neutron guide structure is steel. All structural material in the moderator tank is an aluminum alloy (AlMg-3), which is shown in yellow in Figure 1. The cross sections used in the simulations were from the ENDF/B-VII.1 library. Thermal scattering $S(\alpha,\beta)$ tables (continuous-energy when possible) were used for both moderators and structural materials within the moderator tank [6].

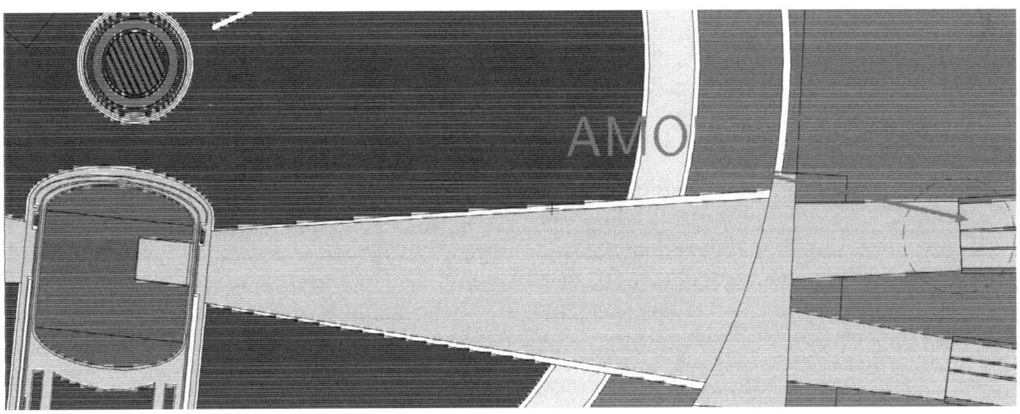

Figure 1. A horizontal slice through the MCNP simulation geometry showing the target (magenta), moderator tank (blue), cold source (purple), nozzle (pink), and neutron guide entrance (grey with circle).

4. Variance Reduction

Two variance reduction methods were used in conjunction to perform the simulations in a reasonable amount of time. First, mesh-based weight windows were generated based on the response of a point detector located inside the AMOR/SANS-I neutron guide. The point detector was positioned 75 cm from the guide entrance, giving it an approximately two degree view towards the cold source. The generated weight windows serve to increase the number of neutrons in regions that make large contributions to the point detector, which in this case is the cold source. In order to not bias the simulation, the weight windows cause neutrons entering the cold source to be split into many neutrons with a lower individual statistical weight.

The second variance reduction method used was a DXTRAN (deterministic transport) sphere. This tool is a non-physical sphere that is placed around the guide entrance that causes neutrons to be generated on its surface. Whenever a neutron undergoes a scattering reaction, the probability of scattering into the sphere's direction is calculated, a line is traced from the scattering location to the sphere's surface, and a new neutron is generated at the surface with a weight reduced by the scattering probability and material attenuation. The generated neutrons are then transported normally inside the sphere [7, 8]. This method serves to preferentially scatter neutrons into the direction of the neutron guide instead of relying on probability to transport them there.

5. Neutron Guide Response

In conjunction with the variance reduction methods described in the previous section, an approximation to the neutron guides was also used. McStas is another Monte Carlo code the has been developed specifically for neutron scattering experiments, and can simulate the neutron reflections that happen with neutron supermirrors. It is solely used to simulate neutron guides and is unable to simulate the detailed physics involved in neutron moderation, however. Since changing the source changes how the neutrons are moderated, MCNP, a code capable of simulating the relevant physics, was used. A patch was developed at Oak Ridge National Laboratory (ORNL) and PSI that implements the McStas guide reflectivity model in MCNPX 2.5.0 and 2.7.0 [1, 9], but the model is analog and is not fully compatible with DXTRAN spheres. If a neutron is sampled to *not* reflect on a guide surface, it is transmitted into the next cell. This means that neutron paths have to be sampled multiple times to account for any reflectivity losses, which slows the simulation down.

Some stability issues were also encountered while using the reflectivity patch in the same simulation as a DXTRAN sphere. In order to ensure stable runs, the simulation was split into two parts. First, a simulation was done where a surface source was written at the guide entrance surface. In this first simulation, the DXTRAN sphere surrounds the guide entrance. Any neutrons entering the guide were recorded and terminated. This source was then used in a second simulation with the guide reflectivity activated and without the DXTRAN sphere. This step transported the entering neutrons to the end of the guide where a tally was scored. Again, this process slowed the simulation down since two MCNP simulations needed to be made for each case. Thus, a guide response approximation was used to eliminate the need for a second run, which simply served to transport the neutrons to the end of the guide.

5.1. Guide Reflectivity

The guide reflectivity model implemented in MCNP is identical to that used in McStas [1]. The reflectivity curve for the current AMOR/SANS-I guide is shown in Figure 2 and the reflectivity equation and parameter values shown in Eq. 1 [10]. The values shown are the same for all the straight guides in the SINQ hall [11]. The dimensions of the guides are slightly different, however, as can be seen from their horizontal cross sections in Figure 1.

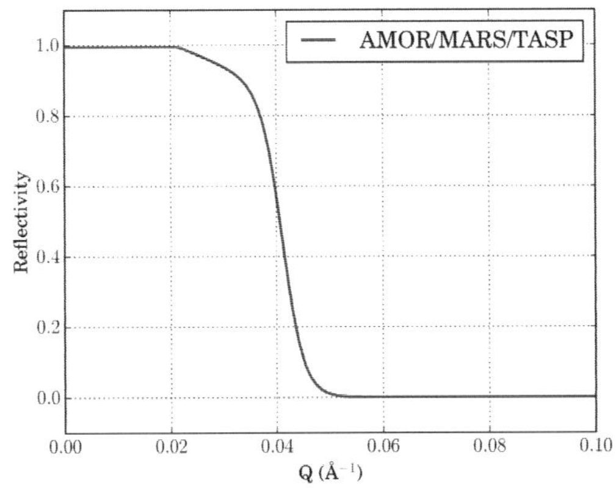

Figure 2. The reflectivity curve of the current SINQ neutron guides.

$$R(Q) = \begin{cases} \text{if } Q > Q_c: \\ \frac{R_0}{2}\left\{1 - \tanh\left(\frac{Q - mQ_c}{W}\right)\right\}\{1 - \alpha(Q - Q_c)\} \\ \text{if } Q \leq Q_c: \\ R_0 \end{cases}$$

$$R_0 = 0.995 \quad Q_c = 2.17 \times 10^{-2} \AA^{-1} \quad m = 1.9$$
$$\alpha = 6.40 \AA \quad W = 4.0 \times 10^{-3} \AA^{-1} \tag{1}$$

Since the neutron reflection probability is based on momentum transfer and the reflections are specular, the probability is determined by the incident neutron angle and energy. If the guide is rectangular, the reflection probability will not change as the neutron travels down the guide since the reflection angle remains constant. These conditions mean that the exiting neutron current should be completely defined by the incident polar angle and energy of a neutron entering the guide. The vertical and horizontal position in the guide entrance determines where neutrons exit the guide, but if the exit current is integrated over exit position, exit position becomes irrelevant and only the transmission probability matters.

5.2. Reconstruction

In order to approximate the spatially-integrated neutron spectrum at the guide exit, a number of fixed-source runs of MCNPX were done as precalculations to the parametric study. An evenly-distributed source was defined at the entrance surface of the guide that emits neutrons in a polar angle bin $1/8°$ wide and logarithmically in energy (to ensure low energies are sampled well). The exiting current at the end of the guide was tallied and divided by the entrance current to yield a transfer function in energy for the polar angle of the source. This was done for 0 to 7.5 degrees of polar angle to yield the set of 60 functions shown in figure 3.

Blue curves are those for incident angles that have an integer value of 0 degrees (e.g. 0.125, 0.250, etc.), green curves for 1, red curves for 2, and so on. The set clearly shows that incident angles above 2 degrees make no contributions to wavelengths shorter than 8 Å in the exit spectrum, and incident angles above 4 degrees make no contributions to wavelengths shorter than 16 Å.

If neutrons are tallied at the guide entrance in the same polar and energy binning as the transfer functions, it can be multiplied by the set and summed over all incident angles to yield the integrated neutron spectrum at the guide exit. This way, the reflectivity model can be eliminated from the transport runs and a good estimate of the exiting current can still be made.

5.3. Self-Consistency Test

In order to verify that the reconstruction method works as expected, a comparison was made to an exit spectrum calculated without using the approximation. To calculate the exit spectrum, a surface source was first written at the guide entrance using a DXTRAN sphere. This surface source was then used in a second run with the guide reflectivity model activated. During the

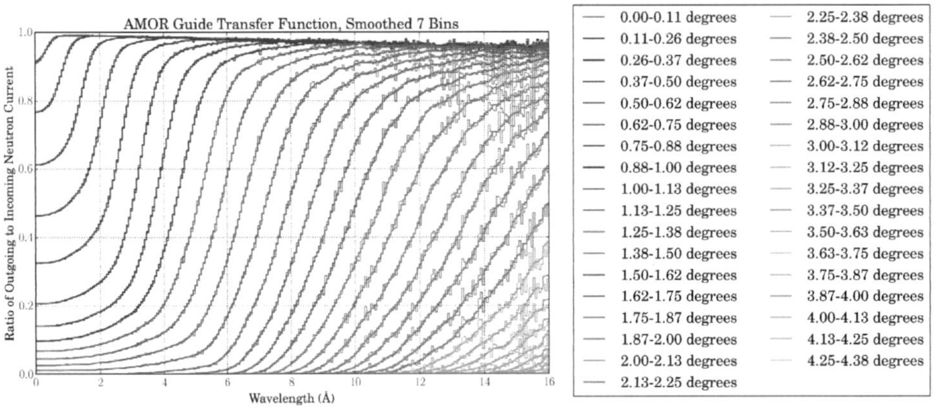

Figure 3. Set of transfer functions for a $m = 2$ guide.

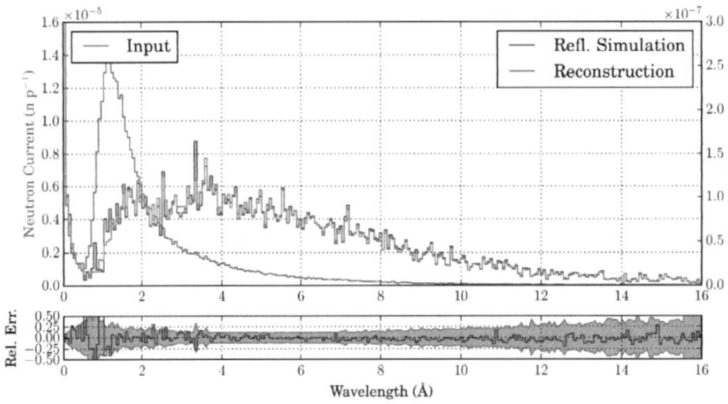

Figure 4. The reconstructed guide exit current compared to the exit current calculated with a MCNP simulation with guide reflectivity activated. A total set of 60 polar bins from 0 to 7.5 degrees was used. The right axis is for the exit spectra and the left axis is for the entrance spectrum. The lower plot shows the reconstruction relative error compared to the reflectivity simulation. The green envelope is the 1-sigma relative error of the reflectivity simulation.

second run, the entry and exit current was tallied. The entry current was binned in polar angle according to the divisions in the precalculated transfer functions and was used to reconstruct the exit spectrum so it could be compared to the exit tally. The results of the comparison are shown in Figure 4. The green line is the total, integrated neutron spectrum at the guide entrance, the blue line is the integrated exit current that was transported by the reflectivity model and tallied at the guide exit (the benchmark), and the red curve is the reconstructed exit spectrum. The entrance spectrum shown in Figure 4 is integrated over all incoming angles, and shows a thermal spectrum due to contributions from the thermal D_2O tank. If only the polar bins up to about 2 degrees (approximately the extent of the D_2 volume as view from the guide entrance) were integrated instead, the spectrum would have a cold peak as well. The maximum error seen is about 70% at 0.55 Å, but decreases with energy and is almost always within the statistical error of the simulation. The average 1-norm value (norm divided by number of bins) of the relative error is 7.8%, and the average 2-norm value is 0.8%.

6. REH Parametric Study in the Horizontal Plane

The REH parametric study was done to investigate a wide range of geometries differing in the horizontal plane, where the guide width is smaller than the hole size and all the guides have different view angles. Four REH depths (14, 10, 6, 2 cm), four opening widths (4, 8, 12, 16 cm), and eleven hole slopes (-0.5, -0.4, -0.3, -0.2, -0.1, 0.0, 0.1, 0.2, 0.3, 0.4, 0.5) were investigated in the study, totaling 176 cases. Figure 5 shows the parameters in relation to the cold source geometry. The REH vertical dimensions were chosen to be the full width of the nozzle with a slightly tapering inward slope. The vertical dimensions were kept constant in this study, but will be the focus of future work.

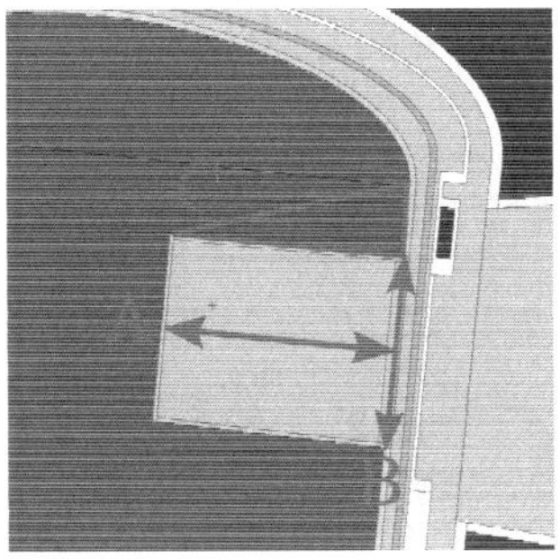

Figure 5. The geometric dimensions varied in the REH parametric study. "A" is the hole depth, "B" is the opening width, and "C" is the hole slope.

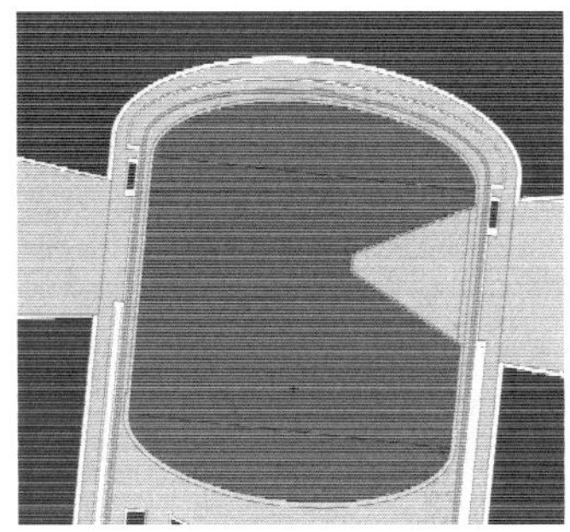

Figure 6. Horizontal cut of the cold source D_2 volume showing the best-performing geometry for the REH.

With all of the variance reduction and approximations implemented, each case took about one hour to run on a 144-core configuration on a computer cluster at PSI. The spectra from all the cases are shown in the upper plot of Figure 7, and their figures of merit are shown in the lower plot. The figure of merit was chosen to be the total current exiting the AMOR/SANS-I guide six meters from the cold source surface from 3 to 6 Å – the wavelength band which is useful for most of the instruments in the guide hall [12]. The parameters were varied in slope-width-depth order, with slope being the fastest varied parameter. Figure 7 has labeled, colored blocks in the lower plot to show which parameters are constant for that block. Slope and opening width have the most effect on the exit current at large hole depths.

The geometry of the best-performing case is shown in figure 6. The depth is 10 cm, the opening width is 12 cm, and the slope is 0.5. This opening is as wide as the nozzle and almost tapers to a point at the maximum depth. The maximum depth is 4 cm from the radial center of the cold source.

Figure 8 shows the gain from the best performing REH geometry over an absent REH. The maximum gain is 1.29 at 5.5 Å and its average value from 3-6 Å is 1.24. Similar gains should be seen by all instruments in the SINQ guide hall. The drop in gain after 5.5 Å is suspected to be from the liquid D_2 cross section decreasing between 5.5 to 25 Å, making it more transparent to colder neutrons and reducing the benefits of the REH.

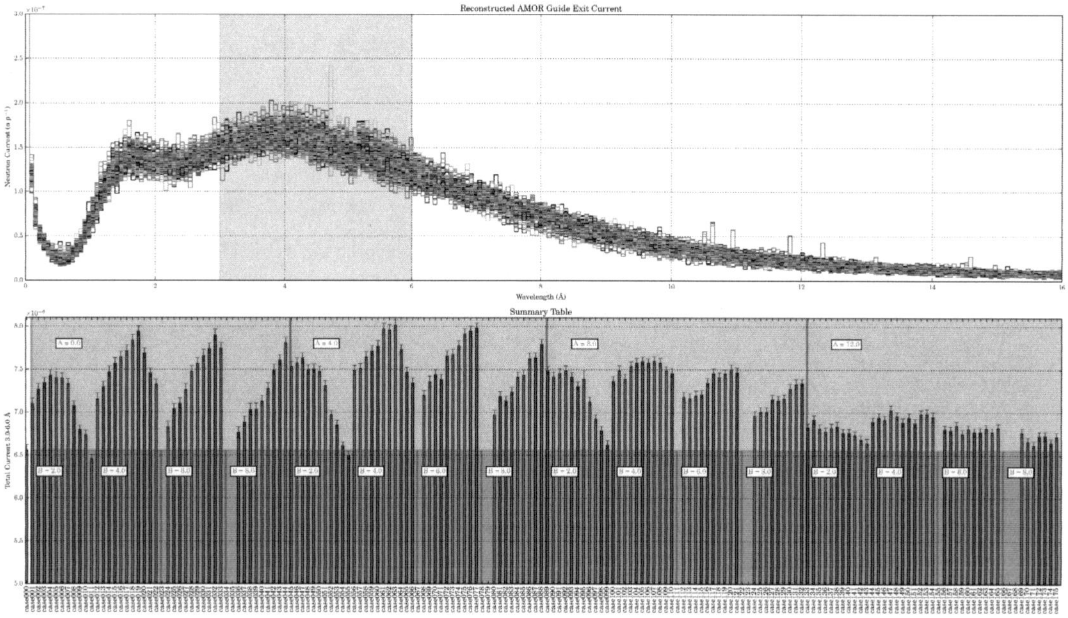

Figure 7. The envelope formed by the spectra from all 176 cases considered (upper) and their figures of merit (lower) which is total current from 3 to 6 Å. The red band denotes the wavelength range which has been summed over to produce the totals shown in the lower plot. Case000 is the case with the REH completely filled.

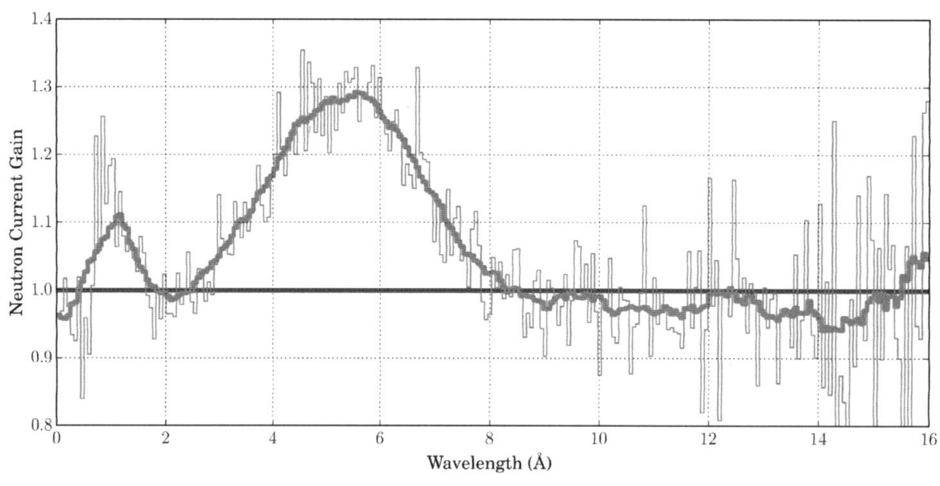

Figure 8. The gain factor of the neutron guide exit current of best-performing REH geometry over the REH being absent. The thick, red line is the simulation data smoothed over 15 bins.

7. Conclusions

From this study, it can be concluded that the flux reconstruction method is at least self-consistent, and can speed up calculations significantly. Furthermore, a near-optimal horizontal REH geometry has been determined. The REH increases total exit current for wavelengths of 3-6 Å by a factor of 1.24 on average. The peak gain is 1.29 at a wavelength of 5.5 Å.

8. Future Work

The optimal horizontal dimensions of the REH have been determined, another study is needed to investigate REH sensitivity to changes in the vertical dimensions. The effect the REH has on the neutron flux on the opposite side of the cold source will also be determined. Different types of reflector materials and configurations will also be investigated. The reflector sits behind the cold source (the side opposite the target), and reducing cold neutron losses in it may increase the guide exit current. Adding a belt of liquid H_2 to produce a "cold spot" for focusing instruments will also be investigated. In these studies, the flux reconstruction method will once again be used.

A guide reflectivity patch for MCNP 6.1 will also be done. Ideally, it will also include a switch to reduce neutron weight upon reflection instead of leaking non-reflected neutrons. This should help speed up in-beam calculations where this signal is of interest and not the noise.

Acknowledgments

This work was supported by Swiss National Science Foundation grant 200021_150048/1.

References

[1] Gallmeier F X, Wohlmuther M, Filges U, Kiselev D and Muhrer G 2009 *Nuclear Technology* **618** 768–772

[2] Rüegg C 2014 Upgrade plans for SINQ – moderators, guides, instruments, and new science *21st Meeting of the International Collaboration on Advanced Neutron Sources (ICANS-XXI)* (Mito, Japan)

[3] Atchison F, Spitzer H and Wagner W 2003 New cold D_2 moderator vessel with re-entrant beam window for SINQ *16th Meeting of the International Collaboration on Advanced Neutron Sources (ICANS-XVI)* (Forschungszentrum Jülich)

[4] Thomsen K 2014 *Physics Procedia* **60** 278 – 293 ISSN 1875-3892 3rd International Meeting of the Union for Compact Accelerator-driven Neutron Sources, UCANS-III, 31 July - 3 August, 2012, Bilbao, Spain & the 4th International Meeting of the Union for Compact Accelerator-driven Neutron Sources, UCANS-IV, 23-27 September 2013, Sapporo, Hokkaido, Japan URL `http://www.sciencedirect.com/science/article/pii/S1875389214005859`

[5] Y Wu and FDS Team 2009 *Fusion Engineering and Design* **84** 1987–1992

[6] Parsons D K and Conlin J L 2012 Release of continuous representation for s(α,β) ace data Tech. rep. Los Alamos National Laboratory Los Alamos, NM, USA

[7] X-5 Monte Carlo Team 2003 *MCNP - A General Monte Carlo N-Particle Transport Code, Version 5* Los Alamos National Laboratory, Los Alamos, NM, USA (Revised 2/1/2008)

[8] Pelowitz D B, Goorley J T, James M R, Booth T E, Brown F B, Bull J S, Cox L J, Durkee J W, Elson J S, Fensin M L, Forster R A, Hendricks J S, Hughes H G, Johns R C, Kiedrowski B C, Martz R L, Mashnik S G, McKinney G W, Prael R E, Sweezy J E, Waters L S, Wilcox T A and Zukaitis A 2013 *MCNP6 USERS MANUAL VERSION 1.0* Los Alamos National Laboratory, Los Alamos, NM, USA rev. 0 ed

[9] Klinkby E, Lauritzen B, Nonbøl E, Willendrup P K, Filges U, Wohlmuther M and Gallmeier F X 2013 *Nuclear Instruments and Methods in Physics Research Section A: Accelerators, Spectrometers, Detectors and Associated Equipment* **700** 106–110 ISSN 0168-9002 URL `http://www.sciencedirect.com/science/article/pii/S0168900212011874`

[10] Willendrup P, Farhi E, Knudsen E, Filges U, Lefmann K and Stein J 2014 *User and Programmers Guide to the Neutron Ray-Tracing Package McStas, version 2.1* Physics Physics Department, Technical University of Denmark 2800 Kongens Lyngby, Denmark

[11] Allenspach P 2000 *Neutron News* **11** 15–18 URL `http://dx.doi.org/10.1080/10448630008233742`

[12] Reiss T, Filges U, Gallmeier F, Talanov V and Wohlmuther M 2013 *Nuclear Instruments and Methods in Physics Research Section A: Accelerators, Spectrometers, Detectors and Associated Equipment* **728** 117 – 125 ISSN 0168-9002 URL `http://www.sciencedirect.com/science/article/pii/S0168900213009479`

Journal of Physics: Conference Series **746** (2016) 012036 doi:10.1088/1742-6596/746/1/012036

High brilliant thermal and cold moderator for the HBS neutron source project Jülich

T Cronert[1], J P Dabruck[2], P E Doege[2], Y Bessler[4],M Klaus[5], M Hofmann[4], P Zakalek[1], U Rücker[1], C Lange[3], M Butzek[4], W Hansen[3], R Nabbi[2] and T Brückel[1]

[1] JCNS - Jülich Centre for Neutron Science, Forschungszentrum Jülich GmbH, Germany
[2] NET - RWTH Aachen University, Germany
[3] Institute of Power Engineering, TU Dresden, Germany
[4] ZEA-1 - Engineering and Technology, Forschungszentrum Jülich, Germany
[5] Bitzer-Stiftungsprofessur für Kälte-, Kryo- und Kompressorentechnik, TU Dresden, Germany

E-mail: t.cronert@fz-juelich.de

Abstract. The proposed High Brilliance Neutron Source (HBS), recognized within the Helmholtz Association of German Research Centres, will optimize the entire chain from particle source through particle accelerator, target, moderator, reflector, shielding, beam extraction, beam transport all the way to the detector, utilizing the nuclear Be(p,n) or Be(d,n) reaction in the lower MeV energy range. A D_2O moderating reflector prototype (MRP) and a cold source were constructed and build according to MCNP parameter studies. The MRP was tested in a feasibility study at the TREFF instrument at MLZ (Garching). Cold beam extraction from the flux maximum within the moderator based on liquid para H_2 and other cold moderators will be tested by energy spectroscopy via TOF-method. Different ratios of liquid ortho/para H_2 will be fed to the cold moderator. The ratio will be controlled by feeding from reservoires of natural liquid H_2 and a storage loop with an ortho/para converter and determined via online heat capacity measurement.

1. Introduction

The continuing shutdown of research reactors in Europe and the simultaneous rise of the European Spallation Source (ESS) will increase the demand for medium flux neutron sources for simple capacity, medium flux experiments, education and instrument development. This demand for a wide fundament of sources in the neutron community can be met by compact accelerator driven neutron sources (CANS) as a network of multiple, local, medium flux sources, taking advantage of the recent developments in accelerator and neutron optics design to produce a brilliance to rival those of current medium flux reactors at a price tag affordable for a national project. The Jülich Centre for Neutron Science (JCNS) has proposed to build the High Brilliance Neutron Source (HBS) on the basis of a CANS design as a project recognized within the Helmholtz Association of German Research Centres as part of the German neutron strategy.

2. HBS - High Brilliance Neutron Source

The High Brilliance Neutron Source (HBS) plans to utilize the nuclear Be(p,n) or Be(d,n) reaction in the lower MeV energy range. Since much lower energies are used, compared to GeV spallation, much more sophisticated methods of neutron extraction can be employed, so that the brilliance at the sample should be able to reach levels comparable to those of medium flux research reactors. While in common neutron sources the neutron density maximum within the thermal moderator is highly delocalized (in the case of research reactors) or not accessible (spallation sources), in CANS one can place the neutron extraction mechanism, that is used to feed a neutron guide, right in the center of the neutron density maximum. In the case of the HBS this will be achieved by an extraction mechanism called (thermal and cold) Finger Moderator taking the principle of low dimensional moderator[6], to be applied at the ESS[3], to the maximum approaching a one dimensional "finger" shape.

2.1. Thermal moderating reflector

In the HBS concept there is no strict difference between a classical thermal moderator and reflector, since the materials in the simulation found to be most efficient (graphite, D_2O and beryllium) have high scattering and low absorption cross sections in the reference energy range, thus fulfilling both tasks at the same time. Classical hydrogen rich thermal moderators like light water or polyethylene were found to be contra productive due to their high absorption cross section, compared to the previously mentioned moderating reflector materials. In current spallation sources, light water premoderators of a few centimeters are used to moderate the incident fast neutrons within a small distance. These premoderators play a role in the upcoming ESS moderator too, where their heat removal capabilities are also very convenient from an engineering point of view. In the HBS, however, the lower neutron energies decrease the necessity of hydrogen based thermal moderators and make it possible to omit them completely, saving absorbing material in the center of neutron density.

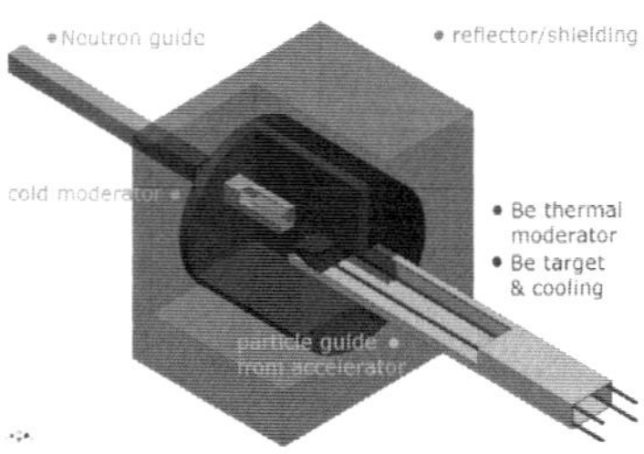

Figure 1. Sketch of target/moderator combination. 125l Be volume, water cooled multipurpose target from LNL[7] and graphite reflector/shielding.

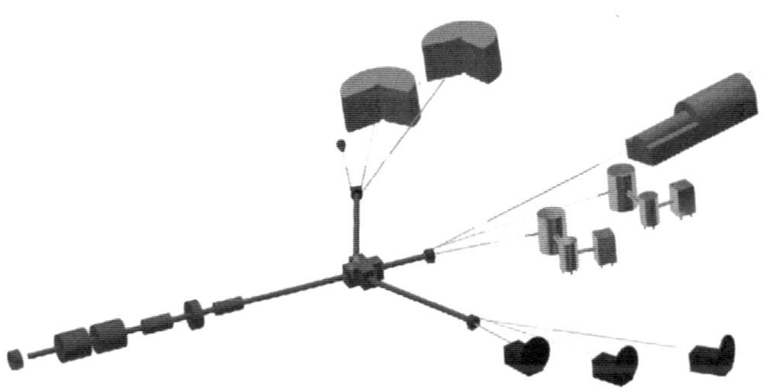

Figure 2. Sketch of a possible HBS facility. One accelerator feeds different target/moderator stations, emitting different puls lengths and neutron energy spectra.

2.2. Cold Finger Moderator

A cold neutron source (Cold Finger Moderator) was designed according to parameter studies of NET and constructed by ZEA 1 - Jülich to be inserted into the Thermal Finger of the moderating reflector. The dimensions are optimized according to the principle of low dimensional moderators proposed for the ESS[3] to be used with liquid para H_2, whose mean free path for energies above 15meV is of the order of 1cm and below 15meV of 10cm[2].

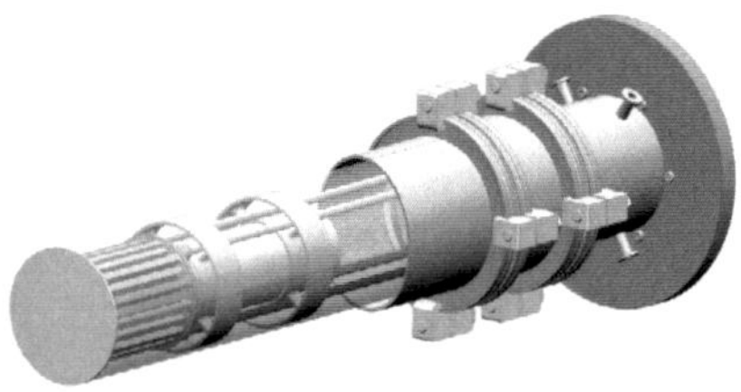

Figure 3. Cold Finger Moderator to be inserted in the moderating reflector prototype.

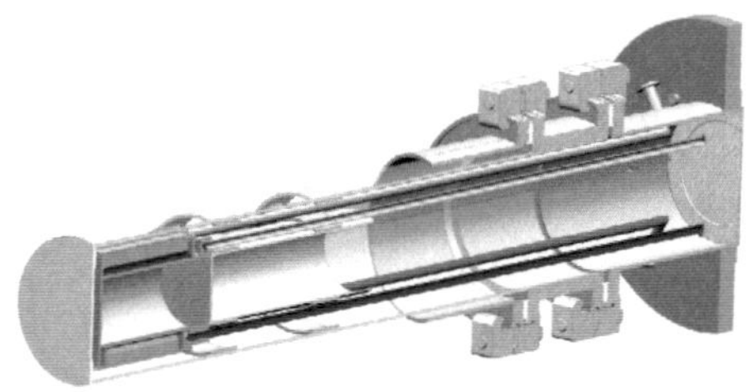

Figure 4. Vertical cut through the Cold Finger. Helium intake (blue) and outtake (red) are located at the bottom.

Fed from all sides with thermal neutrons, a preferred direction for cold neutrons to exit the finger along its alignment axis will be established. Each Cold Finger Moderator fulfills the role of a cold neutron source serving only one beamline and its instruments. Therefore the cold source can be specifically tailored to the needs of a certain instrument and provide neutrons with the needed phase space and energy distribution (e.g. flat vertical liquid H_2 vessel with small divergence in horizontal direction for reflectometry). In addition, neutron lenses can be used directly at the source to decrease divergence and/or enhance the flux.

2.3. Ortho/para configurations

The peak in energy distribution of neutrons emitted from a pure liquid para H_2 moderator is 5meV higher than those of a solid methane moderator, making solid methane more suitable for a majority of scattering applications if only moderating properties are taken into account. By increasing the amount of ortho H_2 in the moderator volume the energy peak can be shifted to lower temperatures at the cost of total flux. In addition, any moderating volume of H_2 will undergo conversion from ortho to para and para to ortho by induced heat from the scattered neutrons and natural conversion at low temperatures, so that in all practical cases one will rarely have a pure, homogeneous para state. However, mixtures of ortho/para H_2 are difficult to simulate using MCNP, so that we will aim to measure these properties experimentally. In addition to feeding the Cold Finger Moderator with pure para and natural H_2, we will also install a storage loop with a catalyst producing pure para H_2 to be mixed with natural H_2 gas, which will be condensed in the cold moderator vessel. The ortho/para ratio of the inserted gas will be controlled by heat capacity measurement of gas samples drawn during the measurements in realtime.

2.4. Time structure

Due to the rather large volume of the moderating reflector the time the neutron cloud stays within the moderator would seem rather large at first sight. However, since the Finger Moderator is extracting neutrons from within the moderator rather than from its surface, neutrons do not spend many collision processes in thermal equilibrium, but are extracted shortly before or after they reach energies equivalent to room temperature, thus shortening the overall pulse. First simulations show that 90% of the emitted thermal and cold neutrons leave the moderator after 0.02-1ms. Further simulations on that matter will be made, when we will be able to verify them at a suitable pulsed neutron source.

3. HBS Moderator test at the AKR-2

The MCNP calculations of NET of RWTH Aachen and MLZ will be validated at the AKR-2 reactor at TU Dresden which provides a source strength of 10^8 n/s with the fission spectrum delivered to our prototype moderator surface. The AKR-2 was chosen due to its extremely well characterized neutron spectrum calculated with MCNP by TU-Dresden (last update 2015) and verified experimentally in the epithermal and fast energy regime by measurements with scintillation technique (800keV - 12MeV), recoil proton proportional counters (20keV - 5MeV) and Bonner spheres (<20keV)[1]. The nominal thermal power of 2W and the relatively low flux guarantee that thermal energy transfer to the cold moderator, activation of the components used, radiation damage and tritium production in the D_2O are negligible.

3.1. Cold source

The Cold Finger Moderator prototype consists of a reservoir of 120ml (r=2.5cm, h=6cm) and a He heat exchanger inside an Al vacuum tube of 3.5cm outer radius. If the radius would have been optimized on brilliance alone, the radius should have been reduced even further at the cost of integral flux. But to create a flux suitable enough for the experimental validation (order of 10^3 cold neutrons per second at the detector position) at the AKR-2 and for manufacturing purposes the radius of 2.5cm was chosen. The final design is suitable not only for liquid H_2 but also for solid methane, mesithylene and comparable candidates for cold moderators. Those cold moderators would be ideal with other geometries (lower height of the cylinder), according to the simulations, but due to comparability of different moderator types by a given geometry and practical purposes this fixed final geometry was chosen and produced. The design profits from the conditions at the AKR-2, where the thermal energy intake by the neutrons scattered in the liquid H_2 is negligible and the whole setup is operated at 1.5bar H_2 and 1bar He pressure, so that the wall thickness can be considerably reduced compared to the supercritical ESS setup. However, the Cold Finger Moderator was tested up to 5bar and tests from the cold source of the Budapest Research Reactor[5], which is similar in size, show, that cooling in the region of 300W cooling power is feasible. Therefore, no engineering changes in the design would have to be done for the current prototype to be used in a medium flux source. The Cold Finger Moderator can be changed within 30min and the condensation time can be significantly reduced if needed.

Figure 5. Cold moderator vessel, leakage test during welding 10^{-9}mbar/s at 3bar He pressure and 500-77K temperature.

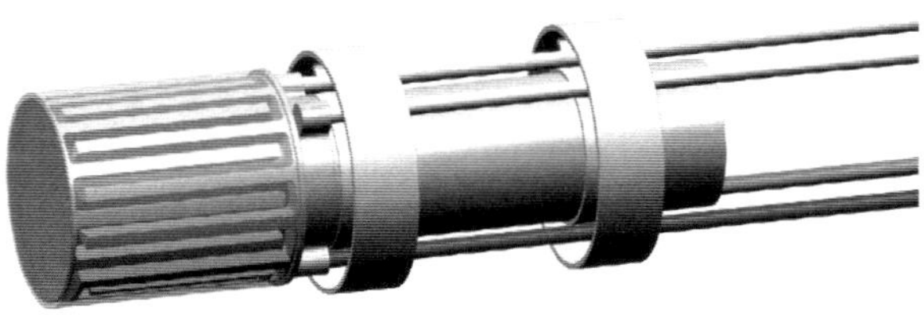

Figure 6. Cold Moderator Vessel with insulating spacer. Helium intake (blue) and outtake (red) at the bottom.

3.2. Experimental validation

A feasibility study was conducted at the TREFF instrument at MLZ (Garching). The cylindrical D_2O moderating reflector prototype (MRP) ($r = 31cm, h = 35cm, V = 100l$), designed for the test at AKR-2, was fed with cold neutrons ($\lambda = 4.75\mathring{A}, \lambda/2 = 2.38\mathring{A}$ and $\lambda/3 = 1.58$ at a ratio of approx $22 : 6 : 5$) which were moderated to thermal equilibrium at $E = 26meV \hat{=} 1.8\mathring{A}$. The Thermal Finger is inclined at 15° with respect to the cylinder axis. The detector was rotated around the intersection of MRP axis and Thermal Finger axis. Results in (Fig. 7) are fitted with a gaussian distribution with a FWHM of 5° and 7° centered close to the exiting angle. It is noted, that the surface of the moderator itself is not emitting any significant amount of neutrons and that all flux is solely generated by the Thermal Finger.

The energy spectra were recorded by TOF method using a three-window-chopper at the Thermal Finger's beam window with 6667rpm providing 3ms pulses at the detector recorded in 100 timechannels. The inset in Fig. 7 shows energy spectra at the three different angles indicated by dotted blue lines. It can be seen that the main range is well spread over the thermal region in addition to a peak disrupting the regular distribution close to the energy of the incoming, undermoderated $\lambda = 4.75\mathring{A}$ beam.

The feasibility study confirms the mechanism of the extraction process from the flux maximum of neutrons in the MRP and can be seen as a first indicator for the working of the Finger Moderator principle until more sophisticated experiments have been made.

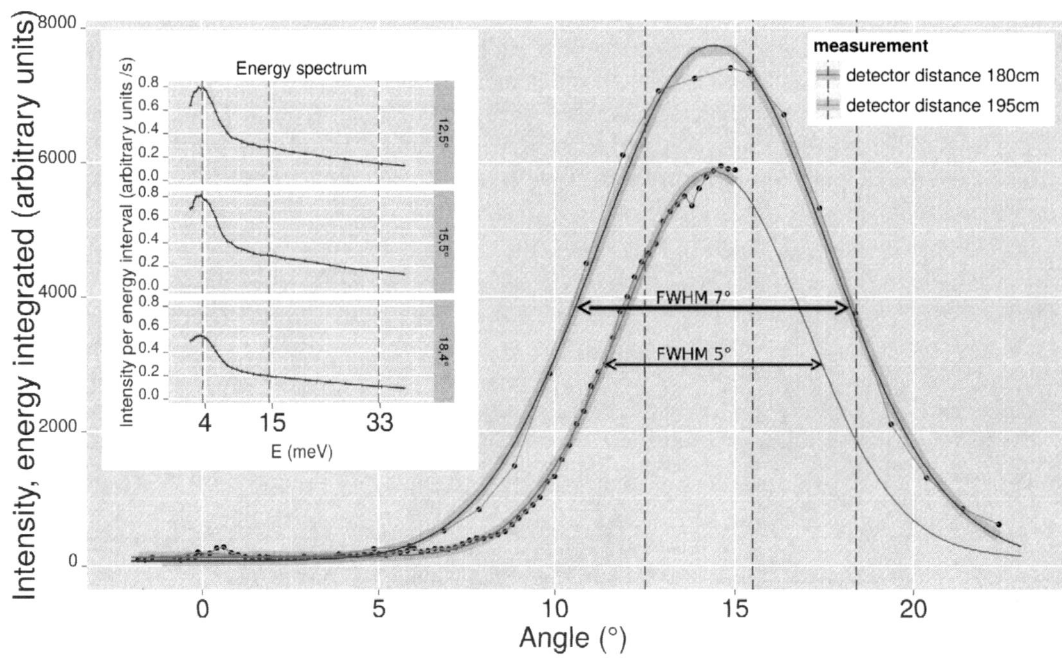

Figure 7. Angle dependent emission of Thermal Finger. The dashed blue lines indicate the position of the energy spectra (inset). The dashed red lines in the inset indicate the energy of incoming neutrons.

4. Conclusion

Making a CANS compatible in brilliance to a medium flux reactor or spallation source is all about gain factors, which are allowed in CANS and forbidden in others sources due to high energies and radiation damage. MCNP simulations have shown[4], that with the Thermal and Cold Finger extraction mechanism in a Be thermal moderator we can already gain orders of magnitude in bridging the gap between spallation and nuclear Be(d,n) reactions without taking the advantages of beam transport into account. The experiments at AKR-2, which will start in October 2015, will not only validate the simulations, but also enable cold moderator tests, which can not easily be simulated and provide a first version of an advanced moderator prototype.

References

[1] Tagaziria H. and Hansen W., Neutron spectrometetry in mixed fields proportional counter spectrometers, *Radiation Protection Dosimetry* Vol. **107**, Nos 13, pp. 7393 (2003) Nuclear Technology Publishing.

[2] Kai T.,Harada M., Teshigawara M., Watanabe N. and Ikeda Y. *Nuclear Instruments and Methods in Physics Research* A **523** 2004 398-414

[3] Batkov K., Takibayev A., Zanini L. and Mezei F. *Nuclear Instruments and Methods in Physics Research* A **729** 2013 500-5

[4] Dabruck J., Cronert T., Rücker U., Bessler Y., Klaus M., Lange C., Butzek M., Hansen W., Nabbi R. and Brückel T. to be puplished in *Il Nuovo Cimento - Colloquia and Communications in Physics*

[5] Grosz T., Mityukhlyaev V., Rosta L., Serebrov A., Zaharov A. and Hargitai T. Thermo-Hydraulic test of the moderator cell of liquid hydrogen cold neutron source for the Budapest reactor

[6] Mezei F., Zanini L., Takibayev A., Batkov K., Klinkby E., Pitcher E. and Schönfeldt T. Low dimensional neutron moderators for enhanced source brightness

[7] Esposito J., Colautti P., Fabritsiev S., Gervash A., Giniyatulin R., Lomasov V., Makhankov A., Mazul I., Pisent A., Pokrovsky A., Rumyantsev M., Tanchuk V. and Tecchio L. *Applied Radiation and Isotopes* **67** (2009) S270S273

The New Sorgentina Fusion Source-NSFS: 14 MeV neutrons for fusion and beyond

A. Pietropaolo[1,*], P. Console Camprini[1], P. Agostini[1], R. Amendola[2], M. Angelone[1], D. Bernardi[1], F. Bruni[3], M. Capogni[1], D. Colognesi[4], R. Faccini[5], A. Filabozzi[6], D. Flammini[1], F. Fiori[7], M. Frisoni[1], F. Grazzi[4], M. Pillon[1], A. Pizzuto[1], L. Quintieri[1], F. Sacchetti[8], P. Valente[9]

1 ENEA, Fusion and Nuclear Safety Technologies Department
2 ENEA, Department of Sustainability of production and territorial systems
3 Università degli Studi Roma Tre , Physics Department
4 Consiglio Nazionale delle Ricerche, Institute of Complex Systems
5 Università degli Studi di Roma Sapienza, Physics Department
6 Università degli Studi di Roma Tor Vergata, Physics Department
7 Università Politecnica delle Marche - Di.S.C.O, Via Brecce Bianche, 60131 Ancona, Italy
8 Università degli Studi di Perugia, Physics Department
9 Istituto Nazionale di Fisica Nucleare, Laboratori Nazionali di Frascati

* **Corresponding author**: antonino.pietropaolo@enea.it

Abstract. The importance of the design for the realization of an intense 14 MeV neutron facility devoted to test and validate materials suitable for harsh neutron environments, such as a fusion reactor, is well established. The "New Sorgentina" Fusion Source (**NSFS**) is a project that proposes an intense D–T 14 MeV neutron source achievable with T and D ion beams impinging on 2 m radius rotating targets. **NSFS** may produce about 10^{15} n/s at the target and has to be intended as an European facility that maybe realized in a few years, once provided a preliminary technological program devoted to the operation of the ion source in continuous mode, target heat loading/removal, target and tritium handling, inventor as well as site licensing. In this contribution, the main characteristics of **NSFS** project will be presented and its possible use as a multipurpose facility outlined.

1. Introduction

The investigation on structural materials to be used on next generation fusion machines, where intense fluxes of neutrons are expected is of paramount importance. Among all the issues to be addressed, the investigation of structural damages induced by intense fluxes of 14 MeV neutrons emerging from D-T reactions is crucial for the design of fusion reactors for the production of energy. The evaluation of the "atomic reactions-induced" damages in metallic materials upon neutron irradiation is given in terms of the so-called displacement per atom (**dpa**), defined as the number of times (N_d) that an atom is displaced for a given fluence: $N_d = N_0\sigma_d\varphi T_r$, N_0 being the number of atoms in the specimen, σ_d its **dpa** cross section, φ the neutron flunce rate and T_r the irradiation time. The interplay of "atomic damages" and "nuclear damages", e.g. transmutation, affects materials in different way: swelling, ductility loss, increased probability of crack propagation and increased creeping rate, just to mention a few examples. These effects and their correct investigation and evaluation have a deep impact in the choice of the materials and thus on the overall costs of a fusion reactor. The availability of an intense, well calibrated 14 MeV neutron source thus becomes a crucial need towards the construction of a fusion reactor like **DEMO**, and in the mid-term is important as well to test neutron diagnostics (detectors and related electronics) to assess their effective use in intense and harsh neutron environments. The capability of such a neutron source to provide beams for **dpa** and in situ investigation of structural damages after irradiation using neutron-based techniques, such as scattering and/or imaging may be also a useful feature for complete and thorough analyses on the irradiated materials. In this context, the **NSFS** project [1,2] is presented: the main characteristics of the facility are briefly described and the possible use of the neutron beams for experimental activities related to and also beyond fusion are outlined.

2. NSFS main features

NSFS design employs a well-proven neutron source type with existing technology in an innovative plant concept which makes this source brighter, following the relevant experience gained by **RTNSI** and **RTNSII** facilities successfully constructed and operated at **LLNL** [3]. **NSFS** design concept improves such a proven approach, utilizing ion source and accelerator stages from neutral beam injector devices utilized at large experimental tokamaks together with a properly scaled rotating target technology. A twofold water-cooled rotating target of about 2 m radius is operated at rotational speed of about 1000 rpm. Inlet fluid flow of about 110 liters/s is inserted at about 50°C bulk temperature to remove 8 MW thermal power per target (16 MW total power) deposited onto the whole target by a 40A current-200keV energy D and T ion beams imping onto the two rotating targets. Three irradiation volumes are foreseen to accommodate: (i) small specimens (0.5 liter) at high flux; (ii) small material/components samples (1.2 liter) at intermediate flux; (iii) larger volume (several liters) for small mock ups at lower flux. The schematic of **NSFS** and the flux available in the central region are illustrated in figure 1

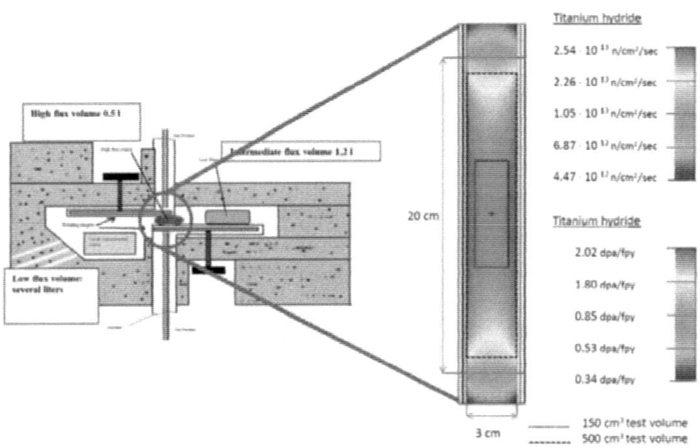

Figure 1: Layout of the NSFS double-wheel target (ion beams on both sides) and neutron flux available in the central region.

3. Possibilities for multipurpose applications

Although **NSFS** is monochromatic, the design of proper moderators could allow at obtaining fluxes of neutrons featuring a wider spectrum. The basic idea relies on the use of a pre-moderator with a high (*n,xn*) cross section at 14 MeV, in order to deplete the 14 MeV neutron population, producing lower energy neutrons. These can be then moderated using "conventional" moderators. Just to clarify this approach, figure 2 shows the results obtained with MCNP5-based Monte Carlo simulation [4], where two pre-moderator materials, namely Cu and Pb, are taken into account as test cases. A 4 cm thick water moderator has been placed behind the 5 cm thick pre-moderator. The 14 MeV neutron source used in the calculations is taken from an external routine describing the **FNG** D-T neutron source [5]. The results of the calculations clearly point out an increase of the intensity of the thermal peak using pre-moderators, with respect to the case of the water moderator only, by a factor 4 and 3.5, for the Cu and Pb case respectively. The effectiveness of the pre-moderator in reducing the intensity of the 14 MeV neutron peak is shown in the inset of figure 2. The 14 MeV peak is reduced by a factor 3 with respect to the sole water moderator. A depopulation of neutrons in the 14 MeV region in favor of lower energy neutrons, make them easier to moderate.

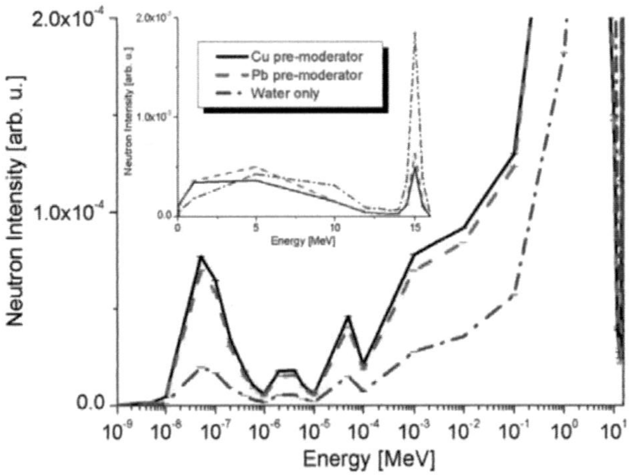

Figure 2: Neutron intensity beyond the moderator system in three different cases: i) only 4 cm water (dashed-dotted line); ii) 5 cm Pb pre-moderator+4 cm water (dashed line); iii) 5 cm Cu pre-moderator+4 cm water (continuous line).

Using a very raw conservative approach, by comparing the neutron production rate of **NSFS** with those of different accelerator-driven neutron sources, and proportionally scaling the fluxes available at the sample position, it is possible to perform an useful exercise providing a first guess of the implementable neutron techniques and their applications for **NSFS**.

3.1. Fusion technology

Fusion reactor materials testing with 14 MeV neutrons is essential before the design of any demonstrator (**DEMO** [6]) can be completed. EUROFER as a breeding blanket structural material, tungsten as the plasma facing component armour, copper alloys for the divertor coolant interface [7] are a few example of main materials. The aim is to deliver 30 **dpa** for structural steels, 10 **dpa** for copper and tungsten. Material investigations for fusion technology concern a broad range of facility components and several property changes under irradiation effects, *e.g.* the influence of nuclear transmutation on the electric characteristics of ceramic insulators, the irradiation effects on window material properties, the selection and test of low activation materials. An important activity will be the assessment of nuclear database and damage cross-section data, as well as the validation of numerical model simulations.

3.2. Electronics and its application

Integrated circuits featuring nanometric dimensions are extremely susceptible to random faults, known as single event effects (**SEEs**), occurring when a highly energetic particle, such as a neutron present in the environment, strikes the sensitive regions of an electronic device [8,9]. Radiation-hardened and radiation tolerant components are often used in military and space applications. Fast neutrons irradiation tests are also important to investigate the nuclear hardness for telecommunication, to assess to which extent the performance of a system, facility, or device is expected to degrade in a given nuclear environment. A potential concern is that mono-energetic neutron-induced single event effects will not accurately represent the real-world effects of broad-spectrum atmospheric neutrons. However, recent studies have indicated that, to the contrary, mono-energetic neutron, and particularly 14 MeV neutrons, can be used to quite accurately measure **SEE** cross sections in modern microelectronics as effectively as white beam, possibly with reduced uncertainties [10].

3.3. Material science

Neutrons provide information on structure and dynamics of materials [11]. Neutron diffraction (ND) and inelastic neutron scattering (INS) provide a thorough description of materials used in many applications ranging from engineering and chemistry, to cultural heritages and life science: the investigation of the structure and physical/chemical properties of materials for fusion reactors is of paramount importance. The measurement of **dpa** described above is just a first aspect of the problem. Irradiation-related defects (of atomic and/or nuclear nature) have to be understood and well characterized. ND at small and wide angles and/or imaging represent the due techniques for these investigations. The identification of stresses, strain, cracks in the bulk of the material as well as morphology is essential in determining the effectiveness of a material for the purpose it was meant for. ND is useful in determining microscopic structure, such as the hydrogen occupancy sites, that in turn may affect macroscopic properties of Pd-based membranes or new and innovative Pd-free compounds (such as electric resistivity [12]) in use for H storage/production. INS represents a very broad spectroscopic area [13], ranging from the study of the slow molecular diffusional regime ($\approx$1 ns) up to that of the fast proton dynamics ($\approx$0.1 ps). Neutron Vibrational Spectroscopy (**NVS**) [14], for example, is a fundamental technique used constantly in educational, research, and industrial laboratories worldwide. It has become an essential tool in medical applications, forensics, environmental compliance, and quality control to cite but a few common uses.
Small-Angle Neutron Scattering (**SANS**) probes structure in materials on the nanometer (10^{-9} m) to micrometer (10^{-6} m) scale. Structure on this length scale is critical to the performance of advanced engineering materials, biological processes in cells, storage of information on magnetic disks, hardness of steels and superalloys, conduction of current in superconductors, and many other materials properties. Possible applications of **SANS** range from biological molecules, polymers and surface properties of catalysts, to colloidal suspensions, metal physics, materials science, nano-crystalline materials, long range spin correlation and flux lines in superconductors [15].

3.4. Fundamental physics

Particle oscillations are familiar phenomena in both classical and quantum mechanics and have provided a wealth of information about the nature of matter and the fundamental forces. The $n-\bar{n}$ oscillation [16] is a unique kind of oscillation phenomenon compared to kaon oscillations as it breaks conservation of baryon number. Another important neutron measurement for fundamental physics is that of the neutron electric dipole moment (**EDM**) [17]. A non-zero electric dipole moment of the neutron (or any fundamental particle) would be a violation of parity (**P**) and time-reversal (**T**) symmetry, implying **CP** violation, already observed in accelerator experiments studying the decays of K and B mesons, but not at a level with can explain the matter-antimatter asymmetry of the universe.
Quantum mechanics fundaments are still subject of debate. Interferometry with massive elementary particles combines particle and wave features in a direct way. In this respect, neutrons are proper tools for testing quantum mechanics because they are massive, they couple to electromagnetic fields due to their magnetic moment, they are subject to all basic interactions, and they are sensitive to topological effects, as well. Recent neutron interferometry experiments based on post-selection methods renewed the discussion about quantum non-locality and the quantum measuring process [18,19].

3.5. Industrial applications

Industrial applications of nuclear analytical methods are useful in metallurgy and coal industries: 14 MeV Neutron Activation Analysis (**NAA**) [20] is capable of determining many light elements that cannot be analyzed using classical **NAA**. One of the most important applications of this technique is the determination of oxygen content in steel during the production phase. Using the reaction $^{16}O(n,p)^{16}N$, where ^{16}N decays with a half-life of 7.13 s, emitting 6.13 MeV, and 7.12 MeV gamma-rays this information can be achieved. Other important applications in metallurgy are the determination of Si, P and Mg in cast iron [21]. Applications can be also found in oil industry (borehole logging), quality assurance and quality controls for nuclear analytical methods, technology for illicit trafficking and anti-terrorism.

3.6. Metrology

Neutron fields can extend over a very wide energy range, this feature setting a challenge for the design of measurement devices and, as a consequence, a parallel challenge of developing measurement standards for their characterization. Neutron standards are required from the lowest energies, (ultra-cold neutrons) up to neutron GeV energies, typical of high-energy accelerators and atmosphere. In the frame of projects such as ITER, a special need is the adequate calibration of neutron diagnostic systems in order to perform accurate neutron emission measurements to determine the fusion power. An intense quite monochromatic neutron reference field, as that expected from **NSFS**, would be of great interest to study instruments in details and asses accurate measurements of neutron emission at 14 MeV. The availability of the intense 14 MeV neutron beam would be a profitable source for metrological concerns, useful to develop improved metrological procedures for fusion applications.

3.7. Neutron radiation biology and therapy and pharmaceutics

In the last decades, studies on the space environment to plan deep space exploration highlighted the the necessity of increase knowledge on the real biological risk of the exposition to low doses of cosmic radiation (mainly energetic protons from solar flares [22] and galactic cosmic rays). These interact with matter producing secondary radiations such as neutrons and gamma rays. Neutrons produced by albedo or interaction with shielding materials could reach 50% of the total radiation [23]. This is a relevant problem since the risks associates to neutron exposure have to be strictly considered for astronauts, beyond the **SEE** discussed above. Cosmic radiation exposure can produce severe damages to astronauts, such as cancer, immune system impairment and neurological damage. The main uncertainty of neutron exposure is the knowledge of the molecular process at low dose exposition [24]. Recently, a molecular approach was performed to investigate the effects of 14 MeV neutron irradiation on living mice [25] using the 14 MeV Frascati Neutron Generator (**FNG**) [26]. It has been shown the capability of neutron radiation, when traversing skin and sub-cutaneous adipose tissue, to induce the release of the Leptin, that in turn reaches the central nervous system likely modifying the lipid metabolism [27]. Radiation therapy is used for the treatment of hyper-proliferative diseases, such as malignant cancer, alone or in combination with surgery and chemotherapy. Neutrons, have potential advantages related to the increased energy deposition in tissues per unit track length. The energy released by neutron delivers a locally multiple damage sites clustered DNA damage far more difficult to repair [28].

In Nuclear Medicine ^{99m}Tc is one of the main radioisotopes currently and extensively used (70% of all radioisotopes diagnostic procedures), with an estimated 10 million medical tests per year. At present, ^{99}Mo, from which the ^{99m}Tc is obtained, is exclusively produced from fission of ^{235}U and in North America this isotope is solely available from Nordion, Canada. A temporary shutdown of two CANDU reactors in Canada in the year 2009 caused so called 'Tecnectium crisis'. **NSFS** represents an interesting tool to investigate the possibility of producing ^{99}Mo/^{99m}Tc through the $(n,2n)$ reaction, that in the range 11-18 MeV of neutron energy has a cross section of 1.5 barn. The cross sections of *(n,α)*, *(n,n'p)* and *(n,p)* reactions on ^{100}Mo are lower and produce very low quantity of impurities, so that large amount (> 100 g) of ^{100}Mo can be safely used and managed for irradiation. Very preliminary calculations showed that irradiating 1 kg of natural-Mo in the high-flux region (see Figure 1), taking into account a realistic irradiation time of about 12 h, an activity of 1 TBq could be profitably obtained.

3.8. Training and neutron science testing

The possibility to exploit the high level capability at the large scale facilities is strictly connected to the growth of a strong community, providing training and instrument testing. The availability of several smaller facilities has been the strength of the European neutron scattering community in the last forty years. Some of these smaller facilities are of high level in terms of source and instrumentation (e.g. **LLB**, **HMI**), but they are expected to progressively shut-down due to their aging. On the other hand, the bigger facilities are progressively becoming less and less adequate for the scientist training and instrument development because they must provide a service for a wide and not expert community of users. The role of the small facilities can become even more important in the near future due to the change of the role of the large facilities and their reduced number. Thus smaller facilities having the possibility of a simple access for young scientists willing to learn the neutron scattering techniques are extremely needed. Properly designed beam lines having the appropriate characteristics to perform instrument component tests will be valuable for training and progress of instrument design. A test beam line can be likely deveoted to detector development and testing, including front-end and acquisition electronics in real conditions. This is a particularly important area both because the reduced production of ^{3}He and the need of new more flexible, faster and cheaper detectors. The crystal monochromator development can be also important. Although the existing monochromator design does not need for specific advances the construction of new devices needs for alignment and testing instruments and some improvement can be also possible for specific applications connected to the future development of new instruments specifically designed for the long pulse sources like **ESS** and the future second target station at **SNS**.

4. An overview on possible instrumentation for NSFS

In this section it is briefly presented a reduced selection (with respect to the overall set of applications outlined above) of possible instruments that might be implemented on **NSFS**, starting from a conservative approach for neutron transport capabilities from target to sample position and taking also into account the techniques to date effectively in use in several small-compact neutron sources worldwide with neutron production rate orders of magnitude lower than that available on **NSFS**.

4.1. 14 MeV neutron irradiation station

NSFS offers different possibilities for 14 MeV neutron irradiation experiments, exploiting the irradiation locations sketched in Figure1, already briefly discussed before. Moreover, a dedicated beam line offering the opportunity of irradiating larger systems can be also foreseen by opening line of sight of the target. At a reasonable distance of about 3 m from the target the neutron fluence rate at 14 MeV might be in the order of 10^{9} n cm^{-2} s^{-1}, with the possibility of irradiating large systems of assemblies of devices enhancing statistical accuracy over a single irradiation run.

4.2. ND, SANS, INS and neutron imaging

A basic instrument implementable on **NSFS** could be a medium resolution diffractometer for the study of disordered systems or powders of high-symmetry crystals. The layout is very simple and an example of such instrument is the D20 beam line at **ILL** or 7C2 at **LLB**. focusing on **NVS** applications, basically the same as those relevant in the spectrometers of the latest generation, such as **TOSCA-II (ISIS)**, **VISION (SNS)**, and **LAGRANGE (ILL)**. **NSFS** characteristics place this source in the so-called "low flux class" [29], together with the reactors of Delft (The Netherlands), Řež

Journal of Physics: Conference Series **746** (2016) 012037 doi:10.1088/1742-6596/746/1/012037

(Czech Republic) and Kjeller (Norway), that is about ten or fifteen times below the **NCNR** reactor of **NIST** (USA) and the **PSI** source (Switzerland). Thus a simple conception, such as the so-called "beryllium filter" (BF) [30] might be implemented on **NSFS**. Among the different BF spectrometers currently in operation, one of the most representative is certainly **FANS** [31], installed at the **NCNR** reactor of **NIST** (USA). Although some details are not available, nevertheless the following three operations could approximately recover the factor 10÷15 of lower neutron intensity compared to **NCNR**: 1) doubling of the secondary sector in the spectrometer so to cover about 230° (incidentally, this is also expected from the future phase II of **FANS**) [gain factor=2]; 2) improving the monochromator [for example by using a doubly focalizing Cu(311)]. This too is required by the future phase II of **FANS** [gain factor=4]; 3) decreasing the monochromator-moderator distance (now about 5.3 m), since **NSFS** will certainly need a biological shielding much lighter than that suitable for a nuclear reactor like **NCNR** [gain factor=1.5÷2]. Considering the neutron production rate of **NSFS** and a conservative approach to the neutron delivery at the sample position as compared to existing sources, a possible configuration for a Small Angle instrument may be that of the SANS-I instrument at the **SINQ** spallation source at **PSI** - Villigen (CH). **SANS-I** at **PSI-SINQ** covers the Q-range from 6 x 10^{-3} to 5.4 nm^{-1} and for the detector displaced laterally by 50 cm up to 10.5 nm^{-1}. This allows the investigation of structures ranging from about 1 to 400 nm. The incoming neutrons are monochromatized by a mechanical velocity selector and collimated on a variable length from 1 to 18 m. The two-dimensional ^{3}He-detector with 128 x 128 elements of 7.5 x 7.5 mm^2 can be positioned at any distance between 1.4 and 20 m from the sample. Additionally it can be displaced laterally by 50 cm inside the large vacuum vessel to increase the accessible Q-range at any detector position. This option is combined with a rotation around its vertical axes to reduce parallaxes effects. The sample support consist of a remotely controlled xyz and rotation table for mounting devices for working in air (e.g. sample changer, high pressure cell, ...) and a vacuum chamber (able to carry an electromagnet).

An imaging systems for **NSFS** is composed of a few items: a pin hole selector to define flus and resolution, a Li/Gd loaded scintillator and CCD-based digital acquisition. Considering the conservative approach used in this discussion for the neutron transport from target to sample e having in mind the typical measuring times for an imaging beam line an exercise was done to roughly determine the possible performances of **NSFS** for this neutron technique. Assuming a thermal neutron flux in the order of 10^6 n $cm^{-2}s^{-1}$ at the sample position, different experimental configurations can be foreseen. In Table 1 only a couple of examples are reported.

Table 1: possible experimental configuration for an imaging beam line at NSFS

Pin hole diameter [mm]	*Maximum Field of View* [mm]	*L/D*	*Resolution* [mm]
15	100	1133	0.122
30	200	567	0.173

In these configurations the typical measuring times are in the order of a few hundreds of seconds, allowing for a medium-to-high level imaging in reasonable times, also allowing tomographic capabilities.

5. Conclusions and perspectives

New Sorgentina Fusion Source (**NSFS**) is a high intensity 14 MeV neutrons source. Its layout and neutron performance were briefly described and discussed. Main applications of **NSFS** using 14 MeV neutrons have been outlined and possible strategies to obtain moderated (thermal) neutron beams were briefly indicated. In a conservative approach, a prevision was attempted to foresee what kind of techniques could be implemented on hypothetical beam lines operating at the **NSFS**. These range, beyond fusion, from irradiation with 14 MeV neutrons for electronics and biology, to neutron diffraction, imaging and neutron vibrational spectroscopy.

References

[1] M. Pillon, M. Angelone, A. Pizzuto, A. Pietropaolo, Fus. Des. Eng. **89**, 2141 (2014).
[2] M. Martone, C. Alessandrini, S. Ciattaglia, R. Coletti, C. Crescenzi, C. Ferro,et al., Feasibility Study of a 14-MeV Neutron Source (Sorgentina), **ENEA Internal Report, 1990**.
[3] H. Sumita, Nucl. Instr. Meth., A **282**, 345 (1989).
[4] X5 MONTE CARLO Team, "MCNP—a general Monte Carlo N-Particle transport code: version5 user's guide", **LANL report LA-CP-03-0245**, (October 2005).
[5] A. Milocco " The D–T Source Routine developed at ENEA and available in SINBAD database" **IJS Internal Report IJS-DP-9988**, (2008).
[6] K. Tobita et al., Fus. Eng. Des. **89** (9-10), 1870 (2014).
[7] D. Stork et al., Fus. Eng. Des. **89** (7-8), 1586 (2014).
[8] E. Normand, IEEE Trans. Nucl. Sci. **43**, 2742 (1996).
[9] C. Andreani, A. Pietropaolo, A. Salsano, G. Gorini, M. Tardocchi, A. Paccagnella, S. Gerardin, C.D. Frost, S. Ansell, and S. P. Platt, Appl. Phys. Lett. **92**, 114101 (2008)
[10] E. Normand and L. Dominik. "Cross Comparison Guide for Results of Neutron SEE Testing of Microelectronics Applicable to Avionics," Radiation Effects Data Workshop (REDW), 2010
[11] C. G. Shull, Rev. Mod. Phys., **67**, 753 (1995).
[12] A. Pozio et al., International journal of hydrogen energy **37,** 7925 (2012).
[13] G. L. Squires, Introduction to the Theory of Thermal Neutron Scattering (Dover, Mineola, 1997)
[14] P. C. H. Mitchell, S. F. Parker, A. J. Ramirez-Cuesta, J. Tomkinson, Vibrational Spectroscopy with Neutrons, World Scientific, Singapore (2005).
[15] https://neutrons.ornl.gov/gpsans/publications
[16] R. N. Mohapatra, Nucl. Instr. Meth. A **284**, 1 (1989).
[17] P. G. Harris, C. A. Baker, K. Green, P. Iaydjiev, S. Ivanov, D. J. R. May, J. M. Pendlebury, D. Shiers, K. F. Smith, M. van der Grinten, and P. Geltenbort Phys. Rev. Lett. **82**, 904 (1999).
[18] Helmut Rauch, Samuel A. Werner, Neutron Interferometry Oxford University Press
[19] M. Iannuzzi, A. Orecchini, F. Sacchetti, P. Facchi, and S. Pascazio, Phys. Rev. Lett. **96**, 080402 (2006); M. Iannuzzi, R. Messi, D. Moricciani, A. Orecchini, and F. Sacchetti, Phys. Rev. A **91**, 020102 (R) (2015).
[20] "Elemental Analysis of Biological Materials": Current Problems and Techniques with Special Reference to Trace Elements, TECHNICAL REPORTS SERIES No . **197**, IAEA (1980).
[21] A. A. Istratov, T. Buonassisi, R. J. McDonald, A. R. Smith, R. Schindler, J. A. Rand, J. P. Kalejs and E. R. Weber, J. Appl. Phys. **94**, 6552 (2003).
[22] LW Townsend, JL Shinn, JW Wilson. Radiat Res.; **126**(**1**), 108(1991)
[23] L Heilbronn, T Nakamura, Y Iwata, Kurosawa, H Iwase and LW Townsend. Rad. Prot. Dos. **4**, 140 (2005).
[24] F Cucinotta, W Schimmerling, JW Wilson, LE Peterson and JF.Dicello NASA / TP – 2002 – 210777.
[25] E Fratini, V Licursi, M Artibani, K Kobos, P Colautti, R Negri and R Amendola, PloS ONE **6**(**4**), e19242 (2011).
[26] http://www.fusione.enea.it/LABORATORIES/Tec/FNG.html.en
[27] M Cestelli-Guidi, C Mirri, E Fratini, V Licursi, R Negri, A Marcelli, and R Amendola. Anal Bioanal Chem. **404**(**5**), 1317 (2012).
[28] L Xue, D Yu, Y Furusawa, R Okayasu, J Tong, J Cao, S Fan., Mutat Res. **670**(**1-2**),15 (2009).
[29] see e.g. https://www.sni-portal.de/kfn/Infos/Neutronenquellen/index-engl.php
[30] A. D. B. Woods, B. N. Brockhouse, M. Sakamoto, and R. N. Sinclair in Inelastic Scattering of Neutrons in Solids and Liquids (IAEA, Vienna, 1961) pages 487-498.
[31] T. J. Udovic et al., Nucl. Instr. Meth. A **588**, 406 (2008) and references therein.

Development of the neutron diffraction method for stress measurements in thick steel samples.

V T Em[1], W Woo[2], B-S Seong[2], P Mikula[3]

[1]*National Research Center «Kurchatov Institute», Moscow, Russian Federation*
[2]*Korea Atomic Energy Research Institute, Daejeon, Republic of Korea*
[3]*Nuclear Physics Institute ASCR, Rez, Czech Republic*

E-mail: vtem9@mail.ru

Abstract. The results of development of neutron method for stress measurements in bulk steel components are considered. Enhancement of the maximum available path length (~85mm) was achieved by using optimized bent perfect silicon crystal monochromators, position sensitive detectors and neutron wavelengths, corresponding to the minimums of neutron total cross section near Bragg edges.

1. Introduction

Thanks to high penetration ability of neutrons, the neutron diffraction method is widely used for nondestructive strain/stress measurements inside bulk polycrystalline materials [1].The method is based on the accurate measurement of the diffraction peak position from a diffracting volume (gauge volume), defined by cadmium slits on the incident and diffracted beams. In a general case, three strain components along the "principal directions" should be measured to calculate stresses. Some components should be measured in "reflection geometry" and some components in "transmission geometry" (Fig 1).

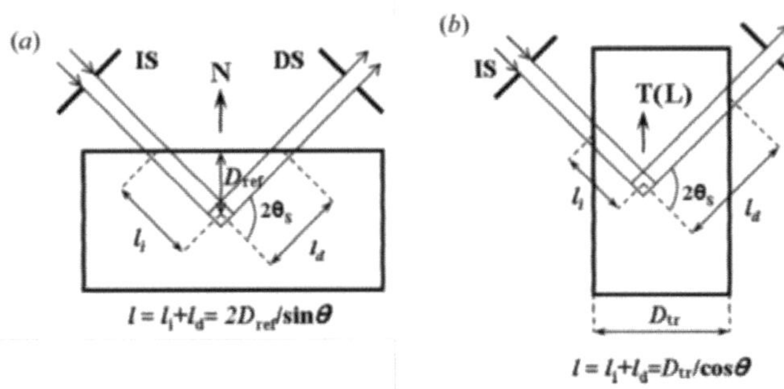

Figure 1. (a) reflection geometry for the measurements of the normal (N) component; (b) transmission geometry for the measurements of the transverse (T) and longitudinal (L) components. The gauge volume is defined by slits in the incident (IS) and diffracted (DS) beams. The neutron path length (l) is the sum of the incident (l_i) and diffracted (l_d) beam path lengths.

Penetration ability of high energy synchrotron radiation is comparable with that for thermal neutrons, but there is difficulty in measuring strain components in "reflection geometry" because of a small (few degrees) scattering angle, θ_S, [2, 3]. Penetration ability of laboratory X-rays is much lower [2]. So, by now, only neutron method allows measuring all three strain components in bulk samples. However, because of low brightness of neutron sources, the maximum feasible path length in ferritic steel, l_m , in modern neutron stress-diffractometers is limited to 60mm [3,4], that corresponds to 40mm thick plate. At the same time, development of heavy industries (e.g. shipbuilding, nuclear industry, chemical industry etc.) demands stress measurements in much thicker steel components.

This paper summarizes our results in development of neutron method for stress measurements in bulk steel components.

2. Experiment

The experiments were carried out at beam port ST-1 of HANARO reactor in KAERI, using the early and upgraded [5-8] versions of the residual stress instrument. The distance between the reactor core to the monochromator was 6.2 m; the monochromator–sample distance, L_{MS}, was ~2.1m; the sample to detector distance, L_{SD}, was ~1.2m. No Soller collimators were used. The open window of $40(w)$x$70(h)$mm^2 at 150mm from the drum center defined the beam impinging on the monochromator.

For testing the resolution and detector signal, the diffraction from the ferritic steel pin of diameter 2mm and height 40mm was measured. The pin was mounted in a sample position at the center of the difractometer and imitated the gauge volume. For comparison of different experimental set-ups for stress measurements, we used the figure of merit, FoM, suggested in reference [9], which for the same acquisition time can be defined as

$$FoM = \frac{I}{(FWHM)^2(\cot\theta_S)^2}, \qquad (1)$$

where FWHM is the full width at half maximum of the diffraction peak in $2\theta_S$. The depth scan in reflection geometry of diffraction peaks 211, 110 of ferritic steel, and 311, 111of austenitic steel, with gauge volume 80mm^3 and measurement time 1h., was conducted to determine the maximal available path length, l_m, defined as corresponding to the 100με (10^{-4}) strain error.

3. Results and discussion

3. 1. Monochromator for a stress-diffractometer

In a stress-diffractometer at stationary reactor, it is necessary to scan only one diffraction peak at $2\theta_S \sim 90^0$ [1]. A bent perfect silicon crystal (BPC) monochromator can be advantageous because its curvature can be adjusted to obtain maximal intensity and resolution of this particular peak [10-12]. Therefore, most of modern stress-diffractometers are equipped with double focusing bent perfect silicon crystal monochromators [13-16].

In experiments for monochromator optimization, we used a four-point type bending device, developed in NPI, Rez. [17], for cylindrical bending in horizontal plane of a perfect Si crystal slabs or "sandwiches" with reflecting area $40(h)$x200mm^2 (w) and thickness up to 6 mm. We did not use vertical focusing because large divergence of the incident beam deteriorates definition of the vertical size of a gauge volume [1].

It was shown [18-22] that BPC monochromator at a take-off angle much smaller than 90^0 can provide a much higher detector signal with excellent resolution in comparison to the mosaic crystal monochromators. Lattice planes (211), (110) of ferritic steel (b.c.c. structure) and planes (311), (111) of austenitic steel (f.c.c. structure) are recommended for strain measurements [1], because they are weakly affected by inter-granular strains. The wavelengths λ~1.6Å for the reflections 211(b.c.c.) and 311(f.c.c.) and λ~2.9Å for the reflections 110(b.c.c.) and 111(f.c.c.) are necessary for the diffraction peak at $2\theta_S \sim 90^0$.

Our experiments with BPC monochromators [8,23-24] showed that for extracting monochromatic neutrons with wavelength λ~1.6Å the best choice (maximal *FoM*) is using BPC Si(220) monochromator with reflecting plane (220) in symmetric diffraction geometry at take-off angle $2\theta_M \approx 50^0$ (Table 1). Using Si(111) monochromator at smaller take-off angle 30^0 significantly increases intensity, however, the broadening of peak results in lower *FoM* [7]. Using Si(113) monochromator at larger take-off angle 60^0 increases resolution (decreases *FWHM*), however, loss in intensity results in lower *FoM* [24]. For comparison with Ge(311) vertical focusing mosaic crystal monochromator ($2\theta_M = 90^0$, λ = 1.83Å), the reflection 220 of austenitic steel pin (Ø2mm, h=40mm) was measured ($2\theta_S = 90.4^0$) using Si(220) monochromator at take-off angle $2\theta_M = 57^0$ for the same wavelength.

Table 1. The figure of merit (*FoM*) for different monochromators, determined from measurements of diffraction peak 211 of ferritic and 220 of austenitic steel pins (Ø2mm, h= 40mm, t =180s.) [8, 23-24].

Monochromator λ = 1.64A	Monochromator size WxHxT (mm^3)	$2\theta_M^0$	*FoM*
Si(111)-BPC	200x40x3.9	30	95
Si(220)-BPC	200x40x3.9	50.6	100
Si(113)-BPC	200x40x4	60	60
Ge(311) mosaic, η = 15', vertical focusing λ =1.83Å	60x220x8	90	38

Neutrons with $\lambda \approx 2.9$Å can be obtained using the Si(220) planes at $2\theta_M \approx 98°$ or the Si(111) planes at $2\theta_M \approx 55°$. The latter is preferable because at $2\theta_M \approx 55°$ the instrument luminosity should be much higher than at $2\theta_M \approx 98°$ [8], while a good resolution can be still achieved.

Thus, the Si(220) monochromator is preferable to extract "short" wavelengths near 1.6Å for the 211(b.c.c.) and 311(f.c.c.) reflections and the Si(111) monochromator to extract "long" wavelengths near 2.9 Å for 110(b.c.c.) and 111(f.c.c.) reflections.

It should be noted that in choosing the optimal resolution, both the *FoM* and the definition of gauge volume should be taken into consideration. Because of the divergence of the diffraction peak, the width of sampling gauge volume, w, is greater than the width of the cadmium slit on the diffracted beam, s. The difference depends on the slit to gauge volume distance, l_{SG}, and FWHM of a diffraction peak [25]. Estimation, made based on the results presented in [25], showed that FWHM= 0.4^0 is necessary for the 20% (w/s=1.2) difference in widths for 2mm slit at l_{SG} = 100mm. Therefore, the diffraction peak width of FWHM=0.4^0 can be considered as reasonable for a stress diffractometer.

3.2. Position sensitive detector for a stress-difractometer

At the scattering angle $2\theta_S = 90^0$, the diffraction cone becomes a plane. Therefore, a position-sensitive detector (PSD), with a possibly high active area, increases the intensity without worsening the spatial and angular resolution [3]. However, a higher active area increases the uncertainty of the direction of the measured strain since the PSD subtends a larger vertical angle. We used a position sensitive detector, specially designed for RS measurements [26-27] with 200(h)x100(w)mm^2 active area, 60% efficiency for λ =1.8 Å, a delay line read-out mode, and spatial resolution of 2.5 mm. However, using 300mm high PSD with 2mm spatial resolution at distance $L_{SD} \sim$ 1m [15-16] is preferable for stress measurements in thick samples, because it can additionally increase intensity, while uncertainty in strain direction $\pm 8^0$ is acceptable.

3.3. Wavelength selection

If a diffraction peak has a Gaussian shape and the background is negligible, the error in the strain determination, *Err* (ε), is [28-29]:

$$Err\ (\varepsilon) = u\cot\theta_S\ /\ I^{1/2}\ , \qquad (2)$$

where u is the standard deviation of the peak profile in θ_S and I is the integrated peak intensity. The integrated peak intensity (I_l) and the peak height (H_l) exponentially decrease with neutron path length, l, in material:

$$I_l = I_0 \exp\ (-\sigma_t n_o l), \qquad (3)$$

$$H_l = H_0 \exp(-\sigma_t\, n_o l), \qquad (4)$$

where I_0 and H_0 are the integrated peak intensity and the peak height for the zero path length, σ_t is the total neutron-cross section per atom for single element material, n_o is the number of atoms per unit volume and l is the sum of the paths for the incident (l_i) and diffracted (l_d) beams ($l = l_i + l_d$, Fig. 1). At a certain depth the peak height becomes comparable with the background (B_l) and the error in strains ($Err(\varepsilon)$) increases [30]:

$$Err(\varepsilon)_l = \frac{\cot\theta_S u_{\theta_S}}{(I_0 e^{-\sigma_t n_0 \cdot l})^{1/2}}\left(1 + 2\sqrt{2}\frac{B_l}{H_0 e^{-\sigma_t n_0 \cdot l}}\right)^{1/2} \qquad (5)$$

The maximum available path length (l_m) can be increased by increasing gauge volume or measuring time. However, from (5) one can see that this is not effective because integral intensity exponentially decreases with neutron path length through the material [27]. A more effective way is to use neutrons with smaller total cross-section, σ_t, because it is in exponent.

In ferritic and austenitic steels, the main contribution to the neutron total cross section gives coherent Bragg scattering [31]. Therefore, the dependence of the total cross section on neutron wavelength has saw-tooth structure with abrupt changes at Bragg edges [32-34] (Fig.2).

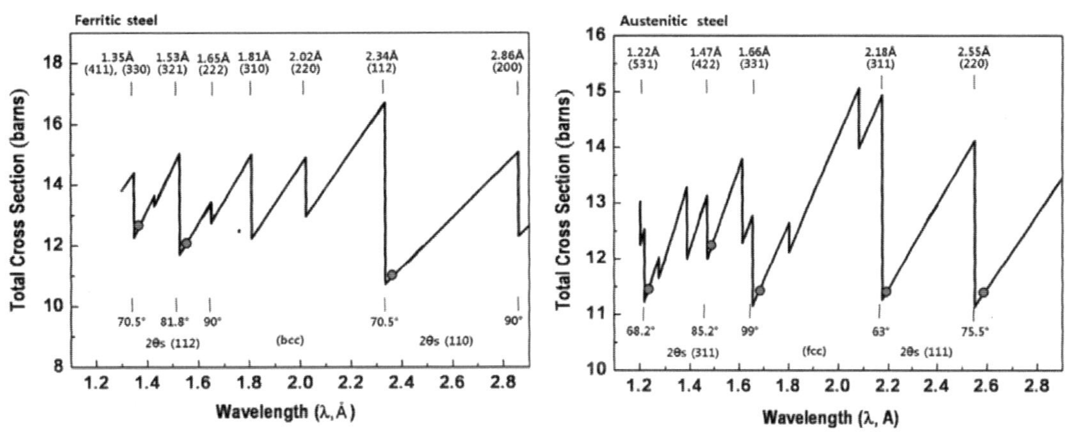

Figure 2. Calculated total neutron cross section of the ferritic steel (low-carbon steel, b.c.c.) and the austenitic steel (stainless steel 304L, f.c.c.) as a function of wavelength. Indices (hkl) of several Bragg edges and corresponding wavelengths are shown at the top. Scattering angles of reflections corresponding to the wavelength at Bragg edges are shown at the bottom. The wavelengths near minimums of cross sections marked with filled circles were tested.

Our experiments [34] showed that the maximum penetration path length in steels can be considerably increased (up to ~85mm) by using neutrons with wavelength λ, corresponding to the local minimum of neutron total cross section near but off the Bragg edges (Table 2).

Table 2. Maximum penetration path lengths (l_m) and depths in reflection (D_{ref}) and transmission (D_{tran}) geometries for different wavelengths in ferritic and austenitic steels [34] (2x2x20 = 80mm^3 gauge volume, 1h measurement time, 10^{-4} precision in strain).

Monochromator	$2\theta_M$(deg)	λ(Å)	Reflection plane	$2\theta_S$(deg)	l_m (mm)	D_{ref}(mm)	D_{tr}(mm
			Ferritic steel (low carbon steel, *bcc*)				
Si(220)	48	1.55	(211)	82.9	77	26	58
Si(111)	45	2.39	(110)	72.1	83	24	67
			Austenitic stel (stainless steel 304L, *fcc*)				
Si(220)	46.5	1.5	(311)	87.8	75	26	54
Si(220)	53.2	1.71	(311)	104	76	30	47
Si(111)	41.2	2.19	(111)	63.8	87	23	74
Si(111)	49.5	2.61	(111)	78	89	26	65

The bent perfect crystal Si monochromator is advantageous for this purpose because the curvature can by optimized for the chosen wavelength in order to simultaneously achieve the maximum detector signal and the minimum peak width for a particular diffraction peak. The stress distribution in 50 mm thick welded ferritic steel plate [35] and dissimilar weld overlay pipe [36] were measured using this method (reflection 110 and wavelength λ = 2.39Å).

3.4. Two reflections/wavelengths method

Further development of this method [37-38] showed that available thickness of stress measurements in ferritic steel can be increased by using two different instrument configurations for measurement of different strain components. From Table 2, one can see that because of difference in scattering angles, the wavelength λ =2.39Å and the reflection 110 of ferritic steel ($2\theta_S$ =72.1^0) are preferable for the measurement of the longitudinal (LD) and transverse (TD) strain components (transmission geometry), and the wavelength λ =1.55Å and the reflection 112 ($2\theta_S$ =82.9^0) are preferable for the measurement of the normal (ND) strain components (reflection geometry).

In general, the calculation of stresses from strains measured with different hkl reflections is not correct because the diffraction elastic constants and inter-granular strains are hkl dependent. However, planes (211) and (110) of α-Fe (bcc) have the same diffraction elastic constants calculated by the Kroner model: $E_{211,110}$ = 225.5GPa, $\nu_{211,110}$ = 0.28 [1]. Furthermore, planes (211) and (110) are less sensitive to the inter-granular strains and evident inter-granular effects in ferritic steel weld could not be observed [39]. The method was testified by measuring stress distribution in 50mm thick weld ferritic steel plate using one reflection 110 and two reflections 110, 211[37-38].

For stress measurements in 50 mm (and more), thick welded ferritic steel plates the following procedure was recommended: wavelength λ =2.39 A and reflection 110 for measurement of TD and LD components at all points through the thickness and ND components up to about 20mm depth. Wavelength λ =1.55Å and the reflection 112 were recommended for measurement of ND component in the remaining deeper locations. Using this "two reflections/wavelengths" method, the stress distribution in 70mm [40] and 80mm [4] thick welded ferritic steel plates were measured.

3.5. Magnetic field for stress measurement

As the neutron beam path increases in a ferritic steel specimen, significant peak broadening occurs, and this reduces the maximal available path length. Peak broadening was observed in ferritic steel and pure iron and was not observed in austenitic steel. Complementary neutron scattering experiments using small-angle neutron scattering (SANS) and wide-angle neutron diffraction (WAND) have been conducted [42] to investigate the reason of peak broadening. SANS experiments showed that applied magnetic field of 1.2T considerably decreases intensity of the small angle scattering in ferritic steel samples. It was supposed that peak broadening is caused by multiple refraction small angle scattering of neutrons on magnetic domain boundaries. The domain sizes calculated based on peak broadening (WAND) support the relationship between magnetic domains and peak broadening. Applied magnetic field of 0.5T increases maximum available path length over 80mm. Therefore it is supposed that the available neutron beam path length in ferritic steel can be considerably increased (~95mm) by applying sufficiently strong (~1.5T) magnetic field.

4. Conclusion

In the last 10 years, the neutron method achieved significant progress in increasing maximal available depth for strain measurements. It was possible thanks to development of dedicated bent perfect crystal silicon monochromators and position sensitive detectors. Further enhancement of available depth in steels was achieved by using neutrons with wavelengths, corresponding to the minimums of the neutron total cross section near but off Bragg edges. Measurement of stress distribution in 80mm thick steel plate is now possible. The maximum available neutron beam path length in ferritic steel can be additionally increased by applying sufficiently strong magnetic field.

Acknowledgements

This research was carried out under the Nuclear Research and Development Programme of the Ministry of Education, Science and Technology of the Republic of Korea. In the Czech Republic, it was supported by the GACR project no. 14-36566G. In Russia, it was supported by the Russian Fund for Basic Research, project № 14-22-01042.

References

[1] Hutchings M T, Withers P J, Holden T M, Lorentzen T 2005 *Introduction to the Characterization of Residual Stress by Neutron Diffractiont* London: Taylor and Francis
[2] Withers P J 2004 *J. Appl. Cryst.* **37** 596-606.
[3] Withers P J 2004 *J. Appl. Cryst.***37** 607 -612
[4] Santistaban J R, Daymond M R, James J A, Edvards L 2006 *J. Appl. Cryst.* **39** 812-825
[5] Lee C H, Moon M K, Em V T, Choi Y H, Cheon J K, Nam U W, Kong K N 2005 *Nucl. Instr. Methods in Phys. Research A***545** 480-489.
[6] Moon M K, Em V, Lee C H, Cheon J K, Choi Y H, Nam U W, Kong K N, 2006 *Physica B* **385–386**, 1038–1039
[7] Em V T, Moon M K, Lee C H 2006 *Neutron News* **12** 4 28-31
[8] Seong B S, Em V, Mikula P, Saroun J, Kang M 2010 *J. Appl. Cryst.* **43** 654-658
[9] Johnson M W, Daymond M R 2002 *J. Appl. Cryst.***35** 49-57.
[10] Popovici M, Yelon W B 1995 *J. Neutron Research***3** 1-25.
[11] Mikula P, Vrana M, Lukas P, Saroun J, Wagner V 1996 *Mater. Sci. Forum* **269** 228-231
[12] Mikula P, Lukas P, Vrana M *1997 Physica B* 234-236
[13] Pirling T, Bruno G., Withers P J. 2006 *Mater.Sci. Forum* **524-525** 217-222
[14] Pirling T., Bruno G., Withers P J.2006 *Mater.Sci. Eng.* A**437** 139-144
[15] Hofman F, Schneider R, Seidl G A, Rebelo-Kornmeier J, Wimproi R C, Garbe U, Brokmeier H G 2006 *Physica.B* **385-386** part2 1035-1037
[16] Kirstein O, Garbe U, Luzin V 2010 *Mater. Sci. Forum* **652** 86-91
[17] Mikula P, Kulda J, Lukas P, Vrana M, Ono M, Sawano J 2000 *Physica B* **174** 276-278

[18] Mikula P, Kulda J, Lukáš P, Vrána M, Wagner V 1994 *Nucl. Instr. Methods in Phys. Research A***338** 18-26.
[19] Kulda J, Wagner V, Mikula P, Saroun J 1994 *Nucl. Instr.Meth.* A**338** 60-64.
[20] Popovici M, Yelon W B, Berliner R, Stoica AD, Ionita I, Law R. 1994 *Nucl. Instr. Methods in Phys. Research.* A **338** 99-110.
[21] Stoica A D, Popovici M, Hubbard C R, Spooner S. 2000 Proc. Intern. Conf. on Residual Stresses (ICRS-6) Oxford UK 1264-1271
[22] Popovici M, Stoica A D 2002 *J. Appl. Phys.* A**74** 273-276.
[23] Moon M K, Lee C H, Em V T, Mikula P, Hong K P, Choi Y N, Cheon J K, Choi Y N, Kim S A, Kim S K, Jin K C 2005 *Physica B* **369** 1-7.
[24] Moon M K, Em V T, Lee C H, Mikula P, Hong K P, Choi Y N, Cheon J K, Nam U W, Kong KN Jin K 2005 *Physica B* **368** 70-75.
[25] Withers P J, Johnson M W, Wright J S 2000 *Physica B* **292** 273-285
[26] Moon M K, Lee C H, Em V T, Kang H D, Nam U W, Gabriel A 2002 *IEEE Trans. Nucl. Sci.* **49** (5) 2411–2414.
[27] Moon M K, Lee C H, Em V T, Kim H J, 2002 *J. Appl. Phys.* A **74** 1437–1439.
[28] Withers P J 2004 *J. Appl. Cryst.* **37** 596-606.
[29] Withers P J 2004 *J. Appl. Cryst.* **37** 607-612.
[30] Withers P J, Daymond M R, Johnson M W 2001 *J. Appl. Cryst.* **34** 737-743.
[31] Sears V F 1992 *Neutron News* **3** 26-37
[32] Sears V F 1989 *Neutron Optics* Oxford University Press.
[33] Hsu T C, Marsiglio F, Root J H, Holden T M 1995 *J. Neutron Research* **3** 27-39
[34] Woo W, Em V, Seong B S, Shin E, Mikula P, Joo J, Kang M 2011 *J. Appl. Cryst.* **44** 747-754
[35] Woo W, Em V T, Mikula P, An G B, Seong B 2011 *Mater. Sci. Eng.* A **528** 4120-4124.
[36] Woo W, Em V, Hubbard C R, Lee H J, Park K S 2011 *Mater. Sci. Eng.* A **528** 8021-8027,
[37] Em V, Woo W, Seong B S, Mikula P, Joo J, Kang M, Lee K H 2012 *J. Phys.: Conf. Series* **340**, 012025
[38] Woo W, Em V T, Seong B S, Mikula P, An G B 2012 *J. Mater. Sci.* **47** 5617-5623
[39] Holden T M, Suzuki H, Carr D G, Ripley M I, Clausen B 2006 *Mater. Sci. Eng.* A **185** 33-37
[40] Woo W, An G B, Kingston E J, De Wald AT, Smith DJ, Hill M R 2013 *Acta Mater.* **61** 3564–3574
[41] Woo W, An G B, Em V T, De Wald A T, Hill M R 2015 *J. Mater. Sci.* **50**, 784-793, DOI 10.1007/s 10853-014-8638-9.
[42] Woo W., Em V T, Shin E, Mikula P, Ryukhtin V 2015 *J. Appl. Cryst.* **48** 350-356

Accumulation of $GdCl_3$ in the feed of a reverse osmosis system during desalination as determined by neutron absorption

D. Schwahn[1] , V. Pipich[2], R. Kasher[3], and Y. Oren[3]

[1] Technische Universität München, Forschungs-Neutronenquelle Heinz Maier-Leibnitz (FRM II), D-85748 Garching, Germany.
[2] Jülich Centre for Neutron Science JCNS-FRM II, Outstation at FRM II, D-85747 Garching, Germany.
[3] Zuckerberg Institute for Water Research, Jacob Blaustein Institutes for Desert Research, Ben-Gurion University of the Negev, Sede Boqer Campus 84990, Israel.

E-mail: d.schwahn@fz-juelich.de, yoramo@bgu.ac.il

Abstract. This article deals with the application of *in-situ* small-angle neutron scattering to investigate wastewater desalination by reverse osmosis. In a first series of experiments we take advantage of the strong neutron absorption of gadolinium (Gd) and use 0.50 g/L $GdCl_3$ in the feed as an indicator for concentration polarization and scaling at the membrane surface. The continuous decline of scattering during the process of desalination indicates an increase of $GdCl_3$ salt concentration which after 15 hours has achieved nearly 100% enhancement with respect to its initial concentration.

1. Introduction

It is well known that the availability of potable water is a serious worldwide problem. There is large effort to further develop technology and to install plants in order to produce potable water from seawater and reclaimed wastewater [1]. An increasing number of desalination plants using reverse osmosis (RO) and nanofiltration (NF) technologies accomplish the strong demand for potable water particularly in arid areas [2]. For those countries that are remote from the sea, recycling of impaired water such as municipal wastewater to a level of unlimited application is a reasonable solution as is done today in several locations. Membrane-based technologies play an essential role in these respects. An ensemble of RO membranes in an operating plant is shown from the photos in Figure 1. Such a plant produces of the order of 500,000 m^3 potable water per day.

A serious problem in RO/NF desalination of municipal wastewater is biofouling and scaling at the membranes, limiting the membrane efficiency and lifetime [3, 4]. Calcium phosphate and organic matter were observed to accumulate on RO-membranes, suggesting that nanoparticles or smaller colloids pass through microfiltration and ultrafiltration membranes, ending up on RO-membranes as a cake-layer fouling [5]. Calcium phosphate scaling in municipal wastewater treatment systems is a severe problem as no efficient antiscalants are currently available [6]. In this respect, scaling is associated to bio-mineralization [7] as biomolecules such as proteins and polysaccharides play an important role in this process . The importance of the interactions of biomolecules and inorganic salts in water treatment processes was shown in recent small-angle neutron scattering (SANS) experiments by the authors [8-10]. It was shown that bio-organic molecules strongly influence mineralization in a model synthetic salt solution. Proteins such as BSA and lysozyme added to a synthetic salt solution

Figure 1. Views of a desalination reverse osmosis plant such as the Ashkelon Desalination Plant in Israel showing stacks of RO-membranes. Such a plant produces of the order of 500,000 m^3 potable water per day.

induced within a few seconds strong precipitation of nearly μm large composite particles composed of organic components as well as calcium phosphate and carbonate, observed by the variation of scattering contrast of the aqueous solution. In another SANS study, gold-nanoparticles were used as templates for studying surface mineralization in the same model solution; it was found that BSA coated gold nanoparticles induces fast formation of precipitates of 0.2 μm size that are composed of 50% - 80% volume fraction of minerals, identified as calcium phosphate and calcium carbonate [9].

This article deals with *in-situ* SANS studies on the accumulation of Gd^{3+} on the membrane surface. This may be caused together with the other salts of the feed by concentration polarization followed by precipitation at the surface of thin film composite (TFC) RO/NF membranes. A special flow-cell was built for these experiments simulating under most realistic conditions as reported in more detail in [11]. In this manuscript we introduce a technique utilizing the strong neutron absorption cross-section of $GdCl_3$ added as an indicator to the synthetic salt solution.

2. Theoretical and experimental background

Gadolinium is a rare-earth element with a strong neutron absorption cross-section as tabulated in [12]. We utilize this property in order to sensitively detect small amounts of Gadolinium (III) chloride ($GdCl_3$) dissolved in a model salt solution simulating a secondary effluent (SSE). The SSE solution represents an aqueous solution for the mineral components of a secondary-treated domestic wastewater in desalination with 80 - 85% recovery of a Shafdan prospective desalination plant in Israel [13].

The corresponding relevant parameters of $GdCl_3$ are compiled in Table 1. The total macroscopic cross-section Σ_{abs} of $GdCl_3$ of thickness D_S determines its absorption neutron transmission T_{abs} according to

$$T_{abs} = \exp\{-\Sigma_{abs} D_s\}. \qquad (1)$$

A general discussion on neutron transmission is found in ref [14]. We added $GdCl_3$ to the SSE solution which corresponds to a volume fraction of $\Phi = 1.1\times10^{-4}$, which according to Eq.(1) corresponds to

Journal of Physics: Conference Series **746** (2016) 012039 doi:10.1088/1742-6596/746/1/012039

Molecule	Molar mass	Mass density	Molar volume	σ_{abs} (λ=1.789Å)	Σ_{abs} (λ=12.8 Å)
Gadolinium (III) chloride	263.61 g/mol	4.52 g/cm^3	58.32 cm^3/mol	49.8×10^3 barn	3.68×10^3 cm^{-1}

Table 1. Neutron relevant parameters of gadolinium (III) chloride ($GdCl_3$)

a transmission of $T_{abs} \cong 0.945$ considering the thickness of $D_s \cong 0.14$ cm of the feed and membrane in the cell. The absorption cross-section is determined as $\Sigma_{abs} = \Phi_{GdCl_3} \times \Sigma_{GdCl_3}$ .Considering the system parameters the volume fraction Φ_{GdCl_3} of $GdCl_3$ is determined according to

$$\Phi_{GdCl_3} = -\ln(T_{abs})/(D_S\Sigma_{abs}) = -1.94\times10^{-3}\times\ln(T_{abs}). \quad (2)$$

The experiments were performed at the VSANS diffractometer KWS3 at the FRM II of the Technical University Munich (TUM) in Garching (Germany) using a specially designed flow-cell [11]. The membrane was a RO thin film composite membrane (TFC) XLE-440 of 140 μm thickness from Dow Filmtech (Minneapolis, MN, USA). Typically, TFC RO-membranes are composed of three layers. An aromatic polyamide layer of 100 to 250 nm thickness is on top of a ~ 40 μm thick micro porous polysulfone layer, both are supported by a third layer – either polypropylene or polyester – consisting of a ~ 100 μm thick non-woven fabric.

Figures 2a to c show views of the cell. Figure 2a shows the feed side of the cell. The lower

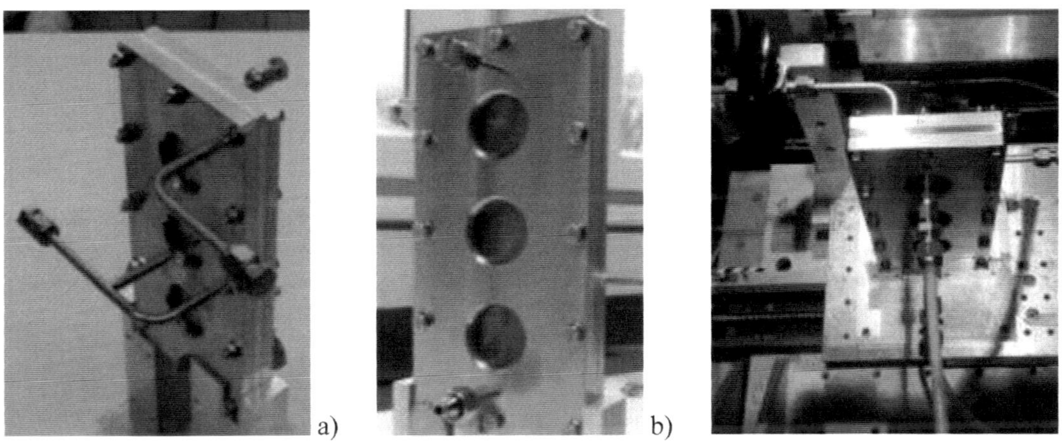

a) b) c)

Figure 2. RO flow-cell for SANS at the (a) feed and (b) permeate side. Figure 2c shows the cell at the sample position of KWS 3. Three sapphire windows of 3 cm diameter are designed for neutrons in order to measure the membrane at three different distances from the inlet of the feed.

and upper capillary tubes are inlet for the feed (i.e. the SSE solution) and outlet for the concentrate (i.e. the concentrated feed), respectively. Figure 2b shows the permeate side with the lower capillary tube as the outlet for the cleaned water which usually has the quality of potable water. The potable water was weighted by a balance as a function of time in order to determine the permeability of the membrane in parallel to the neutron experiment. Figure 2c shows the flow-cell in operation at the sample position of the VSANS instrument KWS3.

3. Experimental Results

The upper Figure 3 shows the time evolution of neutron transmission of the cell during the process of desalination, i.e. of the SSE feed, the TFC membrane and spacer, measured at 6 bar for nearly 20 hours. The permeability and permeate flux measured in parallel are depicted in the lower Figure 3. The SSE feed was prepared on basis of 40% and 60% volume fractions of D_2O and H_2O, respectively, in order to minimize scattering from the membrane. The cavities of the polysulfone and polyester membrane layers are matched under these circumstances, i.e. they do not contribute to scattering [11]. Besides the added salts we added 0.50 g/L $GdCl_3$ to the SSE feed as already mentioned.

An overall decline of permeate flux by more than a factor of 2 is observed which is accompanied by a continuous decrease of neutron transmission. A surprising observation was made after about 10 hours when a minimum of transmission is accompanied by a small hump of permeate flux. Such a

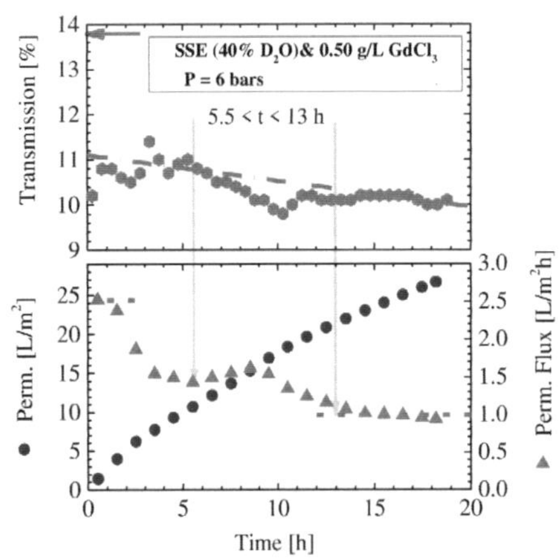

Figure 3. Neutron transmission and permeate flux during RO desalination of a SSE salt solution added with 0.50 g/L of $GdCl_3$.

phenomenon was also observed in a similar experimental without $GdCl_3$ and was interpreted as partial cleaning of the membrane from larger mineral entities [11]. The red arrow in the upper figure represents the transmission coefficient at ambient pressure measured just before starting the RO desalination. Scattering patterns measured at ambient pressure are systematically weaker. The larger intensity, i.e. lower transmission at 6 bar stems from the elasticity of the membrane in particular from the polyester non-woven fabric with its large volume fraction of about 70% [11]. Pressure induced compression of the polyester layer leads to a decline of cavity volume fraction. A decline from 70% down to 50% leads to an enhanced scattering as seen from the prefactor of $d\Sigma/d\Omega(0)$ below in Eq.(4). This effect is quantitatively outlined in ref [11]. The scattering of the new membrane at ambient pressure was always taken as the reference for the 6 bar measurements we will discuss now.

Figure 4a shows scattering patterns measured at the given times during the desalination experiment already presented in context with the transmission data in Figure 3. The scattering data were measured in the Q regime of 10^{-4} Å^{-1} and represent the enhancement of $d\Sigma/d\Omega(Q)$ with respect to the scattering at ambient pressure just before starting the 6 bar experiments. The difference of enhanced scattering, $\Delta d\Sigma/d\Omega(Q)$, was fitted with the Beaucage equation in Eq.(3) function as depicted by the solid lines. Eq.(3) is a function of

$$\frac{d\Sigma}{d\Omega}(Q)=\frac{d\Sigma}{d\Omega}(0)\exp(-u^2/3)+P_\alpha\left[\left(\text{erf}\left(u/\sqrt{6}\right)\right)^3/Q\right]^\alpha \quad (3)$$

the parameter u which is the product of momentum transfer Q and radius of gyration R_g. The momentum transfer is defined according to $Q=(4\pi/\lambda)\sin(\delta/2)$ from the neutron wave length λ and the scattering angle δ. Beaucage's equation represents an approximated scattering law [15]. At small

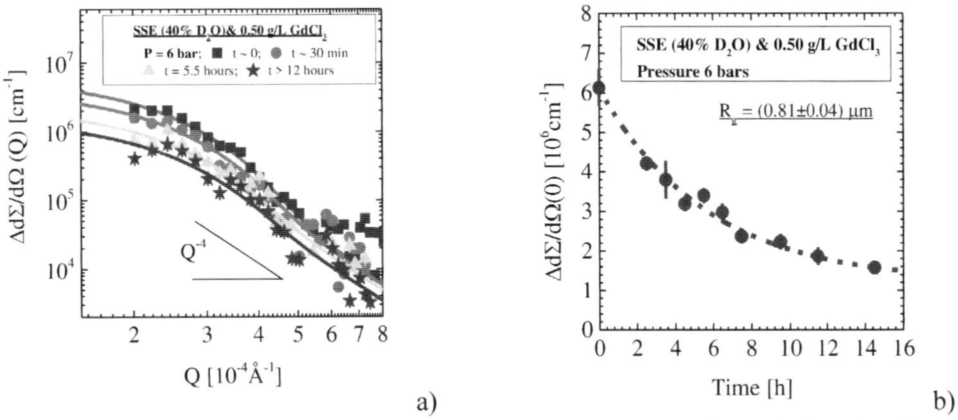

Figure 4. a) Enhanced scattering due to applied pressure which reduces during desalination because of accumulation of $GdCl_3$ concentration in the feed. b) Decline of $d\Sigma/d\Omega(0)$ versus time as derived from fitting scattering law delivering a constant R_g of 0.81 ± 0.04 μm.

Q ($< 1/R_g$) the macroscopic cross-section $d\Sigma/d\Omega$ is determined by the first term of Eq.(3) corresponding to Guinier's approximation, whereas at larger Q ($> 1/R_g$) the scattering usually follows a power law. In many cases one finds a Porod law with an exponent of $\alpha=4$ valid for 3 dimensional particles with a smooth surface. The parameters $d\Sigma/d\Omega(0)$ and P_4 are expressed as

$$\frac{d\Sigma}{d\Omega}(Q{=}0)=\Phi_P\left(1-\Phi_P\right)V_P\left[\rho_P-\rho_S\right]^2 \text{ and } P_4=2\pi N_P S_P\left[\rho_P-\rho_S\right]^2 \qquad (4)$$

Both parameters deliver quantitative structural description of the objects such as mean particle number density N_P, surface S_P, volume V_P, and volume fraction Φ_P [16].

The analysis of the scattering curves in Figure 4a deliver over the time interval a constant radius of gyration R_g of 0.81 μm whereas the enhanced extrapolated scattering cross-section at Q=0, $\Delta d\Sigma/d\Omega(0)$, is declining with time as depicted in Figure 4b. $\Delta d\Sigma/d\Omega(0)$ is following an exponential decline depicted as dashed line as guide for the eye. The constant R_g supports our interpretation that the decline of scattered intensity is due to accumulation of $GdCl_3$ concentration in the feed in front of or at the membrane.

The transmission due to $GdCl_3$ accumulation is evaluated from the ratio of $d\Sigma/d\Omega(0)$ of SSE with and without $GdCl_3$ showing values in the interval from 0.97 to 0.90 as depicted in the upper Figure 5. The change of transmission due to accumulation of $GdCl_3$ in the feed is 7%. Such decline of transmission is not visible in the transmission of the total system which is around 10% thereby inducing a change of only 0.7% over 18 hours. Such change of T is visualized as grey dashed-dotted line in the upper Figure 3.

The volume fraction Φ of $GdCl_3$ is derived from the transmission on basis of Eq.(2) and is plotted as weighted fraction w_{GdCl3} in the lower Figure 5. At the beginning of the desalination process the concentration of $GdCl_3$ is with 0.3 g/L fairly well in agreement with the initially weighted one of 0.5 g/L and approaches a value of about 0.88 g/L after about 14 hours of desalination. This enhancement

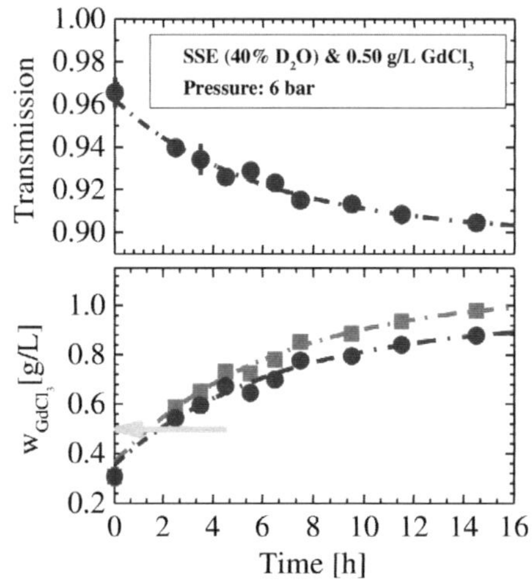

Figure 5. Decline of transmission derived from scattering and the corresponding concentration of $GdCl_3$ versus time (blue symbols). The red symbols show the corrected concentration of $GdCl_3$ in case of a factor 2 increase of scattering due to scaling. The green arrow shows the initially weighted concentration of $GdCl_3$.

corresponds to an increase of about 80% of $GdCl_3$ concentration with respect to the initial value The decay of transmission and the corresponding increase of $GdCl_3$ concentration follow an exponential behavior with a time constant of nearly 6 hours. This statement has only phenomenological character as there is no theoretical basis known at present and is mending as a guide for the eye.

In the analysis so far we assumed that the scattering in Figure 4a is declining only by enhancement of $GdCl_3$ salt. However, from former similar studies in [11] without the strong neutron absorbing salt we know that there is developing enhanced scattering by about a factor of 2 due to scaling and possibly other effects such as concentration polarization. This means that our analysis might underestimate the degree of accumulated $GdCl_3$ content. Assuming enhanced scattering by a factor of 2 over the total time due to scaling mean a slightly larger salt concentration of about 100% $GdCl_3$.

4. Conclusion

We present here one aspect of a larger project on SANS studies on scaling and fouling of RO membranes during desalination. SANS is a new and promising tool in this field as it allows *in-operando* experiments following the process of mineralization on a level of microscopic length scale and distinguishes between inorganic and organic molecules via contrast variation. Results of this work have already been reported in [8-10]. The present experiments are difficult to interpret as neutrons have to pass through several sections of the cell such as feed and the three layers of the membrane. We have to find experimental as well measuring techniques to separate and identify scattering from the different sources. One possible procedure is demonstrated in the present article, namely the determination of the amount of $GdCl_3$ being dissolved together with other salts in the feed forming a concentrated salt solution layer in front of the membrane (concentration polarization) and mineral particles forming a scaling layer at the surface of the membrane. However, from the present experiment we only know the average concentration of $GdCl_3$ at the feed side of the flow-cell. A determination of the amount of $GdCl_3$ of the scaling layer at the surface of the membrane is possible after finishing the desalination experiment and performing the measurement after filling the cell with fresh salt-free water. Another effect needs precaution, namely the enhancement of scattering from the membrane due to scaling. This effect appears much weaker as we have seen from the data in Figure 4b and 5 and can be made even more negligible by slightly increasing the $GdCl_3$ salt concentration.

Acknowledgement

We would like to thank Thomas Starc from Forschungszentrum Jülich for his engaged preparation and participation of the neutron experiments, which were performed at the KWS 3 instrument operated by JCNS at the Heinz Maier-Leibnitz Zentrum (MLZ), Garching, Germany. This work was funded by the German-Israeli Foundation for Scientific Research and Development (GIF), grant no. I-101-307.4-2013.

References

[1] Lattemann S, Kennedy M D, Schippers J C, Amy G, Chapter 2 of *Sustainable Science and Engineering*, Vome 2 (Elsevier, Amsterdam 2010).
[2] Gullinkala T, Digman B, Gorey C, Hausman R, Escobar I C, Chapter 4 of *Sustainable Science and Engineering*, Vome 2 (Elsevier, Amsterdam 2010).
[3] Lee S and Elimelech M, 2006 *Environmental Science & Technology* 40(3), 980-987.
[4] Herzberg M and Elimelech M, 2006 *J. Membr. Sci.* 295(1-2), 11-20.
[5] Ning R Y and Troyer T L, 2007 *Desalination* 208(1-3), 232-237.
[6] Greenberg G, Hasson D and Semiat R, 2005 *Desalination* 183, 273-288.
[7] Mann S, 2001 Biomineralization, Oxford University Press, Oxford 2001.
[8] Pipich V, Dahdal Y, Rapaport H, Kasher R, Oren Y and Schwahn D, 2013 *Langmuir* 29 7607 - 7617.
[9] Dahdal Y, Pipich V, Rapaport H, Oren Y, Kasher R and Schwahn D, 2014 *Langmuir* 30, 15072 - 15082.
[10] Dahdal YN, Pipich V, Rapaport H, Oren Y, Kasher R, Schwahn D, 2016 *Polymer* 85, 77-88.
[11] Schwahn D, Feilbach H, Starc T, Pipich V, Kasher R and Yoram Oren 2016 *J. Membr. Sci.* (submitted).
[12] 1992 *Neutron News* 3 29–37.
[13] Steiner Z, Rapaport H, Oren Y, and Kasher R, 2010 *Environmental Science & Technology* 44 7937–7943.
[14] Palacios J, Schwahn D, Rauch H, 1981 *NDT International* 14, 189-193.
[15] Beaucage G, 1995 *Journal of Applied Crystallography* 28, 717-728.
[16] Roe R J, *Methods of X-Ray and Neutron Scattering in Polymer Science*, University Press, Oxford, 2000.

Investigation of microstructure within metal welds by energy resolved neutron imaging

A S Tremsin[1*], W Kockelmann[2], A M Paradowska[3], Shu-Yan Zhang[2], A M Korsunsky[4], T Shinohara[5], W B Feller[6], E H Lehmann[7]

[1] University of California at Berkeley, Berkeley, CA 94720, USA
[2] STFC, Rutherford Appleton Laboratory, ISIS Facility, Didcot, OX11 0QX, UK
[3] Australian Nuclear Science and Technology Organization, Lucas Heights, NSW 2232, Australia
[4] Department of Engineering Science, University of Oxford, Oxford OX1 3PJ, UK
[5] Japan Atomic Energy Agency, 2-4 Shirakata-shirane Tokai-mura, Naka-gun Ibaraki 319-1195, Japan
[6] NOVA Scientific, Inc., 10 Picker Rd., Sturbridge, MA 01566, USA
[7] Paul Scherrer Institute, CH-5232 Villigen, Switzerland

E-mail: ast@ssl.berkeley.edu

Abstract. The recent development of bright pulsed neutron sources and high resolution neutron counting detectors enables simultaneous acquisition of a neutron transmission spectrum for each pixel of the image. These spectra can be used to reconstruct microstructure parameters within welds, such as strain, texture and phase composition through Bragg edge analysis, and in some cases elemental composition through resonance absorption analysis. In this paper we demonstrate the potential of energy-resolved neutron imaging to study the microstructures of two steel welds, where the spatial distribution of residual strain within the welds, as well as some information on the texture, are obtained with sub-mm spatial resolution. A friction stir weld of two steel plates and a conventional weld of two steel pipes were studied at pulsed neutron facilities, where a $\Delta\lambda/\lambda$ resolution as low as 0.2% can be attained over a wide range of neutron wavelengths ranging from 0.5 Å to 8 Å.

1. Introduction

A number of techniques to study the microstructures within metal welds with sub-mm spatial resolution have been developed to date. Neutron imaging is a rather unconventional method due to the fact that studies can only be performed at a limited number of bright neutron sources. However, the high degree of neutron penetration can provide a unique opportunity to investigate relatively thick samples, where residual strain is not relieved by cutting into thin slices required for more traditional methods, such as X-ray diffraction. It has been demonstrated that thermal and cold neutron radiography is well suited for the investigation of macroscopic internal defects, as well as the distribution of impurities and voids [1]-[5]. With energy-resolved imaging it is possible to investigate some microstructure details within welds, such as the presence of large grains or preferred orientation within certain areas. The analysis of neutron

* To whom any correspondence should be addressed.

transmission spectra can provide information on the microstructure and enable studies of residual strain and texture-related characteristics within the sample. The sharp variation of neutron transmission at certain energies, the so-called Bragg edges, can be used for effective analysis of microstructure within the welds, as long as data is acquired with an adequate energy resolution. The conventional neutron diffraction techniques, already well established, provide data on phase, structure and microstructure, however with limited spatial resolution of a pencil beam scanned across the sample [6],[7]. Measurement of a neutron transmission spectrum in each pixel of the image has recently become available with relatively high spatial resolution of ~100 μm due to recent developments of high resolution neutron detectors capable of measuring both position and time for each detected particle [8], [9].

There are several options to achieve energy-resolution in neutron imaging. Due to the lack of high resolution energy-resolving neutron imaging detectors to this date, there have typically been at least two alternative approaches. 1. The incoming neutron flux is monochromatized by either a velocity selector or by a crystal monochromator and multiple images are measured by scanning through the energy range of interest. 2. Neutrons are produced in short pulses, and the energies of detected neutrons are measured via their time-of-flight. The energy resolution which can be achieved with a velocity selector is typically *dE/E* ~ 15% [10], which allows studies of texture variations and, to some extent, microstructure parameters such as crystallite size and orientation, but is not sufficient for the reconstruction of strain. The energy resolution of a double crystal monochromator setup is typically 3%, providing a higher energy resolution with the expense of neutron flux [11]. Recent developments of neutron imaging facilities at bright pulsed neutron sources [12], [13] enable studies with energy resolution *dE/E* < 0.5% [1], [14], [15] enabling mapping of the strain distributions with sub-mm spatial and ~100 microstrain resolution [16]. The unique combination of bright short-pulse neutron sources and high resolution neutron counting detectors, such as the neutron-sensitive MCP/Timepix used here, enables simultaneous detection of>250000 spectra in one acquisition. In this paper we demonstrate the capabilities of energy-resolved neutron imaging for the studies of microstructures within two steel welds. The distribution of residual strain and some texture variation across the samples are reconstructed with sub-mm resolution by the analysis of measured transmission spectra. Two steel welds are studied with the energy resolved neutron imaging: a butt friction stir weld between two steel plates and a conventional weld between two steel pipes.

2. Experimental setup

The experiments on energy-resolved neutron imaging were conducted at two pulsed neutron beamline facilities: at Engin-X at the ISIS spallation neutron source, Rutherford Appleton Laboratory, UK (friction stir weld) and at the Noboru beamline at the Japanese Proton Accelerator Complex (J-PARC). In both setups we used a high resolution MCP/Timepix neutron counting detector capable of simultaneous detection into multiple neutron energy bins with sub-μs timing resolution and 55 μm spatial resolution [8], [9]. More than 250000 spectra were measured simultaneously by 512x512 pixels, in an array consisting of 2x2 Timepix readout chips mounted behind neutron-sensitive microchannel plates. The energy of each registered neutron was calculated from its time of flight over distances of 50.0 m and 14.17 m for the Engin-X and Noboru experiments, respectively. The set of images, each corresponding to a narrow range of energy (typically dE/E~0.05%) was recorded in each measurement with ~3000 energy slices over the thermal and cold ranges of neutron energy. The measurement time was ~2.5 hours and ~1.5 hours for the friction stir weld and the pipe weld, respectively. It should be noted that the power of the J-PARC source was 200 kW during the latter measurement. However, with a planned increase of power to 1 MW the acquisition time will likely to be reduced to 20-30 minutes for similar neutron counting statistics. The measured spectra were normalized against spectra obtained when there was no sample in the beam. The latter normalization eliminated the effects of the beam spectrum anomalies including beam non-uniformities across the field of view. The samples were mounted at a distance of ~12 mm from the active plane of the detector. The beam divergence was defined by a set of slits and was less than 0.3°, thus limiting the image blur to ~1 pixel of our readout (55x55 μm^2). The active area of the detector used in these experiments was 28 mm in diameter. Multiple readouts per single neutron pulse were implemented in the experiments in order to avoid event pileup. The readout time (limited by our detector deadtime) was 320 μs, resulting in the presence of small gaps in the measured spectrum which were chosen to be away from Bragg edges. Transmission along only one orientation of the sample was measured in the present experiments: for the friction stir weld, it was along the direction of the weld; and for the pipe weld, it was perpendicular to the axis of the pipe. The residual strain was measured as an integral over the entire thickness of the sample. The strain was reconstructed for each pixel of the measured data by fitting an analytical function into the measured Bragg edge transmission [15]-[18].

3. Results and discussion

3.1. Friction stir weld

A butt friction stir weld (FSW) between two 12.7 mm-thick chromium steel plates was characterized with the energy-resolved neutron transmission analysis. The steel composition (in wt %) was 10.5-12.5% Cr, 1.5% Mn, 0.3-1% Ni, 0.03% C, with small quantities of N, P, S and Si, and Fe balance. FSW welds were made at TWI (The Welding Institute, Abington, Cambridge, UK) using refractory metal welding tools of Whorl^{TM} type [19]. Shoulder diameter of 45mm was used with 10-14 mm probe diameter base-tip. During welding the plates were placed on a stainless steel bed and subjected to the down-force of 32 kN. The welding tool was advanced at the rotational speed of 625 r/min and lateral speed of 2mm/s. The dimensions of the completed welds were 203.2×66.2×12.7 mm^3. Infrared thermal imaging was used to monitor the temperature during welding, and revealed that a maximum of ~1000°C. Given the thermal history, the steel phase composition within the weld can be estimated with the help of the Schaeffler diagram [20], revealing that ferrite transforms to austenite during friction-induced heating, and that upon cooling the conversion of ferrite to martensite takes place. A 3mm-thick section out of the same weld has been studied previously by neutron diffraction at FRMII neutron source (Garching, Germany) [21]. However, for the purposes of stress analysis only BCC ferrite peak was used. In the present experiment it was possible to visualise the microstructure within an 11 mm-thick section of the weld (figure 1) using transmission neutron imaging. The contrast formation mechanism can be ascribed to the combination of phase composition and lattice structure, texture (preferred orientation) and residual strain.

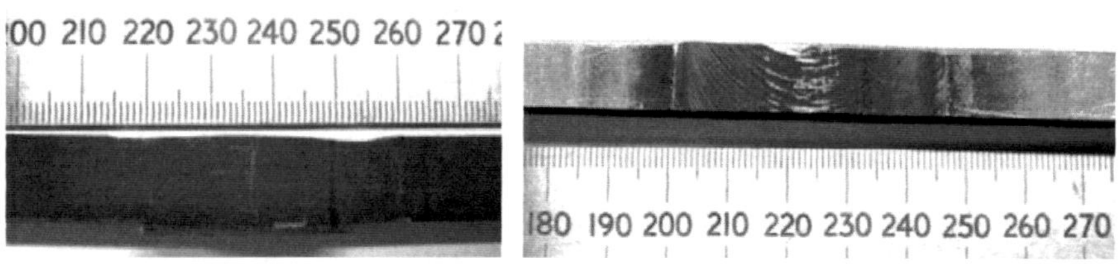

Figure 1. Photograph of the stir-friction weld characterised in the experiment. Left: view on the weld in the direction of neutron flux (longitudinal - direction of welding); Right – top view of the weld (normal direction to the surface of the plate).

The white-spectrum transmission image of the weld showed no features and looked similar to the image of figure 2.f at 4.19 Å. This image has a particular relevance because the wavelengths used in that image are just above the first (110) Bragg edge of BCC steel. It displays no structural features, as Bragg diffraction does not occur at these long wavelengths for BCC steel. The possibility to measure narrow-energy transmission is crucial for studies of phase, crystal and microstructure properties revealed by diffraction of neutrons out of the incident beam. The weld region is clearly seen in the images of figure 2, as well as the parent steel section with horizontal banding corresponding to a variation of texture within the plate due to the rolling manufacturing process.

The neutron transmission spectra obtained in different areas of the weld shown in figure 3 indicate the presence of strong texture within the original steel plates. The images of figure 2, at the same time, reveal the sizes of the textured layers within the plate, while the re-melted weld region shows some large-scale features only around the 110 Bragg edge at 4.06 Å (figure 2.e). The difference between the weld area and the original steel plate is demonstrated in figure 4. A quantitative analysis can be performed by fitting an analytical function to the measured Bragg edges [15]-[18]. Figure 5 illustrates the four fit parameters, which are affected by the microstructure within a sample. The position of the Bragg edge center can be determined with ~10 times better accuracy by fitting, compared to the measured width of the edge, which cannot be smaller than the wavelength resolution of the experimental setup (~0.5% in our case). By fitting the analytical function, the accuracy of strain reconstruction can be improved to a level of ~100 microstrain, depending on the neutron counting statistics in each small area of the sample, down to a single 55 μm pixel.

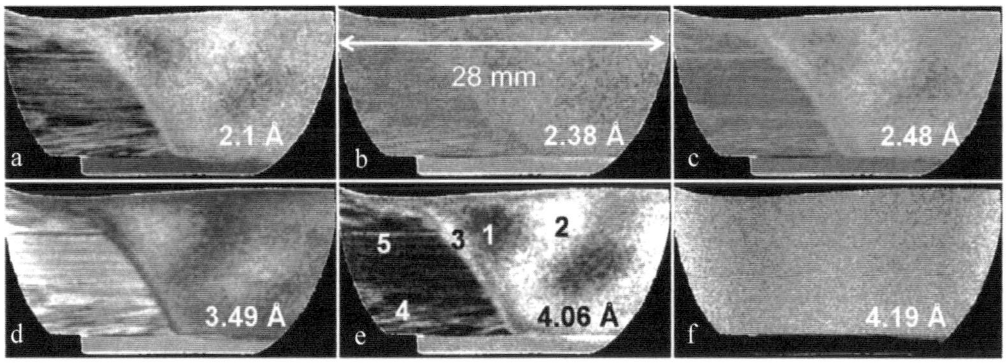

Figure 2. Narrow-energy neutron transmission images of the friction stir weld. The corresponding wavelength of each image is indicated. The labels 1-5 on the4.06 Å image indicate the areas for which spectra are plotted in figure 3.

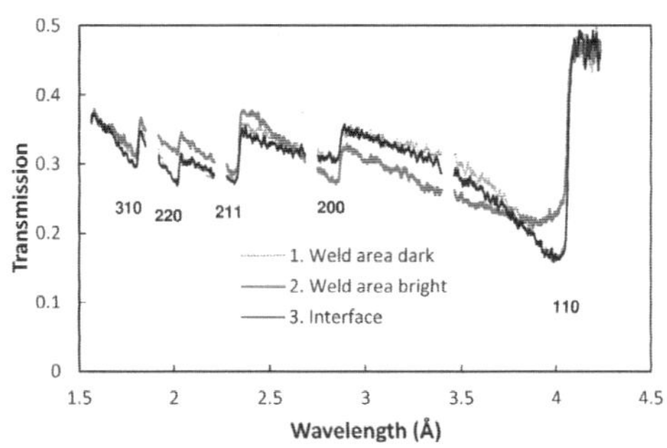

Figure 3.a. The measured transmission spectra of the weld in the areas within the heat affected zone, indicated by the numbers in figure 2.

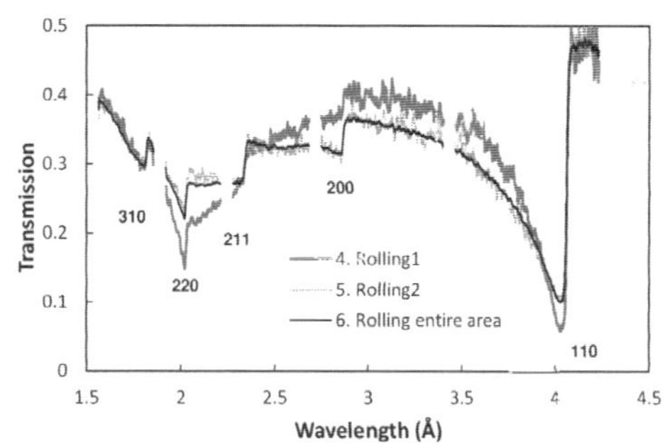

Figure 3.b. Same as figure 3.a, except measured for the areas outside of the weld. Spectra are characterised by strong texture of the rolled material.

The fitting results for the friction stir weld are shown in figures 6-11. The distribution of the residual strain across the sample was calculated from the spectra accumulated over a 1.1x1.1 mm^2 area around each 55 μm pixel, as indicated by the scale bar in figure 6. The '$\lambda_0=2d_0$'-reference value of 4.0534 Å for the strain calculation was taken from data collected in the stirred area of the weld, as demonstrated by the X-ray strain measurements performed on a thin section of the weld [21]. Even after the sample was cut and major macroscopic stresses were nearly relived in the dominant direction (longitudinal) of the welding, the distribution profile can be seen with very high spatial resolution. Tensile strain values are measured in the area adjacent to the stirred zone in the direction along the weld seam. The width of the (110) Bragg edge is larger in the heat affected zone, as seen from figure 7, corresponding to a lower height of the edge, while the (211) edge height is higher in the heat affected zone as seen in figures 4 and 10. The images of figure 2 and spectra of figures 3 and 4 indicate that the rolling texture observed in the original steel plate is reduced in the heat affected region of the weld, while the residual strain distribution of figure 6 shows much lower strain values in the weld region compared to the adjacent area.

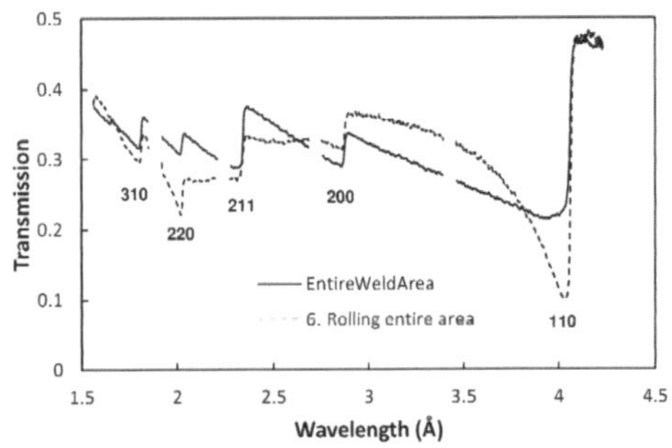

Figure 4. Same as figure 3.a, except shown for the entire weld and entire original steel pate areas.

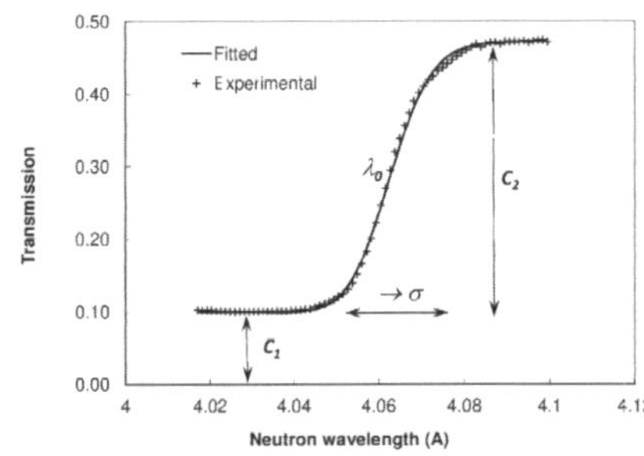

Figure 5. The 110 Bragg edge of measured transmission for the friction stir weld (markers) and the fitted analytical function.

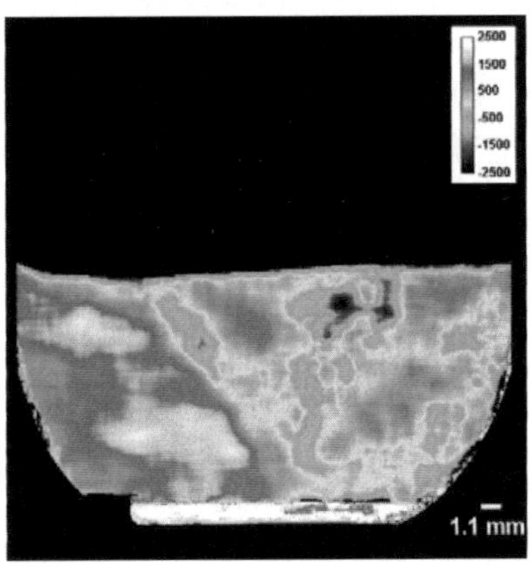

Figure 6. Strain reconstructed from 110 Bragg edge. The legend indicates strain values (με).

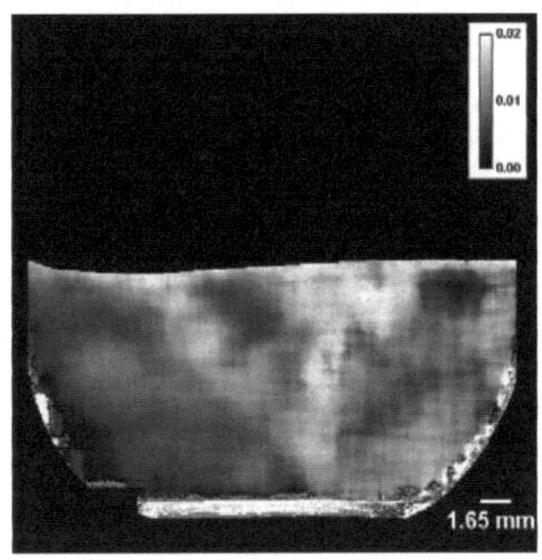

Figure 7. The width of the (110) Bragg edge, reconstructed by a fit to the experimental data.

3.2. Steel pipe weld

A weld between two FCC steel pipes with ~2.5 mm thick walls (figure 12) was investigated by energy resolved imaging at the J-PARC pulsed source facility. The sample was mounted with the incident beam direction perpendicular to the axis of the pipe, as shown in figure 12. The strain values integrated along the beam direction are shown in figure 13.a, with spectra taken in the 1.1x1.1 mm^2 area around each pixel. Different from the friction stir weld described in the previous section higher residual strains are concentrated around the seam area of the welded pipe. The '$\lambda_0=2d_0$'-reference value of 4.1486 Å for the strain calculation was taken as an average value from data collected at a distance from the weld. The height and the pedestal of the Bragg edge shown in figures 13.b and 13.c indicate the presence of large grain orientation variations within the weld region. The narrow-energy band transmission images shown in figure 14 reveal the presence of highly textured areas within the weld, with needle-like structures of ~100 μm thickness. In addition to improved spatial resolution as compared to previously published images of similar welds [2], the improved energy resolution of our experimental setup enables measurements of the transmission spectra within small effective pixel sizes across the entire sample. The transmission spectra shown in figure 15 reflect the presence of the FCC polycrystalline steel structure in the area of the pipe not affected by welding, albeit with indications of considerable preferred orientation (see for example (111) Bragg edge shape of areas 1 and 2). In the region of the weld the spectra reveal the presence of microcrystallites, evidenced by pronounced dips rather than Bragg edges. These dips originating from a group of crystals coexist with the (111) Bragg edge (figure 15.b), indicating the presence of both a random-orientation polycrystalline component and a group of re-solidified large grains within the weld seam.

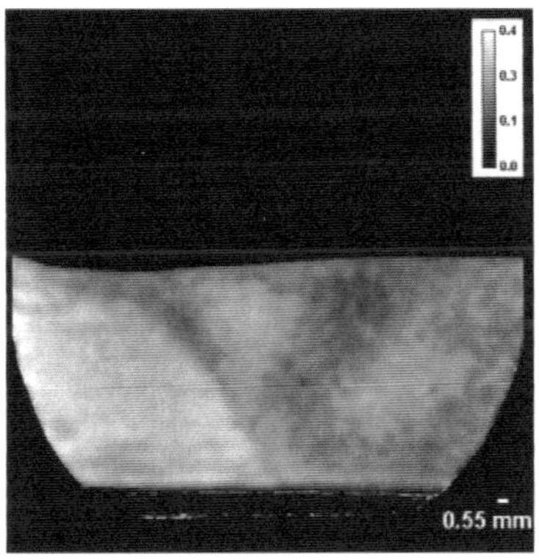

Figure 8. The distribution of height of the (110) Bragg edge (parameter C_2 of figure 5).

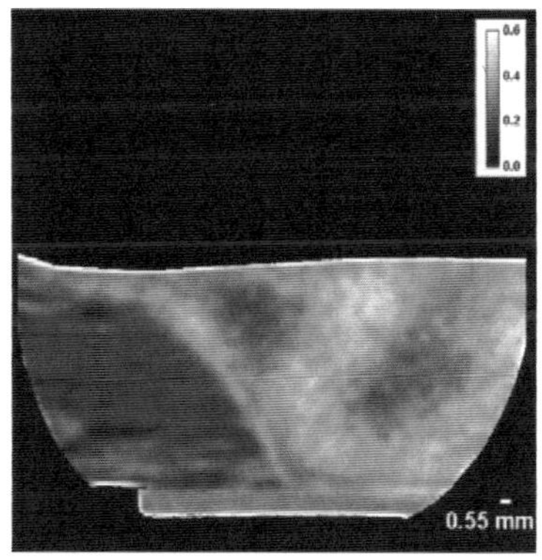

Figure 9. The distribution of (110) Bragg edge pedestal (parameter C_1 shown in figure 5).

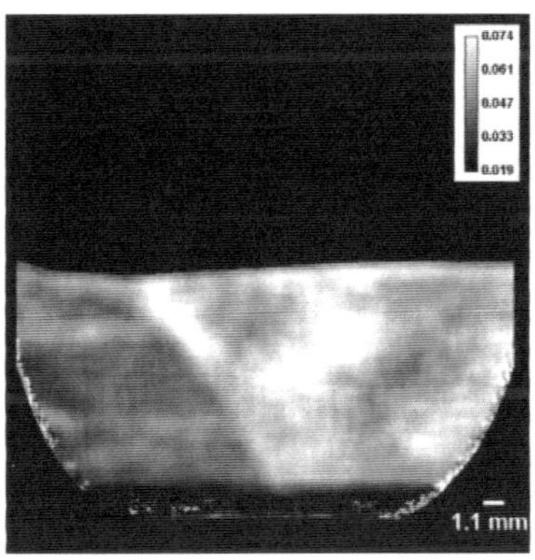

Figure 10. Same as figure 8, except for (211) Bragg edge.

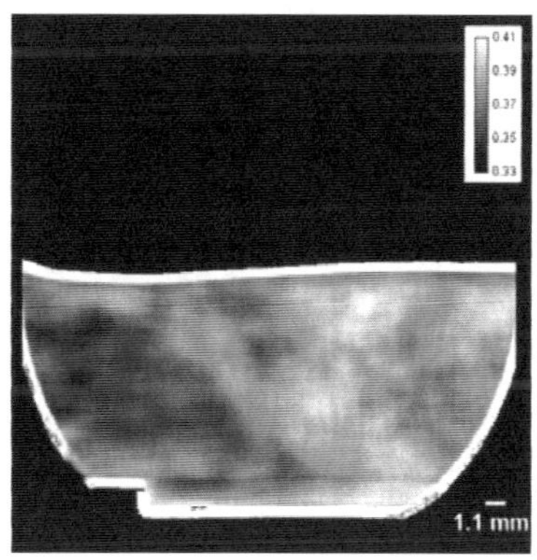

Figure 11. Same as figure 9, except for the (211) Bragg edge.

4. Conclusions

The recent development of bright pulsed neutron sources in combination with high resolution fast neutron counting detectors, enables improved studies of microstructures within weld samples, where strain and texture-related features can be mapped with ~100 μm spatial resolution. The simultaneous measurement of a spectrum in each 55x55 μm^2 pixel provides the potential of recovering some quantitative information on the microstructure through Bragg edge analysis. More advanced Rietveld-type transmission spectrum analysis tools [22] enable mapping of various microstructure characteristics of the welds, throughout their bulk volume, and not merely on the surface. Although the analysis presented in this paper does not provide quantitative information on the orientation distribution function within the sample, as can be obtained with conventional neutron diffraction, the sequence of narrow energy-band images can nevertheless provide valuable and very quick complementary information on residual strain as well as uniformity of texture - with relatively high spatial resolution across the entire sample. It is advantageous that the full region of interest of a sample can be mapped using only a single measurement step, both spatially and as a function of energy.

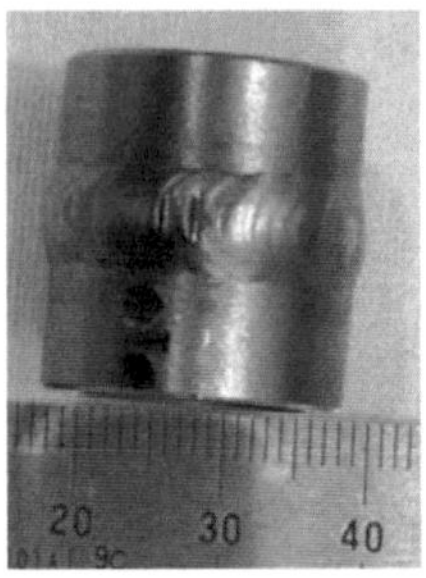

Figure 12. Photograph of the steel pipe weld used in the experiment.

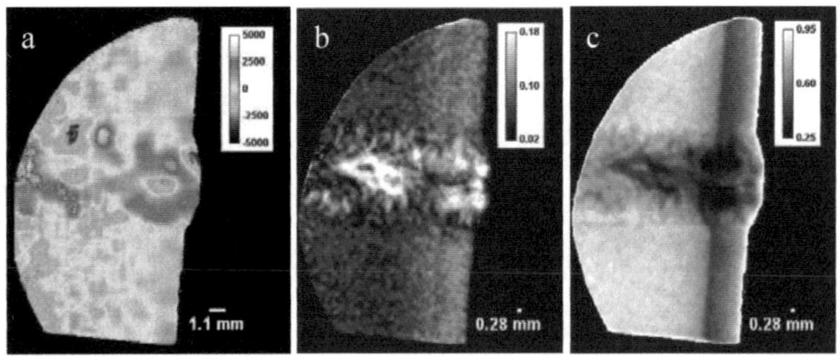

Figure 13. (a) Residual strain distribution obtained from measured spectra around (111) Bragg edge. (b) Height C_2 of the (111) Bragg edge. (c) Pedestal C_1 of the (111) Bragg edge.

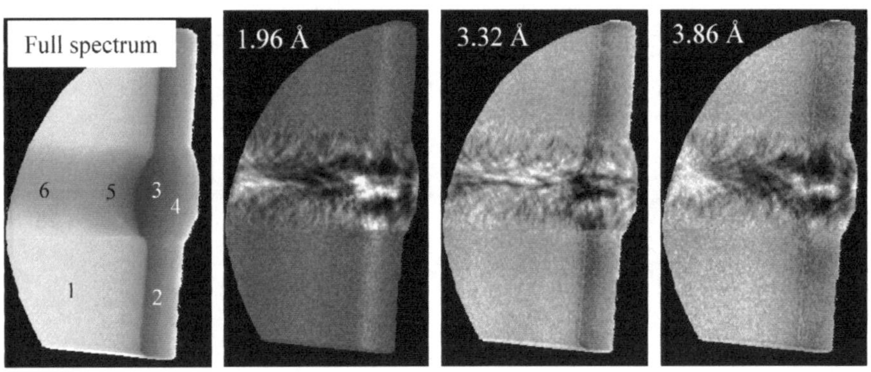

Figure 14. White beam spectrum image and narrow-energy-band images of the pipe weld.

Acknowledgments
The detector used in these experiments was developed within the Medipix collaboration. This work was supported in part by the U.S. Department of Energy under Grants No. DE-FG02-07ER86322, DE-FG02-08ER86353 and DE-SC0009657. The authors would like to acknowledge the generous donation of Vertex FPGA, Vivado design suite and DK-K7-CONN-G connectivity kit by Xilinx Inc. of San Jose, California through their Xilinx University Program.

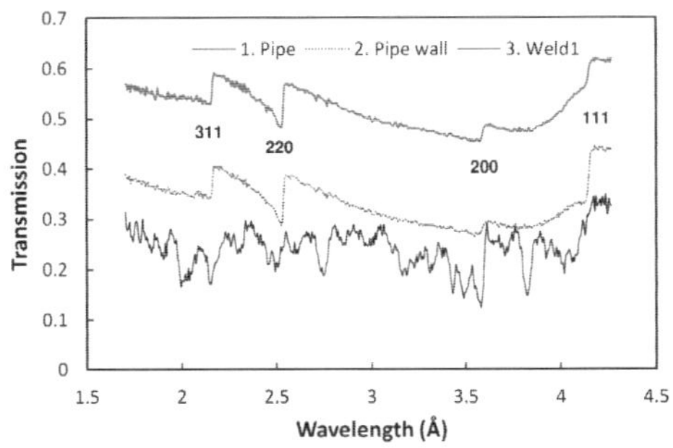

Figure 15.a. The measured transmission spectra of the pipe weld in the areas indicated by the numbers in figure 14.

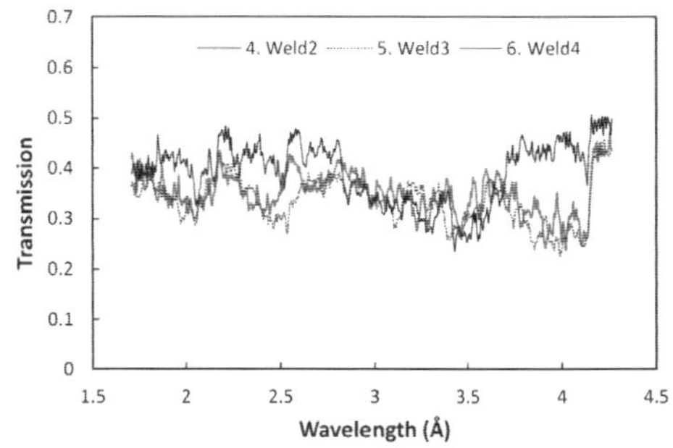

Figure 15.b. The measured transmission spectra of the pipe weld in the areas indicated by the numbers in figure 14.

References

[1] Kockelmann W, Frei G, Lehmann E H, Vontobel, P, Santisteban J R, Energy-selective neutron transmission imaging at a pulsed source, 2007. *Nucl, Instr, Meth. A* **578** 421.

[2] Lehmann E H, Frei G, Vontobel P, Josic L, Kardjilov N, Hilger A, Kockelmann W, Steuwer A., 2009. The energy-selective option in neutron imaging. *Nucl, Instr, Meth. A* **603** 429.

[3] Sato H, Kamiyama T, Iwase K, Ishigaki T, Kiyanagi Y., 2009. Pulsed neutron spectroscopic imaging for crystallographic texture and microstructure. *Nucl, Instr, Meth. A* **651** 216.

[4] Kardjilov N, Manke I, Hilger A, Williams S, Strobl M, Woracek R, Boin M, Lehmann E, Penumadu D, Banhart J, 2012. Neutron Bragg-edge mapping of weld seams. *Int. J. Mater. Res.* **103** 151.

[5] Santisteban J R, Vicente-Alvarez M A, Vizcaino P, Banchik A D, Vogel S C, Tremsin A S, Vallerga J V, McPhate J B, Lehmann E, Kockelmann W, 2012. Texture imaging of zirconium based components by total neutron cross-section experiments. *J. Nucl. Res.* **425** 218.

[6] Withers P J, Webster P J, 2001. Neutron and Synchrotron X-ray Strain Scanning. *Strain* **37** 19.

[7] Woo W, Feng Z, Wang X-L, David S A, 2011. Neutron Diffraction Measurements of Residual Stresses in Friction Stir Welding: A Review. *Sci. Tech. Welding and Joining* **16** 23.

[8] Tremsin A S, 2012. High resolution neutron counting detectors with microchannel plates and their applications in neutron radiography, diffraction and resonance absorption imaging. *Neutron News* **23** 35.

[9] Tremsin A S, Vallerga J V, McPhate J B, Siegmund O H W, 2015. Optimization of high count rate event counting detector with Microchannel Plates and quad Timepix readout. *Nucl, Instr, Meth. A* **787** 20.

[10] Wagner V, Friedrich H, Wille P, 1992. Performance of a high- tech neutron velocity selector *Phys. B: Cond. Matter* **180-181** 938.

[11] Schulz M, Boeni P. Calzada E, Mühlbauer M, Schillinger B, 2009. Energy-dependent neutron imaging with a double crystal monochromator at the ANTARES facility at FRM II. *Nucl, Instr, Meth. A* **605** 33.

[12] Kockelmann W, Zhang S Y, Kelleher J F, Nightingale J B, Burca G, James J A, 2013. IMAT–a new imaging and diffraction instrument at ISIS. *Phys. Procedia* **43** 100.

[13] Harada M, Oikawa K, KasugaiY, Maekawa F, 2011. Shielding Design of a Neutron Beam Line "NOBORU" at JSNS/J-PARC. *Prog. Nucl. Sci. Technol.* **1** 94.

[14] Santisteban J R, Edwards L, Steuwer A, Withers P J, 2001. Time-of-flight neutron transmission diffraction. *J. Appl. Crystal.* **34** 289.

[15] Steuwer A, Santisteban J R, Withers P J, Edwards L, Fitzpatrick M, 2003. In situ determination of stresses from time-of-flight neutron transmission spectra. *J. Appl. Crystal.* **36** 1159.

[16] Tremsin A S, McPhate J B, Steuwer A, Kockelmann W, Paradowska A M, Kelleher J F, Vallerga J V, Siegmund O H W, Feller W B, 2012. High-resolution strain mapping through time-of-flight neutron transmission diffraction with a microchannel plate neutron counting detector. *Strain* **48** 296.
[17] Vogel, S. 2000 *Proc. Math.-Naturwissenschaftliche Fak.* Uni Kiel Kiel. 174.
[18] Santisteban, J R, Edwards L, Fizpatrick M E, Steuwer A, Withers P J, 2002. Engineering applications of Bragg-edge neutron transmission. *App. Phys. A* **74** S1433.
[19] Mishra R S, Ma Z Y, 2005. Friction stir welding and processing. *Mater. Sci. Engn. Reports* **50** 1.
[20] Schaeffler A L, 1949. Constitution diagram for stainless steel weld metal. Metal Prog. **56(11)** 680.
[21] Jun T S, Hofmann F, Hofmann M, Korsunsky A M, 2009, Residual stress characterization in 12%-Cr steel friction stir welds by neutron diffraction. *J. Strain Anal.* **47** 204.
[22] Sato H, Kamiyama T, Kiyanagi Y, 2011. A Rietveld-Type Analysis Code for Pulsed Neutron Bragg-Edge Transmission Imaging and Quantitative Evaluation of Texture and Microstructure of a Welded α-Iron Plate. *Materials. Transactions* **52** 1294.

The New Facilities for Neutron Radiography at the LVR-15 Reactor

J Soltes[1,2], L Viererbl[1], J Vacik[3], I Tomandl[3], F Krejci[4] and J Jakubek[4]

[1] Research Centre Rez Ltd., Husinec – Řež 130, 25068 Řež, Czech Republic
[2] Faculty of Nuclear Sciences and Physical Engineering, CTU in Prague, Břehová 7 115 19 Praha 1, Czech Republic
[3] Nuclear Physics Institute of The Czech Academy of Sciences, Řež 130, 25068 Řež, Czech Republic
[4] Institute of Experimental and Applied Physics, CTU in Prague, Horská 3a/22 128 00 Praha 2, Czech Republic

E-mail: jaroslav.soltes@cvrez.cz

Abstract. Neutron radiography is an imaging method often used at research reactor sites. Back in 2011 a project was started with the goal to build a neutron radiography facility at the site of the LVR-15 research reactor in Rez, Czech Republic. In the scope of the project two horizontal channels were adapted for the needs of neutron radiography. This comprises the HC1 channel which offers an intense thermal neutron beam with a diameter of 10 cm, which can be used for imaging of larger samples, and the HC3 channel which beam is restricted just to 4x80 mm^2, but is highly thermalized, collimated and reduced from gamma background, thus capable of providing better radiograph resolution. Both facilities are equipped with newest Timepix based detectors, with thin ^{6}LiF converters for neutron detection capable of delivering high resolution. Both facilities offer a unique opportunity for non-destructive testing in the Czech region. In 2015 both facilities were put into test operation and several radiographs were acquired, which are presented in the following text.

1. Introduction

The LVR-15 is a multipurpose tank-type light water moderated research reactor operated by the Research Centre Rez near Prague, Czech Republic. The reactor is operated at a maximum thermal power level of 10 MW. The utilization of the reactor is wide, covering material research, industrial and medical radioisotope production, neutron transmutation doping of silicon, neutron activation analysis, experimental boron neutron capture therapy, neutron diffraction experiments, prompt gamma activation analysis, etc. The reactor is equipped with nine horizontal channels and one thermal column providing ten neutron beams for further use (Figure 1). In 2011 in scope of a research project, two of the horizontal channels were chosen to be adapted for neutron radiography facilities. One radiography facility for larger samples was established at the beam exit of the formerly unused horizontal channel HC1. Additionally a second facility for microradiography (radiography with high resolution) of small samples was built on the HC3 channel.

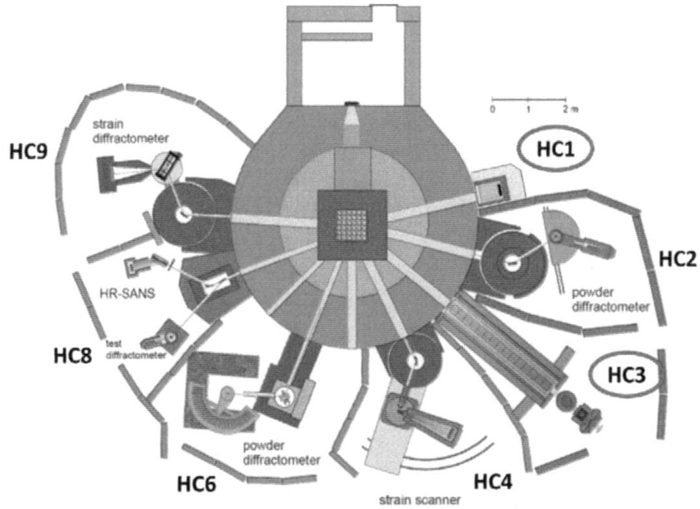

Figure 1. Horizontal channel layout at the LVR-15 research reactor. [1]

2. Neutron radiography on the HC1 channel

Before 2011 the HC1 horizontal channel was unused for many years. For the purposes of the neutron radiography facility several changes had to be done to the channel [2]. One of the key conditions for operating an effective radiography facility is the delivery of a high intensity, parallel, homogeneous and collimated thermal neutron beam at the sample location. Additionally the intensity of fast neutrons has to be kept as low as possible as the fast neutrons may damage the detectors used for neutron imaging. As the spectrum in the empty horizontal channel roughly copies the spectrum in the reactor core, which has a high ratio of fast neutrons, neutron filter components have to be installed inside the channel in order to achieve desired beam parameters. The radial channel HC1 was, therefore, equipped with a silicon single crystal filter-collimator. After the filter instalment it offers a neutron beam with a diameter of 10 cm with an intensity of thermal neutrons of 2×10^8 $cm^{-2}.s^{-1}$ [3]. This channel is designed for the radiography of larger samples (of several cm in size) at the costs of a lower resolution of 100 μm. A shielded irradiation box was installed at the beam exit to house the samples while irradiation. The box is equipped with a positioning device enabling the movement and rotation of the sample (Figure 2).

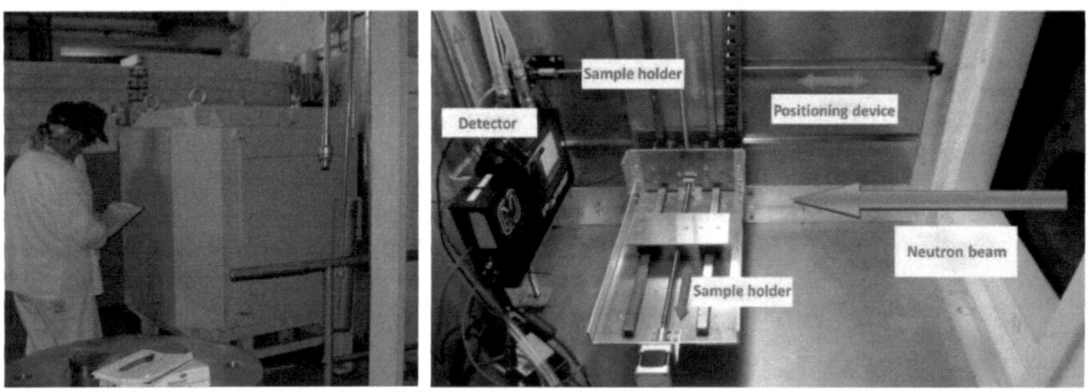

Figure 2. The irradiation chamber of the HC1 channel (left) and its equipment (neutron detector and sample holder) used for neutron radiography.

Several test samples have been recently measured. A first series of radiography measurements was realized before the Timepix detector instalment using a FUJIFILM neutron imaging plate [4]. Radiographs of steel bearings, a 5 cm thick concrete sample and steel samples with cracks are shown in Figure 3. After the first successful series of measurements a Timepix based detector was used for later tests. Several radiographs were gathered using the detector. Two of them showing an amulet and a rose are shown in Figure 4.

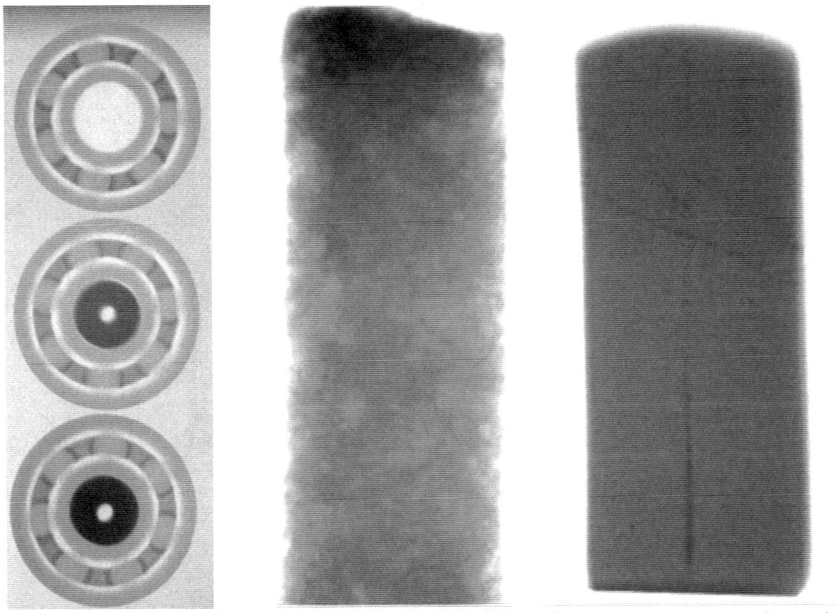

Figure 3. Radiographs from the HC1 facility (ball bearings, concrete sample, stainless steel sample with crack). A neutron imaging plate [4] was used as a detector.

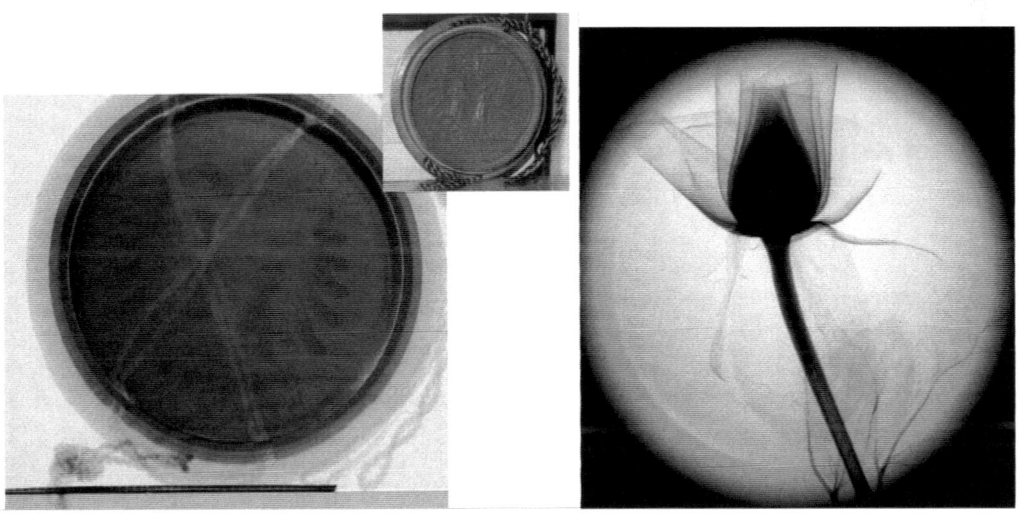

Figure 4. Radiographs of an ancient amulet and a rose acquired from the HC1 radiography facility with a Timepix based neutron detector.

3. Neutron microradiography on the HC3 channel

A second specific neutron radiography facility was built on the channel HC3. In contrast to the HC1 facility its resolution can be up to 10 times higher, however, at the cost of small neutron beam dimensions. The facility uses an old mirror-type neutron guide originally designed for other experimental methods and devices such as prompt gamma activation analysis or neutron depth profiling. The guide itself is composed of 15 coplanar Ni mirror section with the dimensions of 90 x 4 x 380 mm. The total length of the guide is 570 cm. At the exit a beam reduction shielding system containing ^{6}LiF was installed which enables the beam reduction from 4 x 80 mm to any lower value (Figure 5). This way unnecessary sample and construction parts irradiation can be lowered which leads to lowering the gamma background. The guide provides a pure collimated thermal neutron beam with an intensity of 110^{7} $cm^{-2}.s^{-1}$, a high Cd ratio ~ 10^{5} and a very low gamma background. These beam parameters enable a great radiograph resolution of theoretically 10 μm in a contact geometry of the examined sample. The irradiation chamber installed at the beam-exit can be equipped with a rotational sample holder which enable tomographic reconstruction of the examined samples. The results of the first neutron tomography tests carried out at the channel are shown in Figure 5.

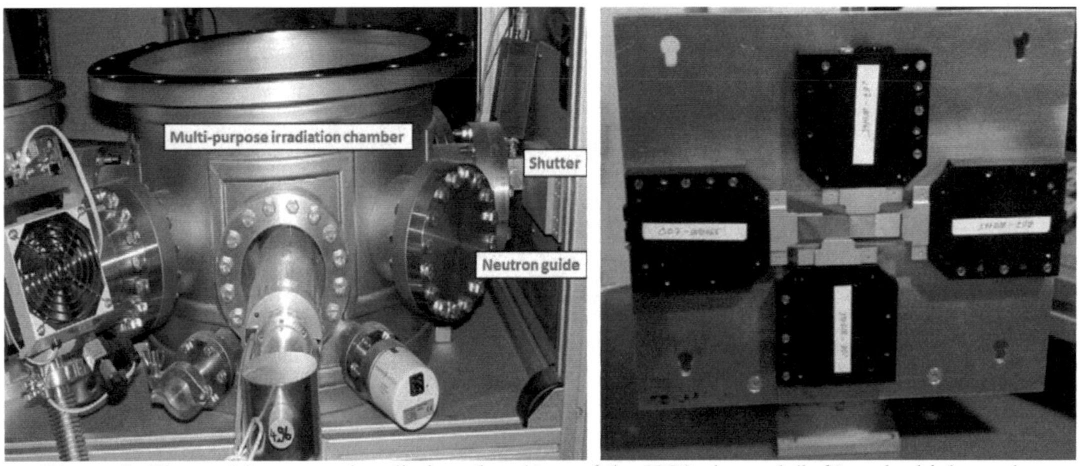

Figure 5. The multi-purpose irradiation chambers of the HC3 channel (left) and which can be equipped with a Timepix based neutron detector (right) for neutron radiography of small samples with better resolution.

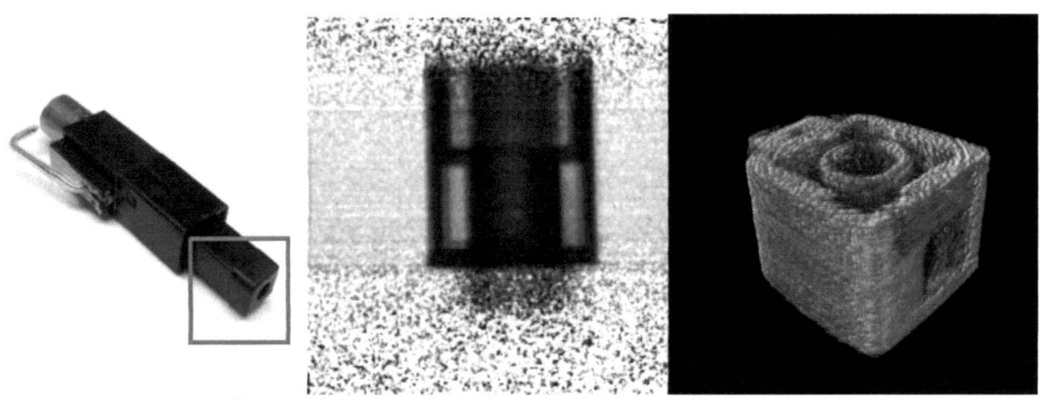

Figure 6. Neutron radiography and tomography reconstruction of a part of a piezoelectric spark generator. The tomograph was created out of 180 projections.

4. Detectors

Both facilities are equipped with Timepix tile-based detectors adapted for neutron imaging by applying a ^{6}LiF conversion layer on the Timepix tile surface.

The facility of the HC1 channel is equipped by large area WIDEPIX$_{4x5}$ [6] detector consisting of 20 Timepix tiles. Its total detection area is 71x57 mm^2 which equals 1.3 mega pixels. The detector detection efficiency for thermal neutrons is about 3 - 4 %. The device is connected to a DAQ computer via a USB 2.0 interface allowing a speed of 2 fps for frames of 1280x1024 pixels. The detector is attached to a positioning device which enables the sample motion and rotation. Both the detector and the device are shown in Figure 7.

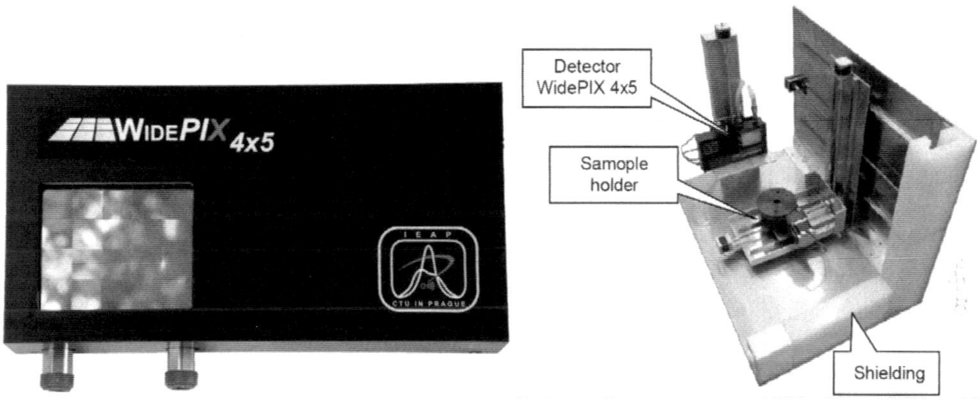

Figure 7. The WIDEPIX4x5 neutron camera (left) and its instalment on the rotational positioning device.

The micro-radiography facility at the HC3 channel is equipped by a modular detector MODUPIX [7] consisting of 4 edgeless Timepix tiles. The modular system allows the assembling of various 2D and 3D detector configurations (Figure 8) while each tile is read-out separately in a parallel mode. The USB 2.0 based FITpix 3.0 interface offers high read-out speed up to 850 frames per second. The detector thermal neutron detection efficiency can reach up to 10 % in the 3D stack mode.

Figure 8. The MODUPIX detector in various detection configurations.

5. Conclusion

Since 2011 two neutron radiography facilities have been built at the site of the LVR-15 reactor in Rez. The first test have shown promising results. Both facilities offer a unique opportunity for non-destructive sample testing in the Czech region. By the beginning of 2015 the facilities has been put to test operation. Through the year more test measurement will be carried out on both facilities leading to their potential minor upgrades. Full operation of both facilities is expected from the beginning of 2016. However, further improvements leading to the improvement of beam parameters (instalment of new neutron guides and collimators) can be expected in near future. As both facilities are equipped with rotational positioning devices neutron tomography is also expected to be introduced.

Acknowledgments
This work was performed within the scope of the research project ALFA No. TA01010237 supported by the Technology Agency of the Czech Republic

References

[1] Nuclear Physics Institute of the ASCR **2012** Neutron Physics Laboratory (NPL) *http://canam.ujf.cas.cz/en/laboratory/npl-neutron-physics-laboratory*
[2] Viererbl L, Soltes J, Lahodova Z, Kostal M, Vins M **2012** Horizontal Channel for Neutron Radiography and Tomography in LVR-15 Research Reactor *Proceedings of RRFM/IGORR 2012*
[3] Soltes J, Viererbl L, Lahodova Z, Koleska M, Vins M **2015** Thermal Neutron Filter Design for the Neutron Radiography Facility at the LVR-15 Reactor *presented at ANIMMA 2015*
[4] FUJIFILM **2000** Imaging plates *http://www.buero-analytik-winden.de/app/download/13602912/Imaging+plates+ausfuhrliche+Beschreibung.pdf*
[5] Vavrik D, Jakubek J, Pospisil S, Vacik J **2011** Neutron imaging with micrometric spatial resolution *Proceedings NSS/MIC* 1285
[6] Jakubek J, Zemlicka J, Jakubek M, Kaestner A, Krejci F, Koester U, Soukup P, Turecek D, Vacik J, Vavrik D **2014** Large area pixel detector WidePIX for neutron radiography *presented at WCNR-10*
[7] Vavrik D, Jakubek J, Kaestner A, Krejci F, Turecek D, Zemlicka J **2014** Modular Pixelated Detector System for Neutron Imaging with Micrometric Resolution *presented at WCNR-10*

Characterization of mono-ethylene-glycol based industrial polyurethanes samples by fast-neutron radiography and neutron tomography

Massimo Rogante[1] **and** Stefan Söllradl[2]

[1] Rogante Engineering Office, Contrada San Michele n. 61, 62012 Civitanova Marche, Italy
[2] Forschungs-Neutronenquelle Heinz Maier-Leibnitz (FRM II), Lichtenbergstrasse, 85747 Garching, Germany

main@roganteengineering.it

Abstract. A complicated structural organization of polyurethanes may have a strong influence on the materials functional properties. Under particular conditions such as mechanical and thermal loading and aging, it leads to the material degradation, even in fresh-prepared bulk polymers and especially if defects are present in the material. Unwanted bubbles can be observed, which form during the expansion of the mixture during its chemical reaction and remain present in the final product. These macro-, micro- and nano-bubbles influence the material's performance. In this work, neutron radiography and tomography have been adopted to characterize at a macro-scale level the bulk of commercially available polyurethane samples, obtained from dissimilar-mixture ratios with different densities and branching levels as well as from different zones of the production mould. The characterisation allowed an estimation of the different dense materials - as they are used, e.g., in soles of shoes - as well as the invisible defects like pores and cracks, responsible for the materials fracture by mechanical loading. The obtained information are expected to be useful for various industrial sectors such as automotive and footwear industry. It will be completed by applying SANS, which has already proved to characterize the microstructure of the bulk-polymer with respect to nano-pores, micro-cracks and their arrangement in the polymer matrix.

1. Introduction

Polyurethanes are polymers containing the urethane linkage in their backbone chain and are made by reacting di-isocyanates with di-alcohols. Since polyurethane chains form easily from the hydrogen bond, these polymers can be very crystalline. This is one reason, why they are frequently used to compose block copolymers (having the properties of thermoplastic elastomers) with soft rubbery polymers. This material is mostly adopted in different industrial sectors (e.g. footwear, automotive, interior design and building). In the footwear industry, e.g. they are used to produce micro-cellular shoe soles. Typical shoe-sole systems utilize a single surfactant. Other novel surfactants are actually under consideration to reduce the soles density while maintaining its physical properties.

The combination of two diverse surfactant chemistries has recently improved the dimensional stability, surface characteristics and, in general, various macroscopic characteristics and physical properties of the final product. A complex structural organization of polyurethanes, including crystalline and amorphous domains, may have a strong influence on the materials functional properties. It leads under particular conditions - such as mechanical and thermal loading, or aging - to the material degradation. This may happen also in the case of fresh-prepared bulk polymers and, in particular, if defects are present in the material [1].

Unwanted bubbles can be detected, which form during the expansion of the mixture while its chemical reaction and remain present in the final product. These macro- (i.e. having size from 0.5 mm and larger), micro- (i.e. having sizes of 1-500 μm) and nano-bubbles (i.e. having sizes of 1-1000 nm) influence the performance of the polyurethanes. Variables as mixture composition, injection speed, injection power, melt temperature, and melt viscosity influence the appearance of those bubbles (e.g. size distribution, interface area, concentration, and volume fraction). In particular, the interface area plays a key role in the polymer material's strength and durability. The spatial distribution of the mentioned bubbles in the bulk material is not homogeneous and depends on the different zones of the mould. A change of one parameter has influences on other production parameters and the knowledge of those parameters is fundamental for the understanding and improvement of the materials performance.

Thermal neutron radiography and fast neutron gauging measurements have been already adopted to evaluate the feasibility of detecting macro voids in polyurethane products and were used to determine its uniformity[2]. A very high citation index exists concerning polymers (for instance, in *Macromolecules, Polymer Content, Physical Review*), but no investigations related to neutron imaging were performed yet. Refs. [1-7] can be considered, e.g. related to neutron beam investigations of different types of polymers.

In the frame of our study, representative samples were selected for the investigation with respect to their attenuation coefficient and inner defects such as bubbles at the instrument NECTAR [8, 9] at FRM-II. Three cylindrical PU samples were investigated and the attenuation coefficients for fission neutrons (1.8 MeV) determined with respect to the density of each sample. To confirm the determined attenuation components from transmission experiments, computed tomography was performed on those samples. In addition, computed tomography was performed at the example of a defective safety shoe to study, down to which size defects such as cavities can be identified in bulk PU materials with fission neutrons. Due to the low interaction of fission neutrons with matter in general, while it has a high interaction with elements like hydrogen it is especially suitable for the investigation of materials like polyurethane, especially if hydrogen-rich samples exceed thicknesses of more than 5 cm compared to cold- and thermal neutron radiography.

2. Materials and methods

Different commercially available polyurethane samples- manufactured with varied known reactant mixture ratios with dissimilar branching levels and obtained from different zones of the production mould - have been investigated. These specimens are made of the following two main polyester based polyurethanes systems: A) for compact parts such as shoe treads; B) for expanded parts such as intermediate shoe soles. The chemical composition is the following: polyol polyester; 4.4-diphenylmethane diisocyanate; tertiary aliphatic amines, and chain extenders. Table 1 reports the main macroscopic (mechanical) characteristics of these polyurethanes.

Tab. 1. Main macroscopic characteristics of the investigated polyurethanes.

Characteristics	**A**	**B**
density (kg/m^3)	1100	420-450
hardness (Shore A)	62-65	46-50
traction (MPa)	15.0	5.7
elongation (%)	600	390

bending at +20°C (kcycles)	30	30
A) for compact (high-density) systems; B) for expanded (low-density) systems		

Fast-neutron radiography and -tomography have been adopted with the main goal to benchmark the capability to identify macro-scale internal bubble structures in bulk material. Fast neutrons with a mean energy of 1.8 MeV were used and the investigation was carried out with a collimated beam (circular collimator of 10 mm) [9], resulting in an $L/D = 233$. A 2.4 mm PP/ZnS scintillator was used with 30% of ZnS to convert the fast neutrons to light. As detector system an Andor DV434-BV CCD camera was used, maintained at a temperature of -50°C with 1024×1024 pixel^2 and a field of view of 317x317 mm^2.

Three homogeneous samples with cylindrical shape (diameters between 50 mm and 60 mm) of PU with different densities were investigated my means of radiography and tomography (see Figure 1).

Fig. 1. PU samples n. 1, 2 and 3 investigated by fast-neutron radiography and neutron tomography.

The density was determined by measuring weight of the samples and the displaced volume. Transmission images of a fast neutron beam through the samples were acquired and the attenuation coefficients determined by measuring the intensity profile (see Figures 1 and 2) over 6 rows (corresponding to the inner part of the rectangles in Fig. 1a and 1b) and averaging this intensity values in each column using evaluation scripts developed in Python [10]. Six rows were selected to obtain the best compromise in noise reduction of a single transmission image with maintaining the representative diameter of the conically shaped sample (see Fig. 1). The minimum transmission I/I_0 was determined from the measured profiles and according to the following equation derived from Lambert-Beer-Law:

$$\frac{\mu}{\rho} = -\frac{\ln\left(\frac{I}{I_0}\right)}{\rho d} \quad (1)$$

where μ is the attenuation coefficient, ϱ is the density, and d is the transmission of a single beam of neutrons through the sample, the mass-attenuation coefficient was determined.

In addition, computed tomography was performed to determine the average-attenuation coefficient for polyurethane with the fission neutron beam available at NECTAR. The tomography was executed with 357 projections, acquired according to the "golden ratio" approach [11] according to equation:

$$\varphi = \left(2\pi\, i\, \frac{1+\sqrt{5}}{2}\right) \bmod (2\pi) \quad (2)$$

where φ is the angle of the i-th projection acquired. The acquisition was stopped after a good compromise between visible details (e.g. bubbles) and acquisition time was reached after regular reconstructions during the data acquisition. All projections are reconstructed using the software Octopus [12] and evaluated using Python [10] scripts. In addition, the tomography of a defective safety shoe was acquired using the golden ratio approach with 325 projections as details in the size of two voxels were visible. The projections were reconstructed using the software Octopus [12] for the reconstruction and VG Studio Max [13] for visualization.

3. Results

The transmission through polyurethane cylinders with different densities was investigated. Fig. 2 shows the radiographs of the investigated cylinders, where Fig. 2a represents the images of Sample 1 (low-density PU) and Sample 3 (high-density PU) in direct comparison and Fig. 2b shows the radiograph of Sample 2 (high-density PU). The black rectangles in the images indicate the origin of the transmission profiles as plotted in Fig. 3. Based on the known thickness of the samples as well as the determined transmission, the attenuation coefficients were calculated depending on the Lambert-Beer Law for attenuation of radiation in matter (Eq. 1).

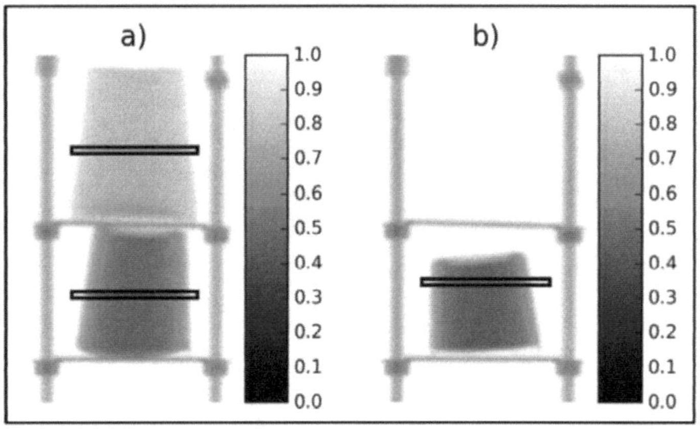

Fig. 2. The radiographs of Sample 1 (upper sample, low density) and Sample 3 (lower sample, higher density) in Fig. 2a and of Sample 2 (higher density) in Fig. 2b). The inside area of the rectangles indicates the region, where the intensity profiles were determined.

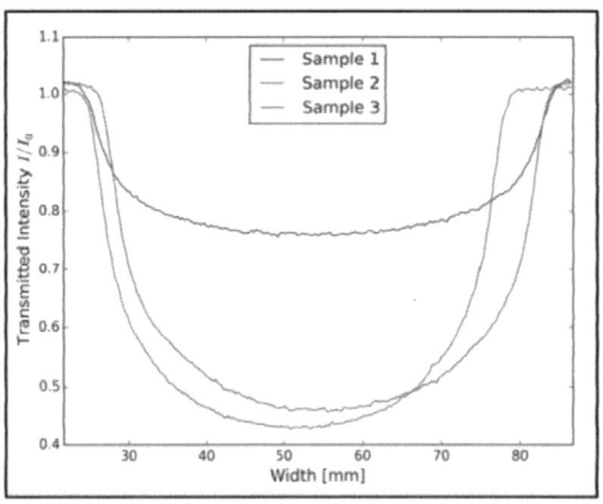

Fig. 3. The transmission profiles averaged over 6 rows of pixels per column. Each profile corresponds to the indicated section from rectangular sections of Fig. 2.

From the obtained data, a clear dependence of the transmission through low-density PU materials (blue line) and high-density PU material (green and red lines) is shown, even if the basic material is the same and the concentration of gas inside the samples determines its density. While a transmission of 76.6% was achieved in Sample 1 though a material thickness of 57 mm, only 42.8% transmission was measured through Sample 2 with a thickness of 50 mm, and 46% in Sample 3 with a thickness of 54 mm.

Based on those data, the attenuation coefficient μ was determined from the radiographs as a function of the investigated material. For all three samples μ was normalized with the density of the material and the mass attenuation coefficient μ/ρ determined as shown in Table 2.

Tab. 2. Measured properties of the cylindrical samples investigated by fast-neutron radiography.

•	Attenuation coefficient (tomograph) μ [1/cm]	Density ϱ [g/cm^3]	Mass attenuation coefficient (tomograph) μ/ϱ [cm^2/g]	Mass attenuation coefficient (radiograph) μ/ϱ [cm^2/g]
Sample 1 (low-density)	0.05 (1)	0.31 (1)	0.132 (3)	0.16 (6)
Sample 2 (high-density)	0.16 (1)	0.96 (1)	0.145 (1)	0.18 (7)
Sample 3 (high-density)	0.14 (1)	0.89 (1)	0.134 (1)	0.16 (6)

The values, determined from the computed tomography of the test samples are shown in Fig. 4. Fig. 4(a-c) illustrates the reconstructed shapes of the investigated PU samples with their corresponding attenuation coefficients. In addition, the histograms corresponding to the inner sections of the white rectangles are plotted in Fig. 4d. The mean attenuation coefficient of each sample and the corresponding mass attenuation coefficient were determined, which were well in accordance with the values determined by the radiograph. However, due to the better statistics of the computed tomography, the uncertainty was significantly reduced.

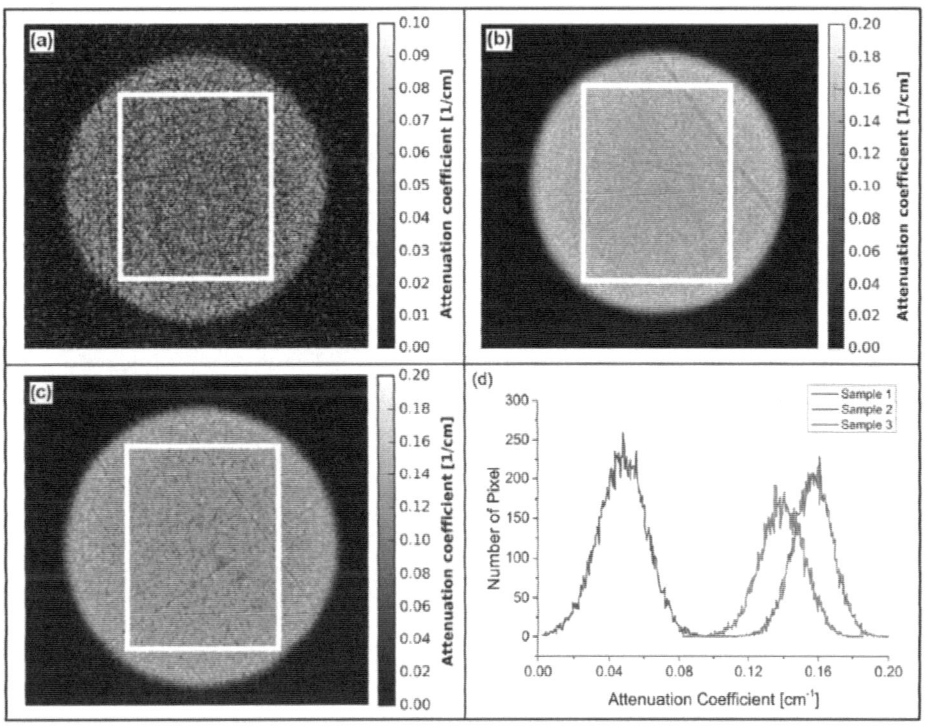

Fig. 4. Reconstructed shapes of the investigated PU samples 1 (a), 2 (b) and 3 (c) with their corresponding mass attenuation coefficients. The rectangles indicates the area, which data are plotted in the histogram shown in (d).

Fig. 5 shows a photograph of the investigated safety shoe in comparison with the reconstructed volume of the computed tomography. Two non-planar sections of the that volume were clipped along the red lines (Fig. 5d) and are illustrated in Fig. 5b and Fig. 5c. The red circles indicate larger defects (more than 2 voxels at a voxel size of 0.29 mm along each edge), observed from the reconstructed volume. Those were confirmed destructively after the experiment by opening the corresponding sections of the sole as shown in Fig. 5a. Based on the determined correlation between attenuation coefficient and density of the samples illustrated in Fig. 4, the density of the different sole components was found to be 0.06 cm^{-1} (0.35 g/cm^3) for the yellowish part and 0.13 cm^{-1} (0.78 g/cm^3) for the greenish part, respectively.

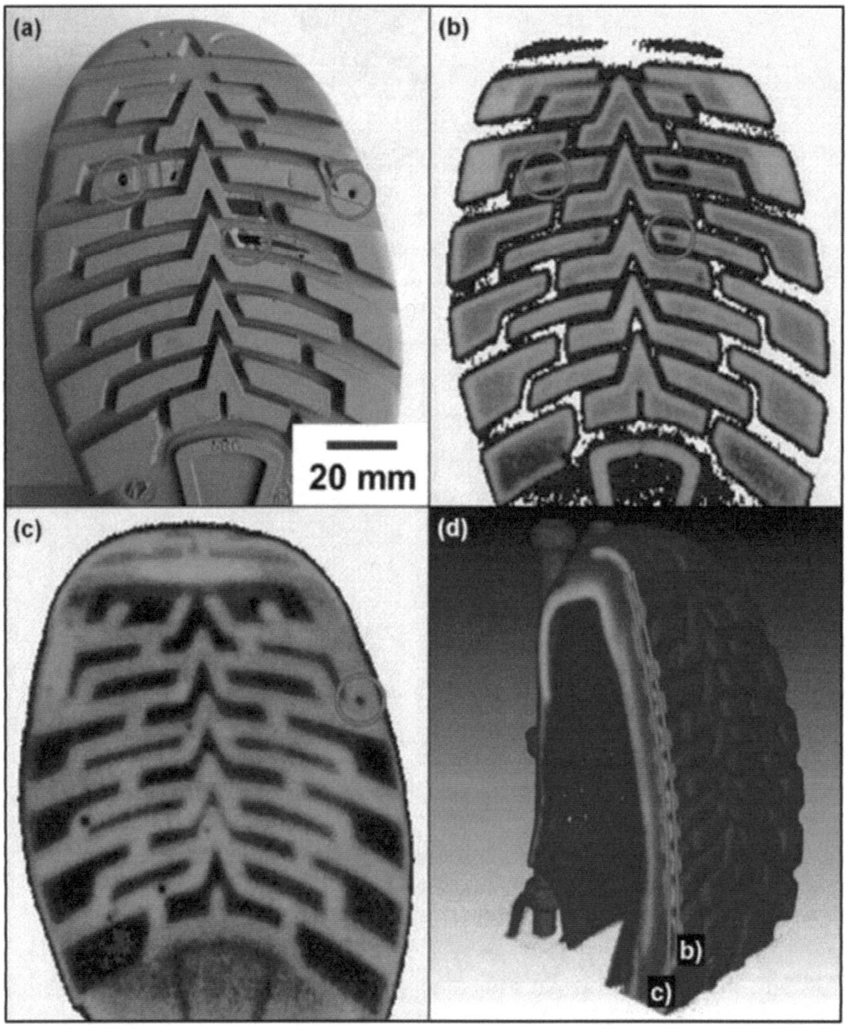

Fig. 5. Photograph of the cut shoe successive to the investigation (physical cut sections to confirm the results) with selected defective sections marked with red circles (a). False colour image of the different attenuation coefficients, cut along a non-planar surface of the sole onto the intersection between the high-density (green) and low-density (yellow) material. Defects can clearly be identified as black sections in the tomography (b). Cut along a non-planar surface parallel to the sole into the mainly low-density part of the sole in a deeper section with defects shown (c). Overview (d) of the shoe with a cut perpendicular to the surfaces illustrated in b) (right line) and c) (left line).

4. Discussion and conclusions

The obtained information about the cavities will be completed by applying SANS, which allows a microstructural characterisation of the bulk-polymer structure with respect to nano-pores, micro-cracks and their arrangement (grouping) in the polymer matrix. SANS, having less Q value than SAXS (~0.005Å^{-1}, or less, if necessary) allows the investigation of objects with larger sizes compared to SAXS. In particular, SANS allows the detection of H inside the bubbles, while SAXS detects only N and C [1]. The sizes of the investigated bubbles can be different, due to the molecular orientation around the same bubbles (internal surface effect). The approximate size of bubbles that can be investigated is in the range 0.5-2 microns. In comparison, fast neutron radiography and tomography is capable of identifying macroscopic defects in the size of 0.58 mm which corresponds to the size of two neighbouring voxels at least. Thus, neutron radiography and tomography, SANS, and SAXS are in this case, complementary.

The macro-, micro- and nano-properties can be analysed and compared to macroscopic characteristics (e.g., strength, density and bending resistance). The results can give an important contribution to the attainment of the main industrial objectives, which are:

- surface improvements with the reduction or the elimination of bubbles
- dimensional stability to control and predict the functional properties, strongly depending on the size and amount of defects as well as on their total area
- to replace expensive components of the mixture with new low cost components, which allow the product to maintain its original characteristics and performances.

The obtained information are expected to be useful for various industrial sectors such as footwear and automotive industry. For the latter, the presence of macro-bubbles in parts such as polyurethane gaskets or seat upholstery may represent starting points of rupture processes, thus information on the internal distribution of bubbles can contribute to the improvement of the production process of these parts.

An industrial application of neutron radiography and tomography has been described, showing the usefulness of this neutron technique in the polymers field. In particular, diverse commercially available polyurethane samples adopted in the footwear sector were successfully investigated, obtaining information on the internal macro-porosity, which could be useful either for the materials improvement, or for the control and estimate of the functional characteristics.

4. References

[1] Rogante M and Lebedev V T 2007 *Polym. Eng. Sci.* **47** 1213
[2] Tsang F Y, Alger D M and Brugger R M 1978 *Nucl. Technol.* **37** 73
[3] Stein R S 1990 *Mol. Cryst. Liq. Cryst.* **180** 119
[4] Hammouda B 1993 *Adv. Polym. Sci.* **106** 87
[5] Higgins J and Benoit H 1994 *Neutron Scattering from Polymers* (Oxford: Clarendon Press)
[6] Scheyer Y, Levelut C, Pelous J and Durand D 1998 *Phys. Rev. B* **57** 11212
[7] Rogante M, Rosta L and Heaton M E 2013 *Meas. Sci. Technol.* **24** 105601
[8] Bücherl T, Söllradl S 2015 *JLSRF.* **1** A19
[9] Bücherl, T., Gostomski, von, C. L., Breitkreutz, H., Jungwirth, M., and Wagner, F. M 2011. *Nucl. Instr. Meth. Phys. Res. A*, **651** 86
[10] Python, Version 2.7.6 [Computer Software] (2013) Retrieved from http://www.python.org
[11] Kaestner A, Münch B, Trtik P, and Butler L 2011 Opt. Eng. **50** 123201
[12] Octopus Reconstruction, Version 8.8.4 [Computer Software] (2014), Retrieved from Inside Matters BVBA, Sylvain Van Der Guchtlaan 24, 9300 Aalst, Belgium
[13] VG Studio Max, Version 2.2.3 [Computer Software] (2013) Retrieved from Volume Graphics GmbH, Wieblinger Weg 92a, 69123 Heidelberg, Germany

Fast and simple method for Goss texture evaluation by neutron diffraction

M Kucerakova[1], K Kolařík, J Čapek, S Vratislav, L Kalvoda
Department of Solid State Engineering, Faculty of Nuclear Sciences and Physical Engineering, Czech Technical University in Prague, Trojanova 13, 120 00 Prague 2, Czech Republic

E-mail: monika.kucerakova@fjfi.cvut.cz

Abstract. Requirement of low power losses is one of the crucial demands laid on properties of electric steel sheets used in construction of various magnetic circuits. For cold-rolled grain-oriented (CRGO) Fe-3%Si sheets used in majority of power distribution transformers, the Goss texture {110}<001> is known to provide the best utility properties (low power loses, high magnetic permeability). Due to the coarse grain size of CRGO steel, neutron diffraction (ND) is dominantly used to characterize the sheets' texture in order to achieve statistically significant data. In this paper, we present a fast and simple method for characterization of Goss texture perfection level in CRGO steel sheets based on monochromatic ND. The method is tested on 8 samples differing in fabrication technology and magnetic properties. Satisfactory performance of the method and its suitability for a detail texture analyses is tested by juxtaposition of the obtained textural and the magnetic characteristics measured by Barkhausen method.

1. Introduction

Highly perfect (100) [001] crystallographic preferential orientation of grains (Goss texture) in cold-rolled grain-oriented (CRGO) Fe-3%Si steel sheets is of crucial importance for achievement of the extraordinary magnetic properties characterized nowadays by typical magnetic losses in range of several tens of W/kg and saturation magnetisation in the easy direction 1.91-1.92 T [1-3]. Still, even for the best materials produced today, there is some level of the Goss texture imperfection that can be reduced in order to achieve even better figures of the resulting magnetic properties. To do this, fine and detail characterization of the Goss texture sharpness must be performed with the angular resolution exceeding the usual limits used in standard texture analyses.

In this paper, a simple neutron diffraction method suitable for fast and highly sensitive evaluation of the ideal Goss texture perfection in large sample series is proposed and tested on 8 samples of CRGO Fe-3%Si steel sheets differing in their fabrication procedures and, consequently, resulting magnetic properties. The later were characterized by Barkhausen noise (BN) measurements, compared with the obtained texture characteristics obtained for the analysed samples and the level of correlation between both types of parameters analysed in order to verify applicability of the proposed method.

2. Experimental

[1] Dr. Monika Kucerakova, CTU in Prague, FNSPE, Brehova 7, 11519 Prague 1, Czech Republic.

2.1. Neutron diffraction texture analysis

Samples were stacked from 6 sheets of square shape, the side being 30 mm long. Total thickness of the resulting specimens was between 1.2 - 2 mm.

The diffraction patterns were collected on KSN-2 neutron diffractometer situated at one of the horizontal channels of the nuclear research reactor LVR-15 at the Nuclear Research Institute, plc. Rez, Czech Republic. The specimen was mounted in the Huber Eulerian cradle 511.5 goniometer and irradiated by the parallel beam of monochromatic neutrons with the wavelength 0.1052 nm obtained by reflection on Cu(200) single crystal. In such setting, the KSN-2 diffraction device offers good intensity and the best resolution value (δd/d) ca. 0.007 within the interval of inter-planar distances d ~ 1.0 ÷ 0.1 nm.

For every specimen, distribution of integral intensity of the (002) reflection (I^0) was recorded at nodes of the angular network (ω, χ) scanned in the range ±10° with the steps Δ ω= Δ χ=1° around the rolling direction (RD). The raw data were then corrected for background, absorption and irradiated volume changes and normalized on the (002) integral intensity of a specimen with randomly oriented crystallites

$$I^{sc}(\chi,\omega) = \frac{1}{I^r(\omega=0)}(I^0(\chi,\omega) - B^0(\chi,\omega))K(\omega) \tag{1}$$

Here $I^{sc}(\chi,\omega)$ is the normalized intensity distribution, $B^o(\chi,\omega)$ is the background intensity, $I^r(\omega = 0)$ is integral intensity of the (002) reflection for the randomly oriented powder sample measured in a symmetrical transmission position ($\omega = 0$) and corrected for the background. The correction factor $K(\omega)$ must be applied used to unify the geometric and material conditions of the measurements performed on the textured specimens and on the reference powder sample

$$K(\omega) = \left(\frac{\rho_p}{\rho_s}\right)\left(\frac{t_p}{t_s}\right)\frac{\cos(\theta+\omega)}{\cos\theta}\exp\left[\frac{\mu_s t_s}{\cos(\theta+\omega)} - \frac{\mu_p t_p}{\cos(\theta)}\right] \tag{2}$$

Rel. (2) is derived for specimens of a cuboid shape, irradiated by the incident beam through only one of the faces. The latter assumption can be easily fulfilled in our case of the small (ω, χ) scanning range. The quantities ρ and t denote the mass density and thickness of the textured (subscript *s*) and the powder sample (p) respectively; Θ is the Bragg angle of the (002) reflection maximum.

The relative percentile volume of crystallites with the [001] crystallographic direction within the angular surroundings Ω of RD is then

$$V^{\%}_{[001]}(\Omega) = j(001)\frac{100}{4\pi}\oiint_{(\Omega)} I^{sc}(\alpha,\beta)\sin\alpha\, d\alpha\, d\beta \tag{3}$$

Here α is the tilt angle between *RD* and the scattering vector, β is the azimuthal angle corresponding to the rotation of *RD* around the scattering vector, and j(001) is the multiplicity of the {001} planes. Within sufficiently narrow angular surroundings of RD, Rel. (3) can be approximated by

$$V^{\%}_{[001]}(\Omega') \cong j(001)\frac{100}{4\pi}\oiint_{(\Omega')} I^{sc}(\chi,\omega)\, d\chi\, d\omega \tag{4}$$

In case of $\alpha \leq 10^o$ and $0 \leq \beta \leq 360^o$, the difference between Ω and Ω' values is less than 6×10^{-4} rad. Rel. (4) was applied to calculate the volume fractions summarized in Table 1.

2.2. Barkhausen noise analysis

Measurements of the Barkhausen effect [4-7] and hysteresis curve characteristics were performed on single sheets with aid of the magneto-elastic analyser Rollscan 300 (Stresstech) equipped with the software MicroScan 600-1 and the standard probe S1-138-15-0. Coercive force Hc, remanence Br, relative permeability μ_r, in-plane angular distribution of the root mean square (rms) of the BN intensity and the angular average of the latter (further referred as RMS) were recorded at four randomly selected places on the tested sheet surface. The final data were obtained as the statistical average of the individual measurements. A sinusoidal modulation of the magnetization field intensity was applied with the frequency 45 Hz and the magnetic probe driving (peak-to-peak) voltage 12.5 V.

3. Results and discussion

The intensity distributions $I^{sc}(\chi, \omega)$ obtained from the experimental data in accordance with Rel. (1) are shown in Figure 1. It is apparent that significant differences exist in character of the distributions concerning both intensity/broadness of the present maxima as well as declination of these from the RD. Integral quantitative representation of the distributions (neglecting such important features as sharpness of the maxima) is provided by the volume fractions presented in the first row of Table 1.

Table 1. Percentile volume fractions $V^{\%}_{[001]}$ of the crystallites having [001] direction oriented within ±10° (χ, ω) angular surroundings of RD.

Sample	S1	S2	S3	S4	S5	S6	S7	S8
$V^{\%}_{[001]}$	49.8	45	38.7	16.3	39.2	27.8	29.2	24.6
Hc (kA/m)	119,5	71,5	71,5	47,4	100,9	70,1	52,8	58,1
Br (mT)	879	639,3	405,4	243,1	589,3	342,8	193,8	178,5
μ_r	8432	10156	6285	5349	6074	5288	4030	3277
RMS	161,5	222,1	136,9	127,8	138,4	111	95,4	81,3

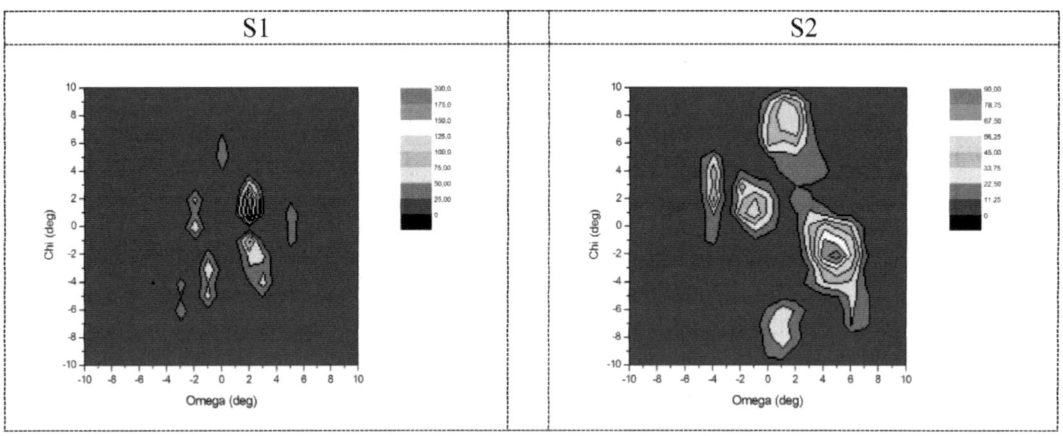

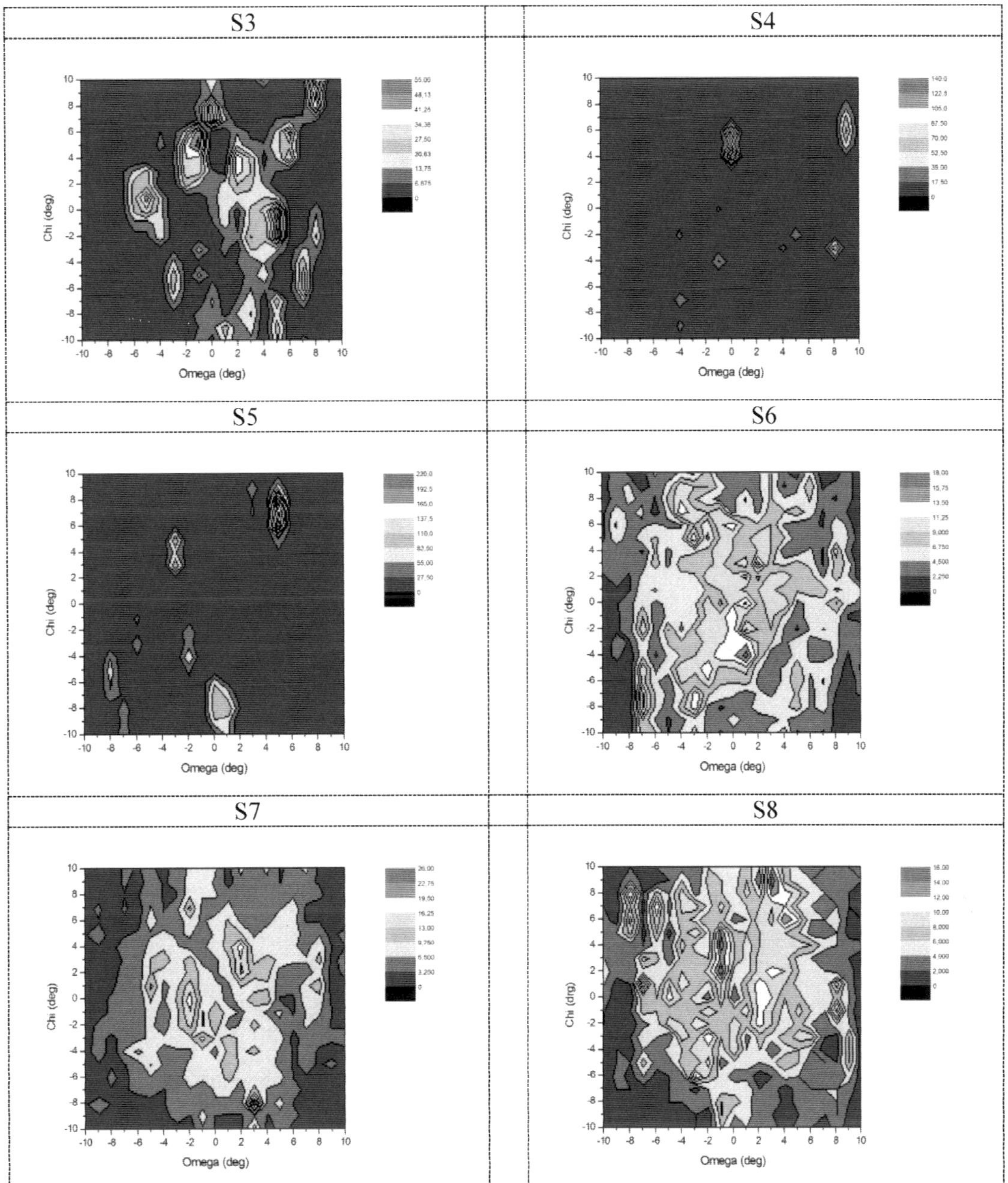

Figure 1. Distribution of [001] direction around rolling direction (RD).

Experimental values of the tested magnetic parameters (Hc, Br, μr, RMS) are summarized in the second part of Table 1. Figure 2 then shows the polar plot of the RMS intensity of BN obtained for the tested samples.

In order to discuss the relationship between the textural and magnetic characteristics, we started with determination of the correlation coefficients linking together the grain volume fractions $V^{\%}_{[001]}$ and the magnetic parameters. The value 0.81, 0.89, 0.77 and 0.68 was obtained for Hc, Br, μ_r and RMS, respectively. Thus, the remanence Br shows the most remarkable overall correlation with the texture

order of the samples. Moreover, some more delicate tendencies can be further noticed. The sample S1 shows the maximal $V^{\%}_{[001]}$ value as well as the highest Hc and Br figures. But, it is not superior in its magnetic permittivity and RMS. These are maximal for sample S2 characterized by a slightly lower $V^{\%}_{[001]}$, but higher sharpness of the texture (c.f. Figure 1). The latter findings seems to be in sound with the expectation that the Bloch walls jumps responsible for the BN signal are easier in case of when the crystallites orientation distribution is more uniform and narrow.
The minimal value of $V^{\%}_{[001]}$ volume fraction provides the sample S4. The latter possess the minimal coercive force value, but the values of Br, μr and RMS are still higher than those observed for samples S6-S8, showing all larger $V^{\%}_{[001]}$ values. Careful look on the detail intensity distributions in Figure 1 reveals that the grain distribution of S4 is apparently sharper than the distributions observed for S6-S8. Thus, the detail sharpness of the grain distribution seems to affect differently the different magnetic characteristics, promoting likely higher μ_r and RMS values and, following the curves in Figure 2 affecting in the same way the angular distributions of the RMS BN intensity, too.

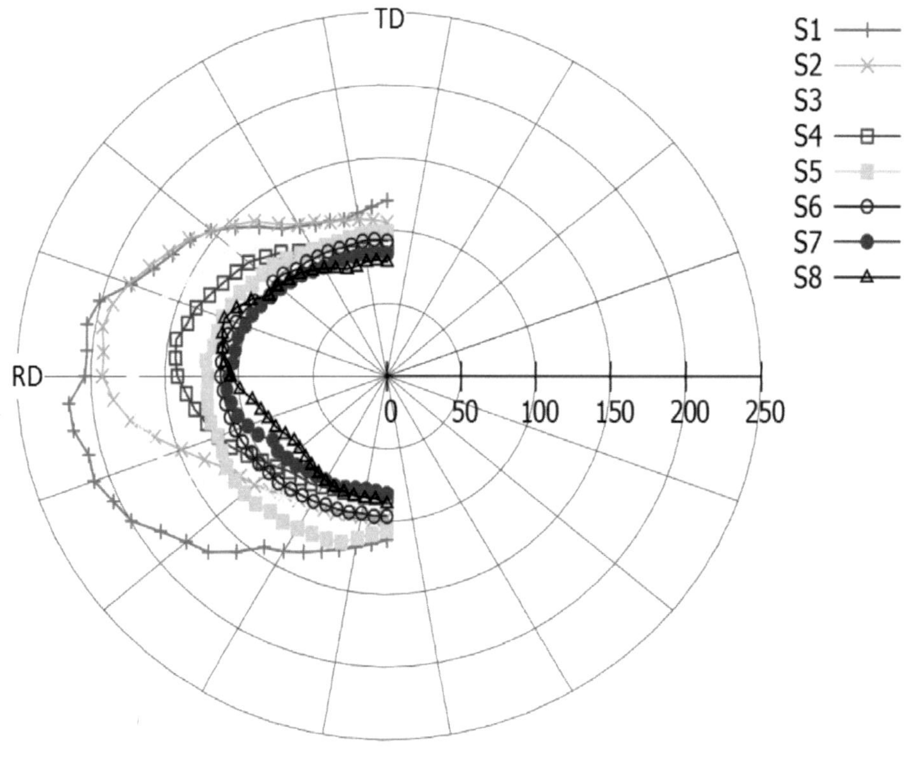

Figure 2. Polar plot of the RMS intensity of BN observed on the tested samples S1 - S8; only the angularly asymmetric part of the signal is shown.

4. Conclusions

Simple method suitable for detailed texture analyses of CRGO Fe3%Si steel sheets is proposed and tested on 8 samples significantly differing in the Goss texture perfection. Comparison of the texture results with the magnetic properties characterized by Barhausen method confirmed that the method provides results that can be well correlated with the main magnetic properties of CRGO sheets. Moreover, combination of the integral view of the investigated grain distribution with analysis of its local variations and sharpness seems to provide a valuable tool for investigation of special relations between different features of the texture distribution and the selected magnetic properties.

5. References

[1] Shimizu R, Harase J, Dingley D J 1990 *Acta Metallurgica et Materialia* **38** (6) 973-978
[2] Lin P, Palumbo G, Harase J, Aust K T 1996 *Acta Materialia* **44** (12), 4677–4683
[3] Harase J, Shimizu R, Dingley D J 1991 *Acta Metallurgica et Materialia* **39** (5) 763-770
[4] Dhar A, Jagadish C, Atherton D L 1992 *Mater Eval* **50** 1139-1141
[5] Krause T W, Clapham L, Pattantyus A, Atherton D L 1996 *J Appl Phys* **79** (8) 4242-4252
[6] Krause T W, Clapham L, Atherton D L 1994 *J Appl Phys* **75** (12) 7983-7988
[7] Dhar A, Clapham L, Atherton D L 2001 *NDT & E International* **34** (8) 507-514

Acknowledgement

This work was supported by the Grant Agency of the Czech Technical University in Prague, grant No. SGS13/219/OHK4/3T/14 and by the Czech Science Foundation, grant No. GACR 14-36566G.

Journal of Physics: Conference Series **746** (2016) 012044 doi:10.1088/1742-6596/746/1/012044

Line profile analyses of a martensitic steel during continuous and stepwise tensile deformations

T Kawasaki[1], **S Harjo**[1], **W Gong**[1] **and K Aizawa**[1]
[1] J-PARC Center, Japan Atomic Energy Agency, Tokai-mura, Naka-gun, Ibaraki 319-1195, Japan

E-Mail: takuro.kawasaki@j-parc.jp

Abstract. Dislocation characteristics in an as-quenched 22SiMn2TiB martensitic steel during tensile deformation were monitored by *in-situ* time-of-flight neutron diffraction combined with the Convolutional Multiple Whole Profile fitting analysis. Two loading conditions, continuous and stepwise followed by unloading, were adopted in the experiments. The diffraction patterns both in the loading (axial) and the transversal directions were measured simultaneously. The dislocation densities obtained from the experiments behaved differently in two loading conditions and in two measured directions, respectively. The different behaviour was mainly due to the increase of intergranular strains with the increase of deformation, and the profiles measured in the axial direction in the loading condition of stepwise followed by unloading gave most reliable dislocation characteristics among the profiles measured in other conditions.

1. Introduction

The density, the arrangement and the character of dislocations are important factors to understand the mechanical properties of materials. These characteristics are often observed directly using a transmission electron microscope (TEM) [1]. Because the TEM observation needs thin film samples and performs only in a limited area, it is difficult to observe the *in-situ* evolution of dislocation characteristics in the bulky material during a mechanical test. Meanwhile, the microstructures of materials are also often evaluated from the peak broadening in X-ray or neutron diffraction profiles [2]. In particular, neutron diffraction is a powerful tool for the microscopic evaluation of bulky materials owing to its high permeability.

The dislocation density in a martensitic Fe-18Ni steel have been studied using X-ray [3] and neutron [4] diffraction through the classical Williamson-Hall plot [5]. These studies were done based on the peak width and concluded that the dislocation densities decreased with the increase of plastic deformation. These results have been judged from the decrease of diffraction peak width observed after the plastic deformation. The Convolutional Multiple Whole Profile (CMWP) fitting method [6, 7] is a state of the art technique for line profile analysis based on physically modelled profile functions for dislocations, crystallite size and planar defects. This technique evaluates the whole peak profile including the tail as well as the width. The *in-situ* neutron diffraction combined with the CMWP fitting method should provide better understanding for the microstructural evolution particularly the dislocation characteristics in steels during deformation.

[1] To whom any correspondence should be addressed.

Recently, many high-resolution time-of-flight (TOF) neutron diffraction instruments built at high-intensity pulsed neutron sources are available for *in-situ* observation of microscopic characteristics, such as lattice strain, texture and phase transformation, of materials during mechanical tests. However, there are only few studies on the dislocation characteristics of steel have been reported using these instruments. In the present study, neutron diffraction patterns of a martensitic steel during tensile deformation were measured, and the evolution of dislocation characteristics were evaluated using the CMWP fitting. Two types of loading conditions were adopted for the deformation, in order to verify which loading condition is more appropriate for the dislocation characterization.

2. Experimental and analyses

2.1. Neutron diffraction measurement

Rod shaped specimens of an as-quenched lath martensitic 22SiMn2TiB steel [8] were used in this study. The diameter and the length of the parallel part were 5.0 mm and 30 mm, respectively. *In-situ* neutron diffraction measurements during tensile tests were performed using the Engineering Materials Neutron Diffractometer TAKUMI [9] at the Materials and Life Science Facility (MLF) of Japan Proton Accelerator Research Complex (J-PARC). The specimen was mounted horizontally in the loading machine sitting on the sample stage of TAKUMI, in which the loading axis was set as 45° to the incident neutron beam, as shown in figure 1. Because TAKUMI has two detector banks that have the scattering angles of +90° and -90°, respectively, the diffracted neutron with the scattering vectors parallel and perpendicular to the loading axis can be measured simultaneously. In figure 1, the geometries of the wave vectors of the incident neutron $\boldsymbol{k}_1$ and the diffracted neutron $\boldsymbol{k}_2$, and the scattering vector $\boldsymbol{Q}$ for both the loading (axial) and the transversal directions are described. The gauge volume was restricted to be 5.0 × 5.0 × 5.0 mm^3 using the incident beam slit and the radial collimators. The instrumental peak resolution was set to be about 0.3 % by the tuning of incident beam divergence. Two tensile loading conditions were adopted in this study; a continuous loading and a stepwise loading. In the continuous loading, the deformation was applied to the specimen until its fracture in a constant crosshead speed with the initial strain rate of 1×10^{-5} s^{-1}. The diffraction patterns measured for the several strain states marked in figure 2 were extracted for the analyses. The time widths for the diffraction pattern extraction were 300 s and 150 s. Because TAKUMI adopted the event-recording technique for the neutron diffraction data acquisition system, the time slicing for extracting intensities in the arbitrary time region from whole data after the measurement was available. In the stepwise loading, the deformation was applied step by step in the same crosshead speed to arbitrary strains followed by unloading for 1800 s. Diffraction patterns related to the unloaded states after deformations were extracted for the CMWP analyses. Figure 2 shows the stress-strain curves of the steel for the different loading conditions.

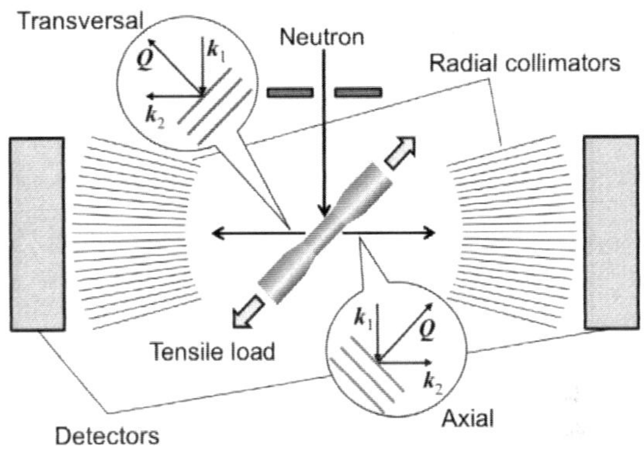

Figure 1. Schematic drawing of the *in-situ* neutron diffraction experiment at the TAKUMI.

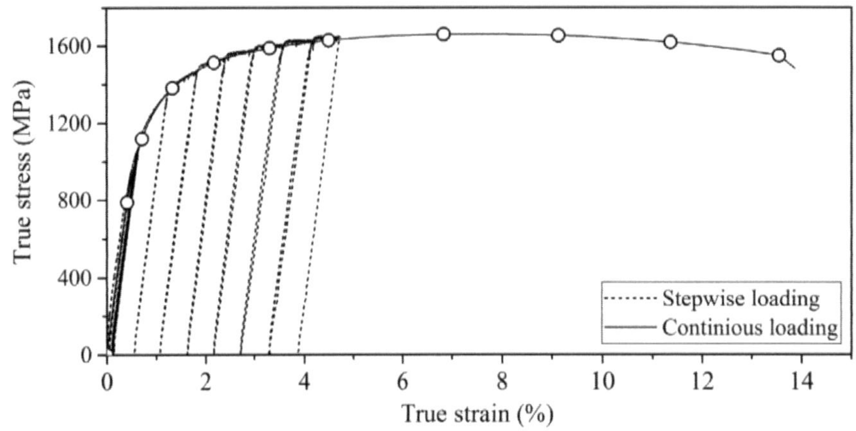

Figure 2. The stress-strain curves of the 22SiMn2TiB steel obtained from tensile deformation tests in the stepwise and the continuous loading conditions for *in-situ* neutron diffraction measurements. The circles on the curve for the continuous loading data indicate the conditions of the data used for the line profile analysis.

2.2. Line profile analysis

Prior to the CMWP fitting, the behaviour of peak width according to the strain was quickly evaluated by the fitting with a simple Gaussian function. The obtained full width at half maximum (FWHM) values of 110 diffraction peaks for both the axial and the transversal directions are plotted in figure 3 as a function of strain. The FWHM values for all the data decrease largely in the strain range between 0% and 2%. For the continuous tensile loading condition, the FWHM values of the data extracted with the time width of 300 s and 150 s are the same within the error bars. These may show that the peak broadening due to a wider range of applied stress that may appeared in the wider extracting time width is negligibly small. The FWHM values of the continuous loading are larger than those of the stepwise loading.

The line profile analyses using the CMWP fitting method were carried out for the data that were extracted with 300 s time width in order to get better statistics for dislocation characterization. The diffraction peak profiles obtained from LaB_6 powder (NIST, SRM 660a) measured in the same instrumental conditions as the *in-situ* neutron diffraction measurements were used as the instrumental resolution function for the analyses.

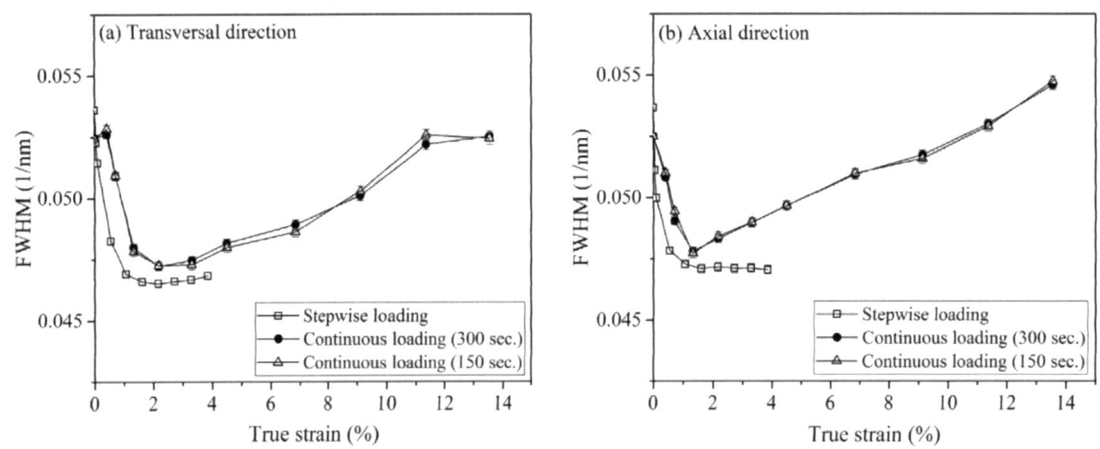

Figure 3. Strain dependences of FWHM values of 110 diffraction peak in (a) the transversal direction and (b) the axial direction.

Figure 4 shows the observed and CMWP fitted diffraction patterns of 22SiMn2TiB steel before deformation. Diffraction peaks that have much smaller intensities in the figure indicate the existence of retained austenite in the specimens. Although these austenite peaks were also fitted, reasonable values could not be obtained due to the small fraction of austenite. The mass fraction of austenite in the specimens estimated by Rietveld refinement was about 3%.

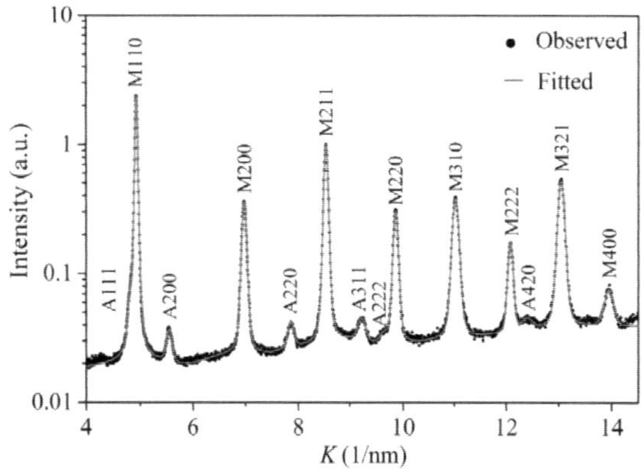

Figure 4. The example of the result of the CMWP fitting. "M" indicates martensite peaks and "A" indicates austenite peaks. 8 diffraction peaks of martensite were used for the present analyses.

3. Results and discussions

From the CMWP fitting, dislocation density (ρ), dislocation arrangement parameter (M^*), dislocation character (edge or screw), crystallite size, etc. were obtained. In this study, results concerning ρ and M^* will mainly be discussed. The value of M^* being larger than 1.0 indicates a random arrangement, and that lower than 1.0 indicates a dipole arrangement or a strongly correlated arrangement of dislocations [10, 11]. The values of ρ and M^* obtained from the CMWP fitting are shown in figure 5. In the transversal direction, almost the same behaviors of ρ and M^* between the stepwise and continuous loadings are found in figures 5 (a) and (c). The ρ values increase gradually from 5.0×10^{15} m^{-2} to 10^{16} m^{-2} order with the increase of plastic strain. The ρ value before deformation is slightly larger than the value from TEM observation of the lath martensitic steel with similar carbon content (0.18 mass %) [12], but considering the physical form (bulk or thin-foil) of the specimens, these values are in good agreement. Meanwhile, the ordering of the dislocation arrangement by the deformation was found from the largely decreasing of M^* according to the increasing strain. The strong correlation of dislocations gives the narrower widths and the sweeping tails of diffraction peaks, i.e., the peak shape changes from the Gaussian to the Lorentzian. [10]. The FWHM values shown in figure 3 are reflecting this phenomenon. On the other hand, different behaviors of ρ for the continuous and stepwise loadings were found in the axial direction as shown in figure 5 (b). Although ρ for the stepwise loading hardly changes during deformation, ρ for the continuous data increases apparently. More detailed discussions about the relationship between the dislocation characteristics and the work-hardening mechanism of the steel can be found in another paper [13]. In the stepwise loading condition, the diffraction peak profiles that were used for the analysis were not affected by the elastic strain because the diffraction intensities during the unloaded states were used for the analysis. In contrast, in the continuous loading condition, the elastic strain caused by the applied load may affect to the observed diffraction peak profiles. If the homogeneity of elastic strain in a specimen is sufficiently high, only the peak shift will be observed in a diffraction pattern. Otherwise, peak broadening due to elastic strain heterogeneity caused by an applied load will appear. The effect of elastic strain heterogeneity to the peak width can be found in the larger FWHM values for the continuous loading data in figure 3. The effects of the elastic strain for the diffraction peak profile should be eliminated in order to evaluate the dislocations more accurately through the line profile analysis. The stepwise loading condition is more appropriate than the continuous one for the *in-situ* diffraction measurement aiming at the microstructural study of the steels during tensile deformation.

The dislocation density ρ in the axial direction is smaller than that in the transversal direction for the stepwise condition. Here, the scattering vector of neutrons measured in the axial direction lie in the loading axis as mentioned above. In other words, the lattice planes that contributed to the diffraction signal were always normal to the loading axis. The crystal grains that contributed to a certain *hkl* diffraction peak received the load in a same crystal orientation. In contrast, scattering vector in the transversal direction is perpendicular to the loading axis. The lattice planes that contributed to the diffraction signal were parallel to the loading axis. Because the orthogonal direction to the scattering vector is not always parallel to the loading axis, the crystal orientations of the loading direction are different for different sets of grain families. The peak profiles that were observed in this direction were highly affected by intergranular strains in this situation. The dislocation densities that were evaluated from the transversal data are larger than that were evaluated from the axial data due to the large elastic strain heterogeneity of the intergranular strain that might cause peak broadening. For this reason, the diffraction pattern measured in the axial direction should be used for the analysis to investigate the behaviour of the dislocation density during tensile deformation.

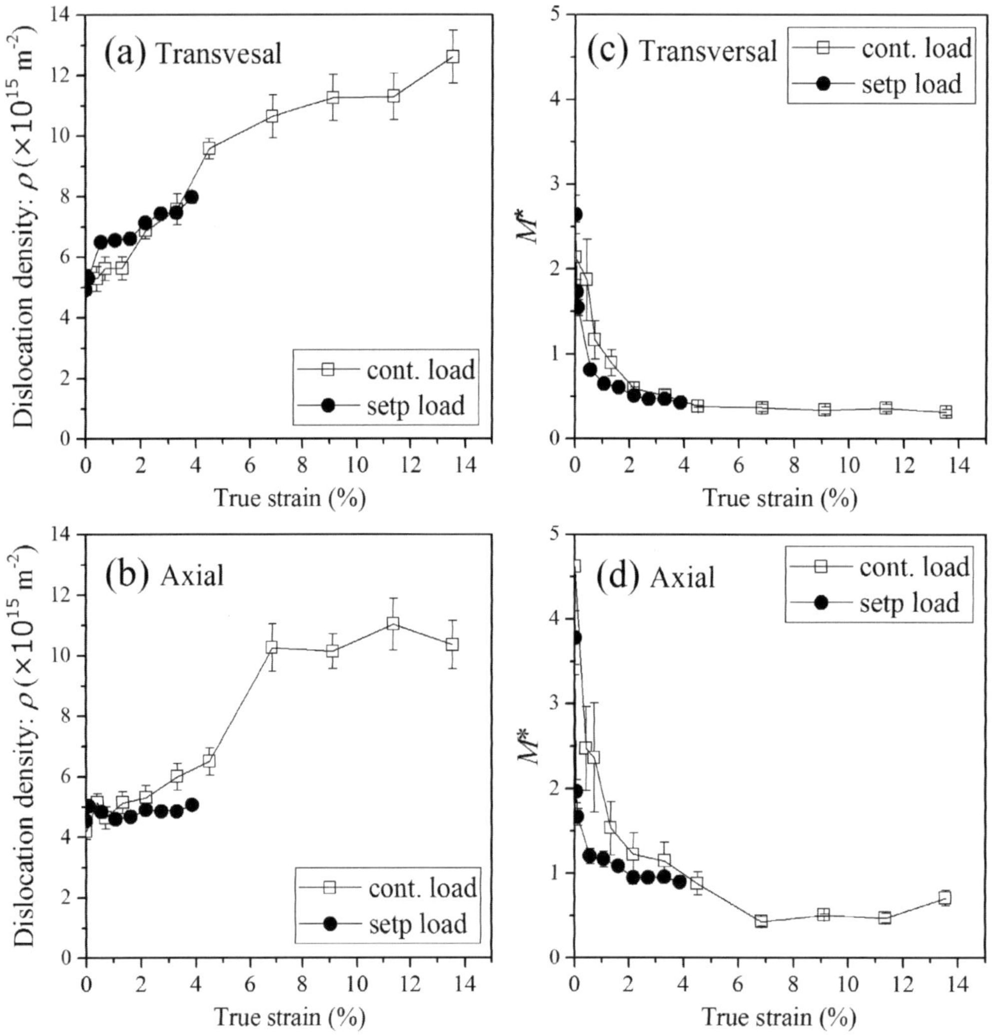

Figure 5. The dislocation density of the 22SiMn2TiB steel in (a) the transversal direction and (b) the axial direction, and the dislocation arrangement parameters in (c) the transversal direction and (d) the axial direction obtained from the line profile analyses using the CMWP fitting method.

4. Summary

Line profile analyses of 22SiMn2TiB martensitic steels during tensile deformations were performed. Both the continuous loading and the stepwise loading conditions were adopted for the *in-situ* neutron diffraction measurements using the high resolution and the high intensity time-of-flight (TOF) neutron diffractometer TAKUMI at the MLF of J-PARC. The analyses of the diffraction profiles measured with these two conditions in the axial direction gave different behaviours of the dislocation densities. The dislocation density ρ and the dislocation arrangement parameter M^* in the axial direction obtained from the continuous loading are larger than those from the stepwise loading, because the elastic strain (lattice strain) heterogeneity was introduced in the diffraction profiles during the data collection under applied stress. The elastic strain heterogeneity is also larger in the transversal direction than in the axial direction leading to the larger observed dislocation density. The diffraction peak profiles measured in the unloaded state and in the axial direction thus should provide more reliable information of dislocation characteristics through line profile analysis.

The line profile analysis for *in-situ* TOF neutron diffraction data using the CMWP fitting method is effective for microstructural study of steels in mechanical tests. Further verification of experimental conditions, such as a search of the standard sample for the instrumental resolution function and an optimization of the incident beam divergence, is needed in order to evaluate dislocation characteristics more accurately.

Acknowledgements

Authors would like to acknowledge Prof. Zhengmin Shi for providing the sample for this study, Prof. Yo Tomota and Prof. Tamas Ungár for fruitful discussion. The neutron diffraction experiments were performed at BL19 in MLF of J-PARC with the proposals of 2014I0019 and 2014P0102.

References

[1] Morito S, Huang X, Furuhara T, Maki T and Hansen N 2004 *Acta Mater* **52** 5405
[2] Williamson G K & Smallman R E 1956 *Phil. Mag.* **1** 34
[3] Nakashima K, Fujimura Y, Matsubayashi H, Tsuchiyama T and Takaki S 2007 *Tetsu-to-Hagane* **93** 459
[4] Morooka S, Tomota Y and Kamiyama T 2008 *ISIJ Int.* **48** 525
[5] Williamson G K & Smallman R E 1956 *Phil. Mag.* **1** 34
[6] Ungár T, Gubicza J, Ribarik G and Borbély A 2001 *J. Appl. Cryst.* **34** 298
[7] Ribárik G, Gubicza J and Ungár T 2004 *Mater. Sci. Eng. A* **387-389** 343
[8] Shi Z, Liu K, Wang M, Shi J, Dong H, Pu J, Chi B, Zhang Y and Jian L 2012 *Met. Mater. Int.* **18** 317
[9] Harjo S, Ito T, Aizawa K, Arima H, Abe J, Moriai A, Iwahashi T and Kamiyama T 2011 *Mater. Sci. Forum* **681** 443
[10] Wilkens M 1970 *Phys. Stat. Sol.* (a) **2** 359
[11] Ungár T 2001 *Mater. Sci. Eng.* A **309-310** 14
[12] Morito S, Nishikawa J and Maki T 2003 *ISIJ Int.* **43** 1475
[13] Harjo S, Kawasaki T, Gong W and Aizawa K, in this issue.

Studies of the radioactive waste confinement matrix using neutron scattering methods

C A Dragolici[1], M Balasoiu[1], L Ionascu[1] and M Nicu[1]

[1] "Horia Hulubei" National Institute for R&D in Physics and Nuclear Engineering, 30 Reactorului Street, P.O. BOX MG-6, Bucharest-Magurele, ROMANIA

E-mail: adrag@nipne.ro

Abstract. The management of the radioactive waste implies the conditioning in a cement matrix as an embedding, stable, disposal material. Cement matrix is the first and most important engineering barrier against the migration in the environment of the radionuclides contained in the packages. Chemically, after mixing with water, cement produces a reacting matrix with a porous microstructure. Knowing how the microstructure develops is therefore desirable in order to assess the compatibility of radioactive streams with cement and predict waste form performance during storage and disposal. This article highlights the work performed in SANS investigation of a new type of cements to be used for conditioning of aluminium waste coming from reactor decommissioning activities. It is well known that the expansion of corrosion products formed on aluminium can lead to cracking of the waste form.

1. Introduction

The set-up of the cement matrix begins as a mixture of water and cement powder. Adjustment of the phase fractions of the individual cement compounds in the anhydrous cement is one of the few means of altering its mechanical properties. The proportions of C_3S and C_2S in the anhydrous cement are important in determining the rate of hardening of the cement, as well as the heat released, which can cause micro cracking and long term durability problems. The main chemical reaction occurs between the water and the C_3S and can be written as:

$$Ca_3SiO_5 + (3 + y - x)H_2O \rightarrow (CaO)x(SiO_2)+(H_2O)y + (3 - x)\ Ca(OH)_2 \qquad (1)$$

where the calcium/silica ratio, x, goes from 0.7 to 1.7 over the course of the reaction [1]. The first term on the right-hand side is the calcium silicate hydrate gel that makes up the binder, usually abbreviated as C-S-H. Its water content, y, remains uncertain. Part of it is in the form of OH- groups in the structure of the gel, and the remainder is immobilized H_2O molecules. The most accepted conditioning process in accordance with the actual international disposal requirements and respecting the waste acceptance criteria for interim storage or/and final disposal is concrete embedding. The use of Ordinary Portland cement (OPC) or modified formula [2] by adding pozzolans as fly ash, silica fume or blast furnace slag developed a pore solution pH as high as 13.5. This research is focused on the development of CEM III and CEM V modified systems based on addition of aluminum sulfate in order to minimize the alkaline attack of cement hydrolysis products on metallic radioactive waste.

2. Materials & Methods

2.1. Materials

To obtain the decreasing of portlandite concentration in shorter time by a direct chemical reaction during the setting time was develop a modified system based on CEM III-A and CEM V-A by addition of aluminum sulfate. The materials used in the modified systems were Portland Cement (CEM-III and CEM-V from Lafarge Romania).

A number of 4 samples were prepared and investigated in a time interval of 232 to 247 hours from their date of production. The samples for this study were made from special types of cements, MKP, CEM V, CEM III and White Portland Cement with specific proprieties. All the samples were mixed with de-ionized water at 0.4 w/c (water to cement ratio) [3] by mass.

2.2. Methods

Electron microscopy studies made on the C-S-H gel phase have revealed the presence of significant fluctuations in the local Ca/Si ratio within the gel phase throughout the specimen. The presence of a bimodal distribution in the local Ca/Si ratio; however, as specimens matured it can be seen that the degree of bimodality decreased (Fig. 1 and 2).

X-ray diffraction was conducted on set and cured mixes (5 days) to ensure the hydration products are formed. The spectra for samples with 50% $Al_2(SO_4)_3$ (CM V-50 and CM III-50) shows that these samples hydrates to form ettringite and gypsum as a primary phases and portlandite as a minor phase.

SANS measurements were done at FLNP from JINR Dubna, using the YuMo instrument from the pulsed reactor IBR-2. As is known from the previous SANS experiments the microstructure of hydrated cement paste contains two fractal objects: volume and surface fractals.

3. Results & Discussion

3.1. Electron microscopy studies

Following the nanophasic model, subcrystalline regions would exist; however, they would be confined to small fragments of Ca-O layers (with attached silicate chain fragments) that are several manometers long. This model can explain many of the crystal chemical observations, macroscopic properties, structural, and compositional trends. In particular, it takes into account the importance of the presence of strong fluctuations in the local Ca/Si ratio. The early presence of bimodality demonstrates that discrete regions exist which initially have different Ca/Si distributions but cannot establish the presence of short-range compositional and/or structural order.

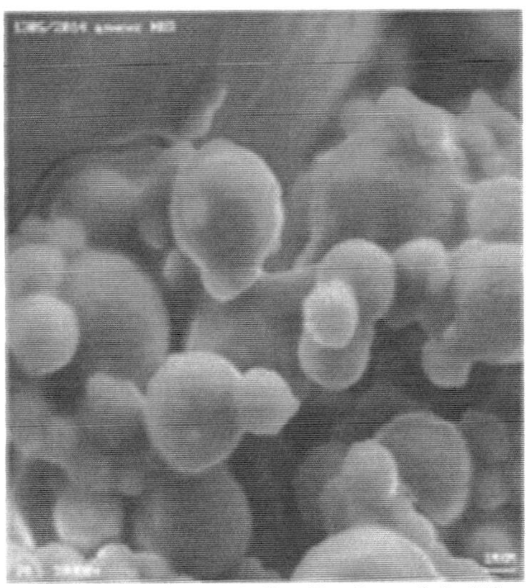

Figure 1. Bimodal distribution of Ca/Si.

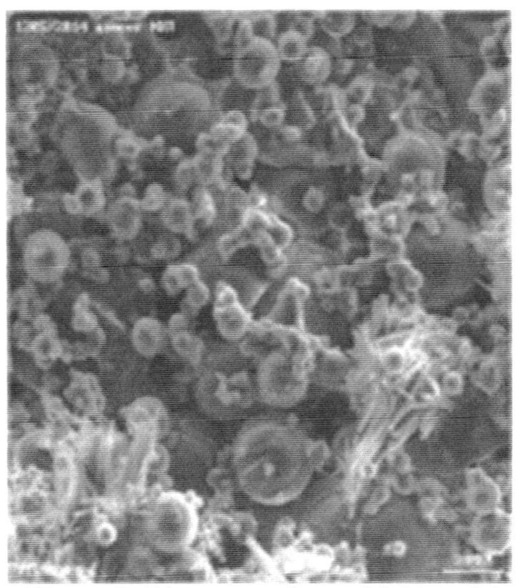

Figure 2. Discrete regions with bimodal distribution of Ca/Si.

Journal of Physics: Conference Series **746** (2016) 012045 doi:10.1088/1742-6596/746/1/012045

3.2. X-ray diffraction studies

The hydration products were studied using a Bruker D8 – Advance x-ray diffractometer with Cu Kα radiation, diffraction angle 2 theta ranged between 5° and 70°, a step size of 0.02° and an acquisition time of 4sec/step. The quantity of gypsum which is formed reduces the rate of the hydration reaction and complicates further the reactions that will occur. The spectra of modified cements with different $Al_2(SO_4)_3$ content (Fig. 3 and 4) show that the diffraction peaks intensity of portlandite decreased gradually with the increase of aluminum sulphate proportion.

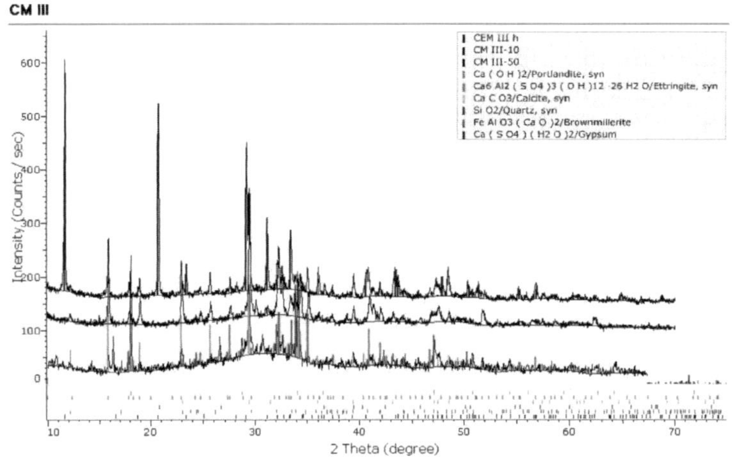

Figure 3. Diffraction spectra for modified cement systems based on CEM III-A.

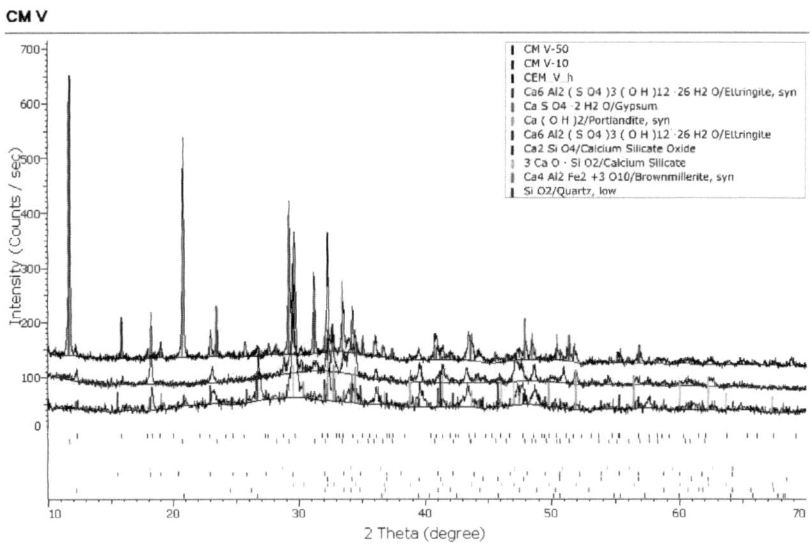

Figure 4. Diffraction spectra for modified cement systems based on CEM V-A

3.3. Small Angle Neutron Scattering studies

SANS measurements (Fig. 5 and Fig. 6) were carried out using the YUMO instrument installed at IBR-2 high pulsed reactor (JINR, Dubna, Russia), equipped with a two- detector system. The coverage interval of scattering vector Q was 0.006 $Å^{-1} < Q < 0.3$ $Å^{-1}$. The measured neutron scattering spectra were corrected for the transmission and thickness of the sample, background scattering on the film substrate and on the vanadium reference sample using SAS software, yielding a neutron scattering intensity in absolute units of cm^{-1}. The volume fractal represents the gel network – (C–S–H) gel clusters consisting of nanometric globules (l~5nm) and the surface fractal is supposed to be a gel deposit on the compact remains of unhydrated clinker cores. The gel clusters [4] contain very fine pores (gel pores with diameters less than about 50nm) which form, however, only a minor part of the total porosity. The major part of the overall porosity consists of capillary pores (diameters larger than 50nm up to millimeters). The fractal dimensions of the gel clusters are strongly dependent on the moisture content and vary from about 2.5 at 100% RH to 3.0 at 55% RH.

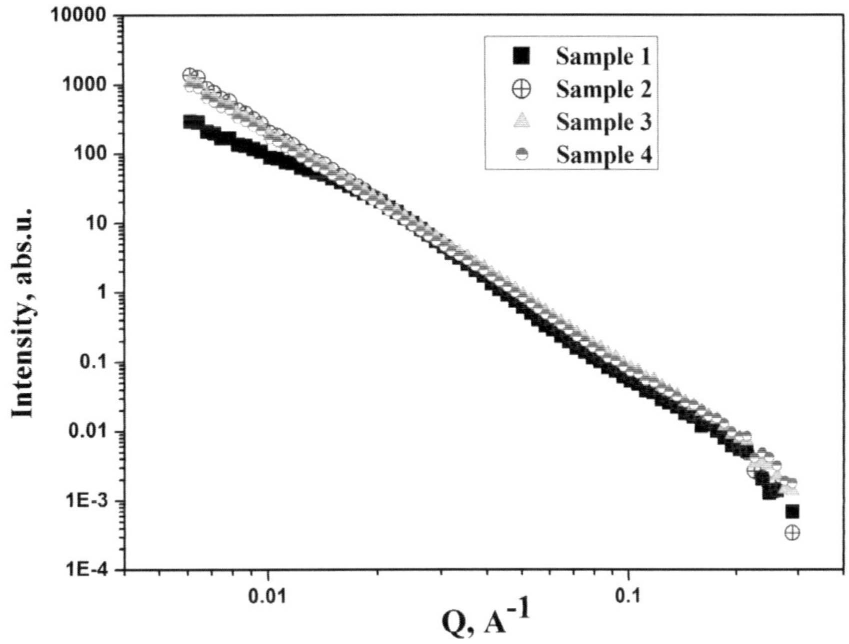

Figure 5. SANS data for MKP, CEM V, CEM III and Portland White Cement.

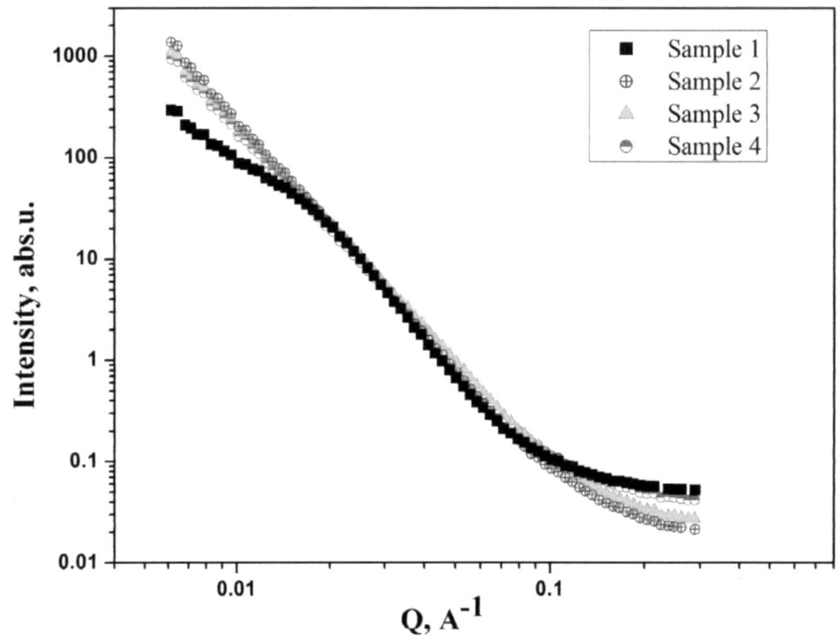

Figure 6. SANS data for MKP, CEM V, CEM III and Portland White Cement.

4. Conclusions

Based on the investigation results the main conclusions are the following:

There is some uncertainty concerning the scaling ratio (l/L) of the volume fractal structure of the gel.

A mix containing 50% $Al_2(SO_4)_3$ which hydrates to form ettringite [5] as a primary phase may be a potential candidate for meeting the aluminium conditioning requirements.

The modified cement grouts have developed a good mechanical structure and the compressive strength values are included in the accepted limits for the matrix durability.

Because of overlap with the surface fractal, the SANS technique is capable to explore the gel volume structure in rather limited interval (1nm, 50nm), the volume clusters may show certain fractal arrangement up to micrometric levels at which our measurements were performed.

References

[1] R.A. Livingston, D.A. Neumann., A. J. Allen, S.A. FitzGerald and R. Berliner 2000 *Neutron News* **11** 4

[2] M. Nicu, L. Ionascu, C. Turcanu, F. Dragolici, Gh. Rotarescu and Gh. Dogaru 2008 *Rom. Journ. Phys.*, **53** 841

[3] Keith Quillin, Geoff Osborne, Amal Majumdar and Bahadur Singh 2001 *Cem. Concr. Res.* **31** 627

[4] Jeffrey J. Thomas and Hamlin M. Jennings, 2006 *Cem. Concr. Res.* **36** 30

[5] M.R. Hartman and R. Berliner 2006 *Cem. Concr. Res.* **36** 364

Dislocation Characteristics in Lath Martensitic Steel by Neutron Diffraction

S Harjo[*,**,1], **T Kawasaki**[*], **W Gong**[*] **and K Aizawa**[*]
[*] J-PARC Center, Japan Atomic Energy Agency, Tokai-mura, Naka-gun, Ibaraki 319-1195, Japan
[**] Elements Strategy Initiative for Structural Materials, Kyoto University, Yoshida-honmachi, Sakyo-ku, Kyoto 606-8501, Japan

E-mail: stefanus.harjo@j-parc.jp

Abstract. In situ neutron diffraction during tensile deformation of an as-quenched lath martensitic 22SiMn2TiB steel, was performed using a high resolution and high intensity time-of-flight neutron diffractometer. The characterizations of dislocations were analyzed using the classical Williamson-Hall (cWH) and modified Williamson-Hall (mWH) plots on the breadth method, and the convolutional multiple whole profile (CMWP) fitting method. As results, the dislocation density as high as 10^{15} m^{-2} in the as-quenched martensitic steel was determined. The dislocation density was found to decrease qualitatively with plastic deformation by the cWH and mWH plots, but hardly to change by the CMWP method. The scanning transmission electron microscopy observation supported the results of the latter method. In the CMWP method, the parameter *M* that represents the arrangement of dislocations was found to decrease rapidly where a very high work hardening was observed.

1. Introduction

The as-quenched lath martensitic steel is one important structural material because it consists fine structures and high dislocation densities showing high strengths. The lath martensitic steel exhibits extremely low elastic limit [1], indicating that the work hardening after the yielding at the beginning of plastic deformation is very high. To understand the deformation behavior, the changes in dislocation density in lath martensitic Fe-18Ni alloys with cold rolling and tensile deformation have been measured using the X-ray [1] and neutron diffractions [2] by employing the classical Williamson-Hall (cWH) plot [3]. The dislocation density was determined to decrease with plastic deformation, though a very high work hardening was observed at the beginning of tensile deformation.

Dislocation density has been evaluated mainly using two primary methods, transmission electron microscopy (TEM) [4] and X-ray diffraction (XRD) peak broadening defined by the breadth method [5,6]. These techniques have been considered to be mutually complementary. The XRD line broadening analysis yields average statistic data and supports the TEM results [7]. Takebayashi et al. [7] have claimed that the dislocation density determined by the cWH plot is not appropriate estimating larger dislocation density values, because the dislocation contrast factor is not considered. Hence, the modified Williamson-Hall plot (mWH) [5,8] coupled with the modified Warren-Averbach method [5,8] taking the contrast factor into account has recently been used to characterize the dislocation density, character and arrangement [7,9]. However, we believe that these analyses are not satisfactory

[1] To whom any correspondence should be addressed.

because the results obtained from thin films (TEM) or surface layers (XRD) are frequently suspected to be smaller than the average values of a bulky whole specimen. Recently, in situ neutron diffraction has been demonstrated to be a powerful tool in various engineering applications [2,10-12], including thermo-mechanically controlled processing [12].

Hence, in the present study, the deformation behavior of a lath martensitic 22SiMn2TiB steel, was investigated using in situ neutron diffraction during tensile test. The characterizations of dislocation were performed using the WH plots and a convolutional multi whole profile fitting (CMWP) method developed by Ungár and his coworkers [13,14], and the obtained results were compared from each other.

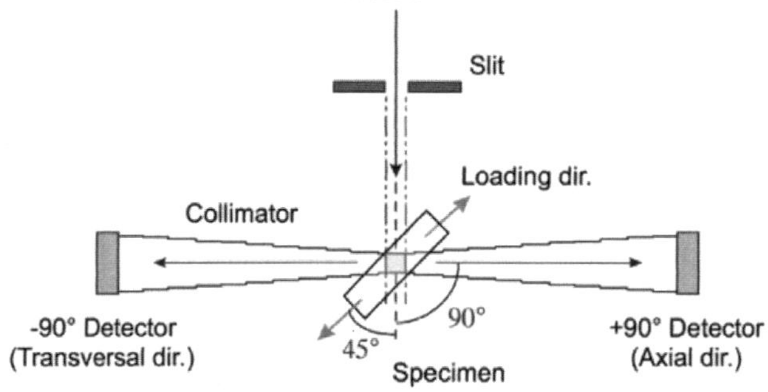

Figure 1. Experimental arrangement of in situ neutron diffraction during tensile deformation on TAKUMI at the MLF of J-PARC.

2. Experimental procedures

The as-quenched lath martensitic steel, 22SiMn2TiB steel [15], was used in this study. The averaged packet and block sizes were 19.6 μm and 3.9 μm, respectively. The rod shape specimen with diameter of 5.0 mm and active length of 30 mm was prepared after the austenitic solution treatment for the in situ neutron diffraction experiment during tensile test. The specimen was mounted horizontally in the loading machine sitting on the sample stage of TAKUMI [16], a high resolution and high intensity time-of-flight (TOF) neutron diffractometer for engineering sciences at Materials and Life Science Facility (MLF) of Japan Proton Accelerator Research Complex (J-PARC). The schematic drawing of experimental setup is shown in figure 1. The neutron diffraction patterns in the axial and transversal directions were measured simultaneously using two detector banks that have the scattering angles of ± 90°. The gauge volume was restricted to be 5.0 × 5.0 × 5.0 mm^3 using the incident beam slit and radial collimators. The TAKUMI instrumental resolution used in this study was tuned to have the instrumental peak resolution of about 0.3 %.

The tensile deformation for in situ neutron diffraction was performed in a stepwise manner. The deformations in the plastic region were increased step by step to arbitrary strains followed by unloading. The neutron diffraction data collection was conducted continuously using an event-recording mode [17] during the tensile deformation. Diffraction patterns related to the unloaded states after plastic deformations were then sliced according to the stress and strain data. The characterizations of dislocation were performed using the WH plots and the CMWP fitting. Diffraction peak profiles of LaB_6 powder measured at the same condition with the in situ neutron diffraction measurement were used to determine the instrumental peak profiles for the dislocation analyses.

The Rietveld refinements on the diffraction patterns were performed using the Z-Rietveld software [18] prior to the dislocation characterization. The specimen contained the retained austenite, and its fraction was determined to be about 3.7 %. The existence of retained austenite was difficult to confirm on the SEM or TEM image, probably due to the tiny size. The lattice constants of martensite and austenite were determined to be 0.28646(0) nm and 0.35912(3) nm, respectively. The austenite peaks still existed after 4.7 % deformation. The inverse pole figures were also determined from the ratios of integrated intensities of (hkl) peaks. The specimen had random orientation before deformation, and a very weak -fiber texture was developed after 4.7 % deformation.

3. Results and discussion

3.1. Stress-strain curve

Figure 2 shows the stress-strain curve obtained during the tensile deformation for the in situ neutron diffraction. The specimen shows a very high tensile strength of about 1.65 GPa. This steel showed a uniform strain of about 6.1 % and fractured at the strain value of about 14 %, when it was tensioned continuously. The tensile deformation shown in figure 2 includes plastic deformation though the highest strain is 4.7 %. The elastic limit measured from the continuous tension test was about 500 MPa, and therefore the rate of work hardening is extremely high.

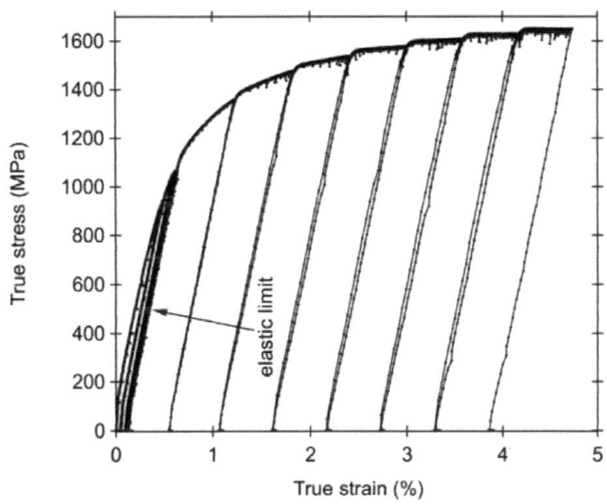

Figure 2. Stress-strain curve of the lath martensitic 22SiMn2TiB steel obtained from the tensile test.

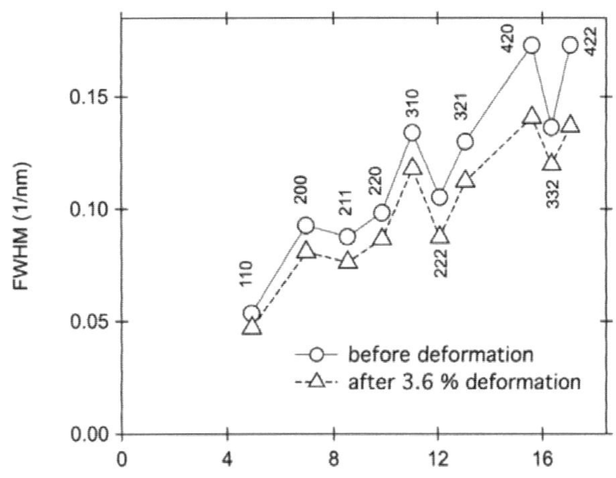

Figure 3. Typical cWH plots obtained for the axial direction before deformation and after 3.6% deformation ($K = 1/d$, where d is the lattice spacing).

3.2. Williamson-Hall plots

The dislocation density is often evaluated from diffraction profiles simply using the cWH plot [3]. Figure 3 shows typical cWH plots for the axial direction. The full width at half maximum (FWHM) data of the specimen is the convolution of the instrumental and physical effects, which are difficult to be separated in physical theories. The scatters of FWHM data of the specimen are large in the cWH plots as shown in figure 3, though the instrumental FWHM data measured using LaB_6 powder was in a good linearity. The dislocation density is estimated from the slope in the cWH plot, and the scatter therefore will decrease the accuracy. The slope of cWH plot for the specimen after 3.6 % deformation is smaller than that before deformation.

Figure 4 shows typical mWH plots for the axial direction at different conditions of deformation. The scatters of FWHM data of the specimen are much improved by introducing the contrast factor (*C*). However, the trends are still similar with those in figure 3. The slope before deformation is the largest, and it becomes smaller with the increase of strain. These trends speculate qualitatively that the high dislocation density induced during martensitic transformation decrease with the increase of strain. The same tendencies have also been reported in the previous studies for Fe-18Ni alloys [1,2] that employed the cWH plot. According to the Taylor equation [19], shown in equation (1), the work hardening can be described by the increase of dislocation density.

$$\Delta\sigma = \alpha b M_T G \sqrt{\rho} \tag{1}$$

Here, $\Delta\sigma$, α, b, M_T, G, and ρ are the amount of hardening, coefficient α, Burger's vector (constant), Taylor factor (constant), modulus of rigidity (constant), and dislocation density, respectively. Thus, the trends in figures 3 and 4 are difficult to understand the origin of the high rate of work hardening shown in figure 2. Note, that the profile does have the FWHM value and the tail. The tails are ignored in the FWHM values. This is one of the reasons why in any breadth method it fails for providing dislocation densities.

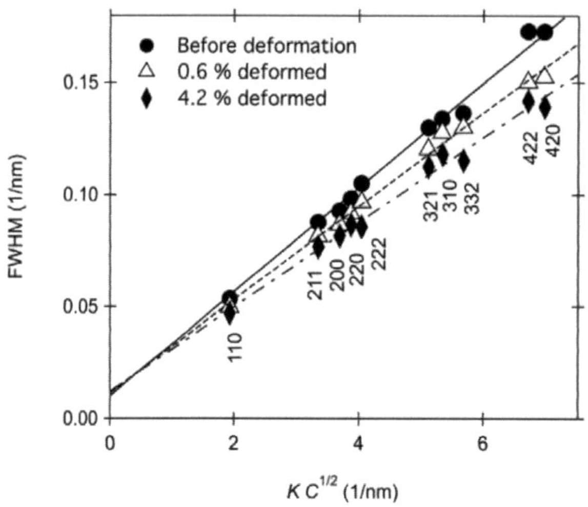

Figure 4. Typical mWH plots obtained for the axial direction at different conditions of deformation ($K = 1/d$, where d is the lattice spacing).

3.3. • CMWP fitting

The CMWP method is an analysis method to characterize defects from the diffraction profiles based on the modified Williamson-Hall plot combining with a convolutional profile fitting method [13,14].

Figure 5 shows the observed and CMWP-fitted neutron diffraction profiles before deformation. In the CMWP fitting, the second phase of the retained austenite was also analyzed to exclude its influence on the results of the main phase of martensite. All martensite peaks are indexed and some of the austenite peaks are also indexed. In the first part of the pattern, the 111 austenite peak can be seen as a shoulder of 110 martensite peak.

The parameters obtained by the CMWP fitting for the axial direction are summarized in figure 5. The value of ρ before deformation is already very high of about 4.0×10^{15} m^{-2}. This value is believed to be consistent with that reported in a lath martensitic steel with similar carbon content (0.18 mass %) studied by Morito et al using TEM [4], by considering that the neutron diffraction observes as bulky whole information while the TEM a limited area of thin film. The value of ρ gradually increases with the large increase of strain though the increase of flow stress is observed. The dislocation densities were also clarified using scanning transmission electron microscopy (STEM) observations on three annular dark field images with the incident beam parallel to <111> and two images <001>. As results, the dislocation density before deformation was averagely determined to be 1.17×10^{15} m^{-2}, and that after 4.7 % deformation to be 1.18×10^{15} m^{-2}, showing no significant different between the two conditions. These results confirm the CMWP fitting results that the increase of dislocation density during deformation is small.

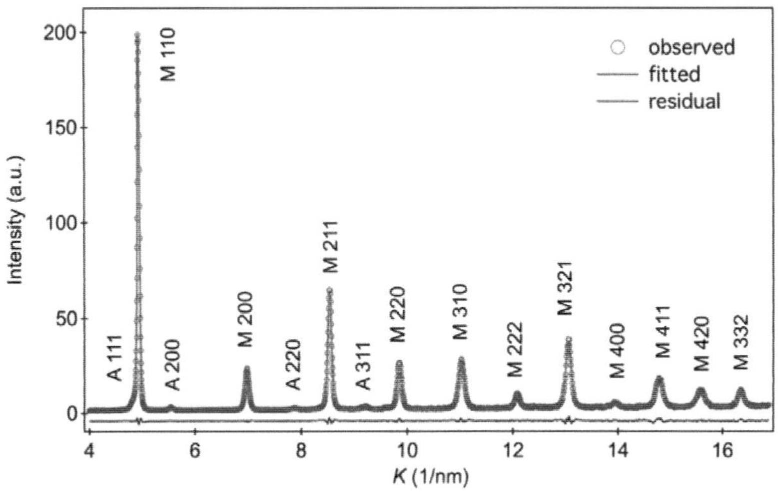

Figure 5. The observed (black-circle symbol) and CMWP-fitted (red line) neutron diffraction profiles before deformation. Blue line is the residual between the fitted and observed profiles.

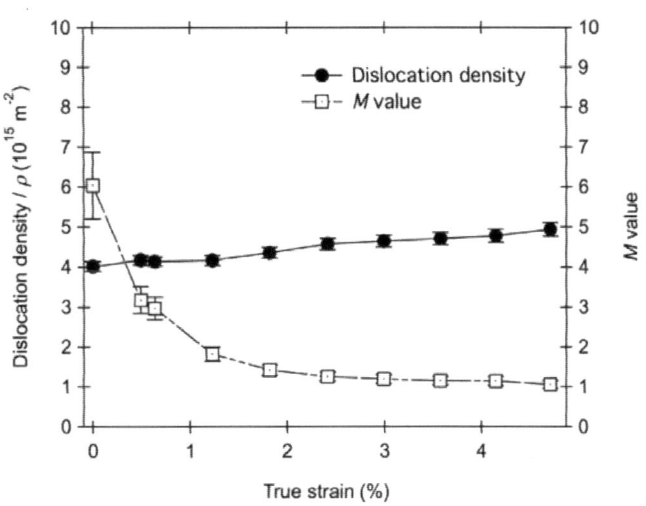

Figure 6. Parameters obtained from the CMWP fitting in the 22SiMn2TiB steel for the axial direction.

The value of contrast parameter q before deformation was determined to be about 1.7, and it was found hardly to change during deformation. The parameter q displays the dislocation character, i.e., the edge or screw component. According to a theoretical computation for iron with slip system of <111>{110}, the value of 1.3 stands for the edge and 2.6 for screw dislocations [20]. The value of q of about 1.7 indicates that the dislocations are mixed with the edge and screw characters. The area-averaged crystallite size was about 57 nm which may display the lath size.

Another parameter M [13], which is a product of the effective cut-off radius of dislocations Re and the square root of ρ ($M = Re\,\rho^{0.5}$ [13]), displays the dislocation arrangement. The value of $M > 1$ stands for random dislocation arrangement, and $M < 1$ for highly correlated arrangement. As is shown in figure 6, the value of parameter M is found to decrease rapidly at the beginning of plastic deformation, then gradually varied with the progress of deformation finally becoming close to 1. This indicates that randomly distributed dislocations in the as-quenched martensite are arranged toward correlated arrangements. Since the increase of dislocation density in the specimen is small during deformation applied in this study, the observed high work hardening is predicted as the result of the increase of coefficient α in the Taylor equation. From comparison between the results in figure 6 and the stress-strain curve in figure 2, the change in parameter M seems to be in inverse proportion to the increase of flow stress. It is speculated that the parameter M has strong relation with the coefficient α in the Taylor equation. However, further discussion to describe the relation is under process.

4. Conclusions

The in situ neutron diffraction during tensile deformation of an as-quenched lath martensitic 22SiMn2TiB steel, was performed using a high resolution and high intensity TOF neutron diffractometer TAKUMI at the MLF of J-PARC. The dislocation density, arrangement and character were analyzed to make clear the work-hardening mechanism of the as-quenched martensitic steel.

(i) The dislocation density in the as-quenched martensitic steel was originally high of 10^{15} m^{-2} order. Although the cWH and mWH plots on the breadth method showed a qualitatively decrease in dislocation density with plastic deformation, the CMWP method revealed little change. The STEM observation supported the results of the latter method.

(ii) According to the results by the CMWP fitting, it is noted that the parameter M related to dislocation arrangement decreases rapidly with plastic deformation keeping little change in dislocation density. The change of parameter M is in the same trend with the change of work-hardening rate.

(iii) The combination of TOF neutron diffraction and CMWP fitting is, therefore, a powerful method to characterize the deformation structures quantitatively using bulky-averaged parameters.

Acknowledgments

Authors wish to acknowledge Prof. Zhengmin Shi for providing the sample for this study, Prof. Yo Tomota, Prof. Tamas Ungar and Prof. Shigeo Sato for valuable discussion. The neutron diffraction experiments were performed at BL19 in Materials and Life Science Facility of J-PARC with the proposals of 2014I0019 and 2014P0102, and got partly financial supports from the Japan Society for the Promotion of Science Grant-in-Aid for Scientific Research (No. 26289264 and 15H05767).

References

[1] Nakashima K, Fujimura Y, Matsubayashi H, Tsuchiyama T and Takaki S 2007 *Tetsu-to-Hagane* **93** 459
[2] Morooka S, Tomota Y and Kamiyama T 2008 *ISIJ Int.* **48** 525
[3] Williamson G K & Smallman R E 1956 *Phil. Mag.* **1** 34
[4] Morito S, Nishikawa J and Maki T 2003 *ISIJ Int.* **43** 1475
[5] Ungár T and Borbély A 1996 *Appl. Phys. Lett.* **69** 3173

[6] Chanda A, De M and Kajiwara S 2000 *Jpn. J. Appl. Phys.* **39** 539
[7] Takebayashi S, Kunieda T, Yoshinaga N, Ushioda K and Ogata S 2010 *ISIJ Int.* **50** 875
[8] Ungár T, Ott S, Sanders P G, Borbély A and Weertman J R 1998 *Acta Mater* **46** 3693
[9] Shintani T and Murata Y 2011 *Acta Mater* **59** 4314
[10] Harjo S, Tomota Y, Lukáš P, Neov D, Vrána M, Mikula P and Ono M 2001 *Acta Mater.* **49** 2471
[11] Tomota Y, Lukáš P, Neov D, Harjo S and Abe Y R 2003 *Acta Mater.* **51** 805
[12] Tomota Y, Xu P G, Oliver E C and Paradowska A 2010 *In-situ Studies with Photons, Neutrons and Electrons Scattering*, ed. T Kannengiesse, S S Babu, Y Komizo and A J Ramirez (Berlin: Springer-Verlag) p 175
[13] Ungár T, Gubicza J, Ribarik G and Borbély A 2001 *J. Appl. Cryst.* **34** 298
[14] Ribárik G, Gubicza J and Ungár T 2004 *Mater. Sci. Eng. A* **387-389** 343
[15] Shi Z, Liu K, Wang M, Shi J, Dong H, Pu J, Chi B, Zhang Y and Jian L 2012 *Met. Mater. Int.* **18** 317
[16] Harjo S, Ito T, Aizawa K, Arima H, Abe J, Moriai A, Iwahashi T and Kamiyama T 2011 *Mater. Sci. Forum* **681** 443
[17] Ito T, Nakatani T, Harjo S, Arima H, Abe J, Aizawa K and Moriai A 2010 *Mater. Sci. Forum* **652** 238
[18] Oishi R, Yonemura M, Nishimaki Y, Torii S, Hoshikawa A, Ishigaki T, Morishima T, Mori K and Kamiyama T 2009 *Nucl. Instrum. Meth. A* **600** 94
[19] Taylor G I 1934 *Proc. R. Soc. A* **145** 362
[20] Ungár T, Dragomir I, Révész A and Borbély A 1999 *J. Appl. Cryst.* **32** 992

VI European Conference on Neutron Scattering (ECNS2015) IOP Publishing
Journal of Physics: Conference Series **746** (2016) 012047 doi:10.1088/1742-6596/746/1/012047

Neutrons as quantum objects

Helmut Rauch
Atominstitut, Vienna University of Technology, Vienna, Austria

Neutrons are proper tools for testing basic laws of quantum physics since they are massive and can be handled and measured with high efficiency. Suitable post-selection experiments demonstrate coherence features of sub-ensembles even when the whole ensemble seems to have lost its coherence. All experiments have the capacity to explain more details by additional pre- and post-selection methods. It will be shown that specific losses are unavoidable in any interaction. Coherence and decoherence are intrinsic quantum effects and can shed light on the measurement problem. Quantum contextuality is a consequence of the entanglement of different degrees of freedom. This makes quantum phenomena more strongly correlated than classical ones. Most experiments have been performed with perfect neutron interferometers and some others by using ultra-cold neutrons and spin-echo systems. An event by event based interpretation can also be brought into agreement with the experimental results. In many cases parasitic beams carry the same information as the main beam and this relates such measurements to "weak" measurements. The coupling and entanglement of various parameter spaces guide us to a more elaborate discussion of quantum effects.

1. Introduction

The ongoing discussion about the interpretation of quantum physics which oscillates between very mystic and more rational interpretations. John Bell contributed to this discussion and made very useful proposal for related test measurements (Bell 1987). In this paper it will be shown that in all experiments there are some kinds of hidden observables mainly because not all pre- and post-selection possibilities are utilised and average values are measured where more detailed investigations would be possible. In this respect we discuss neutron experiments where the particle and wave features are essential and where one can show that much more information can be extracted by using more sophisticated measurement methods. This will be demonstrated by means spatial-, momentum- , time- and polarization post-selection experiments. We will also demonstrate that unavoidable quantum losses may play an important role in the interpretation of the quantum to classic transition. These losses can be used as weak measurements to obtain basic information about the main object without interacting with it. In some sense the analysis follows the pragmatic access of Englert (2013) who showed how quantum phenomena can be explained when one follows strictly the state of knowledge one gains from an experiment starting from a few up to a high number of events and where the averaging procedure becomes more reliable. In this respect one includes the measurement time, or the number of particles used, into the analysis.

Our approach uses neutron interferometer experiments where two coherent neutron beams are produced by dynamical Bragg diffraction from a perfect silicon crystal and they are superposed at the exit crystal plate and exhibit all well-known interference phenomena (Fig. 1; e.g. Rauch and Werner 2015). Since one deals with a stationary situation one can use the time-independent Schrödinger equation and gets for the beams behind the interferometer a superposition of beam I and II which are transmitted-reflected-reflected ψ_{trr} and reflected-reflected-transmitted ψ_{rrt} respectively. From symmetry follows that they are equal in intensity and phase. A phase shift χ can be applied, which is given by the index of refraction of any material and related to a spatial shift $\vec{\Delta}$ $(\chi = \vec{\Delta}.\vec{K})$.

$$I_0 = |\psi_I + \psi_{II}|^2 = |\psi_{trr} + \psi_{rrt} e^{i\chi}|^2 = |\psi_0|^2 \left(1 + |\Gamma(\vec{\Delta})| \cos\chi\right) \quad (1)$$

$\Gamma(\vec{\Delta})$ denotes the coherence function, which defines the coherence lengths $\vec{\Delta}_c$ as its characteristic dimension and is related to the size of the wave packets involved

$$|\Gamma(\vec{\Delta})| = \left|\langle\psi(0)\psi(\vec{\Delta})\rangle\right| \propto \left|\int g(\vec{k}) e^{i\vec{\Delta}\vec{k}} d\vec{k}\right| \quad (2)$$

For Gaussian beams with momentum widths δk one obtains

$$|\Gamma(\Delta)| = \exp\left[-(\Delta.\delta k)^2/2\right] \quad (3)$$

where $g(\vec{k})$ is the momentum distribution. In an experiment Eq. (1) can be approximated by

$$I_{exp} = A(1 + V\cos(\chi + \varphi)) \quad (4)$$

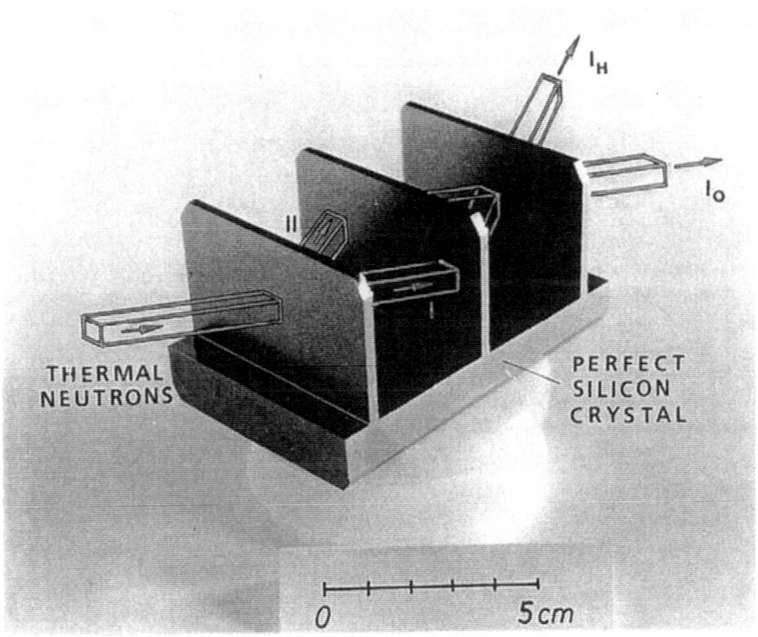

Fig. 1: Photo of a perfect crystal neutron interferometer

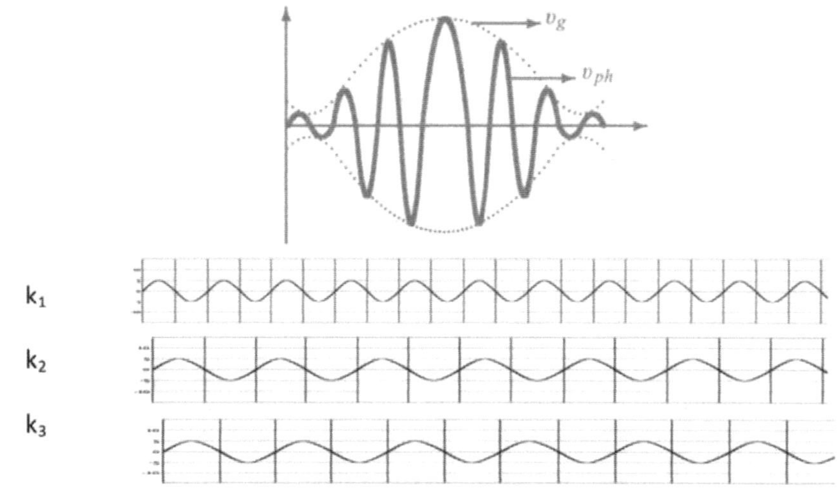

Fig. 2: Wave packet (above) and partial waves (below).

V denotes the visibility of the interference pattern and φ an internal phase caused by some small deviations from the perfectness of the crystal or due to external effects like gravity or magnetic fields. In practice high visibilities (up to 95%) and high order interferences (up to 200th) have been observed.

2. Analysis of the parameters in Eqs. 1-4

All parameters of Eqs. 1-4 are average values over the beam cross section, the momentum distribution and the measurement time. In the following sections we intend to analyse the various parameters of the equations shown above. We try to show in the following sections how they can be specified by various post-selection experiments.

2.1 Wave function – momentum post-selection

The wave function follows from the solution of the time-independent Schrödinger equation and can be written for free space motion as a wave packet centred around k_0 with a width. For Gaussian shaped beams this can be written as

$$-\frac{\hbar^2}{2m}\Delta\psi = E\psi \tag{5}$$

$$\psi(r) \propto \int a(k)e^{ikr}dk \propto \int e^{-(k-k_0)^2/2\delta k^2} e^{ikr}dk \tag{6}$$

which is an eigenvalue solution for the freely moving particle. $a(k)$ denotes the amplitudes of the coherently superposed partial waves and $|a(k)|^2$ the momentum distribution function as used in Eq. 2. Whereas the wave packet is localized within a region compatible with its coherence length

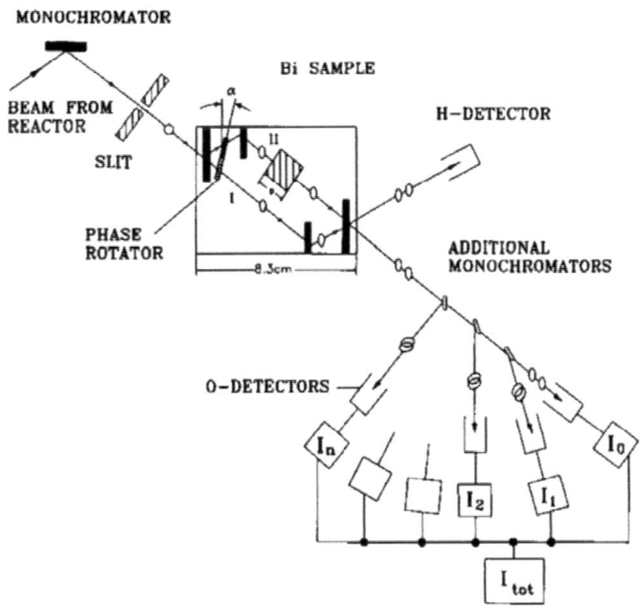

Fig.3: Sketch of a feasible momentum post-selection experiment using various monochromaters (1, ... n) to observe wave features and where the intensities can be summed up to reach I_0, i.e. a situation which permits beam path (particle) information.

$(\Delta_c \delta k \geq 1/2)$ the partial waves are arbitrarily widely spread as shown in Fig. 2. That means that information about the particle exists even far away from the packet. This has to be taken into account when non-locality effects are discussed (e.g. Rauch 1993). Proper momentum post-selection measurements have made these partial waves visible by applying additional monochromatization (Kaiser et al. 1992, Rauch et al. 1996). This shows how more information can be extracted by means of more sophisticated experimentation. Fig. 3 shows an arrangement where some of these far reaching components of the wave functions can be analysed and fringe visibility can be preserved although the beam I_0 without additional monochromator crystals and the artificially summed up intensity ($\sum_1^n I_n$) of all measuring channels do not show interference features at all. This means that one can decide after the measurement whether one is more interested in wave features (individual channel intensities) or particle features (summed up intensity). This has to be discussed on the basis of the Greenberger-Englert duality relation which treats wave and particle properties as a duality system where the particle feature is determined by the path distinguishability (P_D) and the wave features by the visibility (V) of the interference pattern (Greenberger and Yasin 1988, Englert 1996). This indicates that a single particle system can exhibit far reaching features outside its wave packet range and that a two or many body system never separates when it has a common origin.

$$P_D^2 + V^2 \geq 1 \qquad (7)$$

2.2 Beam cross section and wave packets – position post-selection

As mentioned above the size of the wave packet is given by the Fourier transform of the momentum distribution function (Eq. 6) and can be determined from the measurement of the coherence function (Eq. 2), i.e. from the decrease of the visibility at high interference order (Rauch et al. 1996). For a neutron interferometer situation the coherence lengths are different for the various directions due to the different momentum distributions ($\Delta_{longitudinal} \approx 100A$, $\Delta_{vertical} \approx 50A$, $\Delta_{transvers} \approx 20\mu m$), which are much smaller than a typical beam cross section ($\approx 1x2\,cm$). When one considers the available intensity of about 10^4 neutrons/second one notices that one deals with single particle interference and there is no interaction between successive neutrons besides the fact that they are shaped by the same monochromators, collimators and crystal reflections. Thus a situation as shown in Fig. 4 exists where the wave packets of the individual neutron do not overlap at all, but plane wave components do when more than one neutron passes at the time, which is very seldom the case.

The intensity and the coherence features vary across the beam cross section and therefore the parameters of Eq. 4 vary as well, as shown in Fig. 5. This is a position post-selection result when a position sensitive detector is used to scan the intensity, the contrast (V or $|\Gamma(\Delta)|$) and the internal phase (φ) of the interference pattern. One notices that the measured interference pattern depends on the size and position of the aperture used. One can also measure the momentum distribution at any position and will notice some differences. This means that the wave-functions are different at any position of the aperture, but they are similar to each other which permits the definition of a mean wave-function for any beam. One should keep in mind that in any experiment an average is taken over aperture area and various momentum distributions at any position within the beam cross section. This underlines that the wave packet features are determined by the apparatus only but each neutron of the beam experience a similar history which causes common features and determine the coherence properties of the beam.

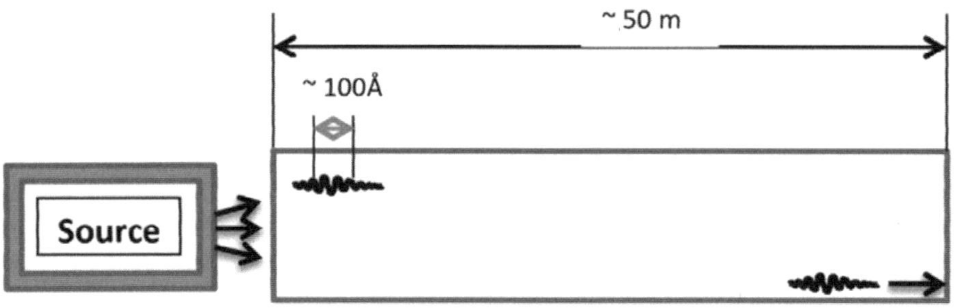

Fig. 4: Sketch of wave packets within the neutron flight path

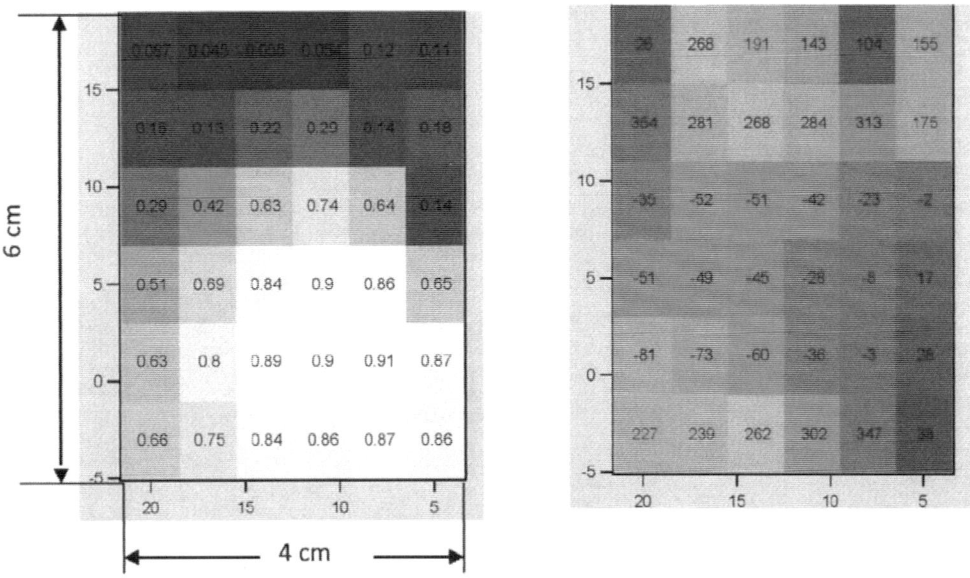

Fig. 5: Result of a position post-selection experiment showing the contrast (left) and the internal phase (right) across the beam cross section.

2.3 Time – post-selection

Neutron choppers can be used to measure the velocity distribution of the beam by means of time-of-flight methods. These methods are similar to the momentum post-selection methods discussed in Sec. 2.1 (Rauch et al. 1992, Jacobson et al. 1996). The energy-resolved interference pattern show a higher visibility than the full beam. A classical analysis of the data is possible as long as the resolution time of the chopper is larger than the coherence time of the beam (5 μs compared to 10 ns).

Another method measures the arrival time of each neutron. For a thermal (Poissonian) beam the probability to measure a neutron within a time interval t after another neutron has arrived is given by (Glauber 1968)

$$W(t) = Ie^{-It} \tag{8}$$

This is shown in Fig. 6 (Zawisky et al. 1994). When measuring two neutrons arriving within short or long time intervals one can achieve a considerable higher contrast and phase sensitivity than analysing the full beam only. Since arrival time measurements can be implemented quite easily this opens a new possibility to improve experimental results. A more complete analysis has been given on the basis of a Bayes estimation by Rehacek et al. 1999. The mean time interval between two arriving neutrons is given by $\bar{\tau} = 1/I$. One notices that for "long" pairs the contrast always reaches nearly 100% and that the interference pattern becomes shifted by π.

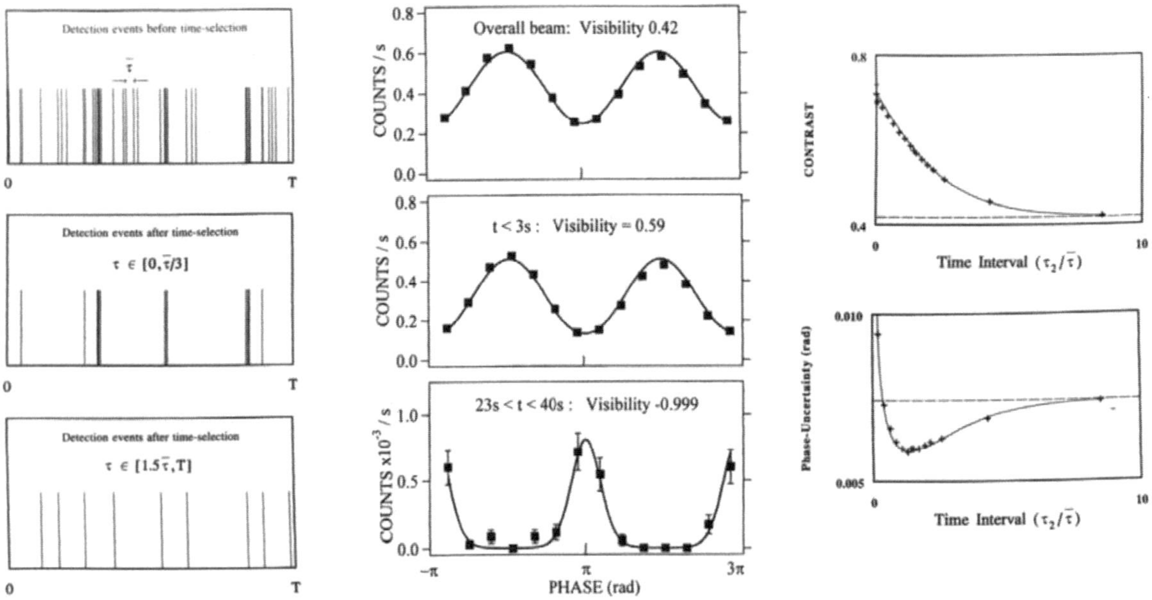

Fig. 6: Arrival times of neutrons and short pair and long pair arrivals (left), measured interference pattern for the overall beam and the short and long pair arrivals (middle) and the measured and calculated contrast and sensitivity (right).

3. Unavoidable losses

In many cases losses during a quantum measurement are neglected and treated as caused by experimental imperfections only. Here we deal with unavoidable losses caused by the theory itself. Such losses may become important for the understanding of quantum decoherence and the quantum measurement process (Zeh 1970, Zurek 1981, Joos et al. 2003). It will be shown that not only dissipative interactions cause an irreversible change of the wave-function, but deterministic ones can cause such a change too. We start with the phase echo experiment where a large positive phase shift should be compensated by a large negative one. This has been verified to a high degree (Clothier et al. 1991), but a closer lpok shows that each phase shifter causes additional back and forth reflections as indicated in Fig.7. When the energy of the particle E is much larger than the height of the barrier ($\overline{V}$) and its thickness produces phase shifts larger than the coherence length of the beam the reflectivity can be written as (e.g. Cohen-Tannoudji et al. 1977)

$$\overline{R} \cong \frac{1}{2}\left(\frac{\overline{V}}{2E}\right)^2 . \tag{9}$$

For thermal neutrons this is very small (10^{-10}) but unavoidable even when specially shaped barriers are taken into account. It should be mentioned that the same information which exists in the main

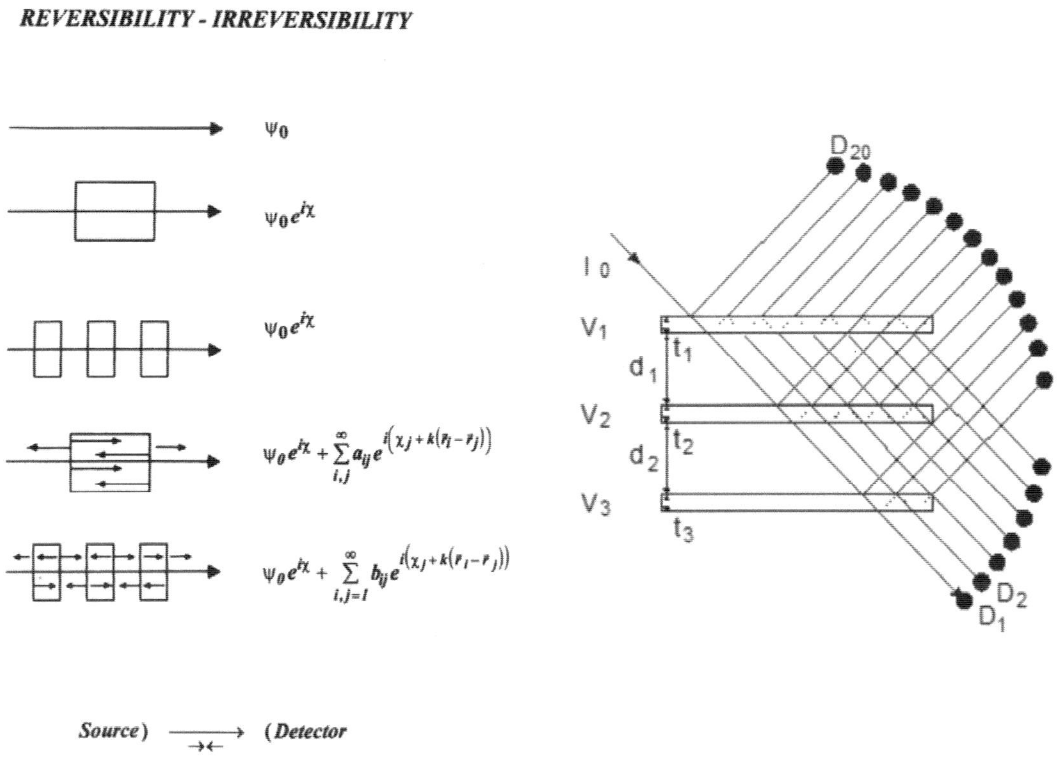

Fig. 7: Approximate and complete wave-functions when differently shaped phase shifters are used (left) and a method to measure these parasitic beams (right).

beam (D_1) is available in the parasitic beams ($D_2 - D_{20}$). This relates such measurements to "weak" measurements (Aharonov et al. 1988, Dressel et al. 2014). The measurement period may be much longer but that does not enter the analysis and there is no method to retrieve all components into the original beam.

4. Discussion

These measurements and their analysis have shown that many coherence properties of a beam can be retrieved by various post-selection methods. Nevertheless a complete retrieval cannot be achieved due to unavoidable losses. These losses need not be caused by dissipative forces but can also be caused by unavoidable quantum losses. Thus irreversibility seems to start with the first interaction the system experiences and can take place at any time scale. In this respect irreversibility and the related measurement process seems to be part of the quantum formalism as stated by Englert (2013). Wave and particle features (interference and beam path) are related according to Eq.7 and they can be taken as a two level system which can be entangled in a contextual sense. This shows that quantum physics involves more entanglements and makes the world and the human being more correlated than classical physics.

References:

Aharonov Y., Albert D.Z., Vaidman L. (1988) Phys. Rev. Lett. 60, 1351

Bell J.S. (1987) „Speakable and unspeakable in quantum mechanics", Cambridge University Press

Clothier R., Kaiser H., Werner S.A., Rauch H., Wölwitsch H. (1991) Phys. Rev. A44, 5357

Cohen-Tannoudji C., Diu B., Laloe F. (1977) "Quantum Mechanics", Vol. 1 (John Wiley & Sons, N.Y.)

Dressel J., Malik M., Miatto F.M., Jordan A.N. (2014) Rev. Mod. Phys. 86, 307

Englert B.-G. (1996) Phys. Rev. Lett. 77, 2154

Englert B.-G. (2013) Eur. Phys. J. 67, 238

Glauber R.J. (1968) *„Fundamental Problems in Statistical Mechanics"*, Ed. E.G.D. Cohen (North Holland)

Greenberger D.M., Yasin A. (1988) Phys. Lett. 128, 391

Jacobson D.L., Allman B.E., Zawisky M., Werner S.A., Rauch H. (1996) J. Jap. Phys. Soc. A65, 94

Joos E., Zeh H.D., Kiefer C., Giulini D., Kupsch J., Stamatescu I.-O. (2003), „Decoherence and the Appearance of a Classical World", 2nd Ed. Springer, Berlin

Kaiser H., Clothier R., Werner S.A., Rauch H., Wölwitsch H. (1992) Phys. Rev. A45, 31

Rauch H. (1993) Phys. Lett. A173, 240

Rauch H., S.A. Werner (2015) „Neutron Interferometry", Clarendon Press, Oxford

Rauch H., Wölwitsch H., R. Clothier, Kaiser H., Werner S.A. (1992) Phys. Rev. A46, 49

Rauch H., Wölwitsch H., Kaiser H., Clothier R., Werner S.A. (1996) Phys.Rev. A53, 902

Rehacek J., Hradil Z., Zawiska M., Pascazio S., Rauch H., Perina J. (1999) Phys. Rev. A60, 473

Zawisky M., Rauch H., Hasegawa Y. (1994) Phys. Rev. A50, 5000

Zeh H.D. (1970) Found. Phys. 1, 69

Zurek W.H. (1981) Phys. Rev. D24, 1516

Error - disturbance uncertainty relations studied in neutron optics

Stephan Sponar [1], Georg Sulyok[1], Bülent Demirel [1], and Yuji Hasegawa[1]

[1] Atominstitut - TU Wien, 1020 Vienna, Austria

E-mail: sponar@ati.ac.at

Abstract. Heisenberg's uncertainty principle is probably the most famous statement of quantum physics and its essential aspects are well described by a formulations in terms of standard deviations. However, a naive Heisenberg-type error-disturbance relation is not valid. An alternative universally valid relation was derived by Ozawa in 2003. Though universally valid Ozawa's relation is not optimal. Recently, Branciard has derived a tight error-disturbance uncertainty relation (EDUR), describing the optimal trade-off between error and disturbance. Here, we report a neutron-optical experiment that records the error of a spin-component measurement, as well as the disturbance caused on another spin-component to test EDURs. We demonstrate that Heisenberg's original EDUR is violated, and the Ozawa's and Branciard's EDURs are valid in a wide range of experimental parameters, applying a new measurement procedure referred to as *two-state method.*

1. Introduction

Heisenberg's uncertainty principle represents, without any doubt, one of the cornerstones of quantum mechanics. In his seminal paper, which was published in 1927, Heisenberg originally introduced a relation between the precision of a position measurement and the disturbance it induces on a subsequent momentum measurement of a particle. This is illustrated in the famous γ-ray microscope thought experiment, which is solely based on the Compton-effect: "*At the instant when the position is determined - therefore, at the moment when the photon is scattered by the electron - the electron undergoes a discontinuous change in momentum. This change is the greater the smaller the wavelength of the light employed - that is, the more exact the determination of the position.*" [1]. A relation is given by $q_1\, p_1 \sim h$, for the product of the mean error q_1 of a position measurement (error) and the discontinuous change (disturbance) p_1 of the particle's momentum. Heisenberg's original formulation can be read in terms modern treatment of quantum mechanics as

$$\epsilon(Q)\eta(P) \geq \frac{\hbar}{2}, \tag{1}$$

for error $\epsilon(Q)$ of a measurement of the position observable Q and disturbance $\eta(P)$ of the momentum observable P. In the subsequent mathematical derivation of Eq. (1), based on the commutator relation $QP - PQ = i\hbar$, Heisenberg showed that the product of the position and momentum standard deviations is given by

$$\Delta Q\, \Delta P \geq \frac{\hbar}{2}, \tag{2}$$

for a class of Gaussian wavefunctions (with standard deviation $\Delta X = \sqrt{\langle X^2\rangle - \langle X\rangle^2}$). Equation 2 was proven shortly afterwards in [2] for arbitrary wavefunctions by Kennard. In 1929 Robertson [3] generalized Eq. (2) to arbitrary pairs of observables A and B as

$$\Delta Q\,\Delta P \geq \frac{1}{2}|\langle\psi|[A,B]|\psi\rangle|, \tag{3}$$

for any state $|\psi\rangle$, with $\Delta Q\,\Delta P < \infty$ and commutator $[A,B] = AB - BA$. Equations 2 and 3 for standard deviations are uncontroversial and have been confirmed by many different experiments of various quantum systems [4, 5, 6].

However, Eqs. 2 and 3 describe the limitation of preparing quantum objects and has no direct relevance to the limitation of measurements on single systems, as suggested by Heisenberg's γ-ray microscope thought experiment [7]. The generalized form of Heisenbergs original error-disturbance relation, as in Eq. (1), would read

$$\epsilon(A)\eta(B) \geq \frac{1}{2}|\langle\psi|[A,B]|\psi\rangle|. \tag{4}$$

The validity of Eq. (4) is known to be limited to specific circumstances [8, 9, 10]. In 2003, Ozawa thus proposed in [11] a new error-disturbance uncertainty relation

$$\epsilon(A)\eta(B) + \epsilon(A)\Delta(B) + \Delta(A)\eta(B) \geq \frac{1}{2}|\langle\psi|[A,B]|\psi\rangle|, \tag{5}$$

and proved its universal validity in the general theory of quantum measurements. Here $\epsilon(A)$ denotes the root-mean-square (r.m.s.) error of an arbitrary measurement for an observable A, $\eta(B)$ is the r.m.s. disturbance on another observable B induced by the measurement, and $\Delta(A)$ and $\Delta(B)$ are the standard deviations of A and B in the state $|\psi\rangle$ before the measurement. Ozawa's inequality Eq. (5) was tested experimentally in matter wave optics using neutrons [12, 13] and later with photons [14, 15].

Though universally valid Ozawa's relations Eq. (5) is not optimal. Recently, Branciard [16] showed that one can improve on the sub-optimality of Ozawas proof and derived the following trade-off relation between error $\epsilon(A)$ and disturbance $\eta(B)$:

$$\epsilon(A)^2\Delta(B)^2 + \Delta(A)^2\eta(B)^2 + 2\epsilon(A)\eta(B)\sqrt{\Delta(A)^2\Delta(B)^2 - C_{AB}^2} \geq C_{AB}^2, \tag{6}$$

with $C_{AB} = {}^1\!/_2|\langle\psi|[A,B]|\psi\rangle|$. For the special case $\langle\psi|A|\psi\rangle = \langle\psi|B|\psi\rangle = 0$, which implies $\Delta(A) = \Delta(B) = 1$, and replacing $\epsilon(A)$ and $\eta(B)$ by $\epsilon(A)\sqrt{1-\frac{\epsilon(A)^2}{4}}$ and $\eta(b)\sqrt{1-\frac{\eta(B)^2}{4}}$, respectively, Eq. (6) can be strengthened yielding the tight relation

$$\epsilon(A)^2\left(1-\tfrac{\epsilon(A)^2}{4}\right) + \eta(B)^2\left(1-\tfrac{\eta(B)^2}{4}\right) + 2\sqrt{1-C_{AB}^2}\,\epsilon(A)\sqrt{1-\tfrac{\epsilon(A)^2}{4}}\,\eta(B)\sqrt{1-\tfrac{\eta(B)^2}{4}} \geq C_{AB}^2, \tag{7}$$

which has been experimentally confirmed in [17, 18] with photons.

2. Measurement theory

For Eqs. (5-7) error $\epsilon(A)$ and disturbance $\eta(B)$ are defined via an indirect measurement model for an apparatus **A** measuring an observable A of an object system **S** as

$$\begin{aligned}\epsilon(A) &= \|(U^\dagger(\mathbb{1}\otimes M)U - A\otimes\mathbb{1})|\psi\rangle|\xi\rangle\|,\\ \eta(B) &= \|(U^\dagger(B\otimes\mathbb{1})U - B\otimes\mathbb{1})|\psi\rangle|\xi\rangle\|.\end{aligned} \tag{8}$$

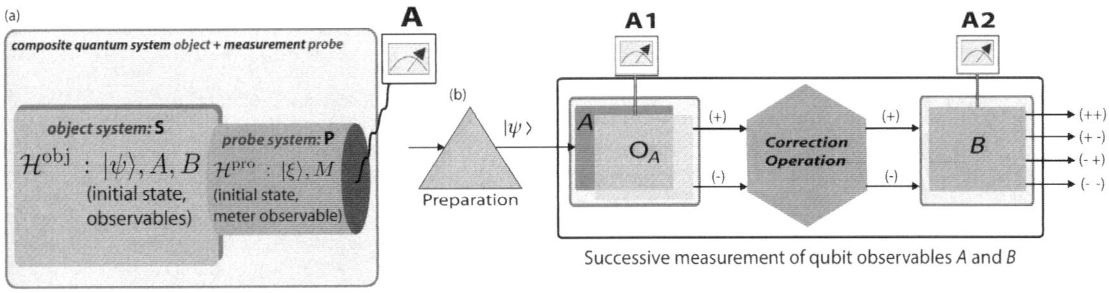

Figure 1. (a) An indirect measurement model for apparatus **A** measuring a general object system **S**, defined on Hilbert space $\mathcal{H}^{\mathrm{obj}}$, is specified by a quadruple $(\mathcal{H}^{\mathrm{pro}}, |\xi\rangle, U, M)$. (b) A successive measurement scheme of qubit observables A and B exploited for demonstration of error-disturbance uncertainty relations. After state preparation (blue) apparatus **A1** carries out a projective measurement of O_A instead of A (magenta), thereby disturbing observable B which is detected by apparatus **A2** (green). Between these two measurements a *correction operation* is performed, in order to minimize the disturbance $\eta(B)$. Error $\epsilon(A)$ and disturbance $\eta(B)$ are quantitatively determined by the four possible outcomes denoted as $(++)$, $(+-)$, $(-+)$, and $(--)$.

Here $|\psi\rangle$ is the initial state of system **S**, which is associated with Hilbert space $\mathcal{H}^{\mathrm{obj}}$. $|\xi\rangle$ is the state of the probe system **P** before the measurement, defined on Hilbert space $\mathcal{H}^{\mathrm{pro}}$, and an observable M, referred to as *meter observable* of **P**. The time evolution of the composite system $\mathbf{S}+\mathbf{P}$ during the measurement interaction is described by a unitary operator U on $\mathcal{H}^{\mathrm{obj}} \otimes \mathcal{H}^{\mathrm{pro}}$. In Eq. (8) the Hilbert-Schmidt norm is used where the norm of a state vector in Hilbert space $X|\psi\rangle$ is given by the square root of its inner product: $||X|\psi\rangle|| = \langle\psi|X^\dagger X|\psi\rangle^{1/2}$. A schematic illustration of a measurement apparatus **A** is given in Fig. 1 (a).

A non-degenerate meter observable M has a spectral decomposition $M = \sum_m m|m\rangle\langle m|$, where m varies over eigenvalues of M, then the apparatus **A** has a family $\{M_m\}$ of operators, called the *measurement operators*, acting on $\mathcal{H}^{\mathrm{obj}}$ and defined as $M_m = \langle m|U|\xi\rangle$. Hence, the error is given by $\epsilon(A)^2 = \sum_m ||M_m(m - A)|\psi\rangle||^2$. If M_m are mutually orthogonal projection operators sum and norm can be exchanged and the error can be written in compact form as

$$\epsilon(A)^2 = \sum_m ||(O_A - A)|\psi\rangle||^2, \tag{9}$$

where O_A is the *output operator* given as $O_A = \sum_m m M_m$. The disturbance can be written as

$$\eta(B)^2 = \sum_m ||[M_m, B|\psi\rangle||^2. \tag{10}$$

All these calculations are elaborated in detail in [10].

2.1. *Threes-state versus two-state method*

Due to occurrence of the term $O_A A + A O_A$ $[X_B B + B X_B]$ when evaluating Eq. (9) [Eq. (10)] error $\epsilon(A)$ [disturbance $\eta(B)$] were claimed to be experimentally inaccessible [19, 20]. However, applying the operator identity $O_A A + A O_A = (A + \mathbb{1})O_A(A + \mathbb{1}) - A O_A A - O_A$ measurement error and disturbance are given by

$$\begin{aligned} \epsilon(A)^2 &= 2 + \langle\psi|O_A|\psi\rangle + \langle\psi|A O_A A|\psi\rangle - \langle\psi|(A+\mathbb{1})O_A(A+\mathbb{1})|\psi\rangle, \\ \eta(B)^2 &= 2 + \langle\psi|X_B|\psi\rangle + \langle\psi|B X_B B|\psi\rangle - \langle\psi|(B+\mathbb{1})X_B(B+\mathbb{1})|\psi\rangle, \end{aligned} \tag{11}$$

where X_B represents the *modified output operator*, given by $X_B = \sum_m M_m^\dagger B M_m$. Consequently error $\epsilon(A)$ and disturbance $\eta(B)$ can be expressed as a sum of expectation values in *three* input states $|\psi\rangle$, $A|\psi\rangle$, $(A+\mathbb{1})|\psi\rangle$ and $|\psi\rangle$, $B|\psi\rangle$, $(B+\mathbb{1})|\psi\rangle$, respectively, which was proposed in [21]. The *three-state method* was used in our previous experiments reported in [12, 13].

Here we introduce the so called *two-state method* using the operator identity $O_A A + A O_A = (A+\mathbb{1})O_A(A+\mathbb{1}) - (A-\mathbb{1})O_A(A-\mathbb{1})$ with error and disturbance denoted as

$$\begin{aligned} \epsilon(A)^2 &= 2 + \langle\psi|(A-\mathbb{1})O_A(A-\mathbb{1})|\psi\rangle - \langle\psi|(A+\mathbb{1})O_A(A+\mathbb{1})|\psi\rangle, \\ \eta(B)^2 &= 2 + \langle\psi|(B-\mathbb{1})X_B(B-\mathbb{1})|\psi\rangle - \langle\psi|(B+\mathbb{1})X_B(B+\mathbb{1})|\psi\rangle, \end{aligned} \tag{12}$$

where $(A \pm \mathbb{1})|\psi\rangle$ and $(B \pm \mathbb{1})|\psi\rangle$ are the projections onto the eigenstates of A and B (multiplied by a factor 2). Note that in the *two-state method* the initial state $|\psi\rangle$ is actually never sent onto the apparatus. Its effect is only enters via the auxiliary states $(A \pm \mathbb{1})|\psi\rangle$ and $(B \pm \mathbb{1})|\psi\rangle$.

2.2. Qubit measurements

In the case of qubit measurements the apparatus illustrated in Fig. 1 (a) has two output ports denoted as (+) and (−). The expectation values of O_A [X_B] from Eqs. (12), necessary for the determination of error $\epsilon(A)$ [disturbance $\eta(B)$], are derived from the intensities at the four possible output ports, depicted the measurement scheme for qubit observables in Fig. 1 (b), denoted as I_{++}, I_{+-}, I_{-+} and I_{--}. The expectation value $\langle\psi|O_A|\psi\rangle$ is obtained from the following combination of count rates

$$\langle\psi|O_A|\psi\rangle = \frac{I_{++} + I_{+-} - I_{-+} - I_{--}}{I_{++} + I_{+-} + I_{-+} + I_{--}}. \tag{13}$$

As already discussed due to the prior measurement of O_A the operator measured by apparatus **A2** is modified from B to X_B, with the corresponding expectation value expressed as

$$\langle\psi|X_B|\psi\rangle = \frac{I_{++} + I_{-+} - I_{+-} - I_{--}}{I_{++} + I_{+-} + I_{-+} + I_{--}}, \tag{14}$$

required to determine the disturbance $\eta(B)$. Consequently all expectation values necessary to determine error $\epsilon(A)$ and disturbance $\eta(B)$ can be derived from the intensities in the two input states $(A \pm \mathbb{1})|\psi\rangle$ and $(B \pm \mathbb{1})|\psi\rangle$, respectively.

3. Experiment

In our experiment the error-disturbance uncertainty relations, as defined in Eqs. (5) and (7), are tested via a successive measurement for qubit observables A and B. The observables A and B are represented by the z- and y-components of the neutron 1/2 spin. A schematic illustration of the setup is given in Fig. 2. The error $\epsilon(A)$ and the disturbance $\eta(B)$ are defined for a joint measurement apparatus, where apparatus **A1** measures the observable $A = \sigma_z$ with error $\epsilon(A)$ for an initial state $|\psi_i\rangle = |+x\rangle$ and disturbs the observable $B = \sigma_y$ thereby with disturbance $\eta(B)$ during the measurement (here σ_z and σ_y denote the Pauli matrices; for simplicity, $\hbar/2$ is omitted in each spin component). Finally apparatus **A2** measures $B = \sigma_y$. This particular combination of observables yields $C_{AB} = 1$ reducing Eq. 7 to

$$\epsilon(A)^2\left(1 - \tfrac{\epsilon(A)^2}{4}\right) + \eta(B)^2\left(1 - \tfrac{\eta(B)^2}{4}\right) \geq 1. \tag{15}$$

Since a precise measurement of A (σ_z) would disturb a posterior B-measurement maximally, an *approximate* measurement of A, referred to as O_A, is performed instead. Thus, apparatus **A1** is designed to actually carry out not the maximally disturbing projective measurement $A = \sigma_z$,

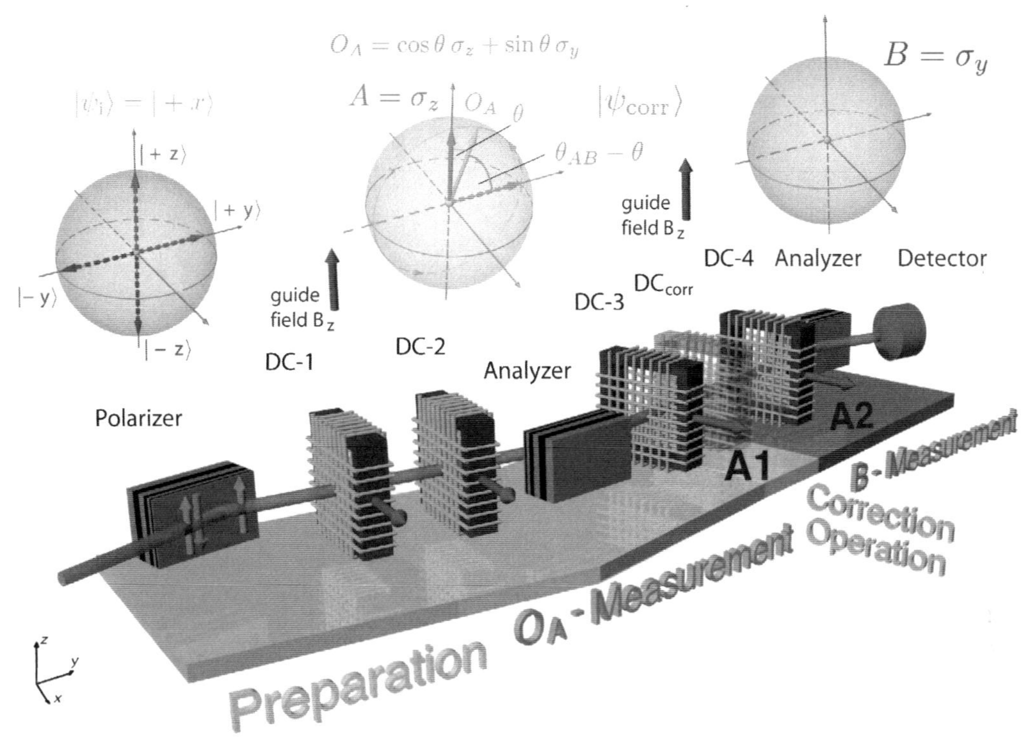

Figure 2. Neutron polarimetric setup for demonstration of the universally valid uncertainty relation for error and disturbance in neutron spin measurements. The setup is divided in four stages: state preparation (blue region), apparatus **A1** carrying out the measurement of observable $O_A = \vec{o}_a(\theta) \cdot \vec{\sigma}$ (magenta region), plus a correction operation (light green region) where the state is rotated by an angle $\theta_{AB} - \theta$, with θ_{AB} denoting the relative angle between the measurement directions of A and B (in our case that is $\theta_{AB} = \pi/2$). Apparatus **A2** performs a subsequent measurement of observable $B = \sigma_y$ (dark green region). The required terms of Eqs. (5) and (7), i.e., error $\epsilon(A)$ and disturbance $\eta(B)$, are determined from the expectation values of the successive measurement of O_A and B (transparent coil $\mathrm{DC_{corr}}$ is virtual, in practice the rotation is also carried out by DC-3).

but instead the projective measurement along a distinct axis $\vec{o}_a(\theta)$, in the yz plane, denoted as $O_A = \vec{o}_a(\theta) \cdot \vec{\sigma}_{y,z} = M_{+1} - M_{-1}$ (where $M_{\pm 1} = 1/2(\mathbb{1} \pm \vec{o}_a(\theta) \cdot \vec{\sigma}_{y,z})$ and $\vec{\sigma}_{y,z} = (\sigma_y, \sigma_z)^T$. Here θ denotes the angle of the measurement direction $\vec{o}_a$ and is our experimentally controlled detuning parameter, so that $\epsilon(A)$ and $\eta(B)$ are determined as a function of θ. Next a unitary operation, as proposed in [16], given by $U_{corr} = e^{-i(\theta_{AB}-\theta)/2\,\sigma_x}$, is performed directly after the projective measurement of O_A, is supposed to reduce the influence of the prior O_A-measurement on the subsequent B-measurement optimally. For qubit-measurements this means that the state is rotated by an angle $\theta_{AB} - \theta$, where θ_{AB} denotes the relative angle between the measurement directions of A and B. When there is no correction operation applied U_{corr} is simply given by $U_{corr} = \mathbb{1}$. In order to detect the disturbance $\eta(B)$ on the observable B, induced by measuring O_A, apparatus **A2** carries out the projective measurement of $B = \sigma_y$ in the state just after the first measurement.

The experiment was carried out at the tangential beam port of the 250 kW research reactor facility TRIGA Mark II of the TU-Wien, Austria. In the preparation section (Fig. 2, blue region)

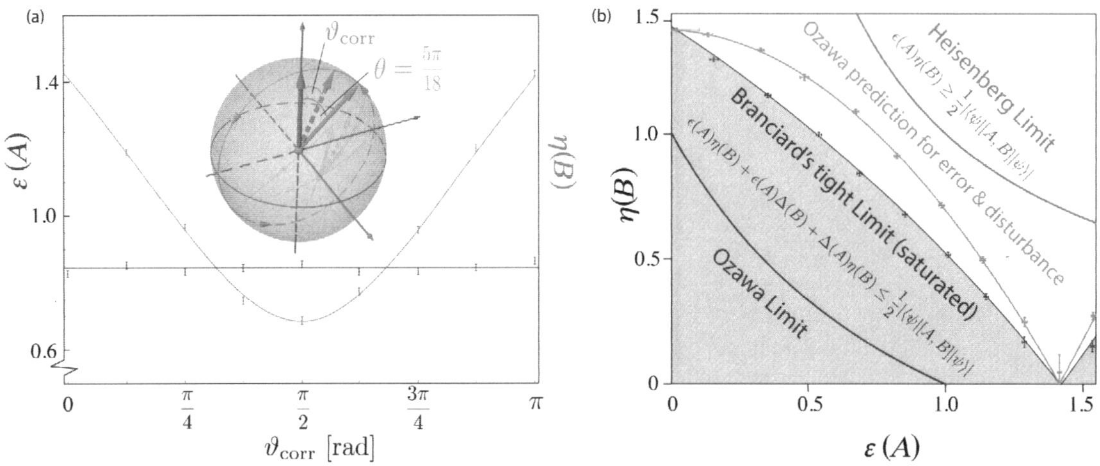

Figure 3. (a) Experimental search of the optimal correction operation: After the projective measurement of O_A (fixed at $\theta = 5\pi/18$) plus unitary rotations U_{corr} the system leaves apparatus **A1** in the state $|\psi(\vartheta_{\mathrm{corr}})\rangle$. A minimum of the disturbance $\eta(B)$ is found for $\vartheta_{\mathrm{corr}} = \pi/2$, while the error $\epsilon(A)$ is not affected by the correction operation, as expected, and thus remains constant. (b) Results of error and disturbance plotted as $\epsilon(A)$ versus $\eta(B)$. Blue curve: the Branciard bound as defined in Eq. (6). Blue marker: Experimental results using the modified apparatus with optimal correction procedure, demonstrating the tightness of this relation. Orange curve error and (sub-optimal) disturbance as in Eq. (5) lhs, without performing any correction operation. Green curve: bound imposed by the Heisenberg's original error-disturbance relation $\epsilon(A)\eta(B) \geq |1/2\langle\psi|[A,B]|\psi\rangle|$, which is violated by our experimental results. Red curve: Ozawas relation Eq. (5) rhs, which is indeed satisfied, but is not saturated.

the auxiliary input states states $(A \pm \mathbb{1})|\psi\rangle = |\pm z\rangle$ and $(B \pm \mathbb{1})|\psi\rangle = |\pm y\rangle$ are generated by spinor rotations within DC-1, due to Larmor precession about the static magnetic field that is produced inside the DC-1 coil, pointing in $+x$-direction. The projective measurement of O_A (apparatus **A1** - magenta in Fig. 2) consist of two sequential operations: first the initially prepared state is projected onto the eigenstates of O_A by DC-2, which rotates the respective spin component of $\vec{\sigma}_a$ belonging to O_A in $+z$ direction. Then, in order to complete the projective measurement the spin, which is pointing in $+z$ after the analyzer, has to be prepared in an eigenstate of O_A. This is achieved by an appropriate adjusted magnetic field of DC-3 (thereby applying the same procedure as for DC-1 in the initial state preparation). The transparent coil $\mathrm{DC_{corr}}$ (light green in Fig. 2) is only for illustrative purpose. In the actual experiment the spinor rotation of $\theta_{AB} - \theta$, ($\theta_{AB} = \pi/2$) about the x-axis is also carried out by DC-3. Finally the B-measurement is performed (apparatus **A2** - dark green in Fig. 2) utilizing DC-4 and the second analyzer. Unlike for the O_A-measurement subsequent preparation of the eigenstates of B is not necessary because the count-rate detection done by a high-efficeny BF_3-detector is spin-insensitive.

3.1. Experimental results

No correction procedure - In the first experimental run no additional correction operation is applied, which means that $U_{\mathrm{corr}} = \mathbb{1}$. The observables $O_A = \cos\theta\,\sigma_z + \sin\theta\,\sigma_y$ and $B = \sigma_y$ are successively measured, resulting in error $\epsilon(A)$ and disturbance $\eta(B)$ as defined for Ozawa's universally valid error-disturbance uncertainty relations from Eq. (5). The obtained results, together with the theoretical prediction and the lower bound, are plotted in Fig. 3 (b), orange curve. Though the theoretical prediction are reproduced evidently and the simple version of Heisenbergs' error-disturbance relation $\epsilon(A)\eta(B) \geq 1/2|\langle\psi|[A,B]|\psi\rangle|$ is violated for all values of our experimentally controlled parameter θ, it can be seen that Eq. (5) is only tight for the trivial cases $\epsilon(A) = 0$ and $\eta(B) = \sqrt{2}$ or vice versa.

Optimal correction procedure - In order to strengthen Eq. (5) (apart from the two points $\eta(B) = 0, \sqrt{2}$) a correction operation can be applied to minimize the disturbance $\eta(B)$. The class of correction operations we study here are unitary transformations in principal realized by an additional spin turner device, denoted as $\mathrm{DC_{corr}}$ (light green in Fig. 2). However, in the actual experiment state preparation after the O_A-measurement and correction operation U_{corr} are both performed using DC-3 alone. In Fig. 3 (a) experimental results are presented, where after the projective measurement of O_A (with detuning fixed at $\theta = 5\pi/18$) the eigenstates of the O_A-measurement is rotated within the zy-plane onto states $|\psi_{\mathrm{corr}}(\vartheta)\rangle = \cos\vartheta/2|+z\rangle + \sin\vartheta/2|-z\rangle$. A minimum of the disturbance $\eta(B)$ is found for $\vartheta = \pi/2$, while the error remains $\epsilon(A)$ remains constant because it is not affected by the correction operation. The angle $\vartheta = \pi/2$ is equivalent to a rotation of $\theta_{AB} - \theta$, with $\theta_{AB} = \pi/2$ after a O_A-measurement with detuning angle θ. As seen in the blue curve of Fig. 3 (b) and proposed in [16] this correction operation is indeed optimal, yielding a tight error-disturbance inequality.

4. Conclusions

We have demonstrated experimental validity of Ozawa's universally valid error-disturbance uncertainty relation as well as its tightened version derrived by Branciard in a neutron polarimetric experiment. The theoretical predictions are reproduced evidently at a high degree of accuracy.

Acknowledgments

This work was supported by the Austrian science fund (FWF) Projects No. P24973-N20 and No. P27666-N20.

[1] Heisenberg W 1927 *Z. Phys.* **43** 172-198
[2] Kennard E H 1927 *Z. Phys.* **44**(4) 326-352
[3] Robertson H P 1929 *Phys. Rev.* **34** 163-164
[4] Shull C G 1969 *Phys. Rev.* **179**(3) 752-754
[5] Yuen H P 1976 *Phys. Rev. A* **13**(6) 2226-2243
[6] Breitenbach G, Schiller S and Mlynek J 1997 *Nature (London)* **387** 471-475
[7] Ballentine L E 1970 *Rev. Mod. Phys.* **42**(4) 358-381
[8] Arthurs E and Goodman M S 1988 *Phys. Rev. Lett.* **60**(24) 2447-2449
[9] Wheeler J A and Zurek W H 1983 *Quantum Theory and Measurement* (Princeton Legacy Library)
[10] Ozawa M 2005 *J. Opt. B: Quantum Semiclassical Opt.* **7** 672
[11] Ozawa M 2003 *Phys. Rev. A* **67**(4) 042105
[12] Erhart J, Sponar S, Sulyok G, Badurek G, Ozawa M and Hasegawa Y 2012 *Nature Physics* **8** 185-189
[13] Sulyok G, Sponar S, Erhart J, Badurek G, Ozawa M and Hasegawa Y 2013 *Phys. Rev. A* **88**(2) 022110
[14] Rozema L A, Darabi A, Mahler D H, Hayat A, Soudagar Y and Steinberg A M 2012 *Phys. Rev. Lett.* **109**(10) 100404
[15] Baek S Y, Kaneda F, Ozawa M and Edamatsu K 2013 *Sci. Rep.* **3** 2221
[16] Branciard C 2013 *Proc. Natl. Acad. Sci. USA* **17** 6742-6747
[17] Ringbauer M, Biggerstaff D N, Broome M A, Fedrizzi A, Branciard C and White A G 2014 *Phys. Rev. Lett.* **112**(2) 020401
[18] Kaneda F, Baek S Y, Ozawa M and Edamatsu K 2014 *Phys. Rev. Lett.* **112**(2) 020402
[19] Werner R F 2004 *Quant. Inf. Comput.* **4** 546-562
[20] Koshino K and Shimizu A 2005 *Phys. Rep.* **412** 191-275
[21] Ozawa M 2004 *Ann. Phys.* **311** 350-416

Journal of Physics: Conference Series **746** (2016) 012049 doi:10.1088/1742-6596/746/1/012049

High-precision neutron spectrometry, using diffraction focusing. Test experiment

I.A. Kuznetsov[1], Ya.A. Berdnikov[2], A.Ya. Berdnikov[2], Yu.V. Borisov[1] Yu.P. Braginetz[1,2], V.V. Fedorov[1,2,3], M.V. Lasitsa[1,2], S.Yu. Semenikhin[1], M.L. Khorina[2], V.V. Voronin[1,2,3]

[1]Petersburg Nuclear Physics Institute NRC KI, Gatchina, Leningrad district, Russia
[2]Peter the Great St.Petersburg Polytechnic University, Saint-Petersburg, Russia
[3]Saint-Petersburg State University, Saint-Petersburg, Russia

E-mail: ikuz@pnpi.spb.ru

Abstract. The effect of double-crystal neutron focusing, using Laue diffraction in large perfect crystals was studied. The observed effect allows reach the angular resolution better than $0.03''$, that is $\sim 10^{-2}$ of the Bragg reflection width. This fact makes it possible to create a new ultra-precise method for neutron spectrometry combining the spin-echo small angle neutron scattering with Laue diffraction.

1. Introduction

This work was motivated by the preparation of new experiment [1] to verify the electrical neutrality of the neutron by a new method, based on spin interferometry technique SESANS (spin-echo small angle neutron scattering), see [2]. The principle of such interferometer is described in [1], where it is shown that this technique allows, in general, to improve the modern precision of the neutron electric charge measurement [3]. However, for this purpose high magnetic fields, large neutron wavelengths of a few tens angstroms, and the several meters dimensions of the setup are required [1]. That makes practically unreacheble the requirements to uniformity and stability of magnetic fields. It should be noticed that increasing the accuracy of the neutron electric charge measurements is an extremely difficult task. The attempts taken over the last 20 years trying to improve the constraint on the neutron electric charge, using, for example, ultracold neutrons [4], have not given any results.

2. Neutron diffraction

As shown in [1] the method can be significantly improved using the neutron Laue diffraction in a perfect crystal. The idea is as follows.

There is a well-known effect of diffraction inhancement when a small variation of the incident neutron beam direction within Bragg width (a few arc seconds) results in a significant change within the Bragg angle (a few tens degrees) of the neutron flux density direction inside the crystal (see. Fig. 1).

The direction of the neutron flux density in the crystal can be described by the angle Ω (measured from the crysallographic plane, i.e. from the direction of neutron flux density along

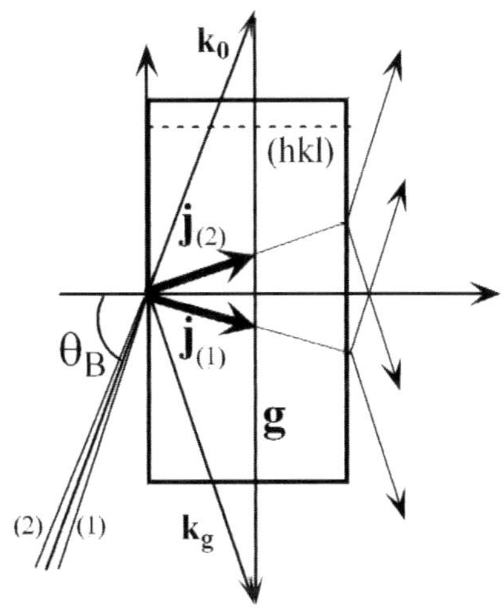

Figure 1. Fig. 1. Symmetric scheme of the neutron Laue diffraction in a perfect crystal. $\mathbf{j}_{(1)}$ and $\mathbf{j}_{(2)}$ are neutron flux density in a crystal for the two directions of the incident beam, slightly different within the Bragg (Darwin) width, $\mathbf{g}$ – reciprocal lattice vector.

the plane for exact Bragg condition) equal to

$$\Omega = \Delta\theta \cdot \frac{E}{v_g} \cdot 2\sin^2\Theta_B, \tag{1}$$

where θ_B is the Bragg angle, $\Delta\theta$ is the deviation from the exact Bragg condition, E is the neutron energy and v_g is the amplitude of g-harmonic of the periodic neutron-crystal interaction potential. This effect was used in the measurement of neutron refraction by the prism in the double-crystal diffraction pattern [5]. A similar effect for a neutron flux density deflection can be observed also due to variation of the neutron energy, because angular deviation from the Bragg condition directly depends on the change in neutron energy:

$$\Delta\theta = \frac{\Delta E}{2E}\tan\theta_B, \tag{2}$$

where θ_B is the Bragg angle, ΔE is the the change of neutron energy. This feature was used for measurements of neutron refraction in the magnetic field [6].

The two-crystal scheme of Laue diffraction with two crystals placed in magnetic fields of opposite directions was considered in [1]. In this scheme, the crystals are used as "amplifiers" of a neutron refraction in the magnetic field. So the magnitude of the effect caused by the influence of an external force acting on a neutron in such scheme is about $K_g = E/v_g$ times greater than that for standard techniques SESANS. Value K_g for the cold neutron diffraction are $\sim 10^5$, becase $E \sim 10^{-2}$ eV and $v_g \sim 10^{-7}$ eV.

However, it is more convenient to use the scheme of experiment based on the effect of the diffraction focusing in Laue diffraction [7]. The experimental scheme is shown in Fig. 2. In such geometry a part of the diffracted beam will be focused on the exit surface of the second crystal [8] so the intensity distribution as well as the neutron polarization distribution at the entrance surface of the first crystal will be reproduced at the exit surface of the second crystal.

Journal of Physics: Conference Series **746** (2016) 012049 doi:10.1088/1742-6596/746/1/012049

We can choose the positions of the coils in such a way (Fig. 2) to make the angles of the neutron spin rotation in the coil K1 equal to those in the coil K2, but those angles will have opposite signs for any neutron trajectory, and so the total angle of rotation will be zero. Calculations of the coils for the geometry under consideration were performed in [9]. The presence of external force acting on a neutron between the crystals will lead to the change of neutron direction and to the focus shift [8]. Due to such a shift a neutron spin will turn by the angle ϕ_x, because some difference arises of the neutron path lengths in opposite magnetic fields. It is easy to show that this angle, corresponding to the focus shift by the value x, will be equal to

$$\phi_x = \frac{2\mu B}{\hbar} \cdot \frac{2x \tan\theta_0}{v}, \tag{3}$$

where v is the neutron velocity, B is the value of magnetic fields in the coils K1 and K2, θ_0 is the angle between the direction of neutron velocity and the normal to the coil boundary, Fig. 2.

If a neutron changes its dirrection in the region (1), Fig. 2, by the anlge α, the shift of the neutron beam spot at the exit surface of the second crystal will be equal to

$$x_\alpha = \frac{\alpha}{\gamma_B} L \tan\theta_B = 2\alpha L \sin^2\theta_B \frac{E}{v_g}, \tag{4}$$

where γ_B is the Bragg width, L is the crystal thickness, θ_B is the Bragg angle. The action of an external force F_n at a distance l will change the direction of the neutron velocity by the angle $\alpha = F_n l/(2E)$ and the shift of the beam focus spot at the exit of the second crystal will be

$$x_F = \frac{F_n l}{v_g} L \sin^2\theta_B. \tag{5}$$

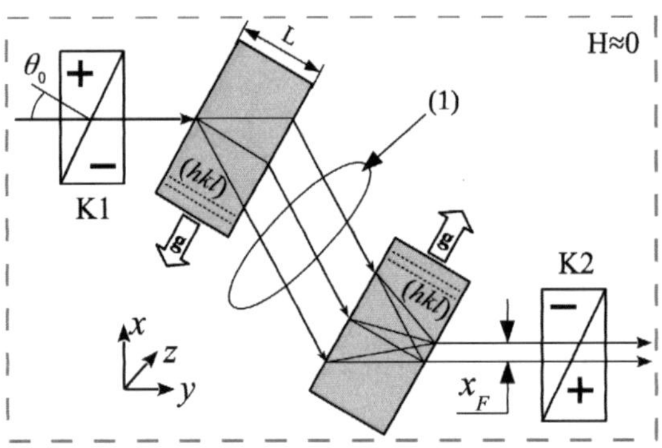

Figure 2. Fig. 2. The idea of the experiment using neutron Laue diffraction. K1, K2 are the coils with the magnetic field B, the signs " + " and " − " correspond to the directions of the magnetic field

The main parameter, which limits the angular resolution of the method, is the spot size of a spatially focused neutron beam. The beam spreading over the surface of the second crystal by the value of Δx will lead to the spreading of angles of neutron spin rotation after the passage the coil K2 by the value of $\Delta\phi$, see (3). To observe the spin rotation effect the condition $\Delta\phi < 2\pi$ should be satisfied.

Physical limit of spatial resolution in the double-crystal diffraction pattern is determined by the value [7]:

$$x_w = \frac{\xi_g \tan\theta_B}{2\pi} = \frac{V}{4F_g d} \tag{6}$$

where $\xi_g = \pi V/(2F_g d \tan\theta_B)$ is the extinction lenght, V is the volume of the crystal unit cell, d is the interplanar spacings, F_g is the structure factor. The values of x_w for cold neutron diffraction are usually of $(10-50)\mu$m. For instance, for the (220) plane of silicon $x_w = 26\mu$m. Such spatial resolution allows to use the magnetic fields up to 1 kGs. As a result, for the silicon crystals with the plane (220) and 10 cm thicknesses, value of the electric field $E_0 = 100\,\mathrm{kV/cm}$ in the area (1), see. Fig. 2, and $l = 1$ m, the neutron spin rotation angle (due to presence of the neutron electric charge e_n) will be, see. (3) and (5)

$$\phi_e \simeq 1 \cdot 10^{18} e_n \tag{7}$$

for the Bragg angle $\theta_B = 70^0$, e_n is in units of electron charge. Thus, for accuracy to measure the angle of the neutron spin rotation at the level of $\sigma(\phi_e) \sim 10^{-5}$, we can expect the sensitivity to measure the neutron charge at the level of $\sim 10^{-23}$ of the electron charge, which is two orders of magnitude better than the moderd constraint [3].

3. Experiment

As the first stage of this project we study the effect of diffraction focusing in double-crystal scheme of Laue diffraction. The experiment was carried out at the reactor WWR-M (PNPI NRC KI, Gatchina). We investigated Laue diffraction in crystals of silicon with the sizes $110 \times 110 \times 100\,\mathrm{mm}^3$ and the working crystallographic plane (220) ($d = 1.92$ Å, $F_g = 32 \cdot 10^{-13}\,\mathrm{cm}$, $v_g = 5 \cdot 10^{-8}\,\mathrm{eV}$). Scheme of the experiment is shown in Fig. 3. Crystal silicon together with the rotating platform placed in an thermostatic volume. The stability of the crystal ltemperature was $10^{-2}\,\mathrm{K/day}$. The distance between the crystals were fixed. Therefore, we had two spots from the direct and forward beam, Fig. 3. An example of the experimentally measured intensity distribution over the exit face of the second crystal is shown in Fig. 4.

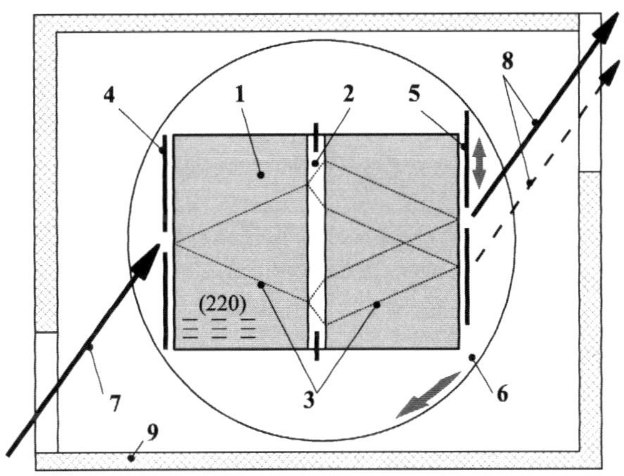

Figure 3. Fig. 3. Experimental scheme. 1 is silicon single crystal , 2 is the gap between the crystals with an additional collimating slit, 3 is example of neutron trajectory in the crystal, 4 is the entrance slit, 5 is the movable exit slit, 6 is the turning table, 7, 8 are incident and outgoing neutron beams, 9 is the thermostat

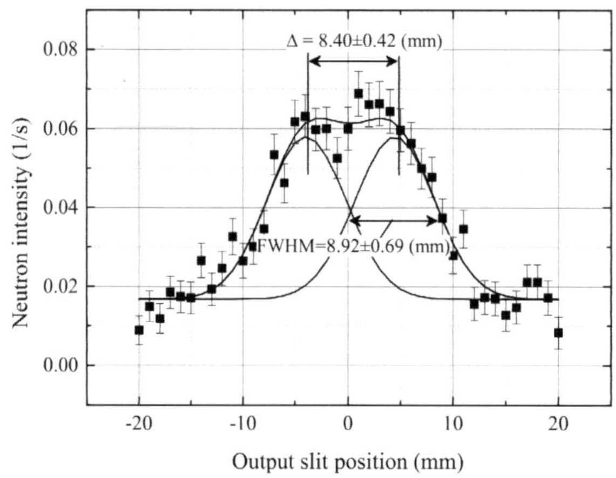

Figure 4. Fig. 4. The intensity dependence on the exit slit position. The sizes of the entrance and exit slits were 4 mm, $\theta_B = 68^0$. The aperture of the beam in the region (2), see Fig. 3 is equal to 110 mm.

The splitting into two components coincided with the calculated value. Spatial line width was 8.9 mm, which was slightly more than the expected value ($\sim$ 6 mm). Apparently, this is due to inperfection of the used crystal, because the reduction of the beam aperture between the crystals, see area (2) in Fig. 3, reduces the width of the line, see Fig. 5. The estimated own spatial resolution obtaned from Fig. 5 was $W_m < 3$. Unfortunately, the low luminosity of the experiment allows us to give only an upper limit for the W_m.

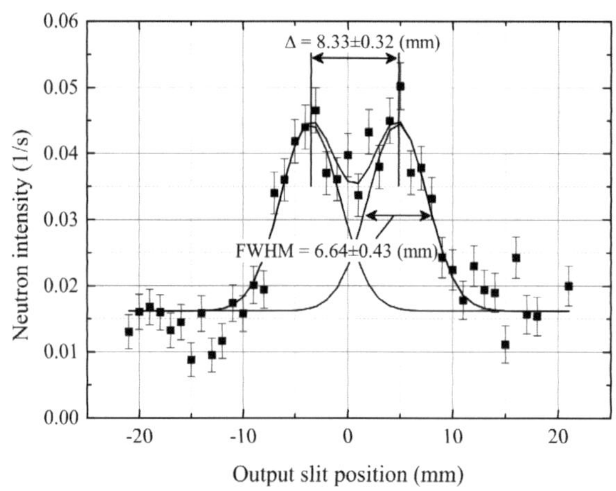

Figure 5. Fig. 5. The intensity dependence on the output slit position. The sizes of the entrance and exit slits were 4 mm, $\theta_B = 68^0$. The aperture of the beam in the region (2), see Fig. 3 is equal to 56 mm.

In the present experiment the value of the beam displacement along exit surface (4) is determined by angular deviation of the neutron in the following way:

$$x_\alpha = 2 \cdot 10^7 \alpha \text{ [mm]}. \tag{8}$$

So, the spatial resolution $W_m < 3$ mm corresponds to the angular one $\alpha_W < 1.5 \cdot 10^{-7} = 0.03''$.

4. Conclusion

The effect of double-crystal neutron focusing was studied. The possibility to create a setup with the spatial resolution better than 3 mm, which corresponds to an angular resolution of $0.03''$ was demonstrated. Numerical calculations show that the statistical sensitivity to measure a neutron electric charge can be $\sigma(e_n) \simeq 1.5 \cdot 10^{-21}$ of the electron charge for 100 days of statistics accumulation for the value of the electric field $E_0 = 100\,\mathrm{kV/cm}$ and the cold neutron fluxes achievable at high flux reactors, such as HFR ILL (Grenoble, France), or PIK reactor under construction (Gatchina, Russia). Further improvement may give approximately two orders better sensitivity, because in priciple the spatial resolution of the proposed experimental scheme can be about two orders better, see. (6). More detailed experimental test at the high flux cold neutron beam should be carried out to answer the question about the experimentally achievable sensitivity.

Authors wish to thank the staff of WWR-M reactor (PNPI, Gatchina) for efforts to maintain performance of this apparatus. This work was supported by Russian Foundation for Basic Research (RFBR 15-02-06257).

References

[1] V. V. Voronin, L. A. Akselrod, V. N. Zabenkin, I. A. Kuznetsov, Physics Procedia **42**, 25 (2013)
[2] E. Knudsen, L. Udby, P. K. Willindrup, K. Lefmann, W. G. Bouwman, Physica B, **406**, 2361 (2011)
[3] J. Baumann, R. Gahler, J. Kalus, W. Mampe, Phys. Rev. D, **37**, 3107 (1988)
[4] Yu.V. Borisov, N.V. Borovikova, A.V. Vasilyev, L.A. Grigorieva, S.N. Ivanov, N.T. Kashukeev, V.V. Nesvizhevsky, A.P. Serebrov, P.S. Iadjiev, Journal of Technical Physics, **58**(5) (1988) 951
[5] S. Kikuta, I. Ishikawa, K. Kohra, S. Hoshino, J. Phys. Soc. Japan, **39**, 471 (1975)
[6] A. Zeilinger, C. G. Shull, Phys.Rev.B **19**, 3957 (1979)
[7] V.L. Indenbom, I. Sh. Slobodetskii, K.G. Truni, JETP, **39**(3) (1974) 542
[8] J. Arthur, C. G. Shull, A. Zeilinger, Phys. Rev., **32**(9), 5753 (1985)
[9] Yu.V. Borisov, A.Ya. Berdnikov, Ya.A. Berdnikov, Yu.P. Braginets, V.V. Voronin, I.A. Kuznetsov, M.V. Lasitsa, S.Yu. Semenikhin, V.V. Fedorov, Technical Physics Letters, **41**(9) (2015) 867–869

A new approach to the experiment intended to test the weak equivalence principle for the neutron

M.A. Zakharov, G.V. Kulin[1], A.I. Frank, S.V. Goryunov D.V.Kustov

Joint Institute for Nuclear Research, Dubna, Russia

E mail: kulin@nf.jinr.ru

Abstract. A new approach to the free fall experiment with UCN is proposed. The idea is that the experiment is performed with neutrons from different equidistant lines of a discrete energy spectrum with a precisely known distance between the lines. Such a spectrum may be formed by neutron diffraction from a moving phase grating. Time-of-flight values of neutrons from different lines of the spectrum must be measured. Neither the initial neutron energy, nor the geometry parameters of the installation are required to be known in this method.

1. Introduction

Apparently, neutrons are the most suitable objects to investigate the gravity interaction of elementary particles. Although gravitational experiments with neutrons have a more than half a century history, the existing experimental data are quite scanty. Almost fifteen years after the first observation of the neutron fall in the Earth's gravitational field [1], the gravitational acceleration neutron was measured in a classical experiment with an accuracy of about 0.5% [2]. However, the fact of gravitational acceleration of the neutron had already been considered earlier as obvious and used by Maier-Leibnitz and Koester [3,4] for precise measurements of the coherent scattering length of neutrons by nuclei.

Later on, there appeared data on coherent scattering lengths for neutron–nucleus interaction that were obtained by measuring cross sections for neutron scattering on atoms. This made it possible to compare scattering length data obtained by two methods and thereby to verify the fundamental principle of the equivalence of the inertial and gravitational masses of the neutron [5]. He obtained the value γ = 1.0016 ± 0.00025 for an equivalence factor that he defined as $\gamma=(m_i/m_g)/(g_n/g_0)$, where m_i and m_g are the inertial and gravitational neutron masses, respectively, and g_n and g_0 are the gravitational acceleration of the neutron and the local gravitational acceleration of macroscopic bodies, respectively. More recently, a similar analysis was performed by Schmiedmayer [6], who obtained the equivalence factor with an accuracy twice as good as that obtained in [5].

The first quantum neutron gravitational experiment was performed in 1975 by Colella, Overhauser, and Werner [7] with a neutron interferometer. They observed the gravitation induced shift of the phase of the neutron wave function. Results of first experiments [7, 8] basically corresponded to theoretical predictions. However, further investigations revealed certain discrepancies. In the latest work [9], the difference between the experimental and theoretical phase shifts was equal to 1% with an error smaller by an order of magnitude. Results of a more recent experiment [10], the accuracy of which was equal to 0.9% do not remove this problem. Very recently, an experiment with a neutron spin-echo spectrometer has been performed [11]. The authors reported that their experimental result for the gravitation induced phase shift agrees within approximately 0.1% with the theoretically expected result, while the overall measurement accuracy was 0.25%.

Nesvizhevsky et al. [12, 13] reported the observation of the quantization of the vertical-motion energy of ultracold neutrons (UCNs) stored on a horizontal mirror. It is possible to hope that detailed

[1] To whom any correspondence should be addressed

investigation of this effect will be very useful for studying the gravitational interaction of the neutron as a quantum particle. Very promising results have been recently obtained by Jenke et al. [14], who observed transitions between quantum states of UCN being stored on a plane mirror in the Earth's gravitational field.

In the experiment [15] the change in the energy of a neutron falling to a known height in the Earth's gravitational field was compensated by a quantum of energy $\hbar\Omega$ that was transferred to the neutron via a non-stationary interaction with a moving diffraction grating. In this experiment the value of the equivalence factor defined as $\gamma = 1 - (m_i g_n / m_n g_{loc})$ was found to be $1 - \gamma = (1.8 \pm 2.1) \times 10^{-3}$. Here m_i is neutron inertial mass, m_n and g_{loc} are neutron table mass and local value of the free fall acceleration, respectively.

In the present paper we discuss the possibility of a new type of experiment for testing the weak equivalence principle for the neutron.

2. New approach to the free fall experiment with the neutron

Let us assume that the weak principle of equivalence is tested in a classical type of the experiment called free-fall or Galileo type of experiments. In such an experiment it is necessary to define the parameters of the equation of the neutron motion at its free falling in the Earth's gravitational field. It is obvious that in order to measure the free-fall acceleration for the neutron it is necessary to know the time of flight, initial velocity and height of falling with an appropriate accuracy.

The original approach to the free fall experiment was proposed by Pokotilovsky [16]. He proposed to use the geometry of a "neutron fountain", in which ultracold neutrons (UCN) following a parabolic trajectory move up and, after passing the turning point, fall down on the annular detector.

In the experiment the time of flight along such a trajectory is measured

$$t = \frac{v_0}{g_n}\left(\sqrt{1 + \frac{2g_n h}{v_0^2}} + 1\right). \tag{1}$$

Here g_n is the neutron free-fall acceleration, v_0 is the vertical component of the initial velocity and h is a difference in height between the source and detector. The most challenging in this approach is measurement of the neutron vertical velocity with high precision. It was proposed to circumvent this problem by a repetitive measurement of time with a set of the heights h_k. It multiplies the number of equations, from the solutions to which the value of g_n may be found. For implementation of this experiment it is necessary to vary the position of the detector in height, thus ensuring stability of its horizontal orientation with high accuracy.

We propose an alternative approach to the Neutron Fountain experiment, which permits avoiding the necessity of the mechanical displacement of any elements of the installation. The idea is to use any quantum non-stationary device, which transforms the initially quasi-monochromatic spectrum to the discrete spectrum with an exactly known value of splitting $\Delta E = \hbar\Omega$, as it is shown in Figure 1. The time-of-flight values of neutrons with a set of energies corresponding to different spectral lines must be measured simultaneously.

The free-fall acceleration g_n can be found from the system of equations

$$t_j = \frac{1}{g_n}\sqrt{\frac{2(E_0 + j\hbar\Omega)}{m}}\left[\sqrt{1 + \frac{m g_n h}{E_0 + j\hbar\Omega}} + 1\right], \tag{2}$$

where E_0 is the initial energy of the vertical motion and j is the order of a satellite line (see Figure 1). The number of equations may be increased further by repetition of measurement with a set of frequencies Ω_k.

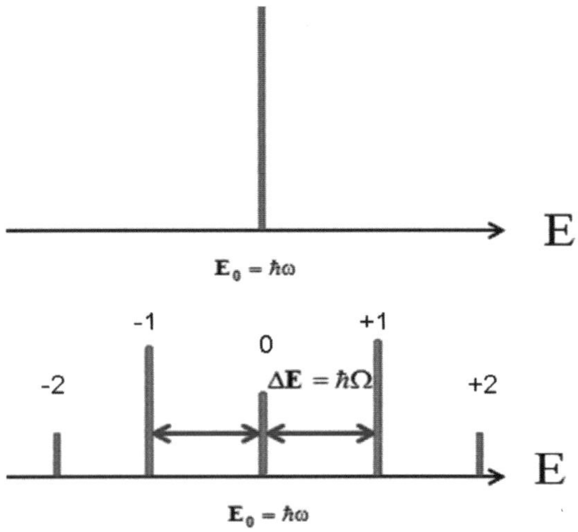

Figure 1. Splitting of the energy spectrum by a non-stationary quantum device.

3. Possibility of experimental implementation

As it is proposed in [16], the experiment must be performed with UCN. The quasi-monochromatic beam will be formed by neutron transmission through the quantum resonant structure - Neutron Interference Filter (NIF) [17-19], which is analogous to the optical Fabry-Perot interferometer. In order to transform the quasi monochromatic spectrum into the spectrum with equidistant lines, a moving diffraction grating [20-22] will be used. The time-of-flight values of neutrons from different diffraction orders will be measures simultaneously by the Fourier spectroscopy method [23].

The frequency of the wave quantum modulation is $\Omega = 2\pi f$, $f = V/d$, where V is the linear speed of the grating and d is its space period. The scheme of the proposed experiment is shown in Figure 2. UCNs from the feeding neutron guide pass sequentially through the NIF, moving grating and Fourier chopper, and after that, moving along a parabolic trajectory, fall down on the annular detector.

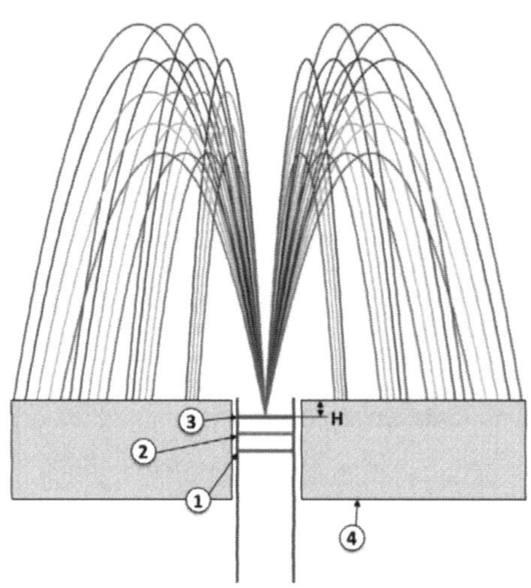

Figure 2. Scheme of the experiment: 1- neutron interference filter (NIF), 2- moving diffraction grating, 3- Fourier chopper, 4- annular detector

To estimate the relation between the accuracy of the measured free-fall acceleration g_n and time of flight experimental errors, a model calculation was done. It was supposed that the spectrum of UCN formed both by NIF with a resonant energy of 110 neV and a moving grating operating with the frequency f. As an illustration, the calculated spectrum formed by a specially designed grating with increased intensities of the high orders [24] is shown in Figure 3. The time-of-flight values of neutrons with energies corresponding to the five spectral lines were calculated in dependence on the modulation frequency for the specified values of g_n, E and h (see Figure 4). For the five values of modulation frequencies pseudo-experimental data for the time of flight were found. It was supposed that the latter were defined by normal distribution with dispersion corresponding to the specified errors in time measurement. Twenty five values of the time of flight found using this procedure (see points in Figure 4) were fitted by the functions

$$t_{j,k}\left(g_n, E_0, h\right) = \frac{1}{g_n}\sqrt{\frac{2(E_0 + j\hbar\Omega_k)}{m}}\left[\sqrt{1+\frac{mg_n h}{E_0 + j\hbar\Omega_k}}+1\right]. \qquad (3)$$

As a result, all three unknown parameters, namely g_n, E_0 and h, were found with the corresponding errors. These calculations show that the error of g_n is proportional to the errors of the time of flight and relative accuracy of $\delta g_n / g_n \approx 10^{-5}$ may be achieved, if the errors of the time measurement are of an order of 10mks.

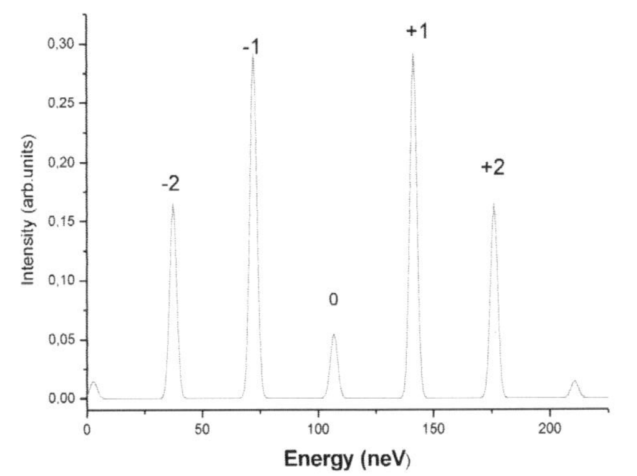

Figure 3. Calculated spectrum of UCN with initial energy of 107 neV after diffraction from the moving grating. Modulation frequency f = 8.4 MHz

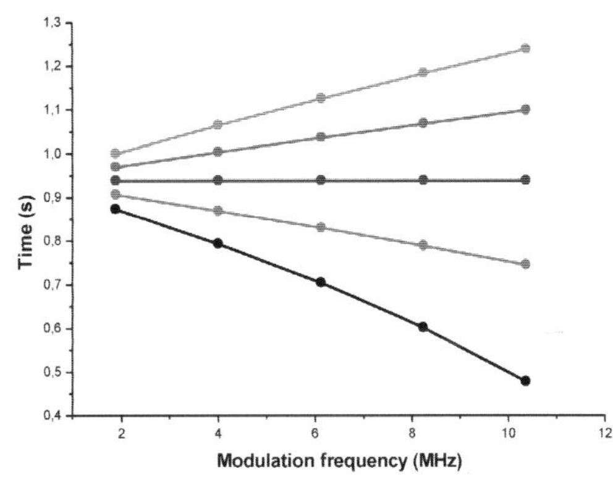

Figure 4. Time of flight of neutrons of the five diffraction orders in dependence on modulation frequency.

4. Conclusions

A new approach to the free fall experiment with UCN is proposed. The idea is that the experiment is performed with neutrons from different equidistant lines of a discrete energy spectrum with a precisely known distance between the lines. Such a spectrum may be formed by neutron diffraction from a moving phase grating. Time-of-flight values of neutrons from different lines of the spectrum must be measured. Neither the initial neutron energy, nor the geometry parameters of the installation are required to be known in this method.

5. Acknowledgements
This work was supported by the Russian Foundation for Basic Research (RFBR grant 15-02-02509).

References

[1] Mc. Reinolds A W 1951 *Phys. Rev.* **83** 172.
[2] Dabbs J W, Harvey J A , Paya D , Horstmann H 1965 *Phys. Rev.* **139** B 757
[3] Maier-Leibnitz H , Angew Z. 1962 *Physik* **14** 738
[4] Koester L 1965 *Z. Phys.* **182** 328
[5] Koester L 1976 *Phys. Rev. D* **14** 907
[6] Schmiedmayer J 1989 *Nuc. Instr. Meth.* A **284** 59
[7] Colella R, Overhauser A W , Werner S A 1975 *Phys. Rev. Lett.* **34** 1472
[8] Staudenmann J.-L., Werner S.A., Colella R., Overhauser A W 1980 *Phys. Rev. A* **21** 1419
[9] Littrell K C , Allman B E , Werner S A 1997 *Phys.Rev. A* 561767
[10] van der Zouw G, Weber M , Felber J, et al. 2000 *Nuclear Instr. Meth. A*, 440 568
[11] de Haan V -O, Plomp J, van Well Ad A et al. 2014 *Phys. Rev.* A **89** 063611
[12] Nesvizhevsky V V, Börner H G , Petukhov A K , et al. 2002 *Nature* **415** 297
[13] Nesvizhevsky V V, Börner H G,. Gagarski A M et al. 2003 Phys.Rev. **67** 102002.
[14] Jenke T , Geltenbort P , Lemmel H and Abele H 2011 *Nature Physics* **7** 468
[15] Frank A I, Geltenbort P, Jentschel M et al. 2007 JETP Letters **86** 225
[16] Pokotilovsky Yu N 1994 *Phys.Atom.Nucl.* **57** 390
[17] Seregin A A 1977 *JETP* **46** 859
[18] Steinhauser K A, Steyerl A, Schechenkofer H and Malik S S 1980 *Phys. Rev. Lett.* **44** 1306
[19] Frank A I, Balashov S V, Bodnarchuk V I, Bondarenko I V et al. 1999 *Proc. SPIE* **3767** 360
[20] Frank A I, Balashov S N, Bondarenko I V et al. 2003 *Phys. Lett. A* **311** 6.
[21] Frank A I, Geltenbort P, Kulin G V et al. 2005 *JETP Letters*, **81** 427
[22] Frank A I , Kulin G V, Bushuev V A . 2016 *Journal of Phys. Conf. Ser. This volume*
[23] Kulin G V , Frank A I , Goryunov S V et al. 2016, *Nuclear Instr. Meth.* (in print)
[24] Bushuev V A , Frank A I, G.V. Kulin 2015 *JETP*, arXiv:1502.04751v1 [physics.optics];
[25] *JETP* **148** in press.

Fundamental physics possibilities at the European Spallation Source

Esben Klinkby[1,2] **for the N-Nbar Collaboration**
Torsten Soldner[3] **for the ANNI Collaboration**

1) European Spallation Source ESS AB, Box 176, S-221 00 Lund, Sweden
2) DTU Nutech, Technical University of Denmark, DTU Risø Campus, Frederiksborgvej 399, DK-4000 Roskilde, Denmark
3) Institut Laue Langevin, CS20156, F-38042 Grenoble Cedex 9, France

E-mail: esbe@dtu.dk, soldner@ill.fr

Abstract. The construction of the European Spallation Source ESS is ongoing in Lund, Sweden. This new high power spallation source with its long-pulse structure opens up new possibilities for fundamental physics experiments. This paper focusses on two proposals for fundamental physics at the ESS: The ANNI instrument and the neutron-anti-neutron oscillation experiment.

1. ANNI

Pulsed beams have tremendous advantages for precision experiments with cold neutrons. In order to minimize and measure systematic effects, they are used at continuous sources in spite of the related substantial decrease in intensity. At the European Spallation Source ESS, such experiments will gain 1-2 orders of magnitude in event rate. Novel concepts become feasible. The cold neutron beam facility for particle physics ANNI [1] is designed to make full use of the pulse structure of the ESS. It has been proposed for construction as one of the public instruments. ANNI is a versatile instrument. In contrast to instruments for neutron scattering experiments, ANNI "only" delivers a neutron beam with well-defined properties (pulse structure, polarization) or, via the included Electron-Proton/Neutron (ep/n) separator, a beam of charged neutron decay products. Spectrometers for measurements will be designed and constructed for specific scientific questions (see section 1.1) by the external visitors. This operation mode is similar to that of existing facilities for particle physics at other neutron sources (see, e.g., [2, 3, 4]).

1.1. Scientific case

Science at ANNI will cover a wide range at the precision frontier of particle physics. The scientific program builds on three pillars: (i) neutron beta decay, (ii) hadronic weak interaction, and (iii) electromagnetic properties of the neutron. Furthermore, similar cold neutron beam facilities are also used for nuclear physics experiments (nuclear spectroscopy for structure and symmetries in nuclei, cross-section measurements), fission studies (understanding of the fission process, cross-section measurements), interferometry (measuring of scattering lengths, tests of the foundations of quantum mechanics), searches for short-range forces by neutron scattering, and technical developments (test and optimization of sources for ultracold neutrons (UCNs), polarization techniques, neutron optical devices, imaging).

(i) Neutron beta decay. Free neutrons decay into proton, electron and anti-electronneutrino. The distribution of the decay products in space and energy can be parameterized by correlation coefficients [5]. In this way, more than 20 observables can be defined [5, 6], for example the beta asymmetry coefficient A or the electron-neutrino correlation coefficient a. In the Standard Model of particle physics (SM), neutron decay is described as pure vector minus axialvector (V–A) interaction, by 2 parameters only. Physics beyond the SM predicts additional interactions, such as V+A, scalar or tensor interactions. Precision measurements of correlation coefficients allow testing the SM and searching for these additional interactions. The theoretical framework only assumes Lorentz invariance but no specific properties of the new physics. Therefore, neutron beta decay is a "broad band" probe for physics beyond the SM [7].

ANNI is optimized for measurements of correlation coefficients in neutron beta decay. Together with instrumentation developed by users, it will improve these measurements by one order of magnitude, unlocking the accuracy range of 10^{-4}. This accuracy tests new physics on a mass scale of 1-100 TeV, far beyond the threshold for direct production at existing or planned particle colliders. See [8, 9, 6] for recent reviews and [10, 11, 7] for theoretical prospects.

(ii) Hadronic parity violation. Weak interaction between quarks in the strangeness-conserving sector is difficult to access experimentally, because of the overwhelming strong interaction. Parity violation is specific to weak interaction. Therefore parity violating observables are used as filters to suppress strong interaction. Hadronic parity violation is traditionally parameterized by the meson exchange model of Desplanques, Donoghue, and Holstein (DDH) [12]. However, quantitative estimates are difficult and come with large uncertainties, due to the strong interaction. Since a few years, this situation is changing qualitatively: Effective field theories allow model-independent parameterizations and QCD on the lattice starts to be able to calculate parameters of the effective theories from fundamental QCD and weak interaction. New lattice gauge theory collaborations have been formed to calculate challenging non-perturbative QCD observables. This is a frontier of computational physics ("exascale computing"). These theoretical developments enable quantitative tests of the non-perturbative limit of QCD, provided that experiments become sufficiently sensitive. See [13, 14] for recent reviews. ANNI will unlock the sensitivity range needed to resolve the predicted tiny effects of hadronic weak interaction experimentally. This will allow a systematic experimental study of these effects which enables quantitative tests of the non-perturbative limit of QCD.

(iii) Electromagnetic properties of the neutron. Electromagnetic properties of the neutron are related to diverse questions of particle physics, see [6] for a review. The electrical neutrality of neutrons and of atoms is not required by the SM [15]. The existing experimental limits can be considered as experimental hint for grand unification [16]. The detection of a finite electric dipole moment (EDM) would indicate a violation of CP (charge conjugation-parity) symmetry beyond the SM, by processes at energy scales of 1 TeV to more than 1000 TeV [17]. If a neutron EDM is found above $10^{-28} e \cdot \mathrm{cm}$, it would be a strong hint for baryogenesis near the electroweak scale; if not, other ways such as leptogenesis or processes close to the inflationary scale are more probable [6].

The pulse structure at ANNI enables novel concepts to measure electromagnetic properties of the neutron. This gives a systematically different access to these fundamental properties. As example, a measurement of the neutron EDM using a pulsed cold neutron beam has been proposed in [18], with an estimated sensitivity of $5 \cdot 10^{-28} e \cdot \mathrm{cm}$ in 100 days of beam time. This would match the sensitivity envisaged by future experiments using stored UCN. We note that ANNI also provides a high flux of neutrons with 8.9 Å which could be used to produce UCN in superfluid He, similar to the planned SNS EDM experiment [19].

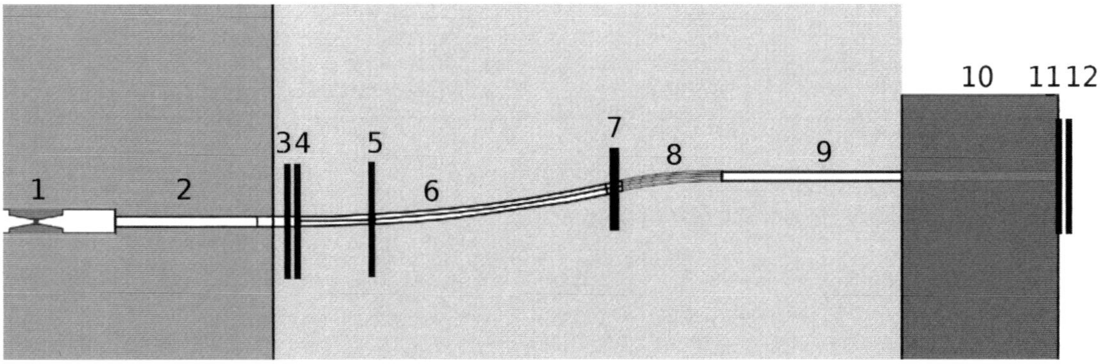

Figure 1: Design of the ANNI guide and chopper system: 1 Flat ESS moderator, 2 in-pile straight guide (horizontal trumpet), 3 first pulse defining chopper, 4 and 5 frame overlap choppers, 6 first bender, 7 shutter, 8 second bender (exchangeable by polarizing bender), 9 straight guide (anti-trumpet), 10 beam definition area, 11 second pulse defining chopper, 12 pulse suppressing chopper. Element 2 is located in the target monolith, 3-9 in the guide bunker, 10-11 in an instrument casemate, and the user experiment or the ep/n separator can be installed in the experimental area, starting at 12.

1.2. Design

The instrument ANNI (see Fig. 1 for guide and chopper system) is designed to provide maximum flux with wavelength information at moderate resolution. These fundamental requirements result in a short neutron guide with two frame overlap choppers. The guide uses the flat moderator (see also section 1.3) and is followed by an area for beam definition. The user experiment can be installed at a distance of 26 m from the moderator. The experimental area has a minimum width of 5 m (corresponding to 11° at 26 m from the moderator), a standard length of 25 m, and is extendable to a length of 50 m (i.e. to 76 m from the moderator) in order to provide sufficient space for specific experiments such as [18]. Several experiments will use the pulse structure in order to define the position of the neutron pulse in space (see example in [20]). For this purpose ANNI includes two pulse defining choppers that can rotate at 70 Hz (to provide short transition times) and a pulse suppression chopper. Pulse multiplication is planned for specific experiments.

Optimization was performed for a series of existing or proposed reference experiments. The best configuration is a guide with S-shape in the vertical plane and a ballistic geometry in the horizontal plane. The final cross-section is $11.2 \times 7\,\mathrm{cm}^2$. In order to go out of direct view twice, for this short guide with large cross-section, a strong curvature is needed. This is achieved by 2 benders, a first with 8 m length and a second with 2.5 m length. The second bender is interchangeable with a polarizing bender of the same geometry, providing moderate polarization (98%) at highest flux. For many precision measurements, a very high beam polarization is required. This is achieved by installing a second bender in the area for beam definition (X-SM geometry, see [21]). The instrument includes equipment for spin flipping, polarization analysis, and a ^{3}He spin filter polarizer.

The ep/n separator serves as a gate for charged neutron decay products for measurements of a large number of correlation coefficients. It delivers a beam of decay electrons and protons under precisely defined and variable conditions and spatially separated from the neutron beam. It also includes interfaces and basic infrastructure for user-supplied secondary spectrometers. The design of the ep/n separator is based on the concept of PERC [22] but features an upstream detector that will further reduce systematic effects and allows operation in a symmetric setup. A first secondary spectrometer is already under construction [23].

Parameter	Value	Comment
Capture flux full spectrum	$5.4 \cdot 10^{10}\,\mathrm{n/(cm^2s)}$	at guide exit
	$1.8 \cdot 10^{10}\,\mathrm{n/(cm^2s)}$	at start of experimental area
Capture flux 2–8 Å(FOCs)	$4.0 \cdot 10^{10}\,\mathrm{n/(cm^2s)}$	at guide exit
	$1.4 \cdot 10^{10}\,\mathrm{n/(cm^2s)}$	at start of experimental area
Particle flux @ 8.9 Å	$5.8 \cdot 10^{8}\,\mathrm{n/(cm^2s\AA)}$	at start of experimental area with guide bridging beam definition area
Divergence distribution	42 mrad (FWHM)	horizontal, full spectrum
	22 mrad (FWHM)	vertical, full spectrum

Table 1: Key performance numbers of ANNI [1]. (FOCs: With frame overlap choppers.)

Furthermore, ANNI includes all infrastructure needed for particle physics experiments, for example monitor detectors, background monitors, a He liquefier, shielding material, electricity and cooling water supply for large electromagnets etc.

1.3. Performance

Simulations of the instrument performance have been carried out using McStas [24] and the flat moderator component "2014" (with corrected solid angle calculation). The final ESS moderator configuration was not known at the time of instrument design. Minor adaptations of the design to one of the two butterfly moderators will be needed; however, no significant changes in the performance are expected.

Simulated key performance parameters are presented in Table 1. These parameters refer to the time average. In order to compare ANNI's performance with that of existing facilities, typical reference experiments have been simulated as benchmarks. Two classes of experiments can be distinguished: (i) Experiments that use pulsed beams in order to control systematic effects gain 1-2 orders of magnitude in event rate at ANNI: Compared to other pulsed sources such as the SNS, the flux at the ESS is one order of magnitude higher and the lower repetition rate allows for a larger wavelength frame at accessible distances from the moderator. At continuous sources, creating the pulse structure costs much more intensity than at ANNI where the peak flux of the ESS can be used. (ii) Experiments that do not use pulsed beams so far will not or only moderately gain in event rate (compared to the most intense continuous beams). However, these experiments may profit in systematics, for example in signal-to-background ratio. It should be noted that the European flagship experiments in particle physics with cold neutrons [22, 25] belong to class (i).

2. Search for $n\bar{n}$ oscillations

2.1. Physics motivation

The lack of discoveries at accelerators exploring the high energy frontier strengthens the motivation to search of new physics elsewhere. The unprecedented proton beam power of 5 MW expected at the ESS poses unique possibilities to search for new physics, including the search for neutron-anti-neutron oscillations.

Deviations from baryon number conservation have not been observed, but yet the Universe consists of baryons rather than anti-baryons. It is an intriguing question, to explain how the asymmetry came to be. Also, to explain the origin of the neutrino mass, interactions with baryon number violation (BNV) are theoretically favored [26]

Even within the SM baryon number is subject only to an approximate conservation law. At the perturbative level baryon number conservation arises due to the specific matter content in the

SM, and corresponds to a so-called *accidental* symmetry. In fact, the SM predicts BNV through rare non-perturbative electroweak instanton processes which violate B and L, but conserves $B - L$ [27, 28].
Furthermore, precision tests of the Equivalence Principle [29] offer no evidence for a long range force coupled to baryon number and thus a local gauge symmetry forbidding BNV. Consequently, BNV arises as a generic feature of many proposed extensions to the SM [30]. A promising means of searching for BNV is via the observation of the $\Delta B = 2$ process, neutron-anti-neutron oscillation [31, 32].

There exists a symbiosis between neutron-antineutron oscillations and neutrino physics via the quantum number $B - L$. A popular model explaining the origin of neutrino mass is the see-saw mechanism [33]. In this framework neutrinos possess a Majorana component and lepton number is violated by two units. Evidence for $\Delta L = 2$ processes are sought in double neutrinoless beta decay searches [34]. Since, howecer, $B - L$ (the true anomaly-free SM symmetry) is also violated by two units it would be natural to expect $\Delta B = 2$ processes. In addition to the complementarity with neutrino physics, neutron-anti-neutron oscillation features in a number other models of physics beyong the SM, for examples R-parity violating supersymmetry [30]. Values of the BNV mass scale for which observable oscillations take place exceed those attainable at colliders. Using a six-fermion BNV operator and dimensional reasoning mass scales of $10 - 1000$ TeV are obtained while other approaches (also leading to an observable signature) predict scales near the grand unified mass [31]. A further motivation for searching for oscillations was recently provided by the observation that such processes violate not only baryon number but also CP [37], thereby addressing two of the Sakharov conditions [36] for baryogenesis. In addition to the theoretical motivation given above a strictly experimentalist consideration of BNV hunting highlights the importance of neutron-anti-neutron oscillation searches. In an oscillation experiment only the violation of baryon number is sought, and not that of other hitherto conserved quantities. Single nucleon decay searches (eg, $p \to \pi^0 e^+$) require lepton number violation in order to ensure angular momentum conservation. Only searches for free neutron oscillation [38, 39] and anomalous nuclear decays, under the neutron oscillation [40] or dinucleon decay-hypothesis [41], offer high precision sensitivity to BNV-only processes.

2.2. Experimental setup

If neutrons can oscillate to anti-neutrons, the probability for the process to occur would be proportional to the free flight time squared, so an experiment to search for neutron-anti-neutron oscillations should aim for highest possible flight time. This requires an intense cold neutron beam and good vacuum. The flight path should be as long as possible and magnetically shielded in order to suppress energy splitting between neutron and anti-neutron energy states inflicted by the (anti)neutron magnetic moment ($\Delta E = \mu B$). To observe the anti-neutrons, an annihilation target (film) is needed and it should be surrounded by tracking detectors in a magnetic field to trace the produced charged pions and by calorimeters to perform energy measurement of the neutral pions. By this approach the invariant mass of the annihilating particle can be reconstructed offline and the experimental background can be severely suppressed. Since this would be a discovery type of experiment, background suppression is essential as any presence of background would reduce the sensitivity. The present limit: $\tau > 0.87 \times 10^8$ s was set by ILL experiment [39], using an approach according to lines of ideas discussed here.
There are a number of reasons to expect that a higher sensitivity can be reached at ESS than was the case for the ILL experiment. The fact that the neutron-anti-neutron oscillation search at the ESS is proposed before the construction of the facility gives rise to a number of advantages when outlining the experiment. Fig. 3 shows the preliminary engineering design of the beam extraction proposed for the neutron-anti-neutron experiment.

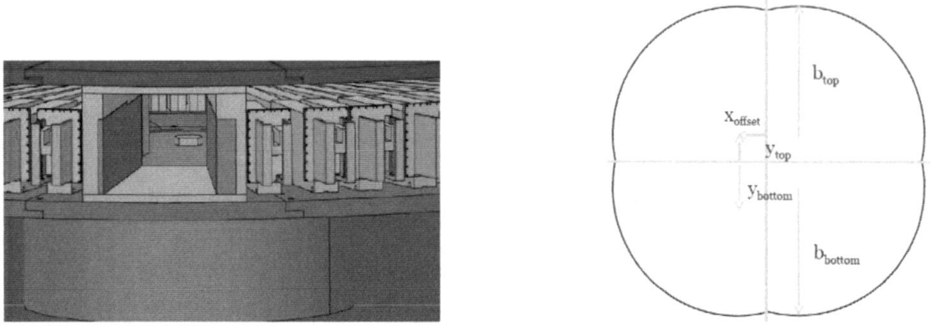

Figure 3: Left: Beam extraction of the neutron-anti-neutron oscillation experiment. Right: Cross-section of the *clover* reflector discussed in the text.

The experiment would temporarily[1] occupy ~15 degree horizontally corresponding to three conventional ESS beam ports. The extended solid angle compared to the scattering instruments is beneficial for the neutron-anti-neutron oscillation experiment due to the loose requirements on beam collimation: any neutron that given one scatter on a focusing neutron optic can be aimed to the annihilation target is useful. Since the annihilation target film is expected to have a size in the order of one meter squared, this means that the acceptance in terms of divergence is far beyond that of conventional beam ports for neutron scattering instruments.
Since the figure of merit of the neutron-anti-neutron oscillation search is proportional to free flight time squared, the instrument is designed to have the longest possible flight path, which given the facility footprint dictates the orientation of the instrument to be perpendicular to the proton driver, where instruments up to 300 meters of length can be accommodated. The highest sensitivity is reached by optimizing for cold neutrons, which given the recent baseline change of the ESS moderator configuration from voluminous para-hydrogen moderators, to flat butterfly shaped ones, constitutes a significant challenge. The newly adapted moderator configuration can be seen in Fig. 4. In the right-hand figure insert the location of the four 'hot-spots' of cold neutron emission are seen to be separated by tens of centimeters. In order to extract as many

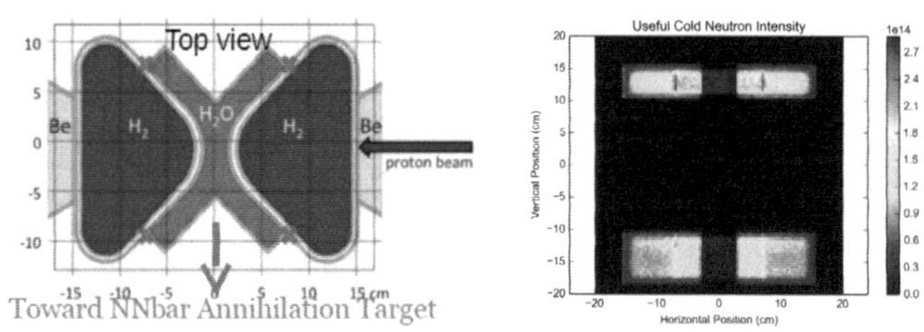

Figure 4: ESS moderator configuration as seen from above (left) and sidewise cold neutron emission map (right).

useful neutrons as possible several options are considered, including the double-cone reflector cut out shown in Fig. 5. To reflect neutrons from all four hot-spots into the annihilation film, a *clover* shaped reflector is proposed (see Fig. 3) and presently being optimized. Assuming a detection efficiency of 50% and that the experiment can be made completely background free, which was the case for the ILL experiment, Table 2 shows the preliminary sensitivity reach. In this table, the *unobstructed* case, refers to the situation where the limitation of the reflector and

[1] The duration of the neutron-anti-neutron experiment is undecided. Possibly 3-5 years.

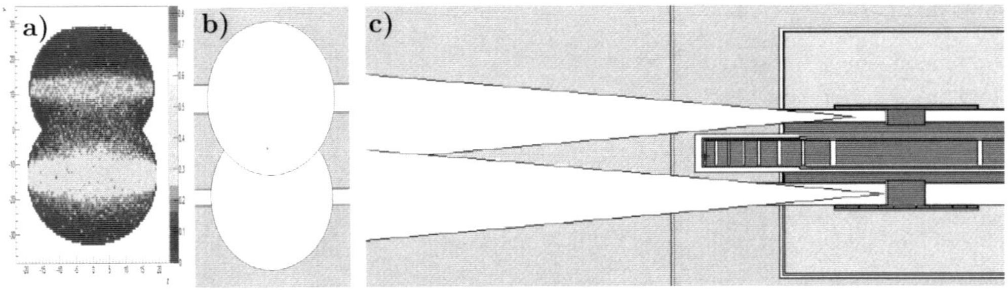

Figure 5: Double cone extraction, considered for the neutron-anti-neutron oscillation experiment, shown in side-view c) and perpendicular b). In a) the position of neutrons arriving at a radius of 2 m from the moderators is shown. The simulations are performed using MCNPX [42, 43], based on the recently adapted butterfly moderator geometry, modified to allow for extended neutron extraction through the double cone shown in the figure.

Source interface	Unobstructed	Beam port entrance	Beam port exit	Be Reflector
Available cold neutron intensity [n/s]	$2.2 \cdot 10^{17}$	$2.00 \cdot 10^{17}$	$1.43 \cdot 10^{17}$	$1.03 \cdot 10^{17}$
Baseline FOM (ILL/year)	211	209	106	64

Table 2: Experimental reach of the proposed neutron-anti-neutron oscillation experiment at ESS and a comparison to the corresponding ILL experiment [39].

beam port are completely neglected, whereas the *beam port entrance* and *beam port exit* refer to the situations taking into account the angular limitations inflicted by the entrance and exit of the beam port respectively (see Fig. 3). In addition to the limitations inflicted by the beam port, the *Be reflector* also takes into account the restrictions of the double cone collimator shown in Fig. 5. In neither case is any attempt made to mirror stray neutrons into useful trajectories - thus the sensitivities listed in the table can be considered conservative.

3. Conclusions and prospects

The ESS provides unique possibilities for experiments concerning fundamental physics. In this paper two such proposals are selected and the concepts of the proposed cold neutron beam facility for fundamental physics ANNI as well as the neutron-anti-neutron oscillation experiment are discussed. ANNI will outperform all existing cold neutron beam facilities for precision measurements of neutron decay, hadronic parity violation, and electromagnetic properties of the neutron, by up to 1-2 orders of magnitude in event rate. Preliminary investigations of a neutron-anti-neutron oscillation experiment suggest that a sensitivity which is 2-3 orders of magnitude better than the present limit should be within reach at the ESS. Recently a collaboration was formed with the intention to propose a neutron-anti-neutron oscillation experiment at ESS [44]. It is expected that a Technical Design Report will be finalized by 2017.

The list of possibilities for fundamental physics at the ESS, however, goes beyond these two examples. The reader is referred to Refs. [45, 46, 47, 48, 49] for additional examples, focussing, but not limited to, ultra cold neutron production and possible use for experiments.

References

[1] Theroine C, Abele H, Konrad G, Märkisch B, Schmidt U and Soldner T, "ANNI a cold neutron beam facility for Particle Physics", ESS Instrument Construction Proposal, unpublished (2015).
[2] Abele H, et al., *Nucl. Instr. and Meth. A* **562**, 407 (2006)
[3] Nico J S, et al., *J. Res. Natl. Inst. Stand. Technol.* **110**, 137 (2005)
[4] Fomin N, et al., *Nucl. Instr. and Meth. A* **773**, 45 (2014)
[5] Jackson J D, et al., *Phys. Rev.* **106**, 517 (1957), *Nucl. Phys.* **4**, 206 (1957).
[6] Dubbers D and Schmidt M G, *Rev. Mod. Phys.* **83**, 1111 (2011)
[7] Cirgliano V and Ramsey-Musolf M J (eds.), *Prog. Part. Nucl. Phys.* **71** (2013)
[8] Severijns N, et al., *Rev. Mod. Phys.* **78**, 991 (2006)
[9] Abele H, *Prog. Part. Nucl. Phys.* **60**, 1 (2008)
[10] Cirigliano V, et al., *Nucl. Phys. B* **830**, 95 (2010)
[11] Bhattacharya T, et al., *Phys. Rev. D* **85**, 054512 (2012)
[12] Desblanques B, et al., *Ann. Phys. (NY)* **124**, 449 (1980)
[13] Haxton W C and Holstein B R, *Prog. Part. Nucl. Phys.* **71**, 185 (2013)
[14] Schindler M R and Springer R P, *Prog. Part. Nucl. Phys.* **72**, 1 (2013)
[15] Foot R, et al., *J. Phys. G* **19**, 361 (1993)
[16] Okun L B, et al., *Phys. Lett. B* **138**, 115 (1984)
[17] Pospelov M and Ritz A, *Ann. Phys. (N.Y.)* **318**, 119 (2005)
[18] Piegsa F M, *Phys. Rev. C* **88**, 045502 (2013)
[19] Ito T M, et al., *J. Phys. Conf. Ser.* **69**, 012037 (2007)
[20] Märkisch B, *Phys. Proc.* **51**, 41 (2014)
[21] Kreuz M, et al., *Nucl. Instr. and Meth. A* **547**, 583 (2005)
[22] Dubbers D, et al., *Nucl. Instr. and Meth. A* **596**, 238 (2008)
[23] Wang X, et al., *Nucl. Instr. and Meth. A* **701**, 254 (2013)
[24] Willendrup P, et al., *Nucl. Instr. and Meth. A* **634**, 150 (2011)
[25] Märkisch B, et al., *Nucl. Instr. and Meth. A* **611**, 216 (2009)
[26] Mohapatra R N and Marshak R E, Phys. Rev. Lett. **44** (1980) 1316
[27] Adler S L, Phys. Rev. **177** (1969) 2426
[28] 't Hooft G, Phys. Rev. D **14** (1976) 3432
[29] Adelberger E G, et al.,Phys. Rev. D **42** (1990) 3267
[30] Barbier R, *et al.*, Phys. Rept. **420** (2005) 1
[31] Mohapatra R N, J. Phys. G **36** (2009) 104006
[32] Phillips D G, II *et al.*, Submitted to: Phys.Rept. [arXiv:1410.1100]
[33] Minkowski P, Phys. Lett. B **67** (1977) 421
[34] Elliott S R, Mod. Phys. Lett. A **27** (2012) 1230009
[35] Babu K S, et al., Phys. Rev. Lett. **97** (2006) 131301
[36] Sakharov A D, Pisma Zh. Eksp. Teor. Fiz. **5** (1967) 32
[37] Berezhiani Z and Vainshtein A, arXiv:1506.05096 [hep-ph]
[38] Fidecaro G *et al.*, Phys. Lett. B **156** (1985) 122
[39] Baldo-Ceolin M, *et al.*, Z. Phys. C **63** (1994) 409
[40] Berger C *et al.* [Frejus Collaboration], Phys. Lett. B **240** (1990) 237
[41] Berger C, *et al.* [Frejus Collaboration], Phys. Lett. B **269** (1991) 227
[42] Waters L S, et al., The MCNPX Monte Carlo radiation transport code. *AIP Conf.Proc.* **896**, 81 (2007)
[43] X-5 Monte Carlo Team, MCNP - A General Monte Carlo N-Particle Transport Code, Version 5, *LA-UR-03-1987* (1987)
[44] G. Broojimans, et al., Expression of interest for a new search for neutron-anti-neutron oscillations at ESS (2015)
[45] Klinkby E, et al., Voluminous D2 source for intense cold neutron beam production at the ESS. [arXiv:1401.6003] (2014)
[46] Lychagin E V, et al., UCN source at an external beam of thermal neutrons, *Advances in High Energy Physics* to be published
[47] Nesvizhevsky V V, An option of "UCN pump" for ESS [arXiv:1405.7495] (2014)
[48] Andreas S, Light Weakly Interacting Particles: Constraints and Connection to Dark Matter. Ph.D. Thesis (Hamburg U.). *DESY-THESIS-2013-024* (2013)
[49] Zimmer O, *Physics Procedia* **51**, 8588 (2014)

Bragg resonance behavior of the neutron refractive index and crystal acceleration effect

Yu.P. Braginetz[1,2], Ya.A. Berdnikov[2], V.V. Fedorov[1,2,3], I.A. Kuznetsov[1], M.V. Lasitsa[1,2], S.Yu. Semenikhin[1], E.O. Vezhlev[1,4], V.V. Voronin[1,2,3]

[1]Petersburg Nuclear Physics Institute NRC KI, Gatchina, Leningrad district, 188300 Russia
[2]Peter the Great St.Petersburg Polytechnic University, 29 Polytechnicheskaya st., St.Petersburg, 195251 Russia
[3]St.Petersburg State University, 7/9 Universitetskaya nab., St.Petersburg, 199034 Russia
[4]Forschungszentrum Jülich GmbH, Germany

E-mail: vvv@pnpi.spb.ru

Abstract. The energy dependence of neutron refraction index in a perfect crystal for neutron energy, close to the Bragg ones, was studied. The resonance shape of this dependence with approximately the Darwin width was found. As a result, the value of deviation from the exact Bragg condition can change during the neutron time of flight through the accelerated crystal and so the refraction index and the velocity of outgoing neutron can change as well. Such new mechanism of neutron acceleration in the accelerating perfect crystal was proposed and found experimentally. This mechanism is march more effective then known one concerning with the neutron acceleration in the accelerating usual media.

1. Introduction

The possibility to control the energy of neutron beams is of a great interest because of the wide neutron applications in various scientific fields from material science to nuclear, particle and astrophysics. The acceleration effect for neutrons scattered by exited isomeric nuclei was first predicted in 1959 [1] and was discovered experimentally in 1980 [2, 3]. The acceleration of neutrons in an inversely populated medium [4, 5] turned out to be very important in processes of stellar nucleosynthesis. In [6] acceleration of neutrons by vibrationally excited nitrogen molecules was observed.

Acceleration of neutrons in the uniform magnetic field due to spin flip by radio-frequency flipper is well known and successfully used in physical experiments (see, e.g., [7]). The phenomenon of neutron acceleration in a strong alternating magnetic field (of amplitude $\sim$ 0.4 T) was observed in [8]. The acceleration of neutrons in a weak alternating magnetic field (of 1–10 G) was measured using anomalous behaviour of the velocity dispersion for neutrons, moving in a crystal close to the Bragg directions [9]. The foundations of the neutron acceleration in a laser radiation field were considered in [10].

Not long ago a new interest arised to the acceleration of neutron, passing through accelerating media [11, 12]. This effect was first observed [13] and described in detail [14]. It was noticed [14] that "the observed effect was a manifestation of quite a general phenomenon – the accelerated medium effect (AME) – inherent to waves and particles of different nature". In

[15] the acceleration and deceleration of neutrons were observed by applying a specific time-of-flight method. In [16] some new special features of the effect for a birefringent medium were discussed with the applications to neutron spin optics and evolution of flavor states of neutrino, propagating through a free space. The acceleration of the samples in the mentioned experiments reached several tens of g units and the value of the energy transfer to a neutron was within the range of $(2-6)\cdot 10^{-10}$ eV, so up to now AME was observed only for ultracold neutrons and by only one research group (see [14, 16]).

In present paper a new much more effective mechanism of acceleration effect is proposed, which is tested and confirmed experimentally for cold neutrons passing through the accelerated perfect crystal.

The essence of the effect is following. Neutron wave function is significantly modifying for neutrons moving through the crystal close to the Bragg conditions. As a result neutrons concentrate on "nuclear planes" or between them [17, 18] ("nuclear planes" mark the positions of maxima of periodic nuclear potential for corresponding system of crystallographic planes). The degree of such consentration depends on the deviation from exact Bragg condition (i.e. on the crystal-neutron relative velocity) as well as the crystal refraction index for neutron. The neutrons enter into accelerated crystal with one mean potential of a neutron-crystal interaction and exit with the other, so the energy change at the crystal boundaries will be different and neutrons will be accelerated or decelerated after passage trough such a crystal, energy transfer to a neutron being at the level of $\sim 4\cdot 10^{-8}$ eV.

2. Neutron crystal optics

Interest to neutron optics in perfect crystals has grown sharply at the last time. It concerns with new possibilities to study fundamental properties and interactions of neutron. For instance, strong electric fields up to 10^9 V/cm, acting on the neutron inside the noncentrosymmetric crystals [18, 19, 20, 21] give new feasibilities to search for neutron electric dipole moment as well as CP-violating pseudomagnetic forces due to exchange of a pseudoscalar axion-like particle, using neutron optics in crystal [22, 23, 24, 25, 26, 27],what can be considered now as a most important tasks in this area.

Admixture of the waves reflected by crystallographic planes to the neutron wave function significantly changes the picture of neutron propagation in the crystal and leads to new phenomena, which manifest sharply defined resonance character with Bragg (Darvin) width. For example, small change of the neutron energy within this width ($\Delta E/E \sim 10^{-5}$ for thermal and cold neutrons) results in significant changing the neutron mean velocity in crystal (the anomalous velocity dispersion), and so the sharp energy dependence arises for the time neutron passes through the crystal [28].

In the present paper we discuss the other phenomenon related to the transformation of neutron wave function in the crystal, namely the resonance dependence of neutron refractive index (i.e. potential energy of neutron inside the crystal) on the incident neutron energy. If neutron passes through the non-absorbing perfect crystal[1] and Bragg conditions are not satisfied for either crystallographic planes, the passage of the neutron through the crystal can be described by refractive index, which depends on the amplitude of zero harmonic V_0 of the neutron-crystal potential (average crystal potential). But when the energy or velocity direction of neutron approaches to the Bragg one, amplitude of the wave reflected by the corresponding plane system will sharply increase. The amplitude of this reflected wave is determined by the corresponding potential harmonic amplitude V_g and the deviations from the exact Bragg condition. When this deviation more than harmonic amplitude we can use the perturbation theory [23, 29]. In this

[1] Under perfect crystal we consider the crystal with the dispersion of the interplanar spacing much less than the intrinsic width of the Bragg reflection.

case if the neutron has initial energy equal to E_0 and wave vector $\boldsymbol{k_0}$ ($E_0 = \hbar k_0^2/2m$), its wave function inside the crystal can be written as

$$\psi = e^{i\boldsymbol{kr}} + \frac{V_g}{E_k - E_{k_g}} e^{i\boldsymbol{k_g r}} \equiv e^{i\boldsymbol{kr}} \left[1 - \frac{1}{\Delta_B} e^{i\boldsymbol{gr}} \right], \tag{1}$$

where $\Delta_B = (E_k - E_{k_g})/V_g = 2(E_k - E_B)/V_g$ is the dimensionless parameter of deviation from exact Bragg condition for the plane system $\boldsymbol{g}$; $\boldsymbol{k}$ and $\boldsymbol{k_g}$=$\boldsymbol{k} + \boldsymbol{g}$ are the wave vectors of incident and reflected waves inside the crystal with the mean refraction index taking into account; E_k and E_{k_g} are the unperturbed neutron kinetic energies in states $|\boldsymbol{k}\rangle$ and $|\boldsymbol{k_g}\rangle$ ($E_k = \hbar^2 k^2/2m = \hbar^2 k_0^2/2m - V_0$, $E_{k_g} = \hbar^2 k_g^2/2m$), $E_B = \hbar^2 g^2/(8m \sin^2 \theta_B)$ is the neutron energy that corresponds to exact Bragg condition.

The presence of reflected wave with the amplitude equal to $1/\Delta_B$ leads to concentration of neutron density in crystal on or between reflecting planes with the degree, depending on value and sign of Δ_B. That in turn leads to additional changing the neutron kinetic energy $\tilde{E_k}$ and, respectively, the value of wave vector and refractive index[2] inside the crystal depending on the deviation parameter Δ_B. So averaging the periodic neutron crystal interaction potential over the wave function (1), one can get for the neutron kinetic energy in crystal

$$\tilde{E_k} = \frac{\hbar^2 \tilde{k}^2}{2m} = E_0 - V_0 + V_g \cdot \frac{1}{\Delta_B}. \tag{2}$$

The last term in equation (2) extremely increases approaching to the Bragg condition ($E_k = E_{k_g}$), so it becomes incorrect (and the perturbation theory as well) already for $E_k - E_{k_g} \sim V_g$. More accurate calculation in two wave approach of the last term (2) gives the following result for kinetic energy of neutron after its entrance into crystal:

$$\tilde{E_k} = E_0 - V_0 + V_g \cdot \frac{\Delta_B}{\Delta_B^2 + 1} \tag{3}$$

The last term in (3) describes the additional potential energy due to neutron localization. It significantly changes with small variation of neutron energy inside the Bragg reflection width $\Delta_B \simeq 1$, i.e. in the narrow energy range within $E_B - V_g < E_k < E_B + V_g$. For thermal and cold neutrons $V_g/E_B \simeq 10^{-5}$. Value V_g is comparable with that of mean crystal potential V_0. So changing the incident neutron energy in the vicinity of E_B, one can observe a well-defined resonance-type energy dependence of neutron refraction index in crystal. For example, for (110) plane of quartz $V_g = 4 \cdot 10^{-8}$eV, $V_0 = 10^{-7}$eV, and $E_B = 3.2 \cdot 10^{-3}$eV for diffraction angle close to $\pi/2$.

3. Experimental setup

Experiment was carried out at the horizontal neutron beam of the WWR-M reactor (PNPI, Gatchina). The energy variations for neutrons passed through accelerated crystal close to the Bragg condition was measured.

If neutron moves through accelerated crystal, the parameter of deviation from the Bragg condition will depend on the time, and so the mean potential of interaction of neutron with the crystal too, see (3). As a result, the refractive index will vary during neutron travel in crystal. But the neutron kinetic energy (wave vectors) inside the crystal cannot change because of homogeneity of the averaged crystal potential. Therefore because of the boundary changes of the neutron kinetic energy at the entrance and exit surfaces of the crystal are different, we should observe acceleration or deceleration of the neutron after passage through such a crystal. It should

[2] The neutron refraction index n is determined as usual $n^2 = 1 - \tilde{E_k}/E_0$.

be noticed that it doesn't matter in what way one changes the deviation parameter at the time of neutron passage through the crystal. For example, one can vary the temperature of crystal or deform it with pressure. Both these ways lead to the variation of the Bragg energy. The motion of the crystal was used due to the convenience of its realization[3]. Numerical estimations show that for the quartz crystal plane (110) the Bragg width in velocity units is equal to $\Delta v_B \simeq 9$ mm/s, i.e. if the crystal velocity changes by 9 mm/s during the time of neutron passage through the crystal, the deviation from the Bragg condition will vary by one Bragg width.

Scheme of experimental setup is shown in Fig. 1. Preselected neutron beam, reflected by the mosaic crystal of pyrolytic graphite (PG) with reflecting plane (002), falls on the monochromator K_1 made from perfect quartz crystal. Reflected by K_1 highly monoenergetic (within Bragg width) neutrons pass through the working crystal K_3 and then are reflected by the crystal-analyzer K_2. The second PG crystal redirects these reflected neutrons to the detector. The quartz crystals K_1, K_2 and K_3 with the same working reflecting planes (110) were arranged to have their plane orientations in parallel directions. The diffraction angle was close to the right one: $\theta_B = 89^0$ ($\lambda \simeq 4.9$Å). Scanning over the Bragg wave length performed by varying temperature difference $T_{21} = T_2 - T_1$ between crystals K_2 and K_1, the temperature T_3 of the crystal K_3 being a reference one. Example of a scanning curve is shown in Fig. 2, when the crystal K_3 is absent. We have scanned the shape of neutron reflex from the K_1 crystal changing T_{21} and so the relative interplanar distance for the K_2 crystal-analyzer. The width of this convolution scanning curve is close to that calculated (solid line) for two perfect crystals.

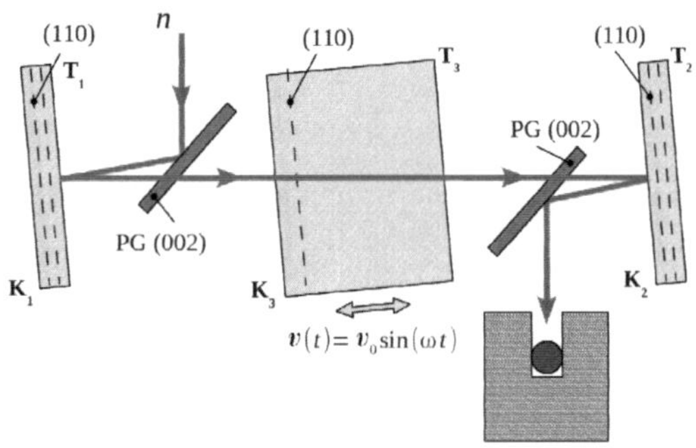

Figure 1. Scheme of experimental setup.

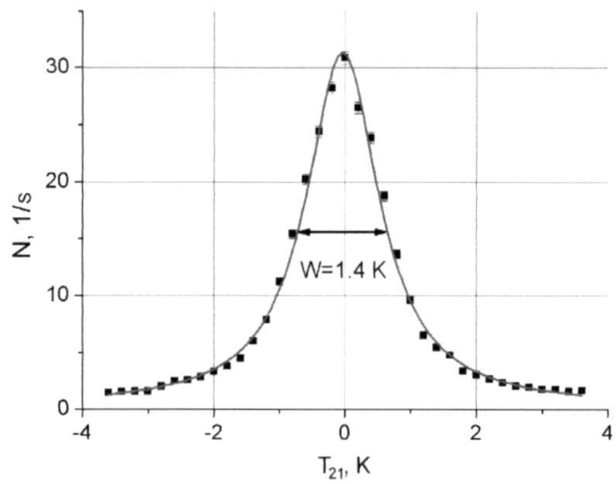

Figure 2. Two-crystal reflection curve (the K_3 crystal is absent). The curve reaches its maximum when interplanar distances for K_1, K_2 crystals coincide ($T_{21} = 0$). Width in units of interplanar spacing is equal to $\Delta d/d \simeq 1.8 \cdot 10^{-5}$.

There was a possibility to vary the temperature of the crystal K_3 and so its interplanar spacing too. Also we could move it in the direction parallel to the reciprocal lattice vector $\boldsymbol{g}$ for working plane. In experiment the crystal was set in harmonic motion by a piezoelectric motor. The frequency of crystal vibration was $\nu_c = 4.5$ kHz and the period $\tau_c = 222$ μs. Vibration amplitude reached 0.15 μm. The crystal length was $L = 5$ cm and neutron time-of-flight through the crystal was $\tau_n = 62\,\mu s$ that was about a quarter of the crystal vibration period.

If the velocity of the working crystal K_3 is depend on time as $v(t) = v_0 \cdot \sin\omega t$, the deviation

[3] Mentioned above the accelerated medium effect [14] in this case is negligible

from the Bragg condition for neutrons moving through that crystal will also depend on time in the same way

$$\Delta_B(t) = \Delta_{B_0} + 4\frac{1}{v_n}\frac{E_B}{V_g}v(t), \ \Delta_{B_0} = 4\frac{E_B}{V_g}\varkappa T_{13}, \tag{4}$$

where v_n is the velocity of incident neutrons, Δ_{B_0} is the deviation from the Bragg condition for resting crystal at $v(t) = 0$, $\varkappa \simeq 1.3 \cdot 10^{-5}$ is the thermal expansion coefficient for a quartz crystal in the direction perpendicular to crystallographic planes. The deviation Δ_{B_0} is determined by the difference of temperatures (interplanar distances) between K_1 and K_3 crystals $T_{13} = T_1 - T_3$. So in further we will use T_{13} as a parameter of deviation from the Bragg condition for neutrons passing the resting K_3 crystal.

The effect of the neutron energy change after passage through the crystal boundaries is determined by variation of the crystal velocity and so averaged potential (3) during the neutron time-of-flight through the accelerated crystal:

$$\begin{aligned} \Delta E(t_0) &= V_g\left(\frac{\Delta_B(t_2)}{\Delta_B^2(t_2)+1} - \frac{\Delta_B(t_1)}{\Delta_B^2(t_1)+1}\right) \\ &= \frac{1-\Delta_{B_0}^2}{(1+\Delta_{B_0}^2)^2}\frac{4E_B}{v_n}\Delta v(t_0), \end{aligned} \tag{5}$$

where $\Delta v(t_0) = v(t_0) - v(t_0 + \tau_n)$, t_0 is the time of neutron entrance into the crystal, τ_n is the neutron time-of-flight through the crystal.

This change of neutron energy (and wavelength too) after accelerated crystal results in shift of the maximum of the scanning curve (Fig. 2). This maximum will be found for some other temperature difference T_{21} of the K_2 and K_1 crystals. Such variations of the scanning curve, depending on temperature and movement of the K_3 crystal, were studied to find that crystal acceleration effect. Time-of-flight technique was used for this purpose.

The main systematic error of this experiment associates with the dependence of the neutron transmission through the K_3 crystal on the deviation from the Bragg condition, that results in the spectrum distortion for neutrons passed through the crystal. Therefore, the position and shape of the scanning curve can change for neutrons passed even through the resting K_3 crystal, when the acceleration effect is absent. Examples of the neutron intensity distribution over the wave length after such passage through the resting K_3 crystal with different deviations T_{31} from the Bragg condition are shown in Fig. 4. It is evident that both intensity of transmitted neutrons and maximum position of the scanning curve can change in different ways. In particular for $T_{31} = 0$ the neutrons after reflection from K_1-crystal cannot penetrate into K_3-crystal. They will be completely reflected (due to exact Bragg condition) and cannot reach the K_2-crystal. So K_2-crystal will reflect only background neutrons. In the other case for $T_{31} \gg 5$K K_3-crystal behaves as a homogeneous medium and practically does not distort the spectrum. In an intermediate case the spectrum will be distorted, because the K_3-crystal reflectivity (and so its transmittance) sharply depends on the neutron wave length.

However, unlike the sought crystal acceleration effect (5), the systematic one is determined only by the deviation from the exact Bragg condition at the time moment t_0 of neutron entry to crystal, but not variation of the deviation during the time-of-flight through the crystal.

So the position $E_s(t_0)$ of the maximum and the maximum intensity $N(t_0)$ of the scanning curve (Fig. 2) in the absence of the crystal acceleration effect will be some functions of deviation $\Delta_B(t_0)$, depending on the crystal velocity $v(t_0)$

$$E_s(t_0) = F(\Delta_B(t_0)), \tag{6}$$

$$N(t_0) = G(\Delta_B(t_0)). \tag{7}$$

For further consideration and comparison with the experimental results expressions (6) and (7) can be expanded in Taylor series over $v(t_0)$ about point $v(t_0) = 0$ (i.e. $\Delta_B(t_0) = \Delta_{B_0}$). Taking into account that the speed of the crystal was significantly less than the typical Bragg widths it is enough to leave expansion terms up to second order over $v(t_0)$

$$E_s(t_0) = A + B \cdot v(t_0) + C \cdot v(t_0)^2, \tag{8}$$

$$N(t_0) = N_0 + N_1 \cdot v(t_0) + N_2 \cdot v(t_0)^2, \tag{9}$$

where A, B, C, N_0, N_1 and N_2 are free parameters depending on Δ_{B_0} to be found from experiment.

As it follows from (5) the crystal acceleration effect contains a term phase-shifted with respect to the false effect (8) by the value $\omega\tau_n/2$. This shift is approximately equal to $\pi/4$ for our experimental conditions. Furthermore, the presence of the acceleration effect does not change the intensity of the line, but only gives its additional shift. Thus there is a phase shift between time dependencies of $N_{exp}(t_0) = N(t_0)$ and $E_{exp}(t_0) = E_s(t_0) + \Delta E(t_0)$ that represents the crystal acceleration effect.

4. Results

An example of experimental dependence of the line position on its maximal intensity $E_{exp}(N_{exp})$ is shown in Fig. 3. In the absence of the acceleration effect it should be observed a bijection between the maximum position and intensity shown by a dotted line. The presence of the neutron energy change after passage through the accelerating crystal leads to dependence $E_{exp}(N_{exp})$ described with a closed curve like Lissajous figure, where the figure square is determined by the crystal acceleration effect. Curved arrows in Fig. 3 show the sweep direction over time. The relation between a line shift in units of the crystal temperature and change of neutron energy is given by the following expression:

$$\Delta E = 2E_B \cdot \varkappa \Delta T. \tag{10}$$

The splitting marked by arrows in Fig. 3 corresponds to $\Delta E_{exp} \simeq 5$ neV.

Dependence of the maximum value for the neutron energy variation (5) due to the acceleration effect on the deviation from the neutron Bragg energy for working crystal (temperature difference T_{13}) is shown in Fig. 5. Measurements were carried out at two different crystal oscillation amplitudes, corresponding to $v_0 \simeq 3$ mm/s and $v_0 \simeq 1.5$ mm/s. Curves show the results of approximating the experimental points by the theoretical curve (5). Thus one can see that the neutron energy change after passage trough accelerating crystal can reach ~ 20 neV.

The mean potential energy of a neutron-crystal interaction, see (3), can be obtained from the experimental dependence shown in Fig. 5, because it is actually a derivative of (3), see (5). One should take into account that far from the Bragg condition the correction to the mean interaction potential due to the presence of g-harmonic V_g tends to zero, see (2), and so neutron refraction will be determined only by the average potential V_0. The result of the interaction potential reconstruction for neutrons moving in a crystal with energies close to the Bragg one is shown in Fig. 6. It is easy to see that the relative change of the incident neutron energy by several units of 10^{-5} leads to the variation of the interaction neutron-crystal potential by $\pm 20\%$.

5. Conclusion

The features of refraction of a neutron, moving in a crystal close to the Bragg condition, has been studied. It is shown that the energy dependence of refractive index has evident resonance shape in a vicinity of the Bragg energy with the corresponding Bragg (Darwin) width (for thermal and cold neutrons $\Delta E/E \simeq 10^{-5}$). The variation of the interaction potential of the neutron with the crystal in this energy range can reach about $\pm 20\%$.

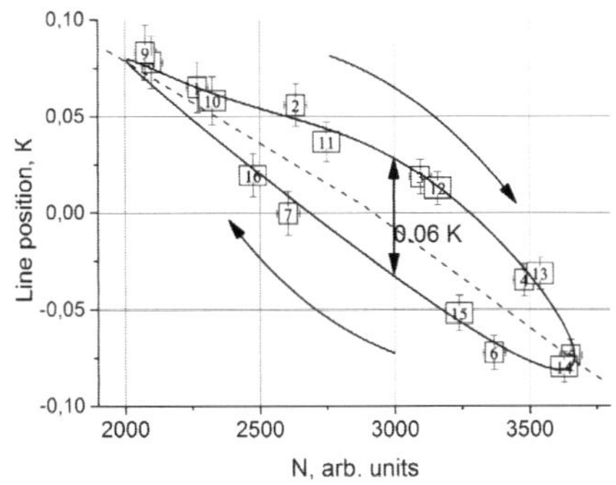

Figure 3. Dependence of the line position on its maximum intensity $E_{exp}(N_{exp})$. Initial deviation from the Bragg condition for the working crystal $T_{13} = +1.5$K. Numbers inside the experimental points correspond to the channel number of time spectrum. Solid curve is the result of fitting the experimental data.

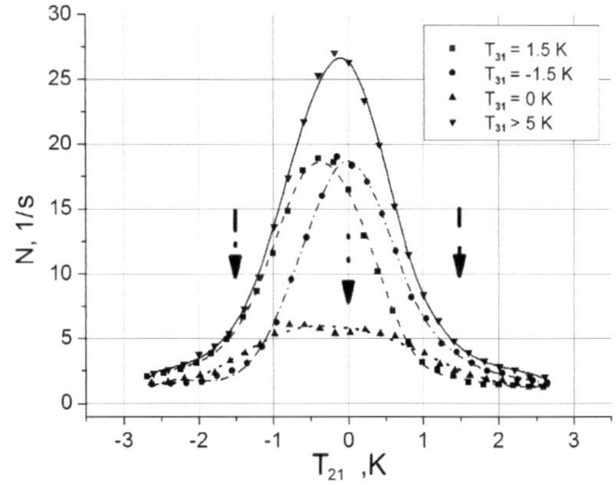

Figure 4. Two-crystal scanning curves for neutrons passed through immovable working crystal with different deviations from the Bragg energy. Exact Bragg positions for K_3 crystal are indicated by arrows.

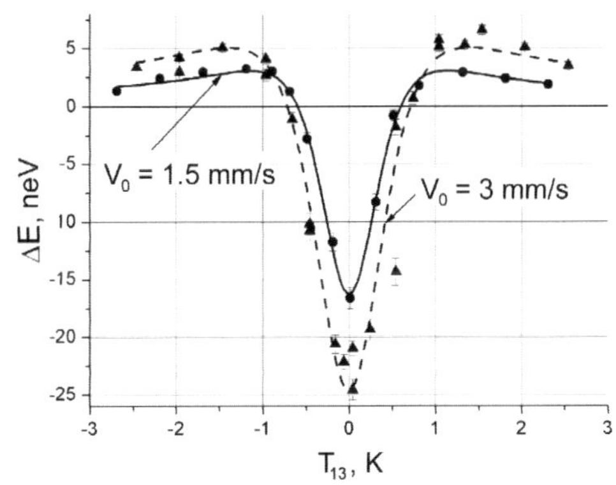

Figure 5. Magnitude of energy variation for a neutron passed through the accelerating crystal as function of the deviation from the Bragg condition for incident beam.

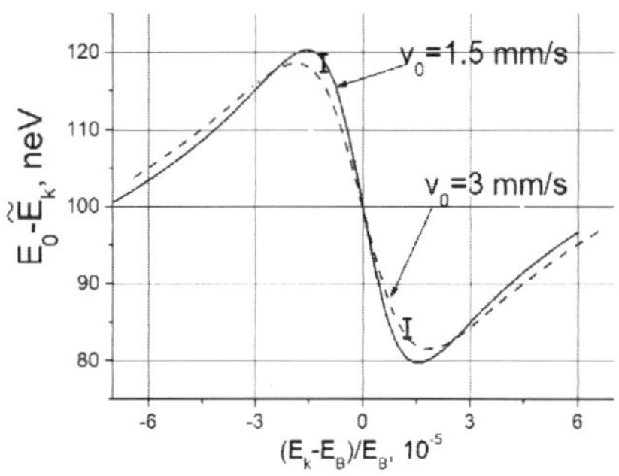

Figure 6. Behaviour of the interaction potential $E_0 - \tilde{E}_k = V_0 - V_g \cdot \Delta_B / (\Delta_B^2 + 1)$ of neutrons with the crystal in a vicinity of the Bragg energy. Calculated and experimental (reconstructed from Fig. 5) curves coincide in error limits. Vertical bars show the scale of the experimental error.

The resonance behaviour of the neutron-crystal interaction potential results in one more new phenomenon. That is neutron acceleration, which is found experimentally for neutrons passed through the accelerated perfect crystal for neutron energies, close to the Bragg one. The effect arises due to change of the parameter of deviation from the exact Bragg condition during the neutron time-of-flight through the accelerated crystal. As a result the refraction index for neutron changes too and so the velocity of the outgoing neutron.

This crystal acceleration effect was observed for the first time. One should take this phenomenon into account in precision neutron optical experiments such as mentioned above, because the neutron refraction index is determined not only by averaged crystal potential, but also by its harmonics, which have the same order of value as the average potential itself.

Acknowledgments

Authors wish to thank the staff of WWR-M reactor (PNPI, Gatchina) for efforts to maintain performance of this apparatus. This work was supported by Ministry of Science and Education of the Russian Federation (program 3.329.2014/K).

References

[1] Yu.V. Petrov, Sov. Phys. JETP **10** (1959) 833.
[2] I.A. Kondurov, E.M. Korotkikh, and Yu.V. Petrov, JETP Lett. 31, 232 (1980).
[3] I.A. Kondurov, E.M. Korotkikh, Yu.V. Petrov, and G.I. Shulyak, Phys. Lett. B 106, 383 (1981).
[4] Yu.V. Pertov and A.I. Shlyalchter, Nucl. Phys. A 292, 88 (1977).
[5] Yu.V. Petrov, Sov. Phys. JETP **36** (1972) 395.
[6] A.B. Laptev, G.A. Petrov, Yu.S. Pleva, and M.A. Yamshchikov, Sov. J. Nucl. Phys. **49** (1989) 204.
[7] H. Weinfurter, G. Badurek, H. Rauch, and D. Schwahn, Z. Phys. B: Condens. Matter **72** (1988) 195.
[8] L. Niel and H. Rauch, Z. Phys. B: Condens. Matter **74** (1989) 133.
[9] V.V. Voronin, Yu.V. Borisov, A.V. Ivanyuta, et al., JETP Lett. **96** (2012) 613.
[10] L.A. Rivlin, Quantum Electron. **40** (2010) 460.
[11] F.V. Kowalski, Phys. Lett. A 182(1993) 335.
[12] V.G. Nosov and A.I. Frank, Phys. At. Nucl. **61** (1998) 613.
[13] A.I. Frank, P. Geltenbort, G.V. Kulin, et al., JETP Lett. **84** (2006)63.
[14] A.I. Frank, P. Geltenbort, M. Jentshel, et al., Phys. At. Nucl. **71** (2008) 1656.
[15] A. I. Frank, P. Geltenbort, M. Jentshel, et al., JETP Lett. **93** (2011) 361.
[16] A.I. Frank, V.A. Naumov, Physics of Atomic Nuclei, **76** (2013) 1423.
[17] P. Hirsch, A. Howie, R. Nicholson, et al. Electron microscopy of thin crystals. Butterworths, London, 1965.
[18] V.L. Alexeev, V.V. Fedorov, E.G. Lapin, et al., Nucl. Instr. and Meth.**A 284** (1989) 181.
[19] V.L. Alexeev, V.V. Voronin, E.G. Lapin, et al., JETP **96** (1989) 1083.
[20] V.V. Fedorov, I.A. Kuznetsov, E.G. Lapin, et al., Nucl. Instr. and Meth. **B 252** (2006) 131.
[21] V.V. Fedorov, V.V. Voronin, Proceedings of the XXXV PNPI Winter School, SPb, 2001, 90–113; Nucl. Instr. and Meth. **B 201** (2003) 230.
[22] V.V. Fedorov, V.V. Voronin, E.G. Lapin, Preprint LNPI-1644, SPb., 1990, 36 p.
[23] V.V. Fedorov, Proceedings of XXVI LNPI Winter School, Leningrad, 1991, 65–118.
[24] V.V. Fedorov, V.V. Voronin, E.G. Lapin, J. Phys. G **18** (1992) 1133.
[25] V. V. Fedorov, M. Jentschel, I. A. Kuznetsov, et al. Nucl. Instr. and Meth. A **611** (2009) 124.
[26] V.V. Fedorov, M. Jentschel, I.A. Kuznetsov, et al., Physics Letters B **694** (2010) 22.
[27] V.V. Voronin, V.V. Fedorov, I.A. Kuznetsov, JETP Letters **90** (2009) 5.
[28] V.V. Voronin , Yu.V.Borisov , A.V.Ivanyuta, et al., JETP Letters **96** (2012) 609.
[29] V.V. Fedorov, V.V. Voronin, Physics of atomic nuclei and elementary particles (Proceedings of XXVI PNPI Winter School, part 1). SPb, 1996, 123–164.

Non-stationary transformation of neutron energy by a moving grating

A.I. Frank[1] [1], G.V. Kulin[1], V.A. Bushuev[2]

[1]Joint Institute for Nuclear Research, Dubna, Russia
[2]Moscow State University, Russia

E mail: frank@nf.jinr.ru

Abstract. The report gives a short review of the studies dedicated to the investigation and applications of non-stationary quantum effects in neutron diffraction from moving periodic structures. The studies were performed in 1994-2015.

1. Introduction

It is known that the neutron is a very appropriate object for demonstration and investigation of wave properties of massive particles. In most performed or proposed experiments a neutron beam or wave are permanent and experimental results may be described with the stationary Schrödinger equation. However, neutron optics acquires new properties, when parameters describing interaction of the neutron wave with an object, depend on time. Non-stationary effect on the wave allows changing considerably such wave properties, as energy spectrum, spin, intensity, phase, direction of propagation, etc.

Apparently, Moshinsky [1] was the first to discuss the problem of the non-stationary quantum effect in the optics of the massive particle. He studied the evolution of the wave function after instantaneous extraction of a perfect absorber from a beam of monochromatic particles. Gerasimov and Kazarnovsky [2] considered the possibility of observation of a number of non-stationary quantum phenomena arising from the interaction of ultracold neutrons (UCN) with a potential barrier oscillating in time. Later on, a number of non-stationary effects were investigated theoretically and experimentally in [3-16].

Under certain conditions neutron diffraction may also be considered as a non-stationary phenomenon. In [17] UCN diffraction by surface Rayleigh waves was regarded as a cause of inelastic neutron scattering resulting in a decrease in the neutron storage time in traps.. Later, the diffraction of cold neutrons by surface acoustic waves generated on the surface of a quartz crystal was observed in [18].

Almost two decades after publishing paper [17], the effect of neutron energy change at diffraction by a moving grating was predicted again in [19]. The starting point of this paper was the idea of a moving grating as of a quantum chopper [5-7] or neutron wave modulator, as moving a periodic structure across the neutron beam modulates the transmitted flux with the frequency V/d. In fact, paper [19], where not only the quantum phenomenon itself, but also the possibility of its observation

[1] To whom any correspondence should be addressed

was discussed, initiated a long period of investigation of neutron diffraction by a moving grating (NDMG). In this report we present a short review of these papers continued up to now.

2. Prediction and first observation of the energy quantization at NDMG

In Ref. [19], theoretical analysis of the NDMG problem was based on several assumptions. Firstly, it was supposed that in the laboratory system of reference the neutron wave is normally incident on the grating surface $\Psi_{in}(z,t)=exp\left[i\left(k_0 z-\omega_0 t\right)\right]$, and the grating moves in the positive direction of the X-axis with the velocity V (see Figure 1). Secondly, the kinematic approach was used, according to which the grating was supposed to be ideally thin and characterized only by a periodic function of transmission $T(x)$.

As in Ref. [17], the diffraction problem was solved in the moving system of reference, where the grating is at rest. The diffraction results in appearance of a set of waves with identical wave numbers $k'_n=\left[k_0^2+k'^2_{xn}\right]^{1/2}$ and amplitudes

$$a_n=d^{-1}\int_0^d T(x)exp(-iq_n x)dx\,, \tag{1}$$

where $k'_{xn}=q_n-k_V$, $k_V=mV/\hbar$, $q_n=nq_0$, and $q_0=2\pi/d$ is the reciprocal lattice vector. Z - component k'_{zn} corresponds to each component k'_{xn}. It is obvious that z-components both in the moving and laboratory systems of reference are identical $k'_{zn}=k_{zn}$. The wave function in the laboratory system is

$$\Psi(z,x,t)=\sum_n a_n exp[i(q_n x+k_{zn}z-\omega_n t)]\,, \tag{2}$$

where

$$k_{zn}=\left(k_0^2+2k_V q_n-q_n^2\right)^{1/2},\quad \omega_n=\omega_0+n\Omega,\quad \Omega=2\pi\left(V/d\right)\text{ and }\Omega<<\omega_0\,. \tag{3}$$

It is seen from equations (2) and (3) that the solution is characterized by a discrete energy spectrum and in a limit of small diffraction angles, when $q_0^2<<k_0^2$, the waves of different diffraction orders differ only by z-components of the wave numbers and correspondent frequencies. In this limit $k_{zn}\cong k_0\left(1+n\Omega/\omega_0\right)^{1/2}$ and the wave function is almost identical to the state formed by the periodical quantum chopper [6, 9].

The cases of the amplitude and phase modulation were both considered in [19]. A special case of the so-called π-grating was analyzed separately. In this case, function $T(x)$ takes the value of 1 or $exp(i\pi)$ altering each half of a period. It is known that at the normal incidence on such a grating the amplitudes of even orders (including zero order) are $a_n=0$, whereas the odd-ordered ones are $a_n=2i/\pi n$ $(n=2s-1)$.

Discussing the possibility of experimental observation of the new quantum phenomenon, it was proposed to use ultracold neutrons passing through two interference filters [20-23], one of which serves as a monochromator and the other as an analyzer. A spectrometer based on this idea was later constructed [24] and the effect of spectrum splitting at UCN diffraction by a moving grating was demonstrated [25, 26] (see Figure 2). Instead of a linearly moving grating, a rotating grating manufactured on the surface of a silicon disk was used. In its peripheral region the radial grooves were made using the lithographic technique. Its angular period was 3.33×10^{-4} rad^{-1} corresponding to a space period of 20 mkm at a diameter of 120 mm. The grooves covered a half of the grating period. The grating could be spinned around the vertical axis. UCN reach the grating through an annular corridor. Owing to the refraction in silicon, neutron waves, which have passed through different elements of the grating, possess different phase shifts $\Delta\varphi=k_0(1-n)h$, where n is the refractive index.

The depth of the grooves h was selected to be approximately 0.14 mkm, which corresponds exactly to a phase shift $\Delta\varphi = \pi$.

In the following experiment [27] UCN spectra appearing due to diffraction at a moving grating were measured at several rotating rates. From the common analysis of these results the intensities of ± 1 diffraction orders were found as $|a_1|^2_{exp} = 0.383(8)$, being in reasonable agreement with the theoretical value $|a_1|^2_{th} = 0.405$ found in kinematic approximation. At the same time it was realized that neglecting the grating form is a very rough approximation, as in the moving system of reference neutrons are incident on the grating at a glancing angle (see figures 3, 4).

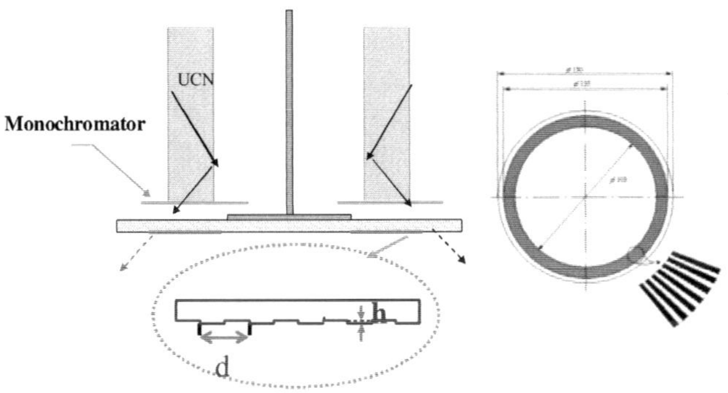

Figure 1. Principle of the experiment. Phase grating is prepared on the surface of the silicon disc, which can be rotated. On the right - geometry of grating and orientation of grooves.

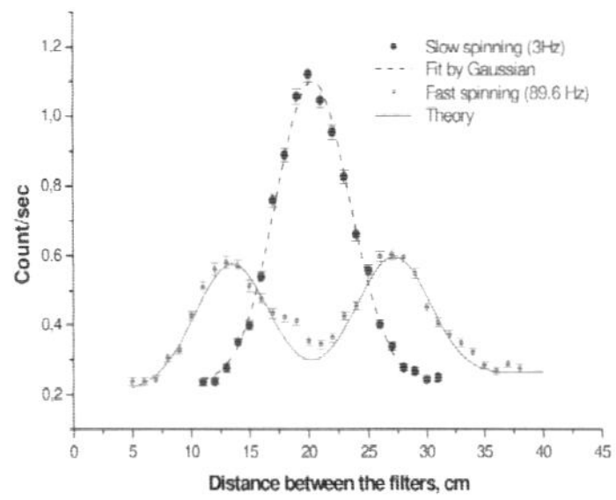

Figure 2. First demonstration of the energy splitting at diffraction of UCN by a moving (rotating) grating [25, 26].

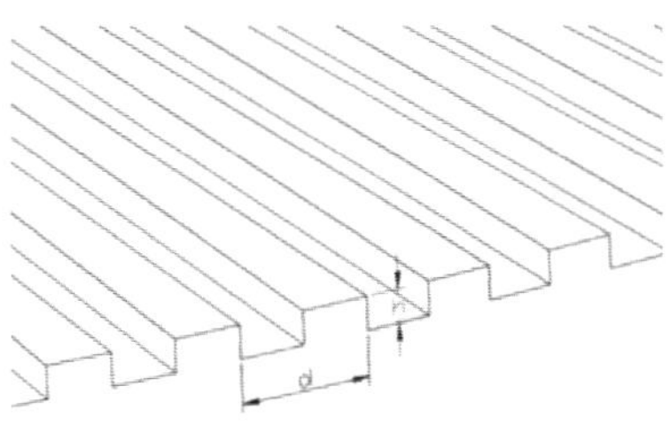

Figure 3. 3D structure of grating.

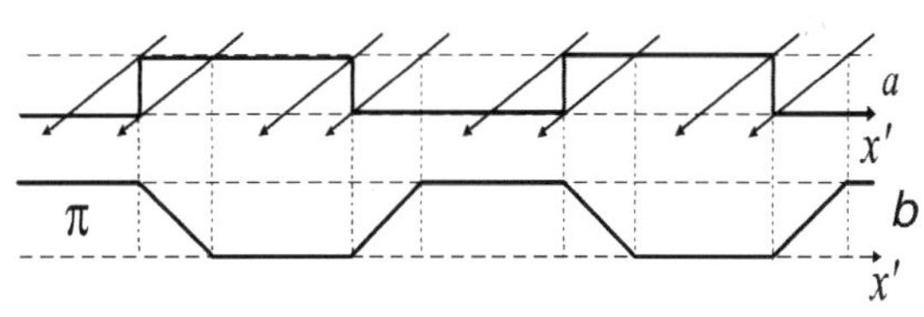

Figure 4. a - grating profile, b - phase profile.

Therefore, in Ref. [28] it was proposed to modify the transmission function of the grating for the phase profile to be of a trapezoidal form, as it is illustrated in Figure 4. In this case, the amplitudes of even and odd orders become equal to $a_n = CB_n$, and $a_n = -B_n/n$, respectively, where $B_n = [1+\exp(-i\pi nC)]/[i\pi(1-n^2C^2)]$, $C = 2hV/dv_{oz}$ and v_{oz} is the neutron velocity.

The intensities of the diffracted waves predicted by this model differ remarkably from the model of a plane grating (see below Figure 8). While the grating velocity V and dimensionless parameter C are both increasing, the first-order intensities of waves should decrease, and in contrast to this, the intensities of the zero and second orders are increasing. However, under the conditions of experiments [25, 26] parameter C was of the order of 0.1 and the effect was insignificant.

3. Applications of neutron diffraction by a moving grating

Concluding paper [26], the authors wrote "We believe that this effect, which has been observed for the first time, may be used for controlled changing of the neutron energy in other experiments. The idea of neutron time-focusing, as proposed earlier, now looks more realistic". Herein they referred to papers [29, 30], where the possibility of application of non-stationary acting at a neutron wave in order to create a neutron time lens was discussed. This idea was discussed in connection with the proposal of pumping UCN produced by a pulse source into a trap [31].

Soon afterwards, the experiment demonstrating the possibility of creation of the neutron time lens was implemented. For the time focusing of neutrons an aperiodic rotating grating was used. Neutrons passed through a small fragment of the grating in such a way that its effective space frequency ξ, i.e. the number of grooves per a length unit, and, accordingly, the transferred energy $\Delta E = 2\pi\hbar V\xi$ varied during the rotation period of the grating. Neutrons of +1 diffraction order were accelerated and those of -1 were decelerated (see Figure 5). The first results of the experiment on neutron time focusing were reported at ECNS-2003 conference [32] and described in greater detail in Ref. [33]. The acting of the neutron time lens is illustrated in Figure 6, which displays the result never published before

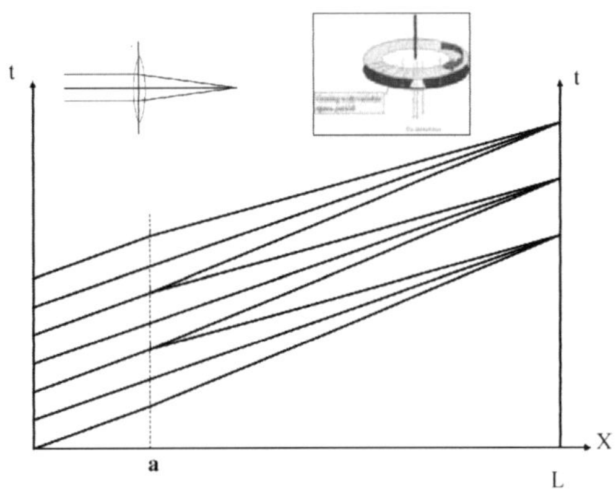

Figure 5. Coordinate versus time scheme of the experiment with the time lens.

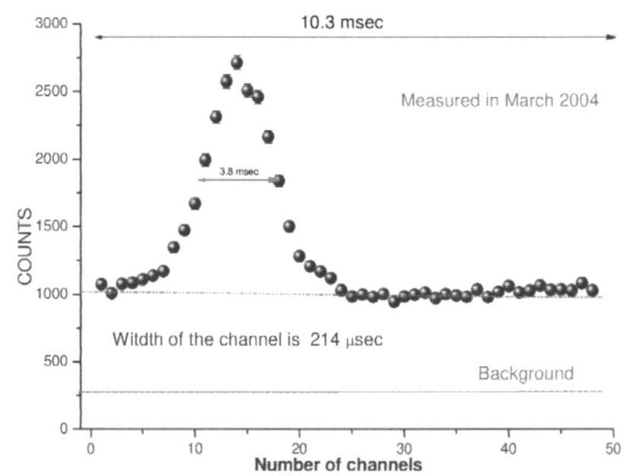

Figure 6. Time focusing peak representing distribution of the neutron arrival times plotted on a time scale equal to the grating rotation period.

Later on, it was realized that the energy quantization effect at NDMG can be used for the investigation of neutron gravity properties [34]. The idea was to compare the gravity energy *mgH* that a neutron gets, when falling from the height *H* with the energy $\Delta E = \hbar\Omega$ transferred to the neutron due to interaction with a non-stationary device. The proper experiment was performed in 2006 [35, 36]. The narrow energy spectrum of neutrons *f*(*E*) was formed by a monochromator located at the height *H* = 0. In order to arrive at the detector and to be detected, neutrons had to pass through an analyzer, which was located below the monochromator by height *H* and had the same spectrometric properties. Neutrons were accelerated in the Earth's gravitational field in the path between the monochromator and analyzer. Since the spectral function *f*(*E*) was narrow, the system already became opaque for a relatively small distance *H*. However, if the neutron energy additionally decreased by $\Delta E = \hbar\Omega$, it was possible to find the position, in which the analyzer transmitted neutrons. In this case, the dependence of the system transmittance on the analyzer position was described by the symmetric function

$$\Phi(H) = \int f(E) f(E + m_i g_n H - \hbar\Omega) dE \,, \qquad (4)$$

where m_i is neutron inertial mass and $m_i g_n$ is the force acting at the neutron in the gravity field of Earth. The maximum of this function or scanning curve was determined by the condition $m_i g_n H = \hbar\Omega$. Neutron interference filters were used as a monochromator and analyzer in this experiment. The rotating grating with the angular period $\alpha = 8.333\times10^{-5}$ rad corresponding to a space period of 5 mkm at a diameter of 120 mm was used in this experiment. It was obtained for the factor of correspondence $\gamma = 1-\left(m_i g_n / m_n g_{loc}\right)$, $\gamma = (1.8\pm2.1)\cdot10^{-3}$ where m_n and g_{loc} are neutron table mass and local value of the free fall acceleration, respectively. The scanning curves (4) measured at different rotating rates are shown in Figure 7.

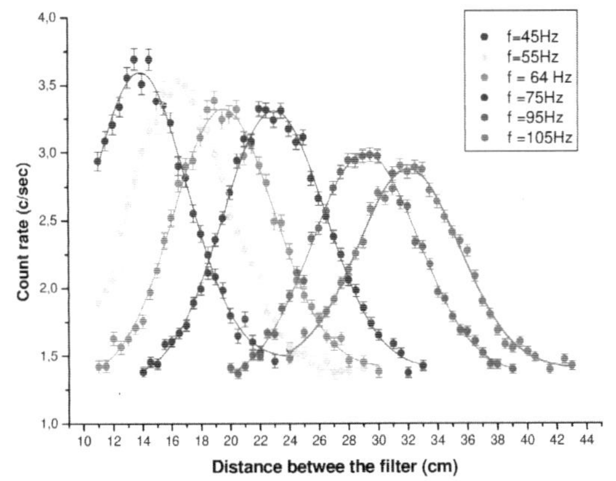

Figure 7. Scanning curves for the line of -1 diffraction order measured for various grating rotation speeds and the fitting Gaussians.

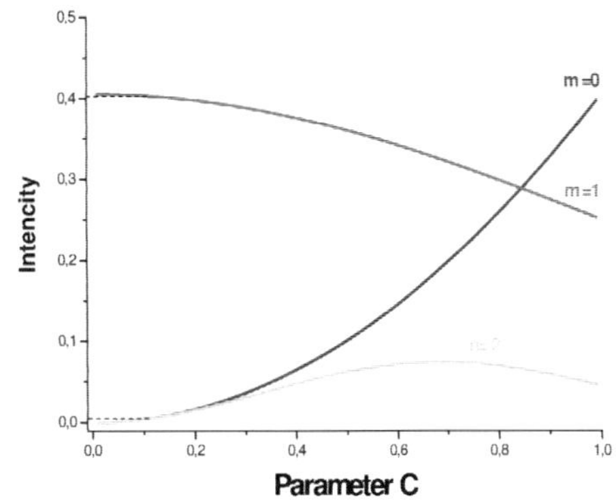

Figure 8. Intensities $J_m = |a_m|^2$ of the waves of the diffraction order m as a function of parameter C

As it is seen in Figure 7, the intensity of the first-order diffraction line decreases remarkably with an increase in the rotation rate. Qualitatively, it corresponds to the prediction in [28] (see Figure 8), whereas quantitative comparison was not performed.

The intension to increase precision of the experiment in order to test the equivalence principle stimulated the improvement of the experimental method. The main idea of comparing the gravity energy mgH with the transferred quantum of energy $\Delta E = \hbar\Omega$ was the same as before, but the approach to the UCN spectrometry changed significantly. The implementation of this project is still in progress; nevertheless, it became evident that the role of even diffraction orders was seriously underestimated in the process of experiment planning. Factor C was four times larger than in [25, 26], and the intensities of even orders rising with the growth of factor C increased remarkably. Due to the fact that the value of the energy splitting $\Delta E = \hbar\Omega$ was much larger than in the previous experiment, it was not possible to measure the spectrum using the existing spectrometer [37]. Moreover, the theory did not give any reliable answer to the question of the relation of different diffraction orders. It became obvious that the phenomenon of NDMG had to be investigated in further detail.

4. Modern stage of NDMG investigation

4.1. Dynamic theory of multiwave diffraction

A significant breakthrough in understanding theoretical aspects of the problem was achieved with appearance of paper [38]. In contrast with Ref. [19, 28] the dynamic approach was developed, taking into account both the three dimensional structure of the grating (see Figures 3 and 4) and the cross-interaction of the diffracted waves at their propagation in the matter of the grating. Further on, we summarize the main results of this work.

We shall consider the problem in the moving system of reference, where the grating is at rest. In this system the wave is obliquely incident on the grating:

$$\Psi'_{in}(x',z,t) = A_{in}(x')exp\left(ik_0 z - i\omega' t\right), \tag{5}$$

where $A_{in}(x')=A_0\exp(-ik_V x')$, $k_V = mV/\hbar$, $\omega' = \omega_0 + \omega_V$, $\omega_V = \hbar k_V^2/2m$.

Neutron propagation in a medium is described by the Schrödinger equation

$$\Delta\psi\left(\mathbf{r},t\right)+\left[k^2 - k_b^2(\mathbf{r})\right]\psi(\mathbf{r},t)=0, \tag{6}$$

where k is the wave number of neutrons in vacuum, $k_b^2(\mathbf{r}) = 4\pi N(\mathbf{r})b(\mathbf{r})$, $N(\mathbf{r})$ is the density of nuclei, $b(\mathbf{r})$ is the coherent scattering length of the medium. Let us assume that the function $k_b^2(x)$, periodic in the region $0 \le z \le h$, has the form

$$k_b^2(x) = \sum_{n=-\infty}^{\infty} \chi_n \, exp(iq_n x), \tag{7}$$

where $q_n = 2\pi n/d$ are reciprocal lattice vectors,

$$\chi_n = \frac{1}{d}\int_0^d k_b^2(x)exp(-iq_n x)dx\,. \tag{8}$$

A zero Fourier amplitude χ_0 determines the value of the average grating refractive index $n_e = (1 - \chi_0/k^2)^{1/2}$ in a layer of thickness h.

We can write the wave function of neutrons in the region $0 \le z \le h$ in the moving coordinate system using (5) as the sum of Bloch functions with amplitudes $\Psi_m(z)$ depending on the vertical coordinate z:

$$\Psi'(x',z,t) = \sum_{m=-\infty}^{\infty} \Psi_m(z)exp(i\mathbf{q}_m\mathbf{r}' - i\omega' t), \tag{9}$$

where the projections of the wave vectors are $q_{mx} = q_m - k_V$, $q_{mz} = k_{0z} = (k_0^2 - \chi_0)^{1/2}$. Herein, it was assumed that the wave number in vacuum is $k' = (k_0^2 + k_V^2)^{1/2}$, and in the layer $0 \le z \le h$ it is $k_{0z} = (k'^2 n_e^2 - k_V^2)^{1/2}$.

By inserting (7) and (9) into (6), we equate the terms with equal exponents and neglect the second derivative of Ψ_m with respect to z. As a result, we obtain a system of differential equations:

$$\frac{d\Psi_m}{dz} = i\alpha_m\Psi_m - i\sum_{n\neq 0}\beta_n\Psi_{m-n}\,, \tag{10}$$

where $\alpha_m = q_m(2k_V - q_m)/2k_{0z}$, $\beta_n = \chi_n/2k_{0z}$. The system of equations (11) should be supplemented by boundary conditions: $\Psi_0(z = 0) = A_0$, $\Psi_{m\neq 0}(z = 0) = 0$. It may be solved digitally.

4.2. Fourier time-of-flight spectrometry

New theoretical results [38] have brought us to the understanding of the need for more detailed experimental investigation of UCN spectra in diffraction by a moving grating. As it was mentioned above, gravity spectrometry with interference filters used in Refs. [23, 24, 37] appeared to be inadequate to the problem. The most appropriate method for the solution of the problem was time-of-flight (TOF) gravity spectrometry [39]. For this purpose the spectrometer [37] was upgraded significantly and converted into a TOF Fourier UCN spectrometer. This spectrometer was used for the new measurements of UCN spectra at diffraction by a moving phase grating. The description of the spectrometer and particular results of diffraction spectrum measurements are reported in [40].

One of the measured spectra compared with theoretical calculations based on the solution of system (10) is shown in Figure 9. It can be observed that the theory corresponds to the results of the experiment quite satisfactorily. Using time-of-flight Fourier spectroscopy, it is possible to measure the

UCN spectra appearing at neutron diffraction by a moving grating in a wide energy range comparable with the initial neutron energy.

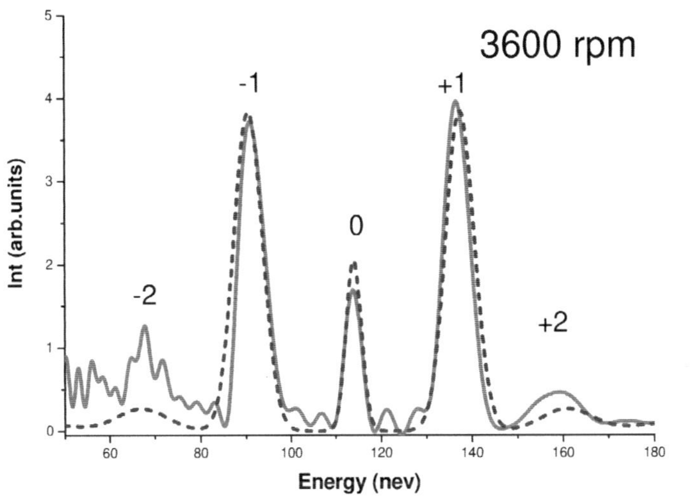

Figure 9. Energy spectra of neutrons diffracted by a moving grating measured by TOF Fourier spectrometer. Experimental (red solid) and theoretical (blue dotted) spectra are shown.

5. Prospects

Presently, it is rather difficult to predict any future application of this relatively new phenomenon. The theory stated in [38] provides us with a wide range of possibilities of formation of diffraction spectra with defined parameters. In particular, it is possible to increase considerably the intensities of higher diffraction orders, for example, the second one, simply by varying the depth of grooves. Using the glancing moving grating [41], it is possible to concentrate the entire flux only in one diffraction order, and consequently, to accelerate and decelerate neutrons without any noticeable widening of the spectrum. The interferometer with a moving grating [42] can have a unique sensitivity to small forces acting at the neutron.

Summarizing the above-said, we have shortly discussed gravity experiments methodically based on the condition $\delta E = \hbar\Omega - mgH = 0$. Let us briefly consider another possibility of using non-stationary neutron diffraction based on the circumstance that the amount of energy $\Delta E = \hbar\Omega$ transferred to the neutron can be accurately defined.

Assumingly, the aim of the experiment is to measure parameters of the motion equation $F(x,E,m,a,t)=0$, where coordinate x, mass m, energy E or velocity $v = (2E/m)^{1/2}$, acceleration a, or force $f = ma$ and time t. It is evident that in order to measure one of the parameters, we have to define the others independently. For example, to measure the free-fall gravity acceleration g of the neutron in the so-called 'free falling experiment', it is necessary to accurately determine the height of falling H and the time of falling t. However, having any periodical non-stationary device such as a moving grating, it is possible to split the initial spectrum and get a state with a set of energies $E_n = E_0 + n\hbar\Omega$, $n = 0, \pm 1, \pm 2..$ with precisely known values $n\hbar\Omega = n\Delta E$. By measuring the time of flight, for instance, for five diffraction orders, we will have not one, but five equations $F_n(x,E_0+n\Delta E,m,a,t_n)=0$ with four unknown values x,E_0,m,a. The accuracy of this common measurement can be improved by overdetermining the equation system applying the method of multiplication of the number of frequencies Ω_j.

The possibility of implementation of this idea to test the equivalence principle for the neutron is considered in the report of Zakharov et al. at the present conference [43].

This work was supported by the Russian Foundation for Basic Research (projects No. 15-02-02367, 15-02-02509).

References

[1] Moshinsky M 1952 *Phys. Rev.* **88** 625
[2] Gerasimov A S and Kazarnovsky M V 1976 *Sov. Phys. JETP* **44** 892
[3] Gähler R and Golub R 1984 *Z. Phys. B* **56** 5
[4] Felber J, Gähler R, and Golub R 1988 *Physica B* **151** 135
[5] Felber J, Müller G, Gähler R and Golub R, 1990 *Physica B* **161** 191
[6] Nosov V G and Frank A I 1991 *J. Moscow Phys. Soc.* **1** 1
[7] Golub R and Lamoreaux S K 1992 *Phys. Lett. A* **162** 122
[8] Nosov V G and Frank A I 1994 *Phys. Atom. Nucl.* **57** 968
[9] Frank A I and Nosov V G 1995 *Ann. New York Acad. Sci.* **755** 293
[10] Nosov V G and Frank A I 1999 *Phys. Atom. Nucl.* **62** 754
[11] Felber J, Gähler R, Golub R et al., 1999 *Foundations of Physics* **29**, 381
[12] Hils Th, Felber J, Gähler R et al., 1998 *Phys. Rev. A* **58** 4784
[13] Summhammer J, Hamacher K A, Kaiser H et al., 1995 *Phys. Rev. Lett.* **75** 3206
[14] Frank A I and Amandzholova D B 1995 *Ann. New York Acad. Sci.* **755**, 858
[15] Kozlov A V and Frank A I 2005 *Phys. Atom. Nucl.* **58**, 1104
[16] Felber J,. Gähler R, Rausch C and Golub R 1996 *Phys. Rev. A* **53** 319
[17] Frank I M 1975 *JINR communication* P4-8851 (In Russian)
[18] Hamilton W A, Klein A G, Opat G I and Timmins P A 1987 *Phys. Rev. Lett.* **58** 2770
[19] Frank A I and Nosov V G 1994 *Phys. Lett.* A **188**, 120
[20] Seregin A A 1977 *JETP* **46** 859.
[21] Steyerl A, Drexel W, Malik S S, Gutsmiedl E 1988 *Physica B*, **151** 36
[22] Frank A I, Balashov S V, Bodnarchuk V I et al., 1999 *Proc. SPIE* **3767** 360
[23] Bondarenko I V, Bodnarchuk V I, Balashov S N et al. 1999 *Phys. Atom. Nucl.* **62** 721
[24] Bondarenko I V, Balashov S N, Cimmino A et al. 2000 *Nucl. Instr. Meth. A*, **440** 591
[25] Frank A I, Bondarenko I V, Balashov S N et al. 2000 *Proceeding of VIII International. Seminar on Interaction of Neutrons with Nuclei ISINN-8*, JINR E3-2000-192, Dubna 448
[26] Frank A I, Balashov S N, Bondarenko I V et al. 2003 *Phys. Lett. A* **311** 6.
[27] Frank A I, Geltenbort P, Kulin G V et al. 2005 *JETP Letters*, **81** 427
[28] Frank A I, Geltenbort P, Kulin G V et al. 2004 *JINR communication P3-2004-207* (in Russian)
[29] Frank A, Gähler R, 1996 *Proceedings of IV International Seminar on Interaction of Neutrons with Nuclei ISINN-4,* JINR E3-96-336, Dubna, 308
[30] Frank A I, Gähler R 2000 *Phys. At. Nucl.* **63** 545
[31] Shapiro F L, 1976 *Neutron Studies.* (Nauka, Moscow) 229. (in Russian)
[32] Balashov S N Bondarenko I V, Frank A I et. al. 2004 *Physica B*, **350** 246.
[33] Frank A I, Geltenbort P, Kulin G V and Strepetov A. N 2003 *JETP Lett.* **78**, 188.
[34] Frank A I, Masalovich S V, Nosov V G 2004 *Proceeding of XII International Seminar on Interaction of Neutrons with Nuclei ISINN-12*. JINR E3-2004-169, Dubna, 215
[35] Frank A I, Geltenbort P, Jentschel M, et al., 2007 *JETP Letters* **86**. 225
[36] Frank A I, Geltenbort P, Jentschel M et al. 2009 *Nucl. Instr. Meth.* A **611** 314
[37] Kulin G V, Frank A I , Goryunov S V et al. 2015 *Nucl. Instr. Meth.* A **792**, 38
[38] Bushuev V A, Frank A I and Kulin G V 2015 arXiv:1502.04751v1 [physics.optics]; *JETP* **148** in press.
[39] Kulin G V, Kustov D V, Frank A I et. al. 2014 *JINR communication P3-2014-72* (in Russian)
[40] Kulin G.V., Frank A.I., Goryunov S.V. et al. 2016 *This volume.*
[41] Frank A I, 2013 *Phys. Atom. Nucl.* **76**, 544
[42] Ioffe A, 1997 *Physica B* **234-236**, 1180
[43] Zakcharov M, Frank A I, Kulin G V et al. 2016 *This volume*

Experiment proposed for the observation of UCN interaction with matter moving with giant acceleration

A.I. Frank[1,1], D.V. Kustov[1], G.V. Kulin[1], S.V. Goryunov[1], D.V. Roshchupkin[2], D.V. Irzhak[2]

[1]Frank Laboratory of Neutron Physics, Joint Institute for Nuclear Research, Joliot-Curie str. 6, Dubna, Russia, 141980

[2] Institute of Microelectronics Technology and High-Purity Materials Russian Academy of Sciences, Academician Ossipyan str. 6, Chernogolovka, Russia, 142432

E-mail: frank@nf.jinr.ru

Abstract

It was assumed that in neutron optics the concept of the effective potential has a limited region of validity, which is unlikely in the case of giant acceleration of matter. In the current work an experimental approach to test this hypothesis has been proposed.

1. Introduction

It is known that the nature of the refractive index is associated with the interference of the incident and scattered waves. The scattering of slow neutrons is isotropic and the scattering amplitude in most cases is constant (taken to be –b). In this case, the phase of the scattered wave differs from the phase of the incident wave by the value $\delta = -kb$, where k is the wave number in vacuum and b is the scattering length. The phase shift δ is responsible for refraction. The refractive index n satisfies the Foldy formula [1, 2]

$$n^2 = 1 - \frac{4\pi\rho b}{k^2}, \tag{1}$$

where ρ is the atomic density of nuclei. The interaction of neutrons with matter can be described by the effective potential

$$U_{eff} = \frac{2\pi\hbar^2}{m}\rho b, \tag{2}$$

where m is the neutron mass.

Representations (1) and (2) are completely equivalent to each other. For this reason, dispersion law (1) is often called 'potential'. The validity of equations (1) and (2) for a uniformly moving medium is indubitable, because the transition to the coordinate system, where matter is at rest for significantly non-relativistic problems characteristic of neutron optics, results only in a change in the phase of wave function [3].

It was predicted in [4, 5] that the energy of neutrons passing through a layer of matter moving at acceleration should be different from the initial energy. The theory was based on the assumption of the

[1] To whom any correspondence should be addressed.

validity of dispersion law (1) in an accelerated medium. The change in energy was really detected in experiments with ultracold neutrons [6, 7], where the acceleration of the sample reached 75 m/s^2 and the energy transferred to neutrons was fractions of neV and was in satisfactory agreement with the calculation. Consequently, the assumption of the validity of the potential dispersion law under the indicated condition is justified.

However, the general conclusion on the validity of the dispersion law in an accelerated medium would probably be erroneous. Indeed, the theory of dispersion, which is, in essence, the theory of multiple scattering of waves, involves a very substantial assumption of sphericity of interfering scattered waves. At the same time, in a non-inertial coordinate system associated with accelerated matter, the concept of spherical waves is invalid, which can affect the condition of their interference [8].

The acceleration, at which a significant deviation from dispersion law (1) is expected, can be estimated as follows [9]:

$$\frac{mad^2}{4E} \ll b, \qquad a \ll \frac{4Eb}{md^2} = W_c, \tag{3}$$

where a is the acceleration of matter and d is the interatomic spacing[2]. According to eq. (3), the critical acceleration Wc is proportional to the energy of neutrons.

Using $E = 100$ neV, $b \approx 5\cdot 10^{-13}$ cm and $d = 5\cdot 10^{-8}$ cm in equation (3), we will obtain $W_c \approx 4\cdot 10^7 cm/s^2$ for ultracold neutrons. This acceleration is achievable in laboratory experiments. Moreover, an experiment on the observation of neutron interaction with an object moving at such acceleration was performed many years ago [10]. However, the results of experiment [10] do not contradict estimate (3) above, as it was carried out not with ultracold neutrons (UCN), but with cold neutrons and in accordance with (3) the value W_c was at least two orders of magnitude greater than the sample acceleration. A similar experiment with UCN would answer the important question of the validity of the dispersion law of neutron waves in a medium moving at giant acceleration.

In this paper we briefly discuss a possible approach to testing the validity of the main equation of neutron optics in the case of giant acceleration of matter.

2. Interaction of neutrons with oscillating potential

Unfortunately, the dispersion law of the neutron wave in accelerating matter is unknown. Nevertheless, certain information about it can be obtained, if we perform an experiment on the observation of neutron interaction with a potential structure oscillating in space and compare the results with the theoretical predictions made in the assumption of the validity of eq.(2). This experiment should be performed both for a relatively low $a \ll W_c$, and for the largest possible acceleration of matter $a \gg W_c$.

Due to the relatively low intensity of UCNs, it is difficult to conduct the experiment in the reflection geometry, as it was done in [10], because such an experiment requires not only monochromatization, but also relatively good beam collimation, which leads to a considerable loss in intensity. Therefore, following ref. [11], we turned to the problem of the neutron tunneling through the oscillating barrier.

The time evolution of the wave function is defined by Schrödinger equation

$$i\hbar\frac{\partial\psi(x,t)}{\partial t} = \mathbf{H}\psi(x,t) \,. \tag{4}$$

The initial wave function is represented by a wave packet defined by

$$\psi(x,p) = \frac{1}{(2\pi\delta^2)^{1/4}}\exp\left(-\frac{(x-x_0)^2}{4\delta^2} - ipx\right), \tag{5}$$

Where x_0 is the initial position of its maximum, p is its momentum, and δ is its space dispersion. The potential is confined to a finite spatial region and is given by

$$V(x,t) = U\left[x - A\cos(\omega t + \varphi)\right], \tag{6}$$

where $U(x)$ is, in the general case, any potential structure moving as a whole

[2] Eq. (3) differs from eq. (6) of ref. [9] by a factor 2 because of mistake in [9].

$$U(x)=\begin{cases} V_0(\xi) & 0\le x\le L \\ 0 & \text{otherwise.}\end{cases} \quad (7)$$

Here $V_0(\xi)$ represents an intrinsic structure of the potential, and for a rectangular barrier it is a constant; L is the total width of the potential, $\xi\in[0;L]$. To solve the time-dependent quantum problem the split operator method [12] has been applied.

The primary calculations were made for the case of oscillating rectangular barrier. It was found that in this case the observable effects in the transmitted state exist only in a narrow region of neutron energy close to the barrier height V_0. Thus we decided to use a more complicated potential structure, which consists of two barriers and a well in between, known as the Neutron Interference Filter (NIF) [13-15]. In a stationary case, such a structure transmits neutrons in a rather narrow energy band due to the resonant tunneling through the potential barriers.

Calculations were made with the following parameters of NIF: the thicknesses of both barriers, separated by 18nm was 25 nm and the value of the barrier potential was 250 neV. This kind of potential structure can be formed by three films of Ni, Ti/Zr alloy and Ni deposed on any appropriate wafer. The potential of the wafer was assumed to be zero. The calculated dependence of NIF transmittivity on the neutron energy is shown in Figure 1.

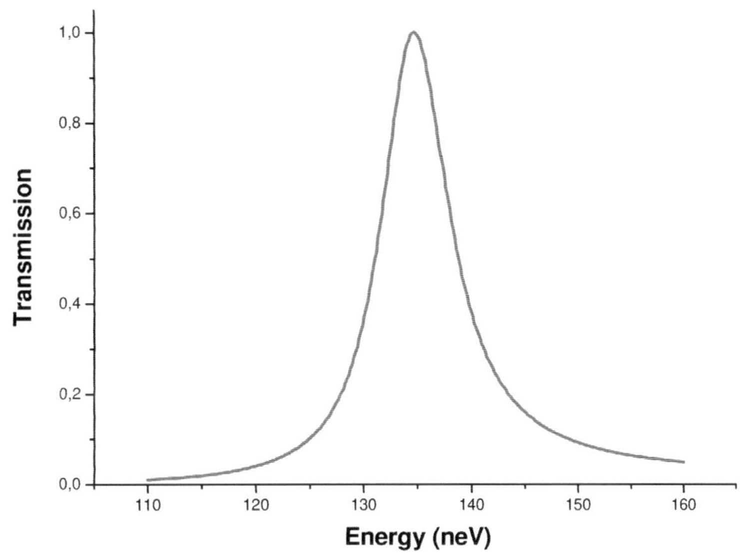

Figure 1. Dependence of NIF transmission on neutron energy

The maximum of the incoming wave packet corresponded to the energy of 135 neV, which was exactly equal to the maximum of NIF transmission, and its dispersion was selected in a way to provide the travelling time through the observation point being much greater than the oscillation period. It corresponded to the degree of monochromatization $\Delta p/p \approx 10^{-3}\div 10^{-4}$.

The calculations were done for two regimes of NIF oscillation. The frequencies and amplitudes of oscillation were, respectively, 100 KHz and 280nm for the low-frequency regime, and 5MHz and 5nm for the high-frequency regime. The maximal acceleration of NIF $a_{max}=A\omega^2$ was $a_{max} = 1.1\times10^7$cm/s^2 and $a_{max} = 5\times10^8$ cm/s^2 for both regimes, which was less and greater than the critical acceleration W_c.

In both cases, the transmitted state was characterized by the line spectrum with the splitting $\Delta E=\hbar\Omega,\ \Omega=2\pi f$, where f is the frequency of oscillation. But, unfortunately, it is very difficult to perform a spectroscopic experiment for the measurement of such spectra. In the low-frequency regime the value of splitting ΔE is of the order of 0.3 neV, which is less than the resolution of the existing spectroscopic methods. In the high-frequency regime the splitting of the spectrum is 20 neV, which looks reasonable, but for realistic values of the oscillation amplitudes the intensities of the first-order line turn out to be very small.

However, it was found that instead of measuring the spectrum it is possible to measure the time dependence of the transmitted flux. In Figure 2 the time oscillation of the flux density of the transmitted neutrons is shown for the case of the low-frequency regime. This value is proportional to the count rate of the detector placed behind NIF.

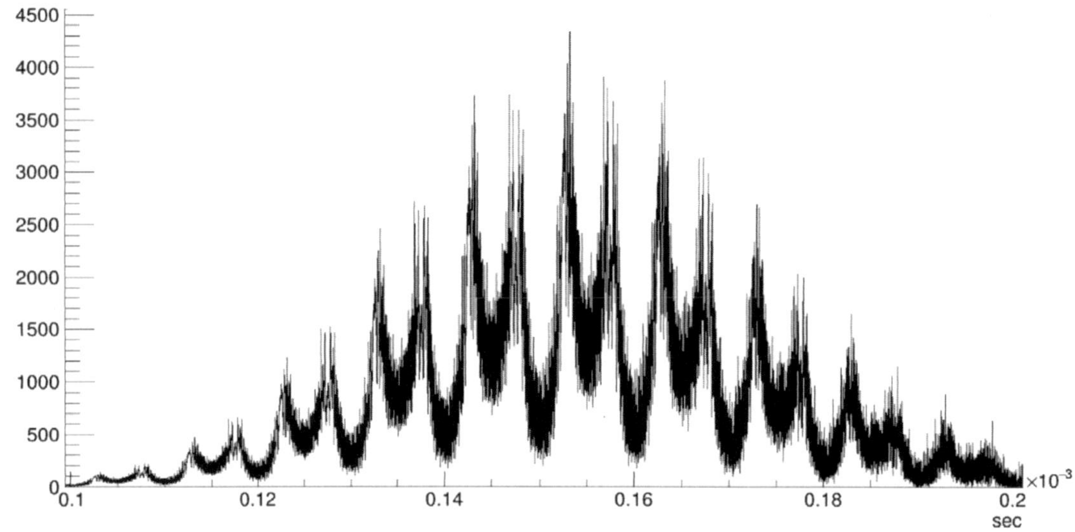

Figure 2. Time oscillation of the flux density at the point behind the potential in the low-frequency regime (see text)

This dependence of the intensity can be easily understood. The smooth flux growing and decreasing during the time of the order of 10-4s demonstrates the travelling of the wave packet (5) trough the point of observation. Oscillations of the flux are conditioned by the space oscillation of the NIF. The periodical variation of NIF velocity $V_{NIF}(t) = V_0 sin(\omega t + \varphi)$ leads to the Doppler shift of the neutron energy in the reference system of NIF. In the case under study, $V_0 \approx 18$ cm/s, which should be compared with the neutron velocity in the laboratory system $v \approx 5$ m/s. The periodical change in the neutron energy $\Delta E/E = 4V_0/v$ in the reference system of NIF is greater than the width of the NIF transmission function and it transmits neutrons only during a certain part of the period, when its velocity is relatively small. As it can be seen in Figure 2, during the oscillation period of 10^{-5}s, as expected, two maxima of intensity appear.

In this case the above-mentioned classical interpretation is correct, as the period of NIF oscillation of 10^{-5} s is two orders of magnitude greater than the time of neutron transmission through NIF, or, equivalently, the life time of neutron quasi-bound state, which is of the order of 10^{-7} s. [16].

In the high-frequency regime (see Figure 3) oscillations are also seen clearly, but their amplitude is much less than in the first case, while the maximal NIF velocity V_0 is of the same order of magnitude as in the low-frequency regime. The point is that the oscillation period of 2×10^{-7} s is now comparable with the transmission time and the classical interpretation becomes inapplicable.

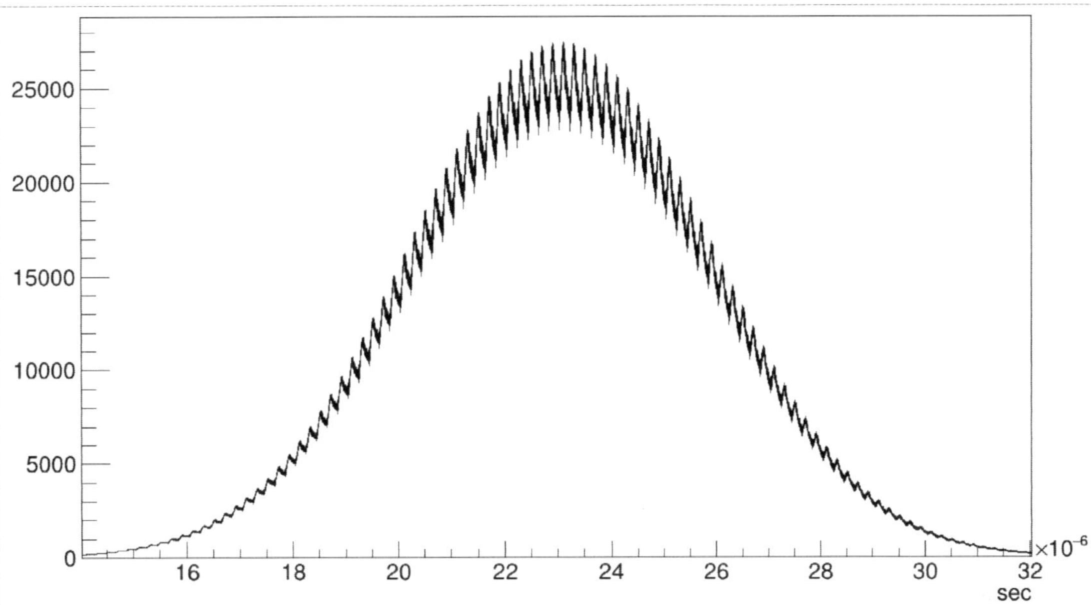

Figure 3. Time oscillation of the flux density at the point behind the potential in the high-frequency regime (see text)

3. Possible experimental implementation

The possible experimental implementation is illustrated in Figures 4, 5. In the low-frequency regime the wafer with the deposed films forming the potential structure of NIF is fixed at the ring-shaped Piezo driver. The advantage of this approach is that neutrons do not pass through the driver and the latter may have the thickness necessary for providing the required vibration frequency. Besides, in this approach it is relatively easy to vary the sample - NIF. The UCN detector must be placed next to the wafer. It may be a silicon detector with a thin (0.2 mkm) layer of ^{10}B.

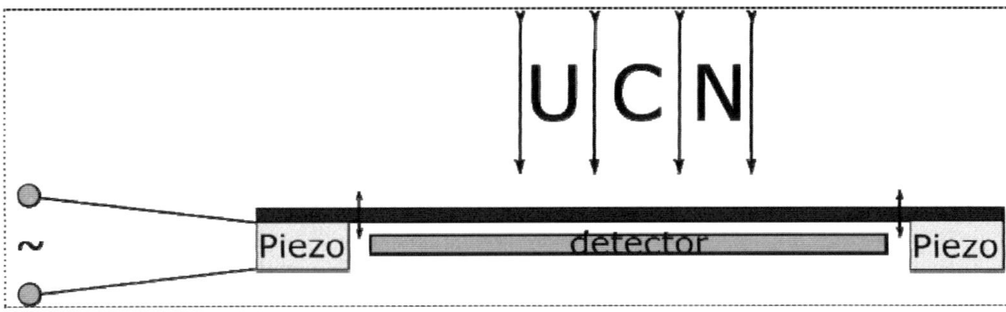

Figure 4. Experimental scheme with the separation of Piezo driver and wafer with NIF. Low-frequency regime with thick piezo.

An alternative approach is shown in Figure 5. In this case the films forming the potential structure of NIF are deposed on the quartz plate, which serves as a resonant Piezo driver. The thickness of the plate must be equal or multiple of a half of ultrasonic wavelength. For the operation at a frequency of 5MHz it must be of the order of 0.5 mm. Such a plate is transmittive enough for UCN. The vibration amplitude may be varied smoothly in certain limits without exceeding the value of the order of 5nm.

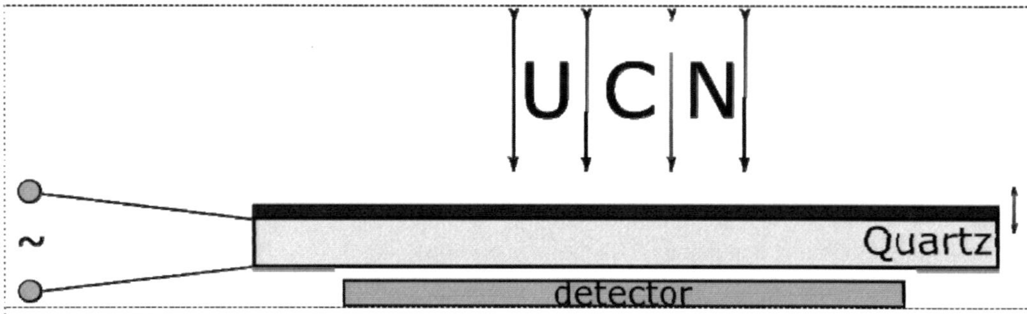

Figure 5. Piezo-driver combined with NIF wafer. High-frequency regime

5. Conclusion

In the current paper it was assumed that in neutron optics the concept of the effective potential has a limited region of validity, which is unlikely in the case of giant acceleration of matter. An experimental approach to test this hypothesis has been proposed.

This work has been supported by the Russian Foundation for Basic Research (grant # 15-02-02367).

References

[1] Foldy L L 1945 *Phys. Rev.* **67** 107
[2] Lax M 1951 *Rev. Mod. Phys.* **23** 287
Lax M 1952 *Phys. Rev.* **85** 621
[3] Horne M A, Zeilinger A, Klein A G and Opat G I 1983 *Phys. Rev. A* **28** 1
[4] Kowalski F V 1993 *Phys. Lett. A* **182** 335
[5] Nosov V G and Frank A I 1998 *Phys. At. Nucl.* **61** 613
[6] Frank A I, Geltenbort P, Jentschel M, Kustov D V, Kulin G V, Nosov V G and Strepetov A N 2008 *Phys. At. Nucl.* **71** 1656
[7] Frank A I, Geltenbort P, Jentschel M, Kustov D V, Kulin G V and Strepetov A N 2011 *JETP Lett.* **93** 361
[8] Frank A I 2009 *Phys. Usp.* **52** 397
[9] Frank A I 2014 *JETP Lett.* **100** 613
[10] Felber J, Gähler R, C Rausch, and Golub R 1996 *Phys. Rev.* **53**, 319.
[11] Pimpale A, Holloway S and Smith R J 1991 *J.Phys.A: Math.Gen.* **24** 3533
[12] Chin S A and Chen C R 2002 *J. Chem. Phys.* **117** 1409
[13] Seregin A A 1977 *JETP* **46** 859
[14] Steinhauser K A, Steyerl A, Schechenkofer H and Malik S S 1980 *Phys. Rev. Lett.* **44** 1306
[15] Frank A I, Balashov S V, Bodnarchuk V I, Bondarenko I V Cimmino A, Geltenbort P. Hoghoj P, Klein A G, Korneev D A, Kozlov AV, Masalovich S V 1999 *Proc. SPIE* **3767** 360
[16] Frank A I, Bondarenko I V, Vasil'ev V V Anderson I, Ehlers G, Høghøj P, 2002 *JETP Lett.* **75** 705

Strain Induced Martensitic Transformation in Austempered Ductile Iron (ADI)

X H Li[1], P Saal[2], W M Gan[3], M Landesberger[2], M Hoelzel[1] and M Hofmann[1*]

[1] *Heinz Maier-Leibnitz Zentrum (MLZ), Technische Universität München, Lichtenbergstr. 1, D-85747 Garching, Germany*
[2] *Institute of metal forming and casting, Technische Universität München, Walther-Meißner-Str. 4, D-85747 Garching, Germany*
[3] *German Engineering Materials Centre at MLZ, Helmholtz-Zentrum Geesthacht, D-85747 Garching, Germany*
**Corresponding author. Tel.:+49 289 14744; fax: +49 89 289 14666*

Keywords:

neutron powder diffraction, retained austenite, phase analysis, texture

Abstract

The strain induced martensitic transformation in austempered ductile iron (ADI) has been investigated using high resolution neutron diffraction on samples compressed ex-situ to different plastic strains. In addition bulk texture measurements using neutron diffraction have been performed to calculate the orientation distribution of ferrite and austenite phases for different strain levels. Combing the detailed texture information with neutron diffraction pattern proved to be essential for quantitative phase analysis and extraction of martensite phase fractions. The martensite content induced by strain in ADI depends on austempering temperature and degree of deformation.

1. Introduction

Austempered ductile iron (ADI) is a nodular ductile iron which has undergone a special heat treatment to enhance mechanical properties to a great extent. The heat treatment process of ADI consists of austenitization followed by quenching to a temperature typically between 250°C and 450°C and isothermal austempering [1]. After such a heat treatment the microstructure of ADI consists of acicular ferrite (α) and high carbon enriched retained austenite (γ). The microstructure strongly depends on the austenitization and austempering temperatures. Details on dependence of microstructure and temperature can be found in [1-6]. In industrial applications alloying elements such as Ni, Mo or Cu are used in order to delay the phase transition kinetics, which improves the austemperability of heavy sections.

The high carbon enriched austenite is metastable and several reports exist indicating a transformation induced plasticity effect (TRIP) when ADI is subjected to deformation [7-9]. During the plastic deformation the high carbon enriched retained austenite can transform to martensite. Austenitization temperature (T_γ), austempering temperature (T_{Aus}), austempering time and alloying

Journal of Physics: Conference Series **746** (2016) 012055 doi:10.1088/1742-6596/746/1/012055

composition can influence the retained austenite fraction, its grain size and stability [3, 4], which in turn will affect the possible martensitic transformation.

The relationship between heat treatment parameters and phase fractions in ADI has been extensively studied by Meier et al. using in-situ neutron powder diffraction on non-textured samples [1]. On the other hand in case of the deformation induced martensitic transformation in ADI quantitative phase analysis is hindered due to texture formation and standard Rietveld analysis fails to produce reliable results. However, the incorporation of texture information into Rietveld refinements to extract quantitative phase information has been already successfully applied in various studies [e.g. 10-12].

A similar approach is therefore used in the current work where the influence of different treatments and composition parameters on the martensitic transformation has been studied using high resolution neutron powder diffraction and texture measurements on samples after compression to different plastic strains. The results from these experiments allow us to relate the martensitic transformation kinetics in ADI with austempering temperature and plastic strains.

2. Experimental

Cylindrical ADI samples which contain 3.6wt% C, 2.8wt%Si and 0.2wt%Mn with dimension of Ø 6mm × 10 mm were prepared using different heat treatment parameters which yield microstructures as shown in table 1 (it should be noted that here the volume fractions relate only to the metallic matrix since the change in nodular graphite fraction during austempering is negligible [1]). T_γ and T_{Aus} where selected according to results published elsewhere [13]. The heat treated ADI samples were compressed to different strain levels ranging from 0% up to 40% with a 10% increment in strain. High resolution neutron powder diffraction data were taken from these samples using the diffractometer SPODI at MLZ [14] within a 2θ range = 10~152° and step width of Δ2θ = 0.05°. The wavelength was set to 1.5483Å using a Ge (551) monochromator.

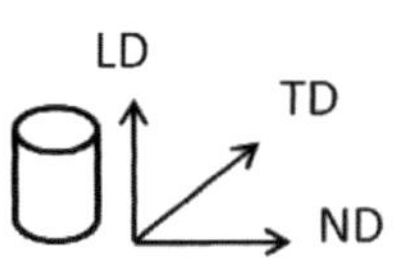

Figure 1. Definition of the sample coordinate system for the pole figures. LD: loading direction; ND: normal direction; TD: transverse direction

Table 1. ADI heat treatment conditions and resulting phase fractions after heat treatment

Austenitization T_γ [°C]	Austempering T_{Aus} [°C]	Austenite [vol%]	Ferrite [vol%]
900	400 350 300	40 34 24	60 66 76

Complete pole figures of ferrite (110) and (200), austenite (111) and (200) of the ADI samples were measured at the materials science diffactometer STRESS-SPEC [15] with a wavelength of 1.68Å using a Ge (311) monochromator. In the measurements the loading direction (LD) of the samples was perpendicularly to the incoming beam direction (figure 1). The orientation distribution functions of crystallites (ODFs) and then the inverse pole figures were calculated from the measured complete pole figures using the harmonic method [16]. For quantitative phase analysis the program MAUD [11, 12] was used in which texture information via the ODF and Rietveld refinement methods can be combined.

3. Results and discussion

3.1 Texture formation in ADI during plastic deformation

Figure 2a shows high resolution diffraction data on a set of samples which were prepared with the heat treatment parameters $T_\gamma = 900°C$/ $T_{Aus} = 350°C$ and then compressed to strain levels of 0%, 20% and 40%, respectively. In case of the unstrained sample the extracted lattice parameters were a = 2.8638(4) Å for ferrite and a = 3.6325(7) Å for austenite. During plastic deformation the carbon enriched retained austenite partly transforms to martensite (at 40% strain: a = 2.880 (2) Å and c = 2.957 (4) Å, c/a-ration = 1.03). The strain induced martensitic transformation is indicated in the diffraction data by a decrease of the austenite peak intensity and an increase of the ferrite peak intensity. The increase in ferrite peak intensity is a consequence of close proximity of ferrite and martensite Bragg reflections, which cannot be separated. Martensite reflections can also be identified as shoulders close to the ferrite peak positions (figure 2b). Extensive peak broadening and overlap of austenite (111), ferrite (110) and martensite (101), (110) reflections make a direct extraction of individual martensite peak intensities difficult using ordinary peak fitting methods.

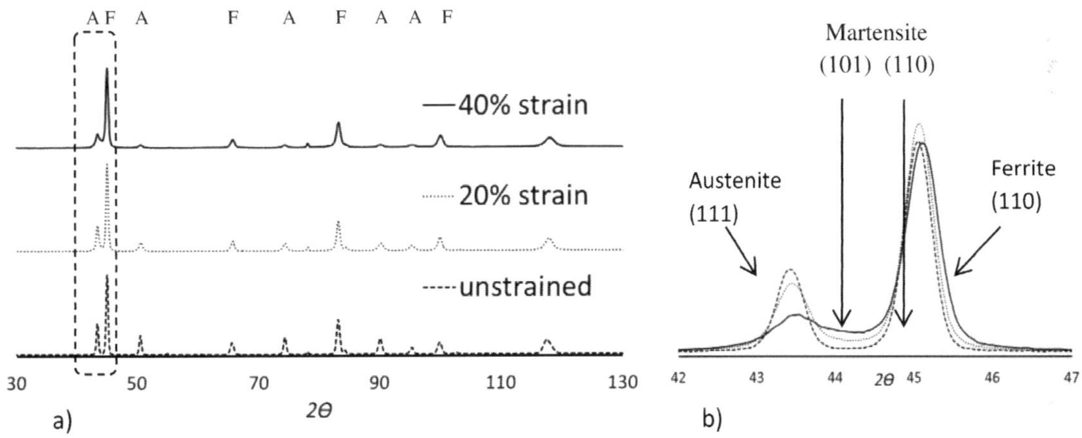

Figure 2a and b. Neutron diffraction pattern of ADI samples, austempered at 350°C and deformed to different strain levels. An enlarged view of the diffraction pattern between 2θ= 42° ~ 47° is shown on the right side. Austenite and ferrite reflections are marked with A and F.

Figures 3 and 4 show the texture evolution of ADI samples (T_γ = 900°C / T_{Aus} = 350°C) with increasing strain. The inverse pole figures have been calculated from measured austenite (111), (200) and ferrite (110), (200) pole figures. The results indicate a weak fibre texture for all samples. Here the typical compression component for austenite [110] (*fcc*) and ferrite [111] (*bcc*) can be readily identified [17]. Comparing the ODFs in figure 4 to the standard ODF of *bcc* and *fcc* materials [18], the texture formation of austenite and ferrite in ADI during the compressive deformation will be identified as follows. The strongest components of austenite phase in the compressed sample are α-fiber // LD spread from (001) $[1\bar{1}0]$ to (110) $[1\bar{1}0]$ plus (111) $[0\bar{1}1]$ and (001) $[\bar{1}10]$. In ferrite the compressive deformation leads to the formation of a texture characterized by two distinct components: (110) $[1\bar{1}2]$ Brass and (112) $[11\bar{1}]$ Copper.

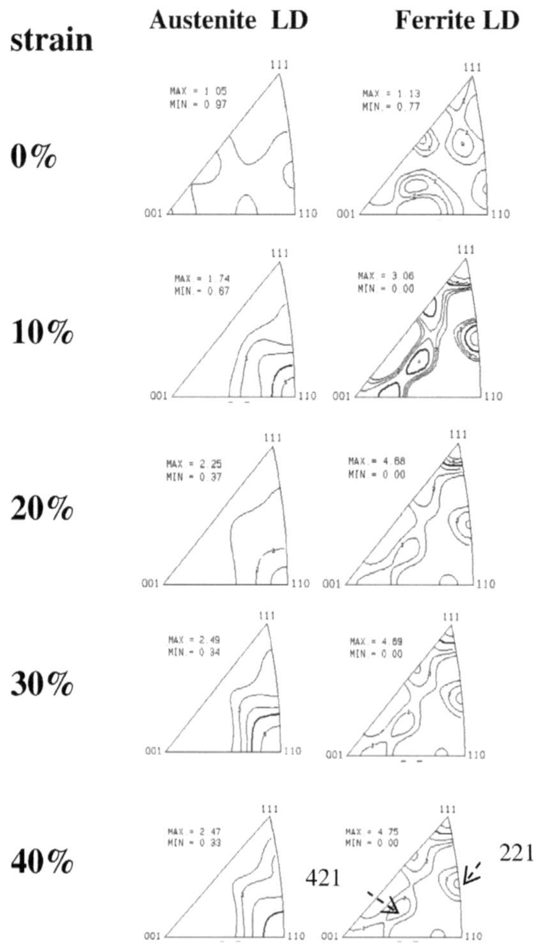

Figure 3. Inverse pole figures of austenite and ferrite phases in ADI (austempered at 350°C) for different strain levels up to 40%.

Comparison of the main texture components - austenite [110] (*fcc*) and ferrite [111] (*bcc*) - of all the differently treated samples (table 1) shows that the texture formation depends only on plastic strains and not on heat treatment conditions. The texture evolution, illustrated by the changes in multiples of random distribution (m.r.d) in the inverse pole figures of the austenite [110] and ferrite [111] directions, which are parallel to the main deformation direction during plastic deformation is shown in figure 5. There is a strong increase in texture during compressive deformation until 20% strain. After reaching 20% strain there is almost no increase with further plastic deformation. This might be attributed to limited activation of dislocation glide systems in *fcc* and *bcc* metals at room temperature. On the other hand the average particle size of austenite and ferrite grains in ADI is in the order of 50 nm [1], nano-crystal sized grain rotation and grain boundary sliding [19] probably take place instead of dislocation gliding during the plastic deformation.

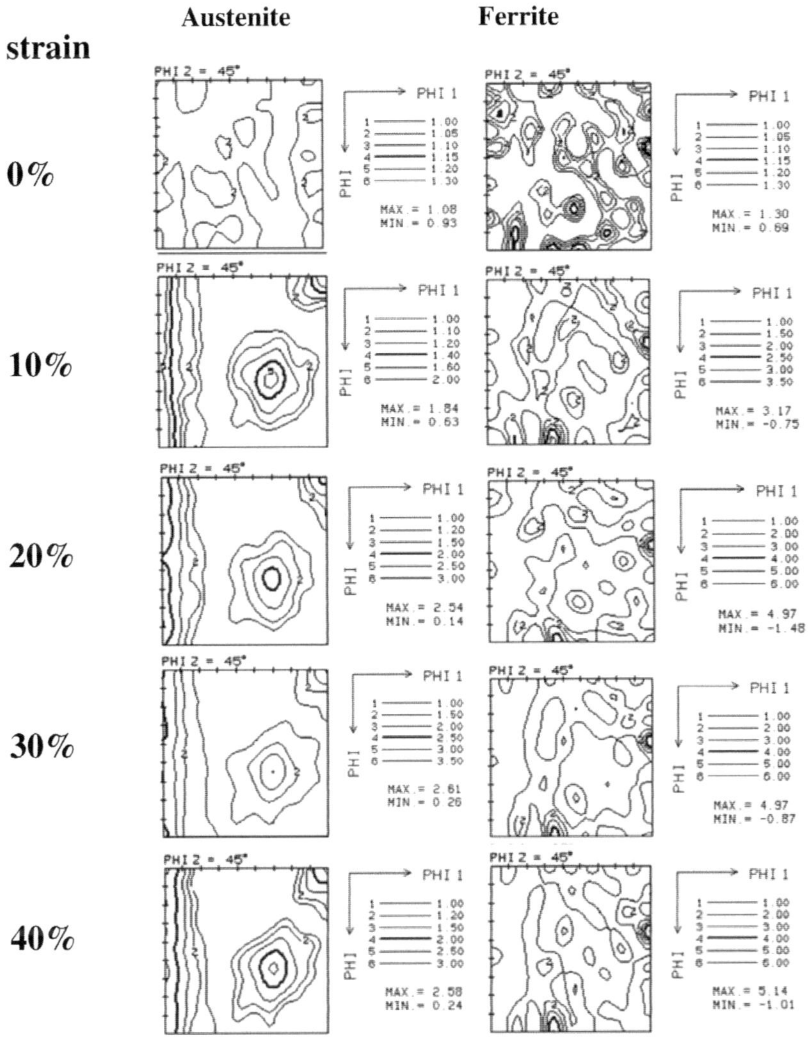

Figure 4. ODFs of austenite and ferrite phases in ADI (austempered at 350°C) for different strain levels up to 40%.

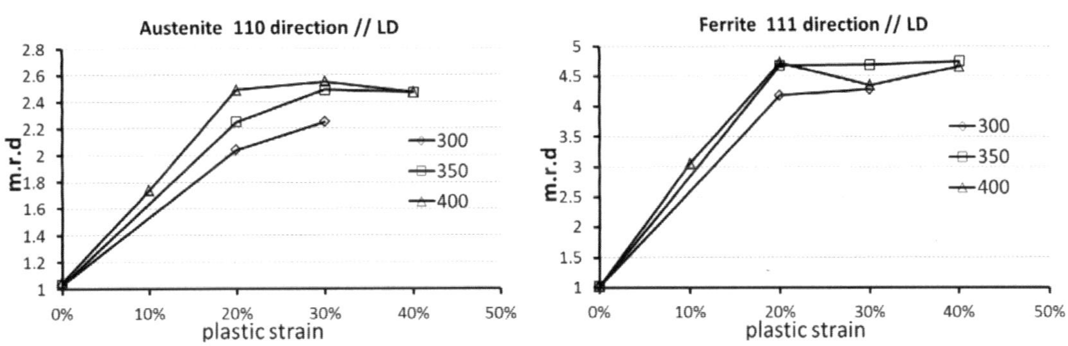

Figure 5. Texture evolution of ADI samples austempered at 300°C, 350°C and 400°C parallel to the main deformation direction during plastic deformation.

3.2 Combining texture and diffraction data for phase analysis

Closer inspection of the inverse pole figures of ferrite reveals additional weak components in [221] and [421] directions (marked in figure 3). These components are usually not found in *bcc* metals (table 2, [18]). Due to the strong overlap of martensite and ferrite reflection positions, intensities for the pole figures cannot be extracted separately and thus these additional components arise most likely from the martensitic phase.

Furthermore this makes it difficult to use the texture results directly when using the ODFs calculated from the (110) and (200) pole figures of ferrite as input in the Rietveld analysis with MAUD. A way around this is to use a single reflection pole figure for the ODF calculation. This is an admissible approach because of the high symmetry of *bcc* crystal structure of ferrite. In addition, originating from the compression deformation martensite as well as ferrite have a distinct orientation relation with the austenite phase. This has been found in most iron based alloys [20-23]. Here the *bcc* ferrite, *bct* martensite and *fcc* austenite are often observed to have a Kurdjumov-Sachs (K-S) and Nishiyama-Wassermann (N-W) orientation relationship. The K-S relation is defined as $(111)\gamma \parallel (110)\alpha$, $[1\bar{1}0]\gamma \parallel [1\bar{1}1]\alpha$, while the N-W relation as $(111)\gamma \parallel (110)\alpha$, $[\bar{1}01]\gamma \parallel [001]\alpha$.

In our case using the (200) pole figure alone yields an ODF and recalculated inverse pole figure in accordance with the standard texture orientations of *bcc* metals (Table 2). As an example figure 6 shows the corresponding ODF and recalculated inverse pole figure for a sample (T_γ = 900°C/ T_{Aus} = 350°C) compressed to 40% strain.

Table 2. Texture orientation // RD or LD in *bcc* and *fcc* metals after different plastic deformation [17].

Cystal structure	tension	compression	rolling
bcc	[110]	[111] and [100]	Mainly from (001)[$\bar{1}$10] to (111)[$\bar{1}$10] Weak component spread from (112)[$\bar{1}$10] to (111)[$\bar{2}$11]
fcc	[111] [100]	Strong component around [110], spread from [110] to [113], plus a weak component [100]	Copper-type (123)[$\bar{4}\bar{1}$2]+(146)[$\bar{2}\bar{1}$1] Silver-type (110)[$\bar{1}$12]

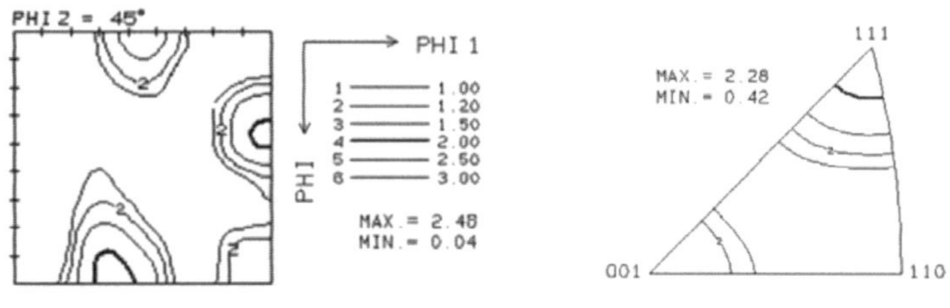

Figure 6. ODF calculated from single component ferrite (200) pole figure of an ADI sample austempered at 350°C and compressed to strain level 40% (left). Corresponding recalculated inverse pole figure (right). Note the absence of the components for the [421] and [221] directions.

3.3 Strain induced martensite transformation

Figure 7 and 8 show the martensite phase fractions in deformed ADI extracted using the recalculated ODFs as input in the Rietveld refinements of the lattice constants, peak shape parameters and phase contents. The strain induced martensite transformation in ADI during plastic deformation is not proportional to the retained austenite content at the beginning of plastic deformation. For the sample treated with T_{Aus} = 400°C and also similar T_{Aus} = 350°C a linear dependence of the martensite phase content with strain is found. Thus at each strain level the same amount of martensite is formed regardless of the initial retained austenite content. However, at around 20% plastic strain the increase in martensite phase fraction deviates from a linear behavior in the sample with the lowest austempering temperature T_{Aus} = 300°C leveling out to about 20% martensite at 40% plastic strain. At this point about 80% of the retained austenite is transformed to marteniste in this sample. To visualize this effect in figure 8 the martensite phase fractions are normalized to the initial retained austenite content from table 1. This shows clearer the dependence on heat treatment parameters, with the highest amount of transformation for the lowest austempering temperature (figure 8). Figure 8 shows that at 40% compressive deformation a total amount of 65% to 80% retained austenite transformed into martensite. Due to sample breakage higher strain levels cannot be achieved using compression and further experiments allowing a higher degree of deformation, for instance cold rolling, are needed to clarify whether the trend seen in the sample austempered at T_{Aus}=300°C becomes apparent in samples heat treated at higher temperatures.

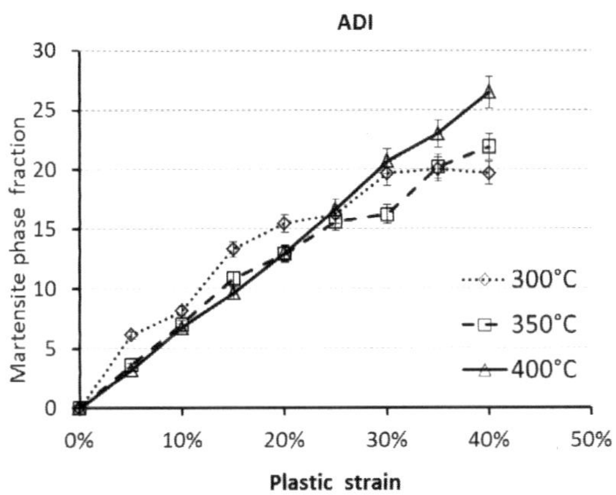

Figure 7. Martensite phase fraction in ADI (austempered at 300°C, 350°C and 400°C) after compressive deformation vs. plastic strain.

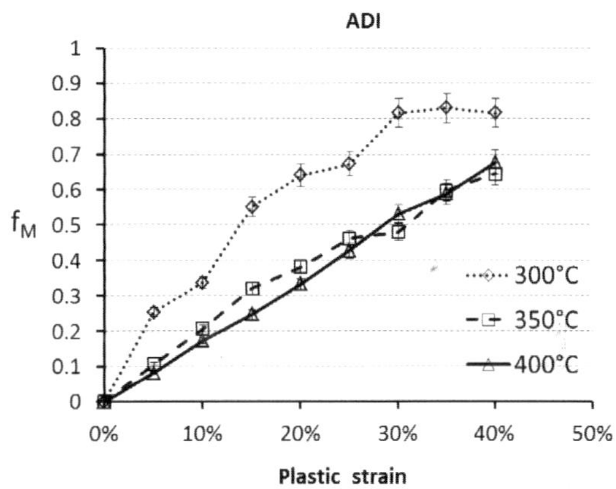

Figure 8. Normalized martensitic phase fraction f_M in ADI after compressive deformation vs. plastic strain.

4. Conclusion

Combining texture information and Rietveld refinements of neutron powder diffraction patterns allows extracting quantitatively phase fractions in ADI. In case of severe overlap of Bragg reflections between marteniste and ferrite in ADI it could be shown that an approach to derive the orientation distribution function for the ferrite phase from just one complete pole figure leads to a more accurate input into the refinement procedure for phase analysis. The results of the quantitative analysis of the martensite phase fractions as a function of deformation shows a dependence of transformation rates with heat treatment parameters with up to 80% of the retained austenite being transformed to martensite for samples heat treated at T_{Aus} = 300°C.

5. References

[1] L Meier, M Hofmann, P Saal, W Volk and H Hoffmann 2013 *Mater. Char.* **85** 124-133
[2] M Daber, P Prasad Rao 2008 *J. Mat. Sci.* **43** 357-367
[3] G Bordin, G C Cecchi and G B Fratucello 1981 *IL Nuovo Cimento* **61B**, N.2
[4] C-H Hsu and T-L Chuang 2001 *Metallurgical and Materials Transactions* A **32A** 2509-2514
[5] J Aranzabal, I Gutierrez, J M Rodriguez-Ibabe and J J Urcola 1997 *Metallurgical and Materials Transactions* A **28A** 1143-1156
[6] S Daber, K S Ravishankar and P Prasad Rao 2008 *J. Mater Sci* **43** 4929-4937
[7] M R Berrahmoune, S Berveiller, K Inal, A Moulin and E Patoor 2004 *Mat.Sci.and Eng.* A **378** 304-307
[8] O Grässel, L Krüger, G Frommeyer and L W Meyer 2000 *Inter. Journal of Plasticity* **16** 1391-1409
[9] W J Dan, S H Li, W G Zhang and Z Q Lin 2008 *Materials and Design* **29** 604-612
[10] M Järvinen 1993 *J. Appl. Cryst.* **26** 525-531
[11] L Lutterotti, S Matthies, H-R Wenk, A J Schultz and J Richardson 1997 *J. Appl. Phys.* **81** 594-600
[12] M Ferrari and L Lutterotti 1994 *J. Appl. Phys.* **76** (11) 7246-55
[13] P Saal, L Meier, X H Li, M Hofmann, M Hoelzel, J N Wagner and W Volk 2015 *Metallurgical Trans.* A, in print
[14] M Hoelzel, A Senyshyn, N Juenke, H Boysen, W Schmahl and H Fuess 2012 *Nucl. Instrum.Meth.* **A667** 32
[15] M Hofmann, G A Seidl, J Rebelo-Kornmeier, U Garbe, R Schneider, R C Wimpory, U Waswuth and U Noster 2006 *Mat. Sci. Forum* **524-525** 211-216
[16] M Dahms and H J Bunge 1989 *J. Appl. Cryst.* **22** 439-447
[17] Hsun Hu 1974 *Texture*, **1** 233-258
[18] U F Kocks, C N Tome and H-R Wenk 2000 *Texture and Anisotropy* (Cambridge University Press)
[19] H van Swygenhoven and Peter Derlet 2001 *Phys. Rev.* **64** 224105
[20] P R Howell, J V Bee and R W K Honeycombe 1979 *Metallurgical Trans.* A, **10A** 1213-1222
[21] G J Mahon, J M Howe and S Mahajan 1989 *Philosophical Magazine Letters*, **59** 273-279
[22] Pat L Mangonon, JR and Gareth Thomas 1970 *Metallurgical Trans.***1** 1577-1586
[23] K Verbecken, L Barbé and D Raabe 2009 *ISIJ International*, **49** 1601-160

Energy-dispersive neutron imaging and diffraction of magnetically driven twins in a Ni_2MnGa single crystal magnetic shape memory alloy

Saurabh Kabra[1], Joe Kelleher[1], Winfried Kockelmann[1], Matthias Gutmann[1], Anton Tremsin[2]

[1] Rutherford Appleton Laboratory, ISIS Facility, Chilton, OX11 0QX, UK

[2] Space Sciences Laboratory, University of California, Berkeley, CA 94720, USA

saurabh.kabra@stfc.ac.uk

Abstract. Single crystals of a partially twinned magnetic shape memory alloy, Ni_2MnGa, were imaged using neutron diffraction and energy-resolved imaging techniques at the ISIS spallation neutron source. Single crystal neutron diffraction showed that the crystal produces two twin variants with a specific crystallographic relationship. Transmission images were captured using a time of flight MCP/Timepix neutron counting detector. The twinned and untwinned regions were clearly distinguishable in images corresponding to narrow-energy transmission images. Further, the spatially-resolved transmission spectra were used to elucidate the orientations of the crystallites in the different volumes of the crystal.

1. Introduction

Large actuation strains, at least an order of magnitude higher than magnetostrictive materials [1], can be achieved in the class of materials known collectively as "magnetic shape memory alloys". The interplay between the high temperature cubic phase and low temperature monoclinic/tetragonal phase (with several twin variants) can be exploited in these alloys to achieve large reversible strains in the stress-temperature-field (magnetic) space. The most successful system in this class of alloys has been the Heusler alloy, Ni_2MnGa, which has been studied as the subject of this article. We have used neutron imaging and diffraction to study twin variants in the low temperature monoclinic phase (so called 'martensite') in single crystalline samples of Ni_2MnGa. The monocrystals have a modulated ordered monoclinic crystal structure at room temperature [2]. The cell angles are close to 90 degrees, making the tetragonal assumption reasonable in many cases.

Owing to a very low twinning stress, the application of an external magnetic field (<0.5 T) is sufficient to twin the crystal into another orientation. As the orientation of the twin variants is different, so is the magnetization direction [3][4]. Therefore, as the field direction is changed, a different twin variant is formed. This is illustrated in figure 1(a) which shows that the twins prefer the crystal c-axis parallel to the externally applied magnetic field. These twins grow and eventually the entire crystal is transformed into the new twinned variant, if a strong enough (>0.65 T) external field is applied.

In this study, we have used single crystal neutron diffraction to find the relative orientations of the twinned crystals and neutron imaging to resolve the morphology of the twins and the untwinned material. Given that these were bulk crystals (several millimetres) neutron methods proved ideal to study them owing to high neutron penetration. Neutron imaging with a high-resolution neutron imaging detector was performed in time of flight (TOF) mode to discriminate the energies of the neutrons. This not only helped in enhancing the contrast between the twins but also the full energy dispersive spectra obtained could be used for further analysis of spatially resolved orientations, strains etc.

2. Experiments and data analysis

2.1. Sample history

Single crystals of Ni_2MnGa with a 5M modulation along the c-axis were obtained from Adaptamat Ltd.[5]. The composition of the samples was 50at.%Ni, 28.5at.%Mn and 21.5at.%Ga and the martensitic transformation temperature was ~50°C. The cuboid crystals were 5 x 5 x 15 mm in size, with the {100} axis along the length of the crystal and the other two crystallographic axis roughly along the remaining edges. Parts of the samples were exposed to an external magnetic field using a Neodymium based permanent magnet. This introduced a region in the crystal which was twinned with respect to the rest of the original crystal. The magnetic field was subsequently removed and the partially twinned system was used for neutron experiments. The removal of the field, did not de-twin the crystal as twinning is not reversible in these materials without the application of an external field or stress.

2.2. Single crystal neutron diffraction

The crystals were measured on the SXD [6] instrument at the ISIS neutron source. The crystals were exposed to the neutron beam for ~20 minutes. The beam size was greater than the sample dimensions and therefore the entire sample including the original crystal and the twin contributed towards the detected signal. SXD comprises of 11 position sensitive detector panels which subtends a large solid angle on to the sample resulting in a large number of diffraction spots to be utilized. Automated peak and cell finding was achieved using the SXD2001 software [7] starting with an initial cell reported in [2]. Two distinct crystalline orientations were found using the automated algorithm. Figure 1(b) shows the orientation relation between the two crystalline orientations that were found using the SXD data. The atomic positions shown in the figure were from the reported crystal structure in reference [2] and were not refined. The orientation of the twin is essentially a ~90 degree rotation about the b-axis of the original crystal thereby swapping the directions of the a-axis and c-axis.

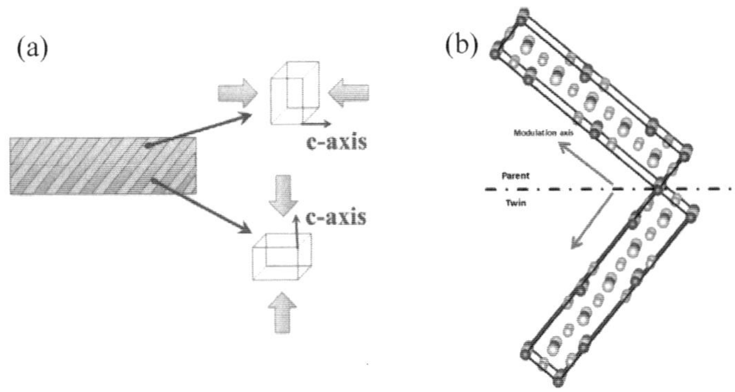

Figure 1. (a) Schematic diagram showing how different parts of the crystal transform to different twin variants as the applied field direction is varied. (b) Schematic diagram showing the orientation relation between the two twin variants obtained from the neutron diffraction analysis.

2.3. Neutron Imaging

Imaging experiments were performed on the ROTAX [8] test beamline at the ISIS neutron source. Simple transmission geometry was used with the sample placed between the beam and the detector (figure 2). The area detector was placed nominally perpendicular to the beam and the sample was placed ~12 mm to the detector active area. The b-axis of the original (untwinned) crystal was approximately parallel to the beam.

The MCP/Timepix detector (figure 2) used for the imaging experiments [9],[10] allowed measurement of the neutron energy for each detected particle. The position of each neutron was detected with the accuracy of 55 μm (current detector pixel size) and timing binning of 20 μs for thermal and cold neutrons. Multiple frames where acquired for each neutron pulse. The acquisition system was synchronized with the trigger of the neutron source, providing the possibility to calculate the energy of each neutron from its time of flight. The active area of the detector was 28x28 mm^2, defined by the 2x2 array of Timepix readout ASICs, developed by the Medipix collaboration [11]. The neutrons arriving at the active area of the detector are converted into an electron cloud by the neutron sensitive microchannel plates, manufactured by Nova Scientific. Each neutron produces ~10^4-10^5 electrons, thus enabling operation with no readout noise. This detector will be used as a standard equipment on the IMAT imaging beamline [12].

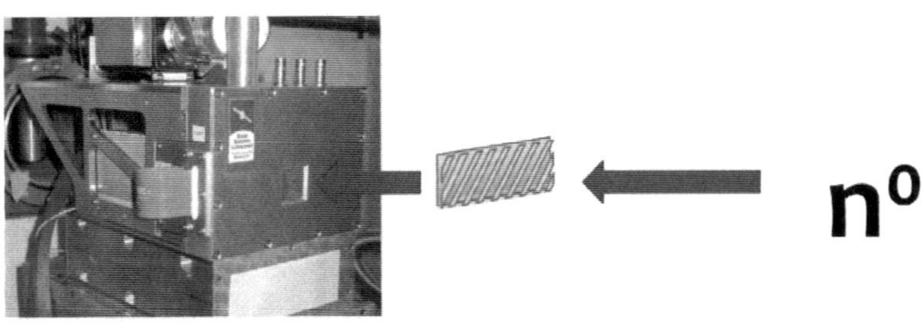

Figure 2. Geometry of the neutron imaging experiment. The sample was mounted close (few mm) to the neutron window of the detector head.

The imaging measurement resulted in a stack of images of dimensions 512 x 512 pixels each. Every image represented a time of flight bin which was later converted to wavelength. Thus the transmission spectra consisting of several thousand energy bins was measured for each pixel of the data set, all in one experiment. An open beam measurement (without any sample) was also carried out under the same setup. This was used to normalize the data both for the flux distribution as a function of TOF and for spatial non-uniformity (combination of beam non-uniformity and pixel sensitivity). Figure 3(a) shows the white beam radiograph of the twinned crystal where all energies have been summed up. No apparent contrast can be inferred. Figure 3(b) shows a corresponding image where only the neutrons in the range of 2.7-2.8 Å have been summed. The twin (yellow-green region) and original crystal (blue region) are easily distinguishable in this image. The neutron intensities as a function of wavelength have been plotted in figure 3(c). The colored dots on 3(b) indicate both locations and the color of the corresponding plots.

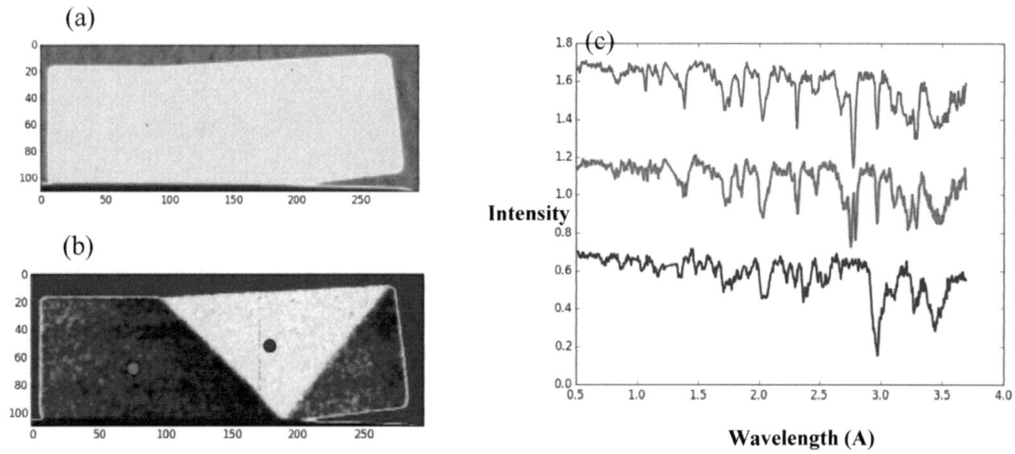

Figure 3. (a) White beam transmission image of the twinned crystal. (b) Radiograph using a limited range of neutron energies showing contrast between original crystal and twin. (c) Transmission spectra from the different regions of the original and twinned crystals (the dots on fig b correspond to regions sampled and colour of the dot corresponds to the colour of the curve)

3. Results and discussion

It is evident from Figure 3 that, by limiting the energy range of the imaging data, contrast can be enhanced between the differently oriented parts of the crystals. The reason for this is that different crystal orientations with respect to the beam results in Bragg condition being satisfied for different sets of crystalline planes. The corresponding neutrons that have diffracted are missing in the transmitted beam and therefore the characteristic 'dips' occur in the transmission spectra (figure 3(c)). Both the intensities and the positions of these dips are related to the orientation of the crystals. Markedly, in figure 3(c), the red and the green plots look quite similar as compared to the blue plot which is quite different from both. This is not surprising, given that the blue plot is from the twinned region of the crystal, while both red and green plot are from the original crystal regions. The two original crystal regions do show some differences because of slight mis-orientation between them arising due to the triangular shape of the twin.

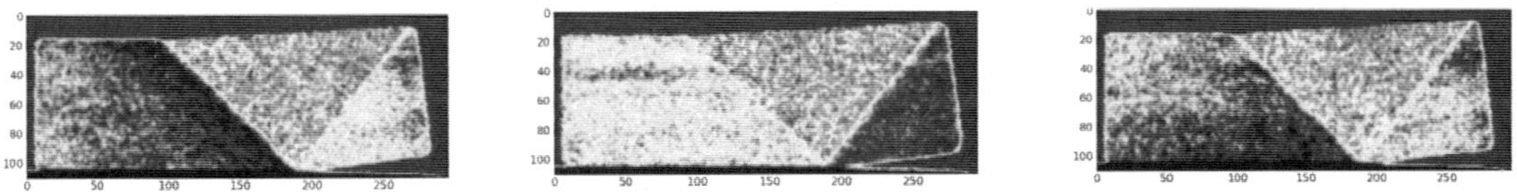

Figure 4. Radiographs of the twinned crystals with various energy ranges showing mosaic features in the original crystal.

Figure 4 shows the radiographs of the twinned crystals at several chosen energy ranges. The figure clearly shows various linear features running along the length of the crystal in the original crystal. These linear features on the single crystal are along the growth direction of the single crystal. It can be speculated that these are probably due to slight misalignments between subdomains during the growth process.

For future work, the authors plan to study the dynamics of the twinning in these crystals by subjecting the crystals to a cyclic magnetic field in situ. These stroboscopic experiments will shed light on the twin kinetics and the dependence of twin morphology on external parameters such as frequency of rotation and the strength of field.

4. Conclusions

Single crystals of Ni_2MnGa can spontaneously twin upon the application of an external magnetic field. Neutrons are an ideal tool to study these twinned crystals due to their high penetration capability and high coherent cross section for this material. Single crystal diffraction shows the orientation relation between the untwinned and the twinned regions while high resolution energy dispersed neutron imaging shows the morphology of the twinned regions. Further, the mosaic microstructure of the original crystal was elucidated during the imaging experiments.

References

[1] Ullakko K, Huang J K, Kantner C, O'Handley R C and Kokorin V V. 1996 Large magnetic-field-induced strains in Ni2MnGa single crystals *Appl. Phys. Lett.* **69** 1966

[2] Righi L, Albertini F, Pareti L, Paoluzi A and Calestani G 2007 Commensurate and incommensurate "5M" modulated crystal structures in Ni–Mn–Ga martensitic phases *Acta Mater.* **55** 5237–45

[3] Karaca H, Karaman I, Basaran B, Chumlyakov Y and Maier H 2006 Magnetic field and stress induced martensite reorientation in NiMnGa ferromagnetic shape memory alloy single crystals *Acta Mater.* **54** 233–45

[4] Lai Y W, Scheerbaum N, Hinz D, Gutfleisch O, Schäfer R, Schultz L and McCord J 2007 Absence of magnetic domain wall motion during magnetic field induced twin boundary motion in bulk magnetic shape memory alloys *Appl. Phys. Lett.* **90** 192504

[5] Tellinen J, Suorsa I, Jaaskelainen A, Aaltio I and Ullkko K 2002 Basic properties of magnetic shape memory actuators *Actuator 2002, Wirtschaftsfoerderung Bremen GmbH, Bremen* pp 566–9

[6] Keen D A, Gutmann M J and Wilson C C 2006 SXD – the single-crystal diffractometer at the ISIS spallation neutron source *J. Appl. Crystallogr.* **39** 714–22

[7] Gutmann M J 2005 No Title *SXD2001,. ISIS Facil. Rutherford Applet. Lab. Oxfordshire, Engl.*

[8] Kockelmann W A, Weisser M, Heinen H, Kirfel A and Schäfer W 2000 Application Spectrum and Data Quality of the Time-to-Flight Neutron Diffractometer ROTAX at ISIS *Mater. Sci. Forum* **321-324** 332–7

[9] Tremsin A S 2012 High resolution neutron counting detectors with microchannel plates and their applications in neutron radiography, diffraction and resonance absorption imaging *Neutron News* **23** 35–8

[10] Tremsin A S, Vallerga J V, McPhate J B and Siegmund O H W 2015 Optimization of high count rate event counting detector with Microchannel Plates and quad Timepix readout *Nucl. Instruments Methods Phys. Res. Sect. A Accel. Spectrometers, Detect. Assoc. Equip.* **787** 20–5

[11] Llopart X, Ballabriga R, Campbell M, Tlustos L and Wong W 2007 Timepix, a 65k programmable pixel readout chip for arrival time, energy and/or photon counting measurements *Nucl. Instruments Methods Phys. Res. Sect. A Accel. Spectrometers, Detect. Assoc. Equip.* **581** 485–94

[12] Kockelmann W, Burca G, Kelleher J F, Kabra S, Zhang S-Y, Rhodes N J, Schooneveld E M, Sykora J, Pooley D E, Nightingale J B, Aliotta F, Ponterio R C, Salvato G, Tresoldi D, Vasi C, McPhate J B and Tremsin A S 2015 Status of the Neutron Imaging and Diffraction Instrument IMAT *Phys. Procedia* **69** 71–8

Neutron diffraction study of metal-matrix composite with fullerite

P A Borisova[1], M S Blanter[2*], VV Brazhkin[3], VA Somenkov[1], V P Filonenko[3]
[1]National Research Centre "Kurchatov Institute", Moscow, Russia
[2]Moscow State University of Information Technologies, Radioengineering and Electronics, Moscow, Russia
[3]Institute for High Pressure Physics RAN, Troitsk, Moscow, Russia

*E-mail: mike.blanter@gmail.com

Abstract. Interaction of amorphous fullerite C_{60} with austenitic Fe-33.2 wt. % Ni alloy at pressures 0-8 GPa and temperatures 600-1100 ^{0}C was studied by neutron diffraction. The amorphous fullerite was obtained by ball milling and mixed with the powder of the crystalline alloy. The interaction at sintering led to the dissolution of carbon in fcc Fe-Ni solid solution and the formation of carbide $(Fe, Ni)_3C$, but the Fe-Ni-C alloy did not undergo phase transformations and preserved the original fcc structure. As a result, the alloy hardened, we could also witness a clear barometric effect: at the pressure of 2 GPa the amount of the dissolved carbon and the microhardness turned out to be significantly higher than those at 8 GPa. During sintering amorphous fullerite is undergoing phase transitions and its microhardness is higher than the microhardness of the metal component. At high temperatures of interaction graphite appears. The presence of Fe-Ni alloy in the composite reduces the temperature of graphite formation in comparison with transformations in the pure amorphous fullerene.

Keywords: neutron diffraction; metal matrix composites; fullerenes; sintering; phase transitions; high-pressure; high-temperatures

1.Introduction

It is widely known that metal-matrix composite materials (MMCs) reinforced by ceramic fibers are superior in terms of the properties to monolithic alloys. They combine the best properties of their two components, such as ductility and toughness of the matrix, as well as a high elastic modulus and strength of the reinforcing elements. These composites are widely used as structural materials in aerospace, automotive, chemical, transport and other industries [1]. Replacing ceramic by carbon fibers or graphitic particles holds much promise because of their high thermal conductivity, low density, good self-lubrication, high damping capacity and a low coefficient of thermal expansion. These MMCs have high dimensional stability and suppress vibration [2,3,4].

With the discovery of fullerenes, carbon nanotubes, and graphene, which all possess a unique set of mechanical and physical properties, scientists started to elaborate on metal-matrix composites with fullerene [5-9], carbon nanotubes [10-11], and graphene [10]. The growing production of carbon nanomaterials and reduction of its cost enhances public interest to such studies. Although using composites with carbon nanotubes brings certain advantages [10-11], a high amount of research has been done on composites with fullerene C_{60} or a mixture of fullerenes as the reinforcing component [5-9].

To develop and produce compositions of such composite materials, one needs to study the interaction of carbon nanomaterials with different metals and alloys under different conditions. In case of fullerenes, most of the performed studies employ crystalline C_{60}. At the same time, it is of interest to investigate the effect of different structural states of fullerenes on such interaction, in particular, the use of amorphous (nanostructured) fullerenes. Previously we have undertaken such studies for the case of Al [12] and Fe [13] and received a different behavior of the metallic component in these composites, associated primarily with the presence of polymorphic transformations in iron. It is therefore of interest to study the interaction of amorphous fullerites with an iron alloy, which has no polymorphic transformations.

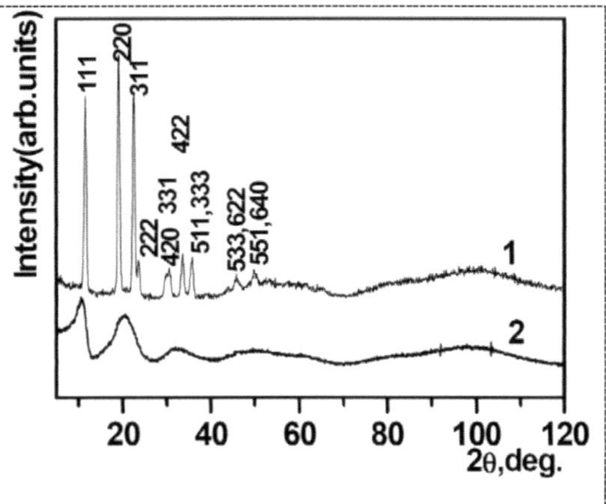

Figure 1. Neutron diffraction patterns of crystalline (1) and amorphous (2) fullerite C_{60}.

In this paper, we selected an austenitic (fcc) alloy Fe-33.2 wt. % Ni as a matrix and investigated structural aspects of its interaction with the amorphous C_{60} in a wide range of pressures (up to 8 GPa) and temperatures (up to 1200 ^{0}C). The method of neutron diffraction was chosen as the main one. This method is the most suitable for localization of light atoms in the presence of heavy atoms as compared with other structural methods.

2. Experimental methods

Our initial samples of C_{60} fullerenes of 99.5% purity were produced by "NeoTechProduct" by high-temperature treatment of graphite, followed by isolation with organic solvents and subsequent chromatographic separation. The neutron diffraction patterns of our initial crystal fullerites correspond to the fcc lattice with a period a = 1.416 nm. Amorphous fullerites (aC_{60}) were prepared by mechanical grinding of crystal C_{60} (~1–1.5 g) in air in a Fritsch type mill with agate cup and balls at low rates over long periods of time (up to 100 hours) fixing the structural changes by neutron diffraction. As a result, at the site of intense diffraction peaks halos were formed (figure 1) corresponding to the amorphization of fullerite.

For the metal component an alloy Fe-33.2 wt.% Ni was melted, it was then turned into shavings by drilling, and the shavings were further ground into a powder in a Fritsch mill. This powder was mixed with the C_{60} amorphous powder (25 at.%) in the same mill. The Fe-Ni alloy retained fcc crystal structure after milling (figure 2).

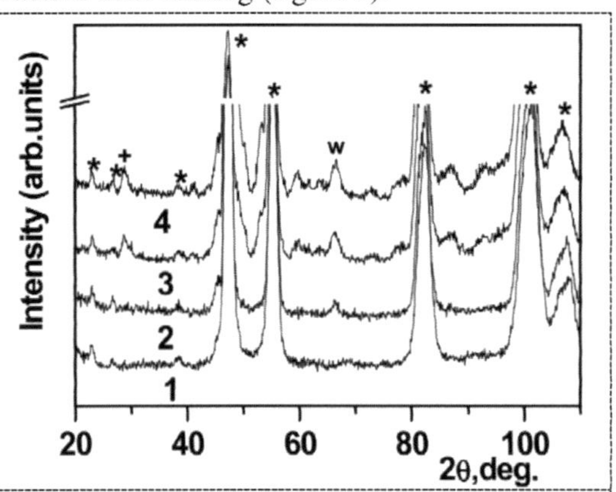

Figure 2. Neutron diffraction patterns of the initial powder mixture of the Fe-Ni alloy and the amorphous C_{60} before (1) and after sintering at a pressure of 8 GPa and temperatures of 600 (2), 800 (3), and 1100 ^{0}C (4). Asterisks show peaks of fcc lattice of metal solid solution, the cross shows the strongest line (002) of graphite. The sign w represents a combined maximum (311) and (132) of cementite.

Annealing of (Fe-Ni)-aC_{60} powder was performed in a high-temperature vacuum furnace at 900 and 1200 ^{0}C for 5 min in each cycle, followed by cooling down to room temperature. Sintering under high pressure occurred in the following way: raw material was pressed to pellets of dimensions 3 mm in height and 5 mm in diameter. Thermobaric treatment of these pellets was carried out by using high pressure toroid-type chambers [14]. The pressure was generated in a cell from lithographic stone placed between hard alloy anvils with a special profile. The pressure accuracy was ±0.2 GPa, and the temperature gradient along the sample height was less than 20 K. The high-pressure treatment was carried out at 2.0, 4.5 and 8.0 GPa in the temperature range 600-1100 ^{0}C. After stabilization of the applied pressure, the samples were heated to the desired temperature at heating rate ≈100 K/min and held for 60 seconds. Then the cells were cooled down rapidly to room temperature (≈100 K/s), before the pressure was released. These sintered samples were in the tablet form, having a diameter of 5 mm and a height of ~ 1 mm. Thus, the samples were studied ex-situ.

The structure of the samples was examined using the neutron diffraction method on the neutron multidetector (224 scintillation detectors) diffractometer DISC [15] with the stationary nuclear reactor capacity of 4-5.5 MW, the wavelength λ of monochromatic heat neutron waves of 0.1664 nm (the part of λ/2 is less than 1%), and neutron flux ~10^{10} $m^{-2}\cdot s^{-1}$. The diffraction pattern processing was performed using a full profile Rietveld refinement with the help of "Fullprof" software. We also determined the microhardness by the device DuraScan20.

3. Experimental results

The investigation of the diffraction patterns (figure 2) showed that at all studied temperatures and pressures the interaction of the Fe-Ni alloy with amorphous C_{60} does not change its crystal structure and, after cooling to room temperature, fcc crystal lattice is preserved. However, with temperature increase the lattice period of the fcc lattice increases significantly (figure 3) and additional lines appear in the diffraction spectrum (curves 2-4 in figure 2).Analysis of the diffraction spectra by the Rietveld method using the package "FullProf" shows the formation of the cementite (Fe, Ni)$_3$C alloyed by nickel [16,17].

Increase in the lattice spacing *Δa* (figure 3) at pressures from 2 to 8 GPa is 0.01-0.048 Å (~ 0.3-1.3% of the lattice spacing of the initial Fe-Ni powder) and grows with the temperature. The value of *Δa* decreases with pressure, i.e. there is a clear barometric effect.

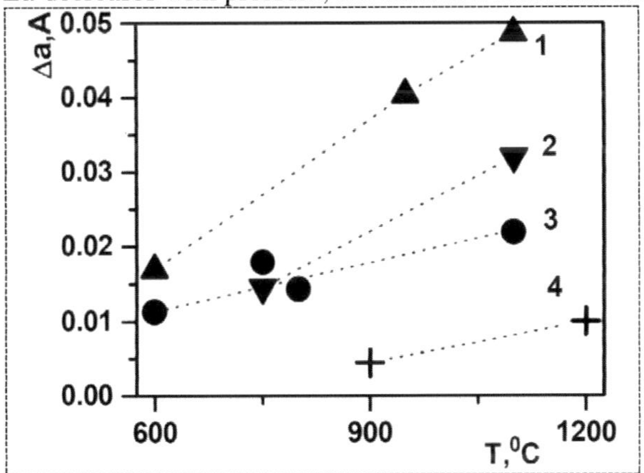

Figure 3. Change in a lattice spacing of the fcc metal lattice after sintering at pressure of 2 GPa (1), 4.5 GPa (2), 8 GPa (3) and in a vacuum furnace (4).

However, after furnace annealing of the powder (consisting of a mixture of Fe-Ni alloy with amorphous fullerite) at high temperatures of 900 and 1200 °C the change in the lattice spacing is much smaller: 0.004 and 0.01 Å respectively.

The change in the amount of cementite was determined from the relative intensity of overlapping lines (311) and (132) (marked with w in figure 2). This maximum was used because it does not overlap with the peaks of the fcc phase. Its relative intensity increases with sintering temperature and is practically independent of the applied pressure (figure 4). In the samples annealed in a furnace without application of a high pressure, the maximum height was significantly less, but also increased with temperature.

The presence of the carbon phase, clearly visible in the microstructure cannot always be traced on the diffraction spectra. In the spectrum of the original mixture *a*C_{60}*- (Fe-Ni)* there are two weak halos at $2\theta \approx 10^0$ and $2\theta \approx 20^0$, which are characteristic for amorphous fullerite (figure 1), but in the annealed or sintered samples the halos cannot be detected on the background of the metal phase spectrum. In sintering at 750 °C and above the most intense graphite peak (002) appears (figure 2). The relative intensity of this peak (amount of graphite) (figure 5) increases with temperature and decreases with pressure. Graphite is not observed when the powder anneals at 900 and 1200 °C in a vacuum furnace.

The properties of the carbon phase do not show such a clear dependence on temperature and pressure, as is the case for the metal component. Microhardness amounts to 0.54-0.73 GPa, and in the majority of the investigated composites it exceeds the microhardness of a metal component by 2-3 times.

4. Discussion

The results demonstrate that at high temperatures and high pressures there is a strong interaction between amorphous C_{60} and the iron-nickel austenitic alloy, which leads to the dissolution of carbon in a fcc metal lattice (a lattice parameter increases) and the formation of carbide (Fe, Ni)$_3$C while retaining the original crystal structure of the metal component. Dissolution of carbon and carbide formation led naturally to the increase of the hardness of the metal component.

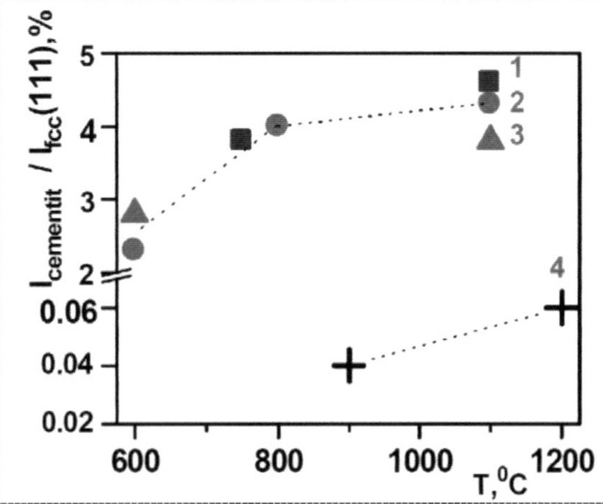

Figure 4. The relative intensity (relative to the intensity of the (111) line of the fcc phase) of the maximum of cementite obtained by imposing of the (311) and (132) reflections. Sintering at pressure of 4.5 GPa (1), 8 GPa (2), 2 GPa(3) and in a vacuum furnace (4).

The observed increase of the lattice spacing in the iron-nickel austenite indicates the dissolution of carbon in austenite. We evaluated the concentration of carbon in solid solution by the known dependence of the austenite lattice parameter on a nickel C_{Ni} and carbon C_C concentration at room temperature [18]: $\Delta a=0.001*C_{Ni} + 0.0316*C_C$, where Δa in Å and the concentration in wt.%. The maximum concentration (after sintering at 2 GPa and 1100 ^{0}C) is ≈1.5 wt.% C and only half of that (≈0.7 wt.% C) at 8 GPa and 1100 ^{0}C. The same reduction of C_C is observed with pressure increase at other temperatures. For example, at 600 ^{0}C $C_C \approx 0.5$ wt.% at 2GPa and ≈0.35 at 8 GPa. The observed decrease in the concentration of dissolved carbon with pressure increase can be explained by the fact that carbon is dissolved by the interstitial type and thus increases the volume of the alloy, which impedes excess pressure. At the same time independence of the amount of cementite of the pressure can be explained by the fact that the cementite density (~ 7.9 g / cm^3 [19]) is close to the density of the Fe-Ni alloy (~ 8.1 g / cm^3).

When we anneal the powder mixture in a vacuum furnace, much less carbon dissolves in the alloy and a smaller amount of cementite is formed. This is apparently due to a less close contact of carbon and metallic phases in the absence of high pressure.

It is difficult to compare our data on the solubility of carbon with the data of other studies. Although there are many publications on the alloys Fe-C and Fe-Ni-C at high pressures [e.g., 16, 17,20-22], most of them are devoted to modeling the structure of the Earth's core, and the studies were conducted at high temperatures close to the melting point. A comparison is also complicated by the fact that different studies have a high variation in the times of the sample exposure at carbon dissolution temperatures, and even sometimes fail to reach equilibrium concentrations, as was the case in the present work. However, some comparisons can still be made. For instance, in the alloy Fe-6 wt.% Ni approximate solubility of carbon is 0.85 wt.% at 2000 ^{0}C and 9 GPa and it decreases with pressure increase up to 20 GPa [17]. The Fe-C system has also demonstrated that the maximum solubility of C in austenite decreases from ≈2 wt.% at 5 GPa to ≈0.5 wt.% at 130 GPa [21].This maximum solubility was observed at the eutectic temperature, which increases with increasing pressure from 1293 up to 2979 ^{0}C. As can be seen, our results do not contradict the data available in the literature.

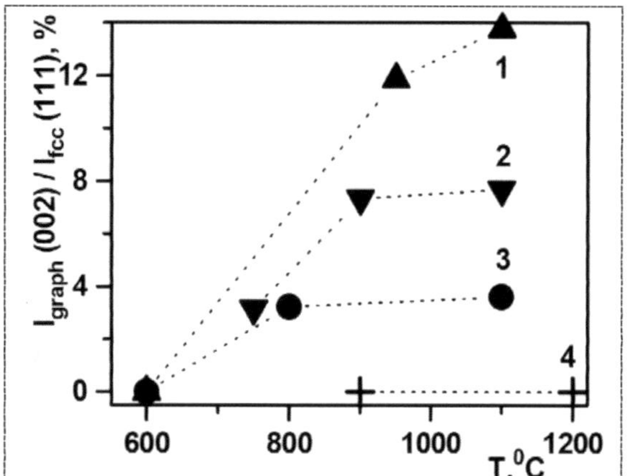

Figure 5. Change of the relative (002) line intensity of graphite after sintering (relative to the intensity of the (111) line of the fcc phase) at pressure of 2 GPa (1), 4.5 GPa (2), 8 GPa(3) and in a vacuum furnace (4).

Preserving the fcc crystal lattice of the iron-nickel alloy with 33.2 wt.% Ni at cooling after sintering is consistent with literature data. In the alloy Fe-31 wt.% Ni-0.02 wt.% C, which is close to our alloy composition, the martensite point is below room temperature (≈205 K) [22]), and even in the alloy Fe-25.3 % Ni-0.73% C (wt.%) with a lower Ni content at room temperature a stable fcc structure was observed [24]. Mechanical activation of the mixture of iron-nickel alloy (50 wt.% Ni) with powder of carbon phase (carbon nanotubes) causes no change in the crystal structure [25].This is consistent with the existence of the fcc structure of our original powder after mechanical activation and after annealing at 900 and 1200 ^{0}C.

The formation of carbide (Fe, Ni)$_3$C, observed in the present work, is consistent with the available data on the Fe-Ni-C alloys at high temperatures and pressures. Thus, according to the cross section of the ternary Fe-Ni-C diagram at 1050 ^{0}C and 10 GPa [16] our composite should contain a metal solid solution and the carbide (Fe,Ni)$_3$C. Although Fe-Ni-C alloys in [16] were sintered from powders of metals and graphite with longer exposures (several hours), while we sintered metal and amorphous fullerite powders over a short time (1 minute) the same structure was registered in our alloy at 1100 ^{0}C and 8 GPa.

During sintering not only the metal component of the composite changes, but also the carbon phase alters. These changes comprise the transformation of amorphous fullerene in amorphous and crystalline graphite [27]. The resulting transformation products of amorphous fullerite have increased microhardness that substantially exceeds microhardness of the metal component. Increased microhardness of the carbon phase enables it as a potential hardening phase in the metal matrix composites.

At high pressures with increasing temperature crystalline C_{60} is known to transform into several polymerized phases, which at higher temperatures are transformed into amorphous and crystalline graphite [26]. In an amorphous fullerite, the stage of formation of polymerized phases is absent and graphite is formed at lower temperatures [27]. In our composite, the line (002) of graphite appears at pressures of 2-8 GPa at 750 ^{0}C or higher, while in the pure amorphous fullerite these temperatures are significantly higher, reaching 1050-1300 ^{0}C [27]. It turns out that the presence of Fe-Ni alloy promotes the conversion of amorphous fullerene to crystalline graphite.

While furnace annealing the powder mixture of amorphous fullerite and the Fe-Ni alloy, we observed no graphite after annealing at 1200 ^{0}C, which agrees well with the data of [28], in which under the same annealing conditions of pure amorphous C_{60} crystalline graphite did not appear even at 1500 ^{0}C. The effect of the Fe-Ni alloy on this process is not observed, apparently due to a poor contact of fullerene and metal powders, and it is consistent with the relatively small change in the alloy lattice spacing (weak carbon saturation).

5.Conclusions

It is shown that between amorphous C_{60} and the austenitic Fe-33.2 wt.% Ni alloy at high temperatures and pressures there is a strong interaction that leads to saturation of the alloy by carbon and formation of carbides (Fe, Ni)$_3$C, while retaining the fcc crystal structure of the alloy. The interaction is weakened with increasing pressure from 2 to 8 GPa. The interaction also reduces the stability of amorphous fullerite against transformation into a stable modification -crystal graphite.

Acknowledgement

This work was performed using the equipment of Unique Scientific Facility "Research reactor IR-8" supported by the Russian Ministry of Science and Education (project code RFMEFI61914X0002). M.S. Blanter, P.A. Borisova and V.A. Somenkov are grateful to the Russian Foundation for Basic Research (projects № 13-02-00208 and 16-02-00193), P.A. Borisova and V.A. Somenkov (neutron diffraction, data analysis) are grateful to Russian Science Foundation (16-12-10065) for the financial support and V.V.Brazhkin (high-pressure synthesis, data analysis) is grateful to Russian Science Foundation (14-22-00093) for the financial support.

References

[1] Miracle D B 2005 Metal matrix composites - From science to technological significance

Compos. Sci. Technol. **65** 2526-40
[2] Tjong S C and Lau K C 2000 Tribological behaviour of SiC particle-reinforced copper matrix composites *Mater. Lett.* **43** 274-280
[3] Rawal S 2001 Metal-matrix composites for space applications *JOM* **53** 14-17
[4] Wu Y W, Wu K, Deng K K et al. 2010 Damping capacities and tensile properties of magnesium matrix composites reinforced by graphite particles *Mater. Sci. Eng. A* **527** 6816-21
[5] Robles-Hernandez F C and Calderon H A 2010 Nanostructured metal composites reinforced with fullerenes *J. Minerals, Metals and Mater. Soc.* **62** 63-68
[6] Choi H J, Shin J H and Bae D H 2010 Self-assembled network structures in Al/C60 composites *Carbon* **48** 3700-7
[7] Khalid F A, Beffort O, Klotz U E et al. 2003 Study of microstructure and interfaces in an aluminum–C60 composite material *Acta Mater.* **51** 4575-82
[8] Tokunaga T, Kaneko K, Sato K, Horita Z 2008 Microstructure and mechanical properties of aluminum-fullerene composite fabricated by high pressure torsion *Scr. Mater.* **58** 735-38
[9] Watanabe H, Fukusumi M, Ishikava K, Shimizu T 2006 Superplasticity in a fullerene-dispersed Mg-Al-Zn alloy composite *Scr.Mater.* **54** 1575-80
[10] Tjong S C 2013 Recent progress in the development and properties of novel metal matrix nanocomposites reinforced with carbon nanotubes and graphene nanosheets *Mater. Sci. Eng. R* **74** 281–350
[11] Bakshi S R, Lahiri D, Agarwal A 2010 Carbon nanotube reinforced metal matrix composites- a review *Int.Mater.Rev.* **55** 41-64
[12] Borisova P A, Blanter M S and Somenkov V A 2014 Neutron diffraction study of interaction between amorphous and crystalline C60 fullerenes and aluminum *Bulletin of the Russian Academy of Sciences. Physics* **78** 1205–1208
[13] Borisova P A, Agafonov S S, Blanter M S and Somenkov V A 2014 Neutron diffraction study of the interaction of iron with amorphous fullerite *Phys. Solid State* **56** 199–202
[14] Khvostantsev L G, Slesarev V N and Brazhkin V V 2004 Toroid type high-pressure device: History and prospects *High Pressure Res.* **24** 371-83
[15] Glazkov V P, Naumov I V, Somenkov V A et al. 1998 Superposition many-detector systems and neutron diffraction of microsamples *Nucl. Instr.Methods.* **A264** 367-374
[16] Rohrbach A, Ghosh S, Schmidt M W et al. 2014 The stability of Fe-Ni carbides in the Earth's mantle: Evidence for low Fe-Ni-C melt fraction in the deep mantle *Earth Planet.Sci.Lett.* **388** 211-221
[17] Narigina O, Dubrovsky L S, Mijajima N, McCammon C A et al. 2011 Phase relation in Fe-Ni-C system at high pressures and temperatures *Phys.Chem.Minerals* **38** 2003-14
[18] Nedolya A V 1997 XRPD application for laser-treated surface of Fe-based alloy study *IL NUOVO CIMENTO* **19 D** 609-615
[19] Bushow K H J, Van Engen P G and Jongebreur R 1983 Magneto-optical properties of metallic ferromagnetic materials *J.Magn.Magn.Mater.* **38** 1-22.
[20] Huang E, Bassett W A and Weathers M S 1988 Phase relationships in Fe-Ni alloys at high pressures and temperatures *J. Geophys. Res.: Solid Earth* **93** 7741-46
[21] Nakajima Y, Takahashi E, Suzuki T and Funakoshi Ken-ichi 2009 "Carbon in the core" revisited *Phys. Earth Planet. Interiors* **174** 2002-2011
[22] Lord O T, Walter M J, Dasgupta R, Walker D and Clark S M 2009 Melting in the Fe-C system to 70 GPa *Earth Planet. Sci. Lett.* **284** 157-167
[23] Guimara˜esa J R C and Rios P R 2015 The mechanical-induced martensite transformation in Fe–Ni–C alloys *Acta Mater.* **84** 436–442
[24] Nadutov V M, Kosintsev S G, Svystunov Ye O et al. 2011 Anti-invar properties and magnetic order in fcc Fe–Ni–C alloy *J. Mag.Mag. Mater.* **323** 2786–91
[25] Azadehranjbar S, Karimzadeh F and Enayati M H 2012 Development of NiFe-CNT and Ni3Fe-CNT nanocomposites by mechanical alloying *Advanced Powder Technology* **23** 338–342
[26] Brazhkin V V and Lyapin A G 2012 Hard and superhard carbon phases synthesized from fullerites under pressure *J. Superhard Mat.* **34** 400-423

[27] Borisova P A, Blanter M S, Brazhkin V V, Somenkov V A and Filonenko V P 2015 Phase transformations in amorphous fullerite C60 under high pressure and high temperature, *J.Phys.Chem.Sol.* **83** 104-108

[28] Agafonov S S, Glazkov V P, Kokin I F and Somenkov V A 2010 Polyamorphic transition in amorphous fullerites *Phys. of the Solid State* **52** 1329-31

Journal of Physics: Conference Series **746** (2016) 012058 doi:10.1088/1742-6596/746/1/012058

Local structures and local exchange interactions in doped magnetic materials of different dimensionality

Albert Furrer[1,2]

[1] Laboratory for Neutron Scattering, Paul Scherrer Institut, CH-5232 Villigen PSI, Switzerland
[2] SwissNeutronics AG, Brühlstrasse 28, CH-5313 Klingnau, Switzerland

E-mail: albert.furrer@psi.ch

Abstract. Defects intentionally introduced into magnetic materials often have a profound effect on the physical properties. Specifically tailored neutron spectroscopic experiments can provide detailed information on both the local exchange interactions and the local distances between the magnetic atoms around the defects. This is demonstrated for manganese dimer excitations observed for the magnetically diluted three-, two- and one-dimensional compounds $KMn_xZn_{1-x}F_3$, $K_2Mn_xZn_{1-x}F_4$ and $CsMn_xMg_{1-x}Br_3$, respectively, with x=0.10. The resulting local exchange interactions deviate up to 10% from the average, and the local Mn-Mn distances are found to vary stepwise with increasing internal chemical pressure due to the Mn/Zn or Mn/Mg substitution. Our analysis qualitatively supports the theoretically predicted decay of atomic displacements according to $1/r^2$, 1/r and constant (for three-, two- and one-dimensional compounds, respectively) where r denotes the distance of the displaced atoms from the defect.

1. Introduction

Atoms in perfect crystals are arranged in a periodic lattice. However, perfect crystals do not exist in nature. All crystals have some inherent defects, *e.g.*, vacancies or substitutional atoms due to the limited purity of the material. Moreover, defects are often intentionally introduced into the crystal in order to manipulate its physical properties. Doping has become an important tool for magnetic materials in the context of high-temperature superconductivity [1] and quantum magnetism [2] as well as for semiconductors to induce ferromagnetism [3]. Defects produce local lattice distortions, *i.e.*, the atoms around the defect are displaced from their crystallographic equilibrium positions, and the exchange interactions around the defects deviate from the average exchange. Experimental information on local exchange interactions and local structures is important to understand the physical mechanisms of the defect crystals. Different experimental techniques exist to unravel local structures, the most prominent methods being x-ray absorption fine structure, nuclear magnetic resonance, nanobeam electron diffraction with a transmission electron microscope, and atomic pair-distribution function analysis, which all yield bulk information; in addition, surface-sensitive imaging methods such as scanning tunneling microscopy and atomic force microscopy are often used. All these methods provide a spatial resolution of typically 0.1 Å, and their performance can hardly be improved. It is therefore desirable to search for alternative techniques pushing the spatial resolution beyond the present limits.

Nature is rich in relationships which connect particular physical properties of materials to their structural details. One such relationship exists for the exchange coupling J of magnetic atoms, which for most materials depends on the interatomic distance R through a linear law dJ/dR as long as dR<<R. Modern spectroscopies measure exchange couplings with a precision of dJ/J≈0.01; thus, spatial resolutions of dR≈0.01 Å can be achieved. This was demonstrated for three-, two- and one-dimensional materials doped with manganese ions [4-6]. More specifically, the singlet-triplet transitions associated with manganese pairs observed by inelastic neutron scattering (INS) are found to exhibit remarkable fine structures as exemplified for the three-dimensional compound $KMn_{0.10}Zn_{0.90}F_3$ in figure 1. The fine-structure lines (indexed by m=0,1,2,...) provide detailed information on both local structural effects and local exchange interactions. As a consquence, the exchange interaction in doped materials is no longer uniformly distributed, but it can markedly deviate from the average exchange.

From the theoretical side, the defect problem was investigated by Krivoglaz [7] who predicted that the atomic displacements in three- and two-dimensional crystals decay asymptotically as $1/r^2$ and 1/r, respectively (r denotes the distance of the displaced atoms from the defect). In one-dimensional crystals the atomic displacements do not decrease with increasing distance *r* at all. INS studies confirmed these predictions by applying statistical models [4-6] as summarized in the present work. In addition, we introduce a phenomenological model to account for the local lattice distortions, and we provide some examples to illustrate the importance of the local exchange interactions around the defects.

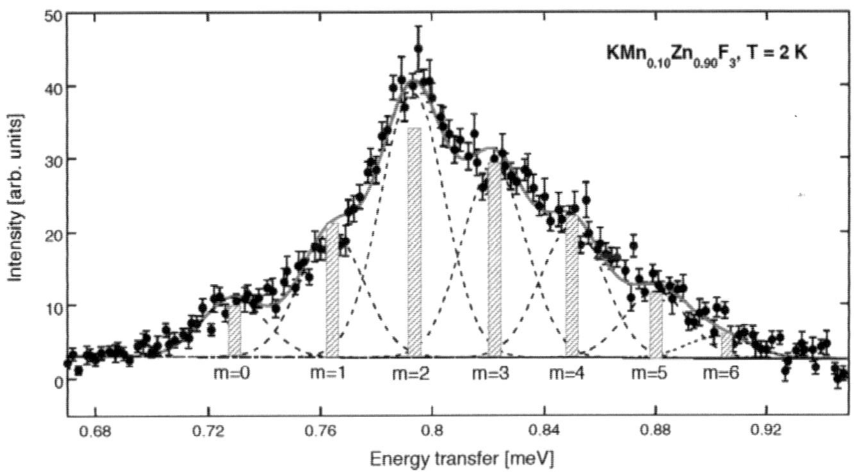

Figure 1. Energy distribution of the Mn singlet-triplet dimer transition observed for $KMn_xZn_{1-x}F_3$ with x=0.10 at T=2 K. The lines correspond to Gaussians resulting from a least-squares fitting procedure. The vertical bars denote the probabilities $p_m(x)$ predicted by (1) with n=26.

2. Theoretical background

Fig. 2 visualizes the atomic positions around isolated Mn dimers in three-, two- and one-dimensional structures relevant for the present work. For x<<1 these positions are mainly occupied by Zn or Mg atoms, which are occasionally replaced by Mn atoms according to statistics. In the following we summarize the corresponding statistical laws compatible with the theoretical predictions [7].

For the three-dimensional case as sketched in figure 2(a) for mixed Mn/Zn structures, the probability for having m Mn atoms at the n nearest-neighbor positions is given by [5]

$$p_m(x) = \binom{n}{m} x^m (1-x)^{n-m} . \tag{1}$$

Because of the short-range $1/r^2$ decay law predicted by Krivoglaz [7] we do not consider further-distant-neighbor positions in (1).

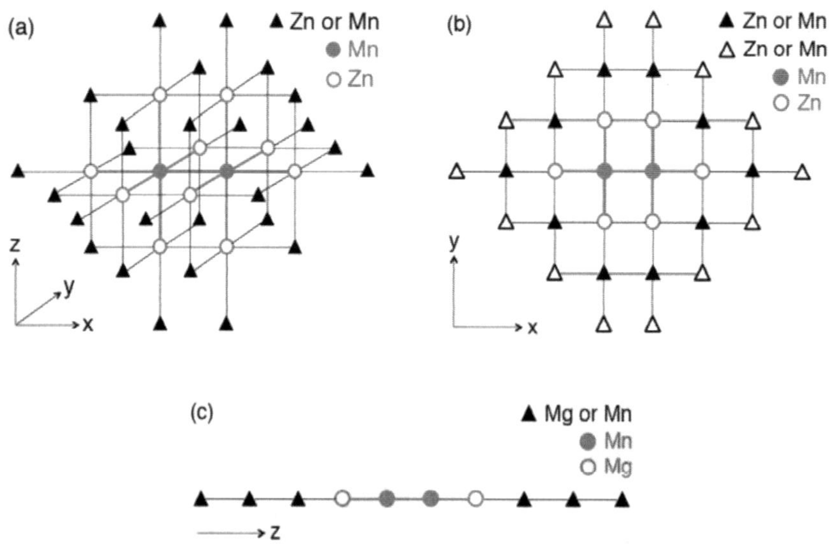

Figure 2. Sketch of neighboring atoms around an isolated Mn dimer marked by spheres. (a) Three-dimensional case studied for $KMn_xZn_{1-x}F_3$. The triangles denote the 26 nearest-neighbor positions. (b) Two-dimensional case studied for $K_2Mn_xZn_{1-x}F_4$. The full and open triangles denote the ten nearest-neighbor and 14 next-nearest-neighbor positions, respectively. (c) One-dimensional case studied for $CsMn_xMg_{1-x}Br_3$.

For two-dimensional systems the 1/r decay law has a long-range nature, so that the nearest-neighbor approximation is not valid. The statistical model was extended to include in (1) also the next-nearest-neighbor positions as sketched in figure 2(b), which turned out to provide a reasonable agreement with the experimental data [5].

For the one-dimensional case as sketched for a Mn_xMg_{1-x} chain in figure 2(c), the probabilities $p_m(x)$ for having m Mn atoms on both sides of the central Mn dimer in a chain of length 2n are given by [4,8]

$$\begin{aligned}
p_0(x) &= (1-x)^{2n} \\
p_1(x) &= 2\binom{n}{1} x(1-x)^{2n-1} \\
p_2(x) &= \left[2\binom{n}{2} + \binom{n}{1}\binom{n}{1}\right] x^2(1-x)^{2n-2} \\
p_3(x) &= 2\left[\binom{n}{3} + \binom{n}{2}\binom{n}{1}\right] x^3(1-x)^{2n-3}
\end{aligned} \tag{2}$$

etc.

The chain length 2n has to be chosen such that the sum rule $\Sigma_m p_m(x)=1$ and the condition $2n \geq m$ are satisfied. These criteria are fulfilled for n=1/x. According to Krivoglaz [7] the decay of atomic displacements is governed by the number m of Mn atoms and not by their specific arrangement in the chain.

Equations (1) and (2) were used to calculate the intensities of the fine structure in the excitation spectra of Mn dimers [4-6]. The positions of the fine-structure lines were analyzed by the spin Hamiltonian

$$H = J\mathbf{s}_1 \cdot \mathbf{s}_2 + D\left[\left(s_1^z\right)^2 + \left(s_2^z\right)^2\right], \tag{3}$$

where $\mathbf{s}_i$ denotes the spin operator of the Mn atoms and D the single-ion anisotropy parameter. From (3) the local exchange parameters J_m can be directly determined. Through the linear law dJ/dR the local intradimer Mn-Mn distances R_m can then be derived as a function of internal chemical pressure induced by the substitutional Mn atoms. Table 1 lists the parameters D and dJ/dR of the compounds considered in the present work.

Table 1. Single-ion anisotropy parameter D and derivative of the nearest-neighbor exchange parameter dJ/dR taken from [6]. κ denotes the local compressibility. ΔR_m corresponds to the consecutive compression of the intradimer Mn-Mn distances R_m as explained in the text.

Compound	D [μeV]	dJ/dR [meV/Å]	κ [GPa^{-1}]	ΔR_m^{obs} [Å]	ΔR_m^{cal} [Å]
$KMn_{0.10}Zn_{0.90}F_3$	5.3(2)	3.3(6)	0.033	0.0044(7)	0.0045
$K_2Mn_{0.10}Zn_{0.90}F_4$	5.2(2)	2.6(4)	0.020	0.0032(6)	0.0027
$CsMn_{0.10}Mg_{0.90}Br_3$	20.7(3)	3.6(3)	0.016	0.0022(4)	0.0026

3. Summary of results

The data obtained for the three-dimensional compound $KMn_{0.10}Zn_{0.90}F_3$ displayed in figure 1 were subdivided into seven individual lines of Gaussian shape with equal linewidths and almost equidistant energy spacings. The probabilities $p_m(x)$ calculated from (1) with 26 nearest-neighbor positions as shown in figure 2(a) are indicated as vertical bars in figure 1. The probabilities first increase with increasing m, have a maximum for m=2, and then decrease until $p_m(x)<1\%$ for $m \geq 7$. There is a rather good agreement with the amplitudes of the individual lines with a goodness of fit $\chi^2=1.9$ [5], thus there is no need to extend the structural model beyond the nearest-neighbor positions. Actually this conclusion is in line with the theoretically predicted $1/r^2$ law for atomic displacements around a defect in three-dimensional crystals [7]. The resulting values of the local exchange parameters J_m and the local Mn-Mn distances R_m are displayed in figure 3(a). When going from m=0 to m=6, the consecutive internal chemical pressure exerted by the Mn substitutions results in a 0.7% compression of R_m and in a 25% increase of J_m. The latter effect clearly shows that the exchange interaction in doped materials is no longer uniformly distributed, but it exhibits marked differences around the defects as will be further commented in section 4.

Similar results were obtained for the two-dimensional compound $K_2Mn_{0.10}Zn_{0.90}F_4$ as summarized in figure 3(b). The data analysis in terms of the probabilities $p_m(x)$ included both the 10 nearest-neighbor and the 14 next-nearest-neighbor positions as shown in figure 2(b), yielding a goodness of fit χ^2=2.6 [5].

Figure 3(c) shows the results obtained for the one-dimensional compound $CsMn_{0.10}Mg_{0.90}Br_3$. The agreement between the amplitudes of the five individual lines and the calculated probabilities $p_m(x)$ defined by (1) turned out to be excellent with a goodness of fit χ^2=1.2 [6], thus claiming quantitative accuracy with the theoretical prediction of a constant decay of atomic displacements for one-dimensional crystals [7].

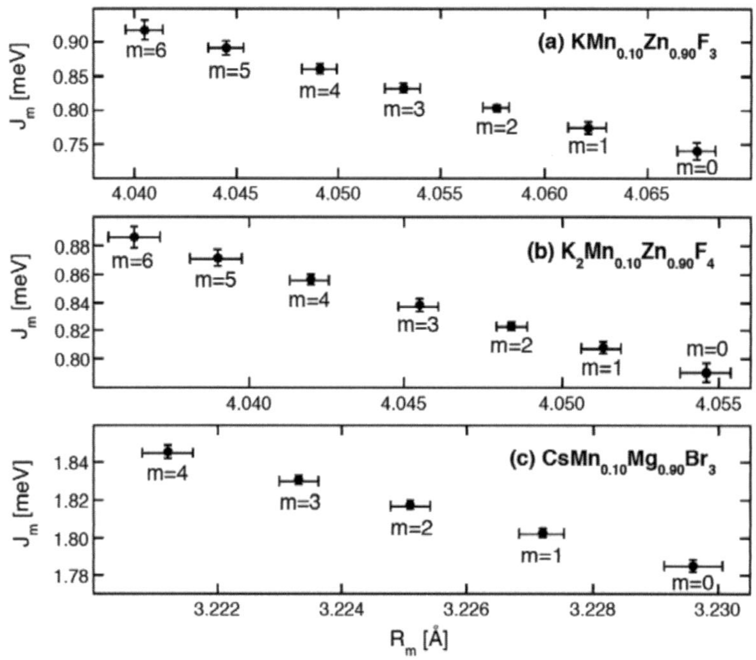

Figure 3. Local exchange interactions J_m and local Mn-Mn distances R_m derived for (a) $KMn_{0.10}Zn_{0.90}F_3$, (b) $K_2Mn_{0.10}Zn_{0.90}F_4$ and (c) $CsMn_{0.10}Mg_{0.90}Br_3$.

4. Concluding remarks

The analysis of the fine structure of Mn dimer excitations supports the theoretically predicted decay of atomic displacements according to $1/r^2$, 1/r and constant resulting from substitutional defects in three-, two- and one-dimensional crystals, respectively. The defects statistically present in the compounds clearly do not produce a continuous local dislocation pattern, but the diplacements lock in at discrete values depending on the number of defects, irrespective of their positions around the isolated Mn dimer. The substitutional atoms create a local internal chemical pressure due to the differences of the ionic radii which are ρ_{Mn}=0.83 Å > ρ_{Zn}=0.74 Å > ρ_{Mg}=0.72 Å for divalent ions with six-fold coordination [9]. Moreover, the nature of the internal pressure is essentially hydrostatic and most likely mediated by the elastic properties of the material. We tentatively describe the consecutive compression of the Mn-Mn distances R_m displayed in figure 3 by the phenomenological law

$$\Delta R_m = R_m - R_{m+1} = \beta \cdot \Delta\rho_{ion} \cdot \kappa \ , \tag{4}$$

where β is a constant prefactor, $\Delta\rho_{ion}=\rho_{Mn}-\rho_{Zn,Mg}$ corresponds to the difference of the ionic radii, and κ is the local compressibility which can be calculated from the elastic constants for the particular dimer geometries sketched in figure 2. The results are listed in table 1, which shows a good agreement between the experimental ΔR_m values taken from figure 3 and those calculated from (4) with a constant prefactor β=1.5 GPa.

Currently, investigations of quantum spin systems are of increasing interest in the search for exotic ground states. A very informative approach has been to study the modifications of the physical properties induced by the controlled introduction of defects in the materials, whose precise behavior depends on the value of the spin, the anisotropy, the dimensionality of the material, and the strengths and signs of the magnetic couplings [10,11]. The present work adds important information on the two latter quantities. We have shown that the exchange interaction in doped materials is no longer uniformly distributed, but it exhibits marked differences around the defects. The variation of the local exchange interactions around the defects manifests itself in a line broadening of the spin-wave dispersion relations as observed, *e.g.*, in INS experiments performed for $Sr_xBa_{1-x}MnO_3$ [12]. Another example concerns the spin-wave dispersion relation observed for $La_{1-x}(Ca_{1-y}Sr_y)_xMnO_3$, which was analyzed by a spin Hamiltonian with a nearest-neighbor and an unusually large fourth-nearest-neighbor exchange parameter [13]. However, the authors of [13] commented that the spin-wave data could also be reproduced by the nearest-neighbor exchange parameter J_1 alone, but with a random spatial distribution of different J_1's resulting from the substitutional effects. The variation of local exchange interactions has also implications on the spin freezing temperature T_g related to impurity induced magnetic order. As demonstrated for the compound $BiCu_{1-x}Zn_xPO_4$, the empiric law $T_g=Jx\varphi/(1+x\varphi)$ works well for small impurity contents x (φ denotes the volume of the correlated magnetic region), but fails for x>3% due to the local variation of the exchange parameter J [11].

The spatial extension of the distorted region around the defects strongly increases with lowering the dimensionality of the spin system. In particular, these effects cannot be neglected in studies of doped spin-chain materials like $SrCuO_2$ [14] and $SrCu_2O_3$ [15] with an exchange interaction of the order of J≈200 meV. Even if the magnetic couplings around the defects are modified only by a few percent, these differences match the relevant energy scale at low temperatures and therefore have to be considered in the data analysis.

References

[1] Bednorz J G and Müller K A 1986 *Z. Phys.* B **64** 189
[2] Sachdev S, Buragohain C and Vojta M 1999 *Science* **286** 2479
[3] Jungwirth T, Sinova J, Masek J, Kucera J and MacDonald A H 2006 *Rev. Mod. Phys.* **78** 809
[4] Furrer A, Strässle Th, Embs J P, Juranyi F, Pomjakushin V, Schneider M and Krämer K W 2011 *Phys. Rev. Lett.* **107** 115502
[5] Furrer A, Podlesnyak A, Krämer K W, Embs J P, Pomjakushin V and Strässle Th 2014 *Phys. Rev.* B **89** 144402
[6] Furrer A, Podlesnyak A, Krämer K W and Strässle Th 2014 *Phys. Rev.* B **90** 144434
[7] Krivoglaz M A 1996 *X-Ray and Neutron Diffraction in Nonideal Crystals* (Berlin: Springer)
[8] Furrer A, Juranyi F, Krämer K W, Schneider M and Strässle Th 2011 *Phys. Rev.* B **83** 024404
[9] Shannon R D 1976 *Acta Crystallogr.* A **32** 751
[10] Alloul H, Bobroff J, Gabay M and Hirschfeld P J 2009 *Rev. Mod. Phys.* **81** 45
[11] Bobroff J, Laflorencie N, Alexander L K, Mahajan A V, Koteswararao B and Mendels P 2009 *Phys. Rev. Lett.* **103** 047201
[12] Pratt D K, Lynn J W, Mais J, Chmaissem O, Brown D E, Kolesnik S and Dabrowski B 2014 *Phys. Rev.* B **90** 140401
[13] Moussa F, Hennion M, Kober-Lehouelleur P, Reznik D, Petit S, Moudden H, Ivanov A, Mukovskii Ya M, Privezentsev R and Albenque-Rullier F 2007 *Phys. Rev.* B **76** 064403
[14] Simutis G *et al* 2013 *Phys. Rev. Lett.* **111** 067204
[15] Larkin M I, Fudamoto Y, Gat I M, Kinkhabwala A, Kojima K M, Luke G M, Merrin J, Nachumi B, Uemura Y J, Azuma M, Saito T and Takano M 2000 *Phys. Rev. Lett.* **85** 1982

Crystal field excitations on $NdFe_3(BO_3)_4$ investigated by inelastic neutron scattering

S. Hayashida[1], M. Soda[1], S. Itoh[2], T. Yokoo[2], K. Ohgushi[3], D. Kawana[1], and T. Masuda[1]

[1]Neutron Science Laboratory, Institute for Solid State Physics, University of Tokyo, Tokai, Ibaraki 319-1106, Japan
[2]Neutron Science Division, Institute of compounds Structure Science, High Energy Accelerator Research Organization, Tsukuba, Ibaraki 305-0801, Japan
[3]Department of Physics, Tohoku University, Sendai, Miyagi 980-8581, Japan

E-mail: hayashida@issp.u-tokyo.ac.jp

Abstract. We performed inelastic neutron scattering experiments on a multiferroic material $NdFe_3(BO_3)_4$ to explore the crystal field excitations, where the magnetic anisotropy of the Nd^{3+} crystal field was responsible for the multiferroicity. Flat modes were observed at 8.5 meV and 17.4 meV in the neutron spectrum. The analysis revealed that these excitations were the transitions from the ground state to the first and second excited states of the Nd^{3+} crystal field.

1. Introduction

Multiferroic compounds, which exhibit both magnetic and dielectric orders, have attracted great attention since the discovery of a huge magnetoelectric (ME) effect in $TbMnO_3$ [1]. Among them, the rare-earth ferroborates $R\mathrm{Fe}_3(\mathrm{BO}_3)_4$ (R = Y and rare-earth metal) are a series of new multiferroic compounds containing R^{3+} ($4f^n$) and Fe^{3+} ($3d^5$ $S = 5/2$) as magnetic ions. They show diverse ME effects as a function of the R^{3+} ions (R = Y, Pr, Nd, Sm, Gd and Tb) [2, 3, 4, 5, 6, 7] because of the variety of magnetic anisotropies and of strong coupling to the Fe^{3+} moments. Recent studies revealed that the mechanism of the magnetoelectricity in these compounds was explained by the spin-dependent metal-ligand hybridization model [8, 9].

We focus on $NdFe_3(BO_3)_4$, where the Nd^{3+} ions ($4f^3$) carry a magnetic moment with $J = 9/2$. The crystal structure of $NdFe_3(BO_3)_4$ is trigonal with space group $R32$. The point symmetry of the NdO_6 octrahedra is D_3; a threefold symmetry along the crystallographic c - axis and three twofold axes in the ab - plane as shown in Fig. 1(a). Studies on the heat capacity, magnetic susceptibility [10], and ME effect [3, 6] revealed that an easy-plane type antiferromagnetic (AF) order and a spontaneous electric polarization simultaneously appeared at T_N = 30 K and a huge electric polarization was induced by a magnetic field. An optical spectroscopy study exhibited five Kramers doublets by the Nd^{3+} crystal field [11]. A neutron diffraction study exhibited the easy-plane type AF order at $T \leq T_N$; the Fe^{3+} and Nd^{3+} magnetic moments aligned ferromagnetically along the a - axis and aligned antiferromagnetically along the c - axis with the propagation vector $\boldsymbol{k} = (0, 0, 3/2)$ [12, 13].

In the present study we explored the crystal field excitations of the Nd^{3+} ion in $NdFe_3(^{11}BO_3)_4$ by means of an inelastic neutron scattering (INS) technique. Very recently it

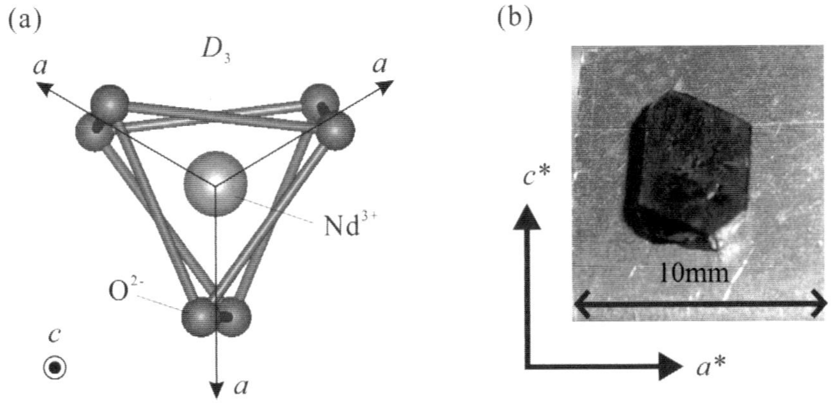

Figure 1. (a) The ab - plane projection of the NdO_6 octahedra in the D_3 symmetry. (b) A single crystal of $NdFe_3(^{11}BO_3)_4$

was revealed that the magnetic anisotropy of the Nd^{3+} crystal field drove the multiferroicity in $NdFe_3(BO_3)_4$ [14]. It is significant to investigate the Nd^{3+} crystal field for the further understanding of the ME effect in the rare-earth ferroborates.

2. Experimental details

Single crystals of $NdFe_3(^{11}BO_3)_4$ were grown by a flux method [15]. Details of the growing method are described in Ref. [14]. Plate-shaped crystals having the crystallographic a^*-c^* plane as shown in Fig. 1(b) were obtained. 22 pieces of the single crystals were coaligned so that the a^*-c^* plane was horizontal. The crystal alignment was performed by a transmission Laue method using a high energy X-ray CCD camera. We placed the crystals on an aluminum holder. The average mass of the crystals was 0.1 g. The total mass of the sample was 2.1 g.

We performed the INS experiment at the High Resolution Chopper Spectrometer (HRC) installed in the Material and Life Science Experimental Facility of J-PARC [16, 17, 18]. White neutrons are monochromatized by a Fermi chopper synchronized with the production timing of the pulsed neutrons at the HRC. The energy transfer $\hbar\omega$ was determined from the time of flight (TOF) of scattered neutrons detected at position sensitive detectors (PSDs). The T_0 chopper was set at 50 Hz, a collimator of 1.5° was installed in front of the sample, and the "S" Fermi chopper with 200 Hz was used to obtain high neutron flux. We used a GM-type closed cycle cryostat to achieve 40 K and 15 K. The energy of the incident neutron beam was $E_i = 51.04$ meV yielding an energy resolution of $\Delta E = 2.5$ meV at the elastic position. We set the incident neutron beam with $\boldsymbol{k}_i // \boldsymbol{a}^*$.

3. Experimental results and discussions

An INS spectrum at 15 K is shown in Fig. 2(a). A well-defined flat excitation is observed at 17 meV and a weak flat one is observed at about 8 meV. Strong intensities at $\hbar\omega \lesssim 6$ meV in the wide Q range are from dispersive Fe-centered spin waves [14]. A weak flat intensity at about 20 meV is from an instrumental artifact [14]. Dispersive spectra in the range of $|Q| > 2$ Å^{-1} are from the aluminum that is used for the sample can and sample holder.

One-dimensional energy cuts obtained by integrating the intensities in the range of 1.5 Å^{-1} $\leq |Q| \leq 1.9$ Å^{-1} are shown in Fig. 2(b). The red triangles and blue circles indicate the data measured at 40 K and 15 K, respectively. The data at 15 K are fitted by the sum of five Gaussian

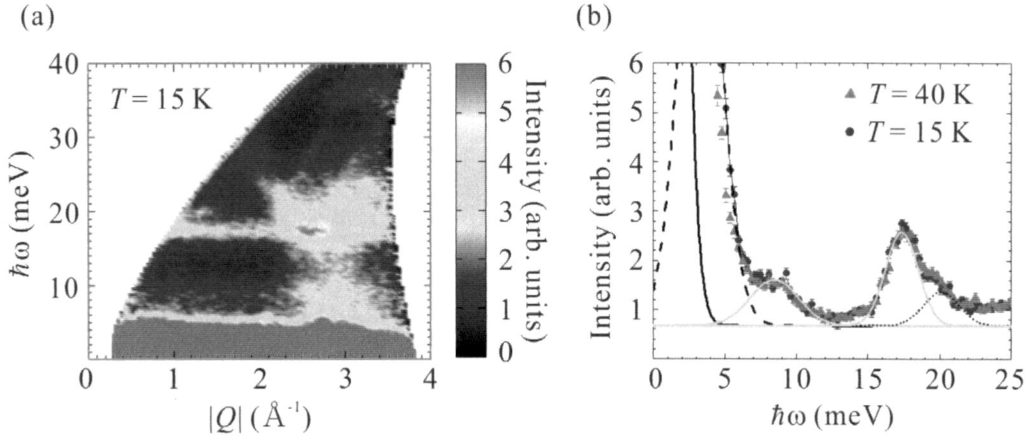

Figure 2. (a) An inelastic neutron scattering spectrum at 15 K. (b) $\hbar\omega$-dependences of the neutron intensities that are obtained by integration in the range of 1.5 $\text{Å}^{-1} \leq |Q| \leq 1.9$ Å^{-1}. The triangles and circles are data at 40 K and 15 K, respectively. Solid, dashed and dotted curves are Gaussian fits to the data at 15 K.

functions; the solid black curve is for an incoherent elastic scattering, the dashed curve is for the Fe-centered spin-wave excitation, the solid green curve is for a couple of intrinsic excitations, and dotted curve is for the instrumental artifact. The peak energies of the green curve are estimated to be 8.5 meV and 17.4 meV, and they agree with the energy of the first and second excited states of the Nd^{3+} crystal field reported in the optical spectroscopy [11]. This suggests that these excitations are the transitions from the ground state to the first and second excited states. A quite small temperature dependence is observed between the data at 40 K and that at 15 K.

We simulated neutron intensities of the Nd^{3+} crystal field. At the Nd^{3+} site with the D_3 symmetry, the crystal field Hamiltonian is defined as follow:

$$\begin{aligned}\mathcal{H}_{\rm CF}(D_3) &= B_0^2C_0^2 + B_0^4C_0^4 + iB_3^4\left(C_3^4 + C_{-3}^4\right) \\ &\quad + B_0^6C_0^6 + iB_3^6\left(C_3^6 + C_{-3}^6\right) + B_6^6\left(C_6^6 + C_{-6}^6\right). \end{aligned} \tag{1}$$

The B_q^p are the crystal field parameters and the C_q^p are the spherical tensor operators. We used the values of the parameters B_q^p reported in Ref. [11].

Then, the neutron intensity is expressed by

$$\frac{d\sigma}{d\Omega dE'} \propto \sum_{n,n'} p_n \left[\left|\langle n'|J_+|n\rangle\right|^2 + \left|\langle n'|J_-|n\rangle\right|^2\right] \delta\left(\hbar\omega + E_n - E_{n'}\right). \tag{2}$$

Here J_+ and J_- are, respectively, the raising and lowering operators of the total angular momentum $\boldsymbol{J}$, and p_n is the product of the probability that $|n\rangle$ state is occupied. E_n and $E_{n'}$ are the energies of the $|n\rangle$ and $|n'\rangle$ levels, respectively. We calculated the excitation energies and the neutron intensities of the transferred states from the ground state to the first and second excited states using *McPhase* software suit [19]. A ratio of the calculated intensities $(I_{\rm 1st}/I_{\rm 2nd})_{\rm cal}$ is 0.71, and that of the experimental intensities $(I_{\rm 1st}/I_{\rm 2nd})_{\rm exp}$ is 0.49. The experimental results are semi-quantitatively consistent with the calculations. Furthermore, we obtained the temperature dependence of the intensity as $(I_{\rm 40K}/I_{\rm 15K})_{\rm cal} = 0.93$ by calculating p_ns at 40 K and 15 K by assuming that the initial state was the ground state. The small temperature dependence is consistent with the data in Fig. 2(b). The observed excitations are, thus, interpreted as the first and second excited states of the Nd^{3+} crystal field.

It is noted that the peak at 8.5 meV is broader than the peak at 17.4 meV in Fig 2(b). This broadening was also reported in the optical spectroscopy [11]. It is known that a spectrum of the crystal field excitation is broadened by coupling of phonon [20] and dynamic Jahn-Teller effect [21, 22]. This broadening of the peak is ascribed to an interaction between the Nd^{3+} moment and lattice vibration since the magnetoelastic coupling was observed in $NdFe_3(BO_3)_4$ [3].

4. Conclusion

The INS measurements were performed on the single crystals $NdFe_3(^{11}BO_3)_4$ to explore the Nd^{3+} crystal field excitations. In addition to the Fe-centered spin waves previously reported, dispersionless excitations are observed at 8.5 meV and 17.4 meV. Our analysis reveals that they are the transitions from the ground state to the first and second excited states of the Nd^{3+} ion in the crystal field.

Acknowledgments

We thank H. Matsuda, and K. Asoh for their contribution to the single crystal growth. The neutron scattering experiment was approved by the Neutron Scattering Program Advisory Committee of IMSS, KEK (Proposal No. 2013S01 and 2014S01). This work was supported by KAKENHI (24340077). S. Hayashida was supported by the Japan Society for the Promotion of Science through the Program for Leading Graduate Schools (MERIT).

References

[1] T. Kimura, T. Goto, H. Shintani, K. Ishizaka, T. Arima, and Y. Tokura, Nature (London) **426**, 55 (2003).
[2] A. K. Zvezdin, S. S. Krotov, A. M. Kadomtseva, G. P. Vorob'ev, Yu. F. Popov, A. P. Pyatakov, L. N. Bezmaternykh, and E. A. Popova, JETP Lett. **81**, 272 (2005).
[3] A. K. Zvezdin, G. P. Vorob'ev, A. M. Kadomtseva, Yu. F. Popov, A. P. Pyatakov, L. N. Bezmaternykh A. V. Kuvardin, and E. A. Popova, JETP Lett. **83**, 509 (2006).
[4] A. M. Kadomtseva, A. K. Zvezdin, A. P. Pyatakov, A. V. Kuvardin, G. P. Vorob'ev, Yu. F. Popov, and L. N. Bezmaternykh, JETP **105**, 116 (2007).
[5] A. K. Zvezdin, A. M. Kadomtseva, Yu. F. Popov, G. P. Vorob'ev, A. P. Pyatakov, V. Yu. Ivanov, A. M. Kuz'menko, A. A. Mukhin, L. N. Bezmaternykh, and I. A. Gudim, JETP, **109**, 68 (2009).
[6] A. M. Kadomtseva, Yu. F. Popov, G. P. Vorob'ev, A. P. Pyatakov, S. S. Krotov, and K. I. Kamilov, Low Temp. Phys. **36**, 511 (2010).
[7] U. Adem, L. Wang, D. Fausti, W. Schottenhamel, P. H. M. van Loosdrecht, A. Vasiliev, L. N. Bezmaternykh, B. Büchner, C. Hess, and R. Klingeler, Phys. Rev. B, **82**, 064406 (2010).
[8] A. I. Popov, D. I. Plokhov, and A. K. Zvezdin, Phys. Rev. B **87**, 024413 (2013).
[9] T. Kurumaji, K. Ohgushi, and Y. Tokura, Phys. Rev. **89**, 195126 (2014).
[10] N. Tristan, R. Klingeler, C. Hess, B. Büchner, E. Popova, I. A. Gudim, and L. N. Bezmaternykh, J. Magn. Magn. Mater. **316**, 621 (2007).
[11] M. N. Popova, E. P. Chukalina, T. N. Stanislavchuk, B. Z. Malkin, A. R. Zakirov, E. Antic-Fidancev, E. A. Popova, L. N. Bezmaternykh and V. L. Temerov, Phys. Rev. B **75**, 224435 (2007).
[12] P. Fischer, V. Pomjakushin, D. Sheptyakov, L. Keller M. Janoschek, B. Roessli, J. Schefer, G. Petrakovskii, L. Bezmaternikh, V. Temerov and D. Velikonov, J. Phys.:Condens. Matter **18**, 7975 (2006)
[13] M. Janoschek, P. Fischer, J. Schefer, B. Roessli, V. Pomjakushin, M. Meven, V. Petricek, G. Petrakovskii, and L. Bezmaternikh, Phys. Rev. B **81**, 094429 (2010).
[14] S. Hayashida, M. Soda, S. Itoh, T. Yokoo, K. Ohgushi, D. Kawana, H. M. Rønnow, and T. Masuda, Phys. Rev. B **92**, 054402 (2015).
[15] L. N. Bezmaternykh, S. A. Kharlamova, and V. L. Temerov Crystallogr. Rep. **49**, 855 (2004), translated from Kristallografiya, **49**, 945 (2004). Original Russian Text Copyright (c) 2004 by Bezmaternykh, Kharlamova, Temerov.
[16] S. Itoh, T. Yokoo, S. Satoh, S. Yano, D. Kawana, J. Suzuki, and T. J. Sato, Nucl. Instr. Meth. Phys. Res. A **631**, 90 (2011).
[17] S. Yano, S. Itoh, S. Satoh, T. Yokoo, D. Kawana, and T. J. Sato, Nucl. Instr. Meth. Phys. Res. A **654**, 421 (2011).

[18] S. Itoh, T. Yokoo, D. Kawana, H. Yoshizawa, T. Masuda, M. Soda, T. J. Sato, S. Satoh, M. Sakaguchi, and S. Muto, J. Phys. Soc. Jpn. **82**, SA033 (2013).
[19] M. Rotter, J. Magn. Magn. Mater. **272 – 276**, e481 (2004).
[20] E. T. Heyen, R. Wegerer, and M. Cardona, Phys. Rev. Lett. **67**, 144 (1991).
[21] S. Kern, C. Loong, J. Faber, and G. H. Lander, Solid State Commun. **49**, 295 (1984).
[22] A. T. Boothroyd, C. H. Gardiner, S. J. S. Lister, P. Santini, B. D. Rainford, L. D. Noailles, D. B. Currie, R. S. Eccleston, and R. I. Bewley, Phys. Rev. Lett. **86**, 2082 (2001).

Zigzag type magnetic structure of the spin $J_{\mathrm{eff}} = ½$ compound α-$RuCl_3$ as determined by neutron powder diffraction

C. Ritter
Institut Laue Langevin, 71 Avenue des Martyrs, 38000 Grenoble, France

E-mail: ritter@ill.fr

Abstract. Using high intensity powder neutron diffraction the magnetic structure of α-$RuCl_3$ has been determined. Following the magnetic propagation vector κ = (½, 0, ½) the J_{eff} = ½ spins of Ru^{3+} adopt a Zigzag type arrangement on the honeycomb lattice of the layered $P3_112$ structure. The magnetic moments are oriented perpendicular to the trigonal axis. Similarities and differences to previously published single crystal data are discussed. The low value of the magnetic moments, μ_{Ru} = 0.5(1) μ_B indicates a possible closeness of α-$RuCl_3$ to the Kitaev spin liquid state.

1. Introduction

The importance of the spin-orbit coupling (SOC) in determining the electronic ground state of transition metal oxides has recently been a focus of both theoretical and experimental efforts [1-6]. On the theoretical side the Kitaev model [7-11]], which describes the interactions of spins J_{eff} = ½ on a honeycomb lattice by including bond dependent exchange interactions resulting from the SOC, is a rare example of exactly solvable models with highly nontrivial properties such as a quantum spin-liquid state and fractional statistics. However, being highly artificial, the model is difficult to realize in a real-world material. A number of promising candidates have been proposed in strongly spin-orbit coupled 5*d* transition-metal oxides (such as Na_2IrO_3 [3] and Li_2IrO_3 [5]), which have the requisite bond-directional interactions of the form in the Kitaev model. The search for the Kitaev spin-liquid has recently been extended to 4*d* transition-metal compounds with α-$RuCl_3$ [12-14] as one of the possible candidates as it crystallizes in a layered structure where the Ru^{3+} ($4d^5$) ions sit on a honeycomb lattice and see a nearly perfect $RuCl_6$ octahedral environment [15]. Under the combined effect of the crystal field and the spin-orbit coupling J_{eff} states having spin and orbital character develop [1]. The on-site Coulomb repulsion leads to the double degenerate J_{eff} = 3/2 bands having lower energy and being fully occupied while the remaining electron occupies a J_{eff} = ½ band and determines the magnetism [12].

Although magnetic susceptibility data had indicated a finite-temperature long-range magnetic order at a magnetic transition temperature of about 14 K [16], the determination of its magnetic structure is of crucial importance in assessing the proximity to the Kitaev quantum spin liquid state. Here we present the results from the first neutron powder diffraction measurements which strongly support the presence of a Zigzag type magnetic order, quite similar to that of Na_2IrO_3. In the low temperature neutron diffraction patterns, we have observed three magnetic Bragg peaks which could be indexed with a propagation vector κ = (½, 0, ½). Using magnetic symmetry analysis, the refinement of these peaks was possible with an allowed irreducible representation and confirmed the three-dimensional nature of the magnetic order despite the fact that neighboring Ru-Cl layers are weakly coupled by van der Waals forces.

2. Experimental

α-$RuCl_3$ crystals were prepared by vacuum sublimation. $RuCl_3$ powder was sealed in a quartz ampoule and heated to 750°C and kept for 2 weeks. Black and thin crystals were grown in the cold zone. For the powder measurements these crystals were powdered. High resolution and high intensity neutron powder diffraction experiments were done on the diffractometers D2B and D20 respectively, both situated at the Institut Laue Langevin at Grenoble, France. The high resolution data were taken at room temperature using a neutron wavelength of λ = 2.39 Å while low temperature data were recorded using a standard Orange cryostat on D20 at 1.8 K and at 50 K with λ = 2.41 Å. On D20 a radially oscillating collimator suppressed any scattering from the sample environment; at both temperatures data were collected for 6 hours. The Rietveld refinement program FULLProf [17] was used to analyse the data, magnetic symmetry analysis was done using the program BASIREPS [18, 19].

3. Results and Discussion

Figure 1 shows the room temperature high resolution D2B data which can be refined within the nuclear spacegroup $P3_112$ with lattice constants of a = 5.9510(4) Å and c = 17.1001(14) Å. The structure is characterized by a strong layer type structure where the coherence between the layers along the crystallographic c-axis is only determined by Van der Waals forces [20].

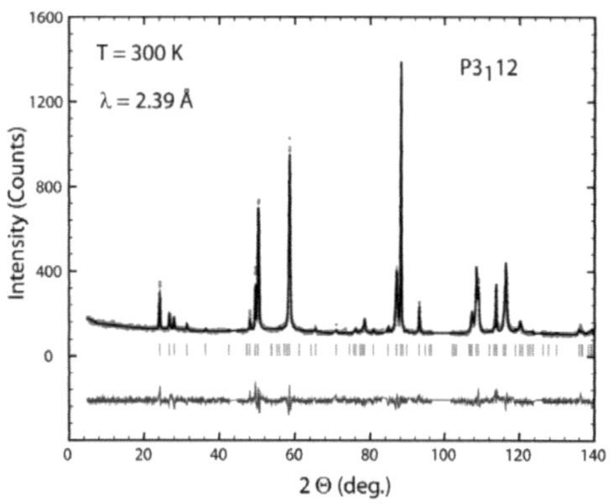

Figure 1. Observed (red dots), calculated (black line), and difference pattern of α-$RuCl_3$ at 300 K refined in $P3_112$. The tick marks indicate the calculated position of the Bragg peaks.

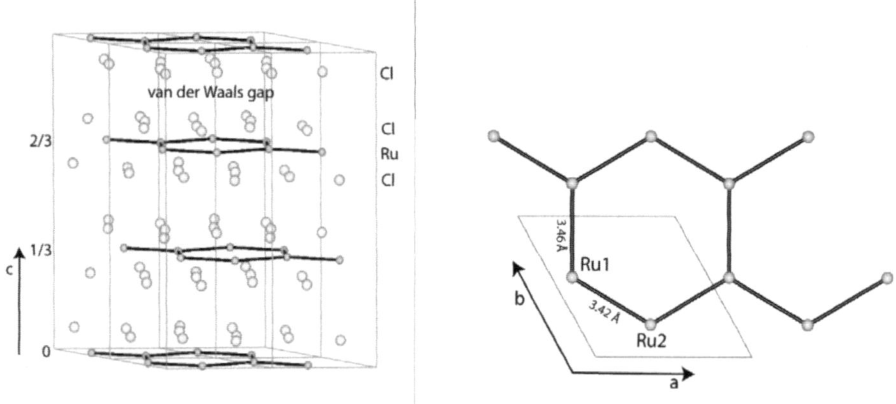

Figure 2. Layer type structure of α-$RuCl_3$ showing the van der Waals gap (left), honeycomb lattice formed by the two Ru sites (right).

Figure 2 shows the structure of α-$RuCl_3$ and the two dimensional honeycomb lattice formed by the two Ru-sites which are both situated on the Wykoff site 3*a* with x = 0.436(8) for Ru1 and x = 0.769(5) for Ru2. The Ru-Ru distances within the honeycomb lattice are not exactly similar; there are two different Ru1 – Ru2 distances amounting to 3.46 and 3.42 Å.

The magnetic scattering developing below T_N is very weak and is only visible in the difference pattern created by subtracting the high intensity data taken at 50 K from those taken at 1.8 K. Figure 3 displays this difference data with the presence of three magnetic reflections. These reflections can be indexed with the magnetic propagation vector κ = (½, 0, ½) or equivalent (0, ½, ½) and (½, ½, ½). The up-down feature visible at 2Θ = 24.5° is created by the slight thermal expansion between 1.8 K and 50 K which leads to an incomplete subtraction of the nuclear 003 reflection.

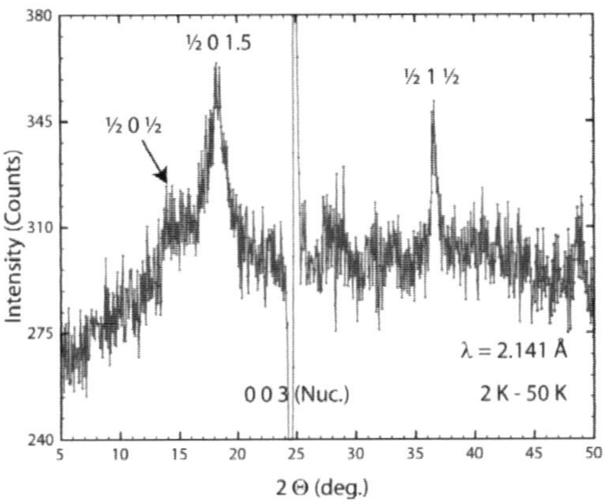

Figure 3. Purely magnetic diffraction pattern created as difference between data taken at 2 K and 50 K. Magnetic reflections are indexed in the nuclear unit cell.

Magnetic symmetry analysis for κ = (½, 0, ½) in spacegroup $P3_112$ shows that the Wykoff site 3*a* is split into two different orbits resulting in four independent Ru-sites in the unit cell. The original 3-fold site is split into a 2-fold site with the atoms on *x*, -*x*, *z* (Ru1(2)_1) and *x*, *x*-*y*, -*z* (Ru1(2)_2) and the unique site at -2*x*, -*x*, 0 (Ru1(2)_3). There are two different allowed irreducible representations (IR) which are listed in Table 1 together with their corresponding basis vectors (BV). For the determination of the magnetic structure of α-$RuCl_3$ one has to use not only the information on the presence and intensity of the measured magnetic peaks but as well on the absence of certain magnetic reflections. Nevertheless one will have to assume certain conditions to be valid in order to find stable solutions for the refinement. As a first condition it will be assumed that all Ru-sites possess equal magnetic moment values. The second condition which limits then sufficiently the number of free refinable parameters is to assume the magnetic spins to adopt a collinear structure. A collinear arrangement perpendicular to the trigonal axis can be described using a combination of BV1 and BV2 of IR1 while a collinear arrangement parallel to the trigonal axis is described by BV2 of IR2. Within each IR one can further choose between an antiferromagnetic coupling between the sites Ru1 and Ru2 defining the so-called Stripe model and a ferromagnetic coupling between the Ru1 and Ru2 sites which gives the so-called Zigzag model.

Table 1. Irreducible representations (IRs) and their basis vectors (BVs) for the Wykoff site 3*a* in space group $P3_112$ with κ = (½, 0, ½). Note that the threefold site is split into two orbits: 1) *x*, -*x*, ⅓ and *x*, 2*x*, ⅔ and 2) -2*x*, -*x*, 0

Ru on *x*, −*x*, ⅓		IR1		Ru on -2*x*, −*x*, 0	IR1
	BV1	BV2	BV3		BV1
x, *y*, *z*	100	010	001	*x*, *y*, *z*	210
x, *x*-*y*, -*z*	110	0-10	00-1		

	IR2				IR2		
	BV1	BV2	BV3		BV1	BV2	BV3
x, y, z	100	010	001	*x, y, z*	0-10	010	001
x, x-y, -z	-1-10	010	001				

Figure 4 displays these two different models in case of IR2, e.g. with spins parallel to the trigonal axis (compare with Figure 2). The last degree of freedom for the refinement of these collinear models consists in the choice between a ferromagnetic or an antiferromagnetic coupling between the Ru1(2)_1 and the Ru1(2)_3 site. This last choice determines the sequence in which the Ru-layers are ordered in direction of the c-axis (Figure 5).

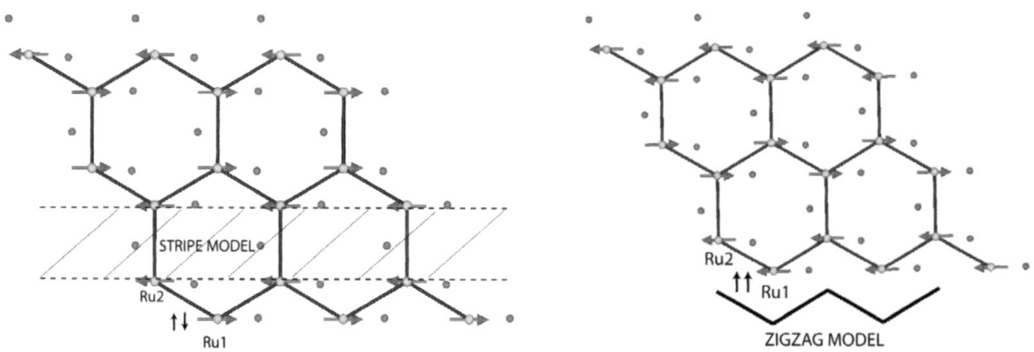

Figure 4. Arrangement of Ru-spins corresponding to the Stripe model resulting from an antiferromagnetic coupling of Ru1 and Ru2 sites (left); corresponding to the Zigzag model resulting from a ferromagnetic coupling of Ru1 and Ru2 sites (right).

Testing the different possible collinear magnetic arrangements it became immediately clear that only IR1 is able to reproduce the measured magnetic reflections: All models using IR2 create a too strong magnetic (½ 0 ½) peak and a too weak (½ 0 3/2) peak. The stripe model of IR2 creates furthermore a strong (-½ 1 3/2) reflection which is not seen in the data. In a similar way it is possible to show that using IR1 the stripe model creates again a strong (-½ 1 3/2) reflection and too weak intensity at the position of the (½ 0 3/2) reflection.

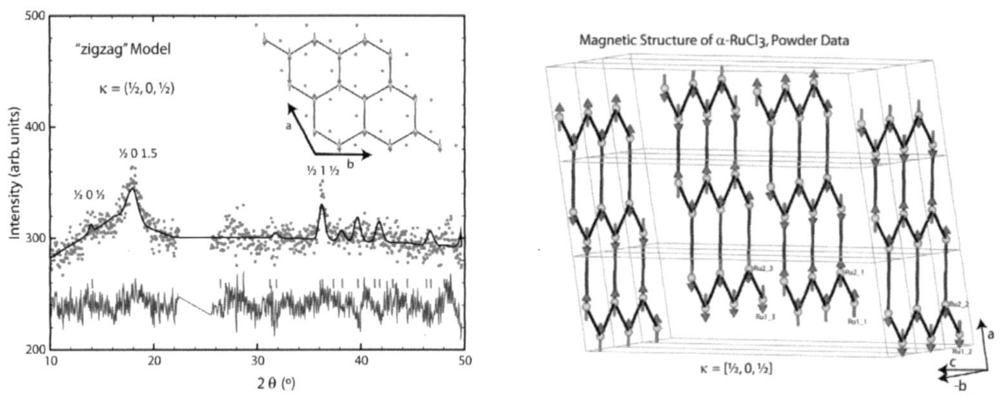

Figure 5. Observed (red dots), calculated (black line), and difference pattern of the purely magnetic diffraction pattern of α-$RuCl_3$ (left). The tick marks indicate the calculated position of the magnetic Bragg peaks. Magnetic structure with a Zigzag type arrangement on the honeycomb lattice (right) and an antiferromagnetic alignment of the Ru1(2)_1 and the Ru1(2)_3 sites.

Only the Zigzag model in IR1 together with an antiferromagnetic arrangement between the Ru1(2)_1, Ru1(2)_2 and the Ru1(2)_3 sites is able to refine consistently the intensities of the measured magnetic reflections. Figure 5 shows the final refinement of the neutron difference pattern 2 K - 50 K together with the arrangement of the magnetic spins within the honeycomb layer. Using the scalefactor as determined from the refinement of the purely nuclear pattern recorded at 50 K and the atom coordinates as determined from the high resolution D2B data the magnetic moment value was determined to $\mu_{Ru} = 0.5(1)\ \mu_B$. The relatively large error bar attached to the refined value is related to the unavailability of a magnetic form factor for Ru^{3+} and the use of the Fe^{3+} form factor instead. The broadening of the magnetic (½ 0 3/2) can be related to a limited coherence length along the crystallographic c-direction induced by the presence of stacking faults.

In the following this result will be compared to previous results on α-$RuCl_3$ obtained by Sears et al. [13] from single crystal neutron diffraction. This group found the presence of two magnetic reflections (½ 0 1) and (½ 0 2) and a magnetic propagation vector κ = (½, 0, 0). The data were not refined and no specific magnetic structure was proposed. Surprisingly single crystal data and powder data find different magnetic propagation vectors. It must be supposed that it is the effect of the stacking faults and the different coherence lengths in a single crystal compared to a powder which leads to this different behaviour. Starting from the magnetic structure determined here from the powder data it is, however, possible to find strong similarities to the data presented by Sears et al.: assuming the same Zigzag type magnetic order and assuming the same orientation of the magnetic moments in direction perpendicular to the trigonal axis one can simulate with κ = (½, 0, 0] the expected magnetic peaks and finds that the by far strongest magnetic reflections are the (½ 0 1) and (½ 0 2) reflections which are exactly the two reflections found by Sears et al. Assuming, on the contrary, a stripe type arrangement in IR1 or a magnetic moment orientation parallel to the trigonal axis corresponding to IR2, reflections not measured by Sears et al. are calculated in the model with κ = (½, 0, 0). This means that apart from the different magnetic propagation vectors both data sets agree in the basic magnetic arrangement of the Ru-spins on the honeycomb lattice. The differences are limited to the arrangement of the Ru-layers in c-direction as can be seen when comparing Figure 6 which displays the simulated magnetic structure for the single crystal data of Sears et al. and Figure 5 which displays the refined solution of the powder data.

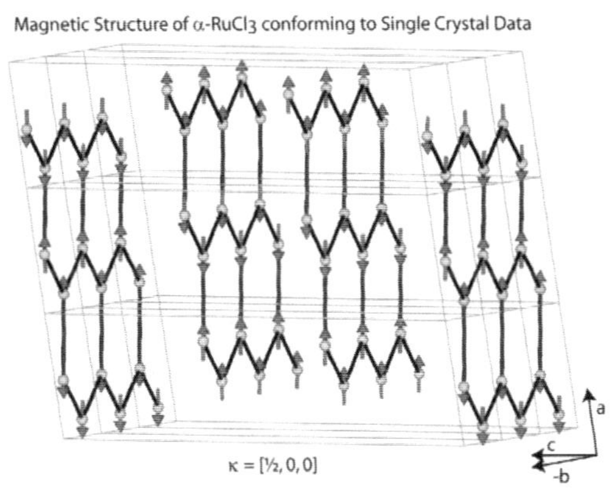

Figure 6. Proposed magnetic structure of α-$RuCl_3$ using the information from the single crystal data of Sears et al. [13] with the magnetic propagation vector κ = (½, 0, 0).

4. Summary

The magnetic structure of α-$RuCl_3$ has been determined from the refinement of high intensity neutron powder diffraction data. The magnetic propagation vector κ = (½, 0, ½) is different from the one found before from single crystal data. A collinear arrangement of the Ru-spins on the honeycomb lattice in a way so as to form Zigzag chains with moments perpendicular to the trigonal axis is, however, able to explain powder and single crystal data. The magnetic moment value amounts to about 0.5(1) μ_B and is therefore strongly reduced compared to the expected value for a J_{eff} = ½ spin indicating possibly the closeness of the system to the Kitaev spin-liquid state [21].

Acknowledgement
I would like to specially thank B.J. Kim and A. Jain (Max Planck Institute for Solid State Research, Stuttgart, Germany) for calling my attention to this topic of research and to N.H. Sung from the same institute for the preparation of the sample.

References
[1] Kim B J, Jin H, Moon S J, Kim J –K, Park B –G, Leem C S, Yu J, Noh T W, Kim C, Oh S –J, Park J –H, Durairaj V, Cao G and Rotenberg E 2008 *Phys. Rev. Lett.* **101** 076402
[2] Kim B J, Ohsumi H, Komesu T, Sakai S, Morita T, Takagi H and Arima T 2009 *Science* **323** 1329-32
[3] Liu X, Berlijn T, Yin W –G, Ku W, Tsvelik A, Kim Y –J, Gretarsson H, Singh Y, Gegenwart P and Hill J P 2011 *Phys Rev. B* **83** 220403(R)
[4] Choi S K, Coldea R, Kolmogorov A N, Lancaster T, Mazin I I, Blundell S J, Radaelli P G, Singh Y, Gegenwart P, Choi K R, Cheong S –W, Baker P J, Stock C and Taylor J 2012 *Phys. Rev. Lett.* **108** 127204
[5] Reuther J, Thomale R and Rachel S 2014 *Phys Rev. B* **90** 100405(R)
[6] Takayama T, Kato A, Dinnebier R, Nuss J, Kono H, Veiga L S I, Fabbris G, Haskel D and Takagi H 2015 *Phys. Rev. Lett.* **114** 077202
[7] Kitaev A 2006 *Ann. Phys.* **321** 2-111
[8] Chaloupka J, Jackeli G and Khaliullin G 2010 *Phys. Rev. Lett.* **105** 027204
[9] Singh Y, Manni S, Reuther J, Berlijn T, Thomale R, Ku W, Trebst S and Gegenwart P 2012 *Phys. Rev. Lett.* **108** 127203
[10] Rau J G, Lee E K-H and Kee H-Y 2014 *Phys. Rev. Lett.* **112** 077204
[11] Trousselet F, Horsch P, Oles A M and You W-L 2014 *Phys. Rev. B* **90** 204404
[12] Plumb K W, Clancy J P, Sandilands L J, Shankar V V, Hu Y F, Burch K S, Kee H-Y and Kim Y-J 2014 *Phys. Rev. B* **90** 041112(R)
[13] Sears J A, Songvilay M, Plumb K W, Clancy J P, Qiu Y, Zhao Y, Parshall D and Kim Y-J 2015 *Phys. Rev. B* **91** 144420
[14] Shankar V, Kim H-S and Kee H-Y 2015 *Phys. Rev. B* **91** 241110(R)
[15] Stroganov E V and Ovchinnikov K V 1957 *Ser. Fiz. Khim.* **12** 152-157
[16] Kobayashi Y, Okada T, Asai K, Katada M, Sano H and Ambe F 1992 *Inorg. Chem.* **31** 4570-74
[17] Rodriguez-Carvajal J 1993 *Physica B* **192** 55-69
[18] Rodriguez-Carvajal J BASIREPS: a program for calculating irreducible representations of space groups and basis functions for axial and polar vector properties. Part of the FullProf Suite of programs, www.ill.eu/sites/fullprof/
[19] Ritter C 2011 *Solid State Phenom.* **170** 263
[20] Fletcher J M, Gardner W E, Fox A C and Topping G **1967** *J. Chem. Soc A* 1038-45
[21] During the final preparation of this manuscript we became aware of related unpublished work on neutron diffraction of a single crystal of α-$RuCl_3$ which reports the existence of (½ 0 1) and (½ 0 3/2) magnetic reflections. Both reflections can be explained by the here proposed Zigzag type magnetic structure with spins pointing in a direction perpendicular to the trigonal axis.

Long Periodic Helimagnetic Ordering in CrM_3S_6 (M = Nb and Ta)

Y Kousaka[1,2], T Ogura[3], J Zhang[4,5], P Miao[4], S Lee[4], S Torii[4], T Kamiyama[4,6], J Campo[7] K Inoue[1,2,8] J Akimitsu[1,2,3]

[1]Graduate School of Science, Hiroshima University, Higashihiroshima, Hiroshima 739-8526, Japan
[2]Center for Chiral Science, Hiroshima University, Higashihirosima, Hiroshima 739-8526, Japan
[3]Department of Physics and Mathematics, Aoyama-Gakuin University, Sagamihara, Kanagawa 252-5258, Japan
[4]Institute of Materials Structure Science, KEK, Tsukuba, Ibaragi 305-0801, Japan
[5]China Spallation Neutron Source, Institute of High Energy Physics, Chinese Academy of Sciences, Dongguan 523803, Guangdong, China
[6]Sokendai (Graduate University for Advanced Studies), Hayama, Kanagawa 240-0115, Japan
[7]Aragón Materials Science Institute (CSIC - University of Zaragoza), Facultad de Ciencias, 50009 Zaragoza, Spain
[8]Institute for Advanced Materials Research, Hiroshima University, Higashihiroshima, Hiroshima 739-8526, Japan

E-mail: koyu@hiroshima-u.ac.jp

Abstract. We report long periodic chiral helimagnetic orderings in ferromagnetic inorganic compounds CrM_3S_6 (M = Nb and Ta) with a chiral space group of $P6_322$. Magnetization in polycrystalline samples and high resolution powder neutron diffraction were measured. Our powder neutron diffraction measurements in CrM_3S_6 successfully separated nuclear and magnetic satellite peaks, having the period of hundreds of angstroms along the c−axis. Therefore, we propose that the magnetic ordering in ferromagnetic CrM_3S_6 is not ferromagnetic, but long periodic chiral helimagnetic ordering.

1. Introduction

Chirality concept is playing a very important role in different areas of the knowledge from applied science such as new materials, nano, spintronic and biosciences to very basic sciences such as particle physics and cosmology. Regarding new materials for spintronic purposes, it crucial to understand how to control the chirality in molecules, crystals and magnetic structures. Recently, the relationship between crystallographic and magnetic chirality has received attention, because the sense of a chiral helical spin structure depends on the right- or left-handed chiral crystal structure that allows a Dzyaloshinskii-Moriya (DM) interaction [1, 2, 3, 4, 5]. When a magnetic field is applied perpendicular to the helical axis, the chiral helimagnetic structure continuously transforms into chiral magnetic soliton lattice in which the period is controlled by the field continuously [6]. Formation of the chiral magnetic soliton lattice is theoretically expected to exhibit attracting new phenomena [7, 8, 9, 10, 11]. In ferromagnetic-based chiral magnetic compounds, it can be shown that a cusp appears at the ordering temperature in the magnetization curves, which is an evidence of the chiral magnetic orderings [12]. However, the pitch angle of the chiral helimagnetic ordering, mainly determined by the ratio of exchange interaction and DM interaction (D/J), is usually very small and the helimagnetic period can be hundreds of angstroms. In some cases, the angular resolution of conventional thermal neutron diffractometers is not high enough to separate fundamental Bragg and magnetic satellite peaks. As a consequence, some compounds with helimagnetic ordering may be easily misinterpreted as collinear ferromagnets.

TM_3S_6 (T = transition metal, M = Nb and Ta), intercalated system of $2H - MS_2$, is described by the chiral space group $P6_322$ [13, 14, 15]. Magnetization measurements of TM_3S_6 show apparently a variety of magnetic behaviors; paramagnetism for $TiNb_3S_6$, VNb_3S_6, $TiTa_3S_6$ and VTa_3S_6, antiferromagnetism for $FeNb_3S_6$, $CoNb_3S_6$, $NiNb_3S_6$, $CoTa_3S_6$ and $NiTa_3S_6$, and ferromagnetism for $CrNb_3S_6$, $MnNb_3S_6$, $CrTa_3S_6$, $MnTa_3S_6$ and $FeTa_3S_6$ [16, 17, 18]. $CrNb_3S_6$ has been recently paid attention due to the observation of the chiral helimagnetism and chiral magnetic soliton lattice. In single crystalline magnetization samples of $CrNb_3S_6$ and $MnNb_3S_6$, the magnetization shows a cusp around the magnetic transition temperature T_C, indicating formation of the chiral helimagnetism and chiral magnetic soliton lattice along the $c-$axis [19]. These incommensurate magnetic structures were directly probed by Lorentz transmission electron microscopy (TEM) [20]. Thermal neutron diffraction studies firstly reported that $CrNb_3S_6$ formed a ferromagnetic structure because the magnetic scattering peaks were observed at the same diffraction angle with nuclear ones [15]. However, small angle neutron scattering experiments probed a magnetic satellite peak with the period of 480 Å[21, 22]. Therefore, some of the apparently ferromagnetic TM_3S_6 compounds can be in fact chiral magnetic compounds forming incommensurate chiral magnetic structures with very long period.

In this paper, we demonstrate that high resolution powder neutron diffraction experiments make it possible to detect incommensurate satellite peaks with the period of hundreds of Å. We report magnetization and powder neutron diffraction experiments in $CrNb_3S_6$ and $CrTa_3S_6$. Our obtained results indicate that these compounds form long periodic incommensurate magnetic structures.

2. Sample preparation

The polycrystalline sample was prepared by the mixture of Cr, M (M = Nb or Ta) and S in the stoichiometric composition in an evacuated silica tube at 1000 °C for one week. The obtained samples were evaluated by X-ray powder diffraction using Cu $K\alpha$ radiation (Multiflex, Rigaku Co.). All the diffraction peaks were successfully indexed using the reported TM_3S_6 structure [13, 14, 15]. The lattice parameters along the $c-$axis were 12.076 Åand 12.126 Å for $CrNb_3S_6$ and $CrTa_3S_6$, respectively. Therefore, all the synthesized samples had no impurity phases.

3. Magnetization measurements

The polycrystalline magnetization measurements were performed using a superconducting quantum interference device (SQUID) magnetometer (Quantum Design MPMS-5). Temperature dependences of magnetization in $CrNb_3S_6$ and $CrTa_3S_6$ are shown in Figs. 1 (a) and (b), respectively. The obtained data indicate that all the compounds show ferromagnetic responses. The magnetic transition temperature T_C in $CrNb_3S_6$ was 130 K, which is several K higher than the reported results [19, 20, 22]. The T_C in $CrTa_3S_6$ was 150 K, which is 35 K higher than the reported results [17]. These increases of the T_C were due to smaller amount of defects in Cr sites, and the details will be discussed elsewhere.

4. Spallation powder neutron diffraction measurements

High resolution powder neutron diffraction experiments were performed at BL08 (SuperHRPD) in the Materials and Life Science Experimental Facility (MLF) of the Japan Proton Accelerator Research Complex (J-PARC). The measurements were performed on a low-angle detector bank

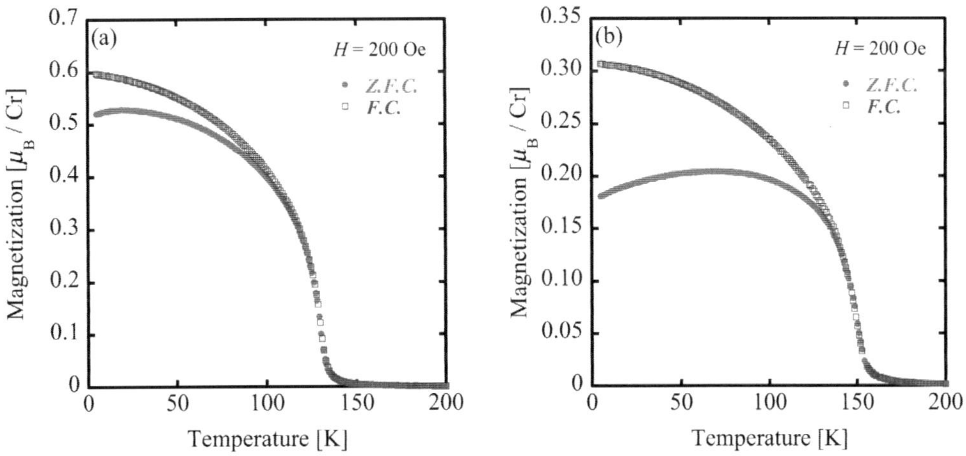

Figure 1. (color online) Temperature dependence of magnetization in (a) $CrNb_3S_6$, (b) $CrTa_3S_6$.

and a backward detector bank. The diffraction peaks were fitted by a pseudo-Voigt function, combining Gaussian and Lorentzian components. As evaluated by the nuclear (0,0,2) peak-width, q resolution of the backward detector bank was 3 times better than that of the low-angle detector bank. Powder diffractograms were collected at magnetic phase at 10 K and paramagnetic phase at 160 K and 180 K for $CrNb_3S_6$ and $CrTa_3S_6$, respectively.

Fig. 2 shows the results around the nuclear (0,0,4) and (1,0,3) reflections of $CrNb_3S_6$ and $CrTa_3S_6$ obtained by the backward detector bank. As indicated by the vertical arrows shown in Fig. 2 (a), we successfully separated the magnetic satellite peaks of $CrNb_3S_6$ around the (0,0,4) and (1,0,3) reflections. All the observed satellite peaks were well indexed by the magnetic propagation vector $k_{mag} = (0, 0, \delta)$ with $\delta = 0.021$. The determined helimagnetic period was slightly longer than the reported value [20, 22]. The fewer defects in Cr sites due to larger T_C may change the helical periodicity governed by D/J. While the thermal powder neutron diffraction measurements of $CrTa_3S_6$ have been reported to be ferromagnetic structure [15], our obtained data shown in Fig. 2 (b) clearly separated the nuclear peaks and magnetic satellite peaks, indicating a helimagnetic structure with the period of 225 Å. We also observed the magnetic satellite peaks around the (0,0,2) reflection on the low-angle detector bank. All the observed satellite peaks were well indexed by the magnetic propagation vector $k_{mag} = (0, 0, \delta)$ with $\delta = 0.054$.

5. Conclusion

In this paper, we succeeded in detecting incommensurate magnetic ordering with the period of hundreds of angstroms in CrM_3S_6 (M = Nb and Ta). While the magnetization measurements of $CrNb_3S_6$ were consistent with the reported results, the magnetic transition temperature T_C of $CrTa_3S_6$ was 35 K higher than that of the reported results[17]. The high resolution powder neutron diffraction of $CrNb_3S_6$ and $CrTa_3S_6$ separated fundamental nuclear Bragg peaks and magnetic satellite peaks with the period of hundreds of angstroms. The magnetic satellite peaks were indexed with the magnetic propagation $k_{mag} = (0, 0, \delta)$, and δ for $CrNb_3S_6$ and $CrTa_3S_6$ were 0.021 and 0.054, respectively. As the crystal structure of CrM_3S_6 is chiral, the observed magnetic satellite peaks in $CrTa_3S_6$ indicate formation of long periodic chiral helimagnetic ordering due to DM interaction.

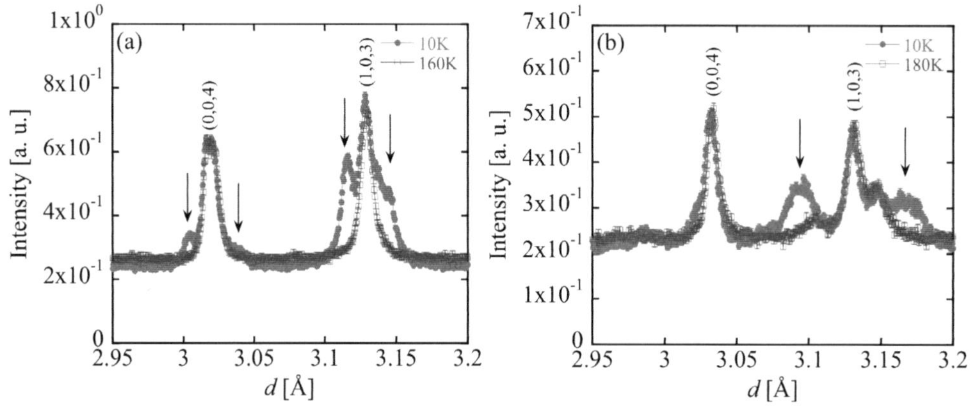

Figure 2. (color online) Neutron powder diffractograms around the nuclear (1,0,3) and (0,0,4) reflections in (a) $CrNb_3S_6$ and (b) $CrTa_3S_6$. The vertical arrows indicate observed magnetic satellite peaks indexed by $k_{mag} = (0, 0, \delta)$ ($\delta = 0.021$ for $CrNb_3S_6$ and 0.054 for $CrTa_3S_6$).

Acknowledgement
The authors express their sincere thanks to Y. Togawa and J. Kishine for fruitful discussions. This work was supported by the Center for Chiral Science in Hiroshima University (the Ministry of Education, Culture, Sports, Science and Technology (MEXT) program for promoting the enhancement of research universities, Japan), and the Japan Society for the Promotion of Science (JSPS) Core-to-Core Program, A. Advanced Research Networks. Y. K. acknowledges a Grant-in-Aid for Scientific Research (C) (No. 26400368) and Scientific Research (B) (No. 15H03680) from the MEXT of the Japanese Government. J. A. and K. I. acknowledge a Grant-in-Aid for Scientific Research (S) (No. 25220803) from the MEXT. J. C. acknowledges the grant No. MAT2011-27233-C02-02.

References

[1] Dzyaloshinskii I E 1964 *Zh. Eksp. Teor. Fiz.* **46** 1420 [Translation: 1964 *Sov. Phys. JETP* **19** 960].
[2] Moriya T 1960 *Phys. Rev.* **120** 91
[3] Tanaka M, Takayoshi H, Ishida M and Endoh Y 1985 *J. Phys. Soc. Jpn.* **54** 2970
[4] Ishida M, Endoh Y, Mitsuda S, Ishikawa Y and Tanaka M 1985 *J. Phys. Soc. Jpn* **54** 2975
[5] Grigoriev S, Chernyshov D, Dyadkin V, Dmitriev V, Maleyev S, Moskvin E, Menzel D, Schoenes J and Eckerlebe H 2009 *Phys. Rev. Lett.* **102** 037204
[6] Kishine J, Inoue K and Yoshida Y 2005 *Prog. Theoret. Phys.* Supplement **159** 82
[7] Bostrem I G, Kishine J and Ovchinnikov A S 2008 *Phys. Rev. B* **78** 064425
[8] Borisov . B, Kishine J, Bostrem I G and Ovchinnikov A S 2009 *Phys. Rev. B* **79** 134436
[9] Kishine J and Ovchinnikov A S 2009 *Phys. Rev. B* **79** 220405 (R)
[10] Kishine J, Ovchinnikov A S and Proskurin I V 2010 *Phys. Rev. B* **82** 064407
[11] Kishine J, Proskurin I V and Ovchinnikov A S 2011 *Phys. Rev. Lett.* **107** 017205
[12] Kousaka Y, Yano S, Kishine J, Yoshida Y, Inoue K, Kikuchi K and Akimitsu J 2007 *J. Phys. Soc. Jpn.* **76** 123709
[13] van den Berg J M, and Cossee P 1968 *Inorg. Chim. Acta* **2** 143
[14] Anzenhofer K, van den Berg J M, Cossee P and Helle J N 1970 *J. Phys. Chem. Solids* **31** 1057
[15] van Laar B, Rietveld H M and Ijdo D J W 1971 *J. Solid State Chem.* **3** 154
[16] Hulligur F and Pobitschka E 1970 *J. Solid State Chem.* **1** 117
[17] Parkin S S P and Friend R H 1980 *Phil. Mag. B* **41** 65
[18] Parkin S S P and Friend R H 1980 *Phil. Mag. B* **41** 95
[19] Kousaka Y, Nakao Y, Kishine J, Akita M, Inoue K and Akimitsu J 2009 *Nucl. Instr. Meth. Phys. Res. A* **600** 250

[20] Togawa Y, Koyama T, Takayanagi K, Mori S, Kousaka Y, Akimitsu J, Nishihara S, Inoue K, Ovchinnikov A S and Kishine J 2012 *Phys. Rev. Lett.* **108** 107202
[21] Moriya T and Miyadai T 1982 *Solid State Commun.* **42** 209
[22] Miyadai T, Kikuchi K, Kondo H, Sakka S, Arai M and Iishikawa Y 1983 *J. Phys. Soc. Jpn.* **52** 1394

Functional crossover in the dispersion relations of magnons and phonons

A Hoser[1] and U Köbler[2]

[1] Helmholtz-Zentrum Berlin, 14109, Germany
[2] Institute PGI, Research Centre Jülich, 52425, Germany

E-mail: hoser@helmholtz-berlin.de

Abstract. Experimental data are presented showing that the dispersion relations of magnons and acoustic phonons can consist of two sections with different functions of wave vector. In the low wave vector range a power function of wave vector often holds over a finite q-range while dispersions for larger wave vector values better approach the atomistic model predictions. In the magnon spectra ~q^x power functions with exponents x=1.25, 1.5 and 2 are identified. The dispersion of the acoustic phonons can be a linear function of wave vector over a surprisingly large range of energy. Since the slope of the linear section agrees with the known sound velocities it can be concluded that the dispersion of the acoustic phonons has got attracted by the linear dispersion of the mass less Debye bosons (sound waves). Due to the different (translational) symmetries of bosons and atomistic excitations (magnons, phonons) the associated dispersions can attract each other. In the same way the different ~q^x power functions in the magnon dispersions indicate that magnon dispersions are attracted by the dispersion of the bosons of the magnetic continuum (Goldstone bosons). This allows evaluation of the otherwise difficult to obtain dispersions of the Goldstone bosons from the known magnon dispersions. Interestingly, the dispersions of Goldstone bosons (Debye bosons) attract magnon dispersions (phonon dispersions) and not vice versa.

1. Introduction

Since development of Renormalization Group (RG) theory it became clear that in crystalline solids one has to distinguish between two translational symmetries: the discrete translational symmetry of the atomistic lattice and the continuous translational symmetry of the infinite or continuous solid [1]. The two symmetries are the generators of distinguished particles with specific excitation spectra. Well investigated excitations of the atomistic solid are magnons and phonons. These excitations can be studied using inelastic neutron scattering. The excitations of the continuous elastic solid, the Debye bosons (=sound waves), are mass less. Mass less bosons are invisible to neutrons. The various bosons of the magnetic continuum we call Goldstone bosons [2]. Because of their different translational symmetries Goldstone bosons (Debye bosons) and magnons (acoustic phonons) can be anticipated to interact weakly only. Interactions between particles of different symmetry can manifest as attraction of the associated dispersion relations. This interaction is mainly for small q-values (large wavelengths). The aim of this phenomenological study is to show that due to this interaction the dispersions of magnons (acoustic phonons) can assume the dispersion of the Goldstone bosons (Debye bosons) for small q-values. Note that the dispersions of freely propagating bosons are simple power functions of wave vector over a large range of energy. In particular, for mass less bosons the dispersion is linear. As a consequence, the dispersion of magnons and acoustic phonons can start as a power function of wave vector over a finite q-range. The power function can be identified as dispersion of the field bosons. This opens the opportunity to evaluate the difficult to obtain dispersions of the Goldstone bosons (Debye bosons) from the known magnon (phonon) dispersions. At the limit of the power function crossover to a different function of wave vector occurs. Note that this type of crossover is different from the better known crossover events between different power functions of temperature either in the temperature dependence of order parameter or of heat capacity [3]. It is evident that atomistic spin wave theory (lattice theory) can give correct description of the magnon dispersions (phonon dispersions) only if interactions with Goldstone bosons (Debye bosons) are negligible. Obviously, in the low wave vector range field theories are necessary to explain the different exponents x of the $\sim q^x$ power functions. When interactions between atomistic excitations and bosons are large, the power function holds over a relatively large q-range. For larger wave vector values the dispersion of magnons and acoustic phonons is determined essentially by inter-atomic interactions. In this q-range atomistic theories appear to be appropriate. In other words, atomistic theories and field theories are complementary and have to be applied to different parts of the observed atomistic excitation spectra.

There is, however, a fundamental difference between Goldstone boson field and Debye boson field. The Debye boson field is a scalar or isotropic field. The dispersion of the Debye bosons is a linear function of wave vector along all crystallographic directions. The sources of the Debye field quanta therefore must be spherical objects. Possibly these are the nearly spherical atomic cores. The Goldstone boson field is a vector field and can have different dispersions along different crystallographic directions. The vector character of the Goldstone boson field is a consequence of the axial generation process of the field quanta by precessing magnetic moments. The Goldstone bosons are essentially magnetic dipole radiation emitted upon precession of the ordered moments [4]. Since integer and half-integer spins precess differently they generate different types of field quanta. The properties of the Goldstone boson fields therefore are different in magnets with integer and half-integer spin. Because emission of magnetic dipole radiation is by stimulated emission the basic boson field is one-dimensional and much resembles the radiation field of a LASER. This is realized in each magnetic domain. The one dimensional field aligns all spins along its axis irrespective of local exchange anisotropies. Stimulated emission of the Goldstone bosons is the origin of perfectly collinear spin structures.

Here we focus on the peculiar phenomenon that in non cubic magnets the dispersion of the Goldstone bosons can be a different power function of wave vector along different crystallographic directions. Formally we have to attribute different field dimensions to the different $\sim q^x$ power functions. In other words, three exponents x can be expected to occur in magnets with integer and in magnets with half-integer spin, respectively (see table 1). Note that for any $\sim q^x$ dispersion relation the heat capacity of the boson field follows universal T^ε power function. In other words, the exponents ε and x are correlated. For the Debye boson field only x=1 and ε=3 occur. It is evident that only one of the observed $\sim q^x$ power functions can be the relevant excitation to define the global dimensionality of the bulk magnet. The global dimensionality can be noticed from the exponent ε observed either in the thermal decrease of the order parameter or in the thermal increase of the magnetic heat capacity [5]. Note that thermal decrease of the magnetic order parameter is controlled by the heat capacity of the Goldstone boson field. Obviously there exists a complicated symmetry selection rule for the global $\sim q^x$ dispersion of the bulk magnet. In RG theory it is customary to call this selection rule relevance [1]. Another difference between the Goldstone boson field and the Debye boson field is that in insulating solids the Debye boson field is the relevant excitation spectrum for temperatures below ~10...15 K only. Relevant means that the heat capacity of the solid is the heat capacity the Debye boson field, and follows universal T^3 function in all solids. Note that universality i.e. independence of lattice structure is the characteristic thermodynamic behavior of a field of freely propagating bosons. Typical for boson dynamics is that the critical power functions hold over a finite distance from critical temperature. Note that the atomistic critical power functions hold asymptotically only. For larger temperatures than crossover at ~10...15 K phonons are the relevant excitations. The dynamics then is lattice specific. Quite generally, boson fields are the relevant excitations in the vicinity of critical temperatures (stable fixed points). In diamagnets there is only one stable fixed point at T=0. In ordered magnets there are two stable fixed points at T=0 and at $T=T_c$. The validity range of the two associated critical power functions is large. The two power functions overlap and give complete description of the magnetic order parameter for all temperatures in the ordered state [3]. As a consequence, in the ordered state the dynamics (of the spins) is entirely determined by the boson guiding field. The magnetic part of the heat capacity is that of the Goldstone boson field. Magnons are non-relevant excitations and do virtually not contribute to the magnetic heat capacity.

2. Interactions between magnons and Goldstone bosons

As we have explained elsewhere [5] the magnon excitation gap is a direct measure of the interaction strength between magnons and Goldstone bosons. Condition for direct interactions between magnons and Goldstone bosons is that the Goldstone bosons have magnetic moment. The bosons therefore must have mass, and the dispersion cannot be linear. Only for the magnets with isotropic boson field and half-integer spin the dispersion of the Goldstone bosons is a linear function of wave vector [6]. These bosons are mass less and cannot have magnetic moment. In fact, there is virtually no magnon gap observed for this class of magnets [3]. It is evident that spin wave theory can give correct description of magnon dispersions only for a vanishing magnon-Goldstone boson interaction. In other words, spin wave theory has problems in explaining a magnon excitation gap, in particular in isotropic magnets with pure spin moments. For magnets with a sizeable magnon gap the continuous part of the magnon dispersion initially also is determined by magnon-Goldstone boson interactions and follows over a finite q-range a single power function of wave vector. The q-range of the power function is the larger the larger the interaction, i.e. the magnon gap is. The observed power function can be identified with the dispersion of the Goldstone bosons. Prominent example to prove this is MnF_2 [7-9]. In MnF_2 the Goldstone boson field is along tetragonal c-axis. Spin order in bulk MnF_2 is as for a single magnetic domain. The one-dimensional boson field aligns all Mn^{2+} moments rigidly along its axis, the crystallographic c-axis. The spin-flop field is as large as ~120 kOe [10]. This anisotropy field is much too large to be explained by atomistic models. Considering the pure spin moment of the Mn^{2+} ion and the isotropic magnon dispersions [8, 9] the spin flop field should be smaller by at least a factor of 100.

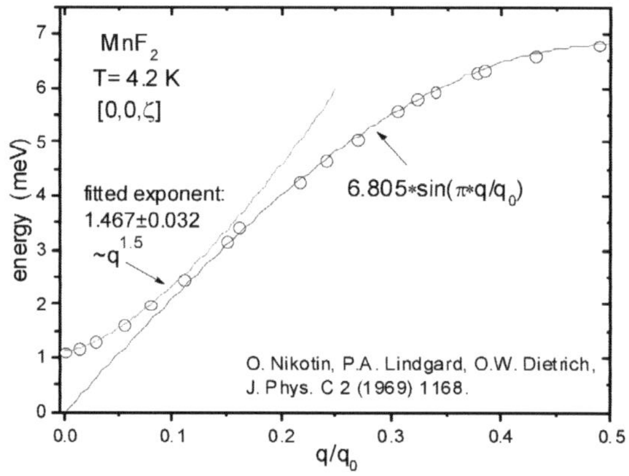

Figure 1. Functional crossover in the magnon dispersion curve of MnF_2 along tetragonal c-axis [8]. For small wave vector values magnon dispersion is given by gap value plus $q^{1.5}$ power function. The $q^{1.5}$ dispersion is that of the one-dimensional Goldstone boson field in magnets with half-integer spin (S=5/2). For larger wave vector values magnon dispersion is excellently described by sine function of wave vector.

At Néel transition only the z-components of the Mn^{2+} moments order [7]. This shows that the magnetic phase transition is executed by the one-dimensional boson field and not by the isotropic exchange interactions. MnF_2 therefore has to be classified as a one-dimensional antiferromagnet. Knowing that the boson field is one-dimensional in MnF_2, the $q^{1.5}$ power function observed in the low q-range of the magnon dispersion along c-axis (figure 1) can be interpreted as dispersion of the one-dimensional boson field in magnets with half-integer spin (S=5/2). This conclusion is consistent with investigations of the dispersion relations of resonating boson states in thin ferromagnetic films [6]. It is important to note that the standing magnetic wave experiments on metallic ferromagnetic films are a completely different and independent experimental method.

The uranium monopnictides nominally are cubic but undergo weak lattice distortions as a function of decreasing temperature. In USb the effective spin is $S_{eff}=1$ according to an observed saturation magnetic moment of $m_s=2.73\pm0.05$ μ_B/U [11]. Integer spin is consistent with $T^{9/2}$ function observed in the temperature dependence of the antiferromagnetic order parameter [12]. However, with decreasing temperature, lattice distortions increase monotonously and, eventually, become relevant. This happens at crossover temperature of $T_{CO}=105$ K where the temperature dependence of the order parameter changes from $T^{9/2}$ function to T^2 function. The T^2 universality class belongs to the three-dimensional but anisotropic boson field in magnets with integer spin. In inelastic neutron scattering measurements performed at T=10 K $\sim q^{1.5}$ and $\sim q^{1.25}$ power functions of wave vector can be identified along different crystallographic directions (figure 2). As systematic analyses of inelastic neutron scattering spectra of many other magnets show [5] the ~$q^{1.5}$ power function is the dispersion of the one-dimensional boson field but the ~$q^{1.25}$ function is the dispersion of the three-dimensional but anisotropic boson field in magnets with integer spin. Observation of T^2 function in the order parameter of USb (at T=10 K) shows that the ~$q^{1.25}$ dispersion is the relevant excitation for $T<T_{CO}$. If ~$q^{1.5}$ function would be relevant the order parameter would follow T^3 function (see table 1). On the other hand it can be expected that for $T>T_{CO}$=105 K where the order parameter follows isotropic $T^{9/2}$ function, ~q^2 dispersion appears in the magnon spectrum. It is typical that the boson defined part of the magnon dispersions depends sensibly on mesoscopic parameters such as weak lattice distortions. The exchange defined part of the magnon spectrum at larger wave vector values is more stable against weak lattice distortions.

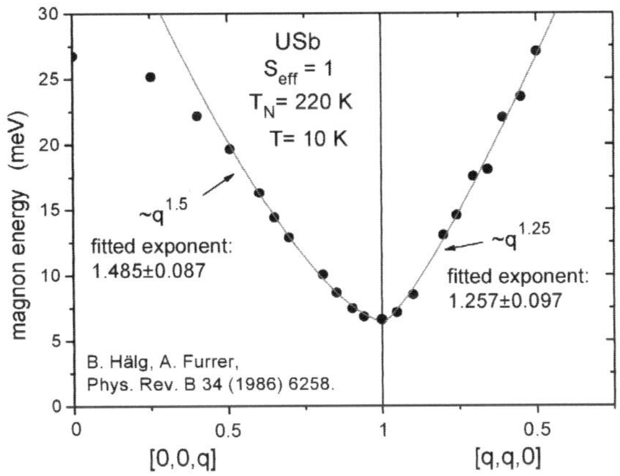

Figure 2. Magnon dispersions of antiferromagnetic USb measured at T=10 K along two crystallographic directions [11]. Along cube edge dispersion of the Goldstone bosons is ~$q^{1.5}$ while along face diagonal ~$q^{1.25}$ dispersion holds. The ~$q^{1.25}$ dispersion is consistent with T^2 dependence observed below T_{CO}=105 K for the magnetic order parameter (see text).

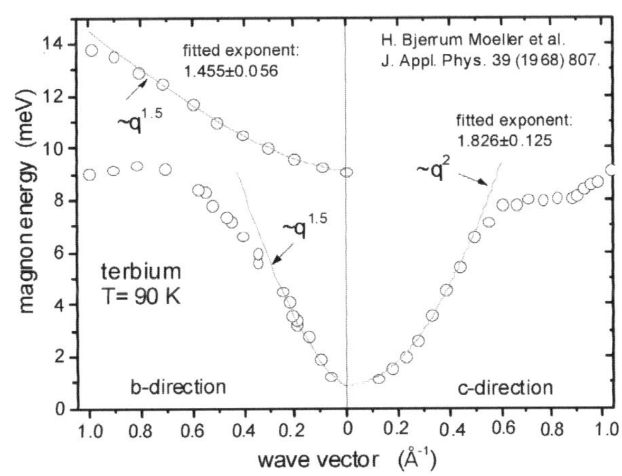

Figure 3. Magnon dispersions of terbium along hexagonal c-axis and transverse to c-axis (along b-direction) [13]. Assuming integer moment of J=6 according to electronic configuration of $^7F_6$ of the Tb^{3+} ion, ~q^2 dispersion is typical for the isotropic boson field but ~$q^{1.5}$ dispersion for the one-dimensional boson field in magnets with integer spin.

Hexagonal terbium has been extensively studied using inelastic neutron scattering [13, 14]. In the magnon spectra of [13] ~q^2 dispersion can be identified along hexagonal c-axis. Transverse to c-axis ~$q^{1.5}$ dispersion holds with high precision (figure 3). Interpretation of the two dispersion functions becomes, however, not clear considering terbium alone. Assuming that crystal field interaction is not relevant and that the full integer moment of J=6 of the configuration $^7F_6$ of the Tb^{3+} ion holds, the two power functions of wave vector are characteristic for magnets with integer spin. In fact, $\sim q^2$ function is the dispersion of the isotropic boson field and $\sim q^{1.5}$ function is the dispersion of the one-dimensional boson field in magnets with integer spin. Final interpretation of the observed $\sim q^x$ dispersion relations must come from magnets with stable magnetic moments and crystal structures that allow definite conclusion on the dimensionality of the boson field [5].

UO_2 is another nominally cubic crystal with integer spin of S=1 that undergoes weak lattice distortion as a function of decreasing temperature. Lattice distortions can be sample dependent. The usually large single crystals used in inelastic neutron scattering studies frequently are strongly strained. As we now know, the Goldstone boson field depends sensitively on lattice deformations. Axial strain in cubic magnets can reduce the dimensionality of the Goldstone boson field from isotropic to one-dimensional. In the magnon dispersions of the large UO_2 single crystal (2x2x4 cm) investigated in [15] dispersions as $\sim q^{1.5}$ can be identified along body diagonal (figure 4). Along cube edge $\sim q^{1.25}$ is observed (figure 5). For perfect cubic lattice symmetry $\sim q^2$ should be observed along all crystal directions. As a consequence, the UO_2 crystal investigated in [15] was not perfectly cubic.

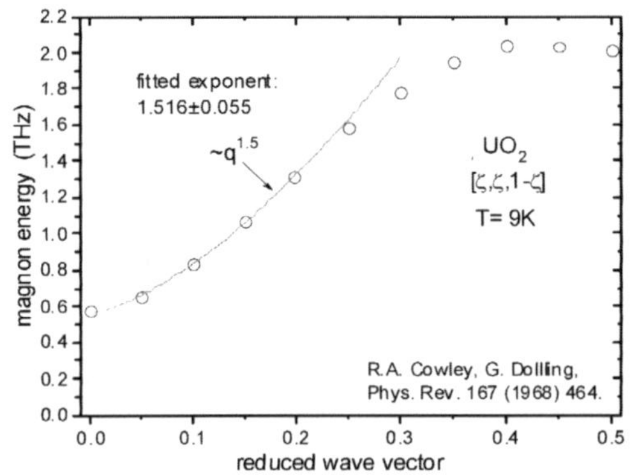

Figure 4. Magnon dispersions of nominally cubic UO_2 along body diagonal [15]. The observed $\sim q^{1.5}$ dispersion is that of the one-dimensional boson field in magnets with integer spin (S=1). This UO_2 sample was not perfectly cubic.

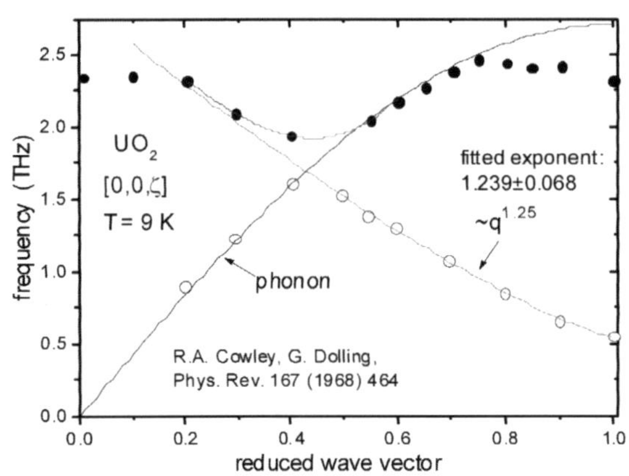

Figure 5. Magnon dispersions of UO_2 along cube edge revealing $\sim q^{1.25}$ dispersion [15]. This dispersion is attributed to the anisotropic 3D boson field in magnets with integer spin (S=1).

In the magnon dispersions of orthorhombic $LaMnO_3$ (figure 6) dispersions as $\sim q^{1.5}$ and q^2 can be identified. According to an observed saturation magnetic moment of m_s=3.87 μ_B/Mn the spin of the

Mn^{3+} ion is S=2 [16]. Since T^3 dependence can be identified in the magnetic order parameter [16] the $\sim q^{1.5}$ excitation spectrum seems to be relevant for the thermodynamics of the three-dimensional solid.

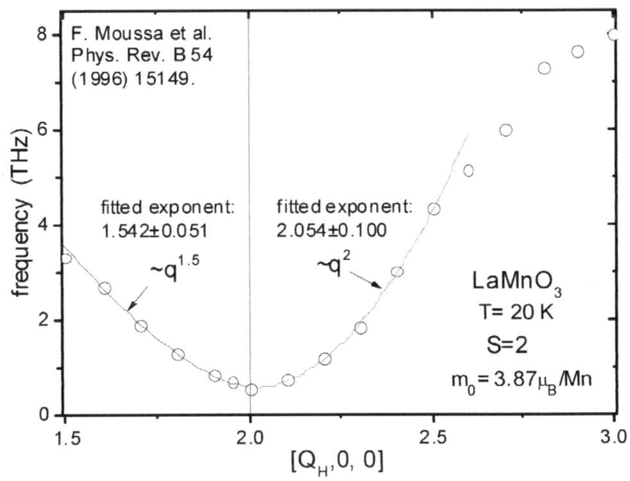

Figure 6. Magnon dispersions of orthorhombic $LaMnO_3$ with integer spin of S=2 [16]. The dispersion $\sim q^2$ is attributed to the isotropic boson field in magnets with integer spin. The dispersion $\sim q^{1.5}$ belongs to the one-dimensional boson field.

As a summary of our studies table 1 compiles the temperature power functions for the heat capacity of the boson fields and the associated dispersion relations in dependence of the dimension of the field (d) and spin quantum number.

Table 1. Power functions of temperature (T^ε) and of wave vector (q^x) for heat capacity and dispersion relation of the Goldstone boson field, respectively, in dependence of field dimensionality (d) and spin quantum number. The T^ε functions apply equally to the thermal decrease of the magnetic order parameter.

field dimensionality	integer spin	half-integer spin
d=3	$T^{9/2}$; q^2	T^2; q
d=2	T^2; $q^{1.25}$	$T^{3/2}$; q^2
d=1	T^3; $q^{1.5}$	$T^{5/2}$; $q^{1.5}$

3. Interactions between Debye bosons and acoustic phonons

For freely propagating sound waves (Debye bosons) the dispersion is a linear function of wave vector for all energies. However, interactions with the atomistic background of phonons and lattice imperfections provide damping to the Debye bosons. Sound velocities and elastic constants therefore depend somewhat on sample quality. Damping shortens the mean free path of the sound waves and decreases their velocity. In fact, sound velocities commonly decrease as a function of increasing temperature [17]. As a consequence the dispersion of the Debye bosons increases slightly weaker than linear.

Interactions between acoustic phonons and Debye bosons usually are stronger for longitudinal polarization than for transverse polarization. In AgBr these interactions are particularly strong. The longitudinal phonon branch initially is a linear function of wave vector up to energy of 2THz (~96 K) (figure 7) [18]. At this thermal energy the dispersion of the Debye bosons is no longer populated and phonons are the relevant excitations. It therefore can be concluded that the acoustic phonons can get attracted by the non populated Debye boson dispersion. This allows verification of the dispersion of the Debye bosons from phonon dispersion up to surprisingly high energies. Since the slope of the linear section agrees with the measured sound velocity [19] it becomes evident that phonon dispersion has been attracted by the dispersion of the Debye bosons. It is evident that the dispersion of the Debye bosons must continue up to much higher energies than the linear section in the phonon dispersion. For transverse polarization phonon data follow with good accuracy sine function of wave vector. However, in order to get correct description of the experimental data it is necessary to add a phase shift to the argument of the sine function. As a consequence, phonon-Debye boson interactions are noticeable all over the Brillouin zone. This seems to be different for the magnon dispersions (figure 1). Unfortunately no phonon data were measured along linear dispersion of v_T.

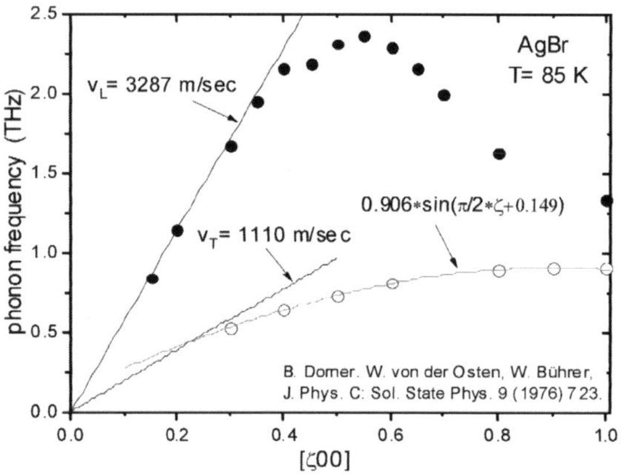

Figure 7. Dispersions of the acoustic phonons with longitudinal and transverse polarization of AgBr [18]. For longitudinal polarization the dispersion of the acoustic phonons agrees with the dispersion of the Debye bosons (sound waves) up to ~2 THz (~96 K). For transverse polarization interaction with Debye bosons necessitates a phase shift in the argument of the sine function.

3. Conclusions

Since development of RG theory it became clear that the continuous translational symmetry of the infinite solid is a particle generating symmetry. Critical dynamics is due to these particles. Sound waves (Debye bosons) are the particles or excitations of the elastic continuum. The particles of the magnetic continuum we have called Goldstone bosons [2]. The excitations of the continuous solid exist in addition to the excitations of the atomistic solid. The two excitations define the dynamics alternatively. Change of dynamics from one excitation spectrum to the other is a crossover. A crossover is a symmetry change. As a consequence, in magnetic solids one has to distinguish between four excitation spectra. Prominent particles or excitations of the atomistic solid are magnons and phonons. In spite of different translational symmetries Debye bosons and Goldstone bosons can interact with the atomistic excitations. These interactions can be surprisingly strong for large wavelengths. In the present communication we have focused on interactions between Goldstone bosons and magnons and on interactions between Debye bosons and acoustic phonons. Debye bosons can interact also with magnons; Goldstone bosons can interact also with phonons. Due to their different symmetries the dispersions of bosons and atomistic excitations can attract each other. Note that the dispersions of particles with identical symmetry avoid each other. It turned out that the dispersions of the bosons attract the dispersion of the atomistic excitations and not vice versa. In other words, for small wave vector values the dispersions of magnons and acoustic phonons assume the dispersion of Goldstone bosons and Debye bosons, respectively. This allows evaluation of the difficult to obtain dispersion relations of the bosons from the known inelastic neutron scattering spectra of the atomistic excitations of magnons and phonons. The Debye boson field is a scalar or isotropic field. There is no preferential field direction. The Goldstone boson field is a vector field and can assume any dimension. The vector character of the Goldstone boson field is a consequence of the axial generation process of the field quanta by precessing spins [4]. Goldstone bosons are essentially magnetic dipole radiation. Essential for the one dimensional Goldstone boson field is that magnetic dipole radiation is generated by stimulated emission. Collinear spin alignment in all magnetic domains is along direction of the boson field. However, coupling of the one-dimensional field components of the individual domains to result into an isotropic field of the bulk magnet is a completely unexplored process.

4. References

[1] Wilson K G and Kogut J 1974 *Phys. Rep.* **12C** 75
[2] Goldstone J, Salam A and Weinberg S 1962 Phys. Rev. **127** 965
[3] Köbler U and Hoser A 2010 *Renormalization Group Theory- Impact on Experimental Magnetism* (Berlin: Springer)
[4] Köbler U and Hoser A 2015 *Acta Phys. Pol.* A **127** 350
[5] Köbler U 2015 *Acta Phys. Pol.* A **127** 1694
[6] Köbler U and Hoser A 2012 *Acta Phys. Pol.* A **121** 1176
[7] Schulhof M P, Nathans R, Heller P and Linz A 1971 *Phys. Rev.* B **4** 2254
[8] Nikotin O, Lindgard P A and Dietrich O W 1969 *J. Phys.* C **2** 1168
[9] A. Okazaki A, Turberfield K C and Stevenson R W H 1964 *Phys. Lett.* **8** 9
[10] Shapira Y and Foner S 1970 *Phys. Rev.* B **1** 3083
[11] Hälg B and Furrer A 1986 *Phys. Rev.* B **34** 6258
[12] Köbler U, Hoser A, Mueller R M and Fischer K 2007 *J. Magn. Magn. Mater.* **315** 12
[13] Bjerrum Moeller H, Gylden Houmann J C and Mackintosh A R 1968 *J. Appl. Phys.* **39** 807
[14] Lindgard P A. 1978 *Phys. Rev.* B **17** 2348
[15] Cowley R A and Dolling G 1968 *Phys. Rev.* **167** 464
[16] Moussa F, Hennion M, Rodriguez-Carvajal J, Moudden H, Pinsard L and Revcolevschi 1996 *Phys. Rev.* B **54** 15149
[17] Hearmon R F S 1979 *Landolt-Börnstein* III, 11, ed K-H Hellwege and A M Hellwege (Berlin: Springer) p 1
[18] Dorner B, von der Osten W and Bührer W 1976 *J. Phys. C: Solid State Phys.* **9** 723
[19] Loje K F and Schuele D E 1970 *J. Phys. Chem. Solids* **31** 2051

Interrupted Magnetic First Order Transitions and Kinetic Arrest probed with In-field Neutron Diffraction

V. Siruguri[†,*], S.D. Kaushik[†], S. Rayaprol[†], P.D. Babu[†], P. Chaddah[$], E.V. Sampathkumaran[+] A. Hoser[‡], and C. Ritter[#]

[†]UGC-DAE Consortium for Scientific Research Mumbai Centre, R-5 Shed, BARC, Mumbai – 400085, India
[$]UGC-DAE Consortium for Scientific Research, University Campus, Khandwa Road, Indore – 452017, India
[+]Tata Institute of Fundamental Research, Dr. Homi Bhabha Road, Colaba, Mumbai – 400005, India
[‡]Helmholtz-Zentrum Berlin für Materialien und Energie, Hahn-Meitner-Platz 1, 14109 Berlin, Germany
[#]Institut Laue-Langevin, 71 avenue des Martyrs, 38000 Grenoble, France

[*]siruguri@csr.res.in

Abstract. In-field neutron diffraction studies were carried out on two compounds that exhibit magnetic first order phase transitions (FOPT). It is shown that the FOPT can be interrupted by an external magnetic field, resulting in a coexistence of kinetically arrested metastable states and equilibrium phases. Use of a novel protocol CHUF (Cooling and Heating under Unequal Fields) helps to determine the coexisting phase fractions and also to observe the devitrification of the kinetically arrested phase into the equilibrium phase, in a manner similar to that found in structural glassy systems.

1. Introduction:

Magnetic first order phase transitions (FOPT) are commonly found in a wide variety of materials like magnetic shape memory alloys, colossal magnetoresistance compounds, magnetocalorics and multiferroics. While in some cases the magnetic FOPT could be of the order-order type, driven by change of magnetic exchange interactions, it could also be accompanied by changes in volume or the symmetry of the crystal lattice. It is well-known that due to the presence of quenched disorder, there is a broadening of the FOPT which results in a distribution of transition temperatures [1-8]. Though temperature is the most commonly used thermodynamic variable for observing FOPT, pressure, magnetic and electric fields could also be useful second thermodynamic variables. It is possible to interrupt the FOPT by using an effective combination of these two variables, resulting in phase coexistence of magnetically ordered metastable states and equilibrium phases, collectively called kinetically arrested state or magnetic glass [1-6]. In such a state, the phase coexistence would persist down to the lowest temperature. The coexisting phase fractions could be quantified by employing the CHUF protocol (Cooling and Heating under Unequal Fields) [7, 8]. It is observed that the kinetically arrested magnetic glass could devitrify into the equilibrium magnetic state at a lower temperature and subsequently re-enter the original high temperature magnetic phase. This is very similar to the case of conventional structurally disordered glasses that devitrify under the influence of temperature or strain. Since neutron diffraction can probe the magnetic and chemical structures of materials, it is capable of detecting the simultaneous presence of ferromagnetic (FM) and antiferromagnetic (AFM) phase fractions in the form of unique magnetic reflections for each phase. In combination with an external magnetic field and the CHUF protocol, this technique would be an ideal probe to differentiate between various coexisting phases as well as observe the devitrification of a kinetically arrested state [1, 2].

We describe here the kinetic arrest phenomenon observed through in-field neutron diffraction measurements in an archetypal half-doped manganite, $La_{0.5}Ca_{0.5}MnO_3$ (LCMO) and a stoichiometric intermetallic compound Nd_7Rh_3. The systems were chosen as examples from a variety of compounds which show the first order magnetic phase transitions (FOPT). In these systems the field-induced magnetic transitions can be manipulated using external magnetic field as a second thermodynamic variable. In LCMO disorder is playing an important role in the phase transition and in Nd_7Rh_3 no chemical disorder is present.

ND experiments were carried out over a large region of field-temperature (*H, T*) space. Upon cooling, $La_{0.5}Ca_{0.5}MnO_3$ (LCMO) undergoes a transition from a paramagnetic state to FM state at ~ 225 K and then, enters a charge-ordered antiferromagnetic (AFM) state at ~ 150 K through a first-order phase transition [1, 9]. In-field ND experiments, employing the CHUF protocol, show that the FOPT in LCMO could be completely interrupted, resulting in a metastable kinetically arrested state which, on subsequent heating in a lower field, devitrifies into its stable low temperature equilibrium AFM phase before reentering the higher temperature FM phase.

Phase coexistence seems quite ubiquitous to materials like LCMO which exhibit disorder-influenced FOPT but similar studies on stoichiometric compounds are quite rare, with notable exceptions like Gd_5Ge_4 and Nd_7Rh_3 [2, 10, 11]. Magnetic studies on Nd_7Rh_3 indicate the existence of a magnetic-field-induced FOPT [10, 11]. Nd_7Rh_3 also undergoes two transitions, the first one which is from a paramagnetic state to an AFM state at 32 K and later at 16 K, it undergoes a FOPT to another AFM state at 16 K, as borne out from magnetization measurements [10, 11]. The magnetic structure at low temperature was shown to be incommensurate and sine-wave modulated along the c-axis and the magnetic moments lie in the hexagonal basal plane [2]. ND measurements performed in presence of an external magnetic field show that the transition to the AFM state could be completely arrested when cooled under a field of 3 Tesla, and upon warming in a field of 0.2 Tesla, there is evidence of partial devitrification of the kinetically arrested state, giving rise to a seemingly complex magnetic order.

2. Experimental details

$La_{0.5}Ca_{0.5}MnO_3$ sample was synthesized by the conventional solid-state route and phase purity was confirmed by XRD. Detailed neutron diffraction (ND) measurements were carried out under various temperature and magnetic field conditions at a wavelength of 1.48 Å using the position-sensitive detector based focusing crystal diffractometer PD-3 installed by the UGC-DAE CSR Mumbai Centre at the Dhruva reactor, Trombay [12]. Room temperature neutron diffraction patterns for both the samples were analysed by Rietveld method for determining single-phase formation. DC magnetization was measured using a commercial 9 T PPMS-VSM (Quantum Design). Further details related to the sample are described elsewhere [1]. Neutron diffraction measurements on Nd_7Rh_3 were carried out as a function of temperature and magnetic field at the diffractometer D1B, Institut Laue Langevin, Grenoble and diffractometer E6, Helmholtz-Zentrum Berlin by making a special arrangement of collimators and direct-beam stop, so that the Q-range could be explored from upwards of $0.027Å^{-1}$ at temperatures below and above the magnetic ordering [2].

3. Result and discussion

3.1. Half-doped charge ordered $La_{0.5}Ca_{0.5}MnO_3$

$La_{0.5}Ca_{0.5}MnO_3$ (LCMO) is one of the widely studied compounds among manganites whose properties have been well documented by several research groups [13-17]. But very few have studied the interesting magnetic field dependent history effects [5, 7]. In an earlier paper [1], we measured magnetization curves in different warming fields after cooling from 300 K in a 6 Tesla field (CHUF protocol), reproduced here in figure 1. It is seen that cooling in a field of 6 Tesla results in a kinetic arrest of the FM phase down to the lowest temperature. Upon application of the CHUF protocol, a devitrification of the kinetically arrested, metastable, ferromagnetic phase to the equilibrium antiferromagnetic state and a subsequent re-entrance to the stable ferromagnetic state at a higher temperature is observed. It is also clear from figure 1 that for the 0.01 T FCW curve, the kinetically arrested ferromagnetic phase devitrifies by about 90 % at 5 K. For higher warming fields, a lesser fraction of the metastable FM phase devitrifies at a higher temperature of 25 K into the stable low temperature antiferromagnetic phase. On further heating, the antiferromagnetic state re-enters the higher temperature ferromagnetic phase at ~230 K, and then finally goes into the paramagnetic phase. For warming fields above 3 T, the kinetically arrested phase no longer devitrifies at any intermediate temperature. ND measurements carried out in the (*H, T*) space reveal the microscopic nature of the coexisting phase fractions.

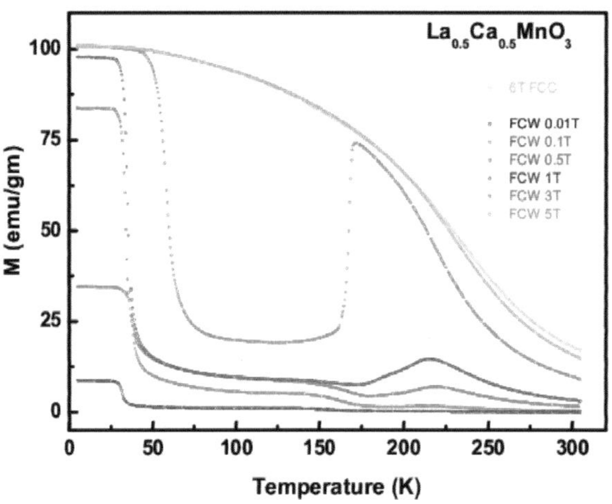

Figure 1: *Magnetization as a function of temperature employing the CHUF protocol (Reproduced from Ref. 1)*

Figure 2 shows the ND patterns of LCMO at 300 K and 9 K in zero field cooled condition and field-cooled in 7 Tesla at 9 K. Magnetic peaks corresponding to the CE-type AFM ordering are observed at 9 K at $2\theta \sim 14.03^{\circ}$ and 15.87°, which are characterized by the two Mn sub-lattices, Mn^{3+} and Mn^{4+} with propagation vectors [0.5 0 0] and [0.5 0 0.5], respectively [15]. The ND pattern at the bottom of the figure corresponds to the condition where a magnetic field of 7 T was applied at 300 K and then the sample was cooled down to 9 K. In this pattern, it is clear that the magnetic peaks pertaining to the CE-type AFM order are completely absent and a strong enhancement in the intensities of the nuclear peaks occurs, indicating a well-defined ferromagnetic ordering. This is a clear indication that the FOPT to the AFM state has been kinetically arrested due to the field cooling of the sample. Refinement of the ND pattern at 9 K cooled in 7 Tesla using the *Pnma* space group along with a magnetic phase with ferromagnetic order gives a moment of 4.16 μ_B on Mn. This is consistent with the value obtained from magnetization measurements. Subsequently, CHUF protocol was employed by reducing the field to 0.5 Tesla at 9 K and then, warming the sample in this field. ND patterns were collected at various temperatures during this warming cycle.

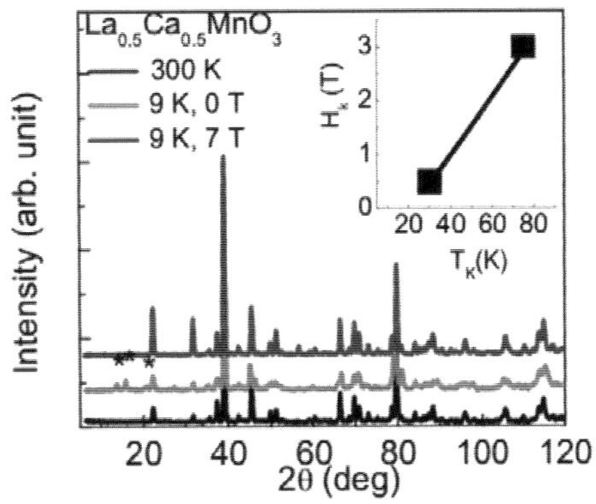

Figure 2: *ND plots of LCMO at (a) 300K, (b) 9 K measured in zero field and (c) in 7T cooling filed.*

The data are plotted in a sequence in figure 3. It is observed that up to 30 K, sample remains in a kinetically arrested, metastable ferromagnetic state as there is no signature of any change in the ND pattern to indicate any AFM ordering. On warming the sample beyond this temperature, the peaks corresponding to the CE-type AFM order (marked as asterisks) appear in the ND patterns and become well defined at 50 K, attaining roughly the same intensity as the ones observed in the zero field cooled case at 9 K. Appearance of these peaks signifies the de-arrest or devitrification of the arrested metastable FM phase and the setting in of the equilibrium AFM order. In the next cycle, the sample was cooled to 9 K in a field of 7 Tesla and then the warming field was set at 3 T (CHUF protocol). The sample remains in a kinetically arrested metastable ferromagnetic state at 9 K when the external field is reduced to 3 Tesla. However, upon warming in a field of 3 Tesla, the devitrification of the kinetically arrested state now occurs only around 75 K, as evidenced by the appearance of weak magnetic peaks attributable to CE-type AFM order. It is further observed that the intensities of the CE magnetic peaks are much lower as compared to the previous case of warming in 0.5 Tesla field, implying a much higher fraction of the kinetically arrested phase remaining throughout this measurement cycle. Therefore, the dotted line shown in the inset of figure 2, connecting the two points at which the CE-type AFM peaks appear is taken to be the devitrification line or the (H_K,T_K) line, separating the kinetically arrested phase from the equilibrium phase. The magnetic moment determined from refinement for the arrested FM phase at 100 K was calculated as 3.83 μ_B, while moments for the devitrified CE-AFM phase come out as 2.02 μ_B for both Mn^{3+} and Mn^{4+}. The AFM peaks disappear upon warming beyond 150 K, and the sample re-enters a fairly strong ferromagnetic phase at 175 K, consistent with the magnetization measurements. Refinement of the pattern at 175 K of this cycle gives a moment of 3.08 μ_B for the re-entrant ferromagnetic phase.

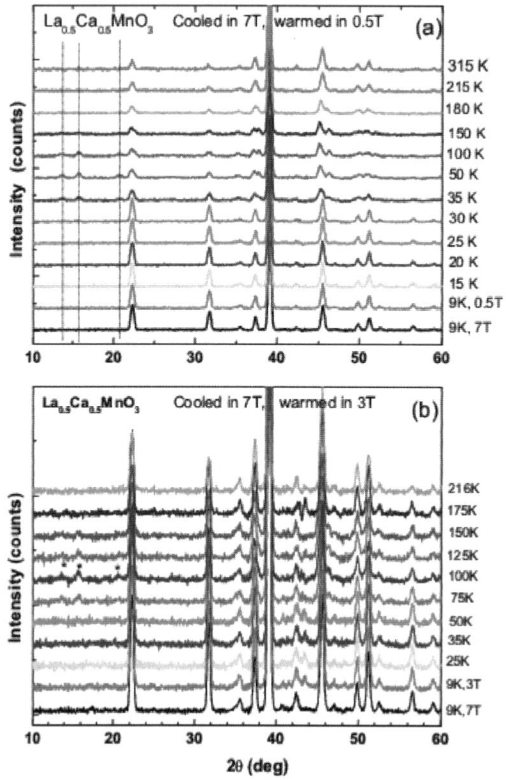

Figure 3: *LCMO ND patterns between 300 and 9 K under CHUF protocol (a) the sample is cooled under 7 T field and field was then reduced to 0.5 T and ND pattern at several temperature were recorded (b) represents ND patterns collected at several temperatures while the sample is cooled under 7 T field and field was then reduced to 3 T.*

In order to understand the microscopic situation in the sample, the Mn-O bond length and bond angles were calculated at various points in the (*H, T*) space and plotted in figure 4. It is well understood that the Mn–O bond lengths in LCMO at room temperature and in the CE-AFM phase are quite different due to Jahn–Teller distortion. It can be reasonably expected that the kinetically arrested FM phase would have almost undistorted MnO6 octahedra, with all the Mn–O distances being approximately the same. Mn–O bond lengths and Mn–O–Mn bond angles at various temperature and field combinations, show some interesting observations. The values for 300 and 9 K show the expected changes in the bond lengths and angles at low temperature. However, for the kinetically arrested FM phase at 9 K, obtained by cooling in a field of 7 Tesla, and also when the field is reduced to 0.5 T (9 K, 0.5 T), the bond angles and lengths do not show any major differences from the 300 K values. A direct effect of the devitrification of the arrested phase at 35 K, 0.5 Tesla is seen in the Mn–O1 bond length and Mn–O1–Mn bond angle, which are comparable to the values of the (9 K, 0 T) set. A similar behaviour is seen at 100 K in the 3 T heating cycle. However, at 175 K warmed in 3 Tesla field, there is a re-entrant transition to the equilibrium FM phase, which again shows an undistorted situation, with the bond angles and lengths associated with O1 and O2 oxygens being similar.

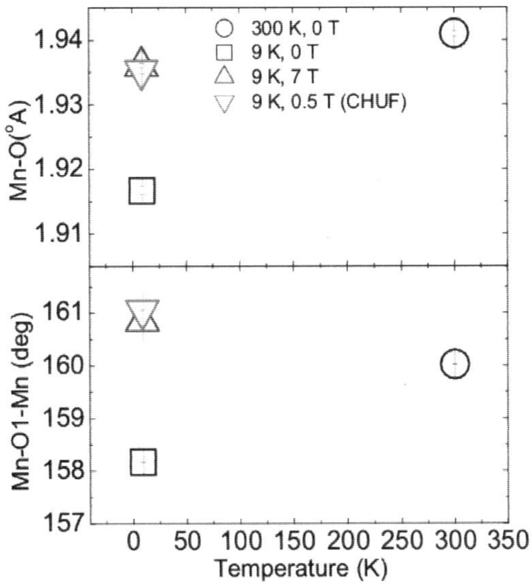

Figure 4: *(a)Mn-O bond length and (b) Mn-O1-Mn bond angles for LCMO between 300 and 9 K under CHUF protocol*

3.2. Nd_7Rh_3

Nd_7Rh_3 exhibits two magnetic transitions, one at (T_{N1} =) 32 K and another around (T_{N2} =) 16 K. The uniqueness of magnetic ordering in this compound has been explained earlier [2]. When the sample is cooled below the magnetic ordering temperatures (T_{N1} and T_{N2}), an intense AFM peak appears along with an enhancement in the intensities of some nuclear Bragg peaks indicating the coexistence of AFM and FM ordering in this system [2]. In figure 5, only the low-Q region, predominantly the AFM part, is focussed upon. The intense AFM peak at 1.6 K (well below the two AFM transitions and cooled in zero field) and centred at Q ~ 0.07Å^{-1}, can be observed (filled squares).

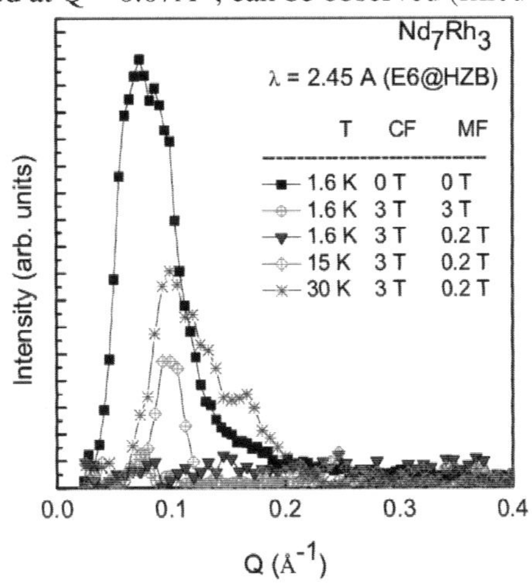

Figure 5: *ND patterns of Nd_7Rh_3 measured employing the CHUF protocol.*

In the next step, the sample was cooled from 70 K to 1.6 K in a field of 3 Tesla and the ND pattern recorded at this stage is shown with red circles in figure 5. It can be clearly observed in this case that the strong AFM peak is completely suppressed, indicating that both the AFM transitions have been interrupted and the system goes into a kinetically arrested metastable state. Then, the field was reduced to 0.2 Tesla after cooling in a field of 7 Tesla and the sample was warmed and ND patterns were recorded at different temperatures around T_{N1} and T_{N2}. As the temperature reaches the vicinity of T_{N2} (~ 16 K), there is a gradual increase in the intensity of the AFM peak suggesting that the sample has partly devitrified into its equilibrium low temperature AFM phase. On increasing the temperature further to 30 K, the magnetic peak becomes broadened and slightly stronger than it was at 15 K. An important observation from this figure is that the magnetic peak position in the devitrified state shifts from Q ~ 0.07 Å^{-1} to ~ 0.1 Å^{-1}, clearly signifying that the underlying magnetic order is quite complex.

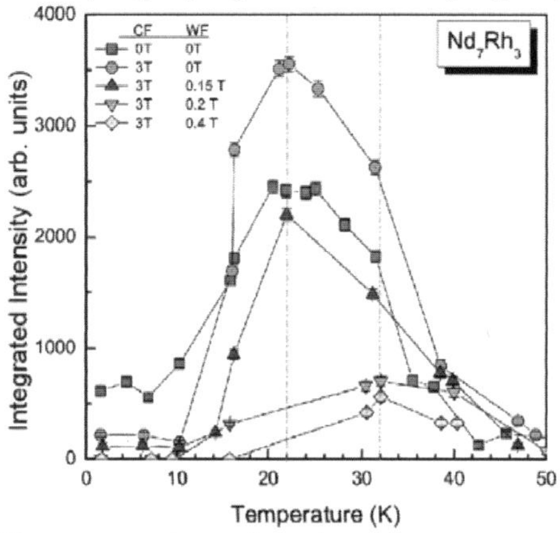

Figure 6: *ND patterns of* Nd_7Rh_3 *measured as a function of temperature in using the CHUF protocol.*

In figure 6, we plot the variation of the integrated intensity of the AFM peaks observed in the Q-range ~ 0.03 - 0.2 Å^{-1}. For the ND measurements carried out in zero field, the intensity of the magnetic peak (red squares) increases in the vicinity of T_{N2}, reaches maximum at around T_{N1} and then reduces to about 1/3rd of its peak value at T = 1.6 K. However, when the sample is cooled in 3 Tesla field (CF), and then warmed in zero field, the devitrification (indicated by increase in the intensity of the AFM peak) happens in the vicinity of 10 K, and at T_{N1}, it reaches its peak value. The process of devitrification shifts to higher temperatures when the sample is cooled in 3 Tesla (CF) and warmed in different fields (WF) as clearly seen in the figure. With the increasing strength of the field under which sample is warmed, the recovery of the magnetic peak intensity happens at higher temperatures, and as can be seen in the case of measurements done while warming in 0.2 Tesla and 0.4 Tesla fields, the full magnetic peak is observed above T_{N2}. The present experimental data clearly show that in-field neutron diffraction measurements using the CHUF protocol throw up some interesting possibilities of manipulating magnetic order. It would be worthwhile to carry out detailed experiments on single crystalline Nd_7Rh_3 to understand the field-induced changes to the magnetic structure and their implications on magnetism of this compound.

4. Conclusion

In-field neutron diffraction measurements performed on two systems that exhibit magnetic FOPT unambiguously demonstrate (i) the phenomenon of kinetic arrest, wherein the FOPT gets interrupted and the system enters a metastable arrested phase, (ii) that the kinetically arrested and equilibrium phases coexist and their phase fractions could be determined using the CHUF protocol, and (iii) de-arrest or devitrification of the arrested metastable phase in to the equilibrium phase in a manner similar to that found in structurally disordered glassy systems [19].

References

[1] Siruguri V, Babu P D, Kaushik S D, Biswas A, Sarkar S K, Madangopal Krishnan and Chaddah P, 2013 J. Phys.: Condens. Matter **25** 496011 (and references therein)
[2] Rayaprol S, Siruguri V, Hoser A, Ritter C, and Sampathkumaran E V, 2014 Phys. Rev. B **90** 134417 (and references therein)
[3] Banerjee A, Dash S, Lakhani A, Chaddah P, Chen X and Ramanujan R V, 2011 Solid State Commun. **151** 971
[4] Schiffer P, Ramirez A P, Bao W, and Cheong S-W, 1995 Phys. Rev. Lett. **75** 3336
[5] Banerjee A, Kumar K and Chaddah P, 2008 J. Phys.: Condens. Matter **20** 255245
[6] Loudon J C, Mathur N D and Midgley P A, 2002 Nature **420** 797; Dagotto E, Hotta T, and More A, 2001 Phys. Rep. **344** 1
[7] Chaddah P, Kumar K and Banerjee A, 2008 Phys. Rev. B**77** 100402(R); Banerjee A, Kumar K, and Chaddah P, 2009 J. Phys.: Condens. Mater **21** 026002
[8] Lakhani A, Banerjee A, Chaddah P, Chen X and Ramanujan R V, 2012 J. Phys.: Condens. Matter **24** 386004 (and references therein)
[9] Radaelli P G, Cox D E, Marezio M and Cheong S-W, 1997 Phys. Rev. B **55** 3015
[10] Sengupta K and Sampathkumaran E V, 2006 Phys. Rev. B **73** 020406
[11] Sengupta K and Sampathkumaran E V, 2006 J. Phys.: Condens. Mater **18** L401
[12] Pimpale A V, Dasannacharya B A, Siruguri V, Babu P D and Goyal P S, 2002 Nucl. Instrum. Methods A **481** 615
[13] Babu P D, Das A and Paranjpe S K, 2001 Solid State Commun.**118** 91
[14] Das A, Babu P D, Chatterjee S and Nigam A K, 2004 Phys.Rev. B **70** 224404
[15] Wollan E O and Koehler W C, 1955 Phys. Rev. **100** 545
[16] Goodenough J B, 1955 Phys. Rev. **100** 564
[17] Rodriguez E E, ProffenTh, Llobet A, Rhyne J J and Mitchell J F, 2005 Phys. Rev. B **71** 104430
[18] Tyson T A, Deleon M, Croft M, Harris V G, Kao C-C, Kirkland J and Cheong S-W, 2004 Phys. Rev. B **70** 24410
[19] Wu W, Israel C, Hur N, Park S, Cheong S-W and de Lozanne A, 2006 Nature Materials **5**, 881

Peculiarities of neutron waveguides with thin Gd layer

Yu Khaydukov[1,2,3], E Kravtsov[4,5], V Progliado[4], V Ustinov[4], Yu Nikitenko[6], T Keller[1,2], V Aksenov[7], and B Keimer[1]

[1] Max-Planck-Institut für Festkörperforschung, D 70569 Stuttgart, Germany
[2] Max Planck Society Outstation at the FRM-II, D-85747 Garching, Germany
[3] Skobeltsyn Institute of Nuclear Physics, 119991 Moscow State University, Moscow, Russia
[4] Institute of Metal Physics, 620990 Ekaterinburg, Russia
[5] Ural Federal University, 620002 Ekaterinburg, Russia
[6] Joint Institute for Nuclear Research, 141980 Dubna, Russia
[7] Petersburg Nuclear Physics Institute, 188300 Gatchina, Russia

E-mail: y.khaydukov@fkf.mpg.de

Abstract. Peculiarities of the formation of a neutron enhanced standing wave in the structure with a thin highly absorbing layer of gadolinium are considered in the article. An analogue of the *poisoning effect* well known in reactor physics was found. The effect is stronger for the Nb/Gd/Nb system. Despite of this effect, for a Nb/Gd bilayer and a Nb/Gd/Nb trilayer placed between Al_2O_3 substrate and Cu layer, it is shown theoretically and experimentally that one order of magnitude enhancement of neutron density is possible in the vicinity of the Gd layer. This enhancement makes it possible to study domain formation in the Gd layer under transition of the Nb layer(s) into the superconducting state (cryptoferromagnetic phase).

1. Introduction

Gadolinium due to the presence of a unique set of magnetic and nuclear properties is widely used in basic research and industrial application (electronics, nuclear energetic, medical and laser technique etc). One further possibly interesting application of Gd is cryptoferromagnetism (CFM) [1]-[3]. A sufficiently thin layer of a weak ferromagnet (F) being in contact with a superconductor (S) is predicted to go into a domain state below the superconducting transition temperature T_C. The domain size in CFM phase is comparable to the superconducting correlation length ξ_S, so that the effective exchange field of the ferromagnet on the size of the Cooper pair is close to zero, thereby providing the coexistence of superconducting and ferromagnetic orders. A phase transition into the CFM phase is possible in ferromagnets with the Curie temperature T_m not much exceeding the superconducting transition temperature T_C. Gadolinium as F layer being in contact with superconducting niobium is predicted to be one of the best candidates for observing this effect [3]. The Curie temperature of bulk Gd is T_m = 293K. However, it can be suppressed in thin films down to 20-40K which is comparable with $T_C \sim$ 10K of Nb layer. Moreover, Gd and Nb are a well known pair of metals which are not solvable, allowing thus creating a sharp Gd/Nb interface [4]. Another distinctive feature of the Gd/Nb systems prepared by magnetron sputtering is high transparency of the S/F interface for Cooper pairs [4],[5].

The most direct method for detecting the CFM phase transition is Grazing Incidence Small Angle Neutron Scattering (GISANS, see e.g. review [6]). However, taking into account the small thickness of the F layer and the small size of domains ($\xi_S \sim$ *10-100nm),* we estimate the GISANS intensity as being $I_{GISANS} \sim 10^{-6}\, I_0$ (I_0 is intensity of the direct beam*)* which is below the typical background limit of neutron instruments. In order to enhance the weak intensity of GISANS scattering from thin F layers a waveguide enhancement of neutron density can be used [5]. To do so, the Nb/Gd system should be placed between layers with high neutron scattering length densities (SLDs). The difficulty of neutron studies of systems with gadolinium is the large neutron absorption cross sections for isotopes ^{157}Gd and ^{155}Gd which have 16% and 15% concentration in the natural mixture of isotopes. The aim of this article is the design of neutron waveguides with Nb/Gd taking into account all the features of Gd interaction with neutrons.

2. Samples design

For optimization procedure a S/F bilayer Nb(d_S)/Gd(d_F) and S/F/S Nb(d_{S1})/Gd(d_F)/Nb(d_{S2}) trilayer systems were considered. While S/F bilayers are simpler to prepare in the technological sense, trilayers S/F/S, due to the presence of superconductors on both interfaces of the F layer may provide more Cooper pair inside it. The thickness of the superconductor must be chosen to be large enough to create superconductivity with T_C close to its bulk value. Following to works [4],[5] we have chosen d_S = 50nm for the S/F bilayer and $d_{S1} = d_{S2}$= 25nm for the S/F/S trilayer. To ensure penetration of Cooper pairs in the whole F layer the Gd layer thickness must be chosen to be of the order of the superconducting correlation length ξ_F in the F layer ($\xi_F \approx$ 1.4nm for Gd, see [4]). On the other hand d_F must be thick enough to guarantee homogeneous deposition of the F layer. Taking into account these considerations we have fixed d_F = 2nm. Next, the investigated bilayer Gd(2nm)/Nb(50nm) and trilayer Nb(25nm)/Gd(2nm)/Nb(25nm) have to be placed between layers with high SLDs to create a waveguide structure. As the substrate we have chosen sapphire since it has high SLD ($\rho = 5.7\times10^{-4}$ nm^{-2}) and allows growing structures with small roughness [9]. As the second layer copper can be used with $\rho = 6.6\times10^{-4}$ nm^{-2}. To protect the copper layer against oxidation a capping layer of Ta(3nm) was considered in calculations.

The presence of the waveguide enhancement in systems with absorbing and ferromagnetic Gd layer can be seen either via dips on the total reflection plateau of non-spin-flip (NSF) reflectivities [10],[11] or as peaks in the spin-flip (SF) channel [7]. In the latter case the vector of magnetization should lie in the sample plane and be normal to the external magnetic field. For our calculations we assumed the Gd magnetization, being equal to its bulk value of 20kG, to be turned normally to the external magnetic field (see insets to Figure 1). To describe the high absorption of Gd, the non-zero imaginary part of the neutron coherent scattering length should be taken into account. We used value b_{Gd} = 6.0-11i fm [13] and bulk density N_{Gd} = 30 nm^{-3} which gives us the scattering length density of Gd $\rho = (1.8 - 3.3i)\times10^{-4}$ nm^{-2}.

To optimize the structures we have calculated the minimum of NSF reflectivity R^{++} and maximum of SF reflectivity R^{+-} varying the thickness of the Cu layer d_l (Figure 1). The thickness dependence of the min(R^{++}) and max(R^{+-}) for the structure Ta(3nm)/Cu(d_l)/Gd(2nm)/Nb(50nm)/Al_2O_3 is shown in Figure.1a. One can see that both min(R^{++}) and max(R^{+-}) have the best values at d_l = 17nm. The thickness of the upper layer of the order of tens nanometers is typical for such waveguide structures [7],[8],[14]. In contrast, for S/F/S structure Ta(3nm)/Cu(d_1)/Nb(25nm)/Gd(2nm)/Nb(25nm)/Al_2O_3 the effect is maximized at a much smaller value d_l = 4nm. The reason for this lies in the large neutron capture cross-section for gadolinium. The effect is somehow similar to the well-known in nuclear reactor technique the *reactor poisoning* effect, when the presence of highly absorbing isotopes in the active zone of a reactor prevents the formation of neutron density sufficient for reactor operation. In our case the neutron capture by Gd atoms prevents appearing enhanced by the waveguide structure neutron density. To check this we calculated the dependence of max(R^{+-}) vs d_1 for the structure with non-absorbing Gd (see dashed line in Figure 1b). Indeed, one can see that for this case a stronger enhancement can be observed for much thicker Cu layers.

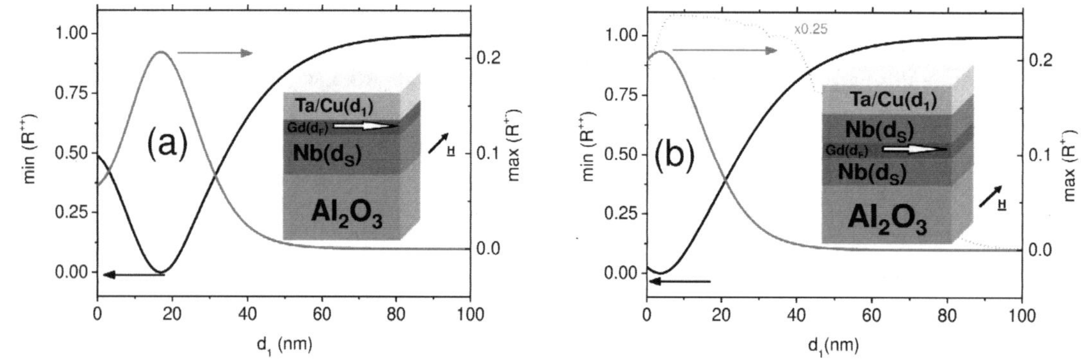

Figure 1. (a) Dependence of the dip on the total reflection plateau for non-spin-flip R^{++} curve (black) and peak maximum for spin-flip R^{+-} curve (red) for the S/F bilayer Ta(3nm)/Cu(d_1)/Gd(2nm)/Nb(50nm)/Al_2O_3 and (b) S/F/S trilayer Ta(3nm)/Cu(d_1)/Nb(25nm)/Gd(2nm)/Nb(25nm)/Al_2O_3. Dashed line is max(R^{+-}) for the same structure but without imaginary part of SLD in Gd. Note that this line is divided by 4. Insets show sketches of the structures.

3. Sample preparation and experimental results

The samples of nominal structure Ta(3nm)/Cu(17nm)/Gd(1-3nm)/Nb(50nm) and Ta(3nm)/Cu(4nm)/Nb(25nm)/Gd(1-3nm)/Nb(25nm) were prepared using an UHV magnetron machine ULVAC MPS-4000-C6 at constant current onto Al_2O_3(1-102) substrates. Before the deposition the substrates were cleaned from organic contaminations with acetone and alcohol. The substrates were further cleaned with reverse magnetron sputtering (2 minutes at argon flow rate of 25 sccm) in the load chamber. The base pressure was lower than 2×10^{-9} mbar. Pure argon gas (99.9998% purity) at flow rate of 25 sccm was used as sputter gas. The deposition was carried out at room temperature (about 25 °C) at magnetron sputtering power of 100 Watts in argon atmosphere of 1×10^{-3} mbar. In these conditions Nb, Gd, Cu, and Ta layers were sputtered at deposition rates of 2.35 nm/min, 6.85 nm/min, 6.45nm/min, and 2.8 nm/min, respectively. The deposition rates were calibrated using test samples with the help of Zygo NewView7300 white light interferometer.

The PNR experiment was conducted on the angle-dispersive reflectometer NREX at the research reactor FRM-II in Garching, Germany. A polarized neutron beam with wavelength 4.26 ± 0.06 Å and polarization 99.99% falls on the sample under grazing incidence angles $\theta_1 = [0.25 - 4]°$. The divergence of the beam was set to $\Delta\theta_1 = 0.025°$ by two slits before the sample. A position sensitive detector was used to measure for every θ_1 intensity of scattered neutrons for a range of outgoing angles θ_2. By switching the spin-flipper state placed before the sample spin-up and spin-down reflectivities were measured. Two samples Ta(3nm)/Cu(17nm)/Gd(3nm)/Nb(50nm) (Gd/Nb) and Ta(3nm)/Cu(4nm)/Nb(25nm)/Gd(3nm)/Nb(25nm) (Nb/Gd/Nb) were measured. Before the measurement the Gd/Nb (Nb/Gd/Nb) sample was cooled down to $T = 9$K (6K) in magnetic field $H = 4.5$kG applied parallel to the surface and normal to the scattering plane. After the cooling down the magnetic field was set to $H = 3$kG (4kG). Temperatures and magnetic fields were chosen to be above T_C of the S layer and the saturation field of the F layer, which were determined in advance with SQUID magnetometry. Both samples have shown difference between the spin-up R^+ and spin-down R^- reflectivities which gives the direct evidence of the presence of ferromagnetism in the Gd layer.

However, since magnetic and superconducting properties are out of the scope of this article, we will discuss them elsewhere.

Figures 2a,b show 2D scattering maps for spin-down neutrons for the Gd/Nb and Nb/Gd/Nb samples correspondingly. Tilted line at the condition $\theta_2 = \theta_1$ corresponds to the specular reflection. The presence of the waveguide enhancement can be seen by the dip on the total reflection plateau at the angle $\theta_{res} = 0.3°$ for both Gd/Nb and Nb/Gd/Nb samples. This dip can be more clearly seen in the insets to Figures 2b,c where the reflectivity curves are presented. Apart of the dip, the resonant enhancement is proved by the presence of channeled beam ($\theta_2 \approx 0$) [12].

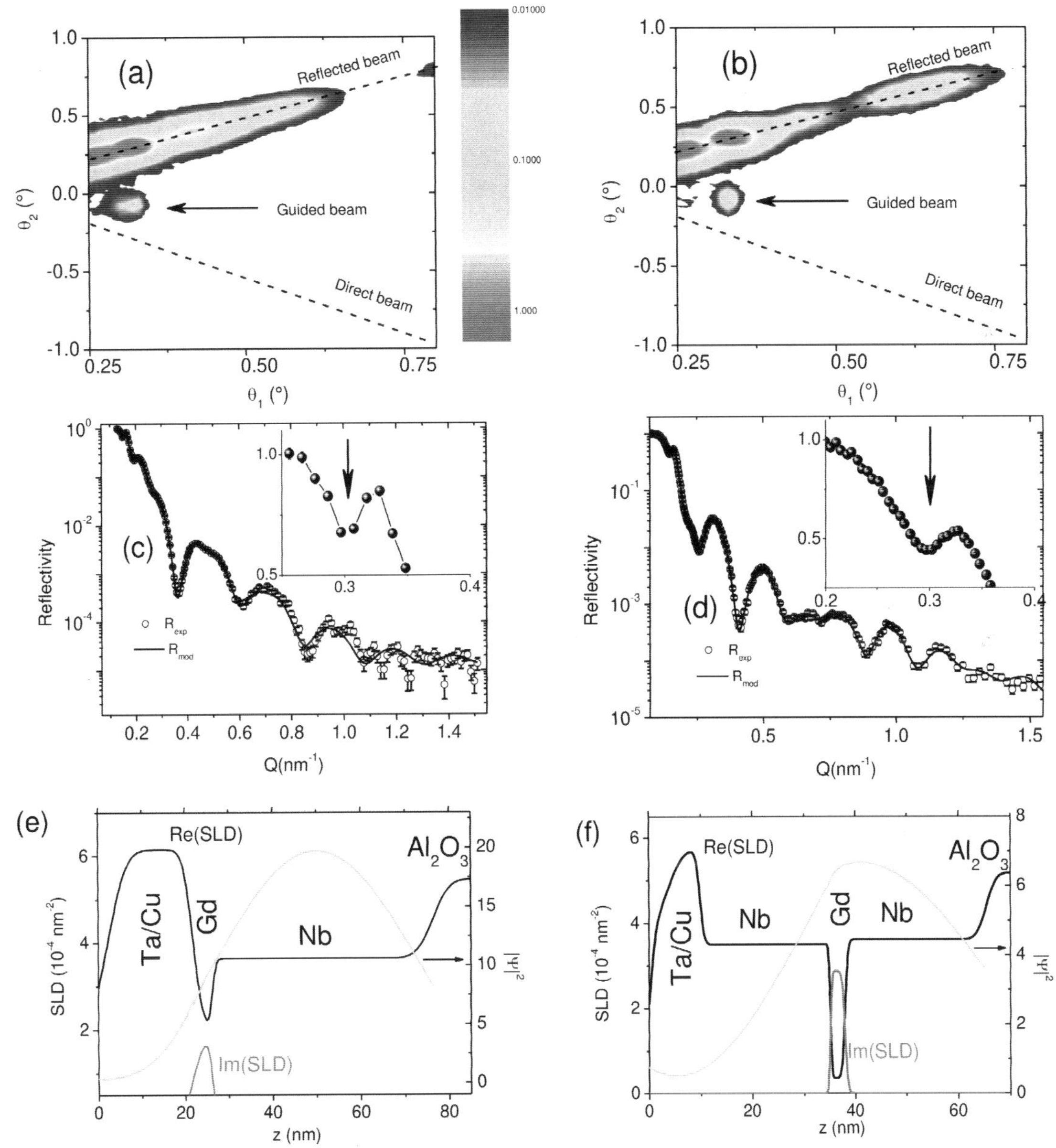

Figure 2. Scattering maps for the Gd/Nb (a) and Nb/Gd/Nb (b) samples. Experimental (dots) and model (solid lines) reflectivity curves for the Gd/Nb (a) and Nb/Gd/Nb (b) samples. Insets show the experimental curves around the resonance (shown by arrows) as a function of θ_1. The depth profiles of SLD and neutron density for the Gd/Nb sample (e) and Nb/Gd/Nb sample (f).

4. Discussions and conclusion

Specular reflectivities were fit to the model curves. Best-fit curves are shown in Figure 2c,d by solid lines. The resulting depth profiles of SLD are shown in Figures 2e,f. From the fit it follows that the real thicknesses of the Nb and Gd layers are close (within 10% maximum) to their nominal values. The fit also shows a rather small rms roughness of the Gd/Nb interface ($\sigma = 0.7$nm for the Gd/Nb and $\sigma = 0.4$nm for the Nb/Gd/Nb sample). The imaginary part of SLD for gadolinium in Gd/Nb and Nb/Gd/Nb samples are 50% and 88% of bulk value, respectively.

Having the SLD depth profile known we can calculate the depth profiles of the neutron density in the waveguide mode (Figure 2e,f). One can see that the neutron density in both cases has a maximum in the center of Nb layer and makes value $|\Psi|^2 = 20$ and $|\Psi|^2 = 7$ for the Gd/Nb and Nb/Gd/Nb samples. This small enhancement is explained by the *poisoning effect* as described in Section 2. However, despite of this effect, the neutron density around the Gd layer in both cases is similar for both systems ($|\Psi|^2 \approx 7$-9). Such an enhancement will allow GISANS intensity from magnetic domains to be increased up to a detectable level of $I_{GISANS} \sim 10^{-5}$-$10^{-4} I_0$.

In conclusion, the features of the formation of the enhanced standing wave in the structure with a thin highly absorbing layer of gadolinium are considered in the article. The analogue of the *poisoning effect* well known in reactor physic was found. Effect is higher when Gd layer is placed in the middle of the waveguide. Despite of this effect for Nb/Gd bilayer and Nb/Gd/Nb trilayer placed between the Al_2O_3 substrate and Cu layer it was shown theoretically and experimentally that one order of magnitude enhancement of neutron density is possible in the vicinity of Gd layer. This enhancement makes it possible to study domain formation in Gd layer under transition of Nb layer(s) into superconducting state (cryptoferromagnetic phase).

The authors would like to thank A.I. Frank, G. Kulin and S. Kozhevnikov for fruitful discussion. This work is based upon experiments performed at the NREX instrument operated by Max-Planck Society at the Heinz Maier-Leibnitz Zentrum (MLZ), Garching, Germany and partially supported by DFG collaborative research center TRR 80 and RFBR (Projects 14-22-01007, 14-22-01063).

References

[1] P.W. Anderson and H. Suhl, Phys. Rev. 116, 898 (1959)
[2] L.N. Bulaevskii, A. I. Buzdin, Zh. Eksp. Teor Fiz 94, 256 (1988)
[3] F.S. Bergeret, K.B. Efetov, A. I. Larkin, Phys. Rev. B 62, 11872 (2000)
[4] J.S. Jiang, D. Davidovic, D. H. Reich, and C.L. Chien, Phys. Rev. Lett. 74, 314 (1995)
[5] J.S. Jiang, D. Davidovic, D. H. Reich, and C.L. Chien, Phys. Rev. B 54, 6120 (1996)
[6] Peter Müller-Buschbaum, Polymer Journal 45, 34-42 (2013)
[7] Yu.N. Khaydukov, Yu.V. Nikitenko, L. Bottyan et al, Cryst. Rep. 55(7),1235-1247, 2010
[8] Yu N Khaydukov, V.L Aksenov et al, J. of Supercond Nov Magn 24, 961 (2011)
[9] Yu. Khaydukov, R. Morari, O. Soltwedel et al, arXiv:1508.02223
[10] H. Zhang, P.D. Galagher, S.K. Satija et al., Phys. Rev. Lett. 72, 3044 (1994)
[11] V.L. Aksenov, L. Cser, N.A. Gundorin et al., Physica B 276-278, 809 (2000)
[12] F. Pfeiffer, V. Leiner, P. Høghøj, and I. Anderson, Phys. Rev. Lett. 88, 055507 (2002)
[13] A.I. Frank, V.I. Bodnarchuk, P. Geltenbort et al., Phys. Atom. Nucl. 66, 1831 (2003)
[14] Yu. N. Khaydukov, G. A. Ovsyannikov et al., Phys. Rev. B 90, 035130 (2014)

Development of a simultaneous SANS / FTIR measuring system and its application to polymer cocrystals

F. Kaneko[1], N. Seto[1], S. Sato[1], A. Radulescu[2], M. M. Schiavone[2], J. Allgaier[3] and K. Ute[4]

[1] Graduate School of Science, Osaka University, Toyonaka, Osaka, 560-0043, Japan
[2] JCNS, Outstation at MLZ, Lichtenbergstraße 1, 85747, Garching, Germany
[3] JCNS and ICS, Forshungszentrum Jülich GmbH, 52425 Jülich, Germany
[4] Dep. of Chem. Sci. and Tech., The University of Tokushima, Tokushima 770-8506, Japan.

toshi@chem.sci.osaka-u.ac.jp

Abstract. In order to provide plenty of structure information which would assist in the analysis and interpretation of small angle neutron scattering (SANS) profile, a novel method for the simultaneous time-resolved measurement of SANS and Fourier transform infrared (FTIR) spectroscopy has been developed. The method was realized by building a device consisting of a portable FTIR spectrometer and an optical system equipped with two aluminum coated quartz plates that are fully transparent to neutron beams but play as mirrors for infrared radiation. The optical system allows both a neutron beam and an infrared beam pass through the same position of a test specimen coaxially. The device was installed on a small angle neutron diffractometer, KWS2 of the Jülich Centre for Neutron Science (JCNS) outstation at Heinz Maier-Leibnitz Center (MLZ) in Garching, Germany. In order to check the performance of this simultaneous measuring system, the structural changes in the cocrystals of syndiotactic polystyrene during the course of heating were followed. It has been confirmed that FTIR spectra measured in parallel are able to provide information about the behavior of each component and also useful to grasp in real time what is actually happening in the sample system.

1. Introduction

Small angle neutron scattering (SANS) has been employed in research that aims at analyzing the mesoscopic scale structures. In combination with partial deuteration technique, SANS is able to characterize the structure of a specific part in a complex system, which makes it unsubstitutable with any other method for structural analysis. SANS has established its unique position as a powerful tool for structural investigation of soft-matter and biological systems. Furthermore, owing to the technological development in instrumentation in recent years such as detectors and neutron guides, SANS has extended its presence not only in the research of static structure but also in the research of time-dependent structure evolution.

The scattering pattern of SANS is determined by the scattering length density (SLD) profile in the object under study. Therefore, it is of great difficulty to derive the sole solution for the arrangement of constituent components only from the scattering pattern of SANS, in particular, when the system of interest is a multicomponent system [1, 2]. This issue increases in importance as the system under research becomes more complicated. If we get any other structural information from the same system, it would reduce greatly the difficulty of analyzing the SANS data.

For this purpose, the simultaneity between data obtained by different kinds of methods is considered to be crucial in the study of time-evolving structure changes of soft materials and biological samples. These systems are so sensitive that it is difficult to reproduce the time course of structural changes completely. Therefore, it is desirable to obtain two or more different kinds of structural information simultaneously from the same sample. Vibrational spectroscopy is a strong candidate for this purpose; indeed it has been employed as a complementary tool of X-ray and neutron diffractometry for investigating the structure of polymer systems. It provides information about the concentration and conformational state of each chemical species in a measuring object. Furthermore, the partial deuteration method for SANS experiments provides another advantage; the vibrational modes of deuterated components shift to low frequencies greatly, eliminating the overlap of the bands due to different chemical species, which makes the analysis and interpretation of spectral data easier. As for synchrotron radiation, simultaneous analyzes by X-ray scattering and Fourier transform infrared (FTIR) spectroscopy have been developed so far [3, 4, 5]. Similarly, the introduction of FTIR spectroscopy to SANS would produce a fruitful methodology for such a simultaneous measurement system.

Owing to a high throughput optical system, a highly sensitive detector, and advanced hardware and software design, it is easy to follow the time evolution of infrared spectra with a compact FTIR spectrometer [6, 7]. Using these advantages of FTIR spectroscopy, we have been developing a simultaneous SANS/FTIR measuring system at the SANS diffractometer KWS2 of the Jülich Centre for Neutron Science (JCNS), outstation at Heinz Maier-Leibnitz Center (MLZ) [8, 9].

In this paper, we are going to present the concept and actual construction of the simultaneous measuring system and also some experimental examples concerning the cocrystals of syndiotactic polystyrene (sPS). sPS is a stereoregular polymer, which has a unique property to form cocrystals with a variety of chemical compounds, by including them into the crystalline lattice. This unprecedented property attracts a wide range of attention, in particular, in the viewpoint of material development [10, 11]. In this research, we studied how the molecular weight of guest molecules affects the structural changes of the sPS cocrystal system on heating process. We will show how the FTIR spectrum measured in parallel with SANS is valuable for interpreting SANS profile changes. It is able to provide how each component in the sample behaves and also monitor chemical structural changes in real time, including cases that happen accidentally, in the sample being measured.

Figure 1. Experimental setup for simultaneous SANS/FTIR measuring system installed at the SANS diffractometer KWS-2 in MLZ (Garching, Germany).

2. Simultaneous SANS/FTIR measurement system

The device for simultaneous SANS/FTIR measuring system consists of a compact portable FTIR spectrometer (PerkinElmer, Spectrum Two) and an optical system of our own making. The device was designed to be installed in the sample chamber of a SANS instrument as show in Figure 1. The concept of the simultaneous system is depicted in Figure 2(a) and the actual construction of the optical system is given in Figure 2(b). The optical system consists of six mirrors (optical elements 1 to 6), which can be grouped into two types by function. Elements 3 and 4, which are irradiated by both the neutron beam (green line) and infrared beam (yellow line), work as a beam mixer and selector, respectively. They transmit the neutron beam efficiently but act as mirrors to reflect the infrared beam. To achieve such functions, elements 3 and 4 are made by aluminum vacuum deposition on optical grade synthetic quartz plates with thickness of 0.5 mm using the characteristics of aluminum and quartz: both are transparent to neutron beam, but the surface layer of aluminum reflects infrared radiation with high efficiency. The other four elements work as mirrors only for the infrared beam. As for the windows (elements 7 and 8) of the sample cell with a brass body wound with a copper tube to temperature control, we used KBr plates of 2 mm thickness because KBr is transparent to in the middle infrared region (4000-400 cm^{-1}) and also has a relatively good transparency to neutron beam.

The two beams in the system are made to run on the same line by element 3 and pass through the same sample position coaxially, and then they are separated from each other with element 4. Each separated beam goes to its own detector system.

Although the large part of the IR path is in open air, the influence of H_2O and CO_2 in the air is efficiently removed through the digital filtering algorithm of the FTIR spectrometer.

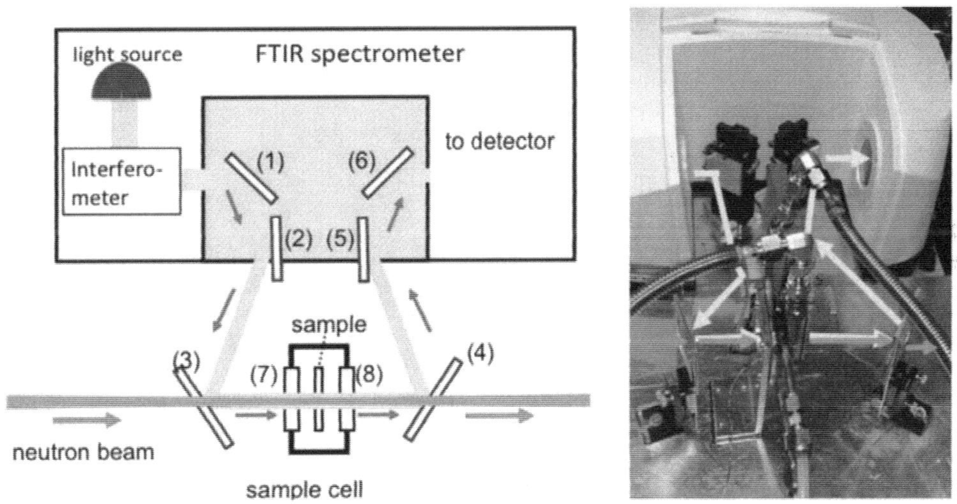

(a) (b)

Figure 2. (a) Schematic drawing of the optical system for simultaneous SANS / FTIR measuring system. Parts 1 to 6 are mirrors and parts 7 and 8 are windows of the sample cell. The infrared and neutron beams are represented with yellow and green lines. (b) The photographic image of the optical system.

3. Experimental

3.1. Sample

In order to confirm the performance of the simultaneous measurement system, the specimens of sPS cocrystal with polyethylene glycol dimethyl ethers (PEGDMEs) were prepared by the following procedures. In order to depress strong incoherent scattering from the sPS matrix, fully deuterated sPS (d-sPS) (Mw=1.1×105 and Mw/Mn = 1.9) was synthesized according to the coordination polymerization developed by Ishihara et al. [12], using fully deuterated styrene with purity more than 98% purchased from Cambridge Isotope Lab. Two kinds of PEGDMEs with different molecular weights were used. One was a PEGDME with molecular weight of 178 (PEGDME178) and the other was a PEGDME with molecular weight of 500 (PEGDME500), both of which were purchased from Sigma-Aldrich and used without further purification.

Uniaxially oriented amorphous d-sPS films about 50 μm thick were prepared by drawing melt quenched amorphous d-sPS films four times in an oil bath kept at 100°C, and then the films were exposed to a vapor of chloroform to obtain sPS/chloroform cocrystal films. The cocrystal films were soaked into liquid PEGDME178 for one day and then kept in a vacuum oven at 40°C for an hour, giving uniaxially oriented sPS/PEGDME178 cocrystal films. The samples of sPS/PEGDME500 cocrystal were prepared by the same procedure, except that liquid PEGDME178 was substituted with a PEGDME500/chloroform mixture at about 1:1 in weight ratio.

3.2. Measurements

All the simultaneous SANS/FTIR measurements were carried out by using the KWS2 diffractometer of the Jülich Centre for Neutron Science (JCNS), outstation at Heinz Maier-Leibnitz Center (MLZ) in Garching, Germany. Scattering data were obtained using a 2D detector with active area of 60 × 60 cm^2 and 128 × 128 channels. A wavelength λ= 0.7 nm ($\Delta\lambda/\lambda$ = 20%) and a sample-to-detector distance of 4 m and 1.35 m were chosen. The typical measured sample area was about 5 × 5 mm^2. The data accumulation time for each data point was 15 min. The temperature of the sample cell was controlled within the accuracy of ±0.2 °C by circulating thermostated oil. The sample 3 kept under a slow flow of air. Transmission IR spectra were taken at a resolution of 2 cm^{-1} and a 10 min interval. The average accumulation time and the number of scans were 10 min and 128, respectively. For measuring time dependence of IR spectra and analyzing them, a commercially available software (PerkinElmer, Timebase) is employed.

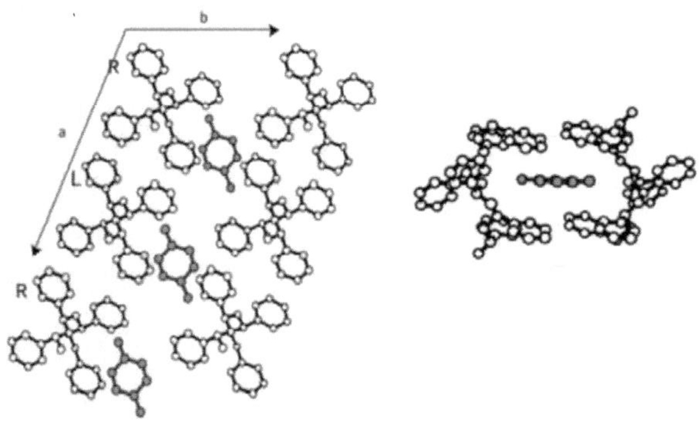

Figure 3. A model of sPS-cocrystal structure (monoclinic δ intercalate cocrystal with toluene) [12]. Left: projection onto the *ab* plane. Right: side view of a cavity containing toluene.

4. Results and Discussion

4.1. Influence of guest molecular weight on the behavior of sPS/PEGDME cocrystals on heating

Although several sPS cocrystal structures with different crystal systems and cavity shapes have been reported so far, the guest molecules are held between regularly arranged sPS helices taking TTGG conformation [10, 11]. The monoclinic δ clathrate structure with two cavities per unit cell (Figure 3 [12]) has been most frequently observed. According to the previous study, the sPS/PEGDME cocrystals prepared by the guest exchange procedure are assumed to maintain this type of structure [13]. The crystalline lamellae of uniaxially oriented sPS cocrystals are stacked in the drawing direction with a repeat period of about 10 nm. The lamellar stacking gives two lamellar reflections in

2D SANS images. Under the present experimental condition that sPS cocrystal film sample consists of fully protonated guest PEGDME and fully deuterated sPS (d-sPS), the distribution of PEGDME molecules principally determines the 2D SANS profile.

SANS profile changes on heating process are obviously different between the two kinds of d-sPS/PEGDME cocrystals with different guest size, as shown in Figure 4. As for d-sPS/PEGDME178 cocrystal, the lamellar reflection remarkably decreases in intensity as temperature increases up to 80 °C, leading to their almost annihilation. A further heating from 100 to 135 °C brings about the reappearance and gradual intensity increase of the two reflections. On the contrary, the lamellar reflection increases in intensity with temperature for d-sPS/PEGDME500 cocrystal.

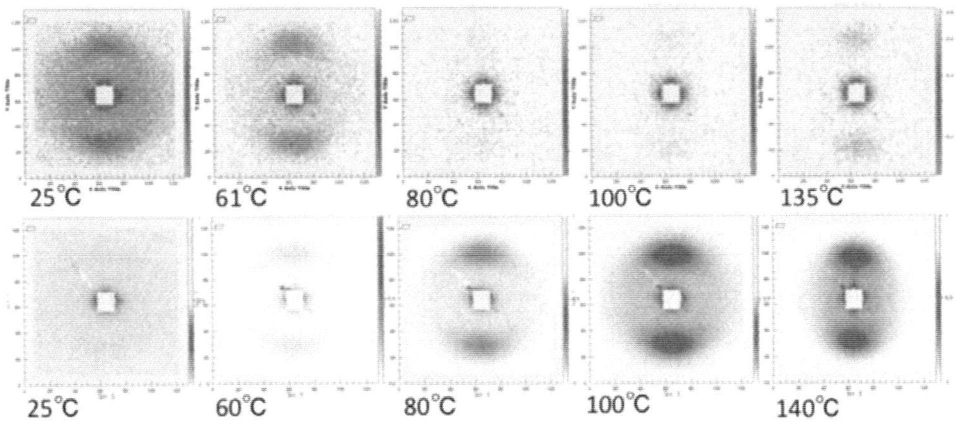

Figure 4. Temperature dependence of SANS 2D images. Top: d-sPS/PEGDME178 and bottom: d-sPS/PEGDME500.

From FTIR spectra measured in parallel with SANS 2D images (Figure 5), it can be seen that the PEGDME content behaves in a completely different manner between the two kinds of sPS cocrystals. As for d-sPS/PEGDME178 cocrystal, the bands due to host d-sPS, such as C-D stretch in the region of 2300 to 2150 cm^{-1} and C=C stretch around 1568 cm^{-1}, do not show any significant intensity changes during the course of temperature change, whereas the bands ascribed to PEGDME178, such as C-H stretch around 2874 cm^{-1} and C-O stretch around 1106 cm^{-1}, decrease in intensity with an increase in temperature. As for d-sPS/PEGDME500 cocrystal, not only the bands due to host d-sPS but also the ones due to guest PEGDME500 do not show remarkable intensity changes during the course of heating, except a slight intensity decrease of the guest bands. These results suggest that there is no significant change as to the d-sPS content in both d-sPS cocrystals and only the PEGDME178 component continues to decrease remarkably during this heating process.

Since the volatility of PEGDME molecules greatly depends on molecular weight, it seems that the nearly the three times difference in molecular weight drastically changes the desorption behavior of PEGDME component from the d-sPS/PEGDME cocrystal film, which consists of two processes, the migration from the crystalline region to the amorphous region and the migration from the amorphous region to the outside of the film. At the moment, we have the following view about the desorption behavior based on the SANS and FTIR data.

As for d-sPS/PEGDME178 cocrystal, the two kinds of migration take place smoothly, which results in a smooth decrease in concentration of the guest molecules both in the crystalline and amorphous region, as schematically shown in Figure 6 (top). Since the distribution of the protonated guest is the main factor for the contrast in scattering length density (SLD) between the amorphous and crystalline regions, the SLD contrast would decrease significantly by the smooth guest desorption, which is consistent with the SANS experimental observation. The slight intensity increase in the highest temperature region can be ascribed to the transformation to the γ phase, whose mass density is higher than that of the amorphous region, leading to a slight resurgence in the SLD contrast between the crystalline and amorphous regions.

On the other hand, the guest migration from the crystallization to the amorphous region takes place principally for d-sPS/PEGDME500 film, and therefore, the guest concentration in the amorphous region monotonously increase with temperature. The marked intensity increase at the high temperatures is attributable to an increase in the SLD contrast caused by this guest, as shown in Figure 6 (bottom).

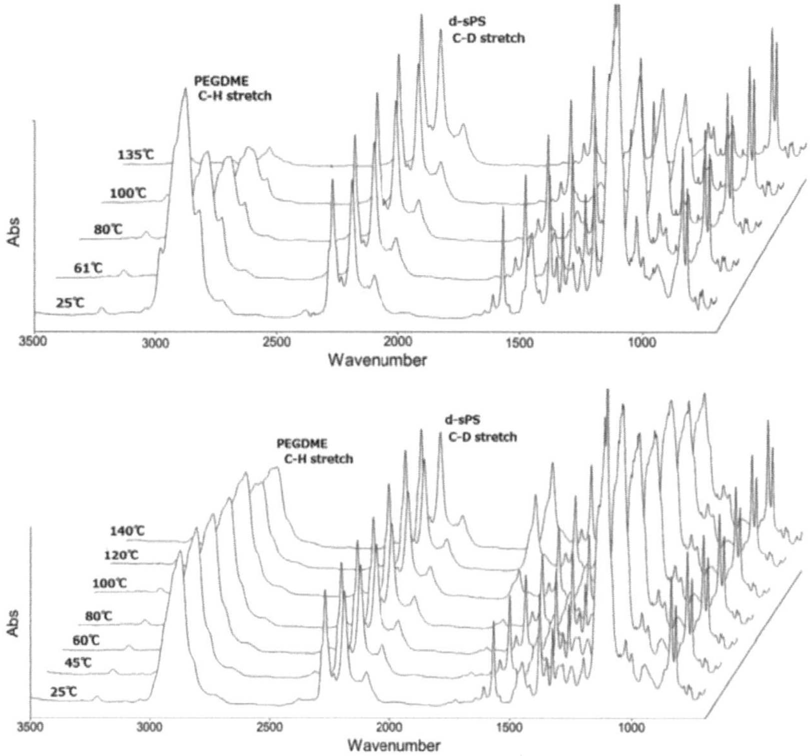

Figure 5. FTIR spectral changes measured in parallel with SANS measurements.
Top: d-sPS/PEGDME178 and bottom: d-sPS/PEG500.

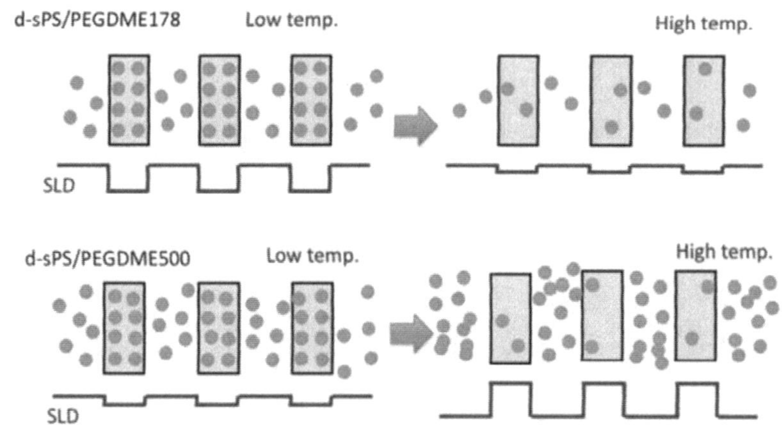

Figure 6. Schematic representation for temperature dependent distribution change of guest molecules. The solid state of d-sPS/PEGDME cocrystal is depicted as a one-dimensional array of crystalline lamellae (yellow box) and interlamellar amorphous regions (white parts). The green circles represent

guest PEGDME molecules. The blue rectangle wave below represents the variation of scattering length density between the crystalline and amorphous regions.

4.2. Decomposition of sPS/PEGDME500 cocrystals on heating

As described in the previous section, the bands due to the guest component show a slight intensity decrease on the heating process of the d-sPS/PEGDME500. In order to look into what happens in the cocrystal sample, the IR spectra measured in parallel with SANS have been checked. A band around 1728 cm^{-1} appears and clearly increases in intensity with temperature, in particular above 80°C, as shown in Figure 7 (right). The 1728 cm^{-1} band can be assigned to the carbonyl stretch mode ν(C=O) due to the aldehyde group of low-mass products by the thermal decomposition of polyethylene glycol, which starts even in relatively moderate conditions [14, 15]. Such low mass products would gradually dissipate into the outside. Therefore, it can be inferred that a small amount of PEGDME500 is lost from the cocrystal film owing the thermal decomposition at elevated temperature, which results in the intensity decrease of the C-H stretch band.

A similar thermal decomposition would occur for PEGDME178 component. However, the most part of the guest molecules dissipate smoothly to the outside because of its low molecular weight before the temperature becomes high enough for the thermal decomposition, and accordingly the band due to aldehyde group did not appear appreciably on the heating process of the d-sPS/PEGDME178 cocrystal, as shown in Figure 7 (left).

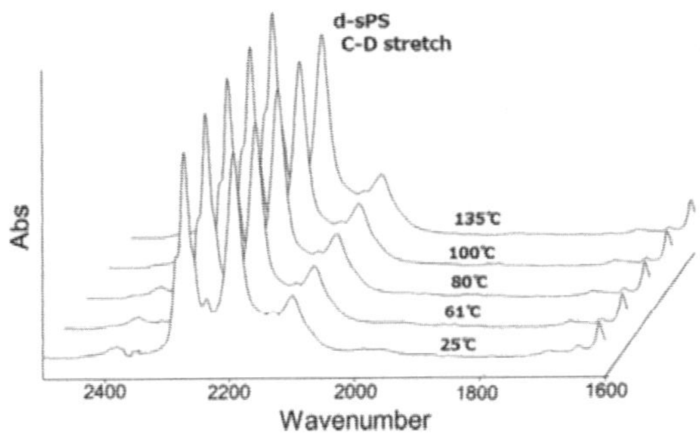

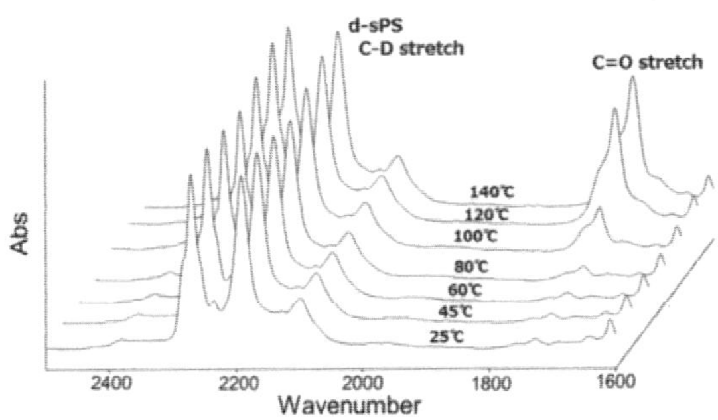

Figure 7. Temperature dependent IR spectral changes of sPS/PEGDME178 (left) and sPS/PEGDME500 (right) cocrystals.

Conclusion

A novel method for the simultaneous time-resolved measurement of SANS and FTIR spectroscopy has been developed. Since a change in chemical composition and chemical structure is sensitively reflected in FTIR spectra, the information obtained by FTIR spectrocopy is helpful to interpret SANS scattering profiles. This system is also able to detect chemical degradation happening in the sample on site, which is difficult to be noticed by SANS measurements. The simultaneous measuring system would be useful for research on a variety of multi-component systems, such as, polymer composites and polymner blends, and other kinds of materials containing non-hydrogeneous parts which are less visible by SANS. Furthermore, it can provide the information about the conformational changes of chemical compounds in different conditions, which would be useful for the analysis of SANS profile.

References

[1] Roe R-J 2000 *Methods of X-Ray and Neutron Scattering in Polymer Science* (New York: Oxford).

[2] Higgins J S and Benoit H C 1997 *Polymers and Neutron Scattering* (New York: Oxford).
[3] Naylor S *et al* 1995. *Nucl. Instrum. Methods Phys. Res., Sect. B Res., Sect.* B **97** 253.
[4] Innocenzi P *et al* 2007 *J. Phys. Chem.* C **111** 5345.
[5] Ratri P J and Tashiro K 2013 *Polym. J.* **45** 1019.
[6] Koenig J L 1999 *Spectroscopy of Polymers* (New York, Elsevior).
[7] Griffiths P R and de Haseth J A 2007 *Fourier Transform Infrared Spectrometry* (New Jersey: Wiley).
[8] Kaneko F *et al* 2015 *Chem. Lett.* **44** 497.
[9] Kaneko F *et al* 2015 *MLZ News* **14** 12.
[10] Guerra G *et al* 2009 *Preparation, structure, properties, and applications of co-crystals and nanoporous crystalline phases of syndiotactic polystyrene* (Syndiotactic polystyrene) ed Schellenberg J (Hoboken, Wiley).
[11] Guerra G *et al* 2012 *J. Polym. Sci. Polym. Phys.* **50** 305.
[12] Chatani Y *et al* 1992 *Polymer* **33** 488.
[13] Kaneko F and Sasaki K 2011 *Macromol. Rapid Commun.* **32** 988.
[14] Bortel E *et al* 1979 *Macromol. Chem.* **180** 2491.
[15] Han S *et al* 1996 *Polymer* **38** 317.

Applications of neutron scattering to heterogeneous catalysis

Stewart F Parker[1*] and David Lennon[2]
[1] ISIS Facility, STFC Rutherford Appleton Laboratory, Chilton, Didcot, OX11 0QX, UK
[2] School of Chemistry, Joseph Black Building, University of Glasgow, Glasgow G12 8QQ, UK

E-mail: stewart.parker@stfc.ac.uk

Abstract. Historically, most studies of heterogeneous catalysts that have used neutron vibrational spectroscopy have employed indirect geometry instruments with a low (<40 cm^{-1}) final energy. In this paper we examine the reasons why this has been the case and highlight the advantages and disadvantages of this approach. We then show how some of these may be overcome by the use of direct geometry spectrometers. We illustrate the use of direct geometry spectrometers with examples from reforming of methane to synthesis gas ($CO + H_2$) over Ni/Al_2O_3 catalysts and an *operando* study of CO oxidation. We conclude with a proposal for a unique instrument that combines both indirect and direct geometry spectrometers.

1. Indirect and direct geometry spectrometers

The first reported example of an adsorbed species (water on silica and alumina [1]) studied by neutron vibrational spectroscopy (inelastic neutron scattering, INS) used a direct geometry spectrometer [2]. However, since then almost all INS studies of catalysts (see [3] and [4] for reviews) have used indirect geometry spectrometers. It is useful to understand why this is the case and also what are the limitations of this approach.

In an INS experiment the aim is to measure the energy transfer E_T (cm^{-1}):

$$E_{\mathrm{T}} = E_{\mathrm{i}} - E_{\mathrm{f}} \quad (1)$$

where the subscripts i and f refer to the incident and final states respectively. The associated momentum transfer Q (Å^{-1}) is given by:

$$Q^2 = k_{\mathrm{i}}^2 + k_{\mathrm{f}}^2 - 2k_{\mathrm{i}}k_{\mathrm{f}}\cos\theta \quad (2)$$

where k is the wavevector ($k = 2\pi/\lambda$), λ is the neutron wavelength (Å) and θ is the scattering angle. Indirect geometry spectrometers for INS spectroscopy all use a small final energy, typically ~30 cm^{-1}. Consequently $k_i >> k_f$ and equation (2) reduces to:

$$Q^2 \approx k_{\mathrm{i}}^2 \quad (3)$$

Energy and wavevector are related (m_n is the mass of the neutron) by:

$$E = \frac{\hbar^2|\boldsymbol{k}|^2}{2m_{\mathrm{n}}} = 16.7|\boldsymbol{k}|^2 \quad (4)$$

* To whom any correspondence should be addressed.

Combining equations (3) and (4):

$$E_{\mathrm{T}} = 16.7\,Q^2 \tag{5}$$

Equation (5) has several important consequences. It states that there is a single momentum transfer value at each energy transfer, thus the instruments have an approximately parabolic trajectory through (Q,E_{T}) space. This means that the resulting spectra resemble vibrational spectra recorded by conventional methods as may be seen in the left part of figure 1, which compares the spectra of $LiOH.H_2O$ [6] (a model compound for adsorbed water on oxide surfaces) recorded by TOSCA (an indirect geometry INS spectrometer [5]), infrared and Raman spectroscopies. Note the excellent resolution in the TOSCA spectrum; the linewidths are comparable to those in the infrared and Raman spectra and this is a major advantage of this type of instrument. For catalysts the region below 2000 cm^{-1} is often difficult to access spectroscopically by conventional means, whether through absorption of the incident radiation by the support for oxide or carbon supported catalysts or few modes being allowed by the surface selection rule for metals. The highly penetrating nature of neutrons and the absence of selection rules make both factors irrelevant for INS spectroscopy. In addition, as there are few options for the user, indirect geometry spectrometers are generally relatively easy to use.

However, there are disadvantages to indirect geometry spectrometers. A major problem is availability: worldwide there are only six operating low final energy indirect geometry spectrometers, whereas there are 10 direct geometry spectrometers [7]. This trend will continue at the European Spallation Source (ESS): four direct geometry spectrometers are planned, whereas there will only be one indirect geometry spectrometer.

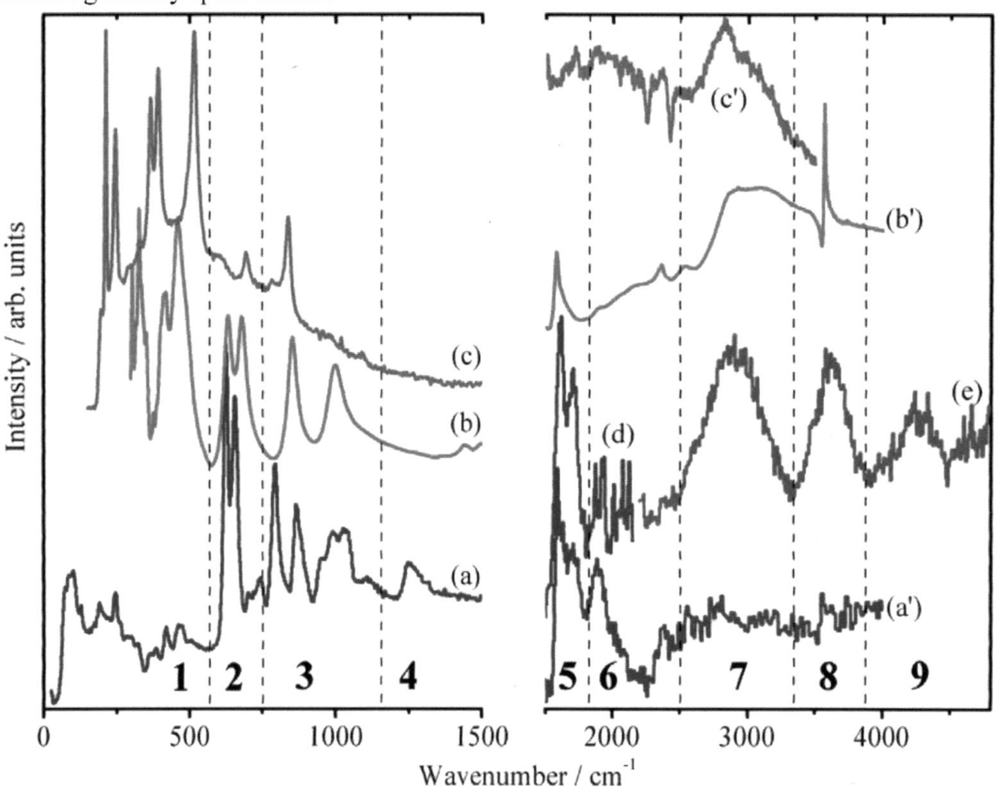

Figure 1. Vibrational spectra of $LiOH.H_2O$. (a), (a') INS, recorded with TOSCA ((a') = (a) × 5 ordinate expansion), (b), (b') infrared, (c), (c') FT-Raman, 1064 nm excitation, ((c') = (c) × 3 ordinate expansion), (d) INS, recorded with HET, E_i = 2420 cm^{-1} and (e) INS, recorded with HET, E_i = 6850 cm^{-1}. The assignments for regions 1 – 9 are given in table 1.

Table 1. Assignment of the vibrational spectra of LiOH.H$_2$O.

Region	cm^{-1}	Description
1	0 - 570	Translations
2	570 - 740	OH$^-$ librations
3	740 - 1160	H$_2$O librations
4	1160 - 1500	1st overtone of OH$^-$ librations
5	1500 - 1840	H–O–H scissors
6	1840 - 2510	Overtones and combinations of H$_2$O librations
7	2510 - 3350	H$_2$O O–H stretch modes
8	3350 - 3900	O–H$^-$ stretch modes
9	3900 - 4800	Combination O–H$^-$ stretch + OH$^-$ librations

Historically, indirect geometry spectrometers have also been relatively insensitive owing to small detector areas, usually <1 m^2, in contrast to state-of-art direct geometry spectrometers that have >10 m^2 of detector area. This is a problem that is becoming less serious for the current generation of spectrometers with high m-number neutron guides (VISION [8] at SNS, TOSCA [5] at ISIS) and/or optimised collection systems (IN1-Lagrange [9] at ILL).

The largest disadvantage of indirect geometry spectrometers is illustrated in the right hand side of figure 1 in the region above 2000 cm^{-1}. While the O–H stretch modes of the water and hydroxyls are clearly seen in the infrared and Raman spectra, figure 1b',c', the TOSCA spectrum is essentially featureless, figure 1a'. The reason for this lies in equation (5): large energy transfer means large momentum transfer; at 2830 cm^{-1} (the maximum in the O–H stretch in the Raman spectrum, figure 1c'), $Q \sim 13$ Å^{-1}. This is a problem because the INS intensity of a mode is given by:

$$S(Q,n\omega_i) \propto \frac{(QU_\omega)^{2n}}{n!}\exp\left(-(QU_{\mathrm{Tot}})^2\right)\sigma \tag{6}$$

where ω_i is the ith mode at transition energy ω, $n = 1$ for a fundamental, 2 for a first overtone or binary combination, 3 for a second overtone or ternary combination *etc*..., U_ω is the root mean square displacement of the atoms in the mode at ω cm^{-1}, σ is the inelastic scattering cross section of the atom and the exponential term is a Debye-Waller factor, U_{Tot} is the total root mean square displacement of all the atoms in all the modes, (both internal and external), and its magnitude is in part determined by thermal motion. (This can be reduced by cooling the sample and so spectra are typically recorded below 30 K). Thus, large Q has two detrimental effects: it suppresses the total intensity *via* the Debye-Waller factor and, worse, the intensity of the overtones and combinations is proportional to Q^4, Q^6... for $n = 2, 3$... and their intensity relative to the fundamental makes the latter unobservable.

Direct geometry spectrometers provide a solution to this problem. Figure 2 shows the $S(Q,E_T)$ map of LiOH.H$_2$O recorded on HET [10]. HET had a low angle (*i.e.* small Q) detector bank at 3 – 29° and a high angle (large Q) at 110 – 140°, (state-of-the art instruments *e.g.* MAPS [11] have continuous Q range, rather than the discrete coverage of HET). By only using the low Q detectors, it is possible to observe the water bend and O–H stretch regions, figure 1d,e. Figure 2 also shows the (Q,E_T) trajectories of TOSCA, which are typical of low final energy indirect geometry instruments. It can be seen that for *any* energy transfer, a direct geometry spectrometer can always access lower Q than an indirect geometry spectrometer.

2. Benefits of access to small *Q*

For molecular spectroscopy in general, and catalysis in particular, access to small *Q* has three major uses: observation of the C–H/N–H/O–H stretch region, quantification of surface species and the study of *operando* reactions. In the next three sections we describe these in more detail.

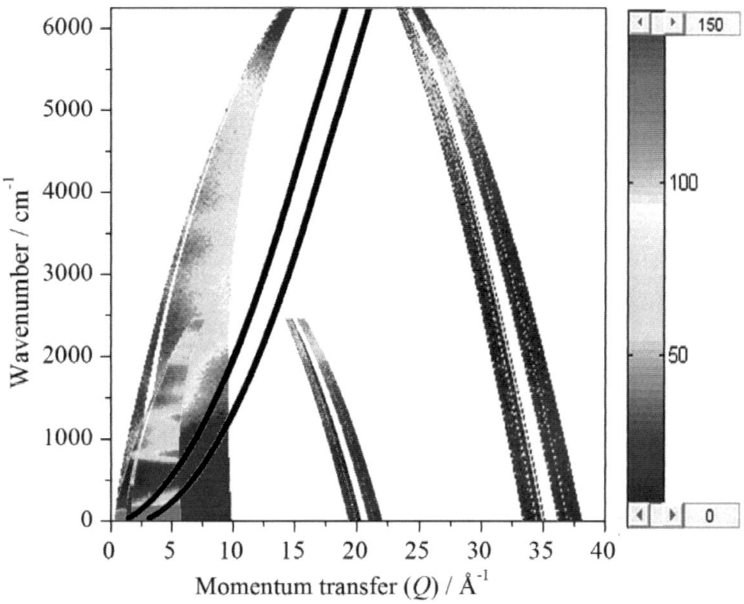

Figure 2. $S(Q,E_T)$ map of $LiOH.H_2O$ recorded on HET [10] with E_i = 2420 cm^{-1} and 6850 cm^{-1}. The black lines are the trajectories for an indirect geometry spectrometer with scattering angles θ = 45° (left) and 135° (right).

2.1. Observation of the C–H/N–H/O–H stretch region

Hydrogen is a vital component in the production of very large scale commodity chemicals such as ammonia, methanol and the Fischer-Tröpsch synthesis of hydrocarbons. Hydrogen is typically produced *via* reforming of methane to synthesis gas (syngas, CO + H_2). Industrially this is done using steam as the oxidant (steam reforming):

$$CH_4 + H_2O \rightarrow CO + 3\ H_2 \tag{7}$$

An alternative method is to use CO_2 as the oxidant (dry reforming):

$$CH_4 + CO_2 \rightarrow 2\ CO + 2\ H_2 \tag{8}$$

Apart from the obvious environmental advantage of CO_2 utilisation, dry reforming is well-suited to hydrogen production from biomass gasification and as the feedstock for Fischer-Tröpsch synthesis. Both methane reforming processes may use a high metal loading Ni/Al_2O_3 catalyst; however, a major problem is catalyst deactivation by coke formation. The nature of the coke is poorly understood and, in particular, the degree of hydrogen incorporation was unknown.

Figure 3 shows INS spectra in the C–H/O–H stretch region recorded with the direct geometry spectrometer MAPS [11] at ISIS for both dry (figure 3a,b) [12],[13] and steam (figure 3c) [14] reforming over Ni/Al_2O_3 catalysts. It can be seen that the spectra are markedly different. Transmission electron microscopy shows physical differences between the overlayers; amorphous carbon in the case of figure 3a and filamentous (graphitic) carbon in the case of figure 3b,c. While the catalysts that were used for dry reforming in figure 3a,b are broadly similar, the α-Al_2O_3 support was obtained from different sources and the calcination conditions were different. In figure 3a, a significant quantity of hydroxyls are present (broad, intense peak at 3335 cm^{-1}) while there is a complete absence of

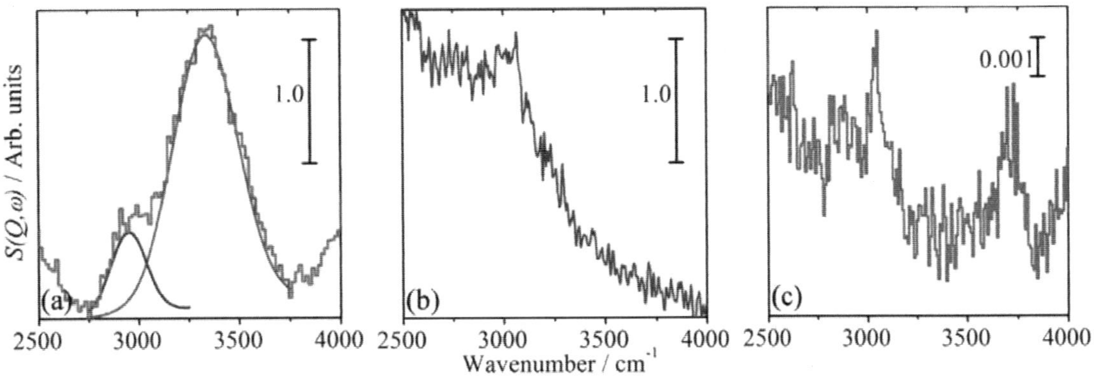

Figure 3. INS spectra of surface species on Ni/Al_2O_3 catalysts after methane reforming recorded on MAPS [11] with E_i = 4840 cm^{-1}. (a) Amorphous hydrocarbonaceous overlayer from dry reforming over Ni(45%)/α-Al_2O_3 [12]. (b) Filamentous hydrocarbonaceous overlayer from dry reforming over Ni(26%)/α-Al_2O_3 [13]. (c) Filamentous hydrocarbonaceous overlayer from steam reforming over Ni(26%)/α-Al_2O_3 [14].

hydroxyls in figure 3b. The hydrocarbonaceous overlayers are also very different: in figure 1a, the overlayer is largely aliphatic sp^3 carbon, as shown by the C–H stretch at 2950 cm^{-1}, while in figure 3b, it is largely aromatic sp^2 carbon, as shown by the C–H stretch at 3040 cm^{-1}. For steam reforming, figure 3c, the catalyst was prepared in the same way as that used for figure 3b, in this case the overlayer is again aromatic (C–H stretch at 3045 cm^{-1}). In this case there is also a small amount of hydroxyls present, which is not surprising in view of the reaction conditions. Thus, ready access to the C–H/O–H stretch region, with reasonable resolution and good sensitivity, enables clear distinctions between the catalysts and shows the dramatic differences that a change of preparation methods or reaction conditions can engender.

2.2. Quantification of surface species

Characterisation of surface species is vital in order to understand a catalytic process, however, being able to quantify the amount of material present is also invaluable and is often difficult. In the case of the hydrocarbonaceous overlayers shown in figure 3, quantification of the amount of carbon present is straightforward by temperature programmed oxidation (TPO: the catalyst is heated in an oxygen-containing gas stream and the amount of CO_2 produced is measured). Qualitatively, it is apparent that the amount of hydrogen present in the spectra in figure 3 is (a) >> (b) >> (c) (note the different scale bars) but quantitative data would be very useful. However, hydrogen at low levels is very difficult to quantify and INS spectroscopy provides a useful capability in this regard.

In the harmonic approximation the term U_ω in equation (6) is given by:

$$(U_\omega)^2 = \frac{\hbar}{2\mu\omega} \tag{9}$$

where μ is the reduced mass. Since hydrogen is the lightest element we can approximate μ as a diatomic, *e.g.* C–H or O–H. Hence U_ω is independent of the nature of the species and all hydrocarbons with a C–H stretch mode at ω cm^{-1} will have the same intensity per C–H oscillator and similarly for hydroxyls. This means that a calibration is transferable. A calibration made with known masses of an appropriate material, in our case we used polystyrene and brucite ($Mg(OH)_2$) for the C–H and O–H regions respectively[12], can be used to quantify the amount of hydrogen present on the surface. The use of a direct geometry is essential for two reasons: the stretch region is relatively clear from interference from other modes and we measure at small Q, in order to minimise the effect of the Debye–Waller term.

By curvefitting the data (as shown in figure 1a), and using the calibration factor the amount of hydrogen can be determined. For the reforming catalysts the amount of carbon-bound hydrogen is very small <100 μmol H (g cat)$^{-1}$. This can be combined with the carbon content derived from TPO to obtain a C:H ratio and these are given in table 2.

From figure 3 and table 2 it can be seen that the nature and quantity of hydrocarbonaceous overlayer is dependent on both the catalyst preparation and the reaction conditions. It is also apparent that the catalyst is very efficient at cycling hydrogen and much less efficient for carbon and it is this that leads to the carbon build-up that results in deactivation.

Table 2. Quantification of hydrocarbonaceous overlayers on Ni/α-Al_2O_3 methane reforming catalysts.

Process	C:H
Dry reforming (figure 1a)	160:1
Dry reforming (figure 1b)	2550:1
Steam reforming (figure 1c)	11690:1

2.3. Operando studies

The current trend in studies of catalytic systems is to study a reacting system under realistic conditions of temperature and pressure while simultaneously maximising the amount of information obtained, so-called *operando* studies. In practice, for gas-solid systems, this means running a reaction with at least one spectroscopic method and with concurrent analysis of the reaction products by gas chromatography or mass spectrometry. Such studies are commonplace at synchrotrons *e.g.* [15] but not at neutron sources and particularly not for INS studies. The reasons are twofold: firstly, the flux is generally insufficient to be able to record data of sufficient quality on the timescale of the reaction and secondly, the need to minimise the Debye-Waller factor (equation (1)) means that the measurements are made at < 30 K. Thus the samples for the work on methane reforming discussed in the previous sub-sections were prepared offline at temperatures of 900 – 1200 K but measured at ~10 K.

With current instrumentation, low flux is becoming less of a limitation due to higher power sources such as SNS and J-PARC and improvements in neutron guides that provide more efficient transport of neutrons. The difficulty caused by the Debye-Waller factor cannot be addressed by indirect geometry instruments, but with a direct geometry instrument, by using only small Q data, the effect of the Debye-Waller factor can be mitigated. This compounds the problem with flux because much of the data is discarded but in favourable cases it is possible to study a system by INS adopting *operando* methods.

The oxidation of CO to CO_2 is a crucial reaction in an automotive catalytic converter. This is accomplished by the Pd component of the three-way (Pt/Pd/Rh) converter. A model system for this component of the three-way converter is hydrous palladium oxide, $PdO.H_2O$. This consists of ~18 Å diameter particles of nanocrystalline PdO with a monolayer or so of hydroxyls capped by 4-7 layers of water [16]. $PdO.H_2O$ has the very useful property that it oxidises CO to CO_2 at room temperature at a rate that is convenient for INS studies. Figure 4a,b show the INS spectra recorded on MAPS together with simultaneous gas analysis by mass spectrometry, respectively.

From figure 4b, it can be seen that the catalyst is initially very effective: there is immediate, almost complete, consumption of CO and concomitant production of CO_2 as soon as the feed (100 cm^3 min^{-1} of He/CO(4%)/O_2(4%)) is switched to the catalyst at time 0 (the flows were set-up using a bypass loop). The heat of reaction results in an increased rate that decreases to an approximately steady-state after an hour or so. This is maintained until ~ 5 hours of elapsed time when CO_2 formation abruptly ends. The reason for this can be seen in the INS spectra recorded on MAPS; (which shows the Pd–O–H bending mode of the surface hydroxyls [16],[17]), as the reaction proceeds there is progressive decrease of the hydroxyls until they are completely consumed. The reason for this is that there are two reactions occurring simultaneously, a catalytic reaction, equation (9), and a direct reaction, equation (10):

$$2CO + O_2 \rightarrow 2CO_2 \quad (9)$$

$$Pd(II)O + CO \rightarrow Pd(0) + CO_2 \quad (10)$$

Both reactions require the presence of hydroxyls and we have shown elsewhere [17] that the direct reaction, equation (10), requires two hydroxyls:

$$CO + 2OH \rightarrow CO_2 + H_2O \quad (11)$$

At room temperature Pd(0) cannot be re-oxidised to Pd(II) and when all of the surface Pd(II) has been reduced, the reaction stops.

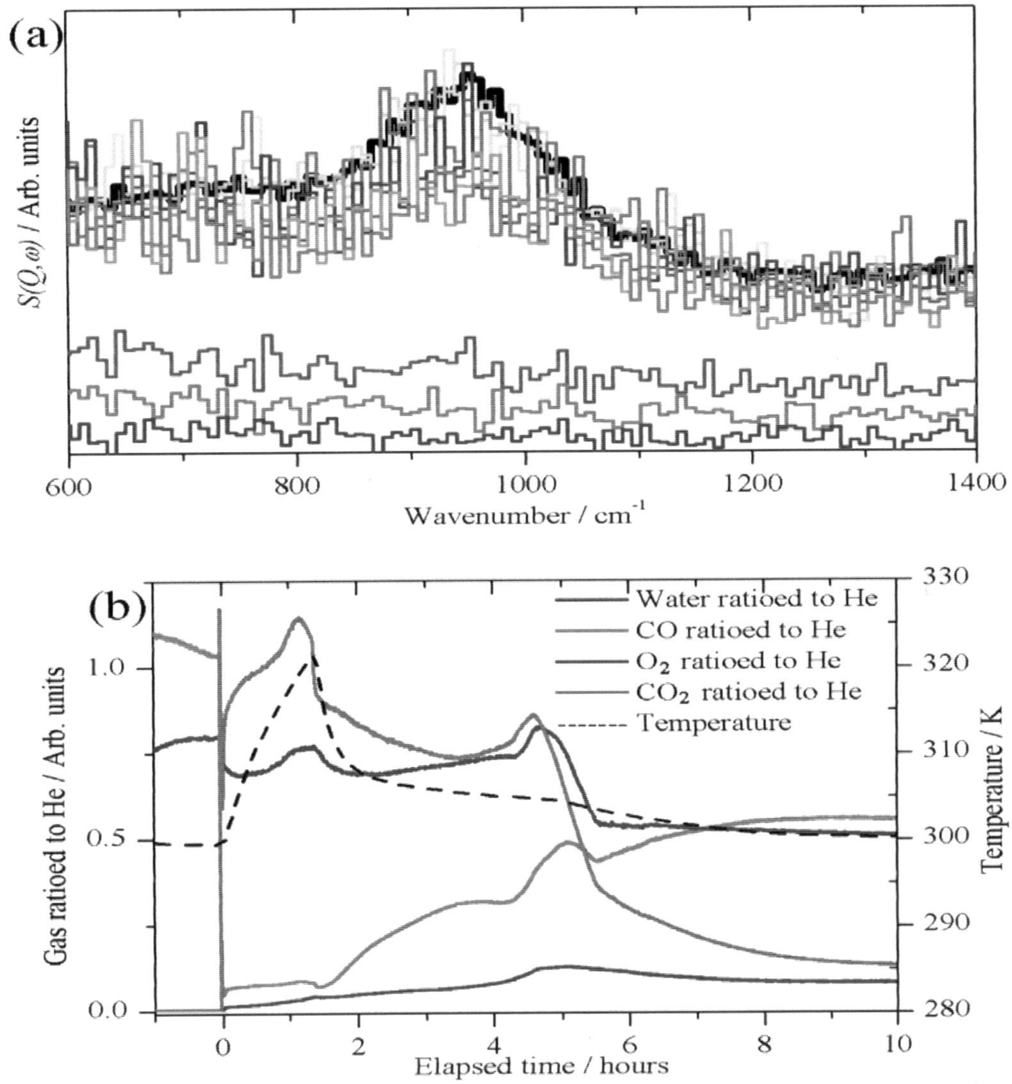

Figure 4. (a) INS spectra recorded on MAPS with E_i = 2420 cm^{-1} of the reaction of 100 cm^3 min^{-1} of He/CO(4%)/O_2(2%) over hydrous palladium oxide at 300 K. The spectra are difference spectra, the spectrum recorded at 24 hours elapsed time has been subtracted from the data. The thick black line is the initial spectrum, the upper series of spectra were recorded over 00:00 to 03:29 elapsed time, the lower three were recorded at 09:45, 12:03 and 14:13 elapsed time. (b) Exit gas analysis by mass spectrometry recorded simultaneously with the INS spectra. The signals from the reactants, CO and O_2, and the products, CO_2 and H_2O, have been ratioed to that of the He carrier gas. The reaction temperature (dashed line) is also shown.

3. Disadvantages of direct geometry spectrometers and the way forward

It is clear that the use of direct geometry spectrometers offers significant advantages for studies of molecular systems and of catalysts. Nonetheless, there are disadvantages to the instruments. The first is that they are complex to operate because there are many possible options regarding flux, spectral range and resolution [11]. The correct choice is not always obvious and even experienced users may require several iterations to arrive at the optimum configuration for a given experiment.

The second disadvantage is that the resolution at <2000 cm^{-1} is generally inferior to that of the best direct geometry instruments, TOSCA and VISION. This is because the resolution is a % of E_i, with the best resolution at $E_T \approx E_i$ and worsening with decreasing E_T. Thus to obtain the best resolution possible across the entire spectral range requires multiple scans (and hence increased beamtime) and merging the scans is not simple.

It is apparent that both direct and indirect geometry spectrometers have significant advantages for studies of catalysts and both are often needed to provide a more complete understanding of a system. Using two different spectrometers often involves delays due to the need to apply for beamtime and repeat the sample preparation. An outline of an instrument, Cerberus [18] figure 5, has been described that offers the capabilities of *both* direct and indirect geometry instruments. The instrument would operate in a sequential mode; with the Fermi chopper out of the beam, an indirect geometry spectrum would be recorded using the indirect geometry modules. These would utilise the focussing analysers developed on VISION to maximise the collection area, while retaining a short secondary flightpath for the best possible energy resolution. On completion of data acquisition, the Fermi chopper would be moved into the incident beam and a direct geometry spectrum recorded. This would usually be at high energy (E_i ~4000 cm^{-1}) to observe the C–H/N–H/X–H stretch region. However, the access to small Q means that time resolved studies at 'catalytically interesting' temperatures up to 473 K would also be feasible with any desired incident energy. Note that in the indirect mode, the detector in forward scattering used for the direct geometry data, would function as a low resolution diffraction detector. While this is not ideal for conventional powder diffraction, it would work well for total scattering studies involving hydrogen, thus providing a complementary structural capability. Cerberus would undoubtedly find much use in the fields of molecular spectroscopy and catalysis.

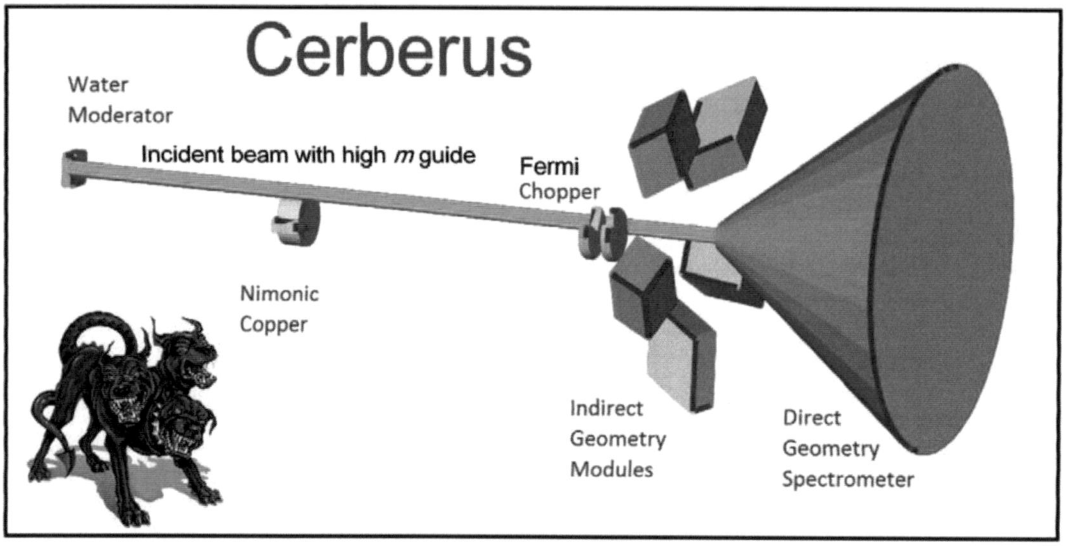

Figure 5. The key components of Cerberus [18] showing the indirect geometry detectors in backscattering (blue) and the area detector in forward scattering for direct geometry and diffraction..

Acknowledgments
The STFC Rutherford Appleton Laboratory is thanked for access to neutron beam facilities. The EPSRC are thanked for project support (grants EP/E028861/1 and EP/E029388/1).

References

[1] Boutin H and Prask H 1964 *Surf. Sci.* **2** 261-66.
[2] Hughes D J, Palevsky H, Kley W and Tunkelo E 1960 *Phys. Rev.* **119** 872-81.
[3] Mitchell P C H, Parker S F, Ramirez-Cuesta A J and Tomkinson J 2005 V*ibrational spectroscopy with neutrons, with applications in chemistry, biology, materials science and catalysis* (Singapore: World Scientific) chapter 10.
[4] Albers P W and S.F. Parker S F 2007 *Advances in Catalysis* **51** 99-132.
[5] Parker S F, Fernandez-Alonso F, Ramirez-Cuesta A J, Tomkinson J, Rudic S, Pinna R S, Gorini G and Fernández Castañon J 2014 *J. Phys. Conf. Ser.* **554** 012003.
[6] Parker S F, Refson K, Bewley R I and Dent G 2011 *J. Chem. Phys.* **134** 084503.
[7] http://neutronsources.org/neutron-centres.html
[8] http://sns.gov/vision/
[9] Ivanov A, Jiménez-Ruiz and Kulda J 2014 *J. Phys. Conf. Ser.* **554** 012001.
[10] Taylor A D, Boland B C, Bowden Z A and Jones T J L 1987 *Rutherford Appleton Laboratory Report* RAL-87–012.
[11] Parker S F, Lennon D and Albers P W 2011 *Applied. Spectroscopy* **65** 1325-41.
[12] Silverwood I P, Hamilton N G, Laycock C J, Staniforth J Z, Ormerod R M, Frost C D, Parker S F and Lennon D 2010 *Phys. Chem. Chem. Phys.* **12** 3102-07.
[13] McFarlane A R, Silverwood I P, Warringham R, Norris E L, Ormerod R M, Frost C D, Parker S F and Lennon D 2013 *RSC Advances* **3** 16577-89.
[14] McFarlane A R, Silverwood I P, Norris E L, Ormerod R M, Frost C D, Parker S F and Lennon D 2013 *Chemical Physics* **427** 54-60.
[15] O'Brien M G, Beale A M, Jacques S D M, Buslaps T, Honkimaki V and Weckhuysen B M 2009 *J. Phys. Chem.* C **113** 4890–7.
[16] Parker S F, Refson K, Hannon A C, Barney E, Robertson S J and Albers P 2010 *J. Phys. Chem.* C **114** 14164–72.
[17] Parker S F 2011 *Chem. Comm.* **47** 1988-90.
[18] Parker S F, Ramirez-Cuesta A J, Albers P W and Lennon D 2014 *J. Phys. Conf. Ser.* **554** 012004.

Highlighting liquid water trapped in Neolithic Chassey flints using Neutron Scattering

L. Noirez[1], Ph. Sciau[2], N. Ratel-Ramond[2], P. Baroni[1]
[1] Laboratoire Léon Brillouin (CEA-CNRS), CE-Saclay, Paris-Saclay University, 91191 Gif-sur-Yvette, France
[2] CEMES, CNRS, Toulouse University, 29 rue J. Marvig, 31055 Toulouse, France

Abstract. A specific lithic manufacturing of flints has emerged in the South of France from 4200 to 3500 ca B.C. This treatment involving a smooth thermal exposure procures an improvement of the ease of detachment and of the cutting properties of the flints. The origin of these new properties remains misunderstood. High resolution large angle neutron scattering (Barotron detector) is used to describe the structural modifications occurring after the thermal treatment. The comparison of the scattering exhibited by native silica flints and flints treated at the Neothilic age does not show a structural change but indicates a significant increase of the quantity of liquid water trapped in the treated flint. It is suggested that the Neolithic treatment was proceeded in several steps including a prior wetting of the flint and then a thermal treatment closing the porosity and trapping the liquid water inside the silica flint.

Introduction

Neutron Scattering is a powerful means particularly adapted to the investigation of cultural heritage objects. Indeed this low energy neutron radiation is non-destructive and non-invasive and provides information on the volume because of its deep penetration length. Offering a unique contrast between elements such as close hydrogen or deuterium, between nuclei of metals or metalloids as silica, neutron scattering can give access to new insights and key information about the structure, the content, the provenance or the history of these precious objects from the past. In this paper, we present the first neutron scattering analysis evidencing that the thermal treatment carried out by the Neolithic societies produces deep physical transformations in the silica flints. Stone tools belongs the first prehistory tools developed by the human activity. These remains of the past have undergone several specific evolutions during millennia. At the end of the 5th millennium (Cal. BC) an important change occurred in the fabrication of the flint tools, which is related to a more pronounced craft and the development of long distance trade. In archaeological terms, this is particularly visible in the southern Chassey culture across the south of France. An exceptional dissemination of specialized lithic flint productions has been identified from the Vaucluse: they are found on nearly 1000 Chassey culture consumer sites and in most cases constitute 99% of the tools. The magnitude of these exchange networks reaches the European scale with the identification of products in Italy and in Catalonia [1]. Recent archaeological researches [2] on Barremo-Bedoulian flints demonstrated that physical transformations take place at a surprisingly moderate temperature around 250°C. The thermal treatment modifies the original gloss of flint, its roughness and its mechanical properties [3]. The first modern flint heating replicative

[1] To whom any correspondence should be addressed.

experiments were performed by Crabtree and Butler in 1964 [4] and subsequent studies have been performed by various authors in order to assess the response and transformations of flint on a wide range of siliceous rocks, with different sizes and heating rates [5]. Recently, studies have tried to clarify the heat-induced transformations occurring in Barremo-Bedoulian flints at rather low temperature (around 250°C) [11-12]. They revealed that the transformation was due to the loss of silanols (Si-OH) and the formation of new Si-O-Si bonds which led to a "closure of the pores" and an improvement of the mechanical resistance at the grain boundaries leading to a more homogeneous distribution of mechanical strength. The production of flint bladelets by a pressure technique is thus facilitated, and the sharpness of the blades (cutting properties) is also improved. The aim of this study is to detail the physical changes occurring in Barremo-Bedoulian flints when

it is heated at low temperature, in particular the role and the presence of the water during these transformations. We therefore sought to detect the presence of water in the Neolithic flints treated by the Chassey culture and to compare with native flints. For this, the study by neutron is particularly relevant since this technique is sensitive to light elements such as liquid water while X-Ray scattering is strongly dependent on heavy elements. With an incoherent cross-section of 80 barns, Hydrogen has the largest scattering interaction with neutrons of all the elements of the periodic table making neutron scattering ideal to detect the presence of natural water (Table 1). In addition the long mean free path of the neutron enables to probe silica samples in the depth.

Table 1. Incoherent neutron cross sections (in barn) of the constituents of the silica flint:

Isotope	Incoherent cross sections:
H	80.26
O	0.0008
Si	0.004

Experimental

Neutron scattering experiments are carried out using a new instrument available at the Laboratoire Léon Brillouin. Initiated about a couple of years ago [6], the concept "Barotron" has evidenced the possibility of using CCD cameras and now of CMOS technology as a mean to explore the reciprocal space by neutron scattering. This two-dimensional position sensitive detector couples the advantages of a solid scintillator in terms of excellent spatial resolution to an excellent efficiency of detection and with the possibility of permanent upgrading of the charge-coupled detection device or of the CMOS technology [7]. This solid-PSD uses a coupling of a photoemission means (LiFZnS solid scintillator) adapted for the neutron radiation, a cooled low-light level charge-coupled detection device and a light amplifier. This setup is advantageous with respect to ICCD (intensified charge coupled devices) and enables a full well capacity of at least 65000 e-/pixel. The LiFZnS solid scintillator is chosen for its very low rate of detection of gamma, its high and fast photoconversion rate of neutrons (about 10^5 photons/neutron) and its fast relaxation time (about 200 ns). The resulting two-dimensional neutron detector displays a large and versatile detection area (260 mm x 260 mm), a high spatial resolution (250000 pixels of 0.35 mm x 0.35 mm), a very low detection threshold (<1 neutron/cm^2/s), a true 16 bits dynamic range that ensures a linear intensity measurement. The performances compete with the best conventional gas chamber detectors in terms of detection efficiency with a spatial resolution of the quality of X-Ray or synchrotron radiation 2D-detectors. Additionally, the compact size of this new neutron detector allows a high versatility including the use in neutronography mode (neutron imaging) and data viewable *in situ* during the acquisition. The detector is coupled to crystalline monochromator delivering a wavelength of $\lambda = 2.36$ Å with bandwidth of less than $\Delta\lambda/\lambda < 2\%$.

The sample of silica flints are cut in thin slices of 1mm thickness for the neutron scattering experiments. They are optically opaque and thus exclude the possibility to use optical means to study the inner structure. The chemical composition of flints originating from Chassey contents more than 99% of SiO_2 in weight, 1% being impurities and a certain amount of water, chemically bound hydroxyl (silanole, SiOH) or free molecular water (0.3-0.6%). The present study describes the results obtained by neutron diffraction on two samples: a native flint and a flint sample originated from Chassey and treated at the Neolithic age. Both are compared at room temperature.

Results- Discussion:

Flint is a chemically precipitated sedimentary silica rock. This cryptocrystalline α-quartz (SiO_2) is made of nanometre sized crystallites [8] assembled to form fibers. The native silica flint contains about 0.3%-0.6% of molecular water retained in voids, crystal defects and grain boundaries [9]. A small amount of water (0.5%-0.8%) is reacting with the silicon forming Si-OH bonds. At temperatures higher than 200°C, the heating is supposed to regenerate the silicon oxide (Si_2O) following the reaction:

$$\text{Si-OH} + \text{Si-OH} \rightarrow \text{Si-O-Si} + H_2O \qquad (1)$$

This chemical reaction, which produces molecular water is responsible for the changes of mechanical properties such as the increase of hardness, an ease of cutting and lowers the fracture ability [10]. The co-product, H_2O, is evaporated from the surface but hardly migrate from the thickness of the mineral. Therefore the question is to know if the crystalline structure is changed during the heating and how much the molecular water remains within the mineral. A first study using infrared spectroscopy has enabled measuring the ability of the samples to rehydrate. The identification of weak chemical transformations during heating has enabled the determination of the working temperatures used at the Neolithic. These latter are estimated between 200 and 250°C [11]. However, the method discriminates hardly the differences between native and archeological flints in terms of hydration. The present neutron scattering experiments are carried out on these native and archaeological minerals and complete the infrared study.

Because of its high cross-section (Table 1), the hydrogen nucleus has a high probability of intersection with the incident neutrons making neutron scattering a sensitive probe to detect this light atom.

Figure 1 displays the two-dimensional pattern produced by a 1mm thick flint sample treated at the Neolithic age. The isotropic scattering (rings) indicates that both treated and the native (not represented here) flints are a powder structure without texture indicating a fine-grained structure. The peak of highest intensity corresponds to the Si-Si distance.

Journal of Physics: Conference Series **746** (2016) 012067 doi:10.1088/1742-6596/746/1/012067

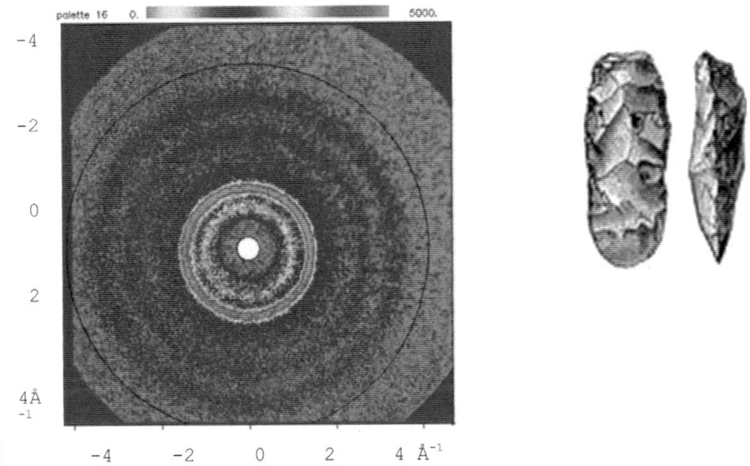

Figure 1: 2D neutron pattern of wide angle neutron spectrum exhibited by silica flints at room temperature (raw data without correction, smoothing or binning. Barotron detector: sample-detector distance: 60mm, λ = 2.36Å, 4mm diaphragm, wavelength dispersion $\Delta\lambda/\lambda \approx$ 2%, acquisition time: 900s, incident flux: 6.10^5 neutrons/cm^2/s). The scattering indicates that both treated and the native (not represented here) flints are a powder structure without texture.

The neutron scattering profiles corresponding to the native and to the treated Neolithic flints are displayed in Figure 2. Both native and treated Neolithic flints display a similar structural pattern with identical peak positions. The thermal treatment has not physically or chemically fundamentally modified the flint; its crystalline structure and the crystallite size are kept (the FWHM corrected from the instrumental resolution of the intense reflection at Q = 1.87 Å^{-1} indicates a short range for the domain of coherence of about 70 Å - no significant structural change is observed before and after the thermal treatment within the accuracy). However the background level is significantly different. The treated sample displays an increase of the background in the whole scattering range with respect to the native one. This increase can be interpreted by a higher proportion of hydrogen in the treated sample; i.e. the Neolithic flint contains a more pronounced quantity of liquid water. From the quantitative comparison of the background levels, the rate of hydrogen is increased of 7% in the sample submitted to the thermal treatment with respect to the native sample; i.e. 7% of about 1.3% of water contained in a native flint. It should be thus concluded that the thermal treatment closes the porosity of the flint and traps a part of the water molecules within the sample. Moreover, since the thermal process (reaction (1)) consists in a release and recombination of hydrogen in water molecules, no increase of the rate of hydrogen should be noticeable. The present observation suggests that a prior wetting or an immersion of the native flint precedes the thermal treatment.

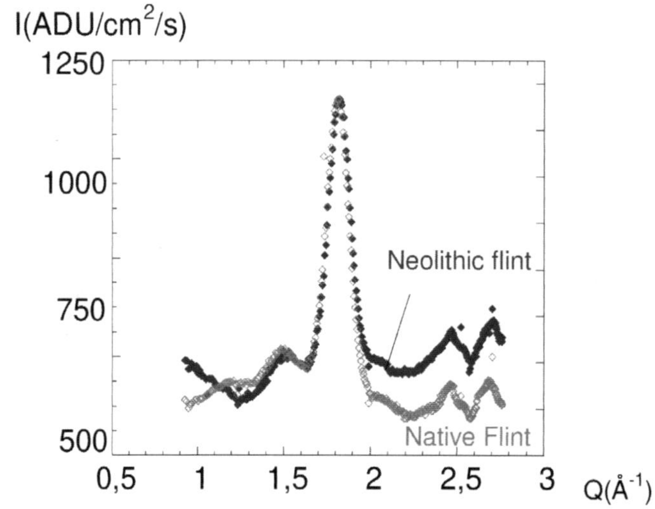

Figure 2: Neutron scattering profile (corrected from the dark and normalized by Vanadium) corresponding to the native and to the treated Neolithic flints. The red points describe the scattering produced by the native flints containing 1.3% of water and the blue points correspond to the scattering exhibited by the treated Neolithic flint. Both samples display a similar structural pattern (identical peak positions) indicating that the thermal treatment did not have chemically modified the flint and that the crystalline structure is kept. The background level is significantly different indicating that the treated sample contains a more pronounced quantity of liquid water.

Conclusions

We present here a pioneering neutron scattering study of the cultural heritage objects: the silica flints of Neolithic age manufactured by the Chassey societies. The present results identifying the presence of molecular water trapped in archaeological crafted objects would not have been obtained by other radiation than neutron scattering. Because of the strong incoherent scattering length of the hydrogen atom, the presence of water (in its liquid or crystalline form) can be easily detected by neutron scattering. Here it is evidenced by two-dimensional high resolution large angle neutron scattering that the thermal treated flints of the Neolithic age by the Chassey civilizations contain a higher concentration of water. The observation of a higher hydrogen concentration in the treated flint compared to the native one, might suggest a wetting of the flint prior the thermal treatment. The thermal procedure then induces a closure of the porosity trapping the liquid water in the mineral. These slight modifications provide important mechanical properties of the treated flints in terms of cutting. Other studies on the Neolithic flints are needed to complete this preliminary study and confirm the suggested scenario.

The present study points out the interest of intermediate and wide neutron scattering angles to characterize condensed matter objects. While conventional X-Ray scattering are available and used for years in many museums and laboratories, offering an accurate spatial resolution, the analysis with neutron scattering is rarely explored. It provides a different contrast particularly for light nuclei as hydrogen. The penetration length of the neutrons is deep in dense and thick materials as minerals. In this respect and in addition to techniques as X-Ray tomography or neutronography (neutron imaging), neutron scattering provides a powerful non-destructive investigation of heritage objects. Moreover, the adaptation of a technology based on optical sensors to two-dimensional neutron scattering procures an unrivaled high spatial resolution together with a fast speed reading (for example CMOS) extends the possibilities to characterize reduced size samples with a high dynamic which is an advantage when the material is rare or precious.

Acknowledgements
The authors thank V. Lea for the archaeological specimens and F. Porchet for crystallography data. This work was supported by the programme ANR-09-BLAN-0324-01 ProMiTraSil.

[1] Lea V., Roque-Rosell J., Torchy L., Binder D., Sciau P., Pelegrin J., Regert M., Cousture M.-P., Roucau C., 2012 *Craft specialization and exchanges during the Southern Chassey culture : an integrated archaeological and material sciences approach, Rubricatum, 5 (Numéro spécial dédié aux actes du colloque de Gava, Networks in the Neolithic. Exchange of raw materials, products and ideas in the Western Mediterranean (*VII-III millennium BC) 2011, M. BORRELL, F. BORREL, J. BOSC, X. CLOP & M. MOLIST Ed.

[2] Binder D., 1984. Systèmes de débitage laminaire par pression : exemples chasséens provençaux. in *Préhistoire de la Pierre Taillée II, économie du débitage laminaire (Table ronde de technologie lithique 3;* Meudon-Bellevue octobre 1982) 71-94. CREP, Paris, Pelegrin J., 1988. *Débitage expérimental par pression;* "du plus petit au plus grand". In: Tixier J. (dir.) Technologie préhistorique (Notes et Monographies Techniques du CRA, n°25), 37-53. Ed. CNRS, Paris.

[3] Lea V 2005 *Antiquity* **79** 51.

[4] Crabtree D.E., Butler B.R. 1964 *Tebiwa* **7** 1.

[5] Mercieca A, Hiscock P 2008 *J. of Archaeological Sci.* **35** 2634.

[6] Baroni P, Noirez L (patent) Two-dimensional detection system for neutron radiation in the field of neutron scattering spectrometry. 2006 PCT WO2006/095013 A.

[7] Baroni P, Noirez L 2014 *Am. J. of Appl. Sci.* **11** 1558.

[8] Rios S, Salje E.K.H., Redfern S.A.T. 2001 *EPJ B* **20** 75.

[9] Schmidt P, Badou A, Fröhlich F, 2011 *Spectrochimica Acta, Part A.: Mol. and Biomol. Spectroscopy,* **81** 552.

[10] Schmidt P, Masse S, Laurent G, Slodczyk A., Le Bourhis E, Perrenoud C, Livage J, Fröhlich F, 2012 *J. of Archaeol. Science* **39** 135.

[11] Schmidt P, Léa V, Sciau Ph, Fröhlich F, 2013 *Archaeometry*, **55** 794.

[12] Schmidt, P., Slodczyk, A., Lea, V., Davidson, A., Puaud, S. & Sciau, P. (2013b). *Phys. Chem. Minerals* **40,** 331-340.

Network structure of molybdate glasses by neutron and X-ray diffraction and reverse Monte Carlo modelling

M Fabian[1,2], E Svab[2], K Krezhov[3]

E-mail: fabian.margit@energia.mta.hu

[1]Centre for Energy Research, 1121 Budapest, Konkoly-Thege st. 29-33, Hungary
[2]Wigner Research Centre for Physics, 1121 Budapest, Konkoly-Thege st. 29-33, Hungary
[3]Institute for Nuclear Research and Nuclear Energy, 1784 Sofia, 72 Tzarigradsko Chaussee, Bulgaria

Abstract. Rare-earth molybdate glasses have been prepared by rapid quench technique, the network structure was investigated by neutron and high-energy X-ray diffraction. For data evaluation the reverse Monte Carlo simulation technique was applied to obtain a possible 3-dimensional network configuration, which is consistent with the experimental data. From the modelling the partial atomic correlation functions $g_{ij}(r)$ and the coordination number distributions CN_{ij} have been revealed. Formation of MoO_4 (55%) and MoO_6 (25%) units was established for the binary $90MoO_3$-$10Nd_2O_3$ glass. The B-O first neighbour distribution show a relatively broad first neighbour distance at 1.40Å, the average coordination numbers show the presents of trigonal BO_3 and tetrahedral BO_4 groups. For $50MoO_3$-$25Nd_2O_3$-$25B_2O_3$ sample mixed MoO_4-BO_4 and MoO_4-BO_3 linkages form pronounced intermediate-range order.

1. Introduction

Rare-earth molybdate phases are known to exhibit a great variety of important physical properties including high ion and electron conductivity of fast oxide ion conductors derived from the parent oxide $La_2Mo_2O_9$ (LAMOX family [1,2], non-linear optical response and luminescent properties ($Sm_2(MoO_4)_3$, $LiLn(MoO_4)_2$, Ln=Nd, Pr, Ho), e.g. [3].

In contrast to the most molybdates with relatively well-characterised crystalline and magnetic structures, e.g. [4-7], the preparation and structural information on amorphous molybdate systems is not ample. MoO_3 is well known as a conditional network former and it is not able to form a glass itself at even rapid cooling rates. The main problems of the preparation of molybdate glasses without modifier ions or participation of classical network formers are connected with the high crystallization tendency of the components.

Recently, the group of Dimitriev [8-10] has undertaken a systematic study to clear up the tendency for glass formation and immiscibility in the molybdate systems containing rare-earth oxides and in the complex mixed boromolybdate glasses. The phase diagram for several ternary oxides, MoO_3-$Nd(La)_2O_3$-B_2O_3 was set up [9] and the basic structural and optical characteristics were investigated by XRD, DTA, SEM, UV-VS spectra and IR spectroscopy [10].

The aim of this work is to get a deeper insight into the structural network characteristics of new series of molybdate conditional glass former containing modifiers Nd_2O_3, MgO and classical glass former B_2O_3. The Nd_2O_3 is appropriate component due to its specific optical properties but it increases the melting temperature. The MgO influences significantly properties as viscosity, glass transition temperature and chemical durability. By introducing of B_2O_3 it is possible to obtain low melting materials.

Journal of Physics: Conference Series **746** (2016) 012068 doi:10.1088/1742-6596/746/1/012068

For this study four new amorphous samples of the glassy MoO_3-Nd_2O_3-(MgO)/(B_2O_3) system have been prepared by rapid quench technique. Both neutron and high-energy X-ray diffraction experiments have been performed, and for structure modelling the reverse Monte Carlo (RMC) computer simulation technique [11] was applied. We expect to obtain new structural data, especially on the characteristics of the partial atomic pair correlation functions, the first neighbour atomic distances, coordination numbers and network former units and their linkages.

2. Experimental details and methods

2.1. Samples

The amorphous materials have been prepared by a conventional melt quenching method. The nominal composition of the glasses synthesized in the present study are binary 90MoO_3-10Nd_2O_3 (hereafter referred to as Mo90Nd10), ternary 80MoO_3-15Nd_2O_3-5MgO (Mo80Nd15Mg5), 75MoO_3-12.5Nd_2O_3-12.5MgO (Mo75Nd12.5Mg12.5) and 50MoO_3-25Nd_2O_3-25B_2O_3 (Mo50Nd25B25) (mol%). Commercial powder of reagent grade MoO_3, Nd_2O_3 and MgO were used as starting materials and were mixed in an alumina mortar. The boron containing sample was prepared from boron oxide, ^{11}B-isotope enriched to 99.6%, to avoid the high neutron absorption of natural boron. The batches were melted at 1200-1300°C for 30 min in air in a platinum crucible. The melting temperature was limited to 1300°C taking into account the volatility of the components. The glasses were obtained by pouring the melts onto an iron plate and by pressing with another iron plate (cooling rate: 10^2 K/s).

2.2. Neutron and X-ray diffraction experiments

Neutron diffraction (ND) measurements have been performed in a relatively broad momentum transfer range, Q=0.4-35Å^{-1}, combining the data measured by the two-axis 'PSD' diffractometer (λ_0=1.068Å) [12] at the 10 MW Budapest research reactor for all samples and by the time-of-flight 'HIPD' instrument [13] at the LANSCE pulsed neutron source for Mo80Nd15Mg5 and Mo75Nd12.5Mg12.5 samples and by the '7C2' diffractometer at the LLB-CEA-Saclay (λ_0=0.726 Å) [14] for Mo90Nd10 and Mo50Nd25B25 samples. The powder specimens of about 3–4 g/each were filled in a cylindrical vanadium sample holder of 8 mm (PSD, 7C2) and 10 mm (HIPD) diameter for the three types of experiments. The structure factors, $S(Q)$s, were evaluated from the raw experimental data independently for the three types of measurements, using the programme packages available at the facilities. The three sets of $S(Q)$s were obtained in the Q=0.4-9.6Å^{-1} (PSD), Q=0.2-15.3Å^{-1} (7C2) and in the 1.5-35Å^{-1} interval using the HIPD experiments. The agreement between the corresponding $S(Q)$ data was within 1% in the overlapping Q range. The statistics of the PSD data is excellent at relatively low Q values, while the statistics and the information content of the 7C2 and HIPD data is unique above 8Å^{-1}. Therefore the $S(Q)$ patterns were combined by normalizing the PSD data to the 7C2 and HIPD data sets in the 4–8Å^{-1} interval by the least-squares method, and the average values of the two spectra were used for further data treatment, while for Q<4Å^{-1} the PSD data and for Q>8Å^{-1} the 7C2 and HIPD data were used. The neutron $S(Q)$ data were obtained with a good signal-to-noise ratio up to Q_{max}=15.3Å^{-1} for Mo90Nd10 and Mo50Nd25B25 and Q_{max}=35Å^{-1} for Mo80Nd15Mg5 and Mo75Nd12.5Mg12.5 samples.

The high-energy X-ray diffraction (XRD) measurements were carried out at the BW5 experimental station [15] at HASYLAB, DESY. The powdered samples were filled into quartz capillary of 2 mm in diameter (wall thickness of ~0.02 mm). The energy of the radiation was 109.5 keV (λ_0=0.113Å). Raw data were corrected for detector dead time, background, absorption, and variations in detector solid angle. The X-ray $S(Q)$ data were obtained with a good signal-to-noise ratio up to Q_{max}=25Å^{-1}.

The overall run of the curves for the two radiations is fairly different, because of the differences in the weighting factors, w_{ij} of the partial structure factors, $S_{ij}(Q)$:

$$S(Q)=\sum_{i,j}^{k} w_{ij} S_{ij}(Q), \quad (1)$$

$$w_{ij}^{ND} = \frac{c_i c_j b_i b_j}{\left[\sum_{i,j}^{k} c_i b_j\right]^2}, \quad (2)$$

$$w_{ij}^{XD}(Q) = \frac{c_i c_j f_i(Q) f_j(Q)}{\left[\sum_{i,j}^{k} c_i f_j(Q)\right]^2}, \quad (3)$$

where c_i, c_j are the molar fractions of the components, b_i, b_j the coherent neutron and $f_i(Q)$, $f_j(Q)$ the X-ray scattering amplitudes, and k is the number of elements in the sample. Note, that the neutron scattering amplitude is constant [16], while the X-ray scattering amplitude is Q-dependent [17]. In order to compare the corresponding weighting factors for the two radiations, Table 1 includes the Q-independent w_{ij}^{ND} together with a characteristic value for $w_{ij}^{XD}(Q)$ at Q=1.06Å^{-1}. It can be seen that the Mo-O and Nd-O atom pairs have a significant contribution for both radiations, while B-O plays a significant contribution only in the neutron experiment. The O-O contribution has a dominant weight in the neutron experiment, in contrast to the X-ray case, where it is much weaker.

Table 1. Neutron and X-ray (at Q=1.06 Å^{-1}) weighting factors, w_{ij}(%) for the investigated glasses.

Composition	**90%MoO$_3$-10%Nd$_2$O$_3$**		**75%MoO$_3$-12.5%Nd$_2$O$_3$-12.5%MgO**		**50%MoO$_3$-25%Nd$_2$O$_3$-25%B$_2$O$_3$**	
Atom pairs	**Neutron**	**X-ray**	**Neutron**	**X-ray**	**Neutron**	**X-ray**
Mo-O	33.79	33.20	28.96	28.24	14.97	16.83
Nd-O	8.44	10.51	10.95	13.45	17.15	24.02
Mg-O	-	-	3.82	1.22	-	-
B-O	-	-	-	-	14.82	1.82
O-O	48.58	10.41	45.81	9.78	38.87	9.35
Mo-Mo	5.87	26.46	4.57	20.38	1.44	7.43
Mo-Nd	2.93	16.75	3.46	19.42	3.30	21.20
Mo-B	-	-	-	-	2.85	1.61
Nd-Nd	0.36	2.65	0.65	4.62	1.89	15.13

On the other hand, the contribution of the metal-metal second neighbour Mo-Mo, Mo-Nd and Nd-Nd atomic correlations appears with a significant weight only in the X-ray experiment.
As a first step of data evaluation the total reduced correlation function, $G(r)$ was calculated by the usual Fourier transformation (see our previous papers [18,19]). The neutron $G(r)$s are shown in Fig. 1.

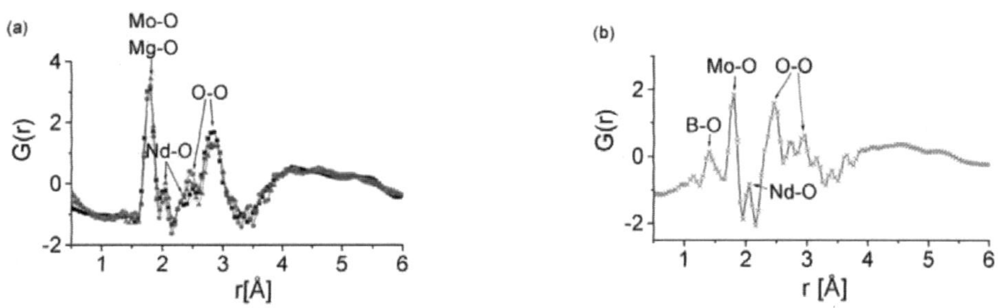

Figure 1. Experimental neutron total reduced distribution function, $G(r)$: a) Mo90Nd10 (square), Mo80Nd15Mg5 (circle), Mo75Nd12.5Mg12.5 (triangle) and b) Mo50Nd25B25 (cross) glasses obtained by Fourier transformation from the corresponding $S(Q)$.

The *G*(*r*) curves show up a sharp peak at 1.8Å for all specimens, which can be attributed to the Mo-O distance. The Mg-O is expected between 1.7-1.9Å, which overlaps with the Mo-O distribution. As far as the first peak is rather sharp, we can conclude that Mg-O first neighbour distance is also at ~1.8Å. The next small intensity but sharp peak appears at ~2.05Å, which may be attributed to the Nd-O first neighbour distance. The O-O distances appear in the 2.5-3Å interval, however, an overlap might be with other second neighbour distributions. In the case of the B-containing specimen, an additional peak appears in *G*(*r*) at ~1.4Å (Fig. 1/b), which can be related to the B-O first neighbour distance. In order to get a deeper insight into the network structure, the RMC simulation technique has been used.

3. Reverse Monte Carlo modelling and results

The experimental *S*(*Q*) data have been simulated by the RMC simulation technique. The RMC minimizes the squared difference between the experimental *S*(*Q*) and the calculated one from a 3D atomic configuration. The RMC algorithm calculates the one-dimensional partial atomic pair correlation functions $g_{ij}(r)$, and they are inverse Fourier transformed to calculate the partial structure factors, $S_{ij}(Q)$. For the RMC starting model disordered atomic configuration was built up with a simulation box containing 10.000 atoms. The density values were 0.0733, 0.0668, 0.0670, 0.0664 atoms·$Å^{-3}$ for Mo90Nd10, Mo80Nd15Mg5, Mo75Nd12.5Mg12.5, Mo50Nd25B25 samples, respectively, and the corresponding RMC half-box lengths were 25.739, 26.548, 26.522 and 26.601 Å.

In the RMC simulation procedure two types of constraints were used, the minimum interatomic distances between atom pairs (cut-off distances) and connectivity constraints. We have used connectivity constraints only for the network former B-O atom pairs. It is reasonable to suppose that boron atoms are surrounded both by three-coordinated and four-coordinated oxygen atoms, forming trigonal BO_3 and tetrahedral BO_4 units as reported for binary neodymium borate glasses [20], for ternary rare-earth boromolybdate MoO_3-Nd(La)$_2O_3$-B_2O_3 [9,10], for B_2O_3-Bi_2O_3-MoO_3 glass [21], similarly, as for binary alkali borate glasses [22 and references therein]. For the starting cut-off distances we have used the characteristic values obtained from the total *G*(*r*)s (see Fig.1), i.e. for B-O 1.28Å, Mo-O 1.65Å , Nd-O 2.0Å, O-O 2.32Å. Several RMC runs have been completed by modifying the cut-off distances in the way, that the results of each run have been carefully checked to obtain reliable data for each $g_{ij}(r)$ and coordination number distributions, $CN_{ij}(n)$. The convergence of the RMC calculations was good and the final *S*(*Q*)s matched very well the experimental ones. (see Fig. 2).

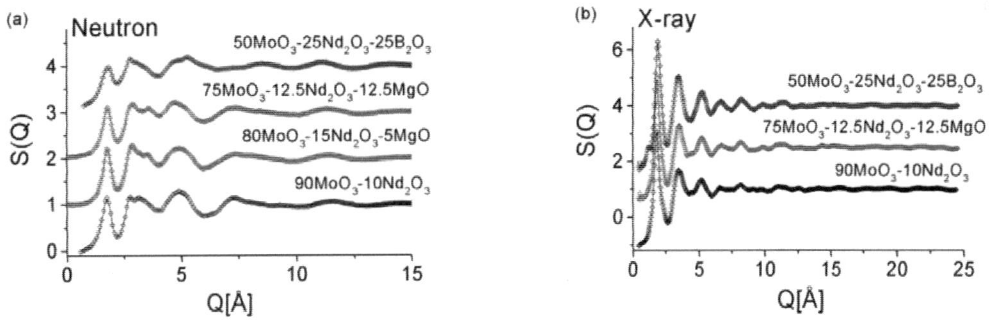

Figure 2. Experimental data (rhombus) and RMC simulation (solid line): a) Neutron and b) X-ray diffraction.

The partial atomic pair correlation functions, $g_{ij}(r)$ have been revealed from RMC simulation with a good reproducibility and acceptable statistics. The oxygen-linked $g_{i\text{-}O}(r)$ and coordination number distributions are shown in Figure 3 and Figure 4.

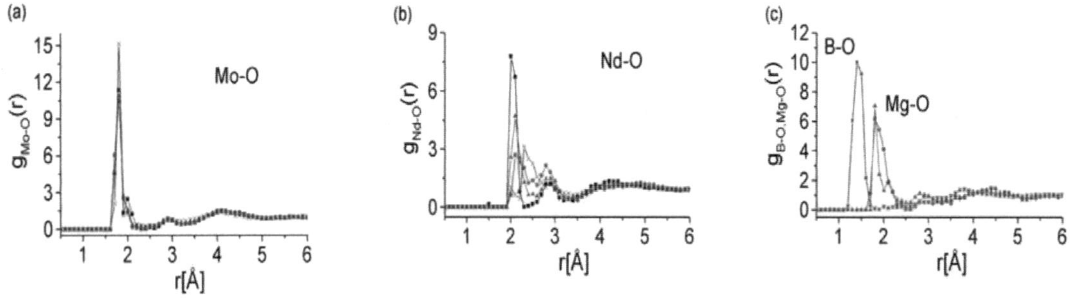

Figure 3. O-linked partial atomic pair correlation functions obtained from the RMC modelling for the: Mo90Nd10 (square), Mo80Nd15Mg5 (circle), Mo75Nd12.5Mg12.5 (triangle), Mo50Nd25B25 (cross).

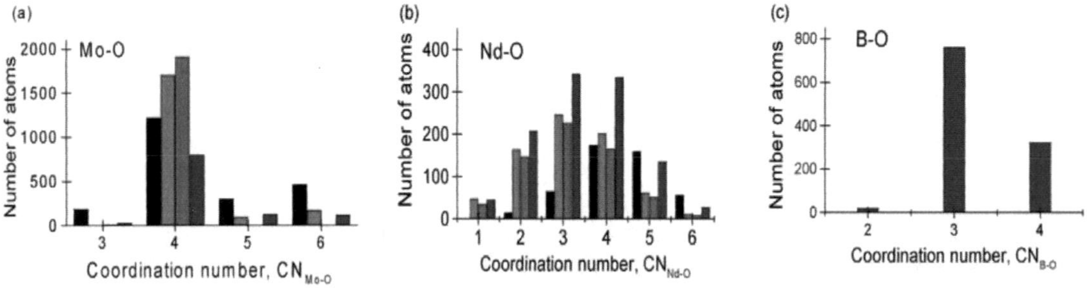

Figure 4. Coordination number distributions obtained from the RMC modelling for the glasses: Mo90Nd10 (black), Mo80Nd15Mg5 (red), Mo75Nd12.5Mg12.5 (green), Mo50Nd25B25 (blue).

The corresponding first neighbour distances are presented in Table 2 and the average CN_{i-O} in Table 3.

Table 2. Oxygen-linked atomic distances, r_{ij} (Å)

Sample	r_{ij} (Å)				
	Mo-O (±0.02)	**Nd-O (±0.05)**	**B-O (±0.02)**	**Mg-O (±0.1)**	**O-O (±0.1)**
Mo90Nd10	1.80/2.0	2.0/2.8	-	-	2.3/2.8
Mo80Nd15Mg5	1.80	2.1/2.8	-	1.8	2.45/2.8
Mo75Nd12.5Mg12.5	1.80	2.1/2.9	-	1.8/2.1	2.4/2.8
Mo50Nd25B25	1.80/2.0	2.0/2.3/2.9	1.4	-	2.4/2.9

Table 3. Average partial coordination numbers, CN_{ij} (atom)

Sample	CN_{ij}				
	Mo-O (±0.2)	**B-O (±0.1)**	**Nd-O (±0.5)**	**Mg-O (±0.5)**	**O-O (±0.5)**
Mo90Nd10	4.9	-	4.3	-	8.0
Mo80Nd15Mg5	4.2	-	3.1	4.3	6.9
Mo75Nd12.5Mg12.5	4.0	-	3.1	3.8	6.2
Mo50Nd25B25	4.2	3.3	4.3	-	5.9

In the studied system the results obtained from RMC modelling have shown significant correlations between several second neighbour atom pairs, as it is shown in Figure 5. This is important, as far as the analysis of these results may give useful information on the linkage of the basic structural units forming the network structure. The second neighbour partial distributions are collected in Figure 5.

Journal of Physics: Conference Series **746** (2016) 012068 doi:10.1088/1742-6596/746/1/012068

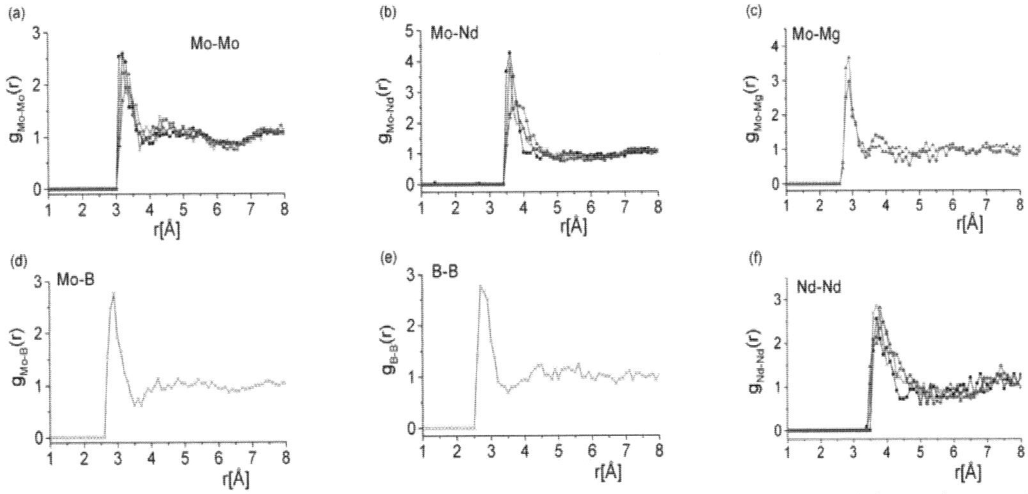

Figure 5. Second neighbour partial atomic pair correlation functions obtained from the RMC modelling for the molybdate glasses, for Mo90Nd10 (square), Mo80Nd15Mg5 (circle), Mo75Nd12.5Mg12.5 (triangle) and Mo50Nd25B25 (cross).

4. Discussion

In the present study for the Mo-O atom pair correlations a well-defined first neighbour distance is obtained at 1.80±0.02Å, and the course of the curves is practically the same for all investigated samples (see Fig. 3/a). This is an indication on the formation of stable Mo-O network structure, even if the sample contains boron oxide network former. The observed Mo-O distance is somewhat greater than the 1.75(5)Å distance reported for glassy $90MO_3$-$10Nd_2O_3$ from EXAFS measurements [23], and shows similarity to the Mo-O 1.73-1.82Å distances in crystalline $LaBMoO_6$ [24] and in Nd_2O_3-MoO_3 [5]. The Mg-O first neighbour distance is obtained also at 1.8±0.1Å, but its accuracy is considerably less than that of the Mo-O distance. This is partly because its relatively low concentration, especially for the Mo80Nd15Mg5 sample, and also due to the low scattering amplitude of Mg for both radiations, resulting in low weighting factors. For the Mo75Nd12.5Mg12.5 sample the Mg-O distributions show a second peak at 2.1±0.1Å, this can overlap with the Nd-O distances. The Nd-modifier shows characteristic Nd-O first neighbour distance at 2.05Å, and a next one at 2.85Å for the binary Mo90Nd10, while for the other compositions the first neighbour distance shifts to slightly higher distances 2.1-2.3Å. The intensity decreases and the peak is less characteristic inspite of the increased Nd-concentration in the ternary samples. The reason for this result may be in the more complicated network due to the increased number of constituent components. The small peak at 2.85Å seems to be reproducible, however, the peak position agrees with the second O-O distance, thus its origin may be artificial. The Nd-O distances reported in the literature are somewhat greater than our results, i.e. 2.51-2.73Å in glassy MoO_3-Nd_2O_3 [8], 2.33-2.5Å in crystalline Nd_2O_3-MoO_3 [5], 2.24-2.32 Å in $Y_{2-x}Nd_xMo_4O_{15}$ [25]. The O-O distribution shows two characteristic peaks at 2.4 and 2.8Å (see Table 2). We assume that the shorter distance belongs to the B-O network, while the longer distance to the Mo-O network. The B-O distribution function shows a relatively broad first neighbour distance at 1.40±0.05Å with a half-width value of 0.25Å, is usually the case for B-O network [18, 22].

Figure 4 shows the coordination number distributions as revealed from the RMC simulation, while the corresponding average coordination numbers, CN_{ij} are tabulated in Table 3.

In the high Mo-containing binary Mo90Nd10 sample the average number of oxygen atoms around Mo is 4.9±0.2 atoms. We have investigated the possibility of a mixed network, containing both MoO_4 and MoO_6 units, by RMC simulation. If we suppose that MoO_4 units similarly to the crystalline $LaBMoO_6$ [24], Nd_2O_3-MoO_3 [5] and usually this is the case for molybdate crystals. For glassy molybdate network the presence of MoO_4^{-2} units were reported [9, 10, 21], while the formation of MoO_6 units was also observed, they are present in the network. In case of binary system it would mean that 55% of the

Mo atoms are 4-fold oxygen coordinated, while 25% are 6-fold coordinated, while the rest 20% are in 3- and 5-fold coordinated states, with an error about 5%. Taking into consideration the $g_{Mo\text{-}O}(r)$ distribution, we can establish that the shorter distance at 1.8Å can be attributed to the MoO_4 units, while in addition two further oxygen neighbours appear at a longer 2.0Å distance forming MoO_6 units.

According to Table 3, with decrease of Mo content the number of oxygen neighbours slightly decreases to 4.0 atoms. For the Mo80Nd15Mg5 and Mo75Nd12.5Mg12.5 ternary samples the average coordination number, as calculated from the RMC modeling, is $CN_{Mo\text{-}O}$=4.2 and 4.0±0.2 atoms, respectively. As far as, the corresponding $g_{Mo\text{-}O}(r)$ distributions contains a characteristic peak at 1.8Å and small shoulder at 2.1Å, we may conclude that 87 and 98% of the Mo atoms form MoO_4 units, while about 8% and 0% form MoO_6 units in case of Mo80Nd15Mg5 and Mo75Nd12.5Mg12.5 samples, respectively. For the Mo50Nd25B25 sample the average coordination number is $CN_{Mo\text{-}O}$=4.2±0.2 atoms. The $g_{Mo\text{-}O}(r)$ distributions besides well known 1.8Å peak show a small "sub"-peak at 2.0Å, than we may conclude that 72% of the Mo atoms form MoO_4 units, while about 10% form MoO_6 units, the rest 18% are in 3- and 5-fold coordinated states, with an error about 5-7%. In case of ternary systems distortion of the tetrahedral MoO_4 units can not be excluded.

The average number of oxygen neighbours of the modifier Nd is 4.3 atoms for both the binary Mo90Nd10 and for the boron-containing ternary sample, while it is significantly smaller for the Mg containing samples, 3.1 atoms. The reason may be that Mg is also a modifier, which plays a similar role like Nd, increasing the number of non-bridging oxygen (NBO) atoms.

The Mg-O coordination number is 4.3 and 3.8 close to form tetrahedral units, for Mo80Nd15Mg5 and Mo75Nd12.5Mg12.5, respectively. The average coordination number suggests that Mg^{2+} cation can act as a conditional network-former. The Mg-O bond distances were in the range 1.8-2.1 Å, in accord with the values experimentally reported for Mg^{2+} cations in tetrahedral coordination [26].

The boron atoms are coordinated mostly by 3 and 4 oxygen atoms, forming trigonal BO_3 and tetrahedral BO_4 units, similarly to other boron containing networks containing modifier atoms as discussed above (see i.e. refs [9-10, 20-22]). The RMC model calculation results not only 3- and 4-fold oxygen coordinated boron atoms, but also 2 oxygen neighbours (Fig. 5/c). This may be an artificial consequence of RMC modelling, as far as the calculation starts from a random distribution of the atoms packed in the simulation box. The relative amount of the different coordinated boron is an important question, especially the number of BO_3 and BO_4 units. Nd_2O_3 acting as modifier causes partial conversion of BO_3 and BO_4 groups leading to the formation of NBO. For Mo50Nd25B25 sample ~68% of B-atoms are 3-coordinated, while ~30% are 4-coordinated, while the rest 2% are in 2-fold coordinated states, with an error about 3%, the average coordination number is 3.3 atoms.

The O-O average coordination number is 8 atoms for the 2-component Mo90Nd10 sample, which is the highest among the investigated samples. The average O-O coordination number reduces to 6 atoms for the samples with increasing number of contributing components, which may be the consequence of mixed structural units.

The RMC modelling also provides information on the second coordination sphere. The basic question is, whether the linkage of the basic structural units show any correlation or not? The analysis of the metal-metal and boron-metal pair correlation functions show pronounced correlations, the corresponding atomic pair distances are shown in Fig. 5. The shortest second neighbour distances are obtained for B-B as 2.7Å, Mo-B 2.9Å which indicate mixed linkages, and Mo-Mg 2.8Å which is related to NBO bond. These results indicate the existence of a pronounced intermediate-range ordering.

5. Conclusions

Rare-earth molybdate glasses have been prepared by melt quenching technique, and the network structure was investigated by neutron and high-energy X-ray diffraction. For data evaluation the RMC simulation technique was applied to obtain a possible 3-dimensional network configuration, which is consistent with the experimental data. The RMC simulation box did contain 10000 atoms and the usual constraints were applied, i.e. atomic density, cut-off distances (nearest neighbour distances for all atom pairs) and connectivity. From the RMC modelling the partial atomic correlation functions $g_{ij}(r)$ and the coordination number distributions CN_{ij} have been revealed. Formations of MoO_4 units were estabilished with 1.80Å Mo-O distance. In the binary glass MoO_4 (55%) and MoO_6 (25%) structural units were revealed. In ternary systems mainly MoO_4 units are present, and with decreasing MoO_3 concentration, the ratio of MoO_6 units roughly decreases. For the B-containing ternary glasses B-O first neighbour distance was obtained at 1.40Å, the B-O network is formed by BO_3 and BO_4 groups. Mixed MoO_4-BO_4 and MoO_4-BO_3 linkages form pronounced intermediate-range order.

Acknowledgement
The authors are grateful to Drs. Gy. Mészáros, L. Alexandrov and Prof. Y. Dimitriev for their helpful contribution in this work. The authors would like to thank Drs. Anna Llobet and Martin von Zimmermann instrument scientists at the neutron- and X-ray diffraction experimental stations at the HIPD (Lujan Neutron Scattering Centre) and BW5 (Hasylab at Desy). The research was supported by the EU-FP7-NMI3 No.283883. Thanks to the financial support of Hungarian OTKA-PD 109384 and to the bilateral cooperation of Hungarian-Bulgarian Academy of Sciences SNK-63/2013.

References

[1] Lacorre P, Goutenoire F, Bohnke O, Retoux R, Laligant Y 2000 *Nature* **404** 856
[2] Georges S, Goutenoire F, Bohnke O, Steil M C , Skinner S J, Wiemhöfer H D, Lacorre P 2004 *J. New Mat. Electrochem. Systems* **7** 51
[3] Kato A, Uchitomi N, Oishi S, Shishido T, Iida S 2006 *Physica Status Solidi (c)* **3(8)** 2709
[4] Naruke H, Yamase T 2003 *Journal of Solid State Chemistry* **173** 407
[5] Barker R S, Evans I R 2006 *Journal of Solid State Chemistry* **179** 1918
[6] Zhao D, Cheng W D, Zhang H, Hang Fang S P 2008 *Dalton Transactions* **28** 3709
[7] Atuchin V V, Chimitova O D, Gavrilova T A, Molokeev M S, Kim S J, Surovtsev N V, Bazarov B G 2011 *Journal of Crystal Growth* **318** 683
[8] Alexandrov L, Iordanova R, Dimitriev Y, Hamda K, Ide J, Milanova M 2008 *Advanced Materials Research* **39-40** 37
[9] Aleksandrov L, Iordanova R, Dimitriev Y 2009 *Journal of Non-Crystalline Solids* **355** 2023
[10] Aleksandrov L, Komatsu T, Iordanova R, Dimitriev Y 2011 *Journal of Physics and Chemistry of Solids* **72** 263
[11] R McGreevy R L, Pusztai L 1988 *Mol. Simul.* **1** 359; http://www.szfki.hu/~nphys/rmc++/opening.html
[12] Sváb E, Mészáros Gy, Deák F 1996 *Materials Science Forum* **228** 247; http://www.bnc.hu/
[13] Proffen Th, Billinge S J L, Egami T, Louca D 2003 *Zeitschrift für Kristallographie* **218** 132
[14] Ambroise J P, Bellissent R 1984 *Rev. Phys. Appl.* **19** 731; http://www-llb.cea.fr/en/
[15] Poulsen H, Neuefeind J, Neumann H B, Schneider J R, Zeidler M D 1995 *J. Non-Cryst. Solids* **188** 63
[16] Hannon A C, ISIS Disordered Materials Database 2006, http://www.isis.rl.ac.uk/disordered
[17] Waasmaier D, Kirfel A 1995 *Acta Crystallogr. A* **51** 416
[18] Fábián M, Sváb E, Proffen Th, Veress E 2008 *Journal of Non-Crystalline Solids* **354** 3299
[19] Fabian M, Proffen Th, Ruett U, Veress E, Svab E 2010 *J. Phys.: Condens. Matter* **22** 404206
[20] Burns A E, Winslow D W, Clarida W J, Affatigato M, Feller S A, Brow R K 2006 *Journal of Non-Crystalline Solids* **352** 2364
[21] Iordanova R, Aleksandrov L, Bachvarova-Nedelcheva A, Ataala M, Dimitriev Y 2011 *Journal of Non-Crystalline Solids* **357** 2663
[22] Fábián M, Sváb E, Proffen Th, Veress E 2010 *Journal of Non-Crystalline Solids* **356** 441
[23] Alexandrov L, Iordanova R, Dimitriev Y 2007 *Phys. Chem. Glasses: European J. Glass Sci. Technol. B.* **48(4)** 242
[24] Wang Z, Zhang L, Wang G, Song M, Huang Y, Wang G 2009 *Optical Materials* **31** 849
[25] Wang H, Peng J, Chen D F, Hu Z 2006 *Solid State Sciences* **8** 1144
[26] Deng F, Yue Y, XiaoT, Du Y, Ye Ch, An L, Wang H 1995 *J. Phys. Chem.* **99** 6029

Journal of Physics: Conference Series **746** (2016) 012069 doi:10.1088/1742-6596/746/1/012069

Target Nanoparticles for Therapy – SANS and DLS of Drug Carrier Liposomes and Polymer Nanoparticles

T Nawroth[1], R Johnson[1], L Krebs[1], P Khoshakhlagh[1], P Langguth[1], N Hellmann[2], G Goerigk[3], P Boesecke[4], A Bravin[4], G Le Duc[4], N Szekely[5], R Schweins[6]

[1] Pharmacy, Chemistry Department, Gutenberg University, D-55099 Mainz, Germany
[2] Biophysics, Biology Department, Gutenberg University, D-55128 Mainz, Germany
[3] Soft Matter and Functional Materials, HZB, Berlin, D-14109 Mainz, Germany
[4] European Synchrotron Facility ESRF, F-38043 Grenoble, France
[5] MLZ, FRM-II Reactor, JCNS outstation, D-38042 Garching, Germany
[6] LSS, Institut Laue Langevin ILL, F-38042 Grenoble, France

nawroth@mpsd.de

Abstract. Target Nano-Pharmaceutics shall improve therapy and diagnosis of severe diseases, e.g. cancer, by individual targeting of drug-loaded nano-pharmaceuticals towards cancer cells, and drug uptake receptors in other diseases. Specific ligands, proteins or cofactors, which are recognized by the diseased cells or cells of food and drug uptake, are bound to the nanoparticle surface, and thus capable of directing the drug carriers. The strategy has two branches: a) for parenteral cancer medicine a ligand set (2-5 different, surface-linked) are selected according to the biopsy analysis of the patient tissue e.g. from tumor.; b) in the oral drug delivery part the drug transport is enforced by excipients/ detergents in combination with targeting materials for cellular receptors resulting in an induced drug uptake. Both targeting nanomaterials are characterized by a combination of SANS + DLS and SAXS or ASAXS in a feedback process during development by synthesis, nanoparticle assembly and formulation.

1. Synopsis and Concept

Heavy metal loaded target Nanoparticles [1-3] shall improve therapy and diagnosis of severe diseases, e.g. cancer, by individual targeting [4,5] of drug-loaded nano-pharmaceuticals towards cancer cells. Specific protein or cofactor ligands, which are recognized by the diseased cells (cancer) or the drug uptake system, are bound to the nanoparticle surface, and thus capable of directing the drug carriers. In the concept a ligand set is coupled by a fast assembly technique (click link) in the very last step of the formulation. In the later clinical cancer therapy application the ligands set (2-5 different) will be selected according to the differential biopsy analysis of the patient tissue (tumor, healthy). The targeting nanomaterials are characterized and optimized by a combination of SANS+DLS [7] and SAXS, ASAXS, and radiotherapy tests with cells in a feedback process during development.

2. Materials and Methods

All materials were at least of analytical grade and obtained from Sigma-Aldrich Schnelldorf and Boehringer Ingelheim. The synthesis was done at the pharmaceutical institute of the Gutenberg University and Nanovel Ltd. & Co KG, Langenlonsheim (www.nanovel.de). The liposomes were prepared by the film method with subsequent vortexing (1 min, 2400 rpm) yielding multilamellar vesicles MLV (160g/l), followed by addition of the drug stock solution and 11-fold extrusion through 100 nm polycarbonate membranes (Millipore). The DLS experiments were done as described earlier [7] at 633 nm (5 mW HeNe laser, Uniphase) with a 5000/EPP correlator, SPIC-II single photon detector (ALV, Langen) and projecting DLS optics devive with 170° backscattering ProSpecD501 (Nanovel). The neutron scattering experiments were done at the ILL at the D11 instrument, and the KWS-2 instrument at the FRM-II reactor, MLZ Garching, with neutrons of 0.6 nm wavelength (9% $\Delta \lambda / \lambda$).

3. Results and Discussion

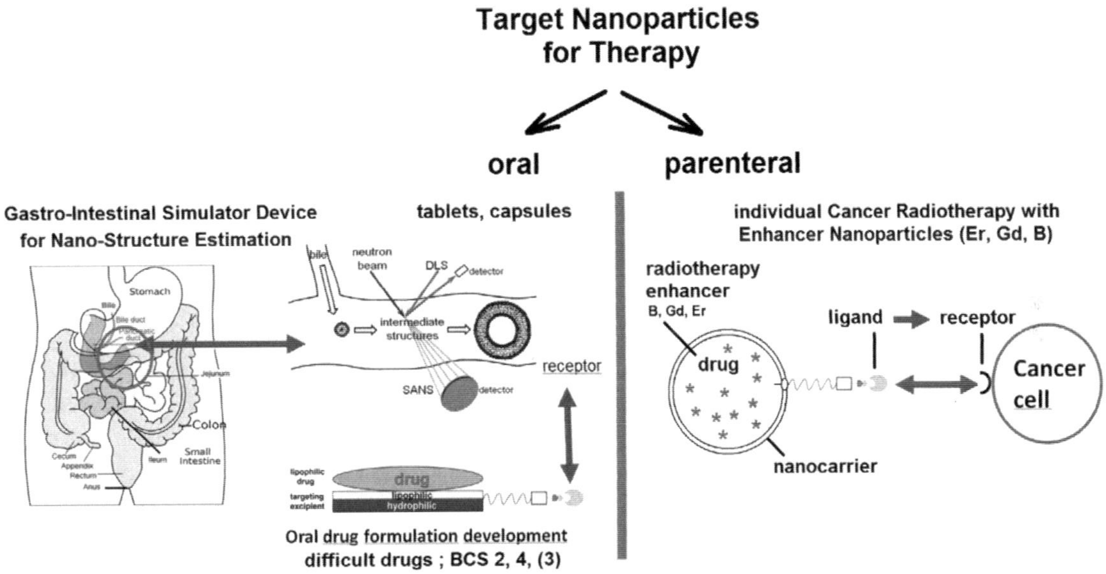

Fig.1: Target Nanoparticles can be applied by oral or parenteral ways. The ligands direct the drug carriers to specific cell receptors. The nanoparticle-ligand structure and dynamics, and the intermediate nano-assemblies with proteins and bile are studied by a combination of SANS and DLS.

We synthesize and combine targeting excipients for oral drug delivery and modifiers of metal drug loaded nanoparticles for radiotherapy, which bear a specific receptor/carrier ligand domain at the end of a linker domain. The other end domain is hydrophobic, i.e. capable of drug binding (drug co-drug complex for oral application), or embedding into the surface of hydrophobic drug carrier nanoparticles for parenteral supply. In both cases the result is a drug carrier directed to specific receptor, as depicted in fig.1. All components are biocompatible and bio-degradable, which limits the selection of chemical structures, but prepares pharmaceutical licensing and medical application. The nanoparticles are biodegradable polymer [3], e.g. PLGA; ferrofluids [6], micelles [5,11,12], lipidic particles [7] or liposomes [8-10]. For oral application the ligand coupling is covalent, while the linkers for parenteral nano-pharmaceuticals are bound by cleavable S-S-bonds as terminal step of the preparation of a person- and case-specific medicine.

The structure of the modified nanoparticles is analyzed by neutron small angle scattering SANS in combination with dynamic light scattering DLS [8,11,12], SAXS and metal specific X-ray scattering ASAXS, while the therapy effect of the drug is investigated with kinetic cell cultures as tumor models [1]. The attached protein shell resembles the protein corona formed by nanoparticles in the blood [13], but it consists of specific receptor targeting proteins, which are bound covalently. The particles in pharmaceutical samples depict a size scale of 1 nm to 100 µm in parallel. The nano-solutions are highly concentrated (up to 30 %, crowded conditions). Thus a combination of SANS, (A)SAXS and DLS is required [8], the latter as back scattering to avoid multiple scattering.

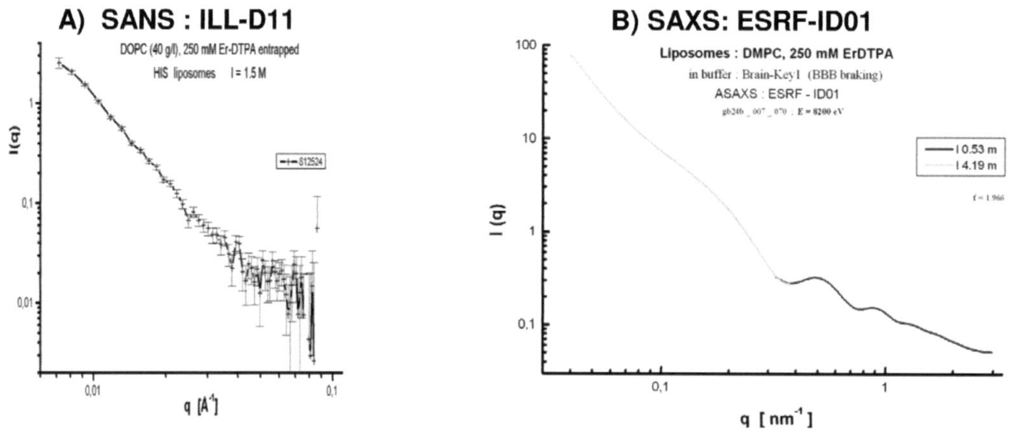

Fig.2: Structure analysis of lanthanide entrapping liposomes (0.25M ErDTPA with (A) neutron small angle scattering and (B) synchrotron X-ray small angle scattering SAXS.

Fig.2 depicts the structure analysis of liposome nanoparticles for radiotherapy with entrapped lanthanides [1-3] by SANS (ILL-D11) and SAXS (ESRF-ID01). The neutron small angle scattering gave radius of gyration ($R_g = 21.3 \pm 0.12$ nm) and membrane span (4 nm) of the metal nanoparticles in highly concentrated D_2O buffer (1.5 M ionic strength). The (A)SAXS investigation revealed a metal-membrane interaction, which is indicated by the oscillation of the scattering curve in Fig.2B. The radiotherapy tests with cells as kinetic tumor model [1] depicted an increased photon induced cell inactivation, when the appropriate drug release was observed ($t_{1/2}$ = 1 h @ c > 0.1 M entrapped drug).

Acknowledgement: We thank for support by the German ministry of science and education BMBF, grant 05KS7UMA.

References

[1] Buch K, Peters T, Nawroth T, Sänger M, Schmidberger H, Langguth P 2012 *Radiation Oncology* **7**, 1-6

[2] Peters T, Grunewald C, Blaickner M, Ziegner M, Schütz C, Iffland D, Hampel G, Nawroth T, Langguth P 2015 *Radiation Oncology* **10**, 52-64

[3] "A particulate system for use in diminishing cell growth / inducing cell killing" 2012 *EU-Patent* 11 007 401.0 ; PCT 13 07 12 keyword "Lanthanide", Johannes Gutenberg-Universität Mainz; inventors: Buch K, Nawroth T, Langguth P, Schmidberger H

[4] Ferrara TA Hodge JW, Gulley JL 2009 *Current Opinion in Molecular Therapeutics* **11**, 37-42

[5] Alemdaroglu FE, Alemdaroglu NC, Langguth P, Herrmann A 2008 Adv. Materials **20**, 899-902

[6] Alexiou C, Schmid R, Jurgons R, Kremer M, Wanner G, Bergemann C, Huenges E, Nawroth T, Arnold W, Parak FG 2006 *Eur Biophys J* **35**, 446-50

[7] Johnson R., Nawroth T., Khoshakhlagh P., Langguth P., Szekely N.K., et al. 2014 *Eur J Lipid Sci Techn* **116**, 1167-1173

[8] Nawroth T., Buch P., Buch K., Langguth P., Schweins R 2011 *Mol Pharmaceutics*, **8**, 2162-72

[9] Nawroth T, Rusp M, May RP 2004 *Physica B* **350**, e635-638

[10] Hofmann AM, Wurm F, Hühn E, Nawroth T, Langguth P, Frey H. *2010 Biomacromolecules* **11**, 568-574

[11] Khoshakhlagh P, Johnson R, Nawroth T, Langguth P, Schmueser L, Hellmann N, Decker H, Szekely NK 2014 *Eur J Lipid Sci Techn* **116**, 1155-116

[12] Khoshakhlagh P, Johnson R, Langguth P, Nawroth T, Schmueser L, Hellmann N, Decker H, Szekely NK 2015 *J Pharm Sci* **104**, 2213-24

[13] Tenzer S, Docter D, Rosfa S, Wlodarski A, Kuharev J, Rekik A, Knauer SK, Bantz C, Nawroth T, Bier C, Sirirattanapan J, Mann W, Treuel L, Zellner R, Maskos M, Schild H, Stauber RH 2011 *ACS Nano* **5**, 7155-7167